Principles of Sedimentary Basin Analysis

Andrew D. Miall

Principles of Sedimentary Basin Analysis

With 387 Figures

Springer-Verlag
New York Berlin Heidelberg Tokyo

Andrew D. Miall
University of Toronto, Department of Geology, Toronto, M5S 1A1
Canada

The figure on the front cover is Fig. 6.5. *Basin fill patterns* p. 282

Library of Congress Cataloging in Publication Data
Miall, Andrew D.
Principles of sedimentary basin analysis.
Bibliography: p.
Includes index.
1. Sedimentology. 2. Stratigraphic correlation.
3. Paleogeography. I. Title.
QE471.M44 1984 551.3 84-1249

Word processing by University of Toronto Secretarial Services Unit
Media conversion by Ampersand Inc., Rutland, Vermont
Printed and bound by Halliday Lithograph, West Hanover, Massachusetts
Printed in the United States of America.

9 8 7 6 5 4 3 2 1

ISBN 0-387-90941-9 Springer-Verlag New York Berlin Heidelberg Tokyo
ISBN 3-540-90941-9 Springer-Verlag Berlin Heidelberg New York Tokyo

For Charlene

Preface

This book is intended as a practical handbook for those engaged in the task of analyzing the paleogeographic evolution of ancient sedimentary basins. The science of stratigraphy and sedimentology is central to such endeavors, but although several excellent textbooks on sedimentology have appeared in recent years little has been written about modern stratigraphic methods. Sedimentology textbooks tend to take a theoretical approach, building from physical and chemical theory and studies of modern environments. It is commonly difficult to apply this information to practical problems in ancient rocks, and very little guidance is given on methods of observation, mapping and interpretation.

In this book theory is downplayed and the emphasis is on what a geologist can actually see in outcrops, well records, and cores, and what can be obtained using geophysical techniques. A new approach is taken to stratigraphy, which attempts to explain the genesis of lithostratigraphic units and to de-emphasize the importance of formal description and naming. There are also sections explaining principles of facies analysis, basin mapping methods, depositional systems, and the study of basin thermal history, so important to the genesis of fuels and minerals. Lastly, an attempt is made to tie everything together by considering basins in the context of plate tectonics and eustatic sea level changes. The vast and important subjects of paleoclimatology, diagenesis, and sedimentary geochemistry are not dealt with here, except for some aspects of diagenetic change that can be related to burial history. None of these areas are crucial to the main theme of the book, although geochemical data are becoming increasingly important in the study of chemical sediments and their depositional environments. The economic applications of basin analysis (e.g., exploring sediment–mineral associations and types of stratigraphic hydrocarbon trap) are not discussed in detail, although some implications for exploration are pointed out at several places in the book.

It is hoped that this book will be useful to those engaged in exploration for the various non-renewable fuel and mineral resources and for students and other geologists carrying out local or regional basin analysis studies, beginning with undergraduates about to leave for their first field school. Chapters 1 to 4 are written mainly at an introductory level, except for the concluding section of Chapter 4, dealing with recent advances in facies model theory. Chapters 5 to 7 are for more advanced students, and Chapters 8 and 9 are addressed to senior undergraduates, graduate students and professional geologists; a working knowledge of the principles of plate tectonics is assumed for this final part of the book. It is intended that the book be used throughout a student's training and professional employment, and it should provide an important reference for graduate and undergraduate courses in stratigraphy, sedimentology, basin analysis, petroleum geology and mineral exploration.

ANDREW D. MIALL

Acknowledgments

The writing of this book has taken about three years, but the ideas and experience on which it is based have been accumulating since I first discovered geology at the age of fifteen, during an optional one-period-a-week course at Brighton Grammar School. I am deeply indebted to the late Arnold Berry, Geography Master at the school, for his lectures and field trips, which rapidly convinced me I had found my vocation.

My introduction to modern sedimentological principles came through the stimulating graduate seminars held by Brian Rust at University of Ottawa. Later, the opportunity for gaining extensive practical experience was made possible by Don Stott, who took me on staff at the Geological Survey of Canada. Writing of this book began shortly after I moved to the fertile academic climate of University of Toronto, an appointment brought about by the efforts of Frank Beales, Geoff Norris and David Strangway.

Many colleagues have helped me formulate my ideas on basin analysis, and I am grateful to them for their stimulating company over the years. Foremost amongst these should be mentioned Don Campbell and John Stuart-Smith at J.C. Sproule and Associates, and my former Survey colleagues Jim Dixon, Ashton Embry, Ulrich Mayr, Ross McLean, Ray Thorsteinsson and Hans Trettin.

For specific advice relating to specialized parts of this book I am indebted to Norm Evensen (radiometric dating), Gary Jarvis (geophysical basin models and plate processes), George Klein (backarc basins and shelf depositional systems) and Geoff Norris (stratigraphic methods).

Photographs were contributed by Jim Dixon, Ashton Embry, Rolf Ludvigsen, Brian Pratt, and the Royal Ontario Museum. My thanks to D. Rudkin at the museum for his assistance with Figure 2.25. Neil Ollerenshaw and the staff at G.S.C., Calgary, are thanked for assistance in obtaining two illustrations for the sections on the dipmeter. About one-third of the figures in the book were drafted by Subash Shanbhag, and the remainder were reproduced from other sources by photographer Brian O'Donovan. Their efforts and their patience during the long gestation of the manuscript are much appreciated.

Some figures and tables are reproduced with permission of the Geological Association of Canada from 'Facies Models', Geoscience Canada Reprint Series No. 1 and from the FAC Special Paper Series.

Geoff Orton and Winston Mottley worked as Research Assistants for short but crucial periods. Typing and word processing were carried out by Diane Gardner and Naomi Frankel, with the final text prepared at the University of Toronto Secretarial Services unit under the supervision of Vera Baker.

Andy Baillie and Tony Tankard undertook the task of critically reading the entire manuscript. Their comments have been most useful, and are much appreciated. In addition, the author is grateful to Roger Walker (who read parts of Chapter 4), Geoff Norris (Chapter 3) and Ray Ingersoll (Chapter 9), who provided some much needed specialized commentary.

Finally I must thank my wife, Charlene, whose assistance, advice and support throughout my career have been irreplaceable. I am particularly grateful for her love and patience during my long obsession with "the book".

Contents

Principles of Sedimentary Basin Analysis

CHAPTER 1

Introduction

1.1 Scope and purpose of book

Most of the world's non-renewable fuel resources and many of its metals and minerals are derived from sedimentary rocks, ranging in age from Archean to Cenozoic. Exploration for such resources requires an understanding of their relationship to host strata, whether they be primary deposits formed at the same time as the sediments, such as coal and placer minerals, or post-depositional deposits whose distribution depends on porosity and permeability trends in the sediments, such as epigenetic ore deposits and petroleum.

The study of such host strata is known by the convenient term **basin analysis**, and its most important product is the documentation of the **paleogeographic evolution** of a sedimentary basin. The work may include many components, amongst which the most important are stratigraphy, structure and sedimentology. The rocks themselves may be examined in outcrop or in wells, or the data may be acquired by geophysical methods.

Basin analysis requires an understanding of many diverse geological specialties and an ability to assess the relationships between varied types of evidence. The skills with which to perform a satisfactory synthesis are rarely taught in any university or industry course, and yet they are demanded by the nature of the work performed by many professionals in the petroleum and mining industry, and by those engaged in regional projects for government surveys. As Baillie (1979) has pointed out, in most such organizations there are now specialists in biostratigraphy, geochemistry, geophysics and, perhaps, petrology, who provide an increasingly complex array of detailed technical information relating to any given research or exploration project, but it is up to the project leaders or district geologists to make sense out of all these services. These individuals must develop the broad overview, and understand enough about the data input to re-examine both their models and the data itself when things cannot be made to fit.

It is also hoped that this book will be useful for advanced undergraduate and post-graduate courses, and that it may provide some alternative ideas for thesis research projects. Too many of these in sedimentary geology deal in great detail with very small areas, ignoring any regional context, or concentrate on developing a particular technique or specialty, without considering the importance of other types of data.

Never was there a greater need for a modern text dealing with basin analysis. The enormous advances made in sedimentology, seismic stratigraphy and geochemistry, the evolution of the depositional systems method of stratigraphic analysis, the documentation of eustatic sea level changes and the impact of plate tectonics, have brought about a revolution in the methods and results of basin analysis during the last twenty years. The last book to tackle the broad questions of stratigraphic method and the relationship between sedimentation and tectonics was that by Krumbein and Sloss (1963). This highly successful book has been a standard text in stratigraphy–sedimentology courses for many years, and still is in some universities, but is now almost completely out of date. Most recent texts in the area have been specialized, dealing with sedimentology (Selley, 1970, 1976; Friedman and Sanders, 1978; Reading, 1978; Blatt et al., 1980), lithostratigraphy (Conybeare, 1979), formal stratigraphic nomenclature (Hedberg, 1976), trend analysis (Potter and Pettijohn, 1977), subsurface log analysis (Pirson, 1977) or seismic stratigraphy (Payton, 1977). The purpose of this book is to show how these subdisciplines fit together into the modern science of basin analysis. Many of the exciting new discoveries about the sedimentary history of the earth were described in a recent short and highly readable book by Hallam (1981), whereas in the present book the focus is on the methods of analysis and synthesis that are being used to make these discoveries.

The various elements of the synthesis, such as facies models, trend analysis, seismic stratigraphy etc., are touched on, but the coverage of these could not attempt to be exhaustive, because the intent is to show how the elements contribute to the whole. Practical methods are emphasized, and theory downplayed. Thus some readers may find coverage of their favorite topic unsatisfactory, and I hope the provision of lists of selected references will serve as an adequate substitute. An example of the practical approach used here is in the treatment of sedimentary structures. Most sedimentology textbooks start by dealing in greater or lesser detail with fluid mechanics, pass on to sediment dynamics and bedform generation, arriving at sedimentary structures at the end of this discourse. This is a "correct" theoretical approach, but most such treatments make only passing reference to the fact that because of sedimentary lag effects (Allen, 1974) and the ubiquitous presence of minor internal erosion surfaces it is usually very difficult to make use of this theoretical material when dealing with a practical problem in ancient rocks. The approach used here is to describe what can actually be seen in the sediments first, and then to discuss what kinds of interpretations can be made from them.

The arrangement of the chapters is from the simple observational details described in Chapter 2, through the problems of stratigraphic and facies analysis, to the complex subject of plate tectonics and its effects on sedimentation, as described in Chapters 8 and 9. However, those engaged in a practical basin problem will, it is hoped, find themselves turning back and forth between the chapters, and the organization of the book is not intended to imply any particular order in which a basin analysis should be performed.

1.2 The modern revolution in stratigraphy

1.2.1 Traditional stratigraphy

The subject of stratigraphy once provided uninspired teachers with unrivaled opportunities to bore their students to distraction. The traditional method of teaching the subject was to choose selected parts of the world and laboriously to ascend the local stratigraphic column, cataloguing the names of the principal stratigraphic units and documenting their lithology and fossil content. This may have been coupled with an attempt to relate stratigraphic style to theories of cratonic and geosynclinal sedimentation, but the main emphasis was on formation and group and member names, and more names. The result was the Layer Cake view of stratigraphy—sediments as geometrically uniform blankets bounded by the sharp vertical lines of the correlation table.

There is no excuse for the continued use of such an approach. Formation names should be left to the local specialist, except for those of particular historic interest or those of units which demonstrate a stratigraphic principle of importance. The many scientific developments touched on above now permit and, indeed, require, a completely different, genetic approach to the subject of stratigraphy. Far from being a dull, descriptive art, stratigraphy should now be the discipline which represents the pinnacle of our achievements in studying sedimentary rocks, based on the most fundamental understanding of global sedimentary and tectonic processes. It is one of the principal aims of this book to outline the elements of this science.

Unfortunately, in the present-day real world of stratigraphy, we have to deal with a wide range of different approaches to the subject. In the heavily populated parts of the developed countries, sedimentary geology is cluttered with myriads of local stratigraphic names built up since the nineteenth century by dedicated workers describing local sequences within the compass of isolated map areas. Old concepts of basin development and a poor understanding of depositional environments and facies relationships meant that earlier stratigraphic units were established in the absence of any real understanding of their origins. Arguments about stratigraphic correlation and terminology have been interminable, since the classic confrontation between Murchison and Sedgwick over the Welsh Lower Paleozoic. P.D. Krynine once said "stratigraphy is the triumph of terminology over common sense". The basic problem has remained unchanged for over a century: the need (or desire) of the geologist to establish a formal stratigraphy, with named units, before he or she has had the opportunity to examine all the evidence. Stratigraphic units ideally should be established on the basis of a basin-wide perspective, but they rarely are. Local terminologies continue to be proposed by geologists studying limited areas, while paying little or no attention to the regional framework.

Until about 1960 geologists had only two, quite different practical tools for studying stratigraphy. These were lithostratigraphy and biostratigraphy (see Chap. 3). The increased pace of deep exploration drilling, particularly since the Second World War, did not at first fundamentally change stratigraphic practices, except that microfossils of all kinds became essential biostratigraphic tools, to complement the older studies of macrofossils, and subsurface logging methods rapidly evolved (Chap. 2).

1.2.2 Modern developments

Beginning gradually in about 1960 five fundamental changes have occurred in sedimentary geology, so that nowadays the discipline of stratigraphy, as practiced in new frontier basins, particularly those offshore, is a completely new science, bearing little relation to the subject described in such classic textbooks as Dunbar and Rodgers (1957) or Krumbein and Sloss (1963). An attempt to summarize the application of these new techniques is shown in Figure 1.1.

The first of these five changes was the evolution of sedimentology into a mature science, capable of explaining the origin of sedimentary rocks through facies studies and facies models. The model concept was first discussed at a conference in 1958 (Potter, 1959) and the first facies models were those of Allen (1963) and Bernard et al. (1962) for fluvial point bars and of Bouma (1962) for sandy turbidites. Nowadays, it is possible to interpret and predict the composition, geometry and orientation of virtually all stratigraphic units using this approach, based on a wide range of studies in modern environments and ancient rock units (Chap. 4).

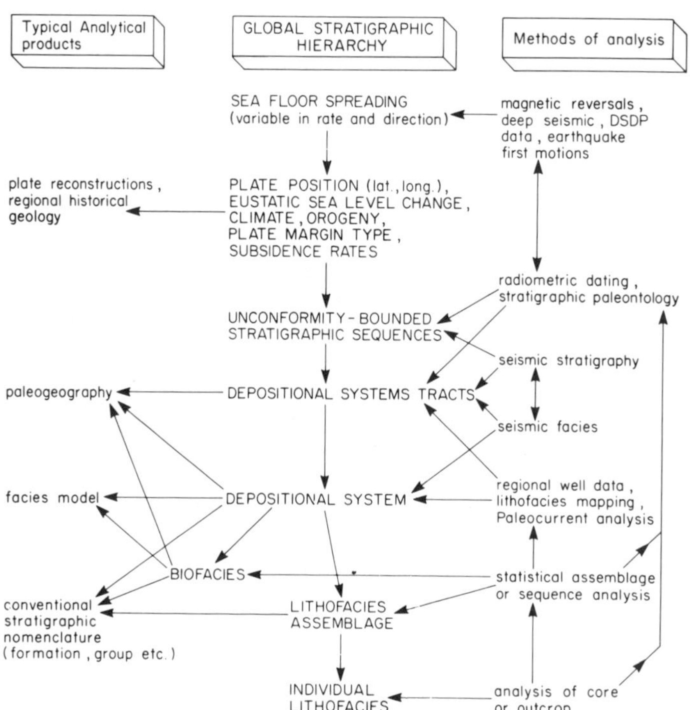

Fig. 1.1. Basin analysis flow chart.

This genetic approach to stratigraphy has been carried a step further in the "depositional systems" method, developed by the Texas Bureau of Economic Geology, and used by Fisher and McGowen (1967), Fisher (1969), Brown (1969), Fisher and Brown (1972), Frazier (1974), Brown and Fisher (1977), Glaeser (1979), Casey (1980), Handford and Dutton (1980), and others. This is the second major development, although the ideas have not yet been widely used outside the Gulf Coast. The concept is an extension of Walther's Law, which states that in a conformable succession the only facies that can occur together in vertical succession are those which can occur side by side in nature (Middleton, 1973). A prograding delta is a good illustration of this. It contains prodeltaic, delta front (bay and distributary mouth) and delta plain (distributary channel and overbank) facies, which occur side by side in that order and the products of which occur together in the same order in vertical succession. A depositional system is this complete package of environments and its sedimentary products. Depositional systems are bounded by unconformities or by facies transitions into adjacent, genetically unrelated systems, such as the passage of a deltaic coast into a clastic sediment-starved carbonate shelf. Use of the depositional systems concept enables predictions to be made about the stratigraphy of large masses of sediment, because it permits interpretations of the rocks in terms of broad environmental and paleogeographic reconstructions (Chap. 6).

The third major development has been the evolution of modern seismic stratigraphic techniques (Payton, 1977). Seismic reflections derived from sedimentary rocks are essentially chronostratigraphic correlation lines (Vail, Todd and Sangree, 1977). Modern acquisition and processing techniques can now enhance seismic reords to such an extent that the fine architectural details of entire basins can be seen and mapped. In this way major depositional systems can be delineated and related to each other. Subsurface geologists can therefore recognize the major stratigraphic subdivisions in a basin before the first well has been drilled, and can describe their composition and make intelligent predictions about internal variation in an early stage of exploration drilling. By being able to see the whole before becoming bogged down in detail, the work of the stratigrapher is made immeasurably easier. Contrast this with the random collection of partially exposed, incomplete sections that the surface geologist has to work with. Application of the seismic stratigraphic approach, particularly in frontier basins, has resulted in many new ideas about stratigraphic relationships (Mitchum et al., 1977) and the effects of sea level changes (Vail et al., 1977a, b). Building on the work of Sloss (1963), Vail and his coworkers demonstrated that the sediments in many basins could be divided into discrete sequences which can be correlated from basin to basin around the world. Global (eustatic) sea level changes are the only possible mechanism for such cyclicity.Documentation of the cycle of sea level changes throughout the Phanerozoic has now been completed, at least in preliminary form, and provides a new tool for regional correlation and paleogeographic interpretation (Chap. 8).

Fourthly, plate tectonic theory has had a profound impact on our understanding of basin evolution and sedimentary styles (Dickinson, 1974; Bally and Snelson, 1980; Miall, 1981). This statement should hardly need explaining to a geologist of the nineteen eighties, and further comment will be deferred to Chapter 9.

The last evolutionary development is the refinement in chronostratigraphy that has come about through developments in radiometric dating techniques and, in the last ten years, the development of magnetic reversal stratigraphy. Chronostratigraphy is the study of the absolute age of rock units, generally expressed in years before the arbitrary but internationally agreed date 1950 A.D. Dating of sedimentary rocks by radiometric techniques is a complex subject. In part, the results depend on the prior establishment of a refined relative age scale using biostratigraphic methods. Harland et al. (1964) edited a milestone compilaton of the research and its results for the Phanerozoic, and it has become a standard reference in the field, although revisions are continuing. The chronostratigraphic framework is now good enough that reasonably accurate ages can be given for at least the major subdivisions of the periods—e.g. for Upper, Middle and Lower Devonian. In some cases, particularly for Mesozoic and Cenozoic rocks, ages of individual stages are now fairly reliable (Chap. 3).

For the post-Triassic a quite different dating technique is rapidly taking shape. This is magnetic reversal stratigraphy. Vine and Matthews

(1963) were the first to demonstrate that the periodic reversal in the earth's magnetic field is recorded in a systematic manner in the new basaltic crust developed at oceanic spreading centers. As the new crust is carried away from either side of these linear spreading axes, it carries the record of the reversals with it as a kind of permanent tape recording. Detailed marine magnetic surveys run across ocean floors in the direction of spreading have produced a documentation of the reversal history that can be dated by biostratigraphic and radiometric means and correlated across all the world's oceans. Distinctive episodes in this reversal sequence are called anomalies, and have been numbered in sequence, starting at 1 for the most recent and extending back to anomaly 33, dated at about 76 Ma and correlated with the Campanian (Late Cretaceous) (Watkins, 1976; LaBrecque et al., 1977). A partial anomaly sequence is available for the rest of the Cretaceous, and has tentatively been extended back to include the oldest undisturbed oceanic crust, of Late Jurassic age (Chap. 3). Attempts to develop similar records in stratigraphic rocks are bedeviled by the problem of non-sequences and erosion surfaces. Few of the intervals within the reversal sequence are distinctive, although some "normal" or "reversed" polarity intervals are markedly longer or shorter than others, and may therefore be recognizable in an incomplete profile. But at present the use of reversal sequences in studying stratigraphic rocks is in its infancy. The method may be used for local correlation, but the older the rocks are, the more difficult it is to compare these results to the standard oceanic sequence.

The oldest undisturbed oceanic crust on earth is Jurassic, and so pre-Jurassic reversal stratigraphy must be derived entirely from sedimentary or volcanic sequences. It remains to be seen whether a reliable magnetic chronology can be developed for these.

1.2.3 The new stratigraphic method

Developments in the five areas discussed above are far from complete, but already they require an entirely new approach to the procedures for defining, correlating and interpreting stratigraphic units.

In the past the formation has been the foundation of stratigraphic procedure, mainly because for geologists working on the ground it was the most convenient scale of unit for mapping purposes (Chap. 3). The proliferation of formations has hindered rather than advanced the development of stratigraphy as a science. Nowadays, in new or frontier areas (where the ground is still uncluttered) the geologist should leave the definition of formations to the end of the work. The first step should be the establishment of the framework of major sequences (Chap. 8), ideally through seismic methods or, if unavailable, through detailed lithostratigraphic correlation and biostratigraphic zonation (Chap. 3). The next step is to interpret each sequence in terms of its component depositional systems (Chap. 6), using sedimentological data, the principles of facies analysis (Chap. 4), and basin mapping methods (Chap. 5). Work in tectonically deformed rocks or in the Precambrian may have to begin at the depositional systems stage. Within each depositional system it should then be possible to recognize units of formation rank on the basis of genetic criteria, using the paleogeographic synthesis developed in the preceding stage. If required these formations can then be defined and named for the convenience of local mappers or prospect developers (Chap. 3), but actual naming becomes a refinement of lower priority using this approach, rather than the culmination of the stratigrapher's efforts.

References

ALLEN, J.R.L., 1963: Henry Clifton Sorby and the sedimentary structures of sands and sandstones in relation to flow conditions; Geol. Mijnb., v. 42, p. 223–228.

ALLEN, 1974: Reaction, relaxation and lag in natural sedimentary systems: general principles, examples and lessons; Earth Sciences Reviews, v. 10, p. 263–342.

BAILLIE, A.D., 1979: The petroleum geologist—scientist or technician?: Bull. Can. Petrol. Geol., v. 27, p. 267–272.

BALLY, A.W., and SNELSON, S., 1980: Realms of subsidence; *in* A.D. Miall, ed., Facts and principles of world petroleum occurrence; Can. Soc. Petrol. Geol. Mem. 6, p. 9–94.

BERNARD, H.A., LEBLANC, R.J. and MAJOR, C.J., 1962: Recent and Pleistocene geology of southeast Texas; *in* E.H. Rainwater and R.P. Zingula, eds., Geology of the Gulf Coast and central Texas; Geol. Soc. Am. Guidebook for 1962 Ann. Mtg., p. 175–224.

BLATT, H., MIDDLETON, G.V., and MURRAY, R., 1980: Origin of sedimentary rocks; 2nd ed.,

Prentice Hall Inc., Englewood Cliffs, New Jersey, 782 p.
BOUMA, A.H., 1962: Sedimentology of some flysch deposits; Elsevier, Amsterdam, 168 p.
BROWN, L.F., Jr., 1969: Geometry and distribution of fluvial and deltaic sandstones (Pennsylvanian and Permian), North-Central Texas; Gulf Coast Assoc. Geol. Soc. Trans., v. 19, p. 23–47.
BROWN, L.F., Jr., and FISHER, W.L., 1977: Seismic-stratigraphic interpretation of depositional systems: examples from Brazilian rift and pull-apart basins; *in* C.E. Payton, ed., Seismic stratigraphy—applications to hydrocarbon exploration, Am. Assoc. Petrol. Geol. Mem. 26, p. 213–248.
CASEY, J.M., 1980: Depositional systems and basin evolution of the Late Paleozoic Taos Trough, northern New Mexico; Texas Petroleum Research Committee, Report No. UT 80-1, Austin, Texas.
CONYBEARE, C.E.B., 1979: Lithostratigraphic analysis of sedimentary basins; Academic Press, New York, 555 p.
DICKINSON, W.R., 1974: Plate tectonics and sedimentation; *in* W.R. Dickinson, ed., Tectonics and sedimentation; Soc. Econ. Paleont. Mineral. Spec. Publ. 22, p. 1–27.
DUNBAR, C.O. and RODGERS, J., 1957: Principles of stratigraphy; John Wiley and Sons, New York, 356 p.
FISHER, W.L., 1969: Delta systems in the exploration for oil and gas; Texas Bur. Econ. Geol. Spec. Publ., 212 p.
FISHER, W.L., and BROWN, L.F., Jr., 1972: Clastic depositional systems—a genetic approach to facies analysis; annotated outline and bibliography; Univ. Texas Bur. Econ. Geol. Spec. Rept., 230 p.
FISHER, W.L., and McGOWEN, J.H., 1967: Depositional systems in the Wilcox Group of Texas and their relationship to occurrence of oil and gas; Trans. Gulf Coast Assoc. Geol. Soc., v. 17, p. 105–125.
FRAZIER, D.E., 1974: Depositional episodes: their relationship to the Quaternary stratigraphic framework in the northwestern portion of the Gulf Basin, Texas Bur. Econ. Geol., Circ. 74–1.
FRIEDMAN, G.M. and SANDERS, J.E., 1978: Principles of sedimentology; John Wiley & Sons, 715 p.
GLAESER, J.D., 1979: Catskill delta slope sediments in the central Appalachian Basin: source deposits and reservoir deposits; *in* L.J. Doyle and O.H. Pilkey Jr., eds., Geology of continental slopes, Soc. Econ. Paleont. Mineral. Spec. Publ. 27, p. 343–357.
HALLAM, A., 1981: Facies interpretation and the stratigraphic record; W.H. Freeman and Company, San Francisco, 291 p.
HANDFORD, C.R., and DUTTON, S.P., 1980: Pennsylvanian–Early Permian depositional systems and shelf-margin evolution, Palo-Duro Basin, Texas; Am. Assoc. Petrol. Geol. Bull., v. 64, p. 88–106.
HARLAND, W.B., SMITH, A.G., and WILCOCK, B., eds., 1964: The Phanerozoic time scale; Geol. Soc. London, Supplement to Quart. J., v. 120, 458 p.
HEDBERG, H.D., ed., 1976: International stratigraphic guide; John Wiley and Sons, New York, 200 p.
KRUMBEIN, W.C., and SLOSS, L.L., 1963: Stratigraphy and sedimentation; 2nd ed., W.H. Freeman, San Francisco, 660 p.
LABRECQUE, J.L., KENT, D.V., and CANDE, S.C., 1977: Revised magnetic polarity time scale for Late Cretaceous and Cenozoic time; Geology, v. 5, p. 330–335.
MIALL, A.D., 1981: Alluvial sedimentary basins, tectonic setting and basin architecture; *in* A.D. Miall, ed., Sedimentation and tectonics in alluvial basins; Geol. Assoc. Can. Spec. Paper 23, p. 1–33.
MIDDLETON, G.V., 1973: Johannes Walther's Law of the correlation of facies; Geol. Soc. Am. Bull., v. 84, p. 979–988.
MITCHUM, R.M. Jr., VAIL, P.R., and THOMPSON, S. III, 1977: Seismic stratigraphy and global changes of sea level, Part two: The depositional sequence as a basic unit for stratigraphic analysis; Am. Assoc. Petrol. Geol. Mem. 26, p. 979–988.
PAYTON, C.E., ed., 1977: Seismic stratigraphy—applications to hydrocarbon exploration; Am. Assoc. Petrol. Geol. Mem. 26.
PIRSON, S.J., 1977: Geologic well log analysis; 2nd ed., Gulf Pub. Co., Houston, 377 p.
POTTER, P.E., 1959: Facies models conference; Science, v. 129, p. 1292–1294.
POTTER, P.E., and PETTIJOHN, F.J., 1977: Paleocurrents and basin analysis; 2nd ed., Springer-Verlag, New York, 425 p.
READING, H.G., ed., 1978: Sedimentary environments and facies; Blackwell, Oxford, 557 p.
SELLEY, R.C., 1970: Ancient sedimentary environments; Cornell Univ. Press, Ithaca, N.Y., 237 p.
SELLEY, R.C., 1976: An introduction to sedimentology; Academic Press, London, 408 p.
SLOSS, L.L., 1963: Sequences in the cratonic interior of North America; Geol. Soc. Am. Bull., v. 74, p. 93–113.
VAIL, P.R., MITCHUM, R.M., Jr., and THOMPSON, S., III, 1977a: Seismic stratigraphy and global changes of sea level, Part three: Relative changes of sea level from coastal onlap; Am. Assoc. Petrol. Geol. Mem. 26, p. 63–82.
VAIL, P.R., MITCHUM, R.M., Jr., and THOMPSON, S., III, 1977b: Seismic stratigraphy and global changes of sea level, Part four: Global cycles of relative changes of sea level; Am. Assoc. Petrol. Geol. Mem. 26, p. 83-98.
VAIL, P.R., TODD, R.G., and SANGREE, J.B., 1977: Seismic stratigraphy and global changes of sea level, Part five: Chronostratigraphic significance of seismic reflections; Am. Assoc. Petrol. Geol. Mem. 26, p. 99–116.
VINE, F.J. and MATTHEWS, D.H., 1963: Magnetic anomalies over ocean ridges; Nature, v. 199, p. 947–949.
WATKINS, N.D., 1976: Polarity subcommission sets up some guidelines, Geotimes, v. 21, p. 18–20.

CHAPTER 2

Collecting the data

2.1 Introduction

A successful basin analysis requires the collection and integration of several, perhaps many, different kinds of data. Direct observation of the rocks themselves may or may not be fundamental to the study. In the case of a surface geological project it will be preeminent, though perhaps supplemented by geochemical and geophysical information, plus laboratory analysis of collected samples. For subsurface petroleum studies actual rock material available for examination may be very limited, consisting of well cuttings from rotary drilling, plus a few short cores. Geophysical well logs and regional seismic lines may provide at least as important a part of the total data base. Investigations for stratabound ores and minerals typically employ networks of diamond drill holes from which continuous core normally is available. This provides a wealth of material for analysis, although certain types of observation, such as analysis of sedimentary structures, may be difficult or impossible in such small-diameter core.

2.2 Types of project and data problems

Data collection and analysis procedures are, of course, determined by the nature of the project. The following are some typical basin analysis problems, with a brief discussion of the data collection potential.

2.2.1 Regional surface stratigraphic mapping project

Work of this nature is one of the primary functions of government surveys, intent on providing complete map sheet coverage of their area of responsibility, both as a service to industry and as a basis for expert advice to government economic planners. Similar regional surveys are commonly undertaken by industry as a preliminary to detailed surface or subsurface exploration, although their studies are rarely as thorough. Many academic theses are also of this type.

Many government surveys are carried out by individuals that are not specifically trained in the analysis of sedimentary basins—the idea being that members of the survey should be generalists, capable of mapping anything. This is the old British tradition, and it is an unfortunate one because it means that the individual survey officer cannot possibly be aware of all the skills that are now available for mapping work in sedimentary rocks. Nor are they encouraged to take the time for the specialized observations which would make their work so much more effective. The argument that this is "left for the academics to do later" is not always satisfactory, for it is commonly the case that the stratigraphy of a succession can only be clarified by those who thoroughly understand its sedimentology. Many fruitless arguments about stratigraphic terminology can be avoided if this is realized at the beginning of a mapping endeavour.

The basis of all surface basin analysis projects is the careful compilation of vertical stratigraphic sections. These are described by the geologist in the field, who also collects samples for subsequent laboratory analysis, taking care to label each sample according to its position in the section. Where should such sections be located? The choice depends on a variety of factors. Firstly they should be typical of the area in which they are found, and should be as free as possible of structural deformation. Obviously they should also be well exposed and, ideally, free of chemical or organic weathering which disturbs or obscures textures and structures.

The best sections for regional correlation are those which include several stratigraphic intervals, but it is rare to find more than a few of these except in exceptionally well exposed areas. Short, partial sections of a stratigraphic unit can provide much valuable sedimentological data such as facies and paleocurrent information, although it may not be possible to locate them precisely within a stratigraphic framework unless they can be correlated by a marker bed or structural interpretation (Chap. 3). Exposures of this type tend to be ignored by the regional mapper, but they should not be ignored by those intent on producing an integrated basin analysis, because they add to the data base and may provide many useful sedimentological clues.

The great advantage of surface over subsurface studies is the potential, given adequate clean outcrop, to see medium to large-scale sedimentary features, such as crossbedding, channels and bioherms, that may be difficult or impossible to identify in a drill hole. These, of course, add immeasurably to any basin interpretation, particularly the paleogeographic and trend analysis aspects. The geologist should also always be searching for lateral variations in lithologies, fossil content or sedimentary structures, as these changes may provide crucial control for paleogeographic interpretations.

The disadvantage of studies carried out exclusively at the surface is that most of the rocks in any given basin are buried, and may be inaccessible to observation over very large areas. Many basins are depositional basins, in the sense that they preserve at the present day essentially the same outline as during sedimentation. The rocks exposed at the surface, especially around the margins of the basin, may have a quite different thickness and facies to those preserved at the center, and may show erosion surfaces and unconformities not present in the center because of the tendency of basin margins to show a greater degree of tectonic instability. This is illustrated in Figure 2.1. A basin analysis carried out under such circumstances might therefore produce very incomplete or misleading results.

2.2.2 Local stratigraphic–sedimentologic mapping project

A common incentive for a detailed local study is the occurrence of some highly localized economic deposit such as an ore body or coal seam. A geologist will examine every available outcrop within a few square kilometers of the deposit, and may also supplement the analysis with logs of diamond drill holes. Another type of local study is an academic thesis project, particularly at the Master's level. A small–scale project is chosen because of the time and cost limits imposed by academic requirements.

Many of the comments given in the preceding section apply to local studies, but there are some additional complications that often arise because of the nature of such projects. A common fault is the emphasis on local features to the exclusion of any real consideration of regional implications. The geologist will erect a detailed local strati-

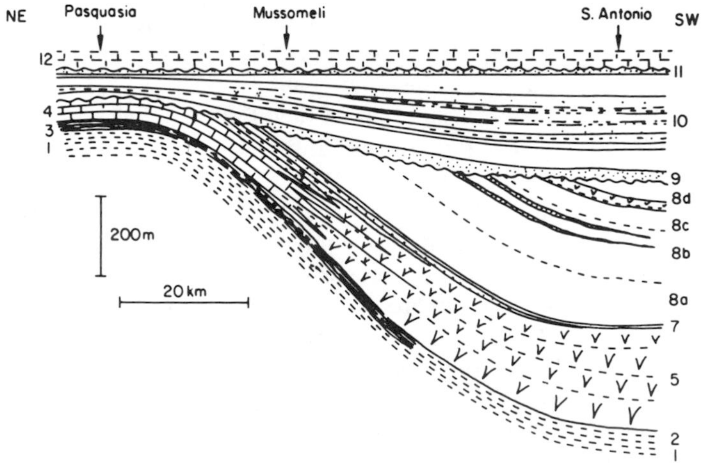

Fig. 2.1. Contrasts between stratigraphic thicknesses and facies at basin center and basin margin. The geology of the deep, hydrocarbon-producing regions of a basin might be quite different from that at the margins, so that surface geology gives little useful information on what lies below. The North Sea Basin is an excellent example. This example is of the Messinian evaporite basin, Sicily (Schreiber et al., 1976).

graphy and fail to show clearly how it relates to any regional framework that may have been established. Or there may be an overemphasis on certain selected parameters chosen, perhaps as a training exercise, to the exclusion of others.

To set against these problems is the advantage a detailed local project may offer for carrying out very complete, sophisticated paleogeographic reconstruction. Rather than relying on selected stratigraphic sections the geologist may be able to trace out units on foot, or study their variation in a close network of diamond drill holes (Fig. 2.2). In this way the detailed variations within, say, an individual reef or channel network or a coal swamp can be reconstructed. Such reconstructions are particularly valuable because they provide a mass of three-dimensional data against which to test theoretical facies models.

Diamond drill holes are produced by a process of continuous coring. They therefore provide a complete lithologic and stratigraphic sample through the units of interest. The small diameter of the core (2–5 cm) does not permit recognition of any but the smallest sedimentary structures (Figure 2.6), but the close drill hole spacing of a few hundred meters or less means that very detailed stratigraphic correlation is usually possible.

Unfortunately, a tradition of retention and curation of diamond drill hole core has rarely emerged in government or industrial organizations except, perhaps, for a few key holes in areas of particular interest. The core normally is discarded once the immediate interest in the area has subsided, and much valuable material is lost to future workers. The mining industry seems to prefer it this way, but what is good for corporate competitiveness is not necessarily the best method for developing a national stratigraphic data base.

2.2.3 Regional subsurface mapping project

Exploration activity in the petroleum industry is now mainly of this type. Companies may initially send surface geological mapping parties into the field, but ultimately the most thorough field studies in a given field area are likely to be done by government survey organizations. Much industry activity is now located in offshore regions for which surface geological information is, in any case, sparse, or unobtainable, or irrelevant. As noted above, the beds exposed at the edge of a basin may bear little relation to those buried near its center (Fig. 2.1).

Regional subsurface work is based initially on geophysical data and subsequently on test drilling. Gravity and aeromagnetic information may provide much useful information on broad structural features, particularly deep crustal structure (Fig. 2.3, Chap. 5). Refraction seismic lines may be shot for the same purpose. More detailed structural and stratigraphic data are obtained from reflection seismic surveys (Chaps. 5, 6, 8), and these provide the basis for all exploration drilling in the early phases of basin development. Seismic shooting and processing is now a highly sophisticated process, and its practitioners like to talk about seismic stratigraphy as if it can provide virtually all the answers, not only about structure, but about the stratigraphic subdivision of a basin, regional correlation, and even lithofacies (Fig. 2.4). However, seismic is only one exploration tool, and its results must be tested against those derived in other ways. For example, test drilling often shows that stratigraphic correlations predicted from seismic interpretation are incorrect. Problems of stratigraphic velocity resolution, the presence of low angle unconformities and the ob-

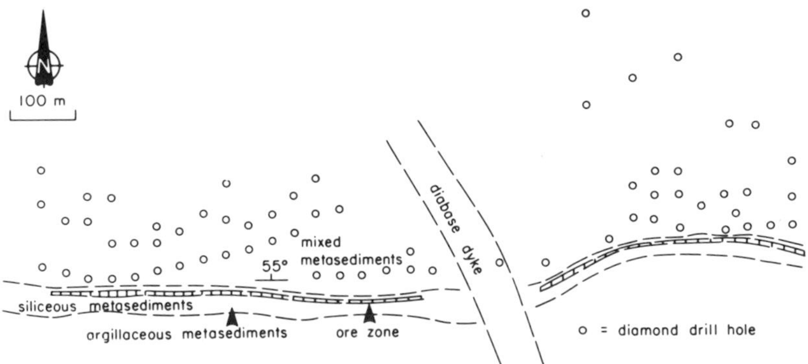

Fig. 2.2. A typical diamond drill hole (DDH) network across a mining property, a gold prospect in Precambrian metasediments, northern Ontario.

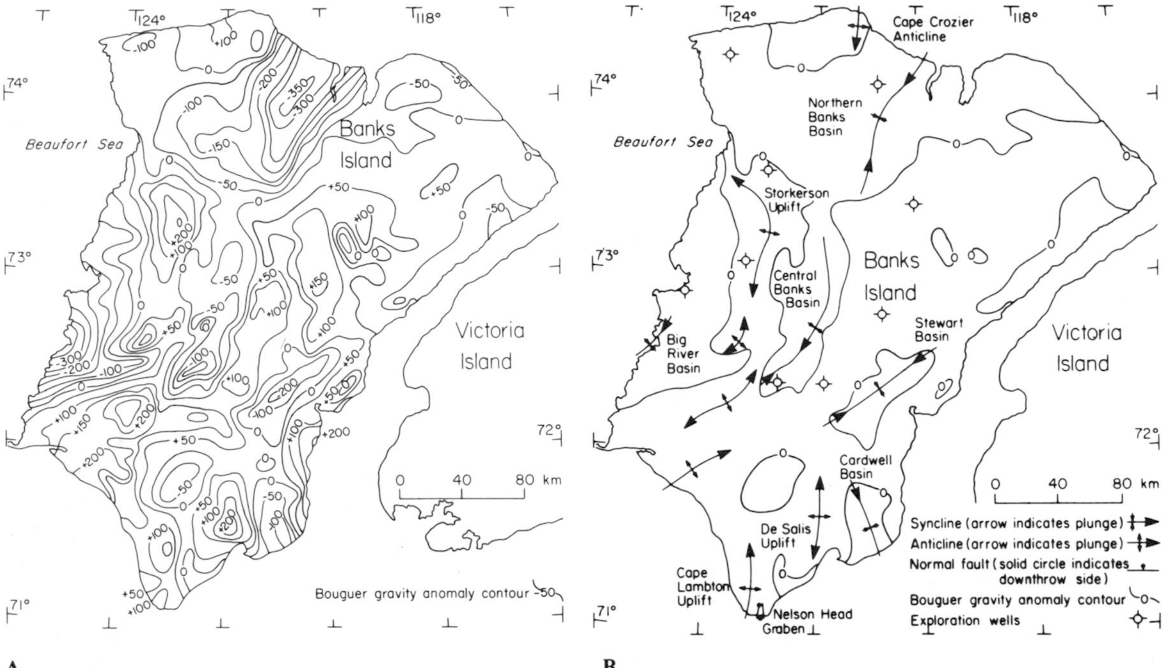

Fig. 2.3. Use of regional gravity data (**A**: Bouguer anomalies) in defining broad structural features (**B**), Banks Island, Arctic Canada. In this case exploratory drilling of the nine wells shown indicated that negative anomalies corresponded closely to the outline of Mesozoic–Cenozoic sedimentary basins up to 2 km thick, resting unconformably on Devonian strata. Gravity data from Stephens et al. (1972), interpretation from Miall (1979).

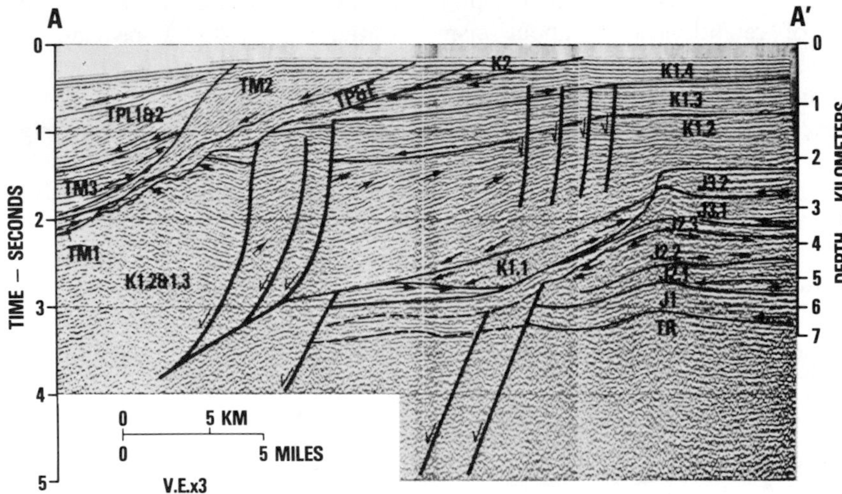

Fig. 2.4. Example of an interpreted seismic line, continental margin of North Africa (Mitchum et al., 1977).

scuring effects of local structure can all introduce errors into seismic interpretation. Dating and correlation of seismic sequences (Chap. 8) depend heavily on biostratigraphic and geophysical log information from exploration wells.

Exploration wells, especially the first ones to be drilled in a frontier basin, are as valuable in their own way as space probes sent out to study the planets, and it is unfortunate that in many countries the resulting data are not treated with the same respect afforded space information, but are regarded as the private property of the organization which paid for the drilling. In a competitive world obviously a company has the right to benefit from its own expenditures but, in the long term, knowledge of deep basin structure and stratigraphy belongs to the people and should eventually be made available to them. In Canadian frontier areas well samples and logs must be deposited with the federal government and released for public inspection two years after well completion. Two years competitive advantage is quite long enough for any company in the fast-moving world of the oil industry, and, after this time is up, the well records become part of a national data repository which anybody can use—with obvious national benefits. Seismic records are released after five years.

The nature of the stratigraphic information derived from a well is both better and worse than that derived from surface outcrops. It is better in the sense that there are no covered intervals in a well section, and such sections are generally much longer than anything that can be measured at the surface (perhaps exceeding 6000 m), so that the stratigraphic record is much more complete. The disadvantage of the well section is the very scrappy nature of the actual rock record available for inspection. Three types of sample normally are available:

1. Cuttings, produced by the grinding action of the rotary drill bit. These generally are less than 1 cm in length (Fig. 2.5) and can therefore only provide information on lithology, texture and mic-

Fig. 2.5. Typical well cuttings. Photo courtesy of J. Dixon.

▷ **Fig. 2.6.** Examples of typical core from a petroleum exploration hole (left) and a diamond drill hole (right).

rofossil content. North American practice is to collect from the mudstream and bag for examination samples every 10 ft (3 m) of drilling depth. When drilling in soft lithologies rock may cave into the mud stream from the side of the hole many meters above the drill bit, so that the samples become contaminated. Also, cuttings of different density may rise in the mud stream at different rates, which is another cause of mixing. It is thus necessary to observe the "first appearance" rule, which states that only the first (highest) appearance of a lithology or fossil type can be plotted with some confidence. Even then, depth distortions can be severe.

2. Full–hole core (Fig. 2.6). The rotary drill bit may be replaced by a coring tool when the well is drilling through an interval of interest, such as potential or actual reservoir bed. However, this type of core is expensive to obtain, and it is rare to find that more than a few tens of meters—perhaps only a few meters—of core are available for any given hole.

The advantage of core is that because of the large diameter (usually in the order of 10 cm) it permits a detailed examination of small to medium-sized sedimentary structures. Macrofossils may be present, and trace fossils are usually particularly well seen. The amount of sedimentological detail that can be obtained from a core is thus several orders of magnitude greater than is provided by chip samples. However, it is frustrating not to be able to assess the scale significance of a feature seen in core. An erosion surface, for example, may be the product of a local scour, or it may be a major regional disconformity, but both could look the same in a core. Hydrodynamic sedimentary structures might be present, but it may not be possible to interpret their geometry and, except in rare instances where a core has been oriented in the hole, they can provide no paleocurrent information. Orientation can sometimes be deduced if the core shows a structural dip that can be determined from regional structural data or dipmeter logs.

3. Side-wall cores. These are small plugs extracted by a special tool from the wall of a hole after drilling has been completed. These cores are rarely available to the geologist because they are used in porosity–permeability tests and are disaggregated for caving-free analysis by biostratigraphers. In any case their small size limits the amount of sedimentological information that they can yield.

In addition to samples and core, each exploration hole nowadays is subjected to an extensive series of geophysical logging methods, which provide records by direct analog tracing or digitizing. The description and interpretation of such logs has itself been the subject of several textbooks and cannot be treated exhaustively here. Log information is discussed further in section 2.4.5 (and the dipmeter is dealt with in sections 5.4.2 and 5.9.6); the following are a few preliminary remarks discussing the utility of logs in a subsurface data collection scheme.

Most geophysical tools measure a single physical property of the rock, such as its electrical resistivity, sonic velocity, gamma radioactivity, etc. (Fig. 2.7). These properties reflect lithology, and can therefore be used, singly or in combination, to interpret lithology. Because measurements with modern tools have depth resolution of a few centimeters they are therefore of great potential value in deriving accurate, depth-controlled lithologic logs, free of the problems of sample caving. The response of a single tool is not unique to each rock type; for example, many different formations will contain rocks with the same electrical resistivity, and so it is not possible to interpret lithology directly from a single log type. However, such interpretations may be possible from a combination of two or more logs, and attempts have been made to automate such interpretations based on computerized calculation routines from digitized log records. Unfortunately, the physical properties of rocks and their formation fluids vary so widely that such automated interpretation procedures can only be successful if they are adapted to the specific conditions of each basin. They thus lose much of their exploration value.

A particularly common use to which geophysical logs are put is in stratigraphic correlation. Log records through a given unit may have a distinctive shape, which a skilled geologist can recognize in adjacent holes. Correlation may therefore be possible even if details of lithology are unknown and, indeed, the establishment of correlations in this way is standard practice in subsurface work (Chaps. 3, 5).

Logs are also of considerable importance in calibrating seismic records—seismic velocities are routinely derived from sonic logs and used to improve seismic correlations.

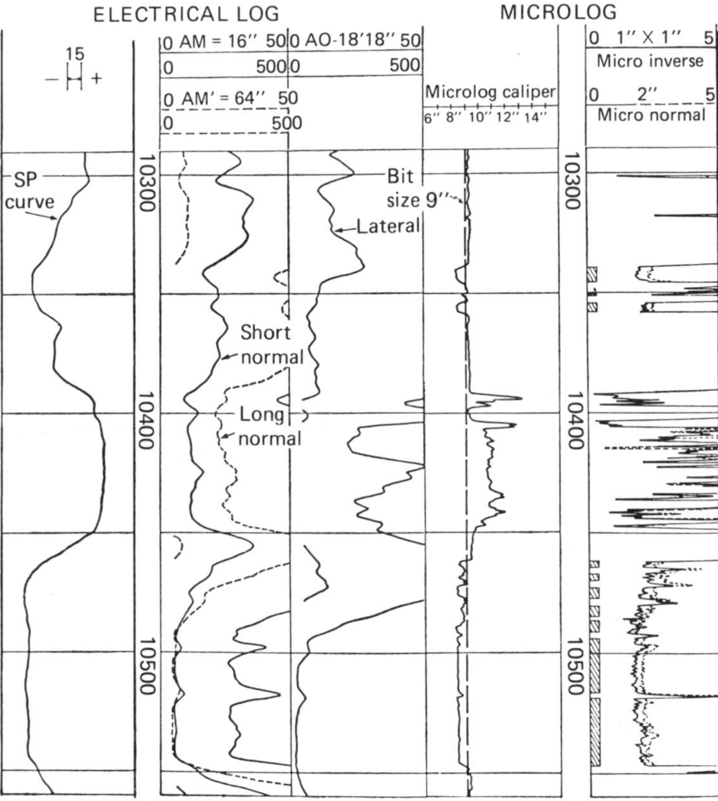

Fig. 2.7. Typical geophysical log suite from a petroleum exploration hole (From Krumbein and Sloss, Stratigraphy and Sedimentation; 2nd ed. W.H. Freeman and Comp, Copyright © 1963).

Regional subsurface work may lead to the development of a list of petroleum "plays" for an area. These are conceptual models to explain the generation–migration–trapping system for petroleum pools; for example, a "Devonian reef play", which might be based on the occurrence in Devonian strata of porous, dolomitized reef masses enclosed in mudrock, providing an ideal series of stratigraphic traps. The development of a petroleum play leads to the evolution of an exploration philosophy, which summarizes exploration experience in choosing the best combination of exploratory techniques and the type of data required to locate individual pools within the play area. At this stage exploration moves from the "play" stage to the "prospect" stage. The first wells drilled in an area might have been drilled on specific prospects, with a specific play in mind, but it is rare for early wells to be successful. Prospect development is the next phase in an exploration program.

2.2.4 Local subsurface mapping project

A network of closely spaced holes a few kilometres (or less) apart may be drilled to develop a particular petroleum or mineral prospect. These may be in virtually virgin territory or they may be step-out wells from a known pool or deposit in a mature petroleum basin or mine. It is at this stage in exploration work that geological skills come most strongly into play. In the early exploration phases virtually the only information available is geophysical; almost all early petroleum exploration wells are drilled on potential structural traps. Once a hydrocarbon-bearing reservoir is located the next problem is to understand why it is there, how large and how porous it is, and in what direction it is likely to extend. Interpretations of depositional environments, applications of facies models and reconstructions of paleogeography may become crucial in choosing new drilling sites.Similar procedures must be followed in order to develop stratabound ore bodies.

The types of data available for this work have been discussed in the previous section. At this point it is pertinent to add a few cautionary notes on the uses of subsurface data. Because of its inherent limitations it is natural that exploration geologists would wish to exploit the data to the full but, paradoxically, this can lead to a very limited approach to an exploration problem. Geologists may be so impressed with a particular interpretive technique that they may tend to use it to the exclusion of all others. This has led to many false interpretations. It cannot be overemphasized that, in basin analysis as, no doubt, in every area of geology, every available tool must be brought to bear on a given problem. Examples of techniques in the area of sedimentology that have been overutilized include the use of vertical profile studies, grain size analysis and grain surface texture analysis to interpret depositional environments.

The use of vertical profiles was first described in detail in print by Visher (1965). It followed from Walther's Law which stated, in effect, that only those environments and facies found side by side in nature could be represented in the same order in a vertical stratigraphic profile (Middleton, 1973; Visher, 1965). Visher and many others since (including this writer) have devoted many pages of print to documenting the various types of vertical profile and demonstrating their environmental significance. Many of these profiles have characteristic log signatures but, unfortunately, few of these signatures are unique, and mistakes will arise (and have arisen) where undue emphasis is placed on "log shape" as an environmental indicator. This is discussed further in Chapter 4.

Grain size analysis of clastic rocks has consumed an enormous amount of research energy. The basic idea is that hydrodynamic sorting of a clastic sediment leaves an imprint on the grain size distribution which varies depending on the nature of the hydrodynamic process. The rock may contain a traction load and a suspension load population or the entire grain mass may have been carried in turbulent suspension. It is thought that these sorting processes have subtly different effects in the different clastic environments such as rivers, deltas or the sea, depending on the nature of the dominant transport process (unimodal currents, waves, wind, etc). Numerous statistical techniques, including Gaussian curve statistics and curve shape, discriminant functions and factor analysis have been proposed that, it was hoped, would provide infallible guides to depositional environment (e.g. Friedman and Sanders, 1978, Chap. 3; Solohub and Klovan, 1970; Glaister and Nelson, 1974). These methods have a natural appeal because they require but a few grams of material, and so can be applied to samples as small as well chips. However, it has been demonstrated in many replication studies that none of the methods is infallible, if only because most clastic rocks consist of detritus that goes through several sorting events before final deposition, including, in many cases, one or more previous cycles of erosion, transport and deposition, to produce the source rock of the unit under study. Diagenesis may also have a major effect on grain size distributions. Nevertheless the temptation is to rely on a sophisticated statistical analysis of a few grain-size samples to the exclusion of other methods of environmental interpretation. The temptation must be resisted!

2.3 Describing the rocks 1. Surface stratigraphic sections

This section deals with the procedures for describing stratigraphic sections in surface outcrops and in subsurface wells and core. The correct observation and recording of these basic descriptive data are crucial to the basin analysis method.

2.3.1 Methods of measuring and recording the data

The most useful record of a surface outcrop is a vertical stratigraphic section. Ideally the location of the section should be chosen to include important stratigraphic features such as formation contacts but, in practice, the location is commonly determined by accessibility—the presence of bars or beaches allowing us to walk along a river cut, or a negotiable gully cutting through a cliff section. Only those geologists who use their profession as an excuse to practice their favorite sport of mountaineering will be able to apply sound geological principles to the choice of section. The rest of us take what we can reach.

In reconnaissance work rapid measurement and description techniques are acceptable. For example a hand-held altimeter (aneroid barometer) may be used in conjunction with dip measurements to reconstruct stratigraphic thicknesses using simple trigonometry. Another method that is commonly described in field handbooks is the pace-and-compass technique, suitable for estimating thicknesses across rela-

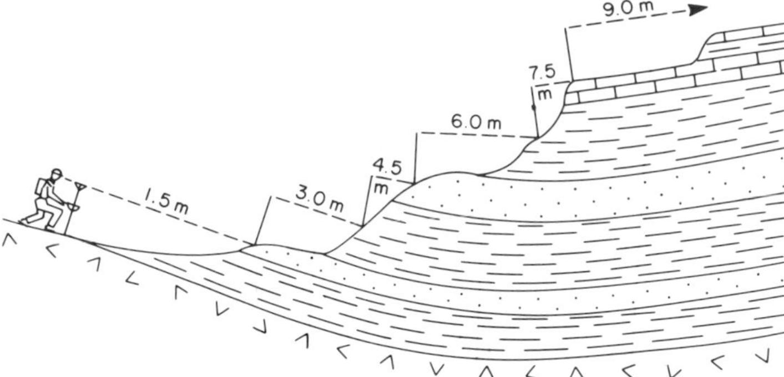

Fig. 2.9. Use of pogo stick in section measurement.

◁

Fig. 2.8. A "pogo stick" for measuring surface sections, showing sighting bar and clinometer on upper 50 cm of rod.

tively level ground, given accurate stratigraphic dip. Or the same distances may be measured from maps or air photographs. Long experience with these methods has shown that they are not very reliable; errors of up to 50% can be expected.

By far the simplest and most accurate method for measuring a section is the use of a Jacob's staff or "pogo stick". The stick is constructed of a 1.5 m wooden rod, 3 cm in diameter, with a clinometer and sighting bar (Fig. 2.8). The clinometer is preset at the measured structural dip and can then be used to measure stratigraphic thickness as fast as the geologist can write down descriptive notes (Fig. 2.9). The best technique is to use two persons. The senior geologist observes the rocks and makes notes while the junior (who can be an inexperienced student) "pogos" his or her way up the section recording increments of 1.5 m on a tally-counter, and collecting samples. The length 1.5 m is convenient for all but the tallest or shortest persons, although it can be awkward to manipulate on steep slopes. The only skill required by the pogo operator is the ability to visualize the dip of the strata in three dimensions across whatever terrain the geologist may wish to traverse. This is important so that the pogo can at all times be positioned perpendicular to bedding with the line of sight extending from the sighting bar parallel to bedding.

It is far preferable to measure up a stratigraphic section rather than down, even though this often means an arduous climb up steep slopes. Many geologists working by helicopter in rugged terrain have made their traverses physically easier by working downhill wherever possible. But not only is it difficult to manipulate a pogo stick downwards in a section, it makes it more difficult for the geologist to comprehend the order of events he or she is observing in the outcrops.

The geologist should search for the cleanest face on which to make observations. Normally this should be weathered and free of vegetation, talus or rain wash. Most sedimentary features show up best where they have been etched out by wind or water erosion, or where a face is kept continuously clean and polished by running water, as in a riverbed or an intertidal outcrop. Such features rarely show up better on fresh fracture surfaces, so a hammer should only be used for taking samples. Carbonates may benefit from etching with dilute acid. The geologist should methodically examine both vertical cuts and the topside and underside of bedding planes; all may have something to reveal, as described later in this chapter. It is also useful, on larger outcrops, to walk back and examine them from a distance, even from a low-flying helicopter or from a boat offshore, as this may reveal large-scale channels, facies changes, and many other features of interest.

In the interests of maximum efficiency the geologist obviously should ensure that all the necessary measurements, observations and sample collections are made during the first visit to a section. It may be useful to carry along a checklist, for reference each time a lithologic change in the section requires a new bed description. Many geologists have attempted to carry this process one step further by designing the checklist in the form of a computer processible data card, or they record the data in the form of a computer code. There are two problems with this:

1. If attempts are made to record every piece of field information on the computer file the resulting file is likely to be very large and cumbersome. Storage and retrieval programming may consume far more time than the original field work, unless the geologist can draw on some preexisting program package. This leads to the second problem:

2. If data are to be coded in the field or if a preexisting program system is to be used, it means that decisions will already have been made about what data are to be recorded and how they are to be recorded before the geologist goes into the field. If the geologist knows in advance what is likely to be found, as a result of some previous descriptions or reconnaissance work, this may be satisfactory, but in the case of isolated field areas such foreknowledge may not be available. In this case there is a certain risk involved in having the observation system designed in advance.

Individual geologists vary in their interests and in the observations they make and, of course, the rocks themselves are highly variable, so that it is not possible to design a single, all-purpose, section-measuring software package. Specialized systems have to be designed for specific projects, and although this leads to expense in programming and debugging, it means that the program can be designed for the specific type of output required. It should not be forgotten, though, that programming, as such, is a technical, not a scientific skill. Students and other workers who write programs for their research get few points for producing a workable program, only for the interesting scientific ideas their programming allows them to test. For example, an obvious application of computerized stratigraphic section data is in the statistical study of facies associations and cyclic sedimentation (Chap. 4).

An example of a successful computerized section description system is that devised by P.F. Friend and his students for studying Devonian nonmarine clastic sequences in East Greenland (Alexander-Marrack et al., 1970; Friend et al., 1976). They designed a form with spaces to be filled in by the geologist in the field (Fig. 2.10). Marks were made in the appropriate tracks with a soft pencil, corresponding to the observed properties. These cards could be fed directly into a computer using an optical-mark page reader. The data form provides a choice of items to be filled in within each small block outlined by a solid line. Areas of the card delimited by heavy horizontal lines across the full width of the page correspond to individual bedding units recognized in the field. Items encoded on the card include thickness of unit, exposed/covered choice, color, grain size, pebbles (present/absent), nature of basal contact, presence of carbonate, internal structures and paleocurrent measurements. The range of possible observations is limited by the card. For example, the method would not be suitable if the section was found to contain thick limestone intervals because, although carbonate content can be indicated, there are no spaces to record typical carbonate sedimentary features such as ooliths, birdseye structures, stromatolites, etc. Also, the method is suitable only for recording vertical stratigraphic changes. The presence of lateral facies changes or the exposure of a large channel could not be recorded in this system.

The particular system designed by Friend worked very well for a team of about a dozen geologists working over three field seasons, but it is unlikely that it could be applied anywhere else without modification. Unless a geologist is committed to doing much the same type of field geology for a long period, the investment of time and resources into designing and setting up such a system may not be worth the effort. In some cases the field records must, in any case, be modified following laboratory work. This is particularly the case with carbonte sediments, which commonly are best studied in polished hand specimens or thin section. In this case much of the effort of the field geologist must be devoted to ensuring that a sufficiently rigorous sample collection routine is observed. In the long run it may be simpler to stay with an old-fashioned field notebook for field observations. Computer files can be built at a later date from corrected data, using a file structure designed for a specific purpose. For example structural data, paleocurrent data and vertical facies succession data lend themselves readily to processing in special purpose computer files

STATION NUMBER

A B C D E F G H J K
000 100 200 300 400 500 600 700 800 900
00 10 20 30 40 50 60 70 80 90
0 1 2 3 4 5 6 7 8 9

SHEET NUMBER

00 10 20 30 40 50 60 70 80 90
0 1 2 3 4 5 6 7 8 9

WRITE STATION AND SHEET NUMBERS ON STUB

0 1 2 3 4 THICK 5 6 7 8 9
·0 ·1 ·2 ·3 ·4 (M) ·5 ·6 ·7 ·8 ·9
·00 ·05 EXPOSED COVERED RED P.RD GN GY OTHER
F.SL M.SL C.SL VF.SS F.SS M.SS C.SS VC.SS CON PEBBLY
GRAD SHP SMTH SCR TOOL MKS. TR.FOS. MD.CR RIPPLES

NO CEM $CaCO_3$ ROCK DOL FLAT SY.P AS.R PL.X TR.X
LIN DEF CONCR - CO_3 OR STUB FOSSILS NO VERT OTHR TR
NO DIR 2 DIR 1 DIR 000 100 200 300
00 10 20 30 40 50 60 70 80 90
STUB ← SPEC RESTART 1 2 SPARE 4 CANC'L

0 1 2 3 4 THICK 5 6 7 8 9
·0 ·1 ·2 ·3 ·4 (M) ·5 ·6 ·7 ·8 ·9
·00 ·05 EXPOSED COVERED RED P.RD GN GY OTHER
F.SL M.SL C.SL VF.SS F.SS M.SS C.SS VC.SS CON PEBBLY
GRAD SHP SMTH SCR TOOL MKS. TR.FOS. MD.CR RIPPLES

NO CEM $CaCO_3$ ROCK DOL FLAT SY.P AS.R PL.X TR.X
LIN DEF CONCR - CO_3 OR STUB FOSSILS NO VERT OTHR TR
NO DIR 2 DIR 1 DIR 000 100 200 300
00 10 20 30 40 50 60 70 80 90
STUB ← SPEC RESTART 1 2 SPARE 4 CANC'L

0 1 2 3 4 THICK 5 6 7 8 9
·0 ·1 ·2 ·3 ·4 (M) ·5 ·6 ·7 ·8 ·9
·00 ·05 EXPOSED COVERED RED P.RD GN GY OTHER
F.SL M.SL C.SL VF.SS F.SS M.SS C.SS VC.SS CON PEBBLY
GRAD SHP SMTH SCR TOOL MKS. TR.FOS. MD.CR RIPPLES

NO CEM $CaCO_3$ ROCK DOL FLAT SY.P AS.R PL.X TR.X
LIN DEF CONCR - CO_3 OR STUB FOSSILS NO VERT OTHR TR
NO DIR 2 DIR 1 DIR 000 100 200 300
00 10 20 30 40 50 60 70 80 90
STUB ← SPEC RESTART 1 2 SPARE 4 CANC'L

0 1 2 3 4 THICK 5 6 7 8 9
·0 ·1 ·2 ·3 ·4 (M) ·5 ·6 ·7 ·8 ·9
·00 ·05 EXPOSED COVERED RED P.RD GN GY OTHER
F.SL M.SL C.SL VF.SS F.SS M.SS C.SS VC.SS CON PEBBLY
GRAD SHP SMTH SCR TOOL MKS. TR.FOS. MD.CR RIPPLES

NO CEM $CaCO_3$ ROCK DOL FLAT SY.P AS.R PL.X TR.X
LIN DEF CONCR - CO_3 OR STUB FOSSILS NO VERT OTHR TR
NO DIR 2 DIR 1 DIR 000 100 200 300
00 10 20 30 40 50 60 70 80 90
STUB ← SPEC RESTART 1 2 SPARE 4 CANC'L

0 1 2 3 4 THICK 5 6 7 8 9
·0 ·1 ·2 ·3 ·4 (M) ·5 ·6 ·7 ·8 ·9
·00 ·05 EXPOSED COVERED RED P.RD GN GY OTHER
F.SL M.SL C.SL VF.SS F.SS M.SS C.SS VC.SS CON PEBBLY
GRAD SHP SMTH SCR TOOL MKS. TR.FOS. MD.CR RIPPLES

NO CEM $CaCO_3$ ROCK DOL FLAT SY.P AS.R PL.X TR.X
LIN DEF CONCR - CO_3 OR STUB FOSSILS NO VERT OTHR TR
NO DIR 2 DIR 1 DIR 000 100 200 300
00 10 20 30 40 50 60 70 80 90
STUB ← SPEC RESTART 1 2 SPARE 4 CANC'L

0 1 2 3 4 THICK 5 6 7 8 9
·0 ·1 ·2 ·3 ·4 (M) ·5 ·6 ·7 ·8 ·9
·00 ·05 EXPOSED COVERED RED P.RD GN GY OTHER
F.SL M.SL C.SL VF.SS F.SS M.SS C.SS VC.SS CON PEBBLY
GRAD SHP SMTH SCR TOOL MKS. TR.FOS. MD.CR RIPPLES

NO CEM $CaCO_3$ ROCK DOL FLAT SY.P AS.R PL.X TR.X
LIN DEF CONCR - CO_3 OR STUB FOSSILS NO VERT OTHR TR
NO DIR 2 DIR 1 DIR 000 100 200 300
00 10 20 30 40 50 60 70 80 90
STUB ← SPEC RESTART 1 2 SPARE 4 CANC'L

0 1 2 3 4 THICK 5 6 7 8 9
·0 ·1 ·2 ·3 ·4 (M) ·5 ·6 ·7 ·8 ·9
·00 ·05 EXPOSED COVERED RED P.RD GN GY OTHER
F.SL M.SL C.SL VF.SS F.SS M.SS C.SS VC.SS CON PEBBLY
GRAD SHP SMTH SCR TOOL MKS. TR.FOS. MD.CR RIPPLES

NO CEM $CaCO_3$ ROCK DOL FLAT SY.P AS.R PL.X TR.X
LIN DEF CONCR - CO_3 OR STUB FOSSILS NO VERT OTHR TR
NO DIR 2 DIR 1 DIR 000 100 200 300
00 10 20 30 40 50 60 70 80 90
STUB ← SPEC RESTART 1 2 SPARE 4 CANC'L

0 1 2 3 4 THICK 5 6 7 8 9
·0 ·1 ·2 ·3 ·4 (M) ·5 ·6 ·7 ·8 ·9
·00 ·05 EXPOSED COVERED RED P.RD GN GY OTHER
F.SL M.SL C.SL VF.SS F.SS M.SS C.SS VC.SS CON PEBBLY
GRAD SHP SMTH SCR TOOL MKS. TR.FOS. MD.CR RIPPLES

NO CEM $CaCO_3$ ROCK DOL FLAT SY.P AS.R PL.X TR.X
LIN DEF CONCR - CO_3 OR STUB FOSSILS NO VERT OTHR TR
NO DIR 2 DIR 1 DIR 000 100 200 300
00 10 20 30 40 50 60 70 80 90
STUB ← SPEC RESTART 1 2 SPARE 4 CANC'L

Fig. 2.10. A form for field recording of stratigraphic sections that can be processed by computer. Heavy horizontal lines define the input for each descriptive unit. Items are entered by making a pencil mark in the dashed line "tracks" (Friend et al., 1976).

which can exploit one of the main advantages of the computer to the full—namely, its ability to carry out complex or repetitive numerical manipulations with great rapidity.

2.3.2 Types of field observation

Subdivision of the section into descriptive units. This is a subjective operation based on the rock types present, the quality and accessibility of the exposure, and the amount of detail required in the description. Very detailed descriptions may require subdivision into units containing (for example) a single mudstone lens or crossbed set, and will therefore be in the order of a few centimeters or tens of centimeters thick. Thicker units can be defined by grouping similar rock types, but sedimentologically useful detail may be lost thereby. For each unit the following observations are made where appropriate.

Lithology and grain size. Lithologic classification of clastic rocks can usually be done satisfactorily by visual observation in the field, without the necessity of follow-up laboratory work. Classification is based on grain size (Table 2.1), which is easily measured on the outcrop. For sand grade rocks it is useful to take into the field a grain size chart (Fig. 2.11), or a set of sand samples each representing one phi class interval through the sand size ranges. These are used for comparison purposes, and permit recognition of the main sand grade subdivisions: very fine, fine, medium, coarse, very coarse. Many tests by the author and others have shown that such observations provide adequate, accurate information on the modal size range of the sandstones. The description should be modified by appropriate adjectives if sorting characteristics require it, for example, pebbly coarse-grained sandstone, silty mudstone, etc. For the purpose of regional facies analysis this is usually the only kind of grain size informtion required. Skewness, kurtosis, and other statistical data are not needed. The use of laboratory grain size analyses is discussed again briefly in section 4.5.1.

Siltstone and mudstone can be distinguished in hand specimen by the presence or absence of a gritty texture, as felt by the fingers or the tongue. This is, of course, a crude method, and should be checked by making thin sections of selected sam-

Table 2.1. Standard grain size scales for carbonate and clastic sediments

Limiting Particle Diameter: m	mm	φ Units	Microns (μm)	CLASTIC ROCKS			CARBONATE ROCKS: Transported Constituents	CARBONATE ROCKS: Authigenic Constituents
				V. Large	Boulders	Gravel	V. Coarse Calcirudite	Extremely Coarsely Crystalline
	2048	−11		Large				
1	1024	−10		Medium				
	512	−9		Small				
	256	−8		Large	Cobbles			
10^{-1}	128	−7		Small				
	64	−6		V. Coarse	Pebbles		Coarse Calcirudite	
	32	−5		Coarse				
	16	−4		Medium			Medium Calcirudite	
10^{-2}	8	−3		Fine				
	4	−2		V. Fine			Fine Calcirudite	Very Coarsely Crystalline
	2	−1		V. Coarse	Sand			
10^{-3}	1	0		Coarse			Coarse Calcarenite	Coarsely Crystalline
	1/2	+1	500	Medium			Medium Calcarenite	
	1/4	+2	250	Fine			Fine Calcarenite	Medium Crystalline
10^{-4}	1/8	+3	125	V. Fine			V. Fine Calcarenite	
	1/16	+4	62	V. Coarse	Silt		Coarse Calcilutite	Finely Crystalline
	1/32	+5	31	Coarse			Medium Calcilutite	
	1/64	+6	16	Medium			Fine Calcilutite	Very Finely Crystalline
10^{-5}	1/128	+7	8	Fine			V. Fine Calcilutite	
	1/256	+8	4	V. Fine				Aphanocrystalline
	1/512	+9	2		Clay/Mud			
10^{-6}								

*(After Friedman and Sanders, 1978; Folk, 1968). Phi (φ) scale is given by $-\log_2$ of particle diameter. Reproduced, with permission, from G.M. Friedman and J.E. Sanders, Principles of Sedimentology, © 1978, John Wiley and Sons, Inc., New York.

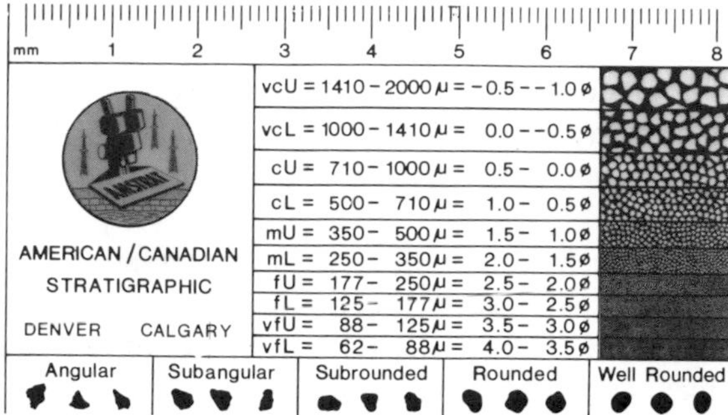

Fig. 2.11. A grain size comparison chart, for use in the field or for logging well cuttings or core (reproduced by permission of American Stratigraphic Company).

ples. However, field identifications of this type commonly are adequate for the purpose of facies analysis.

For conglomerates maximum clast size is often a useful parameter to measure. Typically this is estimated by taking the average of the ten largest clasts visible within a specified region of an outcrop, such as a given area of a certain bedding plane. In thick conglomerate units it may be useful to repeat such measurements over regular vertical intervals of the section. It is also important to note the degree of sorting, clast shape and roundness, matrix content and fabric of conglomerate beds. For example, does the conglomerate consist predominantly of very well-rounded clasts of approximately the same size, or is it composed of angular fragments of varying size and shape (**breccia**)? Do the clasts "float" in abundant matrix, a rock type termed **matrix-supported conglomerate**, or do the clasts rest on each other with minor amounts of matrix filling the interstices—**clast-supported conglomerate**? These features are discussed at length in Chapter 4.

Carbonate rocks commonly cannot be described adequately or accurately in outcrop, and require description from thin sections or polished sections observed under a low-power microscope. Amongst the reasons for this are the ready susceptibility of carbonate rocks to fine-scale diagenetic change, and the fact that weathering behavior in many cases obscures rather than amplifies such changes, as seen in outcrop. Another important reason for not relying on outcrop observation is that some of the types of information required for carbonate facies analysis are simply too small to be seen properly with the naked eye. These include mud content and certain sedimentary textures and biogenic features.

Field geologists traditionally take a dropper bottle of 10% hydrochloric acid with them to test for carbonate content and to aid in distinguishing limestone from dolomite (on the basis of "fizziness"). However, for research purposes the test is quite unsatisfactory and the geologist is advised to abandon the acid bottle (and stop worrying about leakage corroding packsacks). Dolomite commonly can be distinguished from limestone by its yellowish weathering color in the field, but a better field test is to use alizarin red-S in weak acid solution. This reagent stains calcite bright pink but leaves dolomite unstained. Both in hand specimen and in thin section use of this reagent can reveal patterns of dolomitization on a microscopic scale.

Because of the problem with carbonate rocks discussed above the geologist is advised not to rely on field notes for facies analysis of these rocks, but to carry out a rigorous sampling program and supplement (and correct) the field notes using observations made on polished slabs or thin sections. Sampling plans are discussed later in this section. Laboratory techniques for studying carbonates are described by Wilson (1975, Chap. 3).

Evaporites are difficult to study in surface outcrop. They are soft and recessive and commonly poorly exposed, except in arid environments. Like carbonates, they are highly susceptible to diagenetic change, so that field observations must be supplemented by careful laboratory analysis. Only recently have sedimentological methods been applied to the study of evaporites (observations of grain size, textures, bedding, structures, etc.) and this subject offers scope for exciting developments (Chap. 4).

Porosity. Porosity and permeability are of particular interest if the rocks are being studied for their petroleum potential. Observations in surface outcrops may be of questionable value because of the effects of surface weathering on texture and composition, but the geologist should always break off a fresh piece of the rock and examine the fracture surface because such observations commonly constitute the only ones made. Distinguish the various types of porosity such as intergranular (in detrital rocks), intercrystalline (in chemical rocks), and larger pores such as vugs, birdseye texture, moulds of allochems such as oolites or pellets, fossil molds, fracture porosity, etc. More accurate observations may be made in thin section and samples may be submitted to a commercial laboratory for flow tests if required. Porosity types should be reported in terms of the estimated percentage they occupy in the bulk volume of the sample.

Color. Color may or may not be an important parameter in basin analysis. Individual lithologic units may display a very distinctive color which aids in recognition and mapping. Sometimes it even permits a formation to be mapped almost entirely using helicopter observations from the air, with a minimum of ground checking. However, the sedimentological meaning and interpretation of color may be difficult to resolve.

Some colors are easily interpreted—sandstones and conglomerates commonly take on the combined color of their detrital components, pale grays and white for quartzose sediments, darker colors for lithic rocks. As noted above, limestones and dolomites may also be distinguished using color variations. However, color is strongly affected by depositional conditions and diagenesis, particularly the oxidation–reduction balance. Reduced sediments may contain organically derived carbon and Fe^{2+} compounds such as sulfides, imparting green or drab gray colors. Oxidized sediments may be stained various shades of red, yellow or brown by the presence of such Fe^{3+} compounds as hematite and limonite. However, local reducing environments, such as are created around decaying organisms, may create localized areas or spots of reduction color. Color can change shortly after deposition, as shown for example by Walker (1967) and Folk (1976). Moberly and Klein (1976) found that oxidation and bacterial action caused permanent color changes when fresh sediments, such as deep sea cores, are exposed to the air.

Thus the problem is to decide how much time to devote to recording color in the field. Ideally, each descriptive unit in the stratigraphic section should be studied for color using a fresh rock fracture surface and comparisons to some standard color scheme such as the U.S. National Research Council Rock-Color Chart (Goddard et al., 1948). In practice, for the purpose of facies and basin analysis, such precision is not required. Simple verbal descriptions, such as "pale gray", "dark red-brown", etc., are adequate. More precise descriptions may be useful if detailed studies of diagenetic changes are to be undertaken, but recent work has shown that such studies may give misleading results if carried out exclusively on surface exposures, because of the effects of recent weathering (Taylor, 1978).

Bedding. An important type of observation, particularly in clastic rocks, is the thickness of bedding units. Thickness relates to rate of environmental change and to depositional energy. In some cases bed thickness and maximum grain size are correlated, indicating that both are controlled by the capacity and competency of single depositional events. Bed thickness changes may be an important indicator of cyclic changes in the environment and sedimentologists frequently refer to "thinning-upward and fining-upward" or "coarsening- and thickening-upward" cycles. It is important to distinguish bedding from weathering characteristics. For example a unit may split into large blocks or slabs upon weathering, but close examination may reveal faint internal bedding or lamination not emphasized by weathering. Bedding can be measured and recorded numerically, or it can be described in field notes semiquantitatively using the descriptive classification given in Table 2.2.

Table 2.2. Scale of stratification thickness*

Very thickly bedded	Thicker than 1 m
Thickly bedded	30–100 cm
Medium bedded	10–30 cm
Thinly bedded	3–30 cm
Very thinly bedded	1–3 cm
Thickly laminated	0.3–1 cm
Thinly laminated	Thinner than 0.3 cm

*From H. Blatt, G.V. Middleton and R.C. Murray, Origin of Sedimentary Rocks, 2nd edition, © 1980, p. 128. Reprinted by permission of Prentice-Hall, Inc., Englewood Cliffs, New Jersey.

Sedimentary structures produced by hydrodynamic moulding of the bed. Sedimentary structures include a wide variety of primary and post-depositional features. All individually yield useful information regarding depositional or diagenetic events in the rocks, and all should be meticulously recorded and described in the context of the lithology and grain size of the bed in which they occur. The assemblage of structures and, in some cases their orientation, can yield vital paleogeographic information.

Inorganic sedimentary structures can be divided into three main genetic classes, as shown in Table 2.3. These are described briefly and illustrated below in order to aid their recognition in the field, but a discussion of their origin and interpretation is deferred to later chapters. Useful texts on this subject include Allen (1963), Pettijohn and Potter (1964), Michaelis and Dixon (1969), Bouma (1969), Picard and High (1973), Shawa (1974), Lowe (1975), Harms et al., (1975, 1982), Potter and Pettijohn (1977), Hunter (1977), Collinson and Thompson (1982).

Table 2.3. Classification of inorganic sedimentary structures

1. Hydrodynamic molding of the bed
   A. by mass gravity transport
      graded bedding - normal
      - reversed
   B. by noncohesive flow
      lamination
      cross-lamination (amplitude < 5 cm)
      crossbedding (amplitude > 5 cm)
      clast imbrication
      primary current lineation
      fossil orientation
2. Hydrodynamic erosion of the bed
   A. macroscopic
      scours
      channels
      low-relief erosion surfaces
   B. mesoscopic
      intraformational breccias
      hardgrounds
      lag concentrates
      flutes
      tool markings
      rain prints
3. Liquefaction, load and fluid loss structures
      load cast
      flame structures
      ball, pillow or pseudonodule structures
      convolute bedding
      syndepositional faults and slumps
      growth faults
      deformed crossbedding
      dish and pillar structures
      sand volcanoes
      injection features
         dikes
         mud lumps
         diapirs
      synaeresis features
      desiccation cracks
      ice and evaporite crystal casts
      gas bubble escape marks
      teepee structures
      ptygmatic, enterolithic, chicken-wire gypsum/anhydrite

Sediment carried in turbulent suspension by mass gravity transport processes such as debris flows and turbidity currents is subjected to internal sorting processes. When the flow slows and ceases, the sorting may be preserved as a distinct texture termed **graded bedding**. Grading commonly consists of an upward decrease in grain size, as illustrated in Figure 2.12; this is termed **normal grading**. However, certain sedimentary processes result in an upward decrease in grain size, termed **inverse grading**.

Clastic grains can be divided into two classes on the basis of their interactive behavior. **Cohesive grains** are those which are small enough that they tend to be bound by electrostatic forces and thus resist erosion once deposited on a bed. This includes the clay minerals and fine silt particles. A range of erosional sedimentary structures is present in such rocks (Table 2.3), as discussed below. Larger clastic grains, including siliciclastic, evaporite and carbonate fragments, of silt to cobble size, are **non-cohesive**. They are moved by flowing water or wind as a traction carpet along the bed, or by intermittent suspension. The dynamics of movement causes the grains to be

Fig. 2.12. Various types of graded bedding (Pettijohn et al., 1972).

moulded into a variety of **bedforms**, which are preserved as **crossbedding** within the rock.

There are three main classes of bedforms and crossbedding found in ancient rocks:

1. Those formed from unidirectional water currents such as are found in rivers and deltas, and oceanic circulation currents in marine shelves and the deep sea.
2. Those formed by oscillatory water currents, including both wave- and tide-generated features. Although the time scale of current reversal is, of course, quite different, there are comparable features between the structures generated in these different ways.
3. Those formed by air currents. Such currents may be highly variable, and the structure of the resulting deposits will be correspondingly complex. However, examination of ancient wind-formed (eolian) rocks indicates some consistent, surprisingly simple patterns (Chaps. 4, 5).

Recognition in outcrop or in core of the diagnostic features of these crossbedding classes is an invaluable aid to environmental interpretation (as discussed at length in Chapter 4), and therefore crossbedding structures must be examined and described with great care wherever they are found.

The components of crossbedding are illustrated in Figure 2.13. A **foreset** represents an avalanche face, down which grains roll or slump or are swept down by air or water currents. Figure 2.14 shows one of many types of bedform morphology illustrating the way grains advance up the upcurrent (stoss) side of the bedform and are fed to a continuously advancing or prograding downcurrent foreset (lee) surface. In this case the grains are transported by ripples, which themselves contain internal foreset structure. The hydrodynamics of this process are examined further in Chapter 4. Continuous deposition produces repeated foreset bedding or lamination as the bedform accretes laterally, resulting in a crossbed **set** (McKee and Weir, 1953). A **coset** is defined as a sedimentary unit made up of two or more sets of strata or crossbedding separated from other cosets by surfaces of erosion, non-deposition or abrupt change in character (McKee and Weir, 1953). Note that a coset can contain more than one type of bedding.

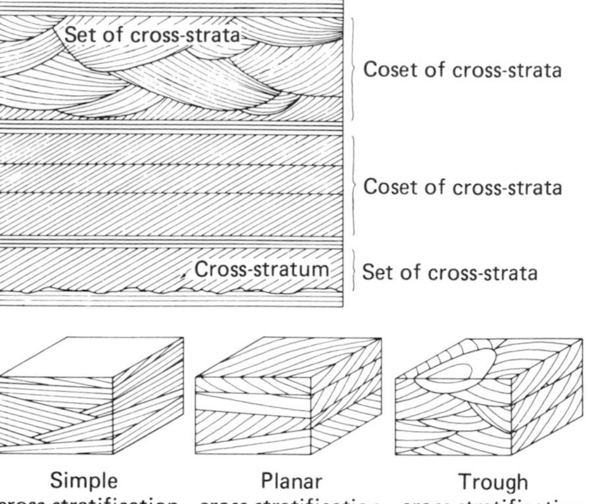

Fig. 2.13. Terminology of crossbedding (Allen, 1963).

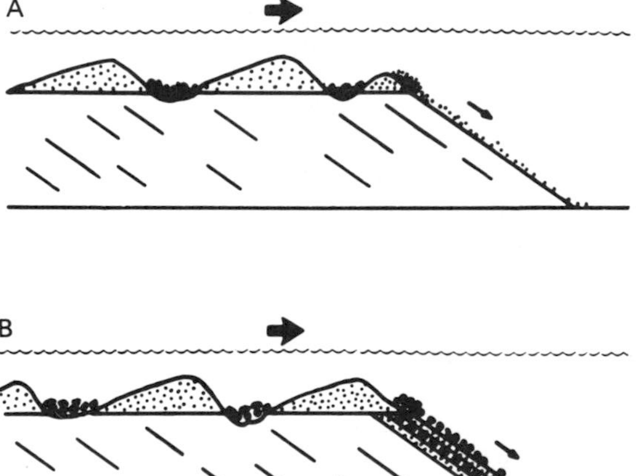

Fig. 2.14. Growth of foresets, advance of a flat-topped bar and development of size sorting by migration of ripples along the bar top (Smith, 1972).

When describing crossbedding attention must be paid to seven attributes, as illustrated in Figures 2.15 and 2.16. All but the last of these were first described in detail in an important paper on crossbedding classification by Allen (1963). In the field crossbeds are classified first according to whether they are solitary or grouped. Solitary sets are bounded by other types of bedding or crossbedding, grouped sets are cosets consisting entirely of one crossbed type. Scale is the next important attribute. In water-laid strata it is found that a bedform amplitude of about 5 cm is of hydrodynamic significance and, accordingly,

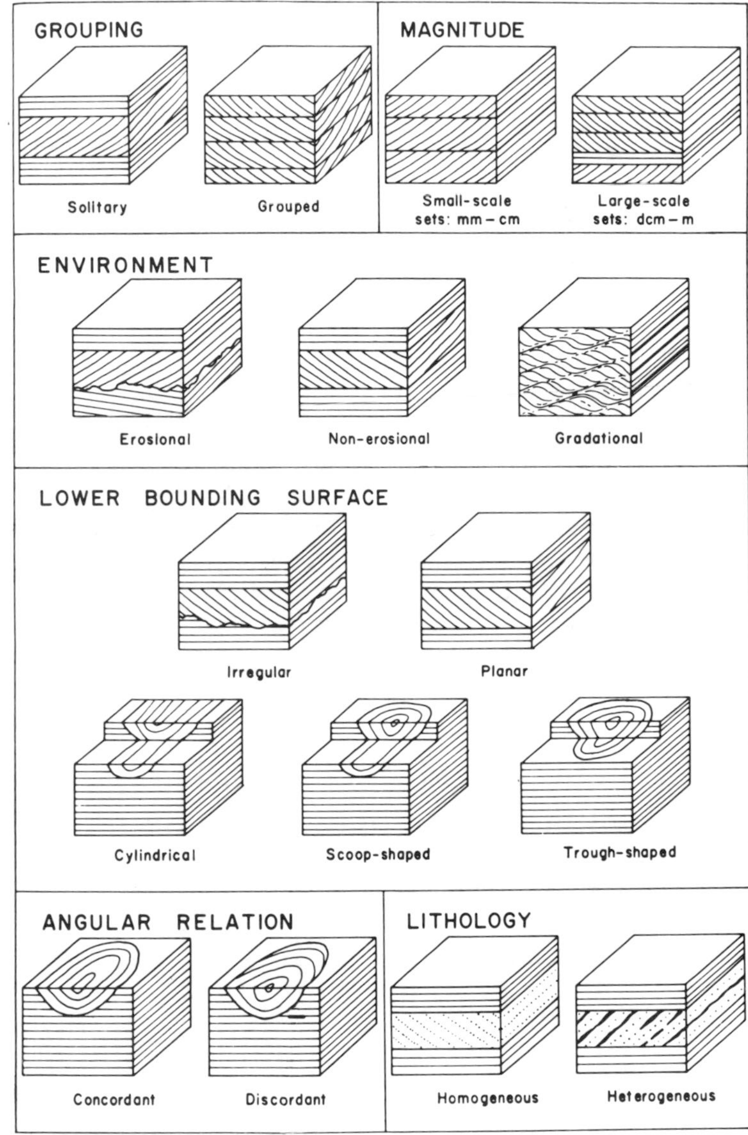

Fig. 2.15. Criteria used in the description and definition of cross-bedding types (Allen, 1963).

this amplitude is used to subdivide crossbeds into small- and large-scale forms. An assumption is made that little or none of the top of a bedform is lost to erosion prior to burial; generally the amount lost seems to increase in approximate proportion to the scale of the bedform or the thickness of the crossbed structure. Forms thinner than 5 cm are termed **ripples**, whereas forms larger than 5 cm are given a variety of names, reflecting in part a diversity of hydrodynamic causes and in part a considerable terminological confusion. This will be discussed further in Chapter 4.

Most crossbed sets contain foresets which terminate at the base of the set, in which case the foresets are said to be discordant. In rare cases where the crossbeds are parallel to the lower bounding surface, as occurs in some sets with curved lower surfaces, the crossbeds are described as concordant.

The crossbeds may show either homogeneous or heterogeneous lithology. Homogeneous crossbeds are those composed of foresets whose mean grain size varies by less than two phi classes. Heterogeneous crossbeds may contain laminae of widely varying grain size, including interbedded sand and mud or sand and gravel, possibly even including carbonaceous lenses.

The minor internal structures within crossbeds are highly diagnostic of origin (Fig. 2.16). The dip

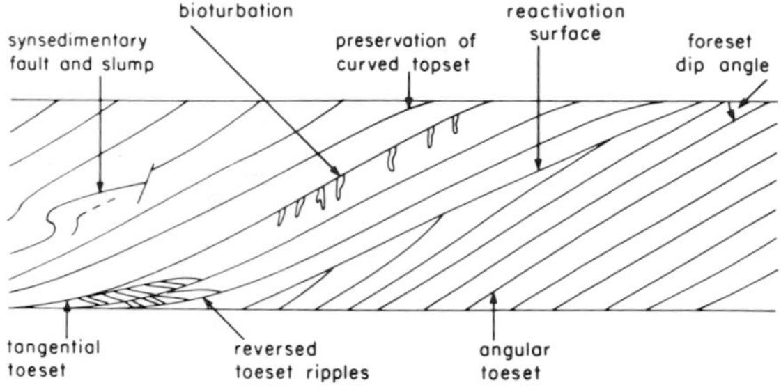

Fig. 2.16. Internal structures of crossbedding.

angle of the foreset relative to the bounding surface is of considerable dynamic significance. Are the foresets curved or linear or irregular in sections parallel to the dip? Is the direction of dip constant or are there wide variations or reversals of dip within a set or between sets? Do the sets contain smaller scale hydrodynamic sedimentary structures on the foresets and, if so, what is the dip orientation of their foresets relative to that of the larger structure? What is the small-scale internal geometry of the foresets—are they tabular, lens or wedge shaped? Do they display other kinds of sedimentary structures such as trace fossils, synsedimentary faults or slumps? Are **reactivation surfaces** present? These represent minor erosion surfaces on bedforms that were abandoned by a decrease in flow strength and then reactivated at some later time (Collinson, 1970).

These seven attributes can be used to classify crossbed sets in the field. It is time-consuming to observe every attribute of every set, but it is usually possible to define a limited range of crossbed types that occur repeatedly within a given stratigraphic unit. These can then be assigned some kind of local unique descriptor, enabling repeated observations to be recorded rapidly in the field notebook. One way to classify crossbed sets is to use a published classification system, of which that erected by Allen (1963) is

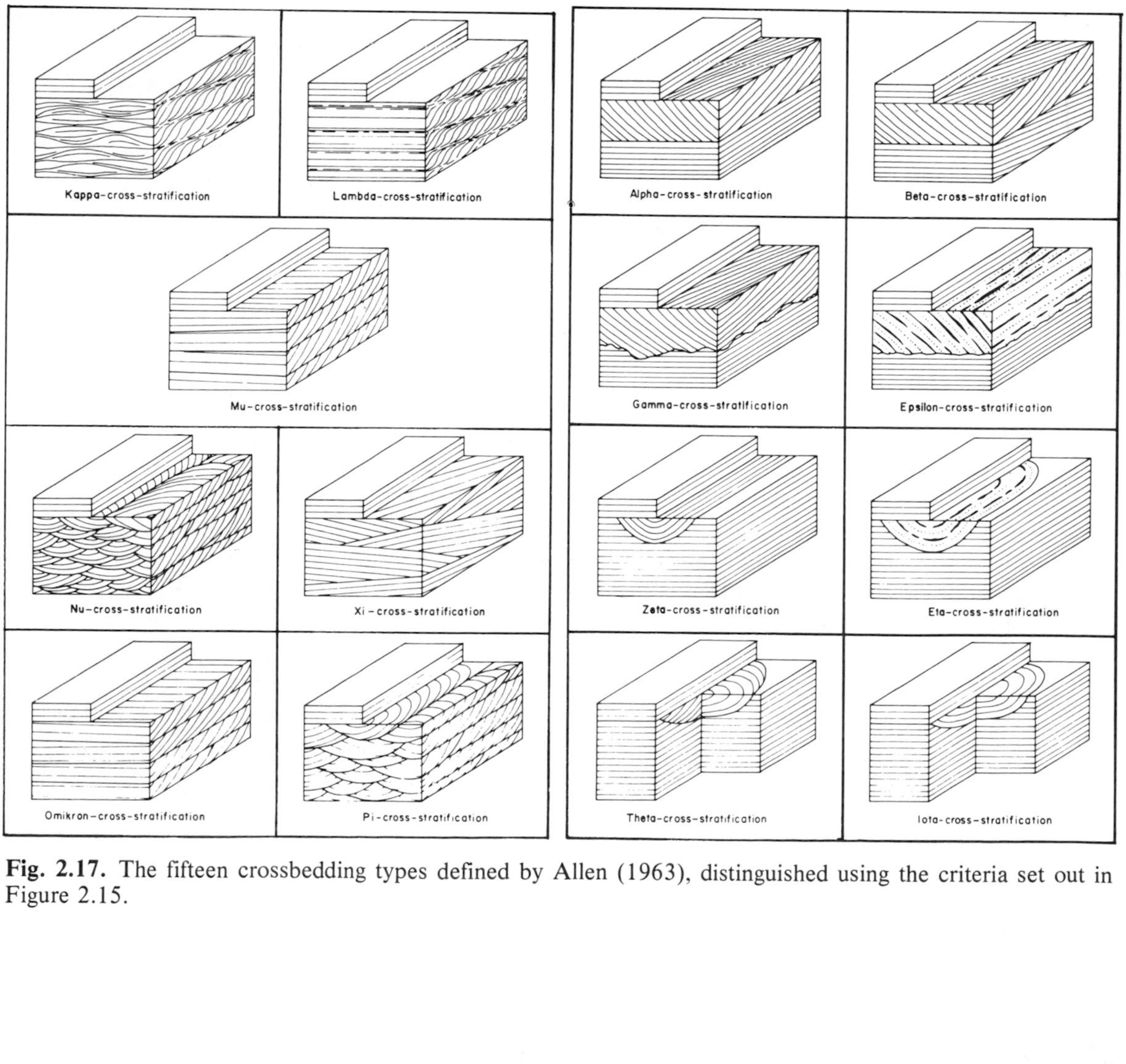

Kappa-cross-stratification
Lambda-cross-stratification
Alpha-cross-stratification
Beta-cross-stratification
Mu-cross-stratification
Gamma-cross-stratification
Epsilon-cross-stratification
Nu-cross-stratification
Xi-cross-stratification
Zeta-cross-stratification
Eta-cross-stratification
Omikron-cross-stratification
Pi-cross-stratification
Theta-cross-stratification
Iota-cross-stratification

Fig. 2.17. The fifteen crossbedding types defined by Allen (1963), distinguished using the criteria set out in Figure 2.15.

the most complete. This is illustrated in Figure 2.17. However, this classification does not encompass any of the internal structures illustrated in Figure 2.16. Allen's classifiction has become widely used, particularly in descriptions of nonmarine rocks; some of his Greek letter names are well known amongst sedimentologists and are used as a convenient shorthand. But care should be taken to supplement this classification with observations of the internal structures. Examples of crossbedding in outcrop are illustrated in Figure 2.18.

A vital component of basin analysis is an investigation of sedimentological trends, such as determining the shape and orientation of porous rock units. **Paleocurrent analysis** is one of several techniques for investigating sedimentary trends based, amongst other things, on studying the size, orientation and relative arrangement of crossbedding structures. Therefore, when describing outcrop sections it is essential to record the orientation of crossbed sets. The procedures for doing this and the methods of interpretation are described in section 5.9.

Crossbedding represents a macroscopic orientation feature, but each clastic grain is individually affected by a flow system and may take up a specific orientation within a deposit in response to flow dynamics. The longest dimension of elongated particles tends to assume a preferred position parallel or perpendicular to the direction of movement, and is commonly inclined upflow, producing an imbricate or shingled fabric. This fabric may be present in sand-sized grains and can be measured optically, in thin section (Martini, 1971), or using bulk properties such as dielectric or acoustic anisotropy (Sippel, 1971). Oriented specimens must be collected in the field for such an analysis. In conglomerates an **imbrication** fabric commonly is visible in outcrop, and can be readily measured by a visual approximation of average orientation or by laborious individual measurements of clasts. Figure 2.19 illustrates imbrication in a modern river bed. As discussed in section 5.9 it has been found that in nonmarine deposits, in which imbrication is most common, the structure is one of the most accurate of paleocurrent indicators (Rust, 1975).

Grain sorting is responsible for generating another type of fabric in sand-grade material, which is also an excellent paleocurrent indicator. This is **primary current lineation**, also termed **parting lineation** because it occurs on bedding plane surfaces of sandstones which are flat-bedded and usually readily split along bedding planes. An example is illustrated in Figure 2.20. Primary current lineation is the product of a specific style of water turbulence above a bed of cohesionless grains, as are the various bedforms which give rise to crossbedding (section 4.5.4). It therefore has a specific hydrodynamic meaning, and is useful in facies analysis as well as paleocurrent analysis.

Rather than the bed itself, objects such as plant fragments, bone or shells may be oriented on a bedding plane. This should be observed if possible, but interpretation commonly is not easy, as discussed in section 5.9.

Sedimentary structures produced by hydrodynamic erosion of the bed. A wide variety of erosional features is produced by water erosion of newly deposited sediment. These result from changes in water level or water energy in response to floods, storms, tides or wind-driven waves and currents. They can also result from evolutionary change in a system under steady equilibrium conditions. Recognition and plotting of these features in outcrop sections is therefore an important component of facies analysis and, with adequate exposure, orientation studies may contribute significantly to trend analysis.

These features range in size up to major **river** and **tidal channels**, **submarine canyons** and **distributary channels** several kilometers across and tens or hundreds of meters deep, but large features such as these can rarely be detected in a single outcrop.They may be visible on seismic sections (Chap. 6), and it may be possible to reconstruct them by careful facies analysis, but this is beyond the scope of our immediate discussion. At the outcrop scale there are two types of erosional features to discuss, those that truncate one or more bedding units and those that scour or pit the bedding plane without significantly disrupting it.

The first type includes **channels**, **scours**, low-relief **erosion surfaces** and **rill markings**, in decreasing order of scale. Typical examples are illustrated in Figure 2.21. These may be classified as macroscopic erosion features (Table 2.3). Channels and scours usually are filled by sediment that is distinctly different in grain size and bedding characteristics from that into which the channel is

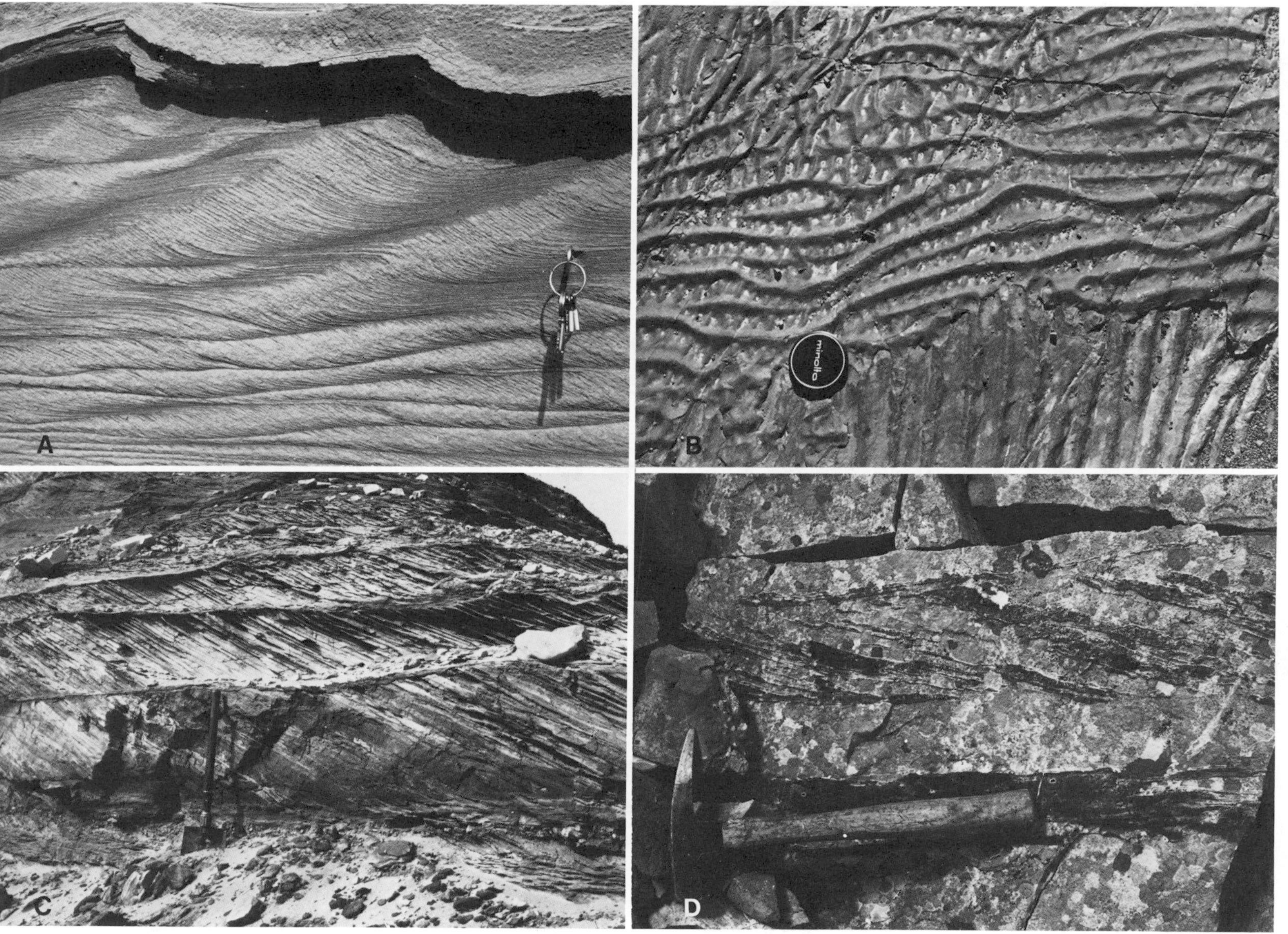

Fig. 2.18 A-D, see p. 29 for legend.

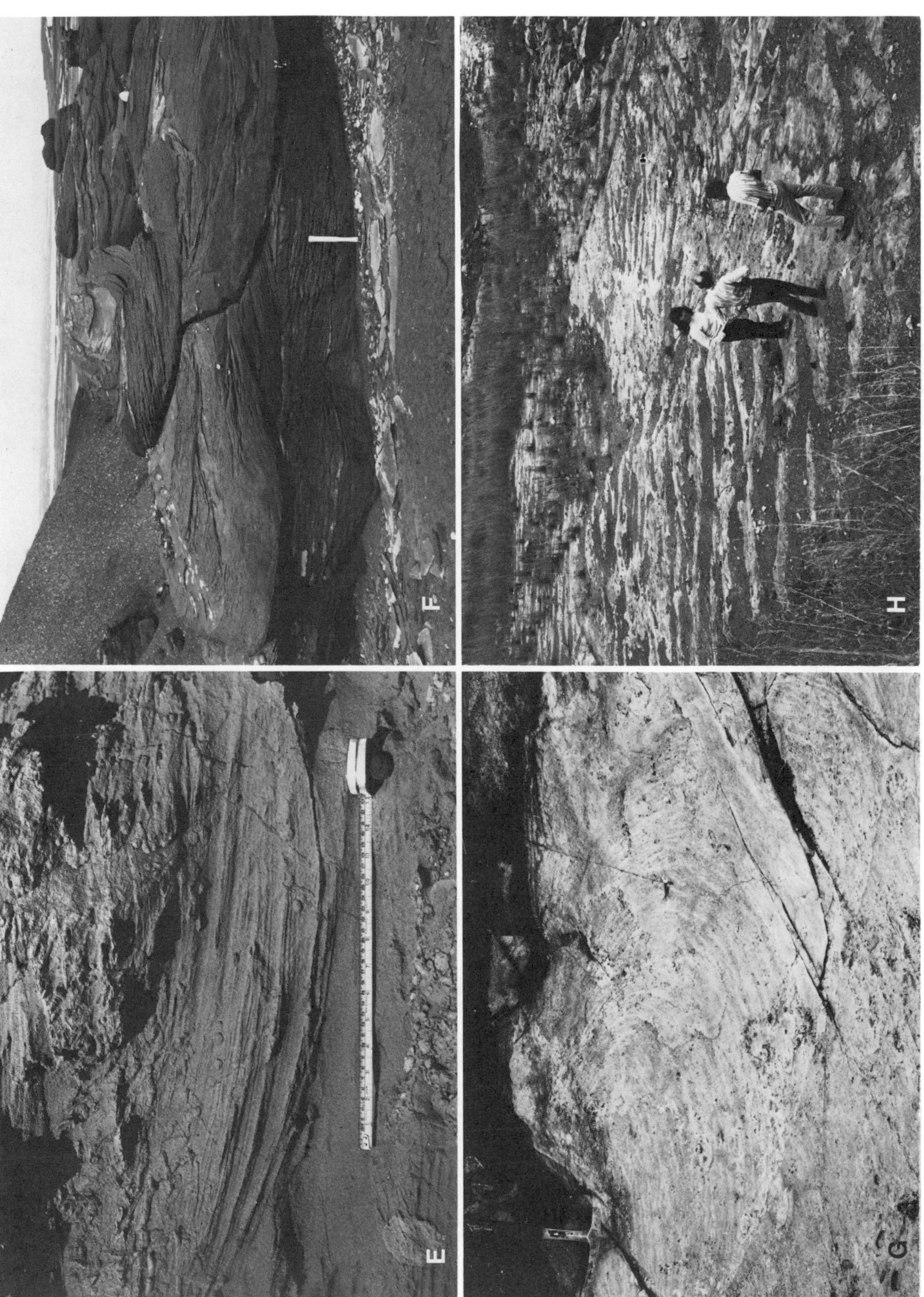

Fig. 2.18 E-H, see p. 29 for legend.

Fig. 2.18 I-L

Fig. 2.18 M,N

Fig. 2.18. Types of crossbedding, as observed in outcrop. **A** Climbing ripple cross-lamination. Note increase in angle of climb, a common feature which indicates decrease in flow strength and increase in detritus settling from suspension. Fluvioglacial, Pleistocene, Ottawa. **B** Ladderback ripples. An older ripple set is visible as short segments of ripple crest preserved in the troughs of a younger set formed in a perpendicular direction. Fluvial, Karoo Supergroup, Beaufort West, South Africa. **C** A coset of planar cross strata (alpha type of Figure 2.17). Note consistency of orientation. Fluvial, Cretaceous, Banks Island, Arctic Canada. **D** Planar crossbed set in a clastic limestone. Foresets are emphasized by chert lenses. Shallow marine, Mississippian, northeast Alaska. **E** Oblique section through a solitary trough crossbed set with trace fossils. Shallow marine, Cretaceous, Banks Island, Arctic Canada. **F** A coset of trough cross strata (pi type of Fig. 2.17). Fluvial, Devonian, Somerset Island, Arctic Canada. **G** Bedding plane view of a trough crossbed. The curved foresets dip toward the left. Fluvial, Huronian, (Proterozoic), near Elliot Lake, Ontario. **H** Bedding plane exposure of straight-crested megaripples. Shallow marine, Karoo Supergroup, near Durban, South Africa. **I** Herringbone crossbedding: five planar crossbed sets showing reversals of flow direction. Shallow marine, Permian, Ellesmere Island, Arctic Canada. **J** Low angle, curved crossbedding, possibly hummocky cross-stratification, with the trace fossil *Ophiomorpha*. Shallow marine, Eocene, Banks Island, Arctic Canada. **K** Large scale crossbedding (epsilon type of Fig. 2.17). Fluvial point bar, Eocene, Ellesmere Island, Arctic Canada. **L** Large scale crossbedding (epsilon type of Fig. 2.17). Fluvial point bar, Carboniferous, Alabama. **M** Large scale eolian crossbedding, Jurassic Navajo Formation, Zion National Park, Utah. **N** Talus slope of a reef, composed of coarse, crinoidal calcarenite and calcirudite, Devonian, Princess Royal Islands, Arctic Canada.

Fig. 2.19. Clast imbrication in a modern river, indicating flow from left to right. Ellesmere Island, Arctic Canada.
◁

◁
Fig. 2.20. **A** Plane bedding in sandstone containing dark, sand sized comminuted carbonaceous debris, Eocene, Ellesmere Island, Arctic Canada. **B** Bedding plane exposure of a sandstone such as that in photo **A**, showing parting lineation, Upper Proterozoic, Banks Island, Arctic Canada.

cut. Almost invariably the channel fill is coarser than the eroded strata indicating, as might be expected, that the generation of the channel was caused by a local increase in energy level. Figure 2.21A illustrates an exception to this, where a channel was abandoned and subsequently filled by fine sediment and coal under low-energy conditions.

It is a common error to confuse trough cross-bedding with channels. Troughs are formed by the migration of trains of dunes or sinuous-crested megaripples. They rest on curved scour surfaces, but these are not channels. The scours are formed by vortex erosion in front of the advancing dunes, and are filled with sediment almost immediately. Channels, on the other hand, may not be filled with sediment for periods ranging from hours to thousands of years after the erosion surface is cut, and so the cutting and filling of the channel are quite separate events.

Erosion surfaces may exhibit little erosional relief, which may belie their importance. In the nonmarine environment, sheet erosion, wind deflation and pedimentation can generate virtually planar erosion sufaces. In subaqueous environments oceanic currents in sediment-starved areas, particularly in abyssal depths, can have the same result. Exposed carbonate terrains may develop **karst** surfaces, with the formation of extensive cave systems. At the outcrop scale careful examination of erosion surfaces may reveal a small-scale relief and the presence of such features as infilled desiccation cracks, basal intraformational or extrabasinal lag gravels (Fig. 2.21E), fissures filled with sediment from the overlying bed, zones of bioclastic debris, etc. In some subaerial environments soil or weathering profiles may have developed, including the development of caliche or calcrete (Fig. 2.21 I, J) and the presence of surfaces of non-deposition. In carbonate environments surfaces of non-deposition commonly develop subaqueously. **Hardgrounds** are organically bored surfaces which may be encrusted with fossils in growth positions. Alternatively they may be discolored by oxidation giving a red stain, or blackened by decayed algal matter. Long-continued winnowing of a surface of non-deposition may leave lag concentrates of larger particles, blackened by algal decay, and possibly including abundant phosphatized fossil material (Fig. 2.21F) (Wilson, 1975, pp. 80–81). In continental slope deposits giant slumps and slides are common and are particularly well exposed as **intraformational truncation surfaces** in deep-water carbonate sediments. An example is illustrated in Figure 2.21K.

It may be difficult to assess the length of time missing at erosion surfaces—some may even represent major time breaks detectable by biostratigraphic zonation. In any case, a careful search for and description of such features in the field is an important part of section description.

Mesoscopic erosional features fall mainly into a class of structure termed **sole markings**. These are features seen on the underside of bedding planes, usually in sandstones, and they represent the natural casts of erosional features cut into the bed below, which is typically siltstone or mudstone. They attest to the erosive power of the depositional event that formed the sandstone bed but, beyond this, most have little facies or environmental significance. However, they can be invaluable paleocurrent indicators.

Flute markings are formed by vortex erosion, typically at the base of turbidity currents. Erosion is deepest at the upcurrent end of the scour and decreases downcurrent, so that in a flute cast the high relief nose of the cast points upcurrent. In rare examples vortex flow lines may be perceived in the walls of the flute (Fig. 2.22A). Flutes generally are in the order of a few centimeters deep.

Tool markings are a class of sole structure formed by erosional impact of large objects entrained in the flow, including pebbles, plant fragments, bone or shell material. The many varieties that have been observed have been assigned names which indicate the interpreted mode of origin. They include **groove-**, **drag-**, **bounce-**, **prod-**, **skip-**, **brush-** and **roll-markings**. A few examples are illustrated in Figure 2.22B, which show the strongly linear pattern on the bed, providing excellent paleocurrent indicators.

Liquefication, load and fluid loss structures. Clay deposits saturated with water are characterized by a property termed *thixotropy*: when subjected to a sudden vibration, such as that generated by an earthquake, they tend to liquefy and lose all internal strength. This behavior is responsible for generating a variety of structures in clastic rocks. Clay beds commonly are interbedded with sand or silt and, when liquefied, the coarser beds have a higher density than the clay

Fig. 2.21 A-D

Fig. 2.21 E,F

Fig. 2.21. Macroscopic erosional features. **A** Abandoned fluvial channel showing levees and fine grained fill, Carboniferous, Kentucky. **B** Tidal channel, filled with calcarenite, cut into finer-grained tidal flat deposits, Carboniferous, Kentucky. **C** A fluvial cutbank, Tertiary, Bylot Island, Arctic Canada. **D** Close-up of a stepped scour surface at the base of a fluvial channel, Permian, southern Poland. **E** Intertidal carbonate mudflat deposits containing several scour surfaces covered by bioclastic debris (above) and a desiccated layer (below) showing incipient brecciation. Compare the latter with photo **G**. Silurian, Somerset Island, Arctic Canada. **F** Discontinuity surface with manganiferous nodules and crusts, Lower Cretaceous, Provence, France. **G** Intraformational breccia produced by break–up and reworking of a desiccated dolomite bed on a tidal flat. Ordovician, Somerset Island, Arctic Canada. **H** Rill marks produced by water seepage from an exposed fluvial bar. Karoo Supergroup, Beaufort West, South Africa. **I** Lowermost bed is a caliche formed by subaerial weathering of a shallow marine carbonate bed. It is followed by a coarse-grained lag deposit formed by a storm (center of view, below overhanging bed). Carboniferous, Kentucky. **J** A caliche breccia, Cretaceous, Provence, France. **K** A slide surface, overlain by a large slumped mass of carbonate sediments. Cretaceous, Provence, France. (Figure 2.21 G-K see p. 34 and 35)

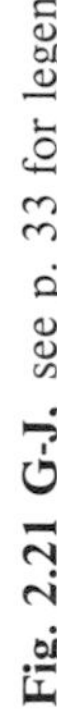

Fig. 2.21 G-J, see p. 33 for legend

Fig. 2.21 K, see p. 33 for legend ▷

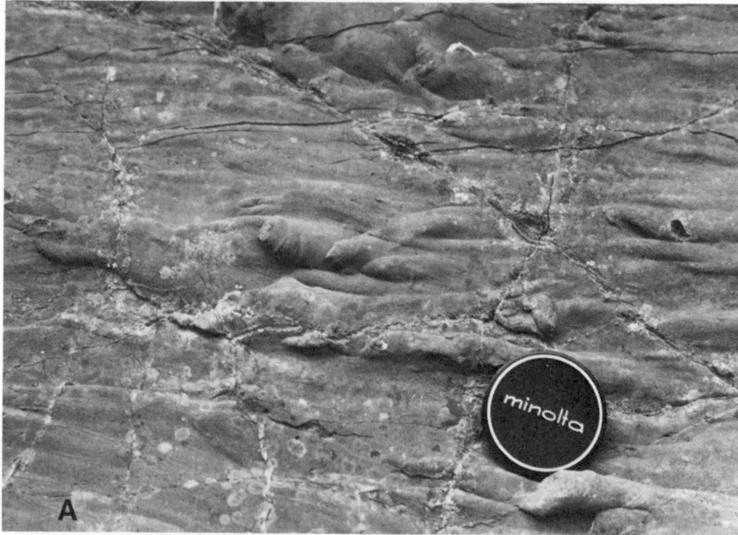

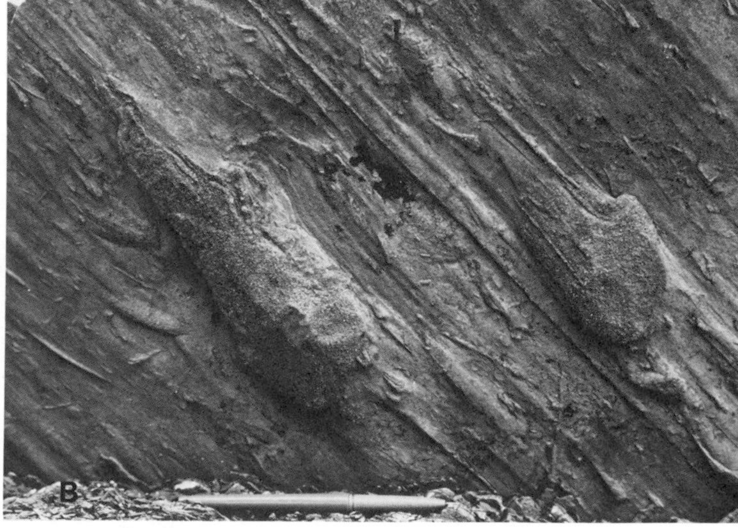

▷

Fig. 2.22. Sole markings **A** Flute markings. Note vortex flow lines on large flute near center of view. Cretaceous, Provence, France.
B Tool markings, mainly groove casts, and load casts. Cretaceous, Provence, France.

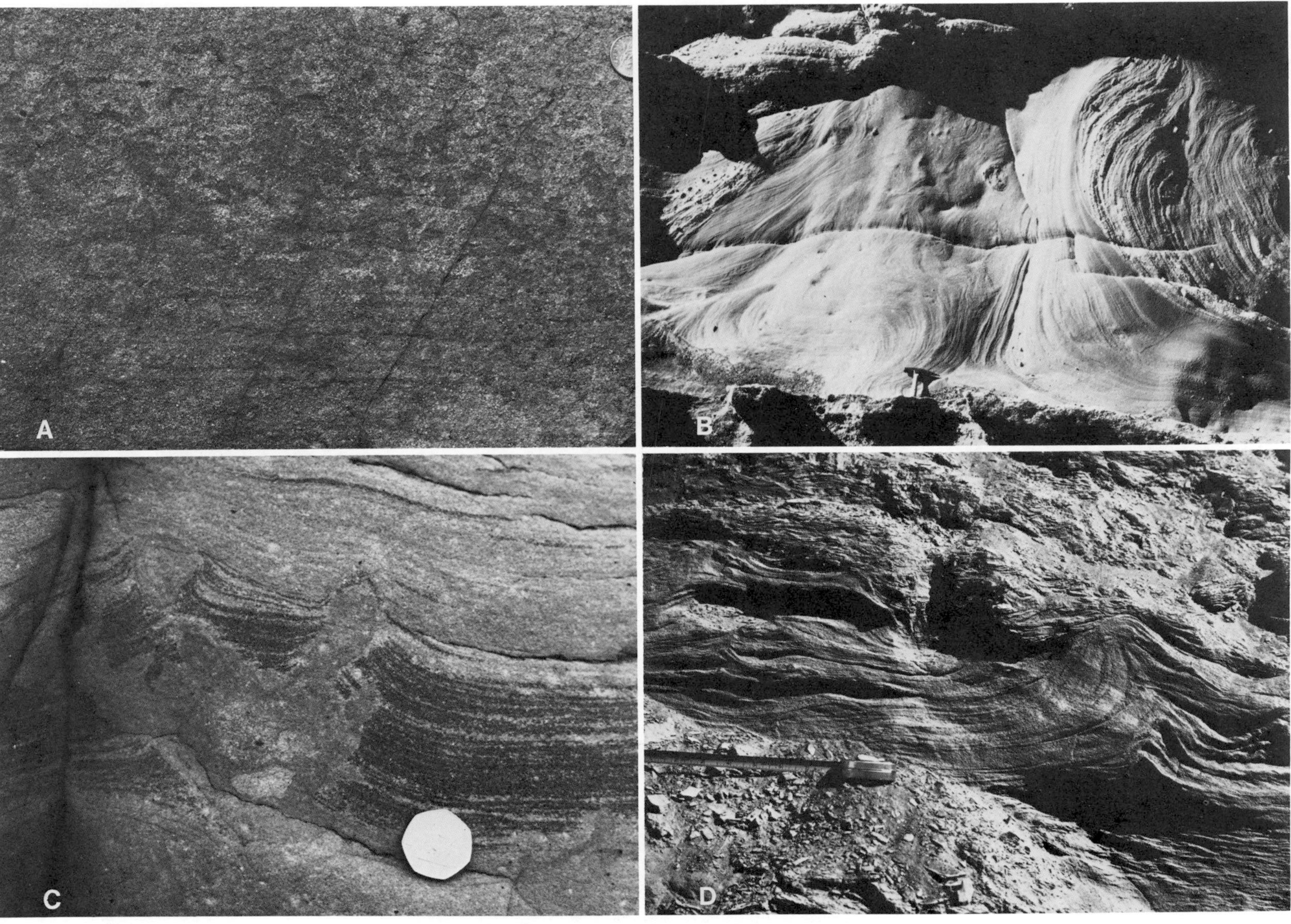

Fig. 2.23 A-D, see p. 39 for legend

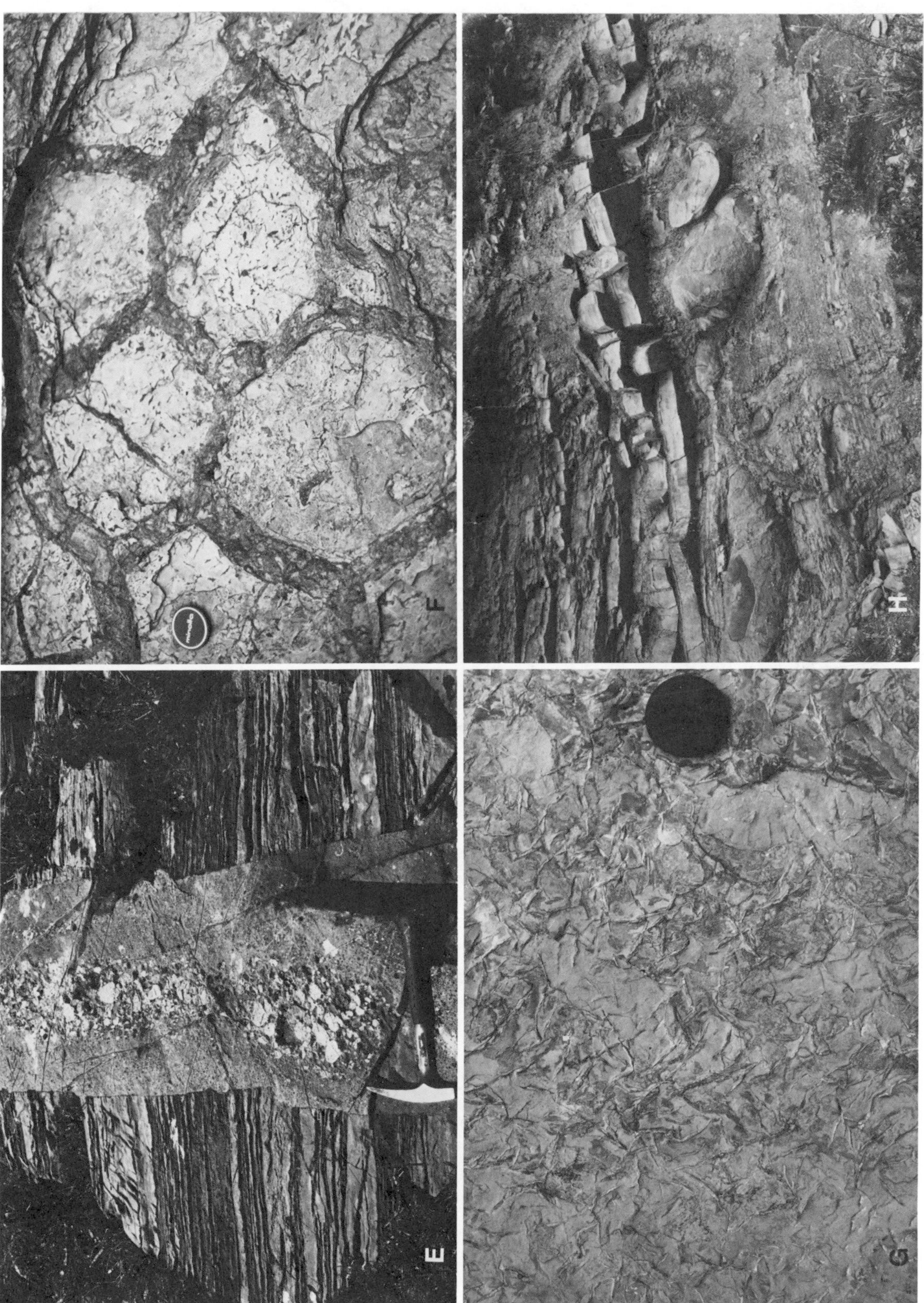

Fig. 2.23 E-H, see p. 39 for legend

Fig. 2.23 I-L

Fig. 2.23. Sedimentary structures produced by liquefaction, load or fluid loss. **A** Dish structures in a grain flow deposit. Cretaceous, Provence, France. **B** Pipe structure produced by penecontemporaneous vertical water escape in a fluvial sandstone. Devonian, northeast Scotland. **C** A set of small water escape pipes in a lacustrine sandstone. Devonian, northeast Scotland. **D** Small sand volcano created by water escape at the depositional surface. Eocene, Banks Island, Arctic Canada. **E** A clastic dyke filled with sandstone and conglomerate. Huronian (Proterozoic), Espanola, Ontario. **F** Desiccation cracks in a lacustrine deposit. Devonian, northeast Scotland. **G** Synaeresis markings. Silurian, Somerset Island, Arctic Canada. **H** Ball and pillow structures in a turbidite unit. Cretaceous, Provence, France. **I** Penecontemporaneous slump produced by failure on a delta front slope, Eocene, Banks Island, Arctic Canada. **J** Load and drape structures caused by presence of massive crinoidal limestone in thin bedded limestones and mudstones. Devonian, Princess Royal Islands, Arctic Canada. **K** Overturned crossbedding, caused by fluid shear of a saturated sandstone. Proterozoic, Banks Island, Arctic Canada. **L** Ice crystal casts in a fluvial floodplain pond. Karoo Supergroup, Beaufort West, South Africa.

and tend to founder under gravity. This may or may not result in the disruption of the sand units. Where complete disruption does not take place the sand forms bulbous shapes projecting into the underlying clay, termed **load structures**. These are usually best seen in ancient rocks by examining the underside of a sandstone bed (Fig.2.22B). They are therefore a class of sole structure, though one produced without current movement. Clay wisps squeezed up between the load masses form pointed shapes termed **flame structures** because of a resemblance to the shape of flames. These are best seen in cross section. Occasionally loading may take place under a moving current, such as a turbidity flow, and load structures may then be stretched, possibly by shear effects, into linear shapes paralleling the direction of movement.

Commonly load masses become completely disrupted. The sand bed may break up into a series of ovate or spheroidal masses which sink into the underlying bed and become surrounded by mud. Lamination in the sand usually is preserved in the form of concave-up folds truncated at the sides or tops of each sand mass, attesting to the fragmentation and sinking of the original layer. Various names have been given to these features, including **ball-**, **pillow-**, or **pseudo-nodule-structures** (Fig. 2.23H). These structures rarely have a preferred shape or orientation and are not to be confused with slump structures, which are produced primarily by lateral rather than vertical movement.

In many environments sediment is deposited on a sloping rather than a flat surface; for example, the subaqueous front of a delta, which is something like a very large scale foreset built into a standing body of water. The difference is that once deposited such material usually does not move again as individual grains because the angle of the slope is too low. Typically, large scale submarine fans and deltas exhibit slopes of less than 2°. However, the sediment in such environments is water saturated and has little cohesive strength. Slopes may therefore become oversteepened, and masses of material may be induced to slump and slide downslope by shock-induced failure. Undoubtedly the thixotropic effects described above facilitate this process. The result is the production of internal shear or glide surfaces and deformed masses of sediment, termed **slump structures**. Failure surfaces may be preserved as **syn-**

depositional faults. Small-scale examples of these features are commonly seen in outcrop, as illustrated in Figure 2.23I and J. Some examples of **convolute bedding** may be produced this way, although others are the result of water escape, as discussed below.

Structures produced by failure and lateral movement commonly retain an internal orientation with a simple geometric relationship to the orientation of the depositional slope. This could include the elongation of slump masses and the orientation (strike) of slide surfaces, both parallel to depositional strike, or the asymmetry, even overturning, of folds in convolute beds. Recognition of these geometric properties in outcrop is important because it helps distinguish the structures from those of different origin, and orientation characteristics obviously have potential as paleoslope indicators.

Very large-scale slumps and syndepositional faults are developed on major deltas. The latter are termed **growth faults** (Chaps. 6, 9). **Olisthostromes** are giant slumps developed on tectonically active continental slopes (Chap. 9).

Deformed or **overturned crossbedding** (Fig. 2.23K) is developed in saturated sand beds by the shearing action of water or turbid flow across the top of the bedform. The upper few centimeters of the crossbedded unit move downcurrent by a process of intragranular shear, and foreset lamination is overturned as a result, producing an upcurrent dip. Obviously, to produce this structure the shearing current must have a similar orientation to that of the current which generated the crossbedding. Deformed crossbedding is common in fluvial and deltaic environments.

As additional sediment is laid on top of saturated deposits, grains within the substrate begin to settle and pack more tightly. Pore waters are expelled in this process, and move upwards or laterally to regions of lower hydrostatic pressure. Eventually they may escape to the surface. This process may take place slowly if sediment is being deposited grain by grain, and the fluid movement leaves little or no impression on the sediments. However, if loading is rapid a much more energetic process of fluid loss takes place, and the sediment itself will be moved around in the process. The result, in sand-grade deposits, is a group of features called **dish** and **pillar structures** (Fig. 2.23A). Dishes are produced by escaping water breaking upward through a lamination and turning up the edges; pillars record the vertical path of water flow moving to the surface. Dishes may be up to 50 cm in diameter. These structures are particularly common in the deposits of sediment gravity flows, such as fluidized flows, in which sediment emplacement is rapid. They are produced by water escape as a flow ceases movement—the loss of the lubricating effect of water itself being one of the main reasons why the flow stops. However, dish and pillar structures have also been observed in fluvial and other deposits, and are therefore not environmentally diagnostic. Obviously they can only be seen in deposits containing lamination, and will not be present if the sand is uniform in texture.

Fluid movement within a bed is an additional cause of convolute lamination. In this case the laminations are folded by internal shear, and may occasionally be broken through by pipes (Fig. 2.23B,C). At the sediment–water interface, escaping water may bubble out as a small spring, building up a miniature **sand volcano** (Fig. 2.23D).

The emplacement of relatively more dense material over a lower density layer was discussed above as the cause of load casts. In addition to the downward movement of the denser material, this situation can be the cause of upward movement of lighter sediment, which is injected, together with contained pore fluids, into the overlying rocks. This can occur on a large scale, producing **diapirs** of evaporite or mud, both of which materials flow readily under overburden weight of a few hundred metres of sediment. These diapirs may be several kilometers across and may extend up for several kilometers through overlying deposits. Evaporite diapirs commonly develop on continental margins (Chaps. 6, 9). Mud diapirs are a characteristic feature of deltas, where coarser deltaic sediment is dumped rapidly on marine mud deposits.

On a smaller scale the same injection process can generate **clastic dykes**, consisting of sheets of sandstone or conglomerate (siliciclastic or carbonate) cutting through overlying or underlying beds (Fig. 2.23E). The host rocks usually are sharply truncated and not internally deformed, indicating that they were at least partially lithified prior to intrusion. Some dykes intrude along fault planes. Some are intensely folded, suggesting deformation by compaction and further dewatering after injection.

Desiccation cracks are readily recognized by

even the untrained eye (Fig. 2.23F). They are one of the best and most common indicators of subaerial exposure in the rock record. They may penetrate as deep as a few meters into the underlying rocks (although a few centimeters is more typical), and normally are filled by sediment from the overlying bed. **Teepee structures** are a variety of large desiccation crack caused by limestone or evaporite expansion on tidal flats. Desiccated beds on tidal flats may peel or curl as they dry, and disrupted fragments commonly are redeposited nearby as an **intraclast breccia** (Fig. 2.21G). A subtly different kind of shrinkage feature has been recognized in recent years, termed a **synaeresis structure** (Fig. 2.23G; Donovan and Foster, 1972). These are formed by volume changes in response to salinity variations in the ambient waters, and superficially resemble desiccation cracks. They may be distinguished by two principal differences: unlike desiccation cracks they do not normally form continuous polygonal networks across bedding planes, but may appear as a loose assemblage of small worm-like relief markings on a bedding surface; secondly, they do not show deep penetration into the substrate, but appear to rest on the bedding plane in which they are found. Synaeresis structures are formed subaqueously, and are common in such environments as lakes, lagoons and tidal pools, where salinity changes may be frequent.

Evaporation and freezing may cause the development of crystals of evaporite salts and water, respectively, on the depositional surface, particularly on alluvial floodplains, supratidal flats and lake margins. Evaporite crystals and nodules may be preserved into the rock record and, of course, major evaporite deposits are common; but individual crystals commonly are replaced by pseudomorphs, or are dissolved and the cavity filled with silt or sand, forming a **crystal cast**. Such structures are a useful indicator of subaerial exposure and desiccation, but do not necessarily imply long term aridity. Gypsum and halite are the two most common minerals to leave such traces. Gypsum forms blade-shaped casts and halite characteristic cubic or "hopper-shaped" structures. Ice casts may be formed in soft sediments during periods of freezing, but have a low preservation potential (Fig. 2.23L).

Particularly distinctive evaporite structures on supratidal (sabkha) flats are termed **ptygmatic**, **enterolithic** and **chicken- wire** structures. These are caused by in-place crystal growth, expansion and consequent compression of evaporite nodules, possibly aided by slight overburden pressures. Enterolithic structure is so named for a resemblance to intestines; chicken wire structure is caused by squeezing of carbonate films between the nodules; ptygmatic folds may be caused largely by overburden pressures (Maiklem et al., 1969).

Lastly **gas bubble escape structures** should be mentioned. These are produced by carbon dioxide, hydrogen sulfide or methane escaping from buried, decaying organic matter. Gas passes up through wet, unconsolidated sediment and forms bubbles at the sediment–water interface, leaving small pits on the bedding surface. These structures have been confused with **rain imprints**, but form subaqueously, as may be apparent from associated features preserved in the rocks. True rain imprints probably are very rare.

Fossils. Body fossils are obviously amongst the most powerful environmental indicators to be found in sedimentary rocks, and should be observed and identified with care. Paleontology and paleoecology are specialized subjects, a detailed discussion of which is beyond the scope of this book. However, those engaged in describing outcrop sections for basin analysis purposes should be able to make use of such information as they can gather. A complete and thorough paleontological–paleoecological examination of an outcrop section may take several hours, days or even weeks of work, involving the systematic examination of loose talus and breaking open fresh material or sieving unconsolidated sediment in the search for a complete suite of fossil types. Extensive suites of palynomorphs or microfossils may be extracted by laboratory processing of field samples. Many apparently "unfossiliferous" or "sparsely fossiliferous" stratigraphic intervals have been found to contain a rich and varied fauna or flora by work of this kind, but it is the kind of research for which few basin analysts have the time or inclination.

We are concerned in this book with reconstructing depositional environments and paleogeography. Fossils can be preserved in three different ways which yield useful environmental information:

1. In-place life assemblages. These include invertebrate forms attached to the sea bottom such

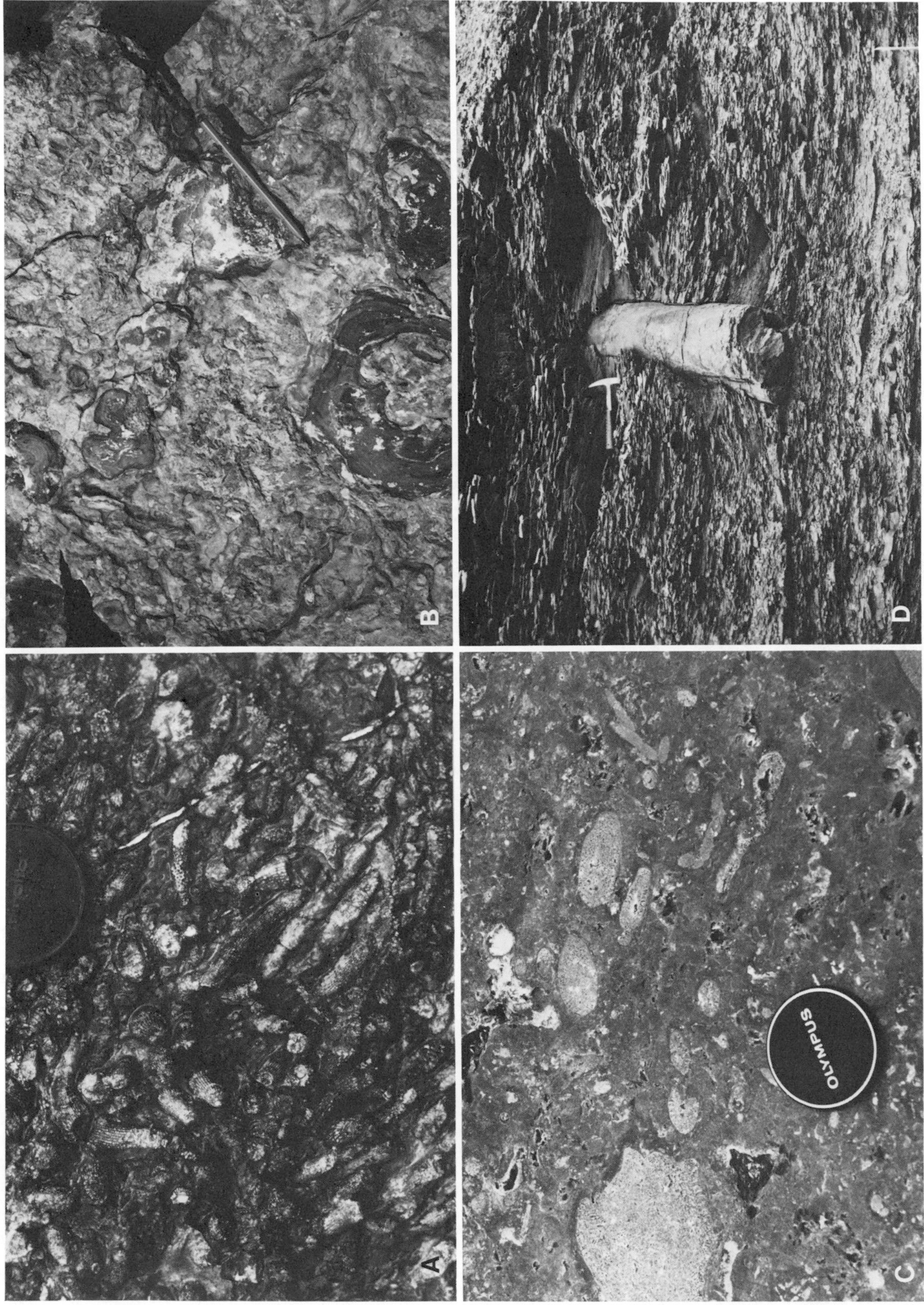
B
D
A
C
OLYMPUS

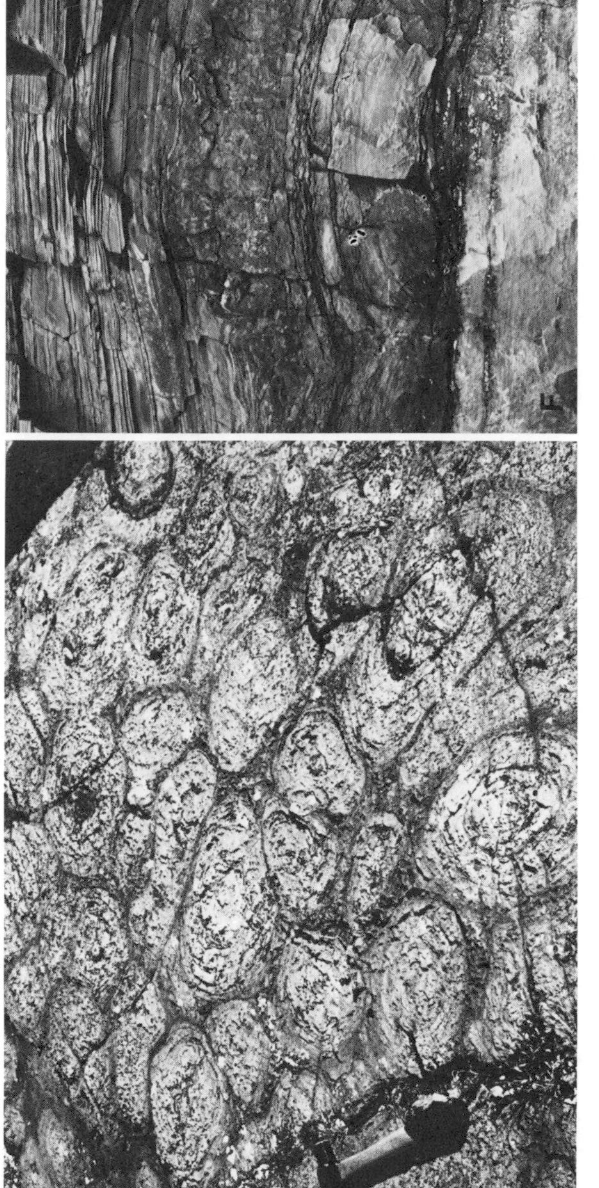

Fig. 2.24. In-place fossils (life assemblages). **A** Biostrome containing Disphyllid and Thamnopora corals, Devonian, N. Spain. **B** Coral biostrome, Devonian, near Norman Wells, N.W.T., Canada. **C** Typical reef facies. Boundstone composed of calcisponges and encrusting stromatolites, Capitan reef, Permian, Guadalupe Mountains, Texas. **D** Tree in growth position, Pennsylvanian, Kentucky. **E** Domal stromatolites, bedding plane exposure, Proterozoic, Dismal Lakes, N.W.T., Canada. **F** Domal stromatolites capped by finger stromatolites, Silurian, Somerset Island, N.W.T., Canada. Photos **A** and **C** courtesy of B. Pratt.

as corals, archaeocyathids, rudists, some brachiopods and pelecypods in growth positions, some bryozoa, stromatoporoids, stromatolites, trees. In-place preservation usually is easy to recognize by the upright position of the fossil and presence of roots, if originally part of the organism. This type of preservation is the easiest to interpret because it permits the drawing of close analogies with similar modern forms, in the knowledge that the fauna or flora almost certainly is an accurate indicator of the environment in which the rocks now enclosing it were formed. Examples are illustrated in Figure 2.24.

2. Almost as good environmental indicators as (1) are examples of soft-bodied or delicately articulated body fossils preserved intact (Fig. 2.25). These indicate very little transport or agitation after death, and preservation in quiet waters such as shallow lakes, lagoons, abandoned river channels or deep oceans. The Cambrian Burgess Shale in the Rocky Mountains near Field, British Columbia, contains one of the most famous examples of such a fossil assemblage, including impressions of soft, non-skeletal parts of many organisms, such as sponges, that are not found anywhere else. The Jurassic Solnhofen Limestone of Germany is another good example.

The bones of vertebrate animals tend to disarticulate after death, due to decay of muscle and cartilage and the destructive effects of predators. Nevertheless, entire skeletons of bony fish, reptiles and mammals are commonly found in certain rock units, indicating rapid burial under quiet conditions. The Solnhofen Limestone is a well-known example, containing, amongst other fossils, the entire skeleton and feather impressions of a primitive bird. Such fossil assemblages must, nevertheless, be interpreted with care because a limited amount of transportation is possible from the life environment to the site of eventual burial. Presumably the bird of the Solnhofen Limestone lived in the air, not at the sediment water-interface where it was deposited!

3. Much more common than any of the foregoing are death assemblages of fossils which may have been transported a significant distance—perhaps many miles—from their life environment. These commonly occur as lag concentrates of shelly debris such as gastropod, pelecypod, brachiopod or trilobite fragments, fish bones or scales, and so on (Fig. 2.26). Such concentrations may be abundant enough to be locally rock-forming, for example, the famous Silurian Ludlow Bone Bed of the Welsh borderlands. They normally indicate a channel-floor lag concentration or the product of wave winnowing, and can usually be recognized readily by the fact that fragment grain size tends to be relatively uniform.

Transported body fossils may not occur as concentrations but as scattered, individual occurrences, in which case each find must be interpreted with care. Did it live where it is now found or was it transported a significnt distance following death? The environmental deduction resulting from such an analysis may be quite different depending on which interpretation is chosen. Evidence of transportation may be obvious and should be sought, for example broken or abraded fossils may have traveled significant distances. Overturned corals, rolled stromatolites (including oncolites), uprooted tree trunks, are all obviously transported.

These problems are particularly acute in the case of microfossils and palynomorphs which, on account of their size, are particularly susceptible to transportation long distances from their life environment. For example modern foraminiferal tests are blown tens of kilometers across the supratidal desert flats of India and Arabia, and shallow-water marine forms are commonly carried into the deep sea by sediment gravity flows such as turbidity currents. Environmental interpretations based on such fossil types may therefore be very difficult, though it may still be possible if the analysis is carried out in conjunction with the examination of other sedimentary features. In fact it was the occurrence of sandstones containing shallow-water foraminifera interbedded with mudstones containing deep-water forms in the Cenozoic of the Ventura Basin, California, that was one of the principal clues leading to the development of the turbidity current theory for the origin of deep-water sandstones.

Because of the great variety of life forms preserved as fossils the subject of paleoecology is a large and complex one. Detailed studies are for the specialist and a complete treatment is beyond the scope of this book. The reader is referred to such texts as Johnson (1960), Fagerstrom (1964), Imbrie and Newell (1964), Heckel (1972), Schafer (1972), Hallam (1973), McKerrow (1978), Dodd and Stanton (1981). A few examples are discussed in Chapter 4.

Biogenic sedimentary structures. Footprints, burrows, resting, crawling or grazing trails and escape burrows are abundant in some rock units, particularly shallow marine deposits; all may yield useful environmental information, including water depth, rate of sedimentation, degree of agitation, and so on. Even the nondescript structure "bioturbation", which is ubiquitous in many shallow marine rocks, can be interpreted usefully by the sedimentologist. Footprints are, of course, best seen on bedding plane surfaces, as are many types of feeding trail and crawling trace (Fig. 2.27). Burrows are better examined in vertical cross section, and are most visible either in very clean, fresh, wetted rock surfaces or in wind- or water-etched weathered outcrops (Fig. 2.28).

Stromatolites are a distinctive component of many carbonate successions, particularly those of Precambrian age, and have been the subject of much detailed study.

Information to record on trace fossils includes morphology, size, attitude, abundance, internal structures and lithofacies associations. Farrow (1975) provided a useful review of field and laboratory techniques for studying trace fossils. Table 2.4 is a classification of some common in-

Table 2.4. Classification of trace fossils*

Rank IV	Rank III	Rank II	Rank I	Examples
A. Track-like trace on bedding plane		a. "Prods" or "scratches"; all alike	(1) Clustered "scratches"	*Paleohelcura*
			(2) Rows of "prods"	*Tasmanadia*
		b. "Prods" or "scratches" of different kinds	Rows of "prods"	*Kouphichnium*
B. Trail-like trace on bedding plane	1. Freely winding	a. Simple trail	(1) No ornament	*Gordia*
			(2) Transverse ornament	*Climactichnus*
		b. Bilobed trail	Transverse ornament	*Cruziana*
		c. Trilobed trail	Transverse ornament	*Scolicia*
	2. Windings in contact with one another; pattern on bedding plane	a. Simple trail	No ornament	*Helminthoidea*
		b. Bilobed trail	Transverse ornament	*Nereites*
C. Radially symmetrical in a horizontal plane	1. Without axial vertical structure	a. Five-rayed	Rays are grooved	*Asteriacites*
		b. Multirayed	Club-shaped rays	*Asterosoma*
	2. With vertical axial structure	a. Circular outline	Conical depression	*Histioderma*
		b. Multirayed	Radial branches	*Lennea*
D. Tunnels and shafts	1. Of uniform diameter	a. Vertical	(1) Isolated	*Tigillites*
			(2) *En masse*	*Skolithos*
		b. Horizontal	Winding	*Planolites*
		c. U-shaped		*Arenicolites*
		d. Regularly branching		*Chondrites*
	2. Variable diameter	Irregular network		*Thalassinoides*
E. Forms having a spreite		a. U-shaped	(1) Vertical plane	*Diplocraterion*
			(2) Horizontal plane	*Rhizocorallium*
		b. Spiral	Inclined plane	*Zoophycos*
		c. Branched	Vertical	*Phycodes*
F. Pouch shaped		a. Smooth surface		*Pelecypodichnus*
		b. Transverse ornament		*Rusophycus*
G. Miscellaneous		Net pattern		*Palaeodictyon*

*From Simpson, 1975.

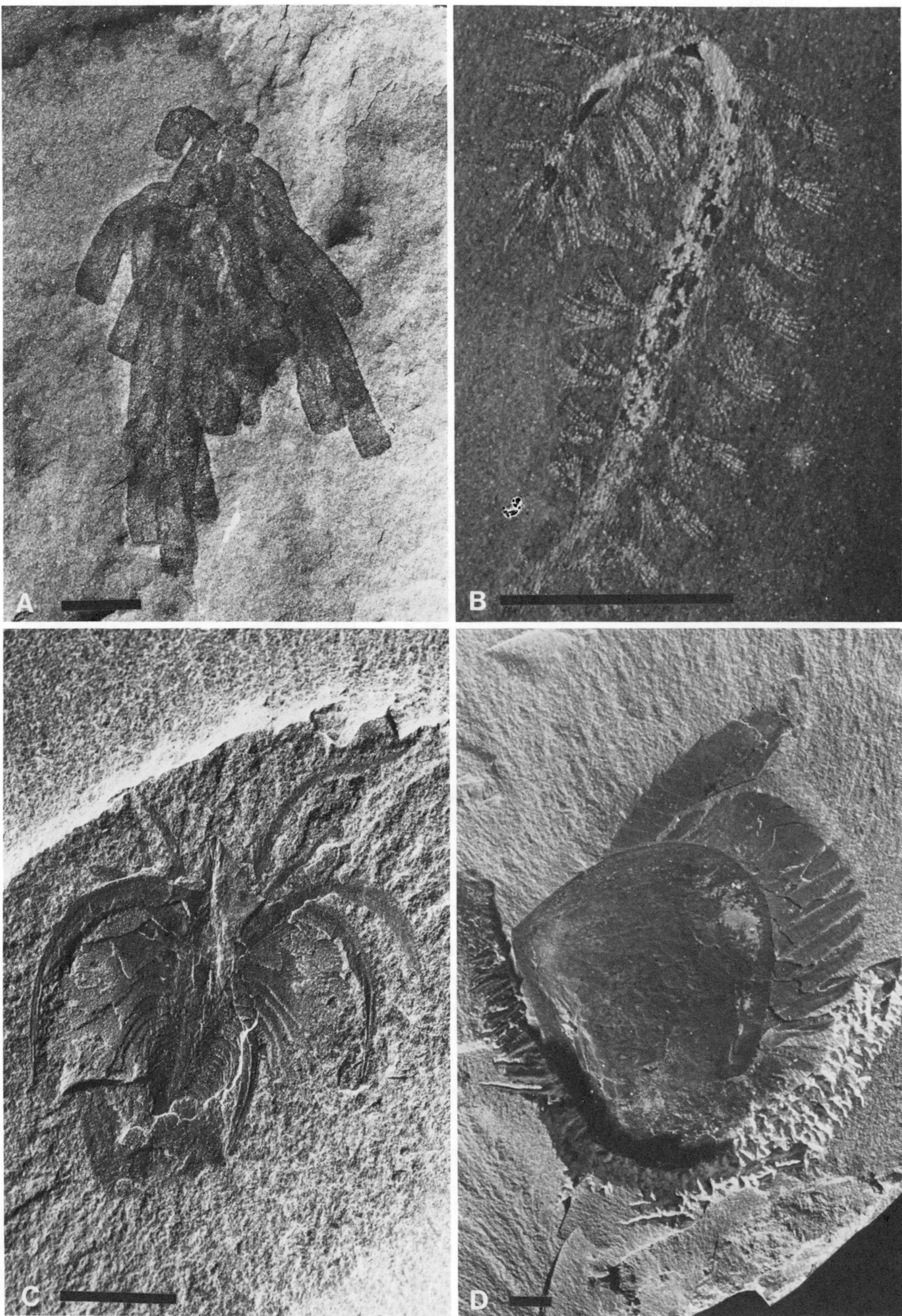

Fig. 2.25. Preservation of delicate structures, including soft parts. **A–D** are from the Middle Cambrian Burgess Shale, Yoho National Park, British Columbia; **E–F** are from the Upper Jurassic Solnhofen Limestone, Bayern, Germany. Bar scales are all 5mm in length. **A** *Vauxia gracilenta*, Walcott, a sponge;

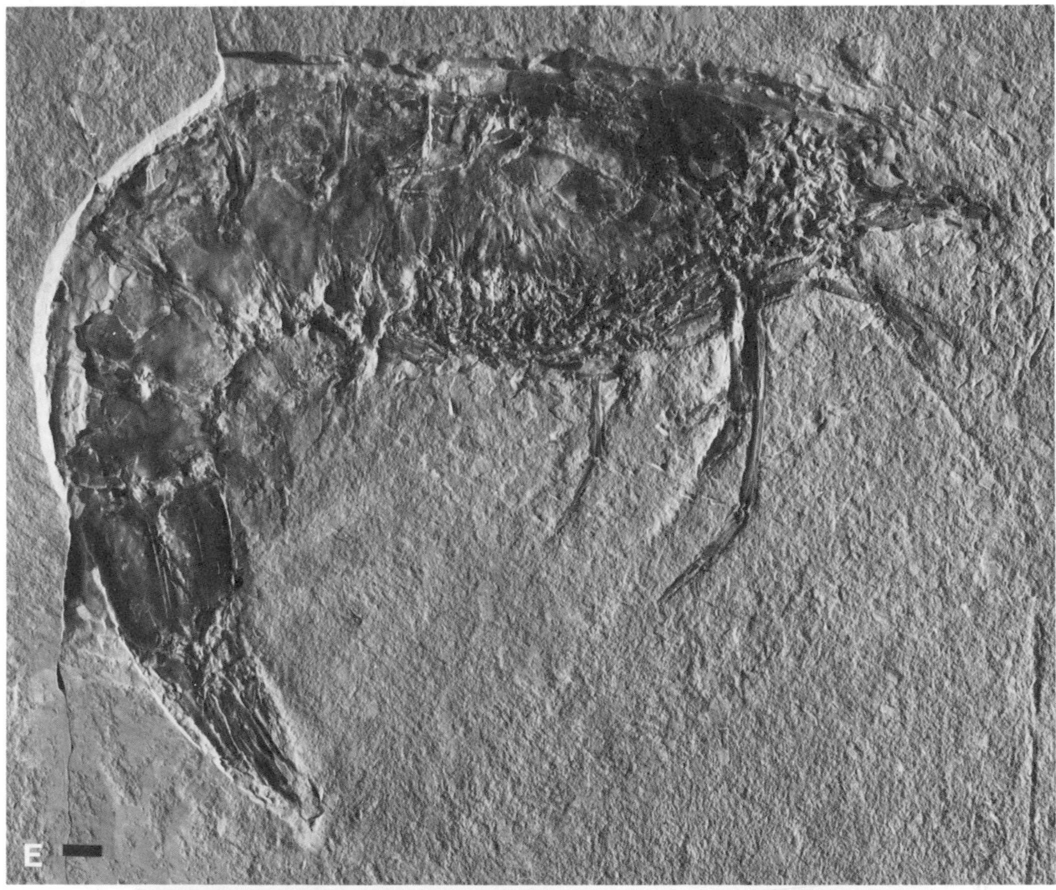

B *Burgessochaeta setigera* (Walcott), a polychaete worm; **C** *Marrella splendens* Walcott, an arthropod; **D** *Canadaspis perfecta* (Walcott), a crustacean; **E** *Antrimpos speciosus* Munster, a shrimp; **F** *Aeschnogomphus intermedius*, a dragonfly. All photos courtesy of D. Rudkin and the Royal Ontario Museum.

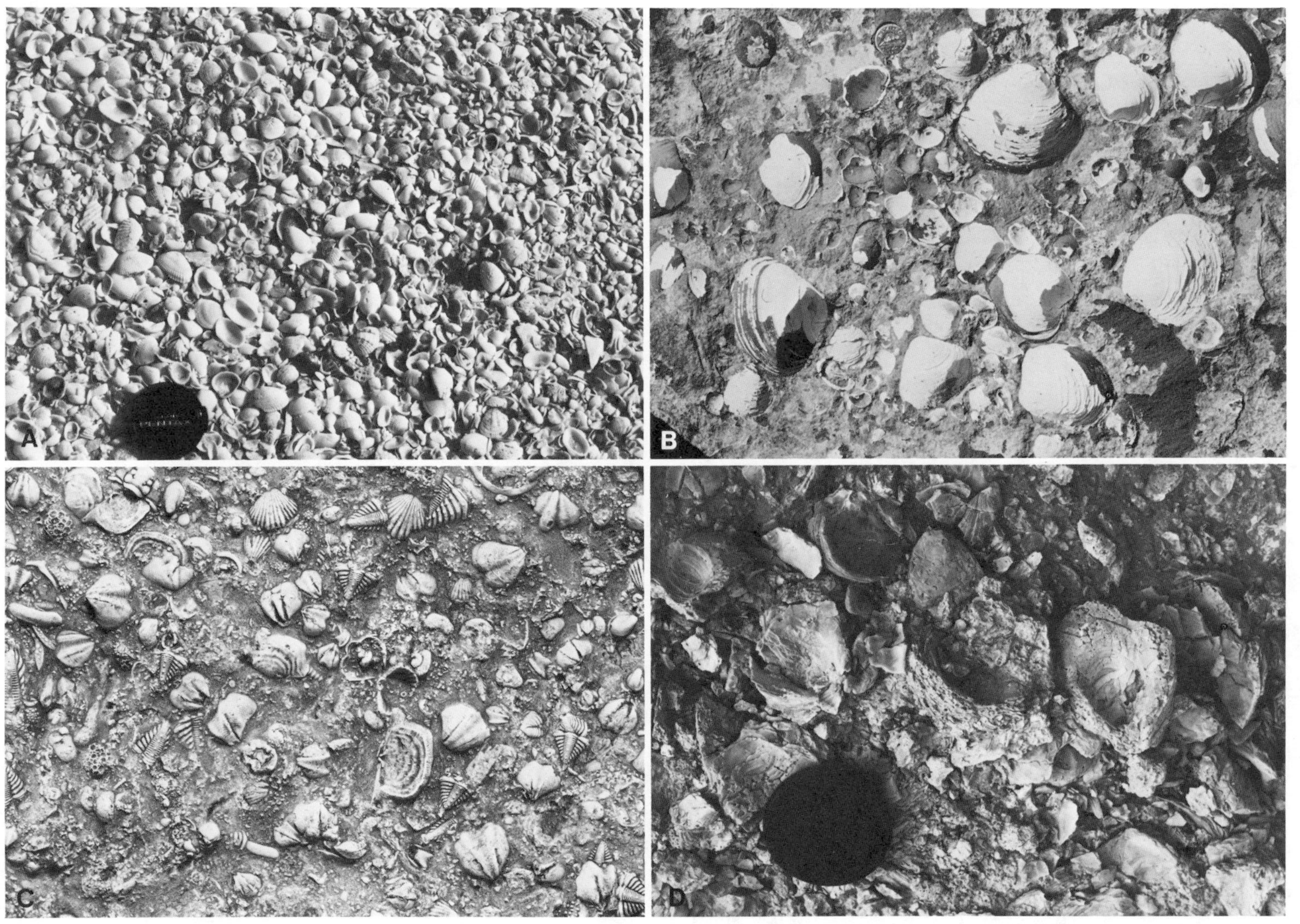

Fig. 2.26. Fossil death assemblages. **A** Modern shell beach, Sanibel Island, Florida. **B** Pelecypods in shallow marine sandstone, Cretaceous, Banks Island, N.W.T., Canada. **C** Trilobites and brachiopods on a bedding plane, Silurian, southern Ontario. **D** Oyster beach deposit, Pleistocene, Mallorca. **E** Crinoidal reef debris, Devonian, Princess Royal Island, N.W.T., Canada. **F** Brachiopods, corals, ostracodes and other bioclastic debris in vertical section, Devonian

lagoonal limestone, southern Ontario. **G** "Starfish" in shallow marine sandstones of the "Molasse Marine", Miocene, near Digne, France. **H** Plant fragments, Pennsylvanian, Kentucky. Photo **C** courtesy R. Ludvigsen; photos **D**, **F** courtesy B. Pratt.

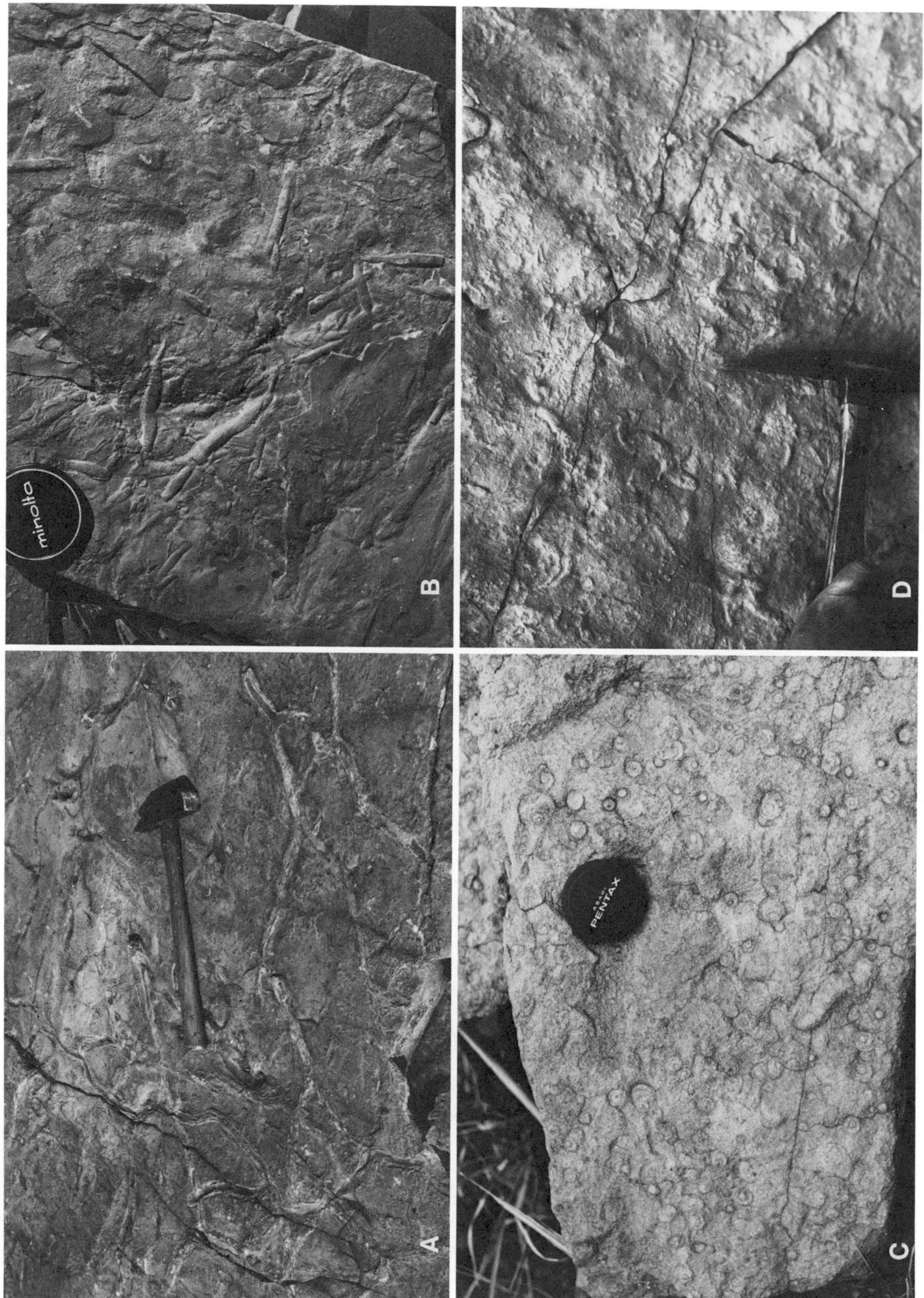

Fig. 2.27 A-D

Fig. 2.27. Trace fossils as exposed on bedding plane surfaces. **A** Interconnected burrows in Ordovician limestone, Ottawa. **B** Short burrows in Ordovician sandstone, Boothia Peninsula, Arctic Canada. **C** Cross sections of vertical, tubular *Skolithus* burrows, Karoo Supergroup, near Durban, South Africa (see Fig. 2.28A). **D** Casts of bird footprints, Cretaceous "flysch", Provence, France. **E** Markings of a resting or swimming crustacean, Karoo Supergroup, Beaufort West, South Africa.

vertebrate trace fossils (from Simpson, 1975). Interpretations of this information are discussed in section 4.5.7. Useful texts on trace fossils include Häntzschel (1962), Goldring (1964), Howard and Skidaway (1971), Howard (1972), Hofmann (1973), Frey (1973, 1975), Crimes and Harper (1977), Basan (1978) and Ekdale (1978).

2.3.3 Sampling plan

The amount of sampling to be carried out in outcrop section depends on the nature of the problem in hand, as discussed in section 2.2. We are not concerned here with sampling of ore, hydrocarbon source beds, or coal to be analyzed for economic purposes, but the sampling required to perform a satisfactory basin analysis. Sampling is carried out for three basic purposes: (1) to provide a suite of typical lithologic samples illustrating textures, structures or distinctive fossils on a hand specimen scale; (2) to provide a set for laboratory analysis of petrography and maturation using polished slabs, the optical microscope, and possibly other tools such as x-ray diffraction analysis and the scanning electron microscope; (3) to provide samples of macrofossils and lithologic samples for microfossil or palynological examination, to be used for studying biostratigraphy.

Illustrative samples. The choice of such samples is usually simple, and can be based on a tradeoff between how much it would be useful to take and how much the geologist can physically carry. Large samples showing sedimentary structures or suites of fossils may be an invaluable aid in illustrating the geology of the project area to the geologist's supervisor or for practical demonstration at a seminar or poster display at a conference. Unless field work is done from close beside a road or is being continuously supported by helicopter it is rarely possible to collect as much as one would like.

Petrographic samples. Before carrying out a detailed sampling program for petrographic work the geologist should think very carefully about what it is the samples are intended to demonstrate. Here are some typical research objectives:

1. Grain size analysis in siliciclastic and carbonate clastic rocks, as a descriptive parameter and as an aid to interpreting depositional environment.

Fig. 2.28. Trace fossils as exposed in cross section. **A** *Skolithus,* Karoo Supergroup, near Durban, South Africa (see Fig. 2.27C). **B** *Diplocraterion,* Cretaceous, Provence, France. **C** Sloping lungfish burrows, Devonian, northeast Scotland. **D** Mottling produced by bioturbation in a dolomite, plus individual small burrows. Ordovician, Boothia Peninsula, Arctic Canada.

2. Petrography of detrital grains, including heavy minerals, as an aid to determining sediment sources.
3. Petrography of carbonate grains as an aid to determining depositional environment.
4. Studies of grain interactions, matrix and cement of both carbonate and clastic grains in order to investigate diagenetic history.
5. Studies of detrital grain fabric in order to determine paleocurrent patterns or, in certain cases, as an aid to interpreting depositional environment.
6. Studies of thermal basin maturity using clay mineral characteristics, vitrinite reflectance, microfossil colour or fluid inclusions.
7. Samples for paleomagnetic study, for use in developing a reversal stratigraphy, a paleopole, or for studying diagenesis.

The geologist must define the scope of the problem before collecting any samples; otherwise it may later be discovered that the collection is unsatisfactory. Work in remote regions is excellent training in such planning exercises, because only rarely is there a chance to return to an outcrop a second time. For the purpose of most regional studies it is useful to examine petrographic variations vertically through a sequence, and areally within a single stratigraphic unit. For example the composition of detrital grains in a sequence may show progressive vertical changes, recording erosional unroofing of a source area or the switching of source areas. Sampling should be adequate to document this statistically. Samples taken say every 10 to 50 m through a sequence will normally suffice. For diagenetic and fabric studies and for paleomagnetic work more detailed sampling may be required. Paleomagnetic research normally requires that several samples be taken at a single locality in order to permit checks of accuracy.

The most detailed sampling program is required for carbonate sequences, particularly those of shallow marine origin which show the most facies variation. Laboratory work on polished slabs or thin sections is required for a reliable facies description of most carbonate rocks, and this may call for sampling every meter, or less, through a section.

For certain purposes it is necessary that the sample be oriented, that is, it should be marked in the field so that its position in space can be reconstructed in the laboratory. For paleomagnetic and fabric studies this need is obvious, but it may also be useful for petrographic purposes, for example where is it necessary to examine cavity-filling detrital matrix to determine time of filling relative to tectonic deformation, as an aid to determining structural top, or for studying microscopic grain size changes related to bedding (e.g. graded bedding). To collect an oriented sample the geologist selects a projecting piece of the outcrop that is still in place, not having been moved by frost heave, exfoliation or other processes, and yet is still removable by hammer and chisel. A flat face on this piece is measured for orientation and marked by felt pen before removal (Fig. 2.29).

Fig. 2.29. Marks to show orientation on a sample before removal from the outcrop.

How much should be collected at each sample station? A few hundred grams is adequate for most purposes. Thin sections can be made from blocks with sides less than 2 cm. Grain mounts of unconsolidated sediment can be made from less than 20 g. Where a particular component is sought, such as disseminated carbonaceous fragments for vitrinite reflectance mesurements, it may be necessary to take a larger sample in order to ensure that enough fragments are included for a statistical study at each sample station. Samples for paleomagnetic study are collected in a variety of ways, including the use of portable drills which collect cores about 2 cm in diameter. Alternatively the geologist may wish to take oriented blocks about 4 x 8 x 15 cm, from which several cores can be drilled in the laboratory. Oriented specimens of unconsolidated sediment are collected by means of small plastic core boxes or tubes pushed into an unweathered face by hand. Textures may be preserved by on-site infiltration with resin.

How to ensure that samples are truly representative? There is a conflict here between statistically valid experimental design and what is practically possible. Statistical theory requires that we take samples according to some specific plan, such as once every 10 m or 30 m (for example) through a vertical section, or by dividing a map area into a square grid and taking one sample from somewhere within each cell of the grid. Only this way can we satisfy the assumptions of statistical theory that our samples are truly representative of the total population of all possible samples. In practice we can never fully satisfy such assumptions. Parts of any given rock body are eroded or too deeply buried for sampling. Exposures may not be available where sampling design might require them, or a particular interval might be covered by talus. An additional consideration is that the very existence of exposures of a geological unit might be governed by weathering factors related to the parameter the geologist hopes to measure. For example imagine a sandstone bed formed at the confluence of two river systems, one draining a quartz-rich terrain and one a quartz-poor terrain. The quartzose sandstone intervals may be quartz cemented, much more resistant to erosion, and therefore much more likely to crop out at the surface. Sampling of such a unit might give very biased petrographic results.

In carrying out a basin analysis we are often dealing with large areas and considerable thicknesses of strata. Petrographic, textural and maturity trends usually are strong enough to show through any imperfections in our sampling program. We collect what we can, taking care that our measurements are controlled by the appropriate geological variables, for example that counts of detrital components are all made on the same grain size range (see Chapter 5).

Biostratigraphic samples. The study of any fossil group for biostratigraphic purposes is a subject for the appropriate specialists who, ideally, should collect their own material. However, this may not be possible and the geologist often is required to do the collecting.

Unless a unit is particularly fossiliferous, the collecting of macrofossils can rarely be performed in a fully satisfactory way by the geologist who is also measuring and describing the section. The search for fossils may take a considerable amount of time, far more than is necessary for the other aspects of the work. In practice what the geologist usually ends up with are scattered bits and pieces, and spot samples of more obviously fossiliferous units, in which it may be fortuitous whether or not any species of biostratigraphic value are present. Given adequate time, for example the two or three field seasons required for Ph.D. thesis research, the geologist may be able to spend more time on collecting and to familiarize him or herself with the fauna and/or flora, but in reconnaissance mapping exercises this is usually impracticable.

The increased sophistication of subsurface stratigraphic analysis by the petroleum industry has led to a greatly expanded interest in fossil microorganisms of all kinds. Groups such as conodonts, acritarchs, foraminifera, palynomorphs, diatoms and radiolaria have been found to be sensitive stratigraphic indicators, and commonly have the inestimable advantage of occurring in large numbers, so that biostratigraphic zonation commonly can be based on statistical studies of taxon distribution. Microfossils are extracted by deflocculation or acid dissolution processes of suitable host rocks. Most useful fossil forms are pelagic or are distributed by wind (palymorphs), so that potentially they may be found in a wide variety of rock types. However, their occurrence is affected by questions of hydrodynamics in the depositional setting, and post-depositional preservation. Palynomorphs may be rare in sandstones because they cannot settle out in the turbulent environments in which sand is deposited, but they are abundant in associated silts, mudrocks and coal. Radiolarians and other siliceous organisms may be entirely absent in mudstones but abundantly preserved in silts, cherts and volcanic tuffs, because they are dissolved in the waters of relatively high pH commonly associated with the formation of mud rocks. Conodonts are most commonly preserved in limestones and calcareous mudstones and may be sparse in dolomites, because dolomitization commonly occurs penecontemporaneously in environments inimical to conodonts, such as sabkha flats.

Armed with advice of this kind from the appropriate specialist, the geologist can rapidly collect excellent suites of samples for later biostratigraphic analysis, and can cut down on the amount of barren material carried home at great effort and expense, only to be discarded. Advice should be sought on how much material to collect. Normally a few hundred grams of the appropriate rock type will yield a satisfactory fossil suite, but more may be required for more sparsely fossiliferous intervals, for example, several kilograms for conodonts to be extracted from unpromising carbonate units.

Samples should be collected at regular vertical intervals through a section, preferably every ten to fifty meters. This will vary with rock type, in order to permit more detailed sampling from condensed units or particularly favorable lithologies. Such sample suites permit the biostratigrapher to plot range charts of each taxon, and may allow detailed zonation. Scattered or spot samples have to be examined out of context, and may not permit very satisfactory age assignments.

2.3.4 Plotting the section

A stratigraphic section can be published in the form of a written description, but this is not a very effective use of the information. It is required for the formal description of type and reference sections of named stratigraphic units, but it is doubtful if much is to be gained by reading such a description. The same information can be conveyed in a much more compact and digestible form as a graphic log, with a central column for lithology and adjacent columns for other features of importance. Such logs have the added advantage that they can be laid side by side, permitting comparisons and correlations between sections from several locations.

It is possible to devise logging techniques with columns and symbols to convey every scrap of lithologic and petrographic information, plus details of fossils, sedimentary structures, paleocurrent measurements, even chemical composition. An example is illustrated in Figure 2.30. Many companies and government organizations print blank logging forms of this type for use by their geologists, to facilitate the logging process and to standardize the results of different workers.

However, such logs have the disadvantage that significant information may be lost in a welter of detail. If a main purpose of a log is to permit

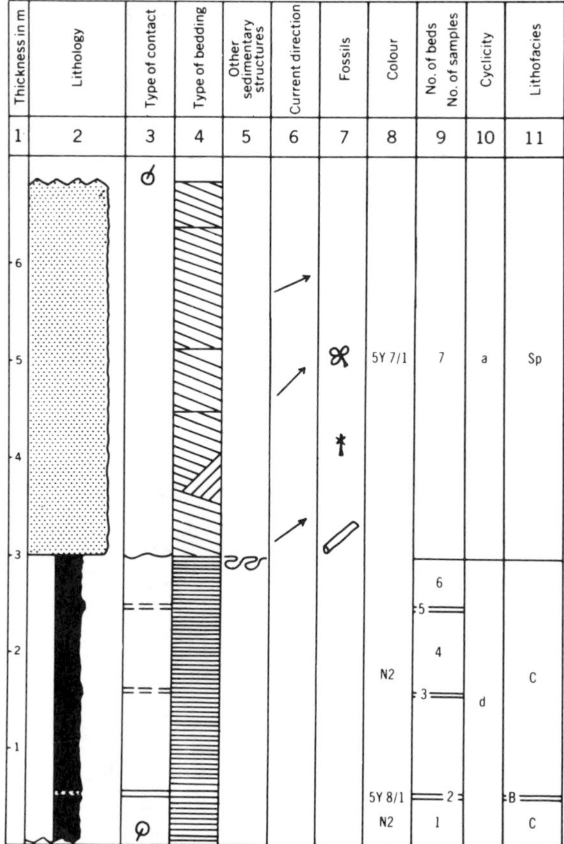

Fig. 2.30. Example of a complex log showing all lithologic information in symbolic form.

visual comparisons between different sections it is advisable to simplify the logs so that the critical features stand out from the page. For the purpose of basin analysis the most important data are those which carry paleogeographic information, such as lithology, grain size, sedimentary structures and fossil content. Paleocurrent data can be added to these, but commonly are treated separately (Chap. 5). As we shall see in Chapter 4 the vertical succession of facies is often of crucial importance in interpreting depositional environment, and so it is helpful to emphasize this in the logs.

For clastic rocks one of the most useful techniques is to vary the width of the central lithology column according to grain size, the wider the column, the coarser the rocks. Examples are illustrated in Figure 2.31. Many other examples are given by Reading (1978). This method imitates the weathering profile of most clastic rocks, as muddy units tend to be less resistant and to form recessive intervals, in contrast to the projecting buttresses and cliffs of coarser sandstones and conglomerates. Drawing the column in this way enables rapid visual comparisons between sections, and also permits instant recognition of gradual trends, such as upward-fining, and sharp breaks in lithology, as at an erosion surface. Subtleties, such as changes in sorting or a bimodal grain size distribution, cannot be displayed in this way, but are rarely as important from a facies perspective and may, in any case, be accompanied by other kinds of facies change such as in sedimentary structure assemblage, which can be readily displayed in visual logs.

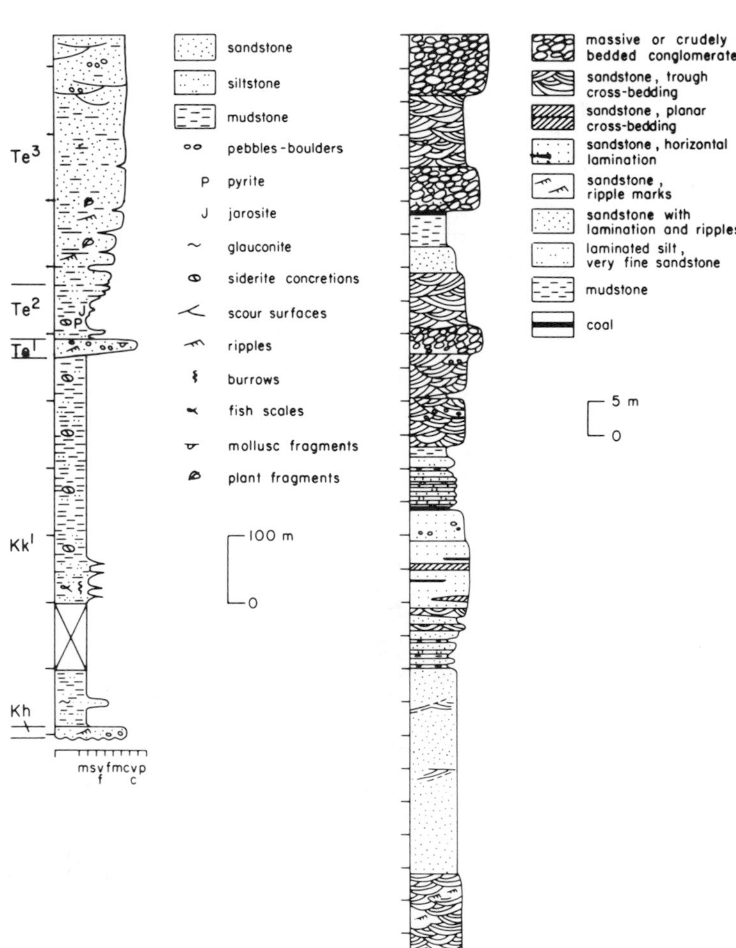

Fig. 2.31. Examples of graphic stratigraphic logs drawn to emphasize clastic sedimentological features. Kh, Kk^1, etc., are formation codes. Grain size is given by width of column, as indicated by device at base of column: m = mud, s = silt, vf, f, m, c and vc = very fine, fine, medium, coarse and very coarse sand, p = pebbles, c = cobbles, b = boulders.

The variable column width technique has not been widely used for carbonate rocks, but there is no reason why it should not be. However, grain size is subject to changes by diagenesis, and this may make interpretation more difficult.

Within the column various patterns can be used to indicate lithology, including symbols for sedimentary structures and fossils. For siliciclastic sequences consisting of interbedded mudstone, siltstone, sandstone and conglomerate, column width conveys most of the necessary lithologic information, and the body of the column can be used primarily for structures and fossils. Some loggers split the column in two, one side for lithology and one for structures, but this is rarely as visually successful. Very little need be placed outside the column, which preserves an uncluttered appearance and increases the graphic impact of the log.

Symbols, abbreviations and other plotting conventions are discussed under subsurface sections.

At what scale should the logs be drawn? This depends on what it is they are intended to demonstrate. Detailed, local sedimentological logs may require a scale of between 1 cm = 1 m and 1 cm = 10 m. Regional stratigraphic studies can be illustrated in large fold-out diagrams or wall charts at scales in the order of 1 cm = 10 m to 1 cm = 50 m, whereas page-sized logs of major stratigraphic sequences can be drawn (grossly simplified) at scales as small as 1 cm = 1000 m. In the petroleum industry the scale 1 in = 100 ft has long been a convenient standard for subsurface stratigraphic work. This translates to a convenient approximate metric equivalent of 1 cm = 10 m, although it is actually closer to 1 cm = 12 m. Metric units should always be used, preferably in multiples of ten.

2.4 Describing the rocks 2. Subsurface stratigraphic sections

2.4.1 Methods of measuring and recording the data

Subsurface sections are logged and described using three types of data—well cuttings, core and geophysical logs (see sections 2.2.3 and 2.2.4). All three typically are available for the large-diameter holes drilled by petroleum exploration companies. Diamond drill holes provide continuous core but nothing else. The logging techniques and the results obtainable are therefore different.

Examination of well cuttings. Samples stored in company and government laboratories are of two types, washed and unwashed. Unwashed cuttings consist of samples of all the material which settles out of the mudstream in a settling pit at the drill site. Unconsolidated mudrocks may disperse completely into the mudstream during drilling, in which case little of them will be preserved except as coatings on larger fragments or occasional soft chips. Washed cuttings are those from which all mud has been removed. The washing process makes the cutting examination process easier, but it further biases the distribution of rock types present if the drill penetrated any unconsolidated muddy units. Stratigraphic well logging is normally carried out on the washed cuttings, whereas palynological and micropaleontological sampling is done on the unwashed material. Stratigraphic logging techniques are described below. A detailed guide and manual was published by Low (1951) and is well worth reading. Many companies also provide their own manuals. McNeal (1959) has some useful comments, and recently the American Association of Petroleum Geologists has issued a logging guide (Swanson, 1981).

As described in section 2.2.3, samples are collected at the well site and bagged every 10 ft (3 m). The bag is labeled according to depth of recovery by the well-site geologist, who makes allowances for the time taken for the mud and cuttings to rise to the surface. Measurements are normally given as "depth below K.B."; K.B. stands for kelly bushing, a convenient measurement location on the drilling platform a few meters above ground level. The altitude above sea level of this point is determined by surveying, so that these drilling depths can be converted to "depths subsea". On offshore rigs K.B. is 25 to 30 m above sea level.

For various reasons not all the cuttings in any given bag may be derived from the depth shown. The problems of caving and variable chip density have been referred to in section 2.2.3. These problems are particularly acute in soft or unconsolidated rocks, and samples from such a sequence may consist of a heterogeneous mixture bearing little relation to such stratigraphic detail as thinly interbedded units of contrasting lithology. The loss of the muds from the cutting suite compounds the problem. There are several ways in which these problems can be at least partially resolved. Caved material may be obvious by the large size of the fragments or by its exotic lithology or fossil content. For example in one well in the Canadian Arctic I logged Jurassic pelecypods and foraminifera 270 m below the unconformable contact of the Jurassic with the Devonian. The geologist will gradually become familiar with the formations under study, and will then readily recognize such obvious contamination.

A more powerful tool is available to the logger and that is to study the cuttings in conjuction with the suite of geophysical logs from the hole. These logs record various physical properties as a measuring tool is slowly run the length of the hole. Modern geophysical logging methods are capable of resolving lithologic variations over vertical intervals of a few decimeters or less, and can therefore be used, with practice, to interpret lithologies in conjunction with the well cuttings. A description of the common geophysical methods and their uses is given later in this chapter. Once the geologist is familar with the geophysical response of the various lithologies in the hole under examination it is possible to use the logs to adjust or correct the sample description. A lithologic log may therefore be drawn up that bears only a loose relationship to the material actually present in the sample bags. Such a log is an interpretive log, and should be clearly labeled as such. It is likely to be more useful in basin interpretation than a log which simply records the cuttings dogmatically, particularly in the case of poorly consolidated beds or those with rapid vertical lithologic variations. Soft muds will be entirely unrecorded in a straight sample log, which may give the geologist a very inaccurate picture of the subsurface stratigraphy. All this can be allowed for in an interpretive log, but interpretations can be wrong, and the geologist must be aware of it when using this type of record.

Figure 2.32 illustrates a comfortable, convenient laboratory arrangement for examining well cuttings. The samples are tipped from the sample bag or vial into a metal tray (Fig. 2.5) and observed under a low-power, reflected light binocular microscope.Immersing the samples in water may aid observation, particularly as dust adhering to the chips can be washed off. The most useful magnification range for such a microscope is from about X5 to X50. The critical geophysical logs should be unfolded to the appropriate depth-interval and placed at one side of the microscope. It is advisable to rapidly scan the samples from several tens of meters of section before beginning the detailed description. Like standing back from an outcrop section this gives the geologist the opportunity to perceive major lithologic variations. These may be correlated with changes in geophysical log response, permitting precise depth control.

As the samples are described the information may be recorded directly on a preprinted logging form, or written out in note form. If the log is to be an interpretive one several tens of meters of section should be examined before plotting the graphic log, so as to give the geologist time to digest what is being observed. It may be necessary for publication or other purposes to produce a written sample description but, as discussed under surface sections, these are difficult for a reader to absorb and are likely to be little used. The Geological Survey of Canada now publishes them in microfiche form to save paper and space.

Many petroleum and mining exploration companies and service companies (such as the American Stratigraphic Company, Canadian Stratigraphic Service Ltd., International Geosystems Corporation) now use computer processible

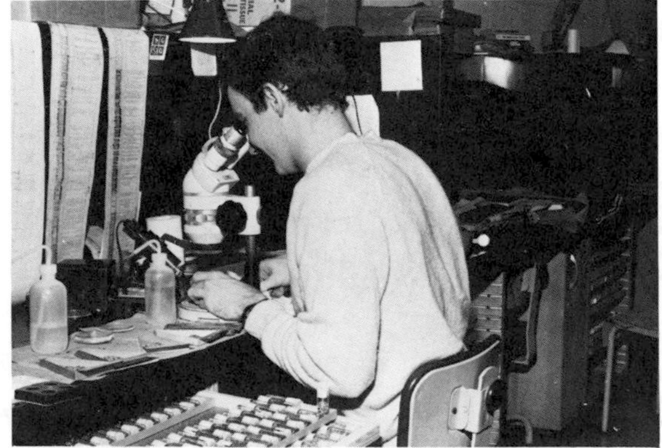

Fig. 2.32. Laboratory arrangements for examining well cuttings (Geological Survey of Canada, Calgary). Photo courtesy of J. Dixon.

logging forms. An example of a form designed for studying outcrops of nonmarine clastic rocks was discussed earlier (section 2.3.1). Similar forms have been designed for well logging, and the data are then stored in digital form in data banks. Retrieval programs may be available that can use these data for automated log plotting, and the same data bank can be exploited for automated plotting of maps and sections, as discussed in Chapter 5.

Examination of core. The large diameter core produced by petroleum drilling is stored either in a company laboratory or an official government repository. North American practice is to divide the core into 2.5 ft (0.8 m) lengths, which are stored side-by-side, two to a box. Top and bottom should be marked on the box. The core usually consists of a series of short pieces, broken from each other by torque during the drilling process. The well-site geologist should number or label each piece with respect to its position in the box and its orientation because, unless this is done, once a piece is moved it may be very difficult to restore it to its correct position, and serious errors may be introduced in reconstructing the vertical lithologic succession. Diamond drill hole (DDH) cores are normally stored in 5 ft (1.6 m) lengths, five to a box. The same remarks apply with regard to the position and orientation of core pieces. DDH cores are rarely brought from the field back into the office, except for crucial holes. They are examined and logged in the field, and then commonly are abandoned. They may even be tipped out of the box to prevent competitors from taking a look. This is a great waste of research material, but the practice seems unlikely to change.

Cores are most conveniently examined in a laboratory specifically designed for this purpose (Fig. 2.33). Core boxes are laid on roller tables, so that they can be readily loaded and unloaded using a forklift vehicle. A movable platform is positioned above the core boxes, on which are placed the microscope, geophysical logs, notebook, etc.

The grinding action of the core barrel during drilling may smear the core surface and obscure lithologic features, and sometimes this can be rectified by washing the core with water or even dilute hydrochloric acid. An even more useful technique is to have the core cut longitudinally with a rock saw, creating a flat section. This should always be wetted with water or etched lightly with dilute acid before examination. The etching technique is particularly useful when examining carbonate rocks, as it tends to generate a fine relief between grains and cement or different carbonate minerals.

Fig. 2.33. A core laboratory (Geological Survey of Canada, Calgary). Photo courtesy of J. Dixon.

Where geophysical logs are available it is important to correlate core lithology with log response. This exercise may reveal that parts of the core are missing, perhaps as a result of fragmentation of soft lithologies. Such information is of importance in attempts to reconstruct a detailed vertical lithologic profile.

For a discussion of description and plotting routines refer to the previous section on well cuttings. Essentially the same methods are used, except that many more features are visible in core, such as bedding features, sedimentary structures and macroscopic trace and body fossils.

2.4.2 Types of cutting and core observation

Large-scale features, including most sedimentary structures and the subtleties of bedding, cannot be identified in well cuttings. They are partly visible in core, but cores usually provide only frustratingly small snapshots of major features, and core research is rather like trying to describe elephant anatomy by examining a piece of skin with a microscope. The following notes are given in the same format as for surface sections, so that the contrasts with the latter can be emphasized. The description of field observation techniques should be referred to where appropriate.

Subdivision of the section into descriptive units. This is best carried out by core and sample examination in conjunction with geophysical logs, in the case of petroleum exploration wells. The combination is a powerful one, and yields good generalized stratigraphic subdivisions with precise depth control. For DDH core the absence of geophysical logs is compensated by the availability of continuous core, and stratigraphic subdivision is simple. Because so many features, such as sedimentary structures, cannot be observed in cuttings, descriptive units in the subsurface tend to be thicker and more generalized than those observed in outcrop. However, for core, the focus on what are really very small samples and the attempt to maximize the use of limited amounts of information tend to lead to very detailed descriptions. Examination of surface sections, sections based on cuttings and those on core, particularly short petroleum cores, require very different concepts of scale. These should be borne in mind when a basin analysis exercise calls for the correlation of all three types of data.

Lithology and grain size. These can be observed satisfactorily in cuttings for all rock types except conglomerates, using the same techniques as described under surface sections. Conglomerates cannot be adequately studied in cuttings where clast size is larger than cutting size. It may not be possible to ascertain which cuttings represent clast fragments and which matrix, and no observations of clast grain size can be made. Remember, also, that unconsolidated silts and muds and evaporites may not be represented in well cuttings. These may require identification using geophysical logs.

Well cuttings commonly contain contaminants which the logger should discard or ignore. Many are easy to recognize, such as metal pipe shavings or bit fragments. Oily substances such as pipe dope or grease may coat some fragment, but can usually be distinguished from natural oil stains by the fact that they coat the cutting and do not penetrate it. Drilling mud may also coat the cuttings, particularly poorly washed samples. Casing cement may appear as a flood of cuttings at certain levels, where the hole was re-entered following the setting of casing. Cement can be easily mistaken for sandy, silty or chalky carbonate. Finally, foreign materials such as feathers, sacking, seeds, cellophane, perlite or coarse mica flakes may be present. These are used in the drilling mud to clog large pores and prevent loss of mud circulation.

Usually there are few restrictions on lithology and grain size determinations in core, except where particularly coarse conglomerates are present.

Porosity. See the discussion in section 2.3.2. Observations and measurements on subsurface rocks are more reliable than in outcrop because of the complications of surface weathering.

Color. See the discussion under the heading of surface sections (section 2.3.2).

Bedding. For core, see the discussion under the heading of surface sections (section 2.3.2). Bedding cannot be seen in well cuttings except for fine lamination, whereas core provides good information on bedding variation. Caution should be exercised in attempts to extract any quantitative information about bed thickness from cores, because of the possibility of core loss, as discussed earlier.

Sedimentary structures. Very few, if any, structures can be observed in well cuttings. However, a wide range of structures is visible in core. Large structures such as major erosion features and giant crossbedding may be difficult to discern because the small sample of the structure visible in the core may easily be confused with something else. For example, thick crossbed sets could be mistaken for structural dip if the upper or lower termination of the set against a horizontal bedding plane cannot be picked out. Dipmeter interpretations may help here, as discussed in sections 5.4.2 and 5.9.6. Erosion surfaces are practically impossible to interpret in small outcrops or core. The break may indicate anything from a storm-induced scour surface to an unconformity representing several hundred million years of nondeposition. The presence of soils or regoliths below the erosion surface is about the only reliable indicator of a major sedimentary break (except, of course, where there is structural discordance).

Ripple marks, and crossbed sets up to a few decimeters thick, are more readily recognizable in core, and the reader should turn to the discussion of these in the section on surface exposures for methods of study. The large-scale geometry of crossbed sets is difficult to interpret in core. For example the difference between the flat shape of a foreset in planar crossbedding and the curved surface of a trough crossbed is practically impossible to detect in core, even the large-diameter petroleum core. Curvature of a typical trough crossbed across such a core amounts to about 2°. Sensitive dipmeter logs have considerable potential for interpreting crossbedding in the subsurface, but the methodology of interpretation has yet to be fully worked out. Dipmeter logs will eventually permit paleocurrent measurements to be made in the subsurface (section 5.9.6), something which cannot now be done without oriented core, and the availability of the latter is practically zero.

Small-scale erosion features such as flutes, tool markings and rain prints, and other bedding features such as desiccation cracks and synaeresis markings, are difficult to find in core because bedding plane sections are rare and the geologist is to be discouraged from creating additional sections by breaking up the core.

Liquefaction, load and fluid loss structures commonly are visible in core and, except for the larger features, should be readily interpretable.

For additional discussion of sedimentary structures observable in core the reader is referred to Michaelis and Dixon (1969) and Shawa (1974).

Fossils. Fossil fragments commonly are visible in well cuttings, but are difficult to recognize and interpret. The best solution is to examine them in thin section, when distinctive features of internal structure may be apparent. A program of routine thin section examination of fossiliferous sequences may be desirable and, particularly if the rocks are carbonates, this can be combined with the lithologic analysis. An excellent textbook by Horowitz and Potter (1971) discusses petrography of fossils in detail.

Macrofossils can rarely be satisfactorily studied in core, except in the case of highly fossiliferous sections such as reefs and bioherms. The reason is that the chance sections afforded by core surfaces and longitudinal cuts do not necessarily provide exposures of a representative suite of the forms present and, unlike sparsely fossiliferous surface outcrops, there is no opportunity to break up more rock in a search for additional specimens. Fragments may be studied in thin section, as described above, but the limitations on quantity of material still apply.

Both cores and cuttings may be used by biostratigraphic specialists, who extract palynomorphs and microfossils from them by processes of deflocculation or acid dissolution. Much useful ecological information may be obtainable from these suites of fossils. For example Mesozoic and Cenozoic foraminifera were sensitive to water depth (as are modern forms), and documentation of foram assemblages through a succession can permit a detailed reconstruction of the varying depths of marine depositional environments (section 4.5.7).

Biogenic sedimentary structures. As in the case of most other mesoscopic features, very little can be seen in cuttings, whereas cores commonly contain particularly well-displayed biogenic sedimentary structures. Those confined to bedding planes may not be particularly easy to find, whereas burrows usually are easy to study, and can provide invaluable environmental interpretation. Refer to the notes and references under "surface sections" and to section 4.5.7.

2.4.3 Sampling Plan

As in the case of surface sections we are concerned here with sampling for basin analysis purposes in three main categories: illustrative lithologic samples, samples for laboratory petrographic analysis using thin and polished sections, x-ray diffraction, etc., and samples for biostratigraphic purposes.

Very limited quantitites of material are available for any kind of sampling in subsurface sections. The cuttings and core stored in the laboratory are all that will ever be available and once used cannot be replaced. Thus they should be used with care and permission always sought from the appropriate company or government agency before removing any material for research purposes.

Core may provide excellent illustrative material for demonstrating lithology, sedimentary structures and facies sequences. The Canadian Society of Petroleum Geologists has established a tradition of holding a "core conference" every year, in one of the government core laboratories in Calgary (other societies are now copying this idea). Each contributor to the conference provides a display of selected core from a producing unit, and uses this as a basis for presenting an interpretation of its geology, with emphasis on depositional environments, diagenesis and petroleum migration history. The educational value of these conferences is inestimable, and they are always well attended. Some have resulted in the production of well-illustrated proceedings volumes, e.g. Shawa (1974), Lerand (1976), and McIlreath and Harrison (1977).

The use of well cuttings as illustrative material of this type clearly is limited. However, cuttings may be sampled routinely for petrographic studies, using etching and staining techniques on the raw, unwashed cuttings, or preparing polished or thin sections. Fortunately very small samples are adequate for this kind of work.

Sampling for biostratigraphic purposes should always be carried out on the unwashed rather than the washed cuttings. Depth control is, of course, better for core, but in petroleum wells there rarely is adequate core for routine biostratigraphic sampling. Very rarely a petroleum exploration well may be drilled by continuous coring for stratigraphic research purposes, and these provide ample sample material free of the problems of sample caving and depth lag. The same is true, of course, for DDH core. The quantity of material required for biostratigraphic purposes depends on the fossil type under investigation. A few hundred grams usually is adequate for palynological purposes, whereas to extract a representative conodont suite from sparsely fossiliferous carbonate sediments may require several kilograms of material. In the latter case samples from several depth intervals may have to be combined.

2.4.4 Plotting the section

Well logging is a routine procedure, and most organizations provide standard forms for plotting graphic logs, with a set of standard symbols and abbreviations. That used by the American Stratigraphic Company and Canadian Stratigraphic Services is typical. An example is illustrated in Figure 2.34. Lithology is shown by color in a column near the center of the log, and accessories, cements, fossil types and certain structures are shown by symbols. To the left of this column are columns for formation tops, porosity type and porosity grade (amount). Formation tops are shown by a formation code. They are interpreted and may be subject to revision. Porosity type is given in a standard code or symbol and porosity grade by a crude graph based on visual estimates. The depth column may be used for symbols to indicate hydrocarbon shows or stains. To the right of the lithology column is a column for crystal or detrital grain size, based on visual estimates or measurements against a grain size comparison set (Figure 2.11). Both to the left and right of the lithology column are spaces for selected geophysical logs. On the right hand side of the log there is a space for typing in an abbreviated description of each lithologic interval. The AmStrat–CanStrat system uses a list of more than 450 abbreviations, covering almost every conceivable descriptive parameter. Remaining columns are used for grain rounding and sorting and for engineering data.

Much of this detail is not necessary for basin analysis purposes. Figure 2.35 illustrates the more limited range of symbols and codes used by Tassonyi (1969) in a study of the subsurface stratigraphy of the Mackenzie Basin in northern Canada. Figure 2.36 illustrates one of his graphic logs. This style emphasizes lithology and other important stratigraphic variables.

2.4.5 Geophysical logs

A wide range of physical parameters can be measured using tools lowered down a petroleum exploration hole. These give information on lithology, porosity, and oil and water saturation. In many cases the measurements are not direct, but require interpretation by analogy, or by correlating values between two or more logs run in the same hole. The subject of petrophysics is a highly advanced one, and beyond the scope of this book. Some approaches are described by Pirson (1977); other are given in the various interpretation manuals issued by the logging companies.

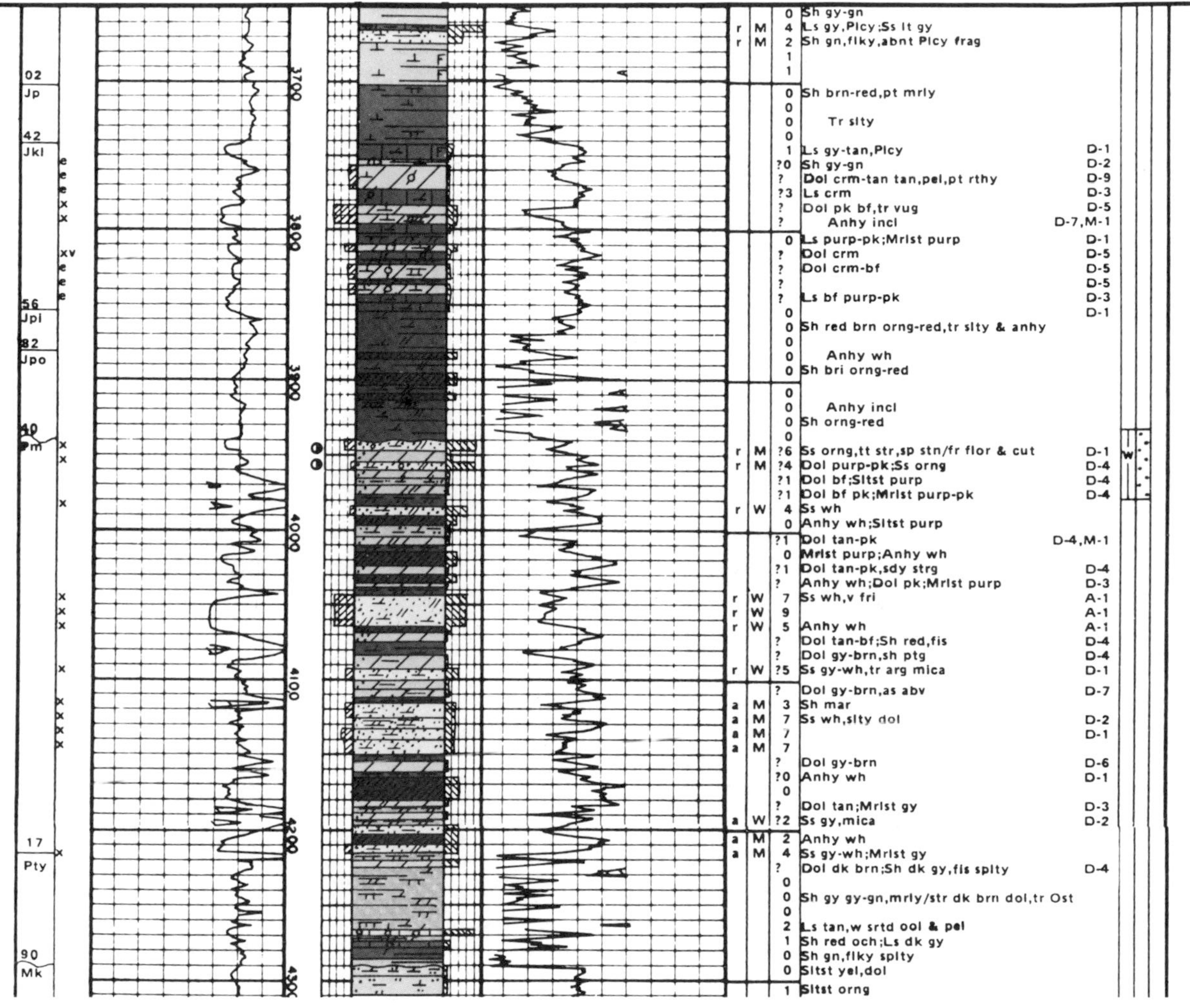

Fig. 2.34. Example of part of a log produced by Am Strat and Can Strat.

Geophysical logs are used routinely by stratigraphers and basin analysts to provide information on lithology and to aid stratigraphic correlation, as discussed in sections 2.2.3 and 3.4.3. In this section some of the principal log types are described briefly and some demonstration of their utility is given.

Gamma ray log (GR). This log measures the natural radioactivity of the formation, and therefore finds economic application in the evaluation of radioactive minerals such as potash or uranium deposits. In sedimentary rocks radioactive elements tend to concentrate in clay minerals, and therefore the log provides a measurement of the muddiness of a unit. Texturally and mineralogically mature clastic lithologies such as quartz arenites and clean carbonate sediments give a low log response, whereas mudstones and certain special sediment types, such as volcanic ash and granite wash, give a high log response. Absolute values and quantitative calculations of radioactivity are not necessary for the stratigrapher. The shape of the log trace is a sensitive lithostratigraphic indicator and the gamma ray log is commonly used in correlation and facies studies. The log has the advantage that gamma radiation penetrates steel, and so the log can be run in cased holes. An example is illustrated in Figure 2.37; Figure 3.7 shows the use of the gamma ray and other log types in stratigraphic correlation. Low GR readings (deflection to the left) correspond to cleaner, sandy parts of the succession. For clastic rocks the variable-width column defined by the two log traces for each well provides a graphical portrayal of grain size variations analogous to the

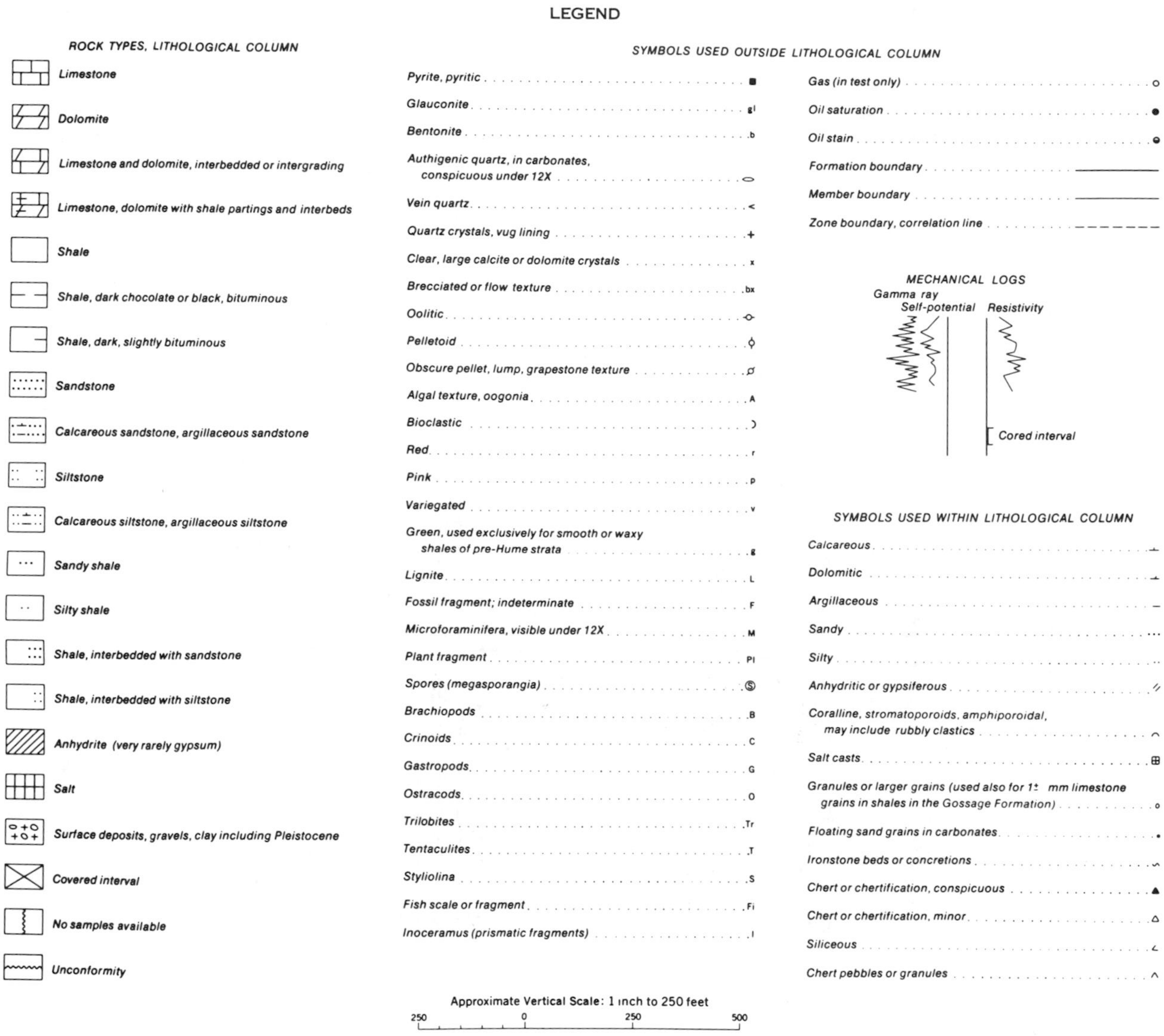

LEGEND

ROCK TYPES, LITHOLOGICAL COLUMN

- Limestone
- Dolomite
- Limestone and dolomite, interbedded or intergrading
- Limestone, dolomite with shale partings and interbeds
- Shale
- Shale, dark chocolate or black, bituminous
- Shale, dark, slightly bituminous
- Sandstone
- Calcareous sandstone, argillaceous sandstone
- Siltstone
- Calcareous siltstone, argillaceous siltstone
- Sandy shale
- Silty shale
- Shale, interbedded with sandstone
- Shale, interbedded with siltstone
- Anhydrite (very rarely gypsum)
- Salt
- Surface deposits, gravels, clay including Pleistocene
- Covered interval
- No samples available
- Unconformity

SYMBOLS USED OUTSIDE LITHOLOGICAL COLUMN

Pyrite, pyritic	■
Glauconite	gl
Bentonite	b
Authigenic quartz, in carbonates, conspicuous under 12X	⬭
Vein quartz	<
Quartz crystals, vug lining	+
Clear, large calcite or dolomite crystals	x
Brecciated or flow texture	bx
Oolitic	-o-
Pelletoid	ϕ
Obscure pellet, lump, grapestone texture	ø
Algal texture, oogonia	A
Bioclastic	)
Red	r
Pink	p
Variegated	v
Green, used exclusively for smooth or waxy shales of pre-Hume strata	g
Lignite	L
Fossil fragment; indeterminate	F
Microforaminifera, visible under 12X	M
Plant fragment	Pl
Spores (megasporangia)	Ⓢ
Brachiopods	B
Crinoids	C
Gastropods	G
Ostracods	O
Trilobites	Tr
Tentaculites	T
Styliolina	S
Fish scale or fragment	Fi
Inoceramus (prismatic fragments)	I

Gas (in test only)	○
Oil saturation	●
Oil stain	◒
Formation boundary	———
Member boundary	———
Zone boundary, correlation line	— — — —

MECHANICAL LOGS

Gamma ray
Self-potential
Resistivity
Cored interval

SYMBOLS USED WITHIN LITHOLOGICAL COLUMN

Calcareous	⊥
Dolomitic	⊥
Argillaceous	–
Sandy	...
Silty	..
Anhydritic or gypsiferous	//
Coralline, stromatoporoids, amphiporoidal, may include rubbly clastics	⌒
Salt casts	⊞
Granules or larger grains (used also for 1± mm limestone grains in shales in the Gossage Formation)	o
Floating sand grains in carbonates	•
Ironstone beds or concretions	∽
Chert or chertification, conspicuous	▲
Chert or chertification, minor	△
Siliceous	∠
Chert pebbles or granules	∧

Approximate Vertical Scale: 1 inch to 250 feet

250 0 250 500

Fig. 2.35. Codes and symbols used in a regional subsurface study of Paleozoic and Mesozoic stratigraphy in the Mackenzie Valley region, Northwest Territories (Tassonyi, 1969).

Fig. 2.36. A subsurface log drawn to emphasize stratigraphically important detail, using the codes and symbols given in Fig. 2.35. Devonian, Mackenzie Valley, Northwest Territories. SP and resistivity logs have been redrawn beside the lithologic column (Tassonyi, 1969).

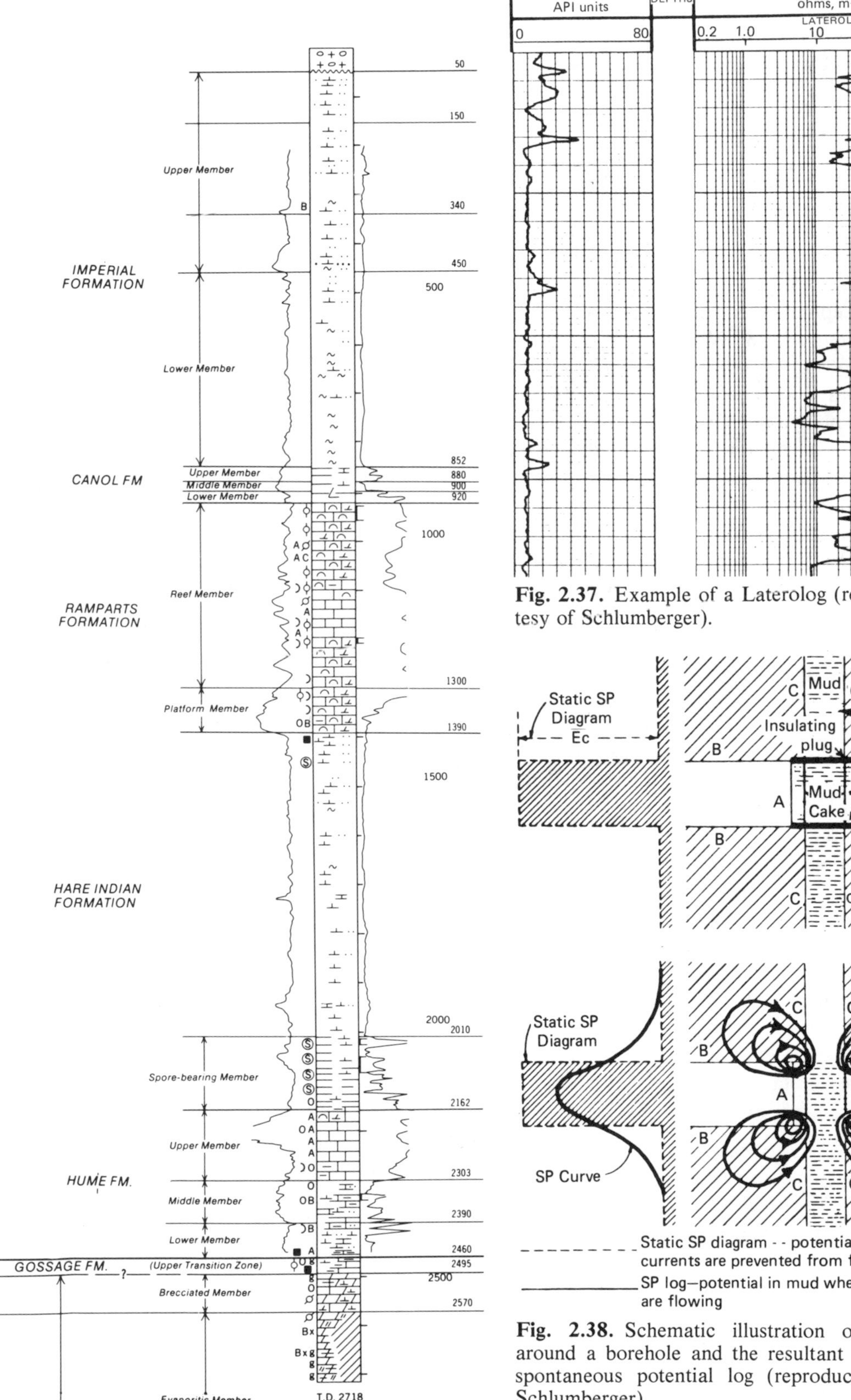

Fig. 2.37. Example of a Laterolog (reproduced courtesy of Schlumberger).

Fig. 2.38. Schematic illustration of current flow around a borehole and the resultant response of the spontaneous potential log (reproduced courtesy of Schlumberger).

variable-width log-plotting methods recommended for drawing surface sections (section 2.3.4). The gamma ray log is therefore widely used for plotting interpretive sample logs of subsurface clastic sections, and for "log-shape" studies.

Spontaneous potential log (SP). The curve is a recording of the potential difference between a movable electrode in the borehole and the fixed potential of a surface electrode. SP readings record mainly currents of electrochemical origin, in millivolts, and are either positive or negative. There are two separate electrochemical effects, as illustrated in Figure 2.38. When the movable electrode is opposite a muddy unit a positive current is recorded as a result of the membrane potential of the mudstone. The latter is permeable to most cations, particularly the Na^{2+} of saline formation waters, as a result of ion exchange processes, but is impermeable to anions. A flow of cations therefore proceeds toward the least saturated fluid, which in most cases is the drilling mud in the hole (log deflection to the right).

Opposite permeable units such as porous limestones and sandstones there is a liquid junction potential which generates a negative potential in the movable electrode. Both anions and cations are free to diffuse into the drilling mud from the more concentrated saline formation waters, but Cl^- ions have the greater mobility, and the net effect is a negative charge (deflection to the left).

The SP log trace, particularly in clastic sequences, is very similar in shape to that of the GR log (examples are given in Figs. 2.39, 3.7, 3.8), and the two may be used alternatively for correlation purposes, as illustrated in Figure 3.7. The SP curve is normally smoother and this log type does not offer the same facility for identifying thin beds. SP deflections are small where the salinities and resistivities of the formation fluids and the drilling mud are similar.

Resistivity logs. Most rock types, in a dry state, do not transmit electric currents and are therefore highly resistive. The main exception consists of those rocks with abundant clay minerals. These transmit electricity by ion exchange of the cations in the clay lattice. In the natural state, rocks in the subsurface are saturated with water or hydrocarbons in pore spaces. Formation waters are normally saline and thus act as electrolytes. The resistivity therefore depends on the salinity and

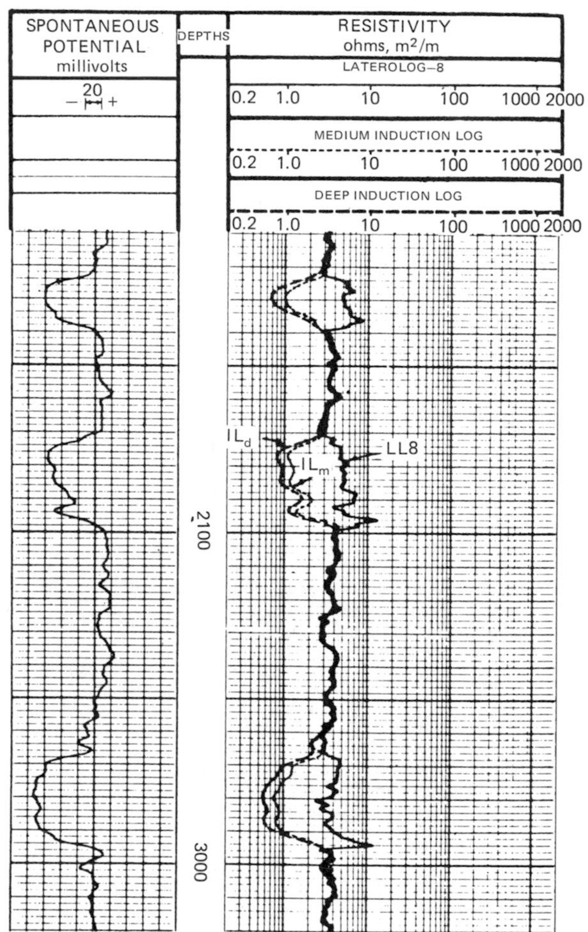

Fig. 2.39. Example of a Dual-Induction Laterolog (reproduced courtesy of Schlumberger).

the continuity of the formation waters. The latter depends, in turn, on porosity and permeability, so that resistivity is lowest in such units as clean sandstone and vuggy dolomite, and highest in impermeable rocks, for example, poorly sorted, "dirty", silty sandstones and "tight" carbonates. Evaporites and coal are also highly resistive. Metallic ores have very low resistivity. Oil is highly resistive and so, under certain conditions, resistivity tools may be used to detect oil-saturated intervals.

A wide variety of resistivity measurement tools has been devised, as listed below. Published logs are commonly identified by the appropriate abbreviation, such as shown in Figure 3.7.

Electrical Survey (ES)
Laterolog[1] (LL)
Induction–Electrical Survey (IES)

Dual Induction Laterolog (DIL)
Microlog[1]
Microlaterolog[1]

The conventional electrical survey was, together with SP, the only logging tool available for many years, in the early days of geophysical logging before World War II. An electrical current is passed into the formation via an electrode, and this sets up spherical equipotential surfaces centered on the electrode–rock contact. Three additional electrodes are positioned on the tool to intersect these surfaces at set distances from the first electrode. The standard spacing is

short normal electrode 40.6 cm (16 in)
medium normal electrode 1.63 m (64 in)
lateral electrode 5.69 m (18 ft 8 in)

The wider the spacing the deeper the penetration of the current into the formation. This permits comparisons between zones close to the hole, permeated by drilling mud, and uninvaded zones further out. The wide spacing of the electrodes also means that the ES tool is not sensitive to thin beds, and so it is not a very satisfactory device for stratigraphic studies.

The Laterolog uses an arrangement of several electrodes designed to force an electrical current to flow horizontally out from the borehole as a thin sheet. A monitoring electrode measures a variable current which is automatically adjusted to maintain this pattern as the tool passes through lithologies of variable resistivity. This device is much more responsive to thin beds. An example, run together with a GR survey and plotted on a logarithmic scale, is shown in Figure 2.37.

Induction logs were developed for use with oil-based drilling muds which, because they are nonconductive, make the use of electrodes unsatisfactory. A high-frequency alternating current is passed through a transmitting coil, creating a magnetic field which induces a secondary current to flow in the surrounding rocks. This, in turn, creates a magnetic field which induces a current in a receiver coil. The strength of the induced current is proportional to formation conductivity.

The DIL survey is a combination of two induction devices and a laterolog device, with different formation penetration characteristics. It is normally run with an SP tool. An example is illustrated in Figure 2.39. Separation between the three resistivity curves occurs opposite permeable units, where the low resistivity of the saline formation waters contrasts with the higher resistivity of the zone close to the borehole, which has been invaded by low-salinity drilling mud. The deep penetration induction log (IL_d) therefore gives the lowest resistivity reading and the shallow penetration Laterolog (LL8) the highest. The presence of these permeable zones is confirmed by the SP curve, which has a pattern very similar to that of the induction logs in the example shown.

The Microlog is the most sensitive device for studying lithologic variations in thin bedded sequences. Its primary petrophysical purpose is to measure the resistivity of the invaded zone. The principle is as follows: during the drilling process mud enters permeable beds and hardens on the surface as a mud cake up to about 1 cm thick. No mud cake is formed opposite impermeable units. The Microlog tool consists of two closely spaced electrodes. Opposite mud cake they record the low resistivity of the mud itself, whereas opposite impermeable units they record the generally higher resistivities of uninvaded rocks. The presence of mud cake is confirmed by a caliper log, which is sensitive to the slight reduction in hole diameter when a mud cake is present. Muddy units commonly cave, and give a very erratic caliper log. The Microlog readings are also likely to be erratic because of the poor electrode contact. All these responses are illustrated in Figure 2.40. The Microlaterolog is a more sensitive version of the Laterolog. Use of both the microresistivity devices permits accurate determinations of permeable sandstone and carbonate thickness, of considerable use in regional subsurface facies studies (Chapter 5).

Sonic log. The sonic tool consists of a set of transmitters for emitting sound pulses and a set of receivers. The fastest path for sound waves to travel between transmitters and receivers is along the surface of the hole, in the rock itself rather than through the mud or the actual tool. The time of first arrival of the sound pulses is therefore a measure of formation density, which depends on lithology and porosity. The sonic log is normally run with a GR tool.

[1]These are registered trade names for tools developed by Schlumberger Limited.

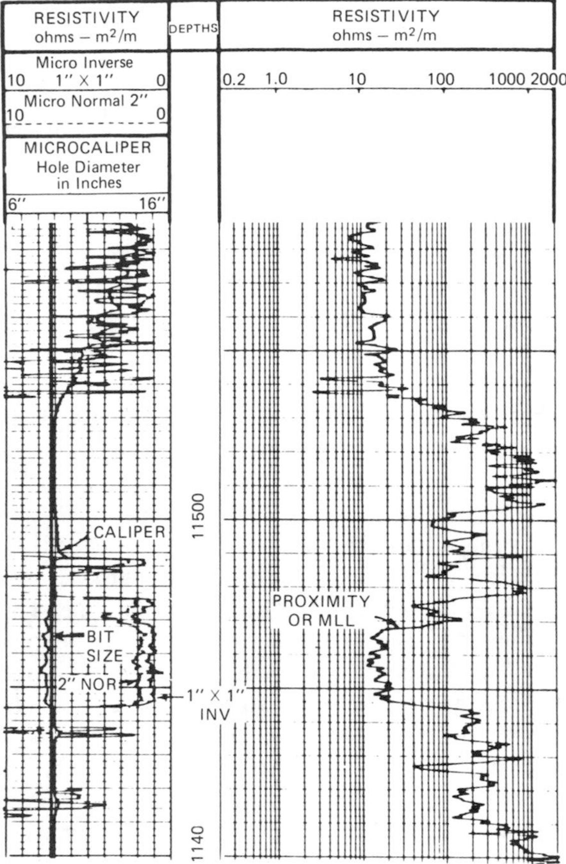

Fig. 2.40. Example of a Microlog (reproduced courtesy of Schlumberger).

The sonic tool has two main uses, the estimation of porosity where lithology is known, and calibration of regional seismic data. For the latter purpose a computer in the recording truck at the wellhead integrates the travel time over each increment of depth as the survey is run. Every time this amounts to one millisecond a small pip appears on a track down the center of the log.

Formation density log. The tool contains a radioactive source emitting gamma rays. These penetrate the formation and collide with it, in a process known as Compton scattering. The deflected gamma rays are recorded at a detector on the tool. The rate of scattering is dependent on the density of the electrons in the formation with which the gamma rays collide. This, in turn, depends on rock density, porosity and composition of the formation fluids.

Neutron log. For this log a radioactive source emitting neutrons is used. These collide with the nuclei of the formation material, with a consequent loss of energy. The greatest loss of energy occurs when the neutron collides with a hydrogen nucleus, and so the total loss of energy depends mainly on the amount of hydrogen present, either in formation waters, hydrocarbons or bound water in clay minerals, gypsum, etc. The detector measures either the amount of scattered low-energy neutrons or the gamma rays emitted when these neutrons are captured by other nuclei.

Crossplots. Where the formation is known to consist of only two or three rock types, such as sandstone–siltstone–mudstone or limestone–dolomite, combinations of two or more logs can be used to determine lithology, porosity and hydrocarbon content. These crossplots are therefore of considerable stratigraphic use where only generalized lithologic information is available from well cuttings.

For clean, non-muddy formations combinations of the sonic, formation density and neutron logs are the most useful. For example, Figure 2.41 and 2.42 show the neutron–density combination. These logs commonly are calibrated in terms of apparent limestone porosity; that is, if the rock is indeed limestone its porosity has the value indicated. For other lithologies the porosity estimate will be in error, but by reading values for both logs it is possible to determine both lithology and correct porosity. The curves in Figure 2.41 give ranges of actual porosity readings for four principal rock types. To show how this graph can be used compare it to the neutron–density overlay given in Figure 2.42. Such overlays may be provided on a routine basis by the logging company, or can be redrawn on request. The thick sandstone interval at the top can be recognized by the distinctly higher readings on the density curve. Values range from about 4 to 12, whereas those on the neutron curve are mostly close to zero. Examination of Figure 2.41 shows that only sandstone can give this combination. The thick limestone at the bottom of Figure 2.42 is indicated by the near coincidence of the two curves. Dolomite or anhydrite would be suggested by relatively higher neutron readings. Mudstones can commonly be recognized by very high neutron readings relative to density, because of the abundant water in the clay mineral lattice. This would be confirmed by the gamma ray response or the SP or caliper log, if available.

Table 2.5 summarizes the use of the sonic, density and neutron logs in lithology identification.

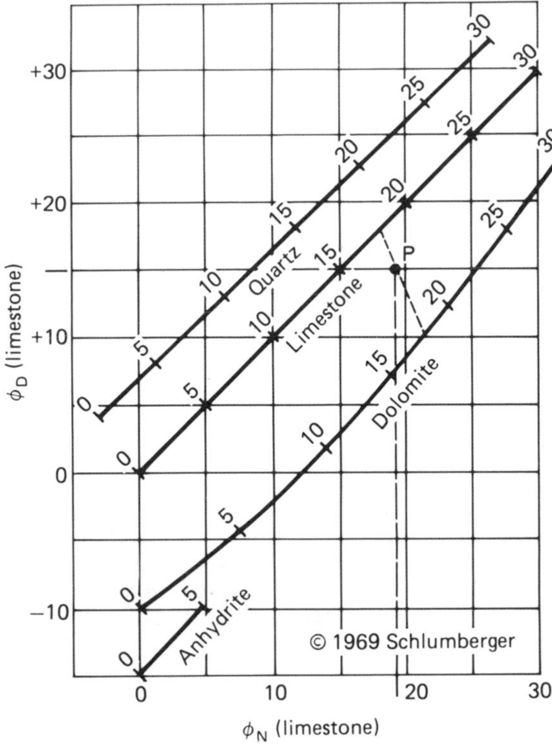

Fig. 2.41. A neutron (ϕ_N):density (ϕ_D) crossplot, showing how readings of the two logs can be used to determine lithology and porosity (reproduced courtesy of Schlumberger).

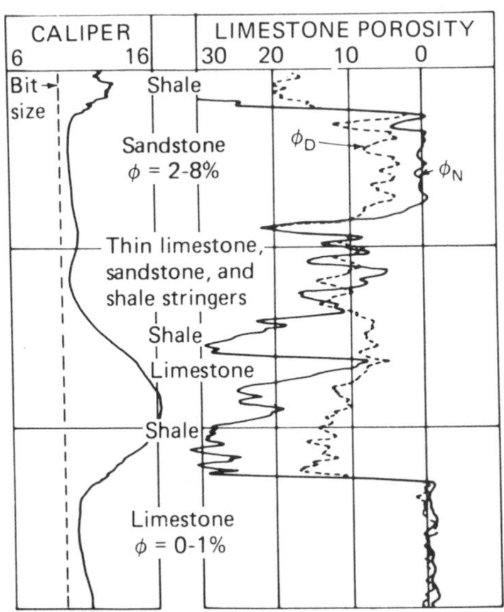

Fig. 2.42. Example of a neutron:density overlay, illustrating how curve separation and deflections can be used to determine lithology (reproduced courtesy of Schumberger).

Table 2.5. Interpretation of density–SNP neutron overlays, assuming single mineral lithologies and liquid filled formations*

Curve relationship	Approximate difference	Additional confirming criteria	Sonic log confirmation (If no secondary porosity)	Probable matrix material	Approx. true porosity
		—At high resistivities—			
$\phi_D \gg \phi_N$	40 p.u.	$\phi_N = 4$, $\phi_D = 43.5$	$\Delta t = 67$	Salt	$\phi \approx \phi_N - 4$
$\phi_D > \phi_{SNP}$	$\phi_{SNP} > 40$ $50 < \phi_D < 80$ $1.3 < \rho_b < 1.5$	Low Gamma Ray High resistivity	$110 < \Delta t < 140$	Coal	
$\phi_D > \phi_N$	5–6 p.u.		$\Delta t - 1.4\ \phi_N \approx 58$	Sandstone	$\phi \cong (\phi_D + \phi_N)/2$
$\phi_D = \phi_N$	0 p.u.		$\Delta t - 1.4\ \phi_N \approx 47$	Limestone	$\phi = \phi_D = \phi_N$
$\phi_D < \phi_N$	8–13 p.u.		$\Delta t - 1.4\ \phi_N \approx 41\text{–}43$	Dolomite	$\phi \cong 0.5\ (\phi_D + \phi_N) + 2.5$ (for $\phi > 8$ p.u.) $\phi \cong 0.7\ \phi_N$ (for $\phi < 8$ p.u.)
$\phi_D < \phi_N$	16 p.u.		$\Delta t - 1.4\ \phi_N \approx 50$	Anhydrite	$\phi \cong \phi_N$
$\phi_D \ll \phi_N$	10–30 p.u.	See Gamma Ray Log		Shale	
		—At high resistivities—			
$\phi_D \ll \phi_N$	28 p.u.	$\phi_N = 49$, $\phi_D = 21$	$\Delta t = 52.0$	Gypsum	$\phi \cong \phi_D - 21$

*Reproduced courtesy of Schlumberger. p.u. = limestone porosity units.

References

ALEXANDER-MARRACK, P.D., FRIEND, P.F., and YEATS, A.K., 1970: Mark sensing for recording and analysis of sedimentological data, *in* J.L. Cutbill, ed., Data processing in biology and geology; Systematics Assoc. Spec. Vol. 3, Academic Press, London, p. 1–16.

ALLEN, J.R.L., 1963: The classification of cross-stratified units, with notes on their origin; Sedimentology, v. 2, p. 93–114.

BASAN, P.B., ed., 1978: Trace fossil concepts; Soc. Econ. Paleont. Mineral. Short Course No. 5, 201 p.

BLATT, H., MIDDLETON, G.V. and MURRAY, R., 1980: Origin of sedimentary rocks; Prentice Hall Inc., Englewood Clifs, 2nd ed., 782 p.

BOUMA, A.H., 1969: Methods for the study of sedimentary structures; John Wiley & Sons, Inc., New York, 458 p.

COLLINSON, J.D., 1970: Bedforms of the Tana River, Norway; Geogr. Ann., v. 52A, p. 31–55.

COLLINSON, J.D., and THOMPSON, D.B., 1982: Sedimentary structures; George Allen and Unwin, London, 194 p.

CRIMES, T.P., and HARPER, J.C., eds., 1977: Trace fossils 2; Seel House Press, Liverpool, 351 p. (Geol. J. Special Issue 9).

DODD, J.R., and STANTON, R.J., Jr., 1981: Paleoecology, concepts and applications; Wiley-Interscience.

DONOVAN, R.N., and FOSTER, R.J., 1972: Subaqueous shrinkage cracks from the Caithness Flagstone series (Middle Devonian) of northeast Scotland; J. Sediment. Petrol., v. 42, p. 309–317.

EKDALE, A.A., ed., 1978: Trace fossils and their importance in paleoenvironmental analysis; Palaeogeog., Palaeoclim., Palaeoec., v. 23, No. 3–4, p. 167–323.

FAGERSTROM, J.A., 1964: Fossil communities in paleoecology: their recognition and significance; Geol. Soc. Am. Bull. v. 75, p. 1197–1216.

FARROW, G.E., 1975: Techniques for the study of fossil and recent traces, *in* R.W. Frey, ed., The study of trace fossils; Springer–Verlag, New York, p. 537–554.

FOLK, R.L., 1968: Petrology of sedimentary rocks; Hemphill's, Austin, Texas.

FOLK, R.L., 1976: Reddening of desert sands: Simpson Desert, N.T., Australia; J. Sediment. Petrol., v. 46, p. 604–615.

FREY, R.W., 1973: Concepts in the study of biogenic sedimentary structures; J. Sediment. Petrol., v. 43, p. 6–19.

FREY, R.W., ed., 1975: The study of trace fossils; Springer–Verlag, New York, 562 p.

FRIEDMAN, G.M. and SANDERS, J.E., 1978: Principles of sedimentology; John Wiley and Sons, New York, 792 p.

FRIEND, P.F., ALEXANDER-MARRACK, P.D., NICHOLSON, J., and YEATS, A.K., 1976: Devonian sediments of east Greenland I: introduction, classification of sequences, petrographic notes; Medd. øm Gronland, Bd. 206, nr. 1.

GLAISTER, R.P., and NELSON, H.W., 1974: Grain-size distributions, an aid in facies identification; Bull. Can. Petrol. Geol., v.22, p. 203–240.

GODDARD, E.N., TRASK, P.D., DE FORD, R.K., ROVE, O.N., SINGEWALD, J.T., Jr., and OVERBECK, R.M., 1948: Rock-colour chart; Geol. Soc. Am., 16 p.

GOLDRING, R., 1964: Trace fossils and the sedimentary surface in shallow-water marine sediments, *in* Develop. in Sedimentology, v. 1, p. 136–143.

HALLAM, A., ed., 1973: Atlas of palaeobiogeography; Elsevier, Amsterdam.

HÄNTZSCHEL, W., 1962: Trace fossils and problematica, *in* R.C. Moore, ed., Treatise on invertebrate paleontology; Geol. Soc. Am. and Univ. Kansas Press, Pt. W., p. W177–W245.

HARMS, J.C., SOUTHARD, J.B., SPEARING, D.R., and WALKER, R.G., 1975: Depositional environments as interpreted from primary sedimentary structures and stratification sequences; Soc. Econ. Paleont. Mineral. Short Course 2, Dallas, 161 p.

HARMS, J.C., SOUTHARD, J.B., and WALKER, R.G., 1982: Structures and sequences in clastic rocks; Soc. Econ. Paleont. Mineral. Short Course 9, Calgary.

HECKEL, P.H., 1972: Recognition of ancient shallow marine environments, *in* J.K. Rigby and W.K. Hamblin, eds., Recognition of ancient sedimentary environments; Soc. Econ. Paleont. Mineral. Spec. Publ. 16, p. 226–286.

HOFMANN, H.J., 1973: Stromatolites: characteristics and utility; Earth Sci. Rev., v. 9, p. 339–374.

HOROWITZ, A.S., and POTTER, P.E., 1971: Introductory Petrography of fossils; Springer-Verlag, New York, 302 p.

HOWARD, J.D., 1972: Trace fossils as criteria for recognizing shorelines in the stratigraphic record, *in* J.K. Rigby and W.K. Hamblin, eds., Recognition of ancient sedimentary environments; Soc. Econ. Paleont. Mineral. Spec. Pub. 16, p. 215–225.

HOWARD, J.D., and SKIDWAY, J.W., 1971: Recent advances in paleoecology and ichnology; AGI short course lecture notes, 268 p.

HUNTER, R.E., 1977: Terminology of cross-stratified sedimentary layers and climbing-ripple structures; J. Sediment. Petrol., v. 47, p. 697–706.

IMBRIE, J., and NEWELL, N.D., ed., 1964: Approaches to paleoecology; Wiley, New York.

JOHNSON, R.G., 1960: Models and methods for analysis of the mode of formation of fossil assemblages; Geol. Soc. Amer. Bull., v. 71, p. 1075–1086.

KRUMBEIN, W.C., and SLOSS, L.L., 1963: Stratigraphy and sedimentation; 2nd ed., W.H. Freeman, San Francisco, 660 p.

LERAND, M.M., ed., 1976: The sedimentology of selected clastic oil and gas reservoirs in Alberta; Can. Soc. Petrol. Geol., 125 p.

LOW, J.W., 1951: Examination of well cuttings; Quarterly of Colorado School of Mines, v. 46, no. 4, 47 p.

LOWE, D.R., 1975: Water escape structures in coarse-grained sediments; Sedimentology, v. 22, p. 157–204.

MAIKLEM, W.R., BEBOUT, D.G., and GLAISTER, R.P., 1969: Classification of anhydrite—a practical approach; Bull. Can. Petrol. Geol., v. 17, p. 194–233.

MARTINI, I.P., 1971: A test of validity of quartz grain orientation as a paleocurrent and paleoenvironmental indicator; J. Sediment. Petrol., v. 41, p. 60–68.

McILREATH, I.A., and HARRISON, R.D., eds., 1977: The geology of selected carbonate oil, gas and lead–zinc reservoirs in Western Canada; Can. Soc. Petrol. Geol., 124 p.

McKEE, E.D., and WEIR, G.W., 1953: Terminology for stratification in sediments; Geol. Soc. Am. Bull., v. 64, p. 381–389.

McKERROW, W.S., ed., 1978: The ecology of fossils, an illustrated guide; Duckworth & Co. Ltd., London, 384 p.

McNEAL, R.P., 1959: Lithologic analysis of sedimentary rocks; Am.Assoc. Petrol. Geol. Bull., v. 43, p. 854–879.

MIALL, A.D., 1979: Mesozoic and Tertiary geology of Banks Island, Arctic Canada: The history of an unstable craton margin; Geol. Surv. Can. Mem. 387.

MICHAELIS, E.R., and DIXON, G., 1969: Interpretation of depositional processes from sedimentary structures in the Cardium Sand; Bull. Can. Petrol. Geol., v. 17, p. 410–443.

MIDDLETON, G.V., 1973: Johannes Walther's Law of the correlation of facies; Geol. Soc. Am. Bull., v. 84, p. 979–988.

MITCHUM, R.M., Jr., VAIL, P.R., and THOMPSON, S., III, 1977: Seismic stratigraphy and global changes of sea level, part two: the depositional sequence as a basic unit for stratigraphic analysis; Am. Assoc. Petrol. Geol., Mem. 26, p. 53–62.

MOBERLY, R., and KLEIN, G., deV., 1976: Ephemeral colour in deep-sea cores; J. Sediment. Petrol., v. 46, p. 216–225.

PETTIJOHN, F.J., and POTTER, P.E., 1964: Atlas and glossary of primary sedimentary structures; Springer-Verlag, New York, N.Y., 370 p.

PETTIJOHN, F.J., POTTER, P.E., and SIEVER, R., 1972: Sand and sandstone; Springer–Verlag, New York, 618 p.

PICARD, M.D., and HIGH, L.R. Jr., 1973: Sedimentary structures of ephemeral streams; Develop. in Sedimentology, v. 17, Elsevier, Amsterdam, 223 p.

PIRSON, S.J., 1977: Geologic well log analysis; 2nd ed., Gulf. Pub. Co., Houston, 377 p.

POTTER, P.E., and PETTIJOHN, F.J., 1977: Paleocurrents and basin analysis; 2nd ed., Academic Press Inc., New York, 296 p.

READING, H.G., ed., 1978: Sedimentary environments and facies; Blackwell, Oxford, 557 p.

RUST, B.R., 1975: Fabric and structure in glaciofluvial gravels, *in* A.V. Jopling and B.C. McDonald, eds., Glaciofluvial and glaciolacustrine sedimentation; Soc. Econ. Paleont. Mineral.Spec. Pub. 23, p. 238–248.

SCHAFER, W., 1972: Ecology and Paleoecology of marine environments; Univ. Chicago Press, 568 p.

SCHREIBER, B.C., FRIEDMAN, G.M., DECIMA, A., and SCHREIBER, E., 1976: Depositional environments of Upper Miocene (Messinian) evaporite deposits of the Sicilian Basin; Sedimentology, v. 23, p. 729–760.

SHAWA, M.S., ed., 1974: Use of sedimentary structures for recognition of clastic environments; Can. Soc. Petrol. Geol.

SIMPSON, S., 1975: Classification of trace fossils, *in* R.W. Frey, ed., The study of trace fossils; Springer-Verlag, New York, p. 39–54.

SIPPEL, R.F., 1971: Quartz grain orientation—1. (the photometric method); J. Sediment. Petrol., v. 41, p. 38–59.

SMITH, N.D., 1972: Some sedimentological aspects of planar cross-stratification in a sandy braided river; J. Sediment. Petrol., v. 42, p. 624–634.

SOLOHUB, J.T., and KLOVAN, J.E., 1970: Evaluation of grain-size parameters in lacustrine environments; J. Sediment. Petrol., v. 40, p. 81–101.

STEPHENS, L.E., SOBCZAK, L.W., and WAINWRIGHT, E.S., 1972: Gravity measurements on Banks Island, N.W.T., with Gravity Map 150; Earth Physics Branch, Energy, Mines and Resources Canada.

SWANSON, R.G., 1981: Sample examination manual; Am. Assoc. Petrol. Geol.

TASSONYI, E.J., 1969: Subsurface geology, Lower Mackenzie River and Anderson River area, District of Mackenzie; Geol. Surv. Can. Paper 68–25.

TAYLOR, J.C.M., 1978: Introduction to state of the art meeting, 1977, on sandstone diagenesis; J. Geol. Soc. London, v. 135, p. 3–6.

VISHER, G.S., 1965: Use of vertical profile in environmental reconstruction; Am. Assoc. Petrol. Geol. Bull., v. 49, p. 41–61.

WALKER, T.R., 1967: Formation of redbeds in modern and ancient deserts; Geol. Soc. Am. Bull., v. 78, p. 353–368.

WILSON, J.L., 1975: Carbonate facies in geologic history, Springer-Verlag, New York, 471 p.

CHAPTER 3

Stratigraphic correlation

3.1 Introduction

Whether the geologist is dealing with roadside outcrops, with subsurface data from a petroleum play, or with regional seismic lines, one of the first problems to be encountered is that of stratigraphic correlation. In order to reconstruct depositional environments and paleogeography or to trace a unit of economic interest the geologist must be able to define a stratigraphy and trace it from one location to another. The procedures for carrying this out are the subject of this chapter.

The three types of data mentioned in the first sentence are examples of the wide range of physical scale and data quality that have to be dealt with by the basin analyst. Formal stratigraphic practices, including the definition of formations and of stages, had their origins in nineteenth century stratigraphy, and have evolved into a set of carefully defined procedures for naming and correlating the various kinds of stratigraphic unit. These methods are based mainly on detailed lithostratigraphic and biostratigraphic information, the collection of which is discussed in Chapter 2. Other important aids to correlation, including radiometric dating and magnetic reversal stratigraphy, are discussed in section 3.6.

A quite different approach to correlation is taken by those dealing with regional reflection seismic data. Seismic work in frontier regions and the deep reflection profiles now being produced by groups such as the Consortium for Continental Reflection Profiling (COCORP) provide sweeping regional cross sections, within which correlations at the detailed level may be far from clear.

With outcrop and well data the problem may be to establish the regional framework from a mass of local detail, whereas with seismic data it is the detail that is hard to see. The ideal combination is, of course, a basin with a network of seismic lines tied to key exploration holes. If there are outcrops too, so much the better, although, as pointed out in section 2.2.1, outcrops around a basin margin may not be very representative.

These differences in data type and scale have led to two different approaches to stratigraphic correlation. In industry exploration work in frontier regions, particularly the great offshore basins, a rather informal, pragmatic approach is taken to such topics as biostratigraphic zonation and the naming of formations. The broad picture can be derived from seismic cross sections, and the details gradually resolve themselves as more well data become available. Application of various basin mapping methods (Chap. 5) and use of the genetic, depositional systems concept of stratigraphy (Chap. 6) are of particular value here.

In the absence of seismic data it is necessary to construct the forest from the individual trees. The stratigraphic framework in most well-explored ("mature") basins was built up this way, and the work has usually been accompanied by considerable controversy, as local specialists have argued about the relationships between the successions in different parts of a basin, or between outcrop and subsurface units.

Whether a basin analysis exercise starts from seismic sections or from outcrop work it is necessary, eventually, to document the fine detail of the stratigraphy by establishing a formal lithostratigraphic and (if in the Phanerozoic) a biostratigraphic framework. Such detail is beyond the needs of most exploration companies, and is an area of research commonly taken over by state geological surveys and by individuals in academic institutions, although the data base and expertise built up in industry may form an essential component. This is one area in which the government core and data repositories referred to in Chapter 2 can prove their usefulness.

Every local biostratigraphic, radiometric and

magnetostratigraphic study potentially can contribute to the long term effort to perfect a global chronostratigraphic (time) scale. Several commissions and subcommissions and numerous working groups of the International Union of Geological Sciences (IUGS) have been tackling this problem for many years, as have many national groups. Some of the results are reported in this chapter, and the interested reader may wish to examine the IUGS journal *Episodes*, which reports on the activities of these various groups and announces important publications.

This chapter is intended to provide an introduction to practical research methods. Chronostratigraphic (including biostratigraphic) research must form an integral part of any ongoing basin analysis project unless it is strictly local in scope. The work is usually performed by specialists. Lithostratigraphic correlation methods are also described here, although the procedures for erecting formal, named units (included here for consistency and completeness) can be left to advanced stages of the analysis. Such naming is best carried out by an individual with sedimentological training, so that the depositional systems concepts described in Chapter 6 can be incorporated into the work.

3.2 Types of stratigraphic unit

Rocks may be described in terms of any of their physical, chemical, organic or other properties, including lithology, fossil content, geochemistry, mineralogy, electrical resistivity, seismic velocity, density (gravity), magnetic polarity or age. Theoretically any of these properties may be used for description and correlation, and most are so used for various purposes. In practice lithology is one of the most important criteria; fossil content is also crucial for rocks of Phanerozoic age. Magnetic polarity is finding increasing application as a correlation tool, particularly for the younger Mesozoic and Cenozoic, and radiometric ages are used to assign "absolute" ages to biostratigraphic, magnetic and other chronostratigraphic units. Other geophysical properties are used in early reconnaissance stages of exploration of a sedimentary basin. Not all these properties will necessarily give rise to the same correlations of a given rock body; for example it is commonly difficult to relate geophysical properties precisely to lithology. Therefore no single type of stratigraphic unit can be used to define all the variability present in nature.

Reflection seismic data are widely used in the exploration of sedimentary basins but, as stated by Sheriff (1976), "the traditional objective... has been mapping of geologic structure without concluding very much about the stratigraphy. Emphasis has been on finding and mapping coherent primary reflections, that is, determining their arrival times and the differences in arrival times with location, and calculating the locations and dips of the interfaces associated with them." This is now changing with the development of the field of seismic stratigraphy (section 5.4.1, Chap. 6, section 8.2). Sheriff pointed out that to be "seen" on seismic data a stratigraphic or structural feature must yield a clear velocity contrast and must be at least the equivalent of a quarter-wavelength in thickness. At shallow depths velocities are in the range of 1.5 to 2.5 km/sec and reflections are relatively high frequency, about 5–100 Hz, so that a quarter-wavelength is in the order of 5–12 m. At greater depths typical reflection wavelength increases considerably. Therefore stratigraphic resolution is fairly coarse (Fig. 8.11). Seismic data are essential for studying large-scale stratigraphic features such as depositional systems (section 5.4.1; Chap. 6) and regional (or global) sequences (section 8.2), but are of less use in the development of the refined stratigraphic subdivisions which are the subject of this chapter. The reader interested in the application of seismic methods to basin analysis should therefore turn to those chapters noted above.

The most important types of stratigraphic unit are as follows:

Lithostratigraphic: based on observable lithologic features including composition and grain size, and possibly also including certain basic sedimentological information such as types of sedimentary structures and cyclic sequences.

Biostratigraphic: based on fossil content. Life forms evolve with time, permitting subdivision on the basis of changes in the fauna or flora.

Chronostratigraphic: this is an interpretive stratigraphy, in contrast to the first two types of unit, which are strictly descriptive. Chronostratigraphy concerns itself with the age of the strata, which may be determined by a variety of means, of which the most important are fossil content, radiometric dating and magnetic polarity.

Both lithostratigraphic and biostratigraphic units tend to be localized in extent. Lithologic character depends on depositional environment, sediment supply, climate, rate of subsidence, etc., all of which can vary over short distances. Lithostratigraphic units may be diachronous, that is to say of different time range in different places, reflecting gradual shifts in environment, for example during transgression or regression. The limits of a stratigraphic unit are either its erosional truncation at the surface or beneath an unconformity, or a facies change into a contemporaneous unit of a different type.

Biostratigraphic units are based on fauna or flora, the distribution of which is ecologically controlled. Also, contemporaneous faunas located in ecological niches that are similar but geographically isolated may show subtly different evolutionary patterns, making comparisons or correlations between the areas difficult. All life forms evolve with time, so that faunas and floras show both spatial and temporal limits on their distribution.

Chronostratigraphy attempts to resolve these difficulties by establishing a global time-based reference frame. However, the accuracy of chronostratigraphic correlation is only as good as that of the time-diagnostic criteria on which it is based.

The evolution of these three types of unit has had a long and complex history (Hancock, 1977) and there is still not univeral agreement on definitions. Hedberg (1976), Hancock (1977) and Harland (1978) discussed some of the practical and philosophical problems. Early geological practice did not distinguish lithology from age, causing severe correlation problems wherever a facies change or a diachronous boundary occurred. More recently there has been controversy over whether the rocks themselves or the more abstract evidence of age which they contain should form the primary basis of chronostratigraphy. The discussions are likely to seem somewhat academic and theoretical to the average basin analyst and will not be discussed at length here.

3.3 Stratigraphic procedures

3.3.1 The International Stratigraphic Guide

Stratigraphic geology was founded by the English geologist William Smith toward the end of the eighteenth century. Since that time a wide variety of local stratigraphies, stratigraphic methods and philosophies has evolved in different parts of the world. Inevitably this has generated much controversy, ranging from disagreements over local correlations to arguments about what criteria should be used to define the composition and limits of global chronostratigraphic units. The first serious attempt to resolve the broader problems was started in 1952 at the 19th International Geological Congress in Algiers. At this time the International Commission on Stratigraphic Terminology was created. Since 1965 this commission, now called the International Subcommission on Stratigraphic Classification (ISSC), has worked under the aegis of the International Union of Geological Sciences (IUGS). Several drafts of a guide to procedures and terminology were produced and published by this commission, based on numerous discussion meetings at International Geological Congresses, plus correspondence with various official national geological bodies and individual specialists. More than 125 individuals from over 40 countries and 39 official national or regional committees have been involved in this work. The culmination of this effort was the publication in 1976 of the *International Stratigraphic Guide: a Guide to Stratigraphic Classification, Terminology and Procedure*, edited by H.D. Hedberg (1976).

The purpose of this guide is to set out those practices and procedures that have been broadly accepted by the international stratigraphic community. It is to be hoped that eventually this guide will be universally adopted, although at present there are many national and regional codes that differ to some degree from that of the ISSC.

Much work remained to be done at the time the ISSC guide was published, particularly in the area of Precambrian and Quaternary stratigraphy and the relatively new field of magnetic reversals. However, the guide provides an excellent basis for stratigraphic work in the Phanerozoic, and is also broadly applicable to the Precambrian. The procedures outlined in this chapter are based mainly on the ISSC guide, to which the reader is referred for further discussion regarding particular points of interest or concern.

Various reports and recommendations concerning particular stratigraphic procedures have been published since the ISSC guide by specialist working groups of IUGS. Some of these are referred to here. A new *North American Stratigraphic Code* has recently been published in the American Association of Petroleum Geologists Bulletin (1983, v. 67, p. 841–875).

3.3.2 Establishing named units

The basis for all stratigraphic documentation consists of named lithostratigraphic and biostratigraphic units. The establishment of a system of names should not be the first objective of the basin analyst, as explained in section 1.2.3, but without the units themselves it is difficult to draw geological maps or write intelligible reports.

Named units are based on the establishment and description of a **type section** or **stratotype**. This may be a well-exposed surface section, a mine section or a well section, chosen to typify the unit. The stratotype should be as complete as possible. It should preferably include the upper and lower contacts of the unit, and should be as nearly as possible continuous and unfaulted. Surface sections should not include long covered intervals or be heavily weathered or only exposed under exceptional circumstances (such as a riverbed exposed during drought). Subsurface sections should be accessible via a record of well cuttings, or (better still) core, stored in a permanent repository. For both surface and subsurface sections as much descriptive documentation as possible should be available in the public records of a state geological survey, museum or university. This should include precise location data, descriptive logs, geophysical logs (for subsurface sections) and as much sample material as can conveniently be curated.

Additional **reference sections** may be designated to illustrate any features not covered in the type section. Where a unit occurs in both the surface and subsurface the type section should preferably be chosen at the surface, but it may be useful to designate a subsurface reference section, especially as this will serve to establish the geophysical log character of the unit.

A named unit is not considered established until it is published in a recognized referred scientific medium. This could include national and international scientific journals and state geological survey publications. Local journals, bulletins and special reports issued by museums or universities are rarely a desirable way to establish a stratigraphic unit, because they commonly do not receive sufficiently wide circulation and may not be rigorously edited. Names used only in company reports or in trade magazines are not considered established, but may, of course, be widely used in internal, informal communication. Abstracts of talks presented at meetings are also not acceptable as a method of formal publication.

The purpose of these publishing rules is to ensure that only the most useful, well-defined names receive public circulation. The literature is full of poorly defined stratigraphic names and much ink has been wasted on controversies over stratigraphic terminology. For this reason it is advisable that before attempting to establish or revise any stratigraphic nomenclature the worker submit the proposals to a formal local or national stratigraphic nomenclature committee (this is a condition of publication for most state geological surveys). Many journal editorial boards include such a committee, and the state geological surveys invariably do. The advice might seem somewhat bureaucratic, but geological surveys are in business to establish the local geological framework and their employees usually include many, if not most, of the experts in the local stratigraphy. Their view of the subject tends to be more regional in scope than that of the average graduate student or mine geologist, and they are usually better acquainted with stratigraphic procedures. (However, the brash young student may delight in an opportunity to overturn the shibboleths of the established order, as personified by a body of aging civil servants. Thus is progress achieved—sometimes.)

The published description of a new or revised stratigraphic unit should include the following:

1. name and rank of unit
2. location of type and reference sections, including a map or air photo
3. detailed description of the unit at the type section, including nature and height in section or well depth of contacts
4. comments on local or regional extent of unit

and its variability

5. graphic log of unit (optional for lithostratigraphic units, only, but desirable, and should include geophysical logs for subsurface sections)
6. statement of location of curated reference material
7. discussion of relationship to other contemporaneous stratigraphic units in adjacent areas.

New or revised stratigraphic names should only be proposed after a careful evaluation of existing terminology and an examination of regional variability beyond the confines of the immediate project area. Both biostratigraphic and lithostratigraphic units commonly have been established on the basis of very detailed work in a very small area, such as a mine or a few closely spaced outcrops. The geologist should always stand back from such detail and attempt to visualize it in a regional context. A local stratigraphic terminology may be useful for in-house company communication but it may not be worth burdening the geological literature with such detail.

Informal units commonly are established in the reconnaissance stage of basin exploration; for example the first successful well in a frontier basin may give its name to an important hydrocarbon reservoir, e.g. the Leduc reef or Parsons sand. Whether such names should be formalized and published depends on further stratigraphic work, and it may not be apparent until a much later stage of exploration exactly what the extent and correlation of the unit should be. Industry geologists rarely have the time for the detailed lithostratigraphic and biostratigraphic follow-up work required, and it may be best left to university or geological survey workers to follow in the wake of such drilling and perform the necessary scientific research.

Seismic records commonly are used to delineate provisional stratigraphic units. An exploration group may establish an informal inhouse convention of coloring prominent reflectors in a consistent way. They then refer (internally) to the "red marker" or the "blue marker". Modern seismic stratigraphy permits a more detailed and rigorous subdivision than this, as explained in Chapters 5, 6 and 8. Seismic sequences and subsequences may be recognized and correlated using very little well control. It is dangerous, however, to assign too much weight to such interpretations until the detailed, scientific follow-up work has been carried out. Too often seismic records are mis-correlated or misinterpreted because of the presence of subtle unconformities, problems of matching reflectors across faults, or simply poor reflector quality. Formal stratigraphic units should never be established without extensive hard lithostratigraphic or biostratigraphic information from a network of wells or outcrops.

3.4 Lithostratigraphy

3.4.1 Recognition of a lithostratigraphic subdivision

In Chapter 2 it was shown what to observe in the rocks and some ideas were given on how to record and present the data. This information can now be used to establish a lithostratigraphic framework. The geologist should have plotted graphic logs of the surface and/or subsurface sections and, for subsurface sections, should have a suite of geophysical logs, if available. In many cases an obvious stratigraphic breakdown will have become aparent as the data collection work has proceeded, and the individual geologist or exploration group may have started to formulate informal units to aid mental comprehension or for internal communication. Where regional seismic records are available the broad regional framework may already be apparent. This process of subdivision should now be carried out in a slightly more rigorous way by comparing the logs and making a preliminary attempt at correlation.

In each section it will be possible to recognize one or more (perhaps many) distinctive lithostratigraphic units based on physical properties such as mineralogical composition and grain size. Bedding characteristics, sedimentary structures, cyclic sequences and fossil content may be used as secondary criteria. Fossils, although primarily of importance as biostratigraphic indicators, may also have importance as facies indicators in lithology studies (Chap. 4). They may be of rock-forming abundance, as in bioherms and coquinas, or may be present as minor but distinctive components of a lithostratigraphic unit. The latter applies to both body fossils and trace fossils.

Lithological variations based on diagenetic changes such as colour changes, dolomitization or emplacement of epigenetic ores should not be used for lithostratigraphic subdivision.

The recognition of a lithostratigraphic unit is a two-part process, requiring subdivision of vertical sections and correlation between sections. The exercise of correlation is one of the most important aspects of basin analysis. Lithostratigraphic correlation ideally should not be divorced from biostratigraphic correlation and considerations of basin structure and the local or regional depositional system, as discussed in section 3.4.3. The geologist is therefore required to perform several mental tasks at once. Omission of these overlapping procedures can lead to serious lithostratigraphic miscorrelations, as the examination of older literature for any well-studied sedimentary basin will demonstrate.

3.4.2 Types of lithostratigraphic unit and their definition

A hierarchy of units has been developed based on the **formation**, which is the primary lithostratigraphic unit:

Group
Formation
Member
Tongue or lentil
Bed

The formation. An important convention has been established that all sedimentary rocks should be subdivided (when sufficient data have been collected) into formations. No other types of lithostratigraphic subdivision need be used, although convenience of description may require them.

What is a formation? This is a question with no final answer. The degree of lithologic variability required to distinguish a separate formation tends to reflect the level of information available to the stratigrapher. Formations may be only a few meters or several thousands of meters in thickness; they may be traceable for only a few kilometers or for thousands of kilometers. Formations in frontier basins usually are completely different in physical magnitude from those in populated, well-explored basins such as much of Western Europe and the United States. As work in frontier basins proceeds some of the larger formations may be subdivided into smaller units and the ranking of the names changed (see below).

The most important criteria for establishing a formation are its usefulness in subdividing stratigraphic cross sections and its mappability. For reconnaissance mapping a thin unit that cannot accurately be depicted at a scale of, say, 1:250,000, is of little use. For more detailed work, mappability at a scale of 1:50,000 or even 1:10,000 may be a more useful criterion. Problems of consistency arise when detailed work is conducted around a mine site within what is otherwise a poorly explored frontier basin (see also Section 2.2.2).

Formations should not contain major unconformities, although minor disconformities may be acceptable. The contacts of the formation should be established at obvious lithologic changes. These may be sharp or gradational. An unconformity is a logical choice for a formation contact. Where lithologies change gradually either vertically or laterally it may be difficult to choose a logical place to draw the contact. For example a mudstone may pass up into a sandstone through a transitional succession with sandstone beds becoming thicker and more abundant upwards. The mudstone–sandstone formation contact could be drawn at the oldest thick, coarse sandstone (with thickness and coarseness carefully spelled out), or at the level where sandstone and mudstone each constitute 50% of the section, or at the youngest extensive mudstone bed. The choice is arbitrary, and it is immaterial which method is selected as long as the same method is used as consistently as possible throughout the extent of the formation.

Other problems of definition arise where there are lateral lithologic changes, requiring definition of a new formation. A simple diachronous contact is not a problem, but where the two units intertongue with one another it may be virtually impossible to draw a simple formation contact. One solution is to give each tongue the same name as the parent formation. A section passing through the transition region may then show the two formations succeeding each other several times. The only problem this causes is if formation contact and thickness data are stored in a data bank and used in automated contouring programs. Without additional input from the operator a computer program might not be able to handle this type of data. Other alternatives are to define the whole transitional rock volume in terms of one of the parent units, or to separate the transitional lithologies as a separate lithostratigraphic entity, or to give separate tongues their own bed, tongue or member names. The ISSC guide does not specify any rigid rules for the solution of such problems. The main criteria should be practicality, convenience and consistency. An illustration of interpreted stratigraphic relationships in the form of a cross section or cartoon will usually serve to clarify any ambiguity. Some examples are discussed in Section 3.4.3.

The group. All other stratigraphic units are based on the formation. A group consists of two or more formations related lithologically. In the past named groups have been established for thick and varied successions without first defining the constituent formations. This is not recommended practice. Groups should not contain major unconformities.

The component formations of a group may not everywhere be the same. Lateral facies changes requiring definition of different formations can occur within a single group. In contrast a component formation may extend laterally from one group to another. Groups are normally defined for regions of complex stratigraphy. Toward the basin margin or basin centre the component formations may lose their individuality, in which case the group may there be "demoted" to a formation, while still retaining the same name.

The terms **supergroup** and **subgroup** are occasionally used to provide an additional hierarchy of subdivision. Usually there are historical reasons for this; some of the higher ranking names may have started out as member, formation or group names, with reclassification and "promotion" being required as additional work demonstrated the need for further subdivision.

The member. This is the next ranking unit below formation. Not all formations need be divided into members, and formal names need be used for only a few, or one, or none of the constituent members, depending on convenience or the level of information available.

There are no standards for the thickness or extent of members, and commonly it is difficult to decide whether to define a given lithostratigraphic unit as a member or a formation. However, recommended practice is that all parts of a succession be subdivided into formations, and so this is the best level at which to start. A member cannot be defined without its parent formation.

For mapping and other purposes it commonly is convenient to establish informal units such as the "lower sandstone member", which do not require formal names.

Tongue or lentil. These are similar to members. Because of their geometric connotations the terms are useful for parts of formations where they interfinger with each other. Formal names may be established for one, several or all such units, depending on convenience and practicality.

Bed. This is the smallest formal, named unit in the hierarchy of lithostratigraphic units. Normally only a few parts of a stratigraphic succession will be subdivided into named beds. Coal seams in mine areas, prominent volcanic tuff horizons and other marker beds are typical examples. Certain stratabound ore-bearing beds such as placer units may also be named.

Choice of names. When establishing a named unit it is standard practice to give it a geographical name, chosen to suggest the location or areal extent of the unit. This may be a river, lake, bay, headland, hill, mountain, town, village, etc. Permanent names are preferable.Subsurface work in frontier basins, particularly in offshore areas, may rapidly use up all the available names, in which case the name of the well chosen as the type section may be used. Failing this, names may have to be invented.

In most cases the geographical name will be followed by the rank designation, e.g. Wilcox Group, Pocono Formation. For beds this is commonly not done, particularly in the case of coal seams, for which a complex mine terminology may have evolved. Many older stratigraphic units use a lithologic term instead of a rank term, e.g. Gault Clay, Austin Chalk, but this is not recommended because the rank of the unit is not clear from the name alone.

Workers should beware of using a geographic name that has already been employed in a different context, or renaming units without justification. Geological survey organizations commonly retain a file of current and obsolete stratigraphic names which the worker may wish to consult.

Changes in rank. Lithostratigraphic units may be changed in rank as the level of knowledge improves. For example the Cornwallis Group started as the Cornwallis Formation, and was raised to group rank when it was realized that it contained three mappable units of formation rank. Conversly the Eureka Sound Group was named for a thick and varied clastic succession, but it was never subdivided into named constituent formations, and was reduced to formation status.

When a unit is raised in rank the original name should not be used for any of the subdivisions, but is best retained for the higher ranking unit or abandoned altogether.

3.4.3 Mapping and correlation procedures

The term **correlation** means to establish the mutual relations existing between stratigraphic units. When used in the context of lithostratigraphic units the term implies that there are physically continuous rock bodies, of more or less constant character, that can be traced from one part of a basin to another. Such units may be diachronous. When dissimilar lithostratigraphic units are said to be correlated, a time connotation is implicit. That is to say, two units may be formed in different environments and contain different lithologic characteristics, but may be said to be correlated because they are known to have been formed at the same time, on the basis of biostratigraphic or other chronostratigraphic evidence.

Correlations based strictly on lithostratigraphic evidence should be viewed with caution, because similar units may be deposited at different times in different areas and have no physical connection with each other. Many mistakes have arisen this way when geologists have attempted to correlate between widely spaced sections.

There are only three conditions under which lithostratigraphic correlation should be attempted without supporting biostratigraphic or chronostratigraphic information:

1. where the beds can be physically traced out on the ground or between very closely spaced subsurface sections
2. where there are distinctive, readily recognizable marker beds, or where facies studies indicate considerable lateral bed continuity
3. in the Precambrian, where only very generalized chronostratigraphic data (radiometric dates) may be available.

Surface mapping. In areas lacking vegetation cover, such as in arid or arctic climates, lithostratigraphic units can commonly be traced in outcrop on the ground, from the air, or from air photographs. In areas of simple to moderate structural complexity photogeological reconnaissance using stereoscopic air photo coverage can produce preliminary maps of large areas very quickly. Standard air photos taken from an elevation of a few thousand meters and printed at a scale of about 1 cm-1 km are the best for this purpose. Satellite images are at too small a scale (although remote sensing technology is developing at a rapid rate, and this statement may soon be out of date). Most of the Canadian Arctic has been mapped by the Geological Survey of Canada using conventional air photographs. The photoreconnaissance maps are used to determine areas of structural complexity or uncertainty, and also locations where the stratigraphy appears to be particularly well exposed. These places are then visited on the ground in order to resolve structural problems and measure detailed stratigraphic sections. Marked facies changes, such as a carbonate–mudstone transition at a shelf edge, may be readily apparent and easily mapped. This is lithostratigraphic mapping in its purest form. Coherent units may be distinguished and mapped, but at this stage there is little or no information on age or possible diachronism of the units. An example of Arctic lithostratigraphic mapping based on air photos is shown in Figure 3.1.

Where there is a significant cover of soil, surficial deposits or vegetation it may still be possible to carry out a similar type of mapping by noting changes in ground texture, soil colour or vegetation type, or subtle changes in relief reflecting varying resistance to erosion of the different units. For example the stratigraphic succession may be studied in detail in good outcrops along river cuts, and mapped out between the rivers by tracing the outcrops into covered regions where the unit shows up as a resistant ridge, or as a belt of distinctively colored subsoil fragments.

In areas of dense vegetation cover, thick surficial deposits or complex structure, detailed surface mapping may not be possible. Lithostratigraphic correlation is then carried out in very much the same way as subsurface work, attempt-

Fig. 3.1. Use of vertical air photographs for lithostratigraphic mapping. An example from southeastern Melville Island, Canadian Arctic (courtesy of A.F. Embry).

ing to match isolated sections by whatever means possible (see next section). Formal lithostratigraphic units such as formations, groups, etc., may be defined, but it may be difficult to plot their distribution accurately on a surface geological map. This is the case with many temperate or tropical regions. In areas such as Western Europe much use is made of soil augurs and temporary exposures, such as trenches dug for pipelines, to supplement outcrop information for mapping purposes.

For further details on mapping procedures the reader is referred to standard field manuals such as that by Compton (1962).

In areas of complex facies changes detailed ground mapping may be a very productive exercise for the basin analyst, because work of this type can provide significant clues for environmental and paleogeographic reconstructions that may be applicable to large parts of the basin. At this stage lithostratigraphic documentation is only a preliminary to detailed facies and trend studies, such as are described in Chapters 4, 5 and 6. An example is illustrated in Figure 3.2. This section is one of many compiled for the Carboniferous, coal-bearing, deltaic sediments of Kentucky and West Virginia by Horne et al. (1978) and Horne and Ferm (1978). They are based on extensive artificial cuts along railroad lines and highways. The series of vertical rectangles outlines detailed sections plotted on the ground across an exposure more than 5 km wide. These are connected by lithostratigraphic correlation lines, most of them traced out on foot in the field. Unornamented areas represent areas not exposed or not studied. The section reveals a series of sandstone beds 5 to 15 m thick, many with irregular, sometimes deeply channeled bases. They are separated from each other by siltstone, mudstone and coal beds. Horne et al. (1978) and Horne and Ferm (1978) discussed the lithostratigraphic succession, lateral variability and facies data of this and other large exposures and interpreted the rocks as the deposits of a coastal fluvial–deltaic plain.

Figure 3.3A illustrates a carbonate example. Jones and Dixon (1975) measured detailed, closely spaced cliff sections at Port Leopold in Northern Somerset Island in the Canadian Arctic. The rocks are Silurian shelf carbonates of the Leopold Formation and show considerable lateral facies changes over distances of only a few tens of meters. Key beds can be traced on foot from section to section permitting a detailed correlation to be documented. Lithostratigraphic variations in this and adjacent sections were used by Dixon and Jones (1978) to erect local members within the Leopold Formation. Extension of this work along well-exposed cliff sections showed that, over a distance of about 130 km, the member boundaries are markedly diachronous. Marker beds consisting of distinctively weathering units of varying lithology could be physically traced along the cliffs, permitting tight correlation of the sections (Fig. 3.3B; the sections in Figure 3.3A correspond to section 11 in Figure 3.3B). In addition, it was found that pulses of quartz sand detritus could be correlated parallel to the marker beds. On regional evidence the line of section is interpreted to be parallel to the line of sedimentary strike, relative to the detrital source, whereas the various shallow shelf environments shifted gradually to the south at the same time (left in Figure 3.3B), generating diachronous lithostratigraphic units.

Section matching. Correlations cannot be walked out or mapped out in the subsurface, and lithostratigraphic work depends on the presence of distinctive units in well records or core that can be recognized and traced from one section to the next. The best correlations are those based on widespread marker beds, because of their chronostratigraphic value. With or without these the geologist must develop visual skills in pattern recognition, placing graphic lithostratigraphic logs or geophysical logs beside each other and comparing all the various attributes present, judging which correlations permit the closest fit. This process is performed routinely by subsurface exploration geologists. Geophysical logs are particularly useful for such purposes. However, without chronostratigraphic markers the geologist must always watch for the possibility of diachronism and lateral facies change. Similar correlation procedures may be necessary for surface work, although here the task tends to be somewhat easier because of the greater range of lithostratigraphic parameters that can be examined in outcrop.

Even the most detailed biostratigraphic zonation usually is of little value for the level of correlation required at the scale of diamond drill hole (DDH) exploration of a mining property, or exploration of a petroleum prospect. The art of sec-

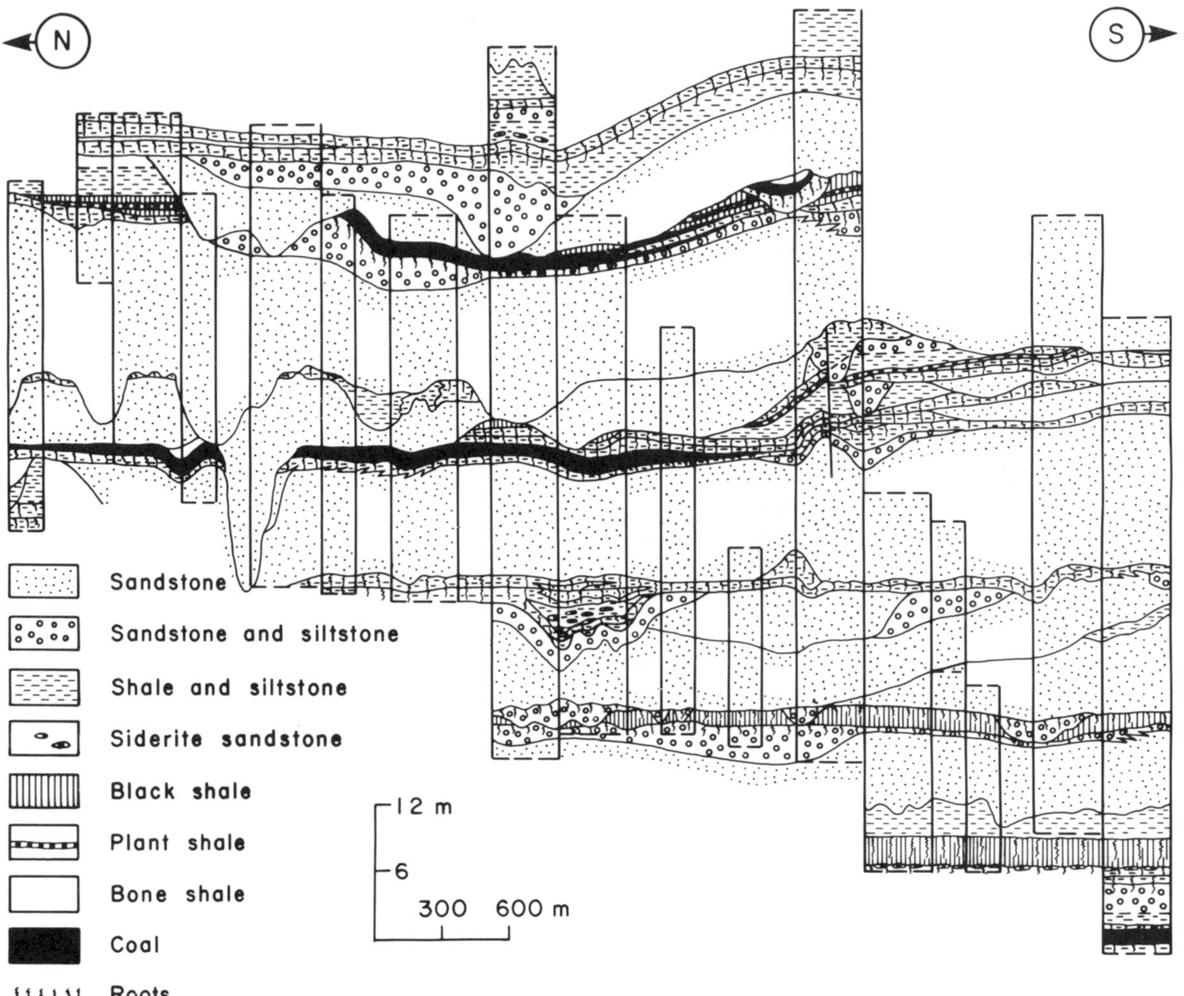

Fig. 3.2. Example of a detailed lithostratigraphic study carried out on well-exposed road side outcrops. Carboniferous, Kentucky (Horne et al., 1978).

tion matching is therefore an important one for the explorationist.

Schultz (1982) pointed out that where a rock body is delimited by marker beds or by well-defined biostratigraphic horizons it has a local chronostratigraphic significance. He proposed the term **chronosome** for such units, and suggested that definition and naming of individual chronosomes would have considerable value for basin analysis work. A chronosome is distinct from a formation in that it can contain lateral lithologic (facies) changes, and is delimited by time planes. It therefore defines a single episode or period of deposition.

An example of the use of a marker bed for correlation is shown in Figure 3.4 (after Hsü et al., 1980). The illustration shows surface and subsurface sections through the deep water conglomerates, sandstones and mudstones of the Ventura Basin, California. An ash bed of Plio-Pleistocene age occurs in all but one of the sections. Hsü et al. (1980) estimated that the sections represent about 15,000 years of sedimentation. The ash bed is the only unit that can be correlated from section to section with any degree of certainty, and without it the relationships between these sections would be impossible to ascertain. The sections are from different parts of an elongate basin dominated by mudstone along the margins (section E, F, G), thin bedded and graded sandstones along the basin axis (transported by longitudinal turbidity currents) (sections A, B, C, H) and occasional slump-derived conglomerate units, such as one in a

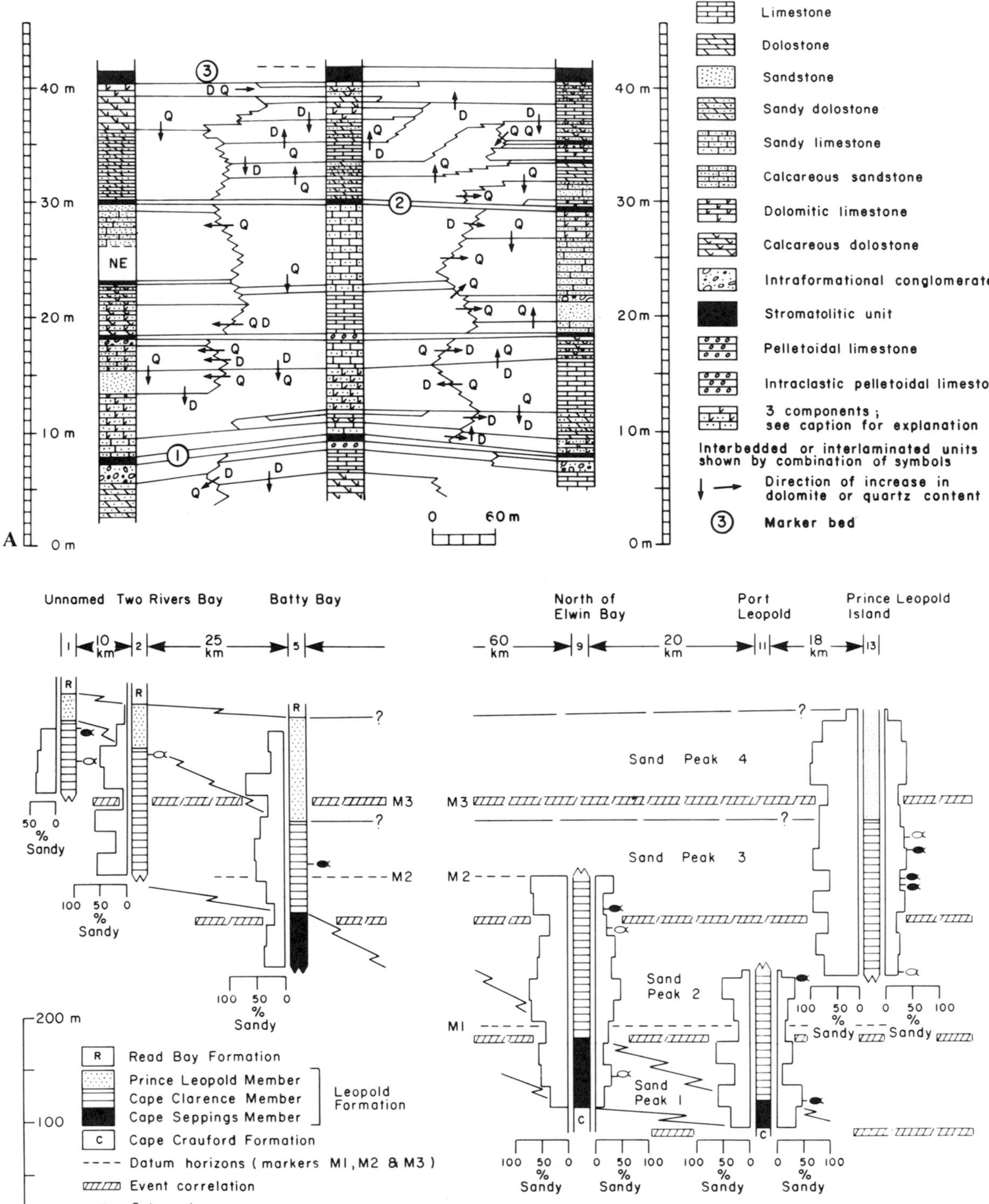

Fig. 3.3A. Use of stromatolitic marker beds (numbered 1, 2, 3) to correlate closely spaced sections showing considerable lateral facies changes. The 3-component lithology consists of interbedded quartz, micritic limestone and dolomite. Silurian, Somerset Island, Canadian Arctic (Jones and Dixon, 1975). **B** Chronostratigraphic correlation of regional stratigraphic sections using weathering markers and correlation of sand pulses (S.P.). The sections shown in Fig. 3.3A are combined here as section 11. Chronostratigraphic correlation demonstrates the diachroneity of the lithostratigraphic units. Silurian, Canadian Arctic (Dixon et al., 1981).

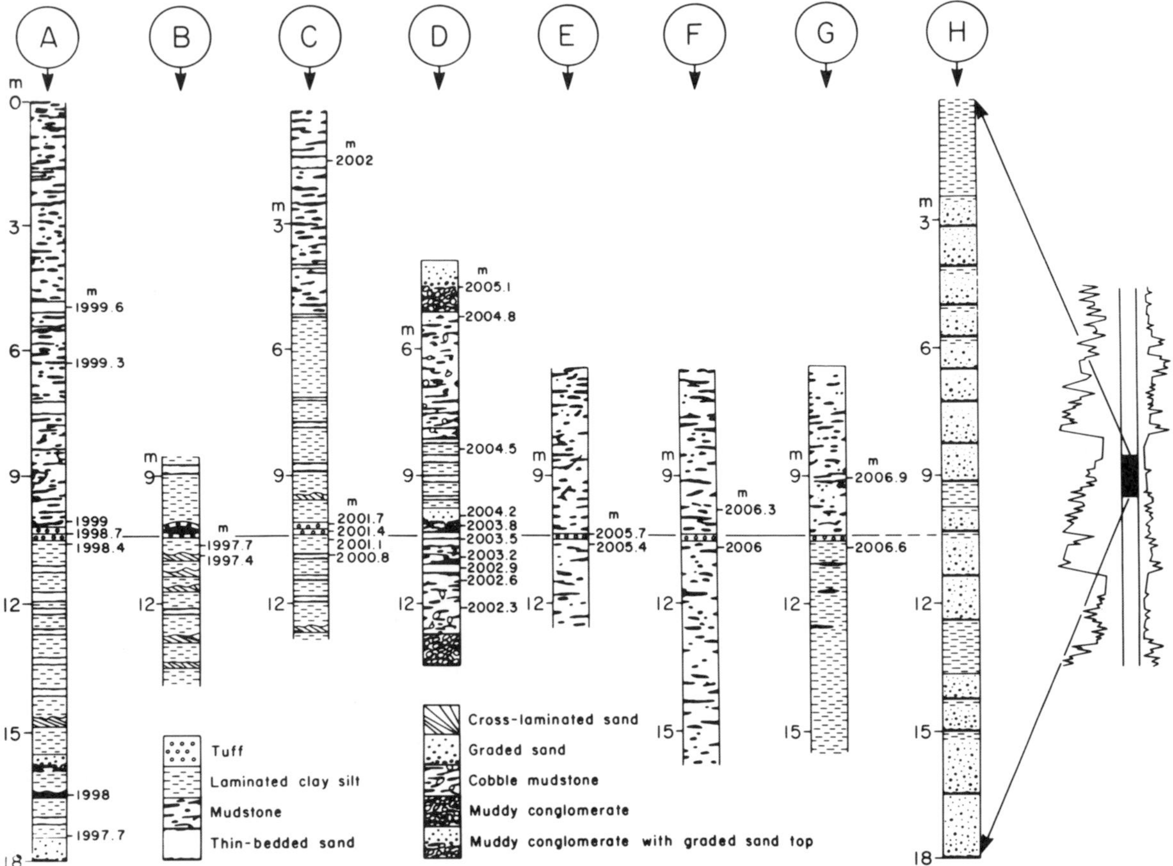

Fig. 3.4. Use of a tuff bed to correlate turbidite units and other deep-water facies. Pliocene, Ventura Basin, California (Hsü et al., 1980).

channel interpreted to have scoured away the ash bed in section D.

Other examples of the use of ash beds for correlation are discussed by Johnson et al. (1979; see section 2.6.5), and Allen and Williams (1981). Problems with this method arise if more than one ash is present, and special laboratory techniques may be required to determine distinctive mineralogical or other features of each ash bed to permit correct correlation (e.g. Westgate, 1980).

A few other lithofacies are suitable for use as lithostratigraphic markers. In carbonate basins abrupt shifts in lithofacies may occur in response to sea level changes, and these may make good markers. Sediment gravity flows are almost instantaneous deposits and therefore have chronostratigraphic potential. The problem is to correlate them correctly between sections (see below). Hardgrounds are surfaces of non-deposition or very slow sedimentation, and may be traceable over many kilometers (Bathurst, 1976; Stoakes, 1980). Transgressive sandstones, such as those occurring above an unconformity, or in the abandonment phase of a delta (Elliott, 1978), may be widespread, relatively thin, glauconitic, mature quartzose units readily recognizable in the sedimentary record. However, they may be markedly diachronous. Miall (1979) used a distinctive bituminous mudstone unit (basal Kanguk Formation) to map the Upper–Lower Cretaceous contact in Banks Island, in the western Canadian Arctic. The bituminous content had oxidized to a bright red, which enabled the contact to be mapped from a low-flying helicopter. The unit also had a distinctive subsurface log signature. This same marker bed is present throughout much of Arctic North America (Miall, 1979). Hallam and Bradshaw (1979) commented on the sedimentary–tectonic interpretation of black mudstones and oolitic ironstones and their stratigraphic significance.

Stratigraphic sequences vary considerably in the ease with which lithostratigraphic correlation may be performed at the member or smaller scale level. In many depositional environments facies changes are numerous, and without marker beds section matching may be impossible. This is the case with fluvial and deltaic deposits and the sediments of most other coastal clastic and carbonate environments. Submarine fan deposits also show considerable facies variation. Environments which may give rise to more lateral consistency include lakes (away from any marginal clastic facies), low-energy carbonate and evaporite shelves, some clastic and carbonate (sabkha) tidal flats, and deep oceanic abyssal plain or trench deposits. In some of these cases practically every sedimentation unit is a marker bed.

Figure 3.5 illustrates carbonate–evaporite correlations made in the Middle Devonian sediments of Saskatchewan (from Wardlaw and Reinson, 1971). The sections were plotted from cores derived from wells extending more than 160 km across the basin. The authors were lucky to locate so many cores through the same thin interval. The consistency of facies and thickness is remarkable. Correlation of some Miocene basin plain turbidites is shown in Figure 3.6. Ricci-Lucchi and Valmori (1980) measured a series of outcrop sections along the axis of a turbidite basin in the northern Apennines. Transport directions were toward the southeast (left to right). A considerable facies change takes place down the basin so that a sandy section such as #1, could not be correlated with a mud-dominated section, such as #12. However, by using the intermediate sections it is possible to trace the lateral facies change of virtually every bed. The amount of fieldwork required to establish this correlation network was considerable. Deep-sea cores in some modern turbidite basins have demonstrated correlations over distances up to 500 km (e.g. Bennetts and Pilkey, 1976; Fig. 6.43).

Many attempts have been made to use statistical techniques for lithostratigraphic correlation. Methods such as Fourier analysis have been used to characterize the cyclicity of certain attributes in a succession, such as petrographic variation or various petrophysical properties (e.g. Preston and Henderson, 1964). Various pattern matching techniques have been used to determine the best fit between two or more sections (e.g. Davies and Ludlum 1973; Smith and Waterman, 1980). However, this approach is not recommended by the author, for the following reasons: in cases such as the turbidite and evaporite examples described above correlation is obvious enough that it can be made manually. Where the correlation is not obvious, statistical techniques are not likely to help, because numerical manipulation cannot easily accomodate thickness and facies changes. These are better dealt with by applying sedimentological principles.

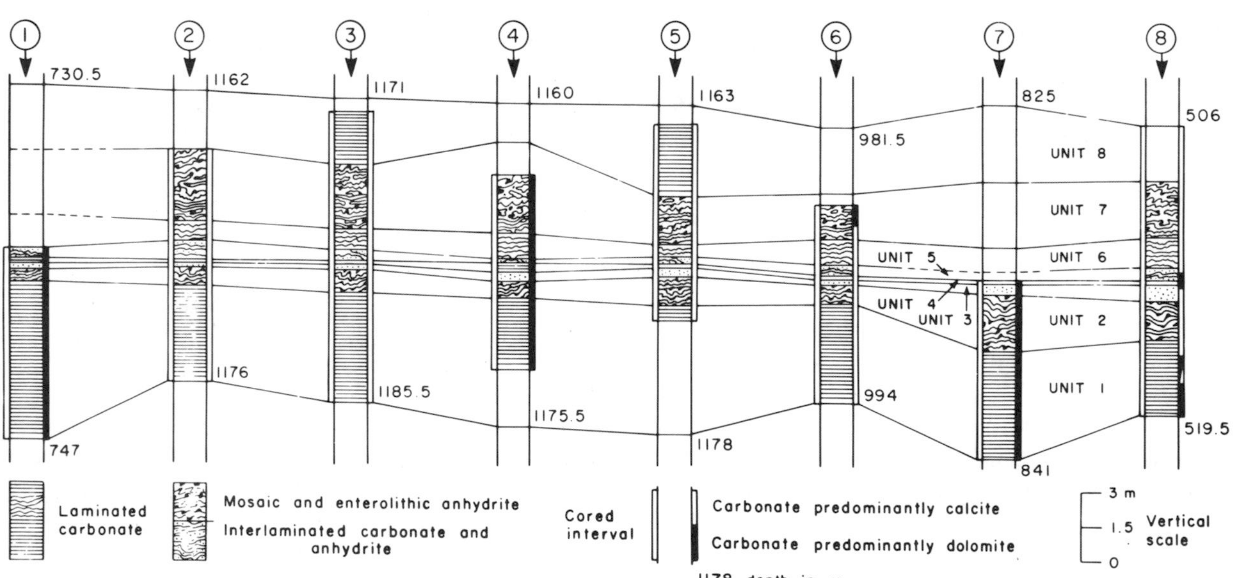

Fig. 3.5. Detailed lithostratigraphic correlation of cored subsurface units consisting of thinly interbedded carbonates and evaporites (Wardlaw and Reinson, 1971).

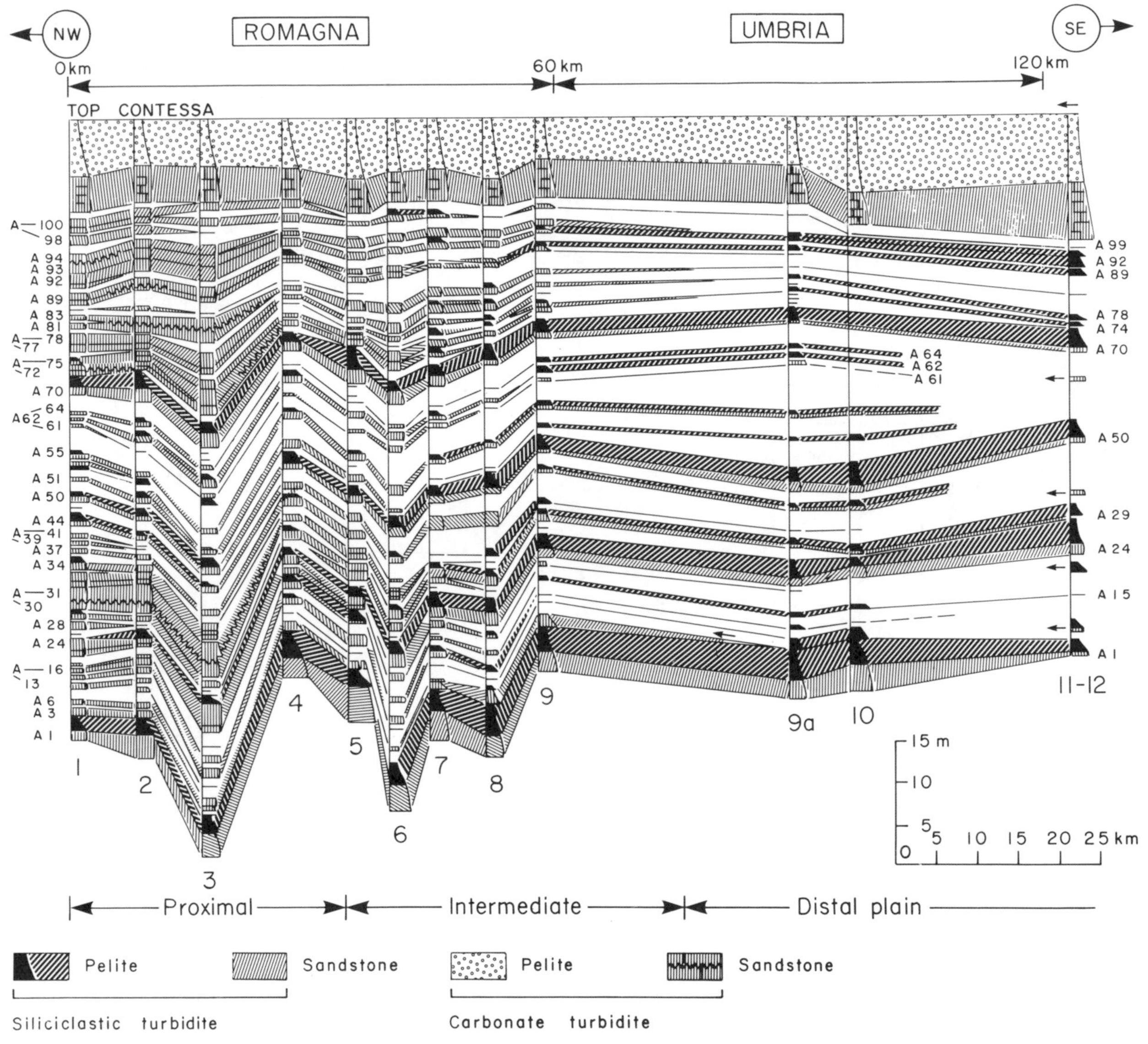

Fig. 3.6. Detailed lithostratigraphic correlation of outcrop sections through basin plain turbidites. The line of section is approximately parallel to paleoslope. Blank intervals are pelites. Cenozoic, Italy (Ricci-Lucchi and Valmori, 1980).

Lithostratigraphic correlation at the formation and member level is routine in surface and subsurface work. Figure 3.7 illustrates the use of geophysical log signature to correlate well sections, an example from the Milk River Formation (Upper Cretaceous) of Alberta (Myhr and Meijer-Drees, 1976). The geologist does not need to know much about the actual lithologies penetrated to carry out such correlations, nor is it necessary to interpret the actual values of the log deflection. Curve shape is sufficiently distinctive to permit precise, reliable correlation until a facies change alters the pattern. Figure 3.8 demonstrates a case were simple log matching cannot be performed reliably (Athabasca Oil Sands, Alberta; from Carrigy, 1971). This is a common problem in channelized fluvial, deltaic and submarine fan deposits. Petroleum geologists frequently use formation pressure data to test interconnectedness of reservoir units. Pressure is plotted against well depth. Continuous (and therefore correctly correlated) units yield data points falling on a straight line. Non-connected units show up as isolated points or as separate but

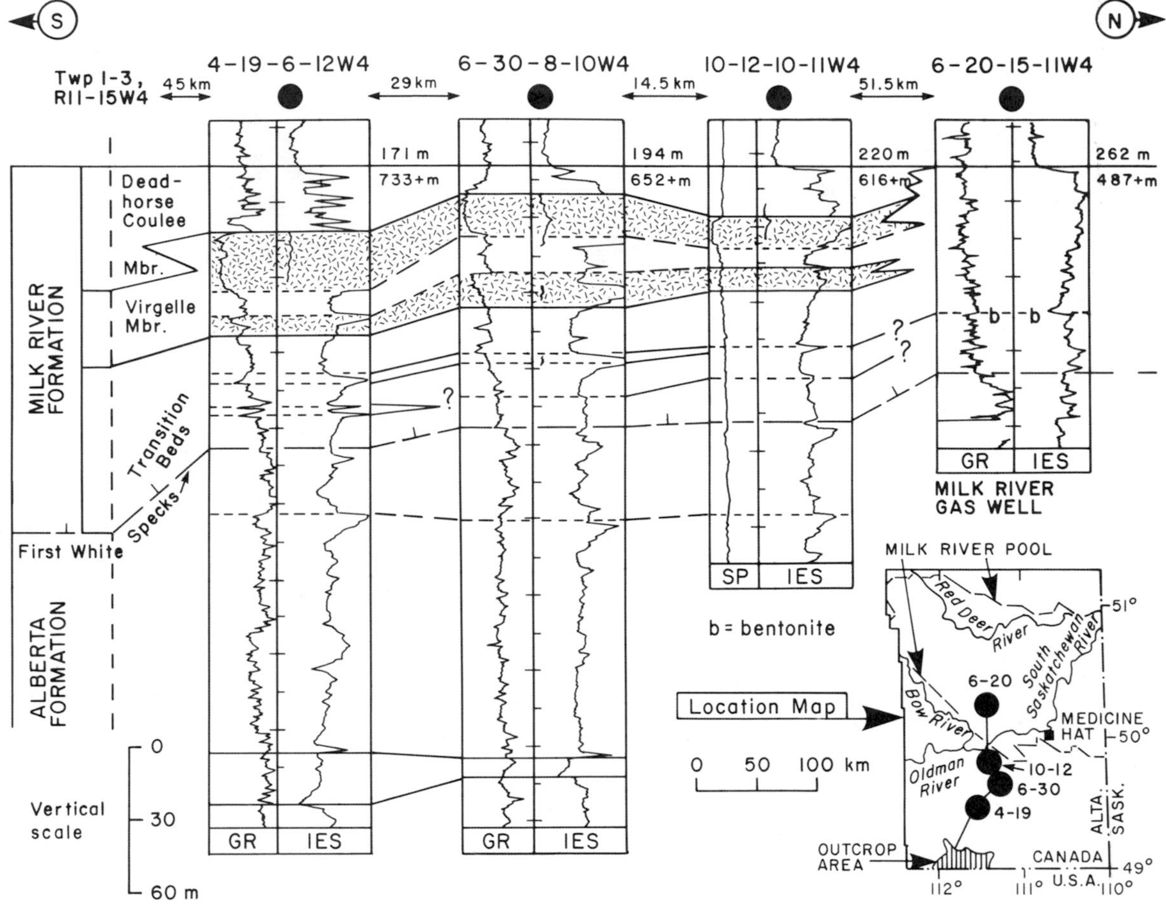

Fig. 3.7. Use of gamma ray (GR) and induction-electrical survey (IES) logs to correlate subsurface sections. This diagram shows that where lithostratigraphic units retain similar thickness and lithology they can readily be correlated using geophysical logs. Cretaceous, Alberta (Myhr and Meijer-Drees, 1976).

parallel straight lines. An example is shown in Figure 3.9. Without such data it is easy to make spurious correlations which may only be revealed as incorrect after many more wells have been drilled.

A particularly ambitious attempt at lithostratigraphic correlation is illustrated in Figure 3.10 (after Young, 1979). The rocks are cratonic and pericratonic sequences of Proterozoic age. A few radiometric dates are indicated and the correlations are therefore loosely controlled by chronostratigraphy. However, virtually all the dashed correlation lines are based on more or less detailed lithostratigraphic section matching. The correspondence between sections 3 and 4 is particularly close (these sections are about 1200 km apart). Imagine a geologist attempting to match Phanerozoic sections, some of them up to 20 km thick, across half a continent, using less than a dozen biostratigraphic control points! The attempt would be greeted with scorn. However, for the Precambrian this represents one of the best examples of regional lithostratigraphic correlation available. Note the recognition of three unconformity-bound sequences, A, B and C, each in the order of 300 Ma in length, and compare these with the Phanerozoic sequences discussed in Chapter 8.

In conclusion, it is worth repeating that the best lithostratigraphic subdivision and correlation is based on an understanding of the depositional systems within which the rocks were laid down. Lithostratigraphy is a descriptive, empirical science, but correlations can be made more comprehensive, and are likely to be more reliable, if they are erected using genetic, sedimentological principles.

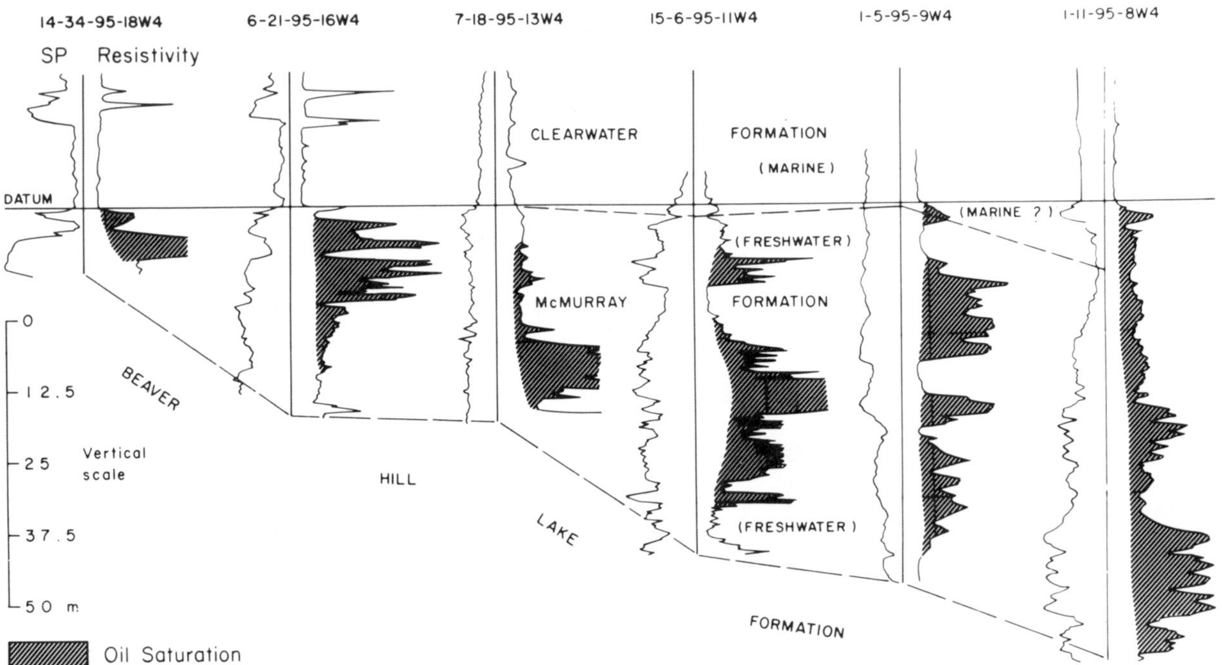

Fig. 3.8. In contrast to Fig. 3.7 this diagram illustrates the difficulty of correlating sections that show marked lateral facies changes. Fluvial deposits of Athabasca Oil Sands, Alberta. Length of section is 200 km (Carrigy, 1971).

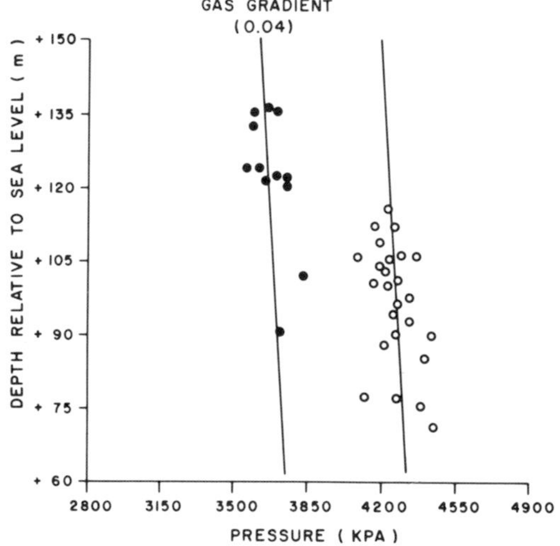

Fig. 3.9. Use of pressure–depth plot to test lithostratigraphic correlation. The points fall into two groups indicating that they represent two channel-fill sandstone bodies isolated from each other by fine-grained units. The data were derived from Lower Cretaceous fluvial sandstones over an area of about 1900 km^2 in Alberta (Putnam and Oliver, 1980).

3.5 Biostratigraphy

Biostratigraphy is the study of the relative arrangement of strata based on their fossil content. Descriptive or empirical biostratigraphy is used in erecting zones for local or regional stratigraphic correlation, and forms the basis for a global system of chronostratigraphic subdivision.

Fossil content varies through a stratigraphic succession for two main reasons: evolutionary changes, and those due to ecological differences such as changes in climate or depositional environment. Biostratigraphy should be based only on evolutionary changes, but it is always difficult to distinguish these from changes that take place in a biostratigraphic assemblage as a result of ecological modifications, and this problem is a cause of continuing controversy for many fossil groups.

Biostratigraphy obviously can only be studied and a classification erected where fossils are present. This rules out much of the Precambrian, although some use has been made of stromatolites for correlation, particularly by Russian geologists, and fossil microbe communities hold promise (section 3.6.7). Even in the Phanerozoic

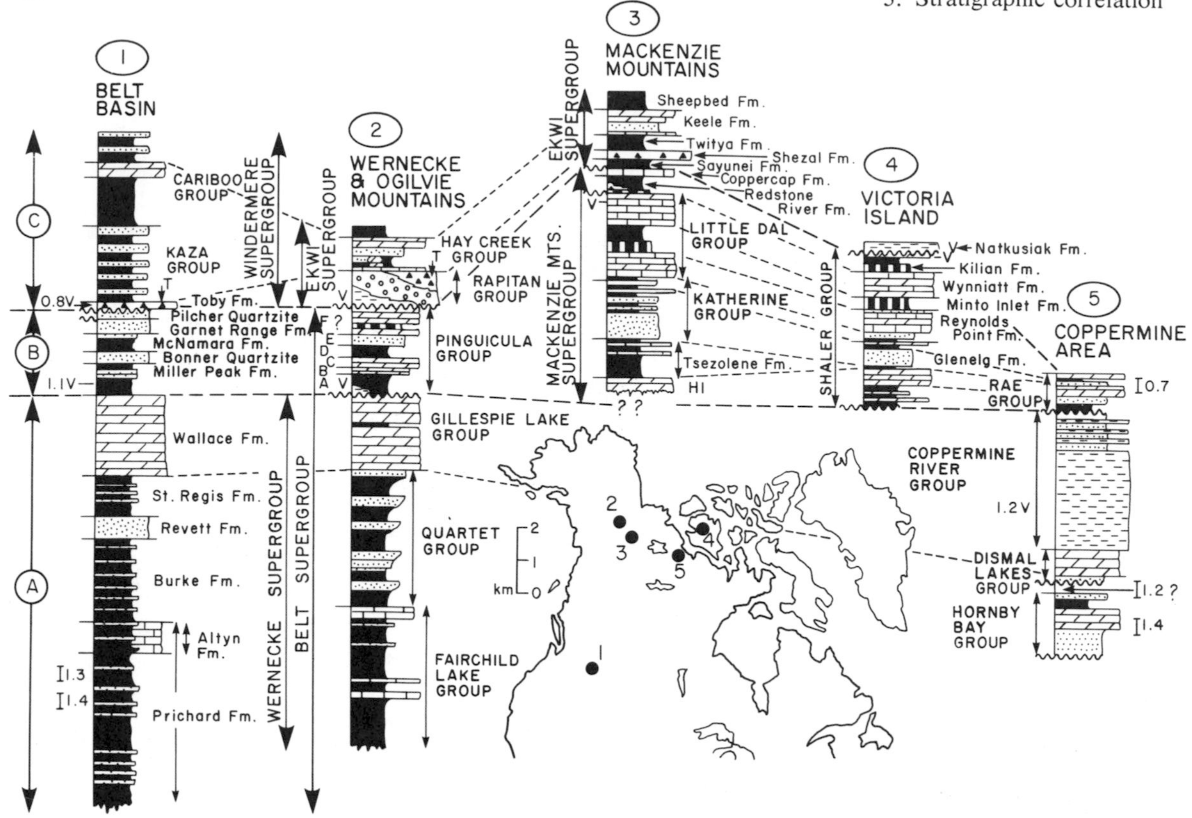

Fig. 3.10. Lithostratigraphic correlation of Proterozoic sedimentary sections. North American craton (Young, 1979).

there are many rock units for which the fossil record is very sparse, and biostratigraphic subdivision is correspondingly crude. This is particularly the case in nonmarine strata or those (particularly carbonates) in which fossil remains have been destroyed by diagenesis.

Biostratigraphy is a study for specialists. Refined work requires intimate knowledge of the phylogeny of a large number of fossil groups, and their regional or global distribution. To accumulate this knowledge may take half a lifetime, and the subject is an excellent example of a science in which the practitioner seems to spend inordinate amounts of time learning more and more about less and less. Some of the leading authorities in a particular fossil group may be able to discuss the cutting edge of their research with only half a dozen other colleagues around the world. This gives them considerable value if one happens to find their kind of fossil, but it may somewhat restrict their scientific scope. Geologists engaged in basin analysis of the Phanerozoic are very rarely such specialists themselves. Biostratigraphers are therefore employed by many organizations to provide these specialized service skills, or they function independently as consultants. They may be engaged much of the time in pursuing paleontological research, but are able to provide biostratigraphic diagnoses for selected fossil types over a specified age range.

Professional biostratigraphic work may take a great deal of field and laboratory time. Sections that a sedimentologist may dismiss as sparsely fossiliferous may yield hundreds or even thousands of specimens to the careful collector. Laboratory extraction of microfossils or palynomorphs may yield similar numbers. It is this kind of work which is necessary for modern, refined biostratigraphic work. Much of the submitted material, particularly that from frontier exploration wells, may itself provide the basis for new biostratigraphic zoning schemes.

Basin analysts should understand what they are getting when they submit their own material to a specialist for identification. Commonly all they are interested in is the age of the enclosing rock, information which can be used for correlation purposes. Age is a chronostratigraphic interpretation based on taxonomic descriptions, but there commonly are problems of fossil identification or interpretation, particularly where the fossil record is sparse or the material is from a new, poorly studied area. The purpose of this section, therefore, is to discuss some of the problems of the biostratigraphic record, and to describe the methods biostratigraphers use in plying their trade.

3.5.1 The nature of the biostratigraphic record

Biostratigraphy and evolution. Biostratigraphy is basically an empirical, descriptive science. Biostratigraphic subdivision and correlation is based on the gradual changes through time of faunas and floras. Such changes reflect evolution, but it is not necessary to understand the evolutionary development (phylogeny) of a group of animals in order to use the faunal information, just as English history is linked to the "age range" of kings and queens, although it is not necessary to know the family relationships of these individuals in order to make use of the chronology. However, it improves one's understanding of biostratigraphic information if it can be related to evolutionary changes.

Biologists and paleontologists have described three types of evolutionary pattern. There remains a controversy over the relative importance of these three styles, but there seems little doubt that all three exist in nature, because well documented examples can be quoted to demonstrate each.

The first was termed **phyletic gradualism** by Eldridge and Gould in 1972 (their 1977 paper is a useful, recent discussion). It refers to long-term evolutionary change, typically in response to geographical, climatic or other environmental pressures. Certain varieties of a species may be favored by these changes, so that there is a gradual adjustment in the stock until a distinctive new species appears. Eldridge and Gould (1977) doubted the reality of this process. They maintained that population groups are genetically conservative and that change will only take place on the fringe of the species range where ecological pressures are most acute. According to this view most species are stable most of the time, and the hope of paleontologists to be able to continually refine biostratigraphic subdivision by perfecting our knowledge of evolutionary lineages is unlikely to be realized. They suggested that unidirectional environmental change rarely continues long enough to "force" evolution, and they quote several examples of supposed lineages that have now been discredited, for example the famous coiling of the mollusc *Gryphaea* through the Jurassic. Trueman (1922) claimed that the shell size, thickness and degree of coiling could be related to position on an evolutionary lineage, and hence age, with a considerable mathematical precision. Later work (Hallam, 1959; Burnaby, 1965; Gould, 1972) showed that the coiling and other changes were related entirely to ecological (environmental) factors, and that similar coiling behavior recurred wherever conditions favoured it. Trueman's lineage and the chronology arising from it turned out to be based on circular reasoning.

However, the disproof of one example does not discredit the entire idea. McKerrow (1971) quoted the even earlier work of Rowe (1899) on the Cretaceous echinoid *Micraster*. Rowe was one of the first to attempt to define a detailed evolutionary lineage; his conclusions are shown in Figure 3.11. The recognition of the succession of species was based on careful stratigraphic collecting in the English Chalk. More recently Kermack (1954) has confirmed the general pattern outlined by Rowe but has placed a somewhat different interpretation upon it. He determined that the changes discovered by Rowe represent a gradual broadening of the ecological niche occupied by *Micraster* over a period of about 15 Ma, without any perceptive environmental change. Some varieties of the genus developed a gradually deeper burrowing habit. However, these are not separate lineages because intermediate (interbreeding) varieties are present at every stage (Fig. 3.12). Nevertheless, this seems to represent a good example of phyletic gradualism, and one that has been of biostratigraphic use for over eighty years.

Kauffman (1977) described two examples of phyletic gradualism in Cretaceous pelecypods. Figure 3.13 illustrates a series of histograms of height–width ratios of the *Inoceramus pictus* lineage, derived from populations collected at about 40 cm intervals (lower graph), and the number of growth ridges in the first 25 mm of shells of *Mytiloides labiatus* (upper graph). The data permit subdivision of the population into species (S), subspecies (SS) and morphological zones (MZ) as indicated in the adjacent columns.

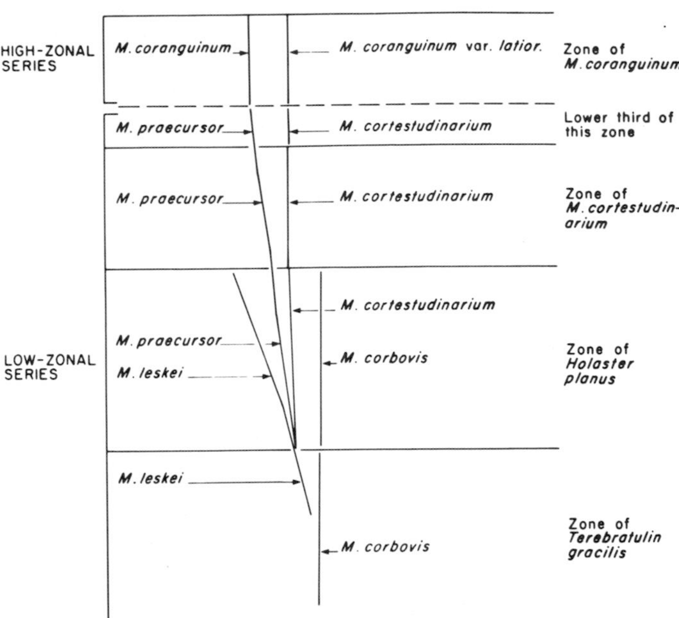

Fig. 3.11. Interpreted lineages of the echinoid *Micraster* in the British Chalk (Upper Cretaceous) (Rowe, 1899).

Two types of environmental adaptation can occur, that which is accompanied by permanent genetic change, and that which can, to some extent (never precisely), reverse itself to recreate the same variety or race of a species more than once, whenever the same environmental conditions are repeated (homeomorphy). Clearly the first type is the only one of use to biostratigraphers, but the literature is replete with ambiguous biostratigraphic determinations that may be falsely based on diachronous environmental change. For example this has been a serious problem with the ammonites (Kennedy and Cobban, 1977), one of the best of biostratigraphic indictors.

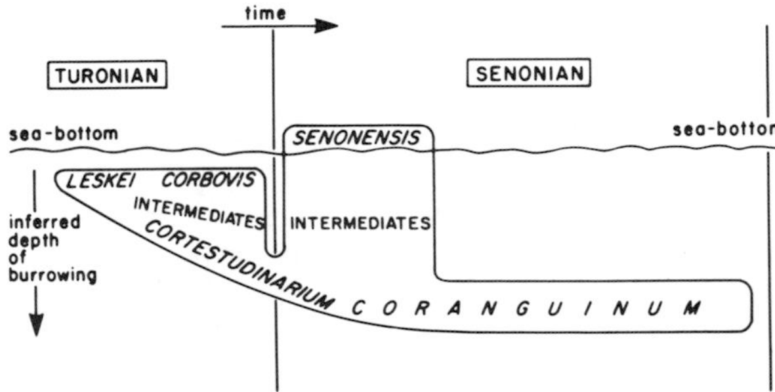

Fig. 3.12. Reinterpretation of *Micraster* phylogeny in terms of a change in its ecological adaptation. It is suggested that a deeper burrowing habit developed with time in some forms, but that intermediate forms were at all times in existence (Kermack, 1954; McKerrow, 1971).

Taxa which evolve by phyletic gradualism have the most potential for refined biostratigraphic zonation, but they require specialist study to recognize the very subtle changes between the varieties. This type of work is usually beyond the abilities of the generalist basin analyst.

The second style of evolution was named **punctuated equilibrium** by Eldridge and Gould (1977). The concept is based on the premise that in a successful, widely distributed taxon the population is genetically conservative. Evolution is thought to occur only where extreme variants are selected by environmental pressures on the fringes of the species range. Rather than a gradual adaptation to an ecological niche or a broadening of a species range by extending slowly into subtly different niches, as in phyletic gradualism, the hypothesis of punctuated equilibrium proposes the spasmodic occurrence of bursts of relatively rapid evolutionary change. Extreme variants of a species can only evolve into a new species if they become isolated by changes in environment, climate or geography, as through the rifting and drifting apart of continental plates. The differences between these two concepts of evolution are illustrated in Figure 3.14.

Sylvester-Bradley (1977) gave an example of punctuated equilibrium based on the work of Howarth (1973) on the Lower Jurassic ammonite genus *Dactylioceras*. Over a few meters of strata the whorl shape and rib pattern show considerable

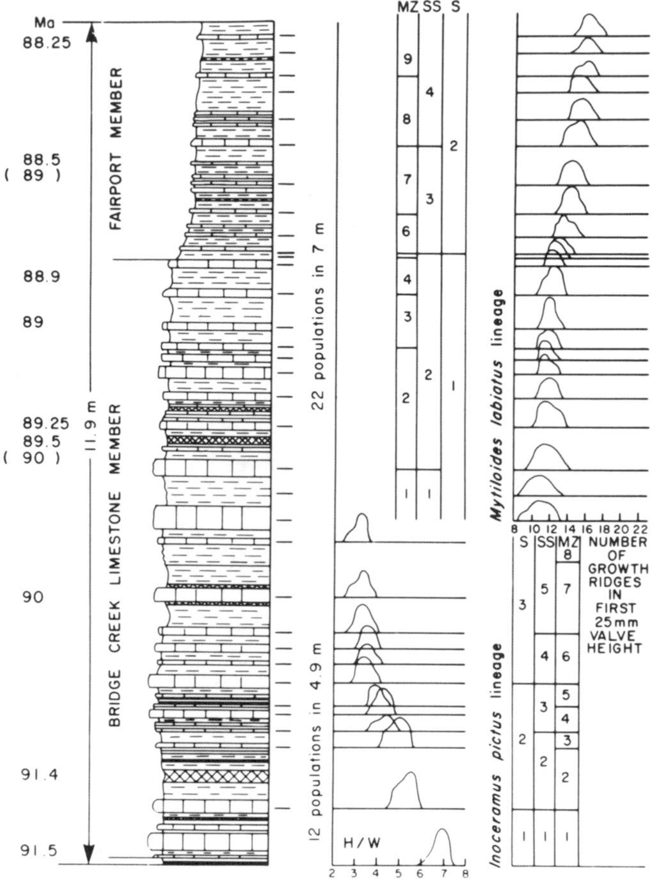

Fig. 3.13. Evolutionary trends in two pelecypod species, Cretaceous of Western Interior. Symbol X in lithologic column indicates bentonite beds used for radiometric age dating. Suggested systematics and zonal subdivisions are shown in the numbered columns: MZ = morphological biozone, SS = subspecies, S = species (Kauffman, 1977). From Concepts and Methods of Biostratigraphy, edited by E.G. Kauffman and J.E. Hazel, p. 124, Fig. 4. Copyright © 1977, Dowden, Hutchinson, & Ross, Inc., Stroudsburg, Pa. Reprinted by permission of the publisher.

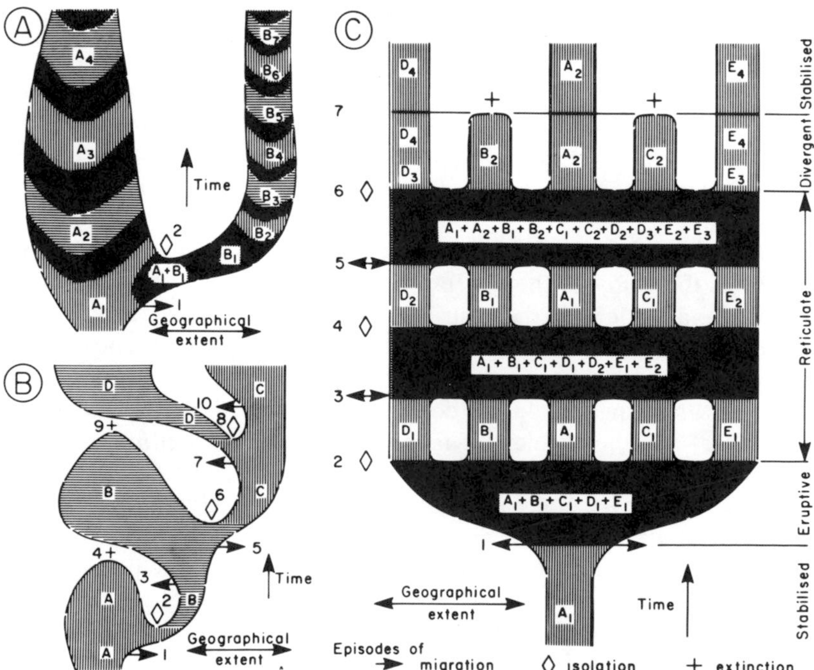

Fig. 3.14. Three models of evolution: **A** phyletic gradualism, **B** punctuated equilibrium, **C** reticulate speciation. **A**,**B**, etc., refer to successive varieties or species; 1, 2, etc., refer to chronology of events. In **B** the same area at left of diagram is successively occupied by three species A, B, C which evolve elsewhere and migrate in (Sylvester-Bradley, 1977). From Concepts and Methods of Biostratigraphy, edited by E.G. Kauffman and J.E. Hazel, p. 43, Fig. 1, p. 45, Fig. 2, p. 46, Fig. 3. Copyright © 1977, Dowden, Hutchinson and Ross, Inc., Stroudsburg, Pa. Reprinted by permission of the publisher.

variation, yet careful collection showed that each range of variation is confined to a distinct stratigraphic interval, indicating sudden small evolutionary changes. Some of the histograms in Figure 3.13 also show sudden changes and, in fact, these pelecypods are thought to show both styles of evolution at different times. Punctuated equilibrium is a style of evolution that tends to produce major but infrequent taxonomic changes (**index fossils**), of the type that can commonly be readily recognized for biostratigraphic purposes without specialist training.

Sylvester-Bradley (1977) proposed a third style of evolution, which he termed **reticulate speciation** (Fig. 3.14). This combines, on a small scale, the mechanisms of both the other two evolutionary styles. Sylvester-Bradley offered the modern common vole as an example of a taxon that has evolved in this way. The vole is distributed virtually globally, and has numerous varieties reflecting adaptation to local variations in climate, vegetation, altitude, isolation on islands, etc. These varieties have evolved in response to rapid global changes following the Pleistocene ice age, and they demonstrate the rapidity with which geographical and ecological changes may bring about evolution. The apparent stability of the species of several taxonomic groups for several million years or more at times during the geological past contrasts with the rapid adaptability of the modern vole. To what extent reticulate speciation will be recognized for fossil groups remains to be determined. However, to recognize this style of evolution would seem to require an immense bank of detailed descriptive data, and therefore it is very much a subject of study for specalists.

Biostratigraphy and biogeography. Biogeography is the study of the geographical distribution of taxa, reflecting the restriction of ranges due to ecological variations, and the geographical isolation of populations. Two topics are included under this heading, the facies control of faunas and floras, and the problem of faunal and floral provinces.

Some taxa are adapted to a benthonic (bottom-dwelling) mode of life and others to a nektonic (swimming) or planktonic (floating) habit. Ideally nektonic or planktonic forms are greatly to be preferred for biostratigraphic purposes because of the likelihood of their being more widely distributed and therefore more practically useful. Benthonic forms tend to be more facies dependent because of their need for certain water conditions or sediment types for feeding and dwelling behavior. However, in practice, benthonic forms are widely used by professional biostratigraphers. Even such static forms as corals, burrowing molluscs or anchored brachiopods have been found to be invaluable for zoning the deposits of the continental shelves. Many benthonic taxa have a planktonic larval stage which ensures wide distribution via marine circulation. Conversely, many planktonic forms, such as the graptolites, are too fragile to survive in agitated, shallow-water environments, and are therefore just as facies-bound as their benthonic contemporaries. In practice, virtually every taxonomic group has some biostratigraphic utility, although considerable problems may arise in attempts to determine the relationships between the various facies-bound faunas, unless environmental fluctuations cause lithofacies of different types, with their accompanying faunas or floras, to become interbedded. Where well exposed, such mixed sequences are of great value in establishing a global chronostratigraphic framework, as discussed in section 3.6.

Classic examples of facies-bound faunas widely used by biostratigraphers are the "shelly" and "graptolitic" faunas of the lower Paleozoic. The shelly fauna actually includes two more or less distinct subfaunas, one in the inner, shallower shelf dominated by brachiopods, and the other on the outer shelf, characterized by trilobites. The graptolitic fauna is confined mainly to low-energy deposits of the continental slope, rise and abyssal plain (Berry, 1977).

A good example of facies control of what might appear at first sight be a recurrent, biostratigraphically controlled fauna, is provided by the brachiopod communities of the Upper Ordovician to Middle Silurian of the Welsh Borderlands. Ziegler et al. (1968) showed that at many localities there is a sequence of assemblages containing, in upward stratigraphic order, *Lingula, Eocoelia, Pentamerus, Stricklandia* and *Clorinda*, followed by a graptolitic fauna. Careful correlation of these sections using graptolites showed that the brachiopod sequence is markedly diachronous. This is shown in Figure 3.15, in which the graptolite zones are shown by horizontal correlation lines labeled A_1 to C_7. One might ask, why are the graptolites trusted more

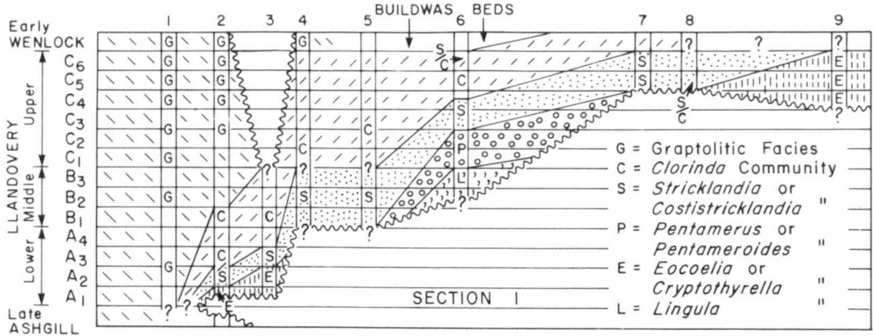

Fig. 3.15. The succession of brachiopod communities in each of these sections is the same (L, E, P, S, C), but use of graptolite biostratigraphy (biozones A_1 to C_7) shows that they are markedly diachronous, and therefore facies controlled. Silurian, Welsh Borderland (Ziegler et al., 1968; McKerrow, 1971).

than the brachiopods for the purposes of chronostratigraphic correlation? The answer is that brachiopods are benthonic organisms known to be prone to facies control, whereas graptolites are tried and tested biostratigraphic indicators. The correlations shown in Figure 3.15 are supported by additional work on two of the brachiopod genera, *Eocoelia* and *Stricklandia*. When examined in detail, evolutionary trends can be detected within the populations of these genera as they are traced from west to east across the line of section shown in Figure 3.15 (and other sections not shown). The overall interpretation of these faunal data is that the sections reveal a gradual eastward transgression and deepening of the water, successive brachiopod communities representing some (unknown) ecological adjustment to increased depths. There is no obvious relationship between brachiopod assemblage and sediment type in this case. (We return to these brachiopod assemblages in section 8.3.2 to discuss their use in unraveling cyclic changes in sea level.)

Faunal provinces are much discussed by biostratigraphers. To the uninitiated such concepts as the Malvinokaffric Province or the Tethyan or Boreal Realms seem nebulous, even mystical. Their description is usually accompanied by global distribution maps showing apparently random dots for each field occurrence of a selected taxon. Rarely are the patterns of dots for different taxa similar, and the innocent reader finds it difficult to determine exactly what it is that defines each province. Clearly the understanding of what constitutes a given faunal province requires a great deal of specialist knowledge, but even the specialists had difficulty before the advent of plate tectonics in comprehending why many of these provinces existed. There was much discussion of the appearance and disappearance of strange, narrow "land bridges" to explain the merging and divergence of provincial variations.

The ammonites provide an excellent example of the various biogeographic styles that can occur in organisms, some offering considerable advantages to the biostratigrapher, others a severe hindrance. Many ammonites underwent a planktonic larval stage which may have lasted from hours to weeks. Where this occurred it would have been of some importance to the distribution of the species. Not all ammonites showed this. Distribution patterns and varying degrees of facies independence show that some ammonites were benthonic in adult life habitat, some were nektonic, and some may have been planktonic. Their facies distribution and provincial tendencies thus varied considerably. Some ammonites may have drifted long distances after death. The modern *Nautilus*, the only living relative of the ammonites, has a buoyant shell after death, and observations in modern oceans suggest that the shell may drift for hundreds, if not thousands of kilometers. Geologically this could be of great importance, but Kennedy and Cobban (1977) suggested that many ammonites in fact became rapidly waterlogged after death and did not float appreciable distances.

Kennedy and Cobban (1977) summarized much of the data for Cretaceous ammonite distribution, and concluded that there were five types of faunal province:

Some genera have a virtually worldwide, or **pandemic** distribution; two examples are shown in Figure 3.16 plotted on a plate tectonic reconstruction of the mid-Cretaceous. These and succeeding illustrations are examples of those dot-plot diagrams, but careful comparison between them reveals some real differences in distribution. Note that in Figure 3.16 the two genera have been recorded in every continent except

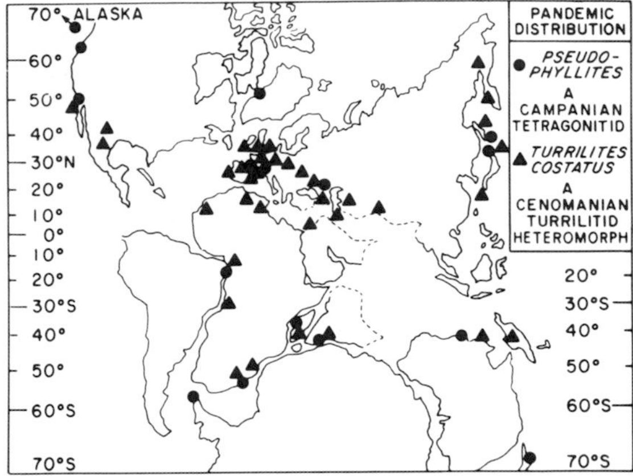

Fig. 3.16. Examples of ammonite genera showing pandemic distributions, plotted on mid-Cretaceous plate tectonic reconstruction of Smith et al. (1973) (Kennedy and Cobban, 1977). From Concepts and Methods of Biostratigraphy, edited by E.G. Kauffman and J.E. Hazel, p. 316, Fig. 2. Copyright © 1977, Dowden, Hutchinson and Ross, Inc., Stroudsburg, Pa. Reprinted by permission of the publisher.

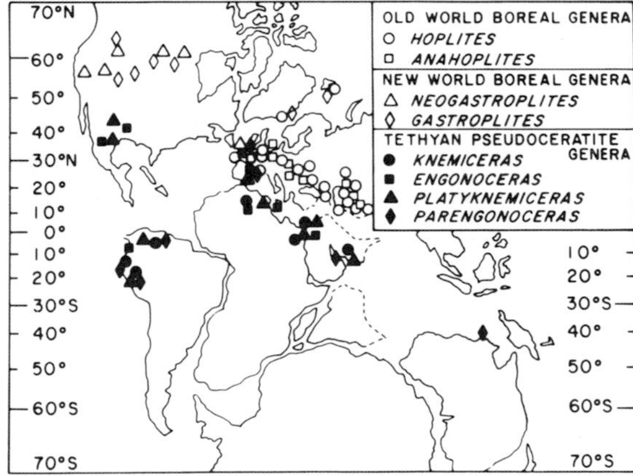

Fig. 3.17. Examples of latitudinally restricted and endemic distributions (Kennedy and Cobban, 1977). From Concepts and Methods of Biostratigraphy, edited by E.G. Kauffman and J.E. Hazel, p. 317, Fig. 3. Copyright © 1977, Dowden, Hutchinson and Ross, Inc., Stroudsburg, Pa. Reprinted by permission of the publisher.

Antarctica (sparse data) and South America. No obvious geographical isolation or latitudinal control seems likely. Pandemic taxa would seem to offer the best possibilities for global correlation. They are relatively facies independent, but it turns out that many are long-ranging forms and thus of limited biostratigraphic usefulness.

Some ammonites have **latitudinally restricted** distributions, reflecting their preference for waters of a certain temperature or salinity and their tolerance of seasonal fluctuations. Some examples are shown in Figure 3.17. They define two provinces, the northern, colder water Boreal province (open dots) and the more tropical Tethyan province (closed dots). In many parts of Europe and North America faunal fluctuations through stratigraphic successions between Tethyan and Boreal (and other) faunas have been cited as evidence of the existence of connecting seaways and transgressions across otherwise barren areas.

Longitudinal restrictions on distribution, such as the presence of land masses or large ocean basins, are a cause of further provincialism. These are added to latitudinal restrictions in the generation of the third type of faunal province: **endemic** distributions. Note that in Figure 3.17 the Tethyan genera show no longitudinal restriction, whereas the four Boreal genera are typical endemic taxa, restricted to either Eurasia or North America (these are examples chosen to illustrate a point and should not be taken to define a universal difference between the Boreal and Tethyan provinces). Endemic ammonites have been shown to have evolved rapidly and thus are of prime biostratigraphic importance, although their provincialism has hindered intercontinental correlation.

Disjunct distributions are those of scattered, but nevertheless widely distributed taxa, three examples of which are shown in Figure 3.18. The distributions are not thought to represent inadequate data or severe facies control, but probably reflect very low population densities.

As noted above, some ammonite taxa may drift in oceanic currents after death. In extreme cases, where an endemic form is involved, such drifted or **necrotic** distributions may prove invaluable for long distance correlation. An example is shown in Figure 3.19.

Many examples of faunal provincialism and facies control could be quoted. The type of province depends on the mobility of the organism and the degree of fragmentation of the continents. For example faunas and floras tended to be less provincial during the period when continents were combined into Pangea (Valentine, 1973). The brief discussion of provinces here is intended to provide only background information on some of the typical problems encounted by biostratigraphers when dealing with faunal and floral distributions. For further examples and more specialized discussion the reader is referred to such texts as Middlemiss et al. (1971), Hallam (1973), Hughes (1973), Kauffman and Hazel (1977), and Gray and Boucot (1979).

Life, death and reworked assemblages. When fossils are used as biostratigraphic indicators it is assumed, if not explicity examined and stated, that they represent the same time period as the rock in which they occur. Some conglomerates contain fossils as clasts indicating a dramatically different age; for example, I have found Permian corals as pebbles in an Early Cretaceous sandstone, and Silurian fossils within limestone boulders in a Paleocene fanglomerate. However, these are obvious examples and easy to recognize. Where the indicated age is not so different the biostratigrapher must be on guard against using such reworked material. For example Barnes et al. (1976) described attempts to relate graptolite faunas in shales of the Ordovician Cow Head Group of Newfoundland to trilobites and conodonts contained in limestone boulders within the shales. The limestone and its shelly faunas were derived from the nearby shelf and slumped into the deep basin as olistostromes (this term is discussed in section 9.3.2). The interbedding of the two faunas is biostratigraphically useful, but it must be assumed that the boulders are slightly younger than the shales and may themselves not always be the same age. Several tests are used by biostratigraphers to distinguish life and death assemblages:

If enough specimens can be collected or measured in place their size distribution may indicate whether the population is a life assemblage (biocoenosis) or has been reworked into a death assemblage (thanatocoenosis). Boucot (1953) argued that many marine invertebrates have high reproductive and initial death rates. This results in a skewed size distribution histogram, with very large numbers of larval and juvenile forms, a moderate number of adults and a much smaller number of large gerontic individuals. In the initial stages the organism may not develop hard parts capable of preservation but, nevertheless, a fossil biocoenosis is expected to show a skewed distribution curve. The action of wave and current sorting on dead shell material will tend to winnow out smaller individuals, and a transported assemblage is likely to be well sorted, with a more nearly "normal", bell-shaped size distribution histogram.

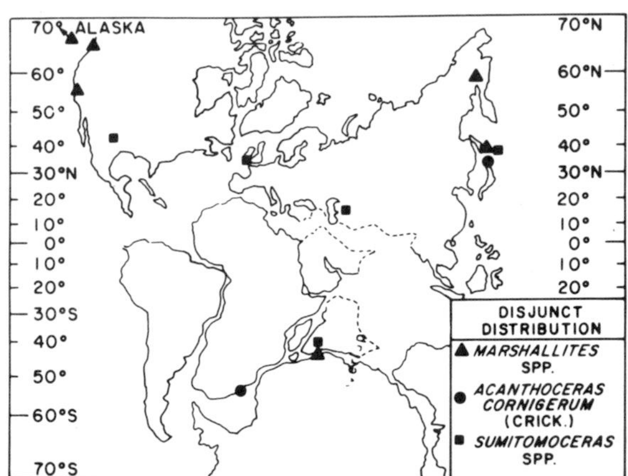

Fig. 3.18. Examples of disjunct distributions (Kennedy and Cobban, 1977). From Concepts and Methods of Biostratigraphy, edited by E.G. Kauffman and J.E. Hazel, p. 318, Fig. 4. Copyright © 1977, Dowden, Hutchinson and Ross, Inc., Stroudsburg, Pa. Reprinted by permission of the publisher.

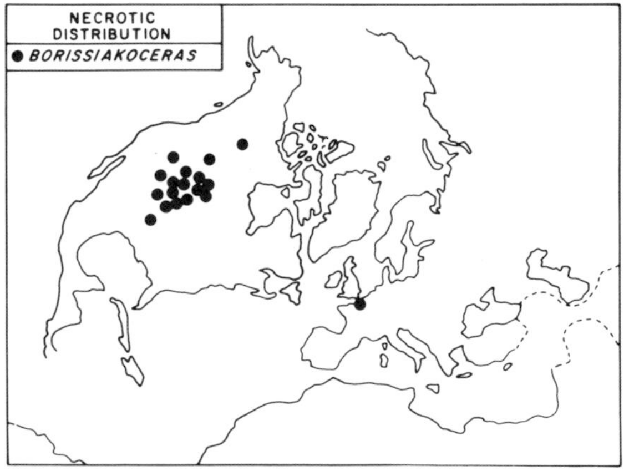

Fig. 3.19. Example of a necrotic distribution. Note lone occurrence in France (Kennedy and Cobban, 1977). From Concepts and Methods of Biostratigraphy, edited by E.G. Kauffman and J.E. Hazel, p. 320, Fig. 5. Copyright © 1977, Dowden, Hutchinson and Ross, Inc., Stroudsburg, Pa. Reprinted by permission of the publisher.

For pelecypods and brachiopods, which consist of two shells held together by soft muscle which decays after death, two additional tests can be made. The ratio of articulated (still joined) to disarticulated shells, and the ratio of opposite valves may be indicative. The chief factors affecting disarticulation are the strength and duration of bottom currents versus the rapidity of postmortem burial. Where the shells have separated they tend to respond differently to currents because, in most cases, opposing valves are of different size or shape. Sorting may therefore tend to build a concentrated deposit consisting preferentially of one or the other of the valves of a given species (Boucot, 1953).

The presence of a death assemblage accumulated by selective sorting, as described above, is not necessarily a biostratigraphic problem. Such assemblages may be produced over a period of only a few weeks or months, or even in a single day during a violent storm—time periods of insignificant biostratigraphic importance. However, at times of slow sedimentation, or during the re-erosion of relatively recent deposits a death assemblage of the "wrong" age may be produced. For example, in areas of negligible sediment supply Kennedy and Cobban (1977) reported that as many as five ammonite zones may be mixed in a single condensed sequence. After burial the process of fossilization involves diagenetic changes such as replacement of the original aragonite by calcite, or by phosphate, pyrite or silica, filling of living chambers by cement, complete dissolution of the shell, etc. Such changes may take place relatively rapidly. Storm re-erosion, deepening of a submarine canyon, or erosion during a regression may re-expose beds bearing such fossils that are now sufficiently resistant to act as clasts in sedimentation processes.

Additional observations can be made to distinguish such death assemblages. The degree of chemical alteration and the amount of breakage, abrasion, rounding, etc., are all likely to be much greater than that of other fossils enclosed with them that were living immediately prior to burial.

Certain microorganisms seem to be particularly susceptible to such reworking. Palynomorphs are highly resistant to destruction (among other ordeals they survive laboratory extraction processes which make use of hydrofluoric acid). For example, modern sediments of the Mississippi Delta contain abundant derived Cretaceous palynomorphs from the mid-Continent; Cenozoic deposits in the Canadian Arctic commonly contain more derived Cretaceous and Devonian taxa than indigenous species. Biostratigraphers may have great difficulty seeing through smoke screens such as these. Of course the taxa commonly show obvious taxonomic differences, but the degree of alteration and abrasion may also be a useful clue. Palynomorphs undergo a color change during burial as a result of progressive oxidation (Staplin, 1969, 1977). The color changes are retained during reworking and can be distinctive, where intermixed with younger, fresher material.

The drilling process itself may produce reworked microorganism assemblages by the process of caving or recycling, as discussed in Chapter 2. Microorganisms may even be present in the drilling mud. Similar techniques to those discussed above can be used to recognize and discount such material during subsurface biostratigraphic evaluation.

Conclusions. The intent of this section has been to show that a single fossil occurrence or a suite or succession of fossils does not necessarily reveal simple or accurate biostratigraphic information. The biostratigrapher may suffer from the geological equivalent of a whiteout—unable to move for lack of a fixed datum. First it must be demonstrated that the assemblage represents the same time period as the host stratum. Next the assemblage or, better still, sequence of assemblages must be sought in more than one location; otherwise the objective of "correlation" is not being achieved. However, even if the same assemblage or sequence is found elsewhere does this prove contemporaneity or are the particular fossils diachronous facies-bound forms? A consistent relationship between fossil type and sediment type might suggest the latter. If the assemblage of interest cannot be located in another section does this mean the rocks are of a different age, or the same age but a different facies, or the same (or different) age and a different faunal province? And how do we correlate a nonmarine vertebrate assemblage with a marine ammonite or foraminiferal assemblage? And what if there is more than one apparently useful biostratigraphic indicator, yielding contradictory correlations? Which one do we trust? Figure 3.15 is a simple example of this in which the answer is clear (the brachiopods are environment-dependent), but far more complex cases could be presented.

The answer to all this is years of painstaking work by many specialists. We will return to some of these questions in section 3.6, when it should become clearer (if it is not already) why few basin analysts are their own biostratigraphers.

3.5.2 Biozones

The basic biostratigraphic unit is the **biozone**, which can be defined as the body of strata containing a prescribed fauna (flora) or faunal (floral) element. Ideally taxa are chosen which have a short time span, are geographically widely distributed, as independent as possible of facies, and are abundant. As discussed below a variety of biozone types have been erected to make the best use of the different ways in which faunas and floras appear and disappear. The historical evolution of much of this nomenclature was entertainingly discussed by Hancock (1977). Further discussions of the various biozone types and biostratigraphic practices are contained in Hedberg (1976), Harland et al. (1972) and other stratigraphic guides. In practice, many biostratigraphic publications do not make completely clear which biozone type is in use. Biozones are named after prominent or otherwise important taxa or are given formal code designations.

Assemblage biozone. A body of strata characterized by a certain assemblage of fossils without regard to their ranges is designated an assemblage biozone. The limits of the zone are drawn within the rock body to encompass all the strata which contain most or all the components of the assemblage. Judgment must be used where some members of the assemblage are absent. The assemblage may consist of whatever fauna or flora is present, but in practice the assemblage normally is restricted to one type of fossil group, otherwise the distribution of the assemblage might be adversely affected by diverse ecological requirements of the individual components.

In fact, the assemblage zone concept is hampered by the fact that no attention is paid to ranges. Assemblages may be entirely ecologically controlled and represent little more than fossil communities which recur whenever conditions are appropriate. Obviously, with time the fauna (flora) will evolve and thus taxonomic differences will become apparent.

Range biozone. The body of strata representing the total range of occurrence (horizontal and vertical) of a particular species, genus, etc., is termed a range biozone. A modification of this is the **acme biozone**, which is defined on the basis of the exceptional abundance of the chosen taxon. Difficulties may arise from dependence on a single species because its occurrence may become sparse at the limits of its range, making it difficult to decide where to draw the boundary. The concept of acme should preferably be defined in some quantitative manner. Both total range and acme may be affected by local ecological (facies) considerations. For these reasons it is better to rely on more than one taxon, as in the concurrent-range zone.

Concurrent-range biozone. This is defined as the body of strata characterized by the overlapping ranges of two or more selected taxa. This is one of the most commonly used types of biozone. Modern work normally makes use of many species and may also incorporate abundance data in order to avoid problems of local facies variation. This is easy to do in the case of microorganisms which tend to occur in great abundance in many rocks.

Decisions must be made as to how many of the selected taxa must be present in a sample in order to define the zone. As Hedberg (1976) pointed out, each taxon will probably have a different distribution in space and time (Fig. 3.20) and so some judgment is required. This is where the extensive knowledge and experience of the biostratigrapher becomes important.

Figure 3.21 is an example of palynological zonation of two wells through a Cretaceous succession in Delaware (from Doyle, 1977). The wells are represented by their gamma ray logs with sample collection depths given in feet. The

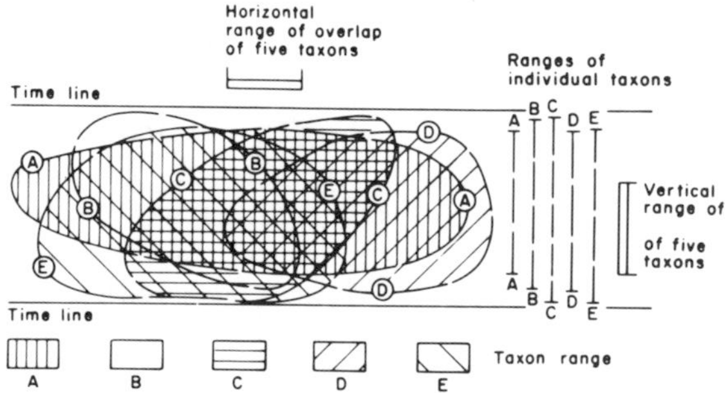

Fig. 3.20. Variations in the extent of a concurrent-range biozone, depending on the number of taxa and degree of concurrence required. From H.D. Hedberg, International Stratigraphic Guide, p. 57, Fig. 7. Copyright © 1976, John Wiley and Sons, New York. Reprinted by permission of the publisher.

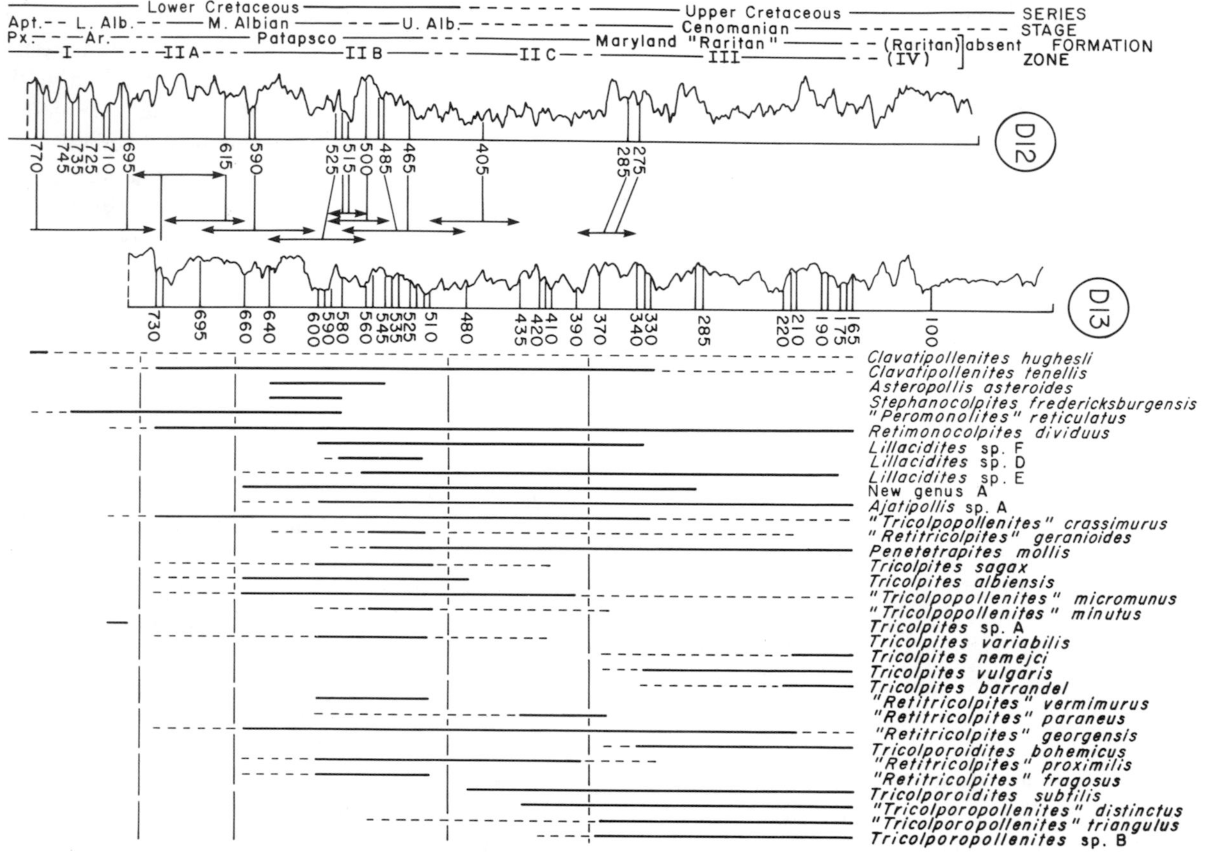

Fig. 3.21. Use of palynological concurrent-range biozones to correlate two subsurface wells (as shown by gamma ray logs). Correlation brackets (double-headed arrows) terminate just above and below samples in the other well that bracket the age of the indicated sample (Doyle, 1977). From Concepts and Methods of Biostratigraphy, edited by E.G. Kauffman and J.E. Hazel, p. 352, Fig. 3. Copyright © 1977, Dowden, Hutchinson and Ross, Inc., Stroudsburg, Pa. Reprinted by permission of the publisher.

ranges of the principal angiosperm pollen types are shown by vertical bars, shown dashed where identification is uncertain. Concurrent-range zones I to IV are shown by horizontal dotted lines and are numbered at the top. It was found that the zones could be most easily defined on the basis of first (oldest) appearance of a taxon, partly because extinct species tended to be reworked, and partly because taxa were found to die out slowly at the upper limit of their range. Work of this type required the counting and documentation of several hundreds of individual pollen grains in each sample. Several or many complete sections through the succession of interest may be required before the data are adequate for the definition of the biozones. Range charts such as that in Figure 3.21 must be prepared for each and carefully compared.

Figure 3.22 shows a series of concurrent range zones that have been erected for the Maastrichtian (Upper Cretaceous) Chalk of northwest Europe, based on belemnites, coccoliths, brachiopods, foraminifera and ammonites (from Surlyk and Birkelund, 1977). Zones are given arbitrary (but permanent) numbers or are named after distinctive species within the zone. This diagram summarizes an immense amount of work. More than 1200 species are present in the Danish Maastrichtian. Surlyk and Birkelund (1977) collected 10 to 30 kg of rock every 2 to 3 m in all major Danish sections. Washing and boiling–freezing treatment allowed the fossils to be picked out clean. About 70,000 brachiopods alone, were obtained in this way. Each 10 kg sample contained 100 to 4000 specimens of all taxa.

A variation on the concurrent-range biozone is

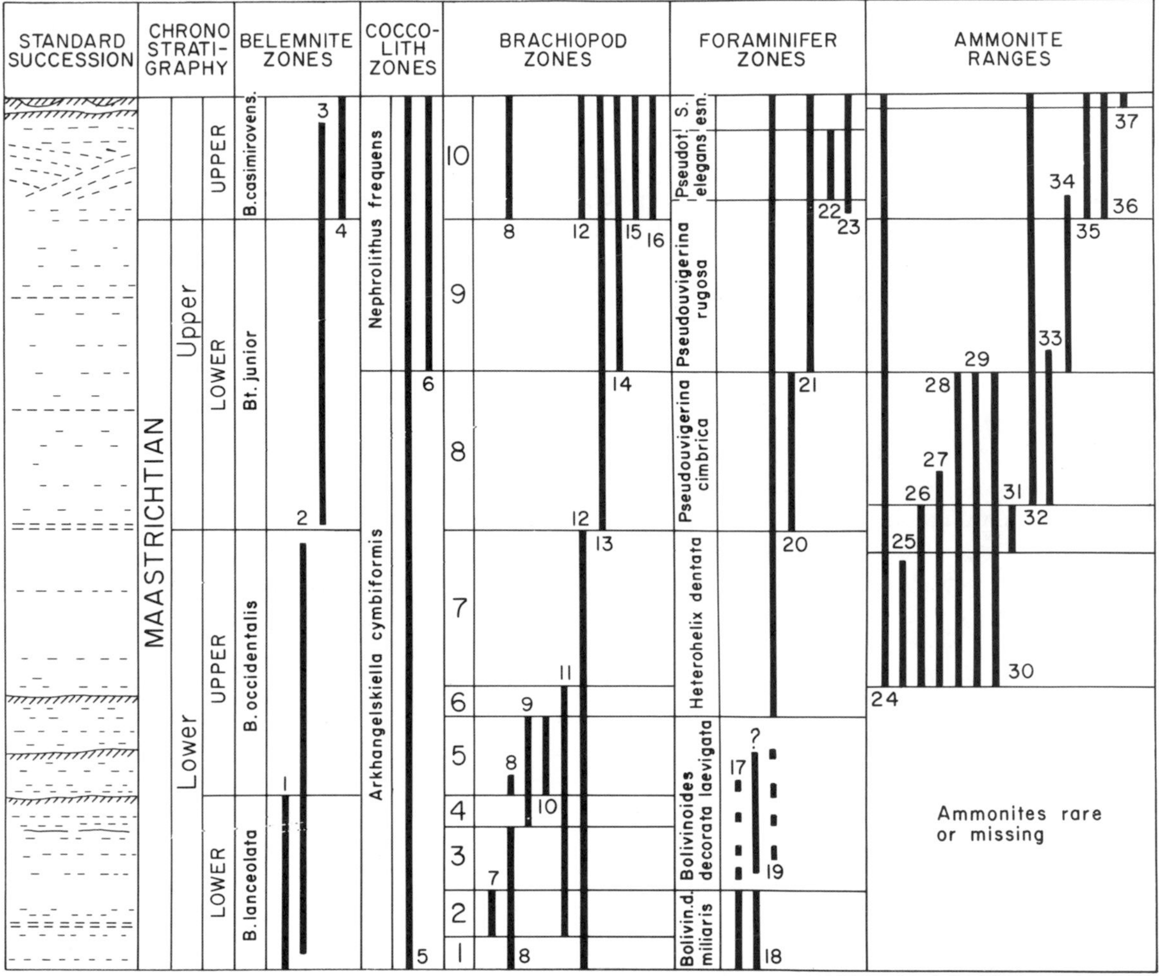

Fig. 3.22. Concurrent-range biozones of Maastrichtian Chalk in northwest Europe. Numbers beside each range bar refer to individual species. (Surlyk and Birkelund, 1977). From Concepts and Methods of Biostratigraphy, edited by E.G. Kauffman and J.E. Hazel, p. 280–281, Fig. 13. Copyright © 1977, Dowden, Hutchinson and Ross, Inc., Stroudsburg, Pa. Reprinted by permission of the publisher.

the **Oppel-zone**, named after the nineteenth-century German geologist who introduced many of the modern ideas about zoning methods. The Oppel-zone is essentially a concurrent-range biozone in which supplementary biostratigraphic criteria are used in its definition, for example abundance data, or the presence or range of some unrelated taxon. Many older zoning schemes were of the Oppel type though they may not have been so identified by the use of his name.

Lineage biozone. This may be defined on the basis of the evolutionary sequence of a particular taxon. It may be a simple linear change, such as the pelecypod shape and growth ridge trends shown in Figure 3.13, or the branching or radiation of a taxon into new varieties or lineages, such as the adaptation of the echinoid *Micraster* to new ecological niches shown in Figure 3.12. Several examples of Cretaceous foraminifera lineage zones are given by Van Hinte (1976b). As discussed in Section 3.5.1, the style of evolution (in particular phyletic gradualism versus punctuated equilibrium) will markedly affect the ease with which lineage zones may be established. The most rapid evolution is most likely to occur on the fringes of the taxon's geographic range. Zonation can best be established there, although it may be of little use in studying conservative stock in a stable environment at the center of the range. Ideally lineage zones should permit the establishment of very refined biostratigraphic subdivisions, but the work of recognizing and discriminating patterns in a series is complex and can be subjective (Hay and Southam, 1978).

Other biozone terms. Many other terms have been used by biostratigraphers for variations on the main zone types described above (George et al., 1969; Van Hinte, 1969; Hay and Southam, 1978; Hedberg, 1976).

For the total range of a taxon the terms range-zone, acrozone and total range zone have been used. Acme-zones have also been called epiboles, peak-zones and flood-zones. Local range-zones may be called teilzones or topozones. Lineage zones are sometimes called phylozones, morphogenetic zones or evolutionary zones.

Some biostratigraphic methods rely on the "first appearance datum" and "last appearance datum" of key, widely distributed taxa. Berggren and Van Couvering (1978) discussed some examples. Statistical treatment of this kind of information is made use of in the graphic correlation technique (see below).

3.5.3 Quantitative methods in biostratigraphy

Over about the last twenty years a rather different approach to biostratigraphic subdivision has evolved, contrasting with the sometimes subjective zoning methods described in the preceding section. The new approach relies on quantitative techniques, including the use of multivariate statistics. It is claimed that these methods are more objective, in that the biostratigraphic subdivisions emerge from normal analysis rather than interpretation, extrapolation, generalization and judgment. Such methods have not received universal acceptance, and there remains the argument widely heard in geology that "if you cannot see a trend in the data no amount of statistics is going to prove one exists." This criticism cannot be applied to the first of the techniques to be described, which is routinely used in some major petroleum companies, and it is clear that statistical techniques are essential to exploit the vast amounts of micropaleontological and palynological data that are being accumulated during routine petroleum exploration drilling.

The graphic correlation technique. This method was first described in a landmark book by Shaw (1964) and has most recently been expounded by Miller (1977), from whom the example used herein has been borrowed.

As with conventional biostratigraphy the graphic method relies on careful field or laboratory recording of occurrence data. However, only two items of data are noted for each taxon, the first (oldest) and last (youngest) occurrence. These define a local range for each taxon. The objective is to define the local ranges for many taxa in at least three complete sections through the succession of interest. The more sections that are used the more nearly these ranges will correspond to the total ("true") ranges of the taxa. To compare the sections a simple graphical method is used.

One particularly complete and well-sampled section is chosen as a **standard reference section**. Eventually data from several other good sections are amalgamated with it to produce a **composite standard reference section**. A particularly thorough paleontologic study should be carried out on the standard reference section, as this enables later sections, for example those produced by exploration drilling, to be correlated with it rapidly and accurately.

The graphic technique, which will now be described, is used both to amalgamate data for the production of the composite standard, and for correlating the standard with new sections. Figure 3.23 shows a two-dimensional graph in which the thicknesses of two sections X and Y have been marked off on the corresponding axes. The first occurrence of each taxon is marked by a circle on each section and the last occurrence by a cross. If the taxon occurs in both sections points can be drawn within the graph corresponding to first and last occurrences by tracing lines perpendicular to the X and Y axes until they intersect. For example the plot for the top of fossil 7 is the coincidence of points $X = 350$ and $Y = 355$.

If all the taxa occur over their total range in both sections, and if sedimentation rates are constant (but not necessarily the same) in both sections the points on the graph fall on a straight line, called the line of correlation. In most cases, however, there will be a scatter of points. The X Section is chosen as the standard reference section, and ranges will presumably be more complete there. The line of correlation is then drawn

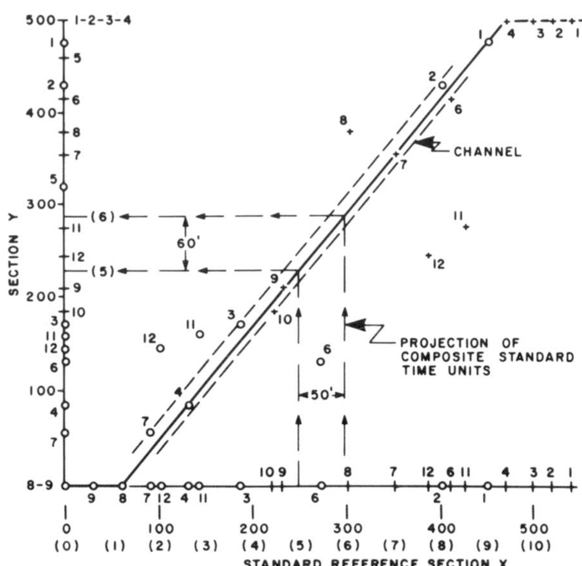

Fig. 3.23. This and the next three figures illustrate Shaw's (1964) graphic correlation method, as discussed by Miller (1977). The plot above shows distribution of first occurrences (open circles) and last occurrences (crosses) in two sections, and positioning of the line of correlation. The channel is the zone on either side of the line of correlation encompassing observation error. From Concepts and Methods of Biostratigraphy, edited by E.G. Kauffman and J.E. Hazel, p. 172, Fig. 3. Copyright © 1977, Dowden, Hutchinson and Ross, Inc., Stroudsburg, Pa. Reprinted by permission of the publisher.

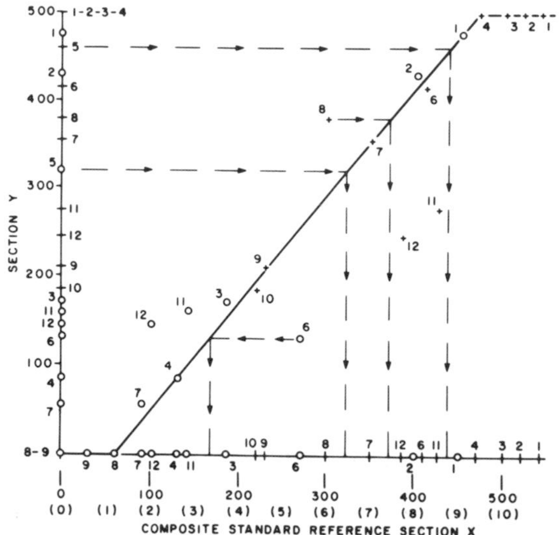

Fig. 3.24. The method used to compile a composite standard reference section. Data from new sections may be used to extend the range of occurrence of taxa that do not show their full range in the standard section (lowest occurrence of species 6, highest occurrence of species 8), and may also be used to transfer data on to the standard section, such as the range of species 5, which does not ocur in the latter (Miller, 1977). From Concepts and Methods of Biostratigraphy, edited by E.G. Kauffman and J.E. Hazel, p. 175, Fig. 4. Copyright © 1977, Dowden, Hutchinson and Ross, Inc., Stroudsburg, Pa. Reprinted by permission of the publisher.

so that it falls below most of the first occurrence points and above most of the last occurrence points. First occurrence points to the left of the line indicate late first appearance of the taxon in Section *Y*. Those to the right of the line indicate late first appearance in section *X*. If *X* is the composite standard it can be corrected by using the occurrence in Section *Y* to determine where the taxon should have first appeared in the standard. The procedure is shown in Figure 3.24. Arrows from the first occurrence of fossil 6 show that in section *X* the corrected first appearance should be at 165 ft. The same arguments apply to the points for last appearances. Corrections of the kind carried out for fossil 6 (and also the last occurrence of fossil 8) in Figure 3.24 enable refinements to be made to the reference section. Combining several sections in this manner is the method by which the composite standard is produced. Data points can also be introduced for fossils which do not occur in the reference section. In Figure 3.24 arrows from the first and last occurrences of fossil 5 in Section *Y* show that it should have occurred between 320 ft and 437 ft in Section *X*.

If the rate of sedimentation changes in one or other of the sections the line of correlation will bend. If there is a hiatus (or a fault) in the new, untested sections (sections *Y*), the line will show a horizontal terrace. Obviously the standard reference section should be chosen so as to avoid these problems.

The advantage of the graphic method is that, once a reliable composite standard reference section has been drawn up, it enables chronostratigraphic correlation to be determined between any point within it and the correct point on any comparison section. Correlation points may simply be read off the line of correlation. The range of error arising from such correlation depends on the accuracy with which the line of correlation can be drawn. Hay and Southam (1978) recommended using linear regression techniques to determine the correlation line, but this approach assigns equal weight to all data

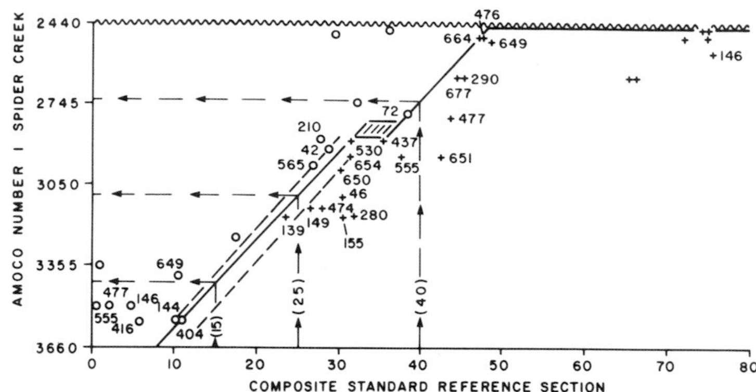

Fig. 3.25. An example of the use of the graphic method, showing plot of data from one well against the composite standard. Break and shaded area on line of correlation are interpreted as an unconformity. Reference section has been divided into arbitrary thickness units (composite standard time units) (Miller, 1977). From Concepts and Methods of Biostratigraphy, edited by E.G. Kauffman and J.E. Hazel, p. 184, Fig. 8. Copyright © 1977, Dowden, Hutchinson and Ross, Inc., Stroudsburg, Pa. Reprinted by permission of the publisher.

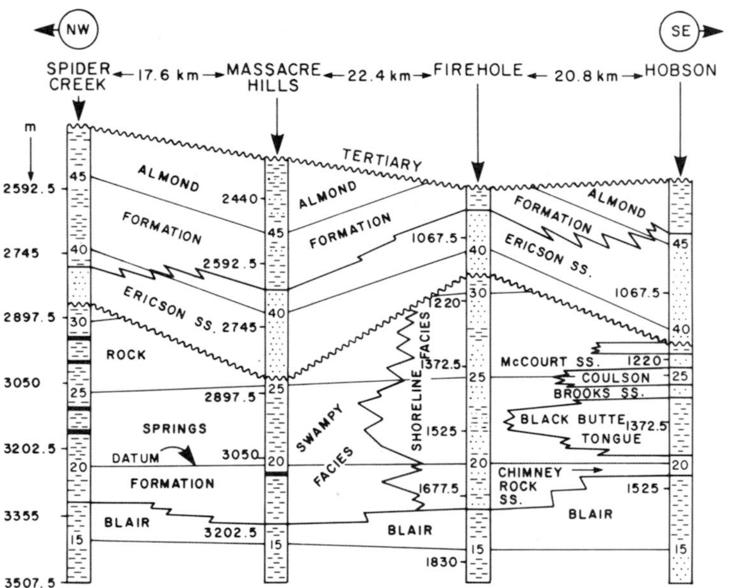

Fig. 3.26. Correlation of four wells in an area of marked lateral facies change using the composite standard time units from Fig. 3.25 (Miller, 1977). From Concepts and Methods of Biostratigraphy, edited by E.G. Kauffman and J.E. Hazel, p. 185, Fig. 9. Copyright © 1977, Dowden, Hutchinson and Ross, Inc., Stroudsburg, Pa. Reprinted by permission of the publisher.

points, instead of using one standard section as a basis for a continuing process of improvement.

Figures 3.25 and 3.26 illustrate an example of the use of the graphic method in correlating an Upper Cretaceous succession in the Green River Basin, Wyoming, using palynological data (from Miller, 1977). The composite standard reference section has been converted from thickness into "composite standard time units", by dividing it up arbitrarily into units of equal thickness. As long as the rate of sedimentation in the reference section is constant these time units will be of constant duration, although we cannot determine by this method alone what their duration is in years. Figure 3.25 shows the method for determining the position of selected time lines on each test section, and in Figure 3.26 the time units are used as the basis for drawing correlation lines between four such sections. Note the unconformity in each illustration and the variation in sedimentation rates in Figure 3.26.

The value of the graphic method for correlating sections with highly variable lithofacies and no marker beds is obvious, and it is perhaps surprising that the method is not more widely used. An important difference between this method and conventional zoning schemes is that zoning methods provide little more than an ordinal level of correlation, whereas the graphic method provides interval data.

Use of multivariate statistics. Many different techniques have been used, including cluster analysis, principal components analysis, principal coordinates analysis and multidimensional scaling (Hazel, 1977). Hay and Southam (1978) discussed a technique for quantifying the relative biostratigraphic "value" of a taxon based on facies independence, and vertical and horizontal range. Other refinements are also covered in their paper. Several other approaches, and useful discussions, are given by Harper (1981), Edwards (1982) and Agterberg and Gradstein (1981).

A single example of these applications is given here, a cluster analysis evaluation of trilobite data from an Upper Cambrian unit in Minnesota (Hazel, 1977). Sixty-five samples were collected from seven sections and were found to contain twenty-four species (Figure 3.27). A Q-mode analysis was performed which compares all the samples with each other on the basis of presence–absence data for each species. A table of similarity coefficient values was produced which enables a dendrogram to be drawn, as in Figure 3.28. Each major cluster corresponds approximately to an assemblage zone, the distribution of which is shown by transferring the ornamentation in Figure 3.28 back to the stratigraphic cross sec-

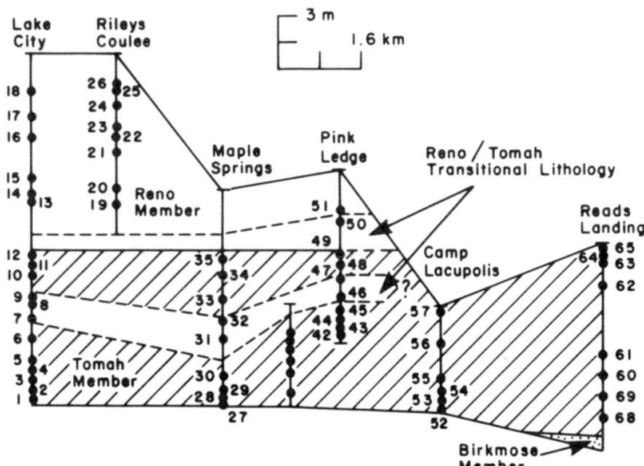

Fig. 3.27. This and the next two diagrams illustrate the use of cluster analysis an an aid in biostratigraphic zonation. The figure above shows lithostratigraphic correlation of some Cambrian rocks in Minnesota and position of numbered trilobite collections (Hazel, 1977). From Concepts and Methods of Biostratigraphy, edited by E.G. Kauffman and J.E. Hazel, p. 194, Fig. 2. Copyright © 1977, Dowden, Hutchinson and Ross, Inc., Stroudsburg, Pa. Reprinted by permission of the publisher.

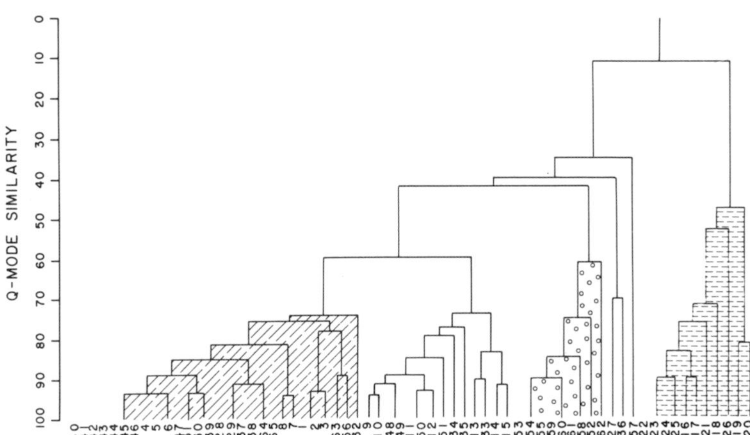

Fig. 3.28. Q-mode dendrogram using 65 samples located as shown in previous diagram and containing 24 trilobite species (Hazel, 1977). From Concepts and Methods of Biostratigraphy, edited by E.G. Kauffman and J.E. Hazel, p. 194, Fig. 2. Copyright © 1977, Dowden, Hutchinson and Ross, Inc., Stroudsburg, Pa. Reprinted by permission of the publisher.

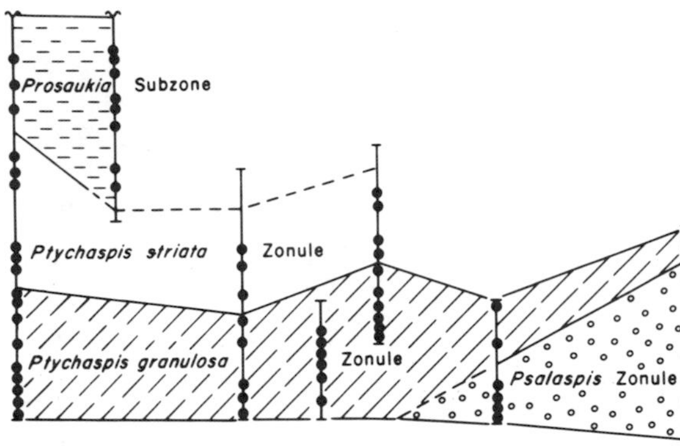

Fig. 3.29. Occurrence in the sections of the clusters outlined in Fig. 3.28, defining four assemblage zones (Hazel, 1977). From Concepts and Methods of Biostratigraphy, edited by E.G. Kauffman and J.E. Hazel, p. 194, Fig. 2. Copyright © 1977, Dowden, Hutchinson and Ross, Inc., Stroudsburg, Pa. Reprinted by permission of the publisher.

tion, as shown in Figure 3.29. Different methods of calculating the similarity coefficients will produce slight differences in clustering, but in general the method can reproduce objectively and very rapidly what would otherwise be a laborious manual procedure.

Quantitative methods are useful for the treatment of large quantities of data, such as the Maastrichtian Chalk study referred to above (Section 3.5.2, Fig. 3.22). They are particularly valuable for studies of microfossils and palynomorphs, which tend to occur in great abundance.

3.6 Chronostratigraphy and geochronometry

Geochronometry is the study of the continuum of geologic time. Geologic events and rock units may be fixed within this time frame by a variety of methods, of which radiometric dating is the most direct.

Chronostratigraphy is the study of the standard stratigraphic scale, comprising the familiar eras, periods and ages (e.g. Paleozoic, Triassic, Campanian). For the Phanerozoic, biostratigraphy is the main basis of the method, but radiometric methods and magnetic reversal stratigraphy are also widely used. Radiometric dating is the only chronostratigraphic tool available for the Precambrian, but is also of considerable importance in providing a calibration scale of biostratigraphic subdivisions in the Phanerozoic (Harland et al., 1964, 1982; Cohee et al., 1978). Magnetic reversals are now of great importance for studying Upper Cretaceous and Cenozoic strata, but are difficult to use in older rocks owing to the difficulty of obtaining complete reversal sequences, and the problems of post-depositional modification (Kennett, 1980).

There has been controversy over the relative meaning of chronostratigraphic and biostratigraphic zone concepts. A biozone can only be erected and used for correlation if the fossils on which it is based are present in the rocks. However, Hedberg (1976) defined a chronozone as a zonal unit embracing all rocks anywhere formed during the range of a specified geological feature, such as a local biozone. In theory a chronozone is present in the rocks beyond the point at which the fossil components of the biozone cease to be present as a result of lateral facies changes. Hedberg (1976) used a biostratigraphic example to illustrate the chronozone concept but, clearly, if the fossil components are not present, a chronozone cannot be recognized on biostratigraphic grounds, and its usefulness as a stratigraphic concept may be rather hypothetical. However, other means may be available to extend the chronozone, including local marker beds or magnetostratigraphic data.

Hedberg (1976) suggested that the **stage** be regarded as the basic working unit of chronostratigraphy because of its practical use in interregional correlation. As Hancock (1977) and Watson (1983) pointed out, Hedberg omitted to mention that all Phanerozoic stages were first defined on the basis of groups of biozones, and they are therefore, historically, biostratigraphic entities. In practice, therefore, so long as biostratigraphy forms the main basis of chronostratigraphy no useful purpose is served by treating them as theoretically different subjects. Stages may be defined by more than one system of biozones, which extends their range and reduces their facies dependence, but, although this improves their chronostratigraphic usefulness, it does not change them into a different sort of unit. Where other methods, such as radiometric dating or magnetostratigraphy, are precise enought to challenge the supremacy of biostratigraphy a case could be made for a separate set of chronostratigraphic units. For the Late Cretaceous to

Recent this point is now being reached, and the global time scale, based on correlations between the three main dating methods, is attaining a high degree of accuracy (section 3.6.7). However, instead of a new set of chronostratigraphic units, this correlation research is being used to refine the definitions of the existing, biostratigraphically based stages. The stage is therefore evolving into a chronostratigraphic entity of the type visualized by Hedberg.

This section deals with four main subjects: the establishment of stages and higher order units from biostratigraphic data, the problem of how to define their boundaries, the use of radiometric data, and the use of magnetic reversals as chronostratigraphic tools. Modern work in seismic stratigraphy indicates that seismic reflections are chronostratigraphic markers. This is touched on later (and in Chapter 8). Finally, a brief discussion of the development of a global time scale is presented, as an introduction to the research literature.

3.6.1 Chronostratigraphic and geochronometric scales

Chronostratigraphy refers to rocks, geochronometry to geological time. Distinction between the two tends to become blurred because the same names are used for both. Table 3.1 lists the terms used for the hierarchy of subdivisions of the two measurements systems, with examples of some names in common use (based on Hedberg, 1976). Schultz (1982) proposed to add the term chronosome to this list, as noted in section 3.4.3. The term is proposed for the body of rock enclosed between two markers that may approximate isochronous surfaces. The markers may be biostratigraphic, magnetostratigraphic, or physical features such as ash beds. A chronosome is therefore analagous to Hedberg's (1976) chronozone, except that, in the original definition of the latter, biostratigraphic criteria were emphasized. Geologists should attempt to make clear whether they are referring to the time scale or the rock record when using any of these terms. One commonly used method is to employ the qualifiers early, middle and late when referring to geologic time, and lower, middle and upper, when discussing the rocks. Geologists should never use lithostratigraphic names in a time sense, e.g. Catskill time, meaning the time during which the Catskill delta was deposited, because lithostratigraphic units are diachronous and have no precise time meaning.

The higher-ranking terms (periods, eras and eons) have mostly been with us since the nineteenth century, and it is assumed that most readers are familiar with at least some of the colorful and controversial history behind their definition (e.g. see Berry, 1968). Many of the period names are British in origin and, it turns out, could hardly have been defined in a worse place because of gaps in the section, atypical facies, and so on. The Silurian–Devonian boundary is the classic example. The original definition of this demarcation point within a marine to nonmarine transition in the sediments of the Welsh borderlands has led to the creation, during about the last thirty years, of almost an entire geological industry or subdiscipline devoted to deciding where the Silurian–Devonian boundary should be drawn stratigraphically, and which section to use as the stratotype (McLaren, 1970). This particular problem will be returned to later when discussing "boundaries".

Of more immediate practical importance to basin analysts is how the lower ranking units—the stages—are erected, and what their use by biostratigraphers means. This will now be discussed.

Table 3.1. Hierarchy of Chronostratigraphic and Geochronometric terms

Chronostratigraphic	Geochronometric	Examples
Eonothem	Eon	Phanerozoic
Erathem	Era	Mesozoic
System	Period	Cretaceous
Series	Epoch	Upper (Late) Cretaceous
Stage	Age	Campanian
Chronoze	Chron	*Orbitoides tissoti*

3.6.2 The Stage

The evolution of the stage concept was a confused one. D'Orbigny and Oppel in the mid-nineteenth century were the first to use the term with essentially its present meaning (Hancock, 1977).

Stages are essentially convenient groupings of biozones. Stage boundaries may be drawn at the top or base of a particularly well or widely developed biozone, or at a prominent faunal change. Many of our modern stage names were rather loosely defined, perhaps on only a handful of taxa when originally established, but have subsequently been refined by more and more detailed study using many different life forms. Biozones are nowadays usually only established using the members of a single phylum (except, perhaps, Oppel-zones), whereas many stages have now been defined in many different ways. For example Devonian stages are based mainly on brachiopods, corals, trilobites, fish, conodonts, and palynomorphs; Cretaceous stages are based mainly on ammonites, pelecypods, brachiopods, foraminifera, nannofossils and palynomorphs.

Most fossil groups tend to be facies-bound, and the key to successful stage correlation is to locate sections where more than one useful fossil group is present, or where repeated facies variations cause ecologically incompatible taxa to occur close together by interdigitation. This is largely a matter of chance. Much depends on lucky exposure of the right section, and the literature is replete with obscure geographic localities that have attained a specialized kind of fame because of the excellence of the biostratigraphic work carried out there. Examples are such small Welsh towns as Llanvirn and Llandeilo (Ordovician), the Eifel district, Belgium (Devonian) and the Barrandian area, Czechoslovakia (Siluro–Devonian boundary). Most stages are named after such places. The end result of this extended effort is that many stages can now be recognized globally, with varying degress of confidence. Plate tectonic movements cause faunal provincialism to vary in many taxonomic groups simultaneously, and so the perfection of this correlation varies from place to place and time to time. For example Berry (1977) reported that correlation of Early and Middle Ordovician graptolitic faunas between Europe and North America has been fraught with controversy, because of faunal provincialism. At that time the proto-Atlantic Ocean (Iapetus) was at its widest.

As an example of modern work in establishing stages by means of detailed faunal work on several animal groups a brief discussion is presented below of the Pridolian and Lochkovian stages, as defined in the Barrandian area of Czechoslovakia. Chlupáč (1972) published a detailed description of the faunas and revised the earlier biostratigraphic subdivisions of these rocks. His faunal list for the two stages includes over 300 species, of which graptolites, conodonts and trilobites constitute the most important biostratigraphic indicators. Other groups providing subsidiary control include eurypterids, phyllocarids, ostracodes, echinoids, cephalopods, gastropods, pelecypods, and brachiopods. Figure 3.30 illustrates one of several short but critical sections measured through the Pridolian-Lochkovian boundary near Klonk. Limestone beds are numbered 1 to 53. This section has now been established as the stratotype of the Silurian–Devonian boundary, and therefore also serves as a good example of a chronostratigraphic boundary, as discussed in the next section.

The Pridolian–Lochkovian succession in this area consists mainly of thinly interbedded, grayish black, calcareous mudstone and grayish black to dark gray, fine-grained (micritic), skeletal, platy weathering limestone, with subordinate beds of pale limestone and coarser, detrital limestone (calcarenite to calcirudite). Graptolites are abundant in the mudstones; trilobites and conodonts occur sparsely in the limestones, particularly in the paler, less muddy units, and become more abundant north of Klonk where the rocks undergo a facies change into a predominantly pure carbonate succession. This interbedding of different facies with their contrasting faunas is one of the most important features of the Czechoslovakian sections from the point of view of biostratigraphic stratotype definition.

Two graptolite biozones have been recognized in the Pridolian and form the main basis for the definition of the unit. They are the *Monograptus ultimus* zone below and the *M. transgrediens* zone above. The latter does not reach to the top of the Pridolian, but is followed by a graptolite "interregnum" containing only sparse, nondiagnostic forms (Fig. 3.30). The base of the Lochkovian is defined by the sudden widespread appearance and abundance of *Monograptus uniformis* in bed

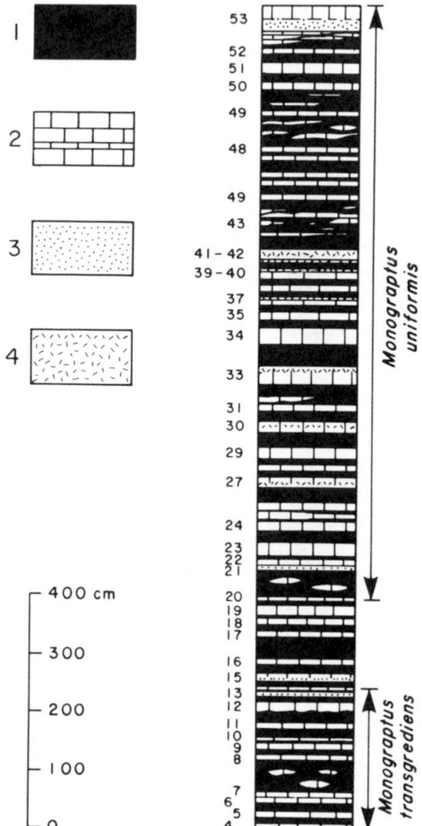

Fig. 3.30. The Silurian–Devonian section at Klonk, Czechoslovakia, used as the boundary stratotype for the Pridolian–Lochkovian stages and of the Silurian–Devonian. Limestone beds have been assigned numbers for reference. Ranges of two key graptolite species shown at right; 1. grayish-black calcareous shales; 2. grayish-black and dark gray, fine-grained, platy limestones (micritic, skeletal); 3. paler gray, granular, fine-grained limestones (skeletal, micritic); 4. coarse, detrital, crinoidal limestones (skeletal, micritic, with biorudite component) (Chlupáč, 1972).

20. Other species of *Monograptus* and of *Linograptus* appear in the upper part of the lower Lochkovian, while in the upper Lochkovian *M. hercynicus* is typical. Some of these species occur in the pure carbonate facies, permitting close correlation with the shelly fauna. It is interesting to note that in spite of the effort biostratigraphers have made to formalize their biozone types Chlupáč's work is typical of many in that no attempt is made to state what kind of biozone is in use. The older of these graptolite biozones appear to be single-taxon range biozones, with concurrent-range biozones for the upper part of the lower Lochkovian and the upper Lochkovian.

The trilobite *Warburgella (Podolites) rugulosa rugosa* is of primary importance in delineating the lower boundary of the Lochkovian. In the Klonk section it appears in limestone bed 21, immediately above the first appearance of *M. uniformis* in the upper part of bed 20 (Fig. 3.30).

The conodont *Icriodus woschmidti* defines a range biozone corresponding approximately with the lower part of the Lochkovian, although in the Barrandian area it ranges down through the graptolite interregnum into the top of the *M. transgrediens* biozone. Conodonts are not common in the somewhat argillaceous facies at Klonk.

This discussion could be extended considerably into a consideration of other faunal groups and some of the subsidiary species that define concurrent-range zones. It is to be hoped, however, that by this time the reader can discern the main threads of a procedure that has been followed innumerable times by many different workers.

Barnes et al. (1976) discussed the correlation between graptolites, conodonts, trilobites and brachiopods for the Ordovician rocks of Canada. This is an interesting case in that the benthonic fauna was used to establish a North American stage nomenclature in the craton (the Richmond, Maysville, etc. in Fig. 3.31), whereas graptolites occur mainly in deeper water deposits and were correlated with the classic British stages (Ashgill, Caradoc, etc.), in spite of the difficulties of faunal provincialism across Iapetus Ocean referred to earlier. To be of any regional use these biostratigraphic schemes must be integrated with each other, and this depended on finding locations, such as a Klonk, where the different biofacies are interbedded. In order to cover the entire Ordovician System it was necessary to study partial sections in the Canadian Rocky Mountains, Nevada, Texas, Newfoundland, the St. Lawrence Lowlands and parts of the Canadian Arctic Islands. One of the key sections was the Lower to Middle Ordovician Cow Head Group of Newfoundland. Here, trilobites and conodonts occur in limestone boulders slumped from the shelf into the deep water basin, where graptolite-bearing shales buried them. Integration of the biozone schemes had to allow for the fact that the boulders were probably slightly younger than the enclosing shales. Elsewhere, transgressions and regressions caused intermingling of faunas from different facies and different faunal provinces.

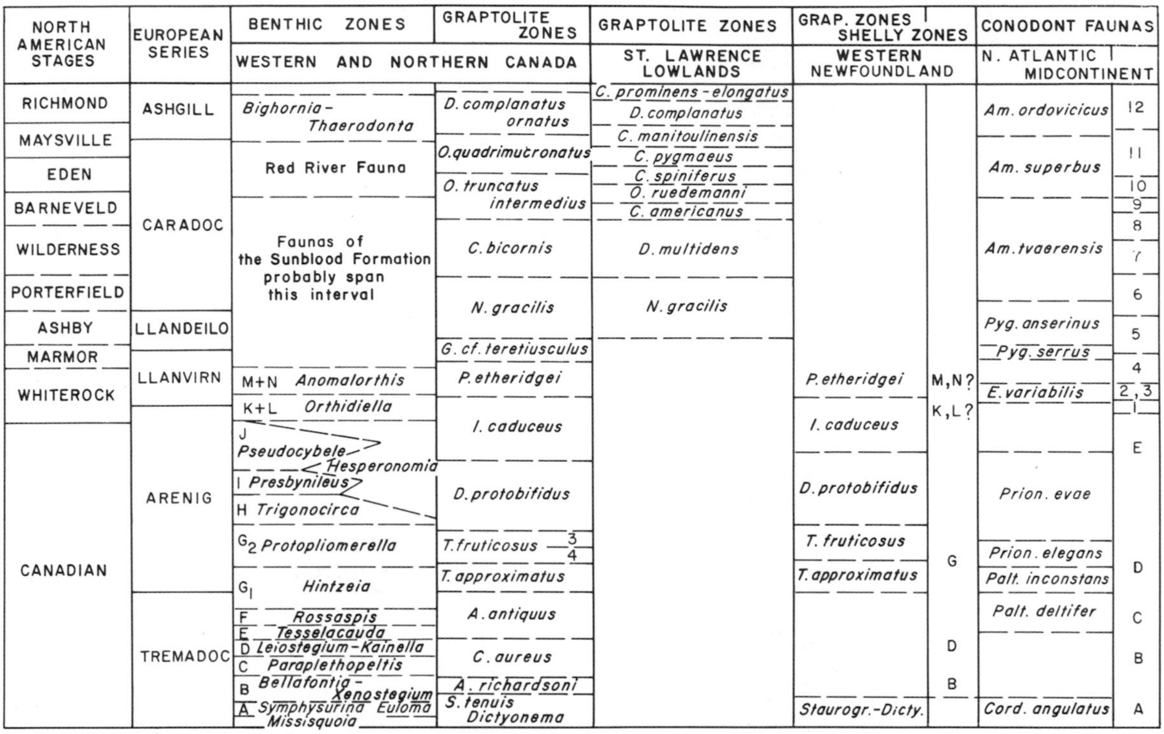

NORTH AMERICAN STAGES	EUROPEAN SERIES	BENTHIC ZONES (WESTERN AND NORTHERN CANADA)	GRAPTOLITE ZONES (WESTERN AND NORTHERN CANADA)	GRAPTOLITE ZONES (ST. LAWRENCE LOWLANDS)	GRAP. ZONES (WESTERN NEWFOUNDLAND)	SHELLY ZONES (WESTERN NEWFOUNDLAND)	CONODONT FAUNAS (N. ATLANTIC)	CONODONT FAUNAS (MIDCONTINENT)
RICHMOND	ASHGILL	*Bighornia–Thaerodonta*	*D. complanatus ornatus*	*C. prominens-elongatus*; *D. complanatus*			*Am. ordovicicus*	12
MAYSVILLE				*C. manitoulinensis*				
EDEN	CARADOC	Red River Fauna	*O. quadrimucronatus*	*C. pygmaeus*; *C. spiniferus*			*Am. superbus*	11
BARNEVELD			*O. truncatus intermedius*	*O. ruedemanni*; *C. americanus*				10; 9
WILDERNESS		Faunas of the Sunblood Formation probably span this interval	*C. bicornis*	*D. multidens*			*Am. tvaerensis*	8; 7
PORTERFIELD			*N. gracilis*	*N. gracilis*				6
ASHBY	LLANDEILO						*Pyg. anserinus*	5
MARMOR			*G. cf. teretiusculus*				*Pyg. serrus*	4
WHITEROCK	LLANVIRN	M+N *Anomalorthis*	*P. etheridgei*		*P. etheridgei*	M,N?	*E. variabilis*	2,3
		K+L *Orthidiella*	*I. caduceus*		*I. caduceus*	K,L?		1
CANADIAN	ARENIG	J *Pseudocybele*; *Hesperonomia*						E
		I *Presbynileus*; H *Trigonocirca*	*D. protobifidus*		*D. protobifidus*		*Prion. evae*	
		G_2 *Protopliomerella*	*T. fruticosus* 3 / 4		*T. fruticosus*	G	*Prion. elegans*	D
		G_1 *Hintzeia*	*T. approximatus*		*T. approximatus*		*Palt. inconstans*	
	TREMADOC	F *Rossaspis*; E *Tesselacauda*	*A. antiquus*				*Palt. deltifer*	C
		D *Leiostegium–Kainella*; C *Paraplethopeltis*	*C. aureus*			D		B
		B *Bellafontia–Xenostegium*	*A. richardsoni*			B		
		A *Symphysurina* *Euloma* *Missisquoia*	*S. tenuis* *Dictyonema*		*Staurogr.-Dicty.*		*Cord. angulatus*	A

Fig. 3.31. An attempt to provide a chronostratigraphic framework for the Ordovician of Canada, by relating various biozone schemes to each other. Note the use of two stage nomenclatures, one erected for North American cratonic sediments, the other the classic European, mainly graptolite-based nomenclature (reproduced with permission, from Barnes et al., 1976).

3.6.3 Chronostratigraphic boundaries

It would seem to be a logical assumption that the best place to draw a boundary between stages and higher-ranking chronostratigraphic units would be at a hiatus where the fauna changes abruptly. Many units have, in the past, been so defined, using major unconformities or even lithostratigraphic changes for boundary demarcation. Others have been defined on the basis of the extinction of particular faunal groups, causing great consternation when these fossils were later found in much younger rocks. McLaren (1970) discussed the Frasnian–Famennian boundary within the Late Devonian. Although this boundary has not been well defined worldwide it seems that there was wholesale replacement of faunas across it, including the disappearance of tentaculitids, the near extinction of stromatoporoids, and the complete extinction of many coral, brachiopod and trilobite groups. Other well-known mass extinctions occurred at the end of the Permian and the end of the Cretaceous. These are obviously significant events and good places to draw boundaries, so long as they are real and not the product of biostratigraphic generalization from inadequate data. There has been much literature discussion of these extinctions and of possible catastrophic causes, such as meteorite impact or supernovae (Silver and Schultz, 1982).

There are few dramatic boundaries of this type, however, and most stratigraphers now agree that chronostratigraphic boundaries should be "defined whenever possible in an area where 'nothing happened' " (McLaren, 1970). Hedberg (1976) stated that the boundary stratotypes of a stage should be within sequences of continuous deposition—preferably marine—and both should be associated with distinct marker horizons such as biozone boundaries that can be readily recognized and widely traced as isochronous horizons. The problem with using a hiatus as a chronostratigraphic boundary is that a hiatus represents a break in time that is bound to be represented by sediments and their contained fossils somewhere else. When these sediments are eventually discovered, how are they to be classified chronostratigraphically?

Even an apparently continuous sequence might later be discovered to contain a time break. For this reason there is now general (but not universal) agreement that boundary stratotypes should serve as the base of the unit above, rather than the top of the unit below (e.g. see Carter, 1974). If it is subsequently realized that the boundary stratotype is a hiatus, sediments spanning the time break are automatically assigned to the older unit. No revision of the stratotype itself is therefore necessary. This is now known colloquially as the concept of the "topless stage", an easy term to commit to memory.

Once international agreement has been reached a boundary stratotype should not need to be revised, because the precise position of the boundary is less important than that there is agreement on where it happens to be. All possible biostratigraphic means may be used to trace the boundary into different regions. To physically fix the boundary for future study the practice has recently been started of emplacing a permanent marker in the stratotype outcrop (e.g. Ager, 1964). These have come to be termed "golden spikes", although some other metal, causing less of a drain on research funds (and less prone to theft) are used in practice.

The first golden spike was inserted in bed 20 of the section at Klonk (Fig. 3.30). By international agreement the first appearance of *Monograptus uniformis uniformis* was chosen as the base of the Lochkovian stage and as the base of the Devonian System (McLaren, 1973).

It is particularly apt that the first golden spike should have been inserted at such a place as Klonk. This happy accident of onomatopoeia celebrates two events—the establishment of a better method for resolving boundary problems and the willing self-termination of the Silurian–Devonian boundary industry.

As noted by Carter (1974) much of Phanerozoic stratigraphy is now being revised and formalized using the boundary stratotype and topless stage concept. Virtually all this work is heavily based on biostratigraphy, although radiometric age data are now providing invaluable supplementary information in the form of an approximate numerical time scale in years. For the Cenozoic and perhaps the younger Mesozoic magnetostratigraphy has also assumed great importance. Only in these cases are chronostratigraphic units in any sense conceptually different from biostratigraphic units. Some of the recent work in this area is discussed in section 3.6.7.

Problems remain in applying the topless stage approach to well cuttings. As discussed in sections 2.2.3 and 2.4.1 the nature of the drilling process is such that "tops" are the only reliable kind of stratigraphic data.

3.6.4 Radiometric dating

Radiometric methods. Several naturally occurring isotopes of common elements are radioactive and, by a process of radioactive decay, produce one or a series of daughter products. The decay process ceases when a stable isotope is produced. The rate of decay for each radioactive series is constant—it is conventionally expressed by the "half-life" of the radioactive isotope. This is the length of time required for half the atoms in a given quantity of the element to decay. Half-lives range from fractions of a second to hundreds of millions of years.

Minerals containing the radioactive isotopes are formed by crystallization in an igneous melt or by authigenesis in a sediment. At this point their isotopic composition is "frozen-in", so that any subsequent changes are, in principle, the result only of radioactive decay through time. By measuring the relative proportions of radioactive isotope remaining and of new daughter products produced it is possible to calculate the age of the crystal. In practice there are many problems that have to be resolved, including the possibility of contamination by weathering, outward diffusion and escape of daughter products (yielding ages that are too young), and resetting of the clock by later thermal or tectonic events. Folding or fracturing of the rock, or metasomatism, can cause such changes, and so the measurements are best carried out in cratonic areas or on fold belts that have not been subsequently disturbed. Igneous minerals are set when they cool past their "Curie point", which is a temperature of a few hundred degrees. For large plutonic bodies this may be a considerable period of time after the initial crystallization. Very deep burial, regional or contact thermal metamorphism can change the apparent age. Radiometric ages commonly record either the time of primary cooling, a subsequent homogenizing event or an intermediate date between the two.

In most cases the amount of isotope material available for analysis is extremely small, and laboratory methods permit a reproducibility of only about 3% at best (Lambert, 1964). McDougall (1978) reported that a 3% error between the results of different laboratories working on the same material is not uncommon. A more fundamental source of error is disagreement about the decay constants of the various radioactive isotopes. Armstrong (1978) showed that calculated ages may change by at least 3%, depending on the decay constant used. Steiger and Jager (1978) reported on an attempt to standardize these values. Because of these sources of error the decay constant used and the calculated experimental error should always be quoted, and should be borne in mind when using a radiometric age in stratigraphic work. A 3% error is equal to $\pm$1.5 Ma at 50 Ma B.P. or 15 Ma at 500 Ma B.P. These are significant quantities, as we shall see in a later section when we discuss the global time scale.

Two main radiometric methods are used in sediments, the potassium–argon (K–Ar) method, using the isotopes ^{40}K and ^{40}Ar, and the rubidium– strontium (Rb–Sr) method, using the isotopes ^{87}Rb and ^{87}Sr. The methods are outlined in detail in Harland et al. (1964), and are discussed in more detail by Moorbath (1967), Goldich (1968) and Dalrymple and Lanphere (1969). Several papers in Cohee et al. (1978) discuss more recent developments, and Harland et al. (1982) provided an up-to-date review. ^{14}C methods are used for sediments back to about 70,000 a old. For dating the primary crystallization of orogenic materials uranium–lead methods are preferable, and are widely used in studies of the Precambrian (e.g. see Stockwell, 1973). The K–Ar method is somewhat better for younger rocks up to less than 1 Ma, and the Rb–Sr method for rocks older than about 10 Ma, but both are widely used throughout the Precambrian and Phanerozoic. Both may be used either for "whole rock" analyses or for selected minerals, particularly biotite, muscovite, amphibole, pyroxene, chlorite, feldspars and zircon, plus a few sedimentary minerals, as discussed below. Carrying out individual mineral determinations and whole rock analyses provides a useful cross-check of the results. For example, because of the varying resistance to ^{87}Sr diffusion Moorbath (1964) reported the following empirical age relationships for typical specimens of granitic igneous rocks: whole rock age > microcline age > muscovite age > biotite age.

Uses in stratigraphy. Sedimentary rocks can be dated radiometrically in three different ways:

1. By dating igneous intrusions within the sediments. Although widely used, this method can never provide very accurate ages because of uncertainties about the stratigraphic relationship between the igneous rocks and the sediments. The intrusions are obviously younger than the rocks they intrude, but this provides only a minimum age for the sediments. If truncated at an unconformity, the igneous rocks provide a maximum age for any sediments which follow. Many Precambrian successions have been dated by bracketing relationships of this type.

2. By dating volcanic flows and pyroclastic beds, including ignimbrites, tuffs, agglomerates, bentonites and volcaniclastic rocks interbedded with the sediments. Most volcanic materials provide several of the most readily dated mineral types listed above. This is an area of active research at the time of writing, the details of which are beyond the scope of this book. Interbedded volcanics provide much more precise stratigraphic information than do intrusive rocks because it can be safely assumed that their indicated age is very nearly the same as that of the enclosing sediments. The method is widely used, as indicated by a few examples discussed below. However, many successions lack any evidence of volcanism.

3. By use of a few authigenic sedimentary minerals. Glauconite is by far the most widely used mineral, although illite, sylvite, bone, limestone, fluorite, in fact any rubidium- or potassium- bearing mineral, may be used within the limits of analytical accuracy when dealing with very minute quantities. Numerous glauconite dates have now been published and work is continuing to improve the measurement technique (e.g. see Berggren et al., 1978; Odin, 1978). The mineral is prone to argon loss with mild heating or overburden pressure, and so glauconite dates are most reliable for the Cenozoic and, with care, the younger Mesozoic, particularly for cratonic se-

quences or others that have not been deeply buried. Odin (1978) claimed that even Precambrian sediments preserved under these conditions may yield reliable K–Ar dates on glauconite. Glauconite is rare as a detrital mineral, but this should always be checked lest spurious results are obtained. Both Rb–Sr and K–Ar methods may be used, but the latter is preferred.

Glauconite is widely distributed in shallow marine deposits, which is also a facies in which many of the best biostratigraphic indicators occur, and so glauconite dates have been widely used to provide "absolute" ages of biozones and stages.

Another technique that seems to hold considerable promise is Rb–Sr dating of mudrocks. Cordani et al. (1978) reported on an exhaustive study of Brazilian samples, in which they were able to recognize and correct for various forms of diagenetic modification and derive meaningful radiometric ages.

It should be obvious from the above that precise radiometric dating of sediments can only be carried out in certain marine facies or where volcanic rock are present. This includes only a small proportion of the total volume of sediments. It is therefore unlikely that radiometric methods will ever replace biostratigraphy. However, they are proving of inestimable value in placing ages in years on the biostratigraphic time scale.

Efforts to quantify geological time are as old as geology itself. Early attempts are described in several papers in Harland et al. (1964), which is the proceedings volume of a major symposium held to bring together modern work in the field. The symposium was held in honor of Arthur Holmes, the main pioneer in the use of radiometric methods in studying the geological time scale. At the time of this symposium geologists were fairly confidently assigning ages to series or even stages, although many of these values have since been revised and the literature is full of minor disagreements and inconsistencies that have yet to be eliminated. This has not prevented geologists from making elaborate correlations, calculations of sedimentation rates, and so on, based on radiometric dates without due regard for their experimental error. Jeletzky (1978) and others have argued forcefully against this practice. A single example chosen at random will suffice to illustrate the point.

Berggren (1972) published a major paper on the chronostratigraphy of the Cenozoic using radiometric dates and magnetostratigraphy to supplement biostratigraphic data. In it the Eocene–Oligocene boundary is placed at 37.5 Ma B.P. The main evidence for this is a single determination of 37.5 ± 3 Ma on a "latest Eocene or earliest Oligocene" sand in the Belgian Basin. Odin (1975) dated the boundary as 35.0 Ma B.P. In the scale published by the Polarity Time Scale Subdivision (Watkins, 1976) this same boundary is given (without supporting data) as 38.5 Ma. Berggren et al. (1978) placed it at 37.0 Ma B.P. Wendt (1980) suggested that the *maximum* age of this boundary is 34.8 ± 2 Ma, based on studies in the Atlantic Coastal Plain of North Carolina and Virginia. Lowrie and Alvarez (1981) gave a "revised age" of 38.0 Ma. These ages are within each others' indicated ranges of experimental error, but they reveal a total range of possible error of 8.3 Ma! Foraminifera are the best biostratigraphic indicators at this time, and Berggren's (1972) data indicate that the average length of foraminiferal biozones is about 2 Ma. Thus, even in young sediments where the data should be excellent, radiometric dating is still a crude tool relative to biostratigraphy.

However, radiometric dates are not used in isolation from each other or from other types of data. A sequence of dates from several ash or bentonite beds interbedded with sediments can provide better information through averaging, and by checking the dates against thickness of bracketed sedimentary rocks, assuming constant sedimentation rates. An example of this approach is illustrated in Fig 3.13. Kauffman (1977) obtained dates from most of 10 bentonite beds occurring over 12 m of section, and tied the dates to 34 pelecypod collections. Checks of internal consistency of this type give the procedure a much greater degree of reliability. The use of magnetostratigraphy as an additional aid is discussed in section 3.6.5. None of these additional aids is available for the Precambrian, and chronostratigraphy for this major part of earth history remains relatively crude (see section 3.6.7). Harland et al. (1982) showed that even in the Phanerozoic chronometric calibration of some age boundaries may be in error by more than 30 Ma because of lack of rigorously defined radiometric tie points.

3.6.5 Magnetostratigraphy

Development of the technique. The study of the magnetic properties of rocks is one of the most recent additions to the range of tools available to stratigraphers for correlating sedimentary successions. The first major development, the erection of a satisfactory geomagnetic polarity scale for the last 4.5 Ma, was published by Cox (1969), and in the decade since then magnetostratigraphy has become almost equal in importance to biostratigraphy and radiometric dating in establishing a time scale for the last 80 Ma of earth history. However, it seems at present unlikely that the method will be of more than local use for pre-Cretaceous rocks.

The principle of magnetic stratigraphy is that during deposition of sediments or cooling of molten rock, magnetic iron oxide minerals align themselves with any magnetic field that then exists. This preferred alignment of grains results in a bulk property of the rock referred to as **remanent magnetism**, which in essence acts as a geomagnetic field recorder. Initially it was observed in sequences of volcanic rocks on land that the north and south poles of the earth's magnetic field have reversed fairly frequently. For instance our present polarity began 7 x 10^5 years ago, and before that for about one million years, the poles were reversed except for several short episodes of normal polarity. Magnetic stratigraphy, or magnetostratigraphy, is based on the reversals in the earth's magnetic field recorded in sediments, and the utilization of these changes for stratigraphic and chronological purposes (Kennett, 1980).

Magnetic properties have also been used in two other ways that are not our immediate concern in this book: (1) for reconstructing ancient pole positions in order to track plate trajectories across the globe relative to an assumed stationary geomagnetic axis; (2) for analyzing sea floor spreading patterns by using the polarity reversal pattern as it is imprinted on newly formed oceanic lavas.

The advantage of the polarity reversal sequence to stratigraphers is that the reversals are synchronous, worldwide events. The disadvantage is that there is little to distinguish one polarity episode from another except its duration, which makes it difficult to correlate the reversals to each other, especially as time can be distorted in sediments by variations in sedimentation rate. To resolve this problem magnetostratigraphy has become closely linked to biostratigraphy and radiometric dating for geochronological purposes.

The work on the reversal chronology has followed two major trends. Vine and Matthews (1963) and Morley and Larochelle (1964) discovered that linear magnetic anomalies on the sea floor could be correlated across mid-ocean ridges, showing that they recorded the pattern of growth of oceanic crust. Heirtzler et al. (1968) developed a polarity time scale from this data for the last 79 Ma by plotting the width of the anomalies and assuming constant sea-floor spreading rates. Later, more accurate work on deep-sea drilling cores showed that this assumption was incorrect, and the polarity scale has required revision.

The other approach to reversal chronology has been to carry out detailed sampling and reversal studies in volcanic and sedimentary rocks. Ini-

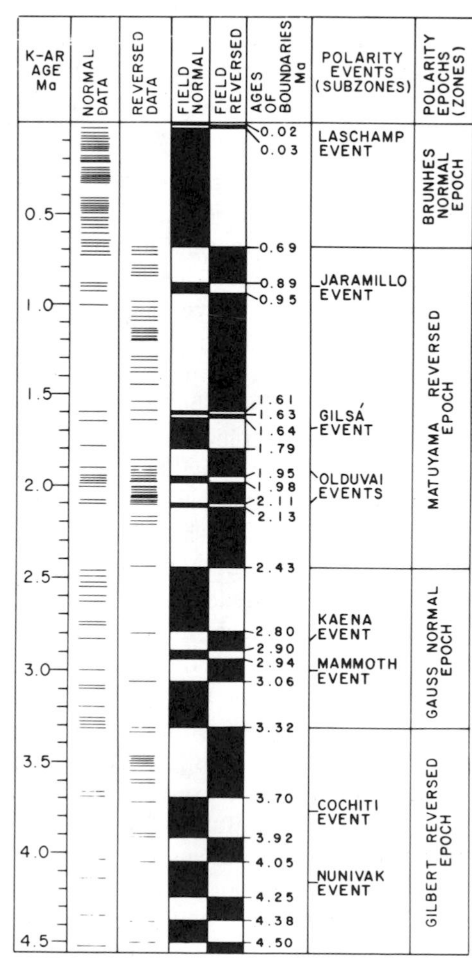

Fig. 3.32. The sequence of magnetic reversals for the last 4.5 Ma, based on studies of radiometrically dated lavas (Cox, 1969, © by the AAAS). The ages of some boundaries and the occurrence of some of the reversal subzones have been revised in recent years (see Watkins, 1976, and text).

tially the reversals were dated using radiometric methods. Cox (1969) published a revised scale for the last 4.5 Ma based mainly on measurements of volcanic lavas (Fig. 3.32). This work is still, for the most part, generally accepted (Opdyke, 1972). Named polarity events in this table are based on various outcrop localities. The first use of deep-sea cores predated the Deep Sea Drilling Project (DSDP) and the cores were very short (see review by Kennett, 1980). Harrison and Funnell (1964) were the first to combine biostratigraphy and chronostratigraphy in an attempt to correlate a reversal event. Later Opdyke et al. (1966) studied some longer cores, up to 12 m in length, and were able to correlate reversal events with the land-based lava sequence. Radiolarian zones closely parallel the reversal correlations (Fig. 3.33). The DSDP started in 1968 and began to have an important effect on chronostratigraphy. However, difficulties were encountered in establishing magnetic stratigraphy directly from the cores because of drilling disturbance and bioturbation. The practice developed of calibrating polarity and biostratigraphic data by correlating deep marine with exposed on-land sections and dating the latter radiometrically. This work has not been without its problems, because of the variations in biofacies between various oceanic areas due to water mass differences, and variations between oceanic and on-land sections. The latter commonly contain a shallower water fauna. Extension of a reliable magnetic chronology to older Cenozoic and Cretaceous sequences has depended on the study of a few well-preserved sections in Italy, New Zealand and a few other areas (e.g. see Ryan et al., 1974; Sclater et al., 1974; Lowrie and Alvarez, 1981; Harland et al., 1982). These times scales will be discussed in section 3.6.7.

In 1972 a Subcommission on a Magnetic Polarity Time Scale (SMPTS) was established as part of the International Commission on Stratigraphy. The deliberations of this group (SMPTS, 1973), plus those of the Subcommission on Stratigraphic Classification, have resulted in a formal nomenclature for magnetostratigraphy. The new hierarchy of reversal units and their approximate duration is given in Table 3.2.

Uses in stratigraphy. Magnetostratigraphy can be used by basin analysts in two main ways. Firstly it may provide part of the data used to correlate local sections with the global chronostratigraphic standard. Secondly, it may be used to correlate local sections with each other without regard to the global scale.

The first use requires detailed knowledge of the chronostratigraphic data base. This is now excellent for the Cenozoic but decreases in quality rapidly back into the Mesozoic. A recent publication of SMPTS extends the reversal scale only as far back as 79 Ma B.P., in the Santonian

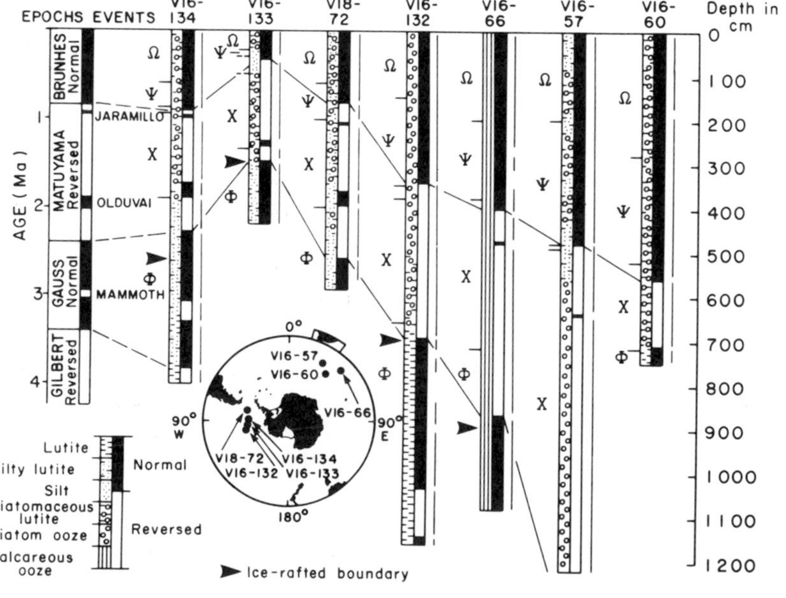

Fig. 3.33. Correlation of magnetic stratigraphy and radiolarian zones in seven cores from the Antarctic. From N.D. Opdyke, B. Glass, J.D. Hays and J. Foster, Science, v. 154, p. 349–357. Copyright 1966 by the American Association for the Advancement of Science.

Table 3.2. Nomenclature of magnetostratigraphy

Magnetostratigraphic units	Old chronologic terms (abandoned)	New geochronologic units	Approximate durations (a)
polarity subzone	polarity event	subchron	10^4–10^5
polarity zone	polarity epoch	chron	10^5–10^6
polarity superzone	polarity period	superchron	10^6–10^7

(Watkins, 1976). Compilations for earlier in the Cretaceous and Jurassic are available (see section 3.6.7) but are based on combining scattered marine and on-land sections. The oldest undisturbed oceanic crust yet described is Middle Jurassic (Larson and Pitman, 1972; Gradstein, 1981). Continuous anomaly sequences and virtually hiatus-free sedimentary cover are available from the oceans back to this time period, but for earlier rocks it is necessary to turn to incomplete on-land sequences. No reliable reversal sequence is therefore available for pre-Middle Jurassic time, and it is likely to require many years of work before one becomes available.

For the purpose of local stratigraphic correlation no knowledge of the global scale is required. However, because reversal events are not unique it is not possible to correlate them by matching sequences from different stratigraphic sections unless they contain particularly distinctive long or short polarity intervals. Some supplementary criteria may be required to assist in matching, such as marker beds or biostratigraphic zonation. Picard (1964) and Irving (1966) were among the first Western workers to use paleomagnetic correlation for sediments on the continents. The technique has become widely used for nonmarine sediments because of the scarcity of other means of precise correlation. However, problems of post-depositional modification of the magnetic remanence have introduced complications (Turner, 1980).

A single example of the use of magnetostratigraphy in a practical field problem is described here briefly. For several years a team has been exploring the nonmarine Siwalik Group (Oligocene–Quaternary) of Pakistan, in part because of its rich vertebrate fauna and in part because of the information the sediments yield about the tectonics of the Himalayas, from which the sediments were derived. Magnetostratigraphy, coupled with radiometric dating of several ash beds, has provided a useful means of local correlation between sections that show marked lateral facies changes (Keller et al., 1977; Barndt et al., 1978; Johnson et al., 1979). It was also possible to propose a correlation with the global scale. Figure 3.34 illustrates the correlation of three closely spaced sections in the Pabbi Hills area, near Jhelum. From three to five oriented rock specimens were collected from each of 113 sites within the sections. These were subjected to laboratory tests to determine stability of the field and absence of magnetic overprinting, and the pole positions obtained corrected for structural dip. The reversal zones were correlated with the standard scale of Opdyke (1972) using the following argument. The oldest remains of *Equus* (horse) were found at level 400 m in the composite section (Fig. 3.34). The oldest occurrence of *Equus* in North America is dated 3.5 Ma and in Asia 2.5 Ma. It is considered unlikely that the Pabbi Hills fossils are older than 3.5 Ma. Note that the fossil locality is in a short normal polarity sequence within a long reversed interval. Only two such dominantly reversed intervals are present in the magnetostratigraphic scale, the Matuyama and the Gilbert zones. The Gilbert zone extended from 5.1 to 3.3 Ma B.P. and the Matuyama from 2.41 to 0.70 Ma B.P. (revised ages of Opdyke, 1972; modified from Cox, 1969; see his chart in Fig. 3.32 for comparison). The evidence of *Equus* suggests that this is the Matuyama zone. The two short normal events then correlate with the Olduvai and Jaramillo subzones. The Gilsá event (subzone) of Fig. 3.32 is not universally recognized and does not appear on Opdyke's (1972) chart.

Four composite sections have been correlated using the presence of two tuff horizons (Visser and Johnson, 1978) and the polarity zones, as shown in Figure 3.35.

The importance of this work is that it permits precise local and global correlation of vertebrate localities, permits accurate calculations of sedimentation rates, and provides accurate control for

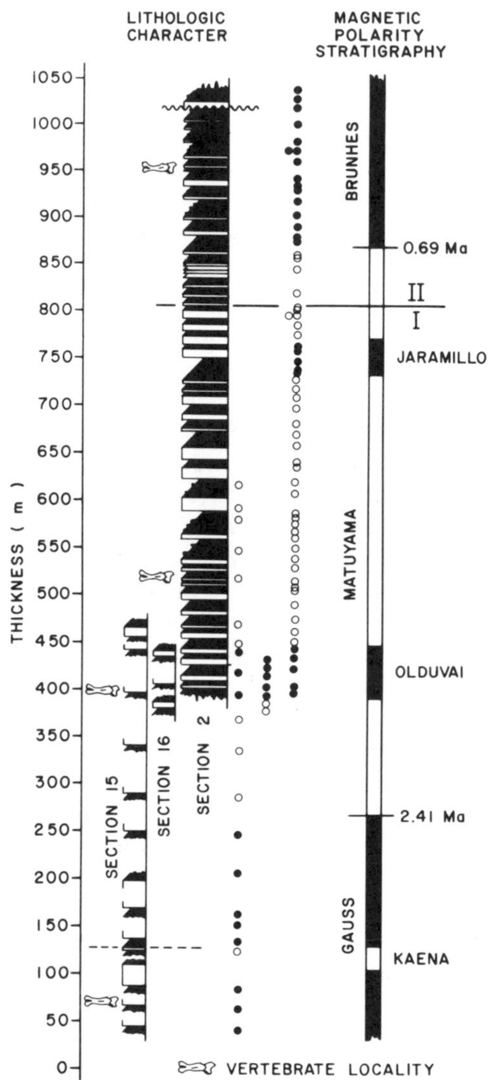

Fig. 3.34. Schematic lithology and magnetic reversal stratigraphy, composite Siwalik section, Pabbi Hills area, Pakistan. Most of the section consists of fluvial fining-upward cycles. Correlation of the reversals with the standard sequence of Cox and others is discussed in the text. (Johnson et al., 1979).

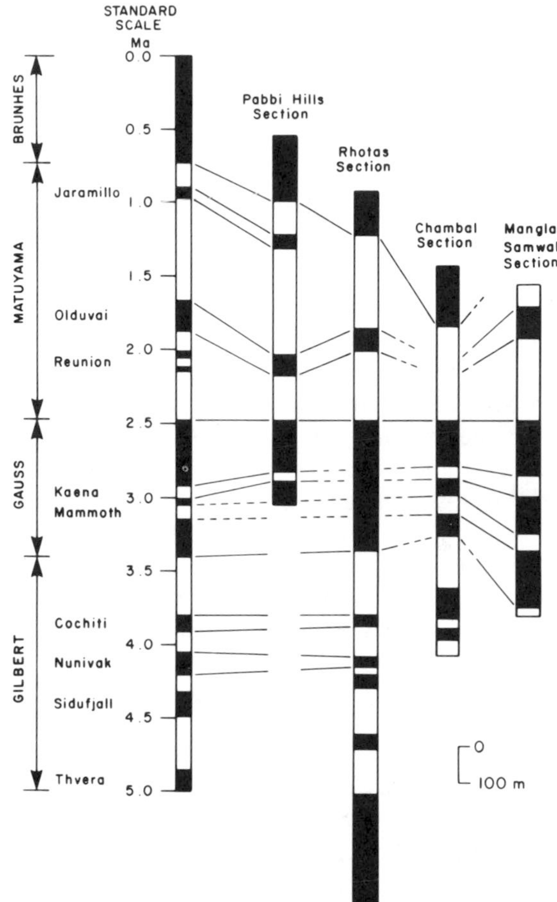

Fig. 3.35. Correlation of local reversal sequences in four composite Siwalik sections, with the standard sequence at left. Two tuff horizons were also used for correlation (redrawn from Johnson et al., 1979).

studying sedimentological characteristics, basin architecture and tectonic events.

3.6.6 Other correlation techniques

Improvements in seismic techniques have led to the realization that seismic reflectors are chronostratigraphic markers. Vail et al. (1977) pointed out that "there is no continuous physical surface that follows the top of a time transgressive unit." Such units are marked by the local variations in lithofacies, but each rock body is composed of individual bedding units. Bedding planes which delimit these units are essentially synchronous, and it is these planes which provide seismic reflectors. Seismic stratigraphy is therefore important for local subsurface correlation and has been used to construct a global sequence stratigraphy, as shown in some detail in Chapter 8. The use of seismic stratigraphy in analyzing depositional systems is described in Chapters 5 and 6.

Several other techniques have been used for dating and correlating Pliocene and Quaternary sediments, for example the fluctuations in oxygen isotope content in response to temperature changes (Shackleton and Opdyke, 1976), and the variations in amino acid content (Nelson, 1981).

These techniques have yet to be applied to pre-Pliocene strata. Detrital composition of sandstones may have local chronostratigraphic value, as shown in a detailed case study by Dickinson and Rich (1972).

3.6.7 The global time scale

Basin analysts engaged in any kind of project, be it local or regional in scope, using surface or subsurface data, require the most accurate dating available for their stratigraphic sections. This is vital for local, internal correlation within a project area, when relative ages may suffice. It is even more essential if attempts are to be made to relate local events to a regional, continental or global chronology. The aim of this chapter has been to discuss the principles behind the various techniques used in stratigraphic correlation. In this section a brief survey will be presented of the time scales that have been established for the geological column. Use of any of these scales requires "professional help", to borrow the words of the advice columnists. They summarize an enormous amount of interdisciplinary effort extending back, in the case of biostratigraphy, for nearly two hundred years. This discussion can therefore focus only on some of the most recent review articles in which the data have been collated and synthesized. Those interested in the details of a particular time period must search through the appropriate research literature themselves. Harland et al. (1982) provided a comprehensive and readable synthesis, with many references.

The Phanerozoic. The first basis on which the Phanerozoic time scale was erected was the relative scale of biostratigraphy. More recently radiometric dating has provided ages in years for biostratigraphically based subdivisions. These ages are now assigned, with varying degrees of confidence, to most Phanerozoic stages. One of the first complete compilations of this work was that published by Harland et al. (1964). Chapters in this book deal individually with every period from the Cambrian to the Tertiary. Discussions in Cohee et al. (1978) are also invaluable. For the Cambrian to Triassic the dates assigned in Harland et al. (1964) have undergone only minor modifications. Van Eysinga's (1975) chart which, because of its compactness and comprehensiveness should be (and very nearly is) on every geologist's office wall, contains more recent information. In addition to radiometric ages this chart also shows the relationship of various systems of biostratigraphic stage nomenclature to each other. Armstrong (1978) compiled a useful summary of pre-Cenozoic Phanerozoic dates, and Harland et al. (1982) compiled all available data in a useful book with many detailed charts.

For post-Triassic time a great deal of chronostratigraphic information is now available, based on information from oceanic sediments and including the relatively new technique of magnetostratigraphy. Van Hinte (1976a, b) has published a detailed biostratigraphic, radiometric and magnetostratigraphic compilation for the Jurassic and Cretaceous which, at the time of writing, serves as the standard work. For example it was used in the calibration of global seismic stratigraphic sequences, which are discussed in detail in Chapter 8. Van Hinte's Jurassic scale is reproduced as Figure 3.36. The long-established ammonite biozone scheme is shown in the center. Coccolith zones, based on more recent work, are shown on the right. A compilation of magnetostratigraphic reversals is shown on the left, and the entire scheme is calibrated against a radiometric age scale.

This is an authoritative-seeming scale, but an examination of the data base shows how tentative it really is. This point cannot be overemphasized, because there is a tendency to use such scales as precise time indicators, assuming that every biozone has been accurately radiometrically dated. This is far from being the case. Howarth (in Harland et al., 1964) considered only 12 of all published radiometric dates for the Jurassic to be accurate. Lambert (1971) argued that none of them is reliable. Van Hinte (1976a) based his scale on only three radiometric control points, for the Triassic–Jurassic and Jurassic–Cretaceous boundaries, and a mid-Jurassic age. Each of these is associated with a ± 4 to 5 Ma error. To derive a linear scale from these three calibrations requires the assumption that the stages (Harland et al., 1964) or the ammonite biozones or the coccolith biozones are equal in duration, and arbitrarily assuming equal time slices for each. For no clear reason Van Hinte chose to use the ammonite biozones, and subdividing the Jurassic time scale in this way assigns each biozone a duration of approximately 1 Ma. As Van Hinte (1976a) stated,

Fig. 3.36. A Jurassic time scale (reproduced, with permission, from Van Hinte, 1976a).

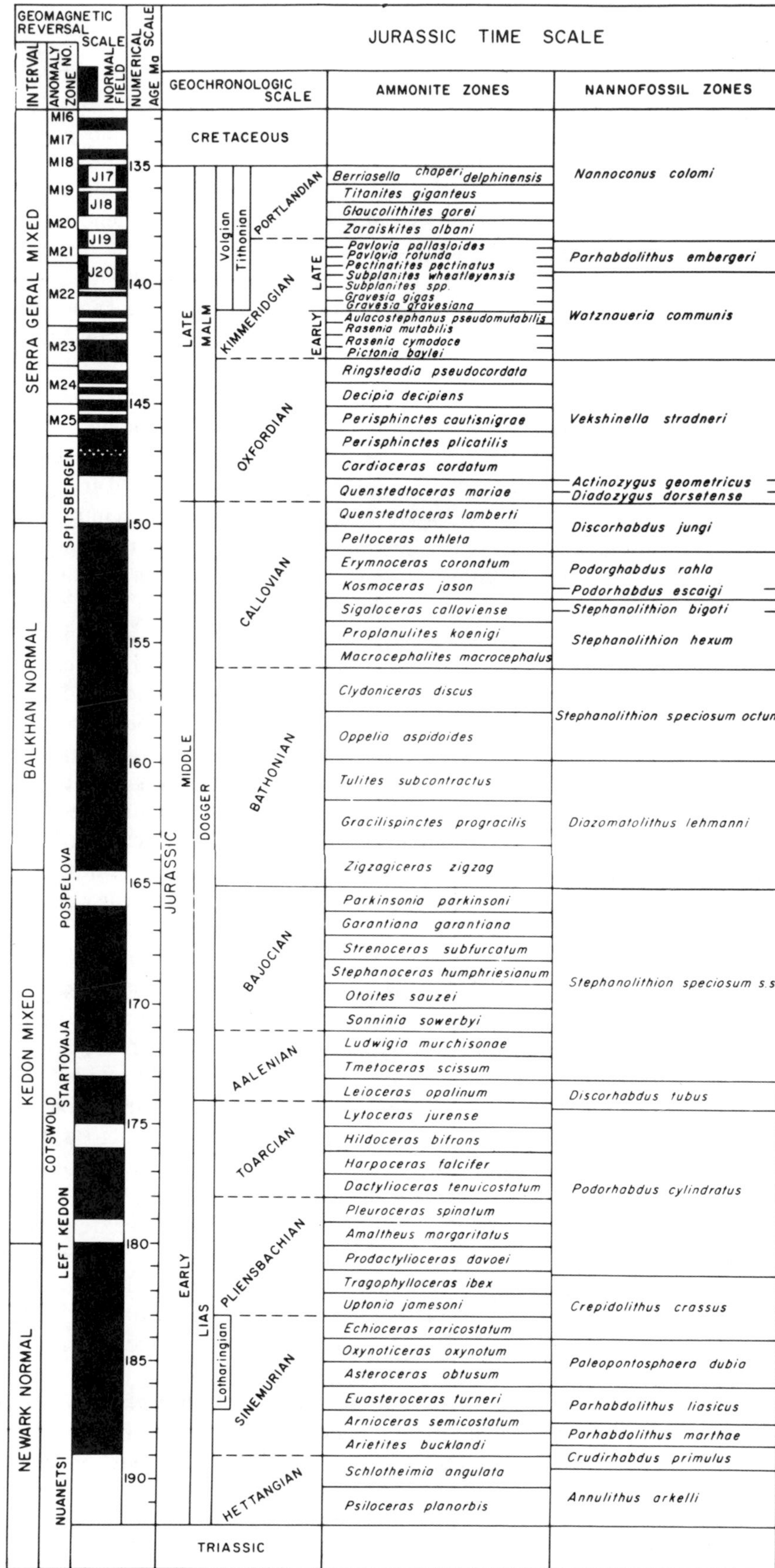

this is "highly artificial". The younger part of the scale is modified to accord with a few other scattered radiometric ages. The scale is correlated with DSDP data by comparing sedimentation rates between biostratigraphically based control points. Van Hinte (1976a) quoted several such exercises which confirm the general accuracy of the scale. However, several authors have demonstrated that biozones and stages are rarely of equal duration (e.g. Armstrong, 1978; Odin, 1978), and so further modification of this scale seems likely.

Considerably more data were available to Van Hinte (1976b) for the Cretaceous. A simplified version of his scale is shown as Figure 3.37. The biostratigraphic framework is based on ammonites and other macrofossils, benthic and planktonic foraminifera, nannofossils and radiolaria. Three main radiometric control points were used, namely: beginning Cretaceous, Early/Late Cretaceous boundary and end Cretaceous. Within this control the scale follows the Harland et al. (1964) convention of equal length stages, with exceptions made to accomodate reliable radiometric dates and DSDP data. No universally accepted polarity scale is available for the Jurassic and Cretaceous. Van Hinte's was a compilation from several sources. The revised scale drawn up by Harland et al. (1982) for the Middle Jurassic to Paleocene is given in Figure 3.38.

W.A. Berggren was one of the first to carry out these interdisciplinary correlation exercises on a major scale, with his work on the Cenozoic. Several of his earlier attempts led to a milestone paper (Berggren, 1972) which has formed the basis for all subsequent Cenozoic correlation (Fig. 3.39). Many modifications have been published, both for the Paleogene (Sclater et al. 1974; Berggren et al., 1978) and the Neogene (Van Couvering and Berggren, 1977) which have resulted in numerous refinements (see also several papers in Cohee et al., 1978; Harland et al., 1982), but the 1972 paper stands as establishing both the basic framework and the methodology.

The accuracy of Cenozoic correlation is an order of magnitude better than that of the Mesozoic because of the availability of numerous reliable radiometric glauconite dates (glauconite is commonly not a very reliable age indicator for pre-Cretaceous time). Nevertheless, as discussed in section 3.6.4, substantial errors are still apparent when comparing radiometric ages carried out by different workers on different sections (see in particular the discussion of the Eocene–Oligocene boundary). The revisions by Berggren et al. (1978) incorporated a considerable amount of new information from nonmarine mammal-bearing sequences, and revisions of the ages of linear oceanic magnetic anomalies. There is still considerable disagreement about the latter. Figure 3.40 compares the estimated ages of the numbered anomalies from three sources. Those of Sclater et al. (1974) are based on a linear extrapolation of an assumed constant rate of sea-floor spreading between control points at the Cretaceous–Paleocene and Oligocene–Miocene boundaries. Tarling and Mitchell (1976) based their scale on intra-Paleogene extrapolations from European radiometric data published by Odin (1975). Berggren et al. (1978) did not accept the validity of Odin's work and three alternate ages of anomalies 21, 24 and 26 are shown in Figure 3.40, based on their own correlations. At the time of writing no standard Cenozoic magnetic polarity scale had been accepted by a majority of workers. Dawes (1982) reported an attempt to reach a consensus amongst a group discussing Arctic and North Atlantic geology. The attempt was not successful, because of the existence of at least seven competing scales, each with its adherents. In general, recent work has shown that ages formerly assigned to Late Cretaceous and Paleogene reversals are about 10% too high. Further refinements seem inevitable, and any basin analyst planning to use reversal stratigraphy clearly should work closely with the appropriate experts. Harland et al. (1982) provided what is probably the most up-to-date compilation of biostratigraphic, radiometric and magnetostratigraphic data. Their magnetic polarity scale for the Cenozoic is shown in Figures 3.38 and 3.41.

For space reasons Neogene biostratigraphic scales are omitted, and the reader is referred to Berggren (1972), Van Couvering and Berggren (1977), Cohee et al. (1978), Lowrie and Alvarez (1981), and Harland et al. (1982).

The Precambrian. There is, as yet, no global time scale for the Precambrian. There are three main reasons for this:

Fig. 3.37. A Cretaceous time scale, simplified after Van Hinte (1976b). ▷

CRETACEOUS TIME SCALE

Geomagnetic reversal scale (Interval / Anomaly zone no. / Normal field) and **Numerical age scale (Ma)**:

Interval	Anomaly zone no.
BERINGOV MIXED	29, 30, 31, 32a, 32, 33a, 33
MERCANTON NORMAL	AKOH; SITE 263 MIXED ZONE; GATON
SERRA GERAL MIXED	M1, M5, M11, M15, M20

Numerical age scale (Ma): 65, 70, 75, 80, 85, 90, 95, 100, 105, 110, 115, 120, 125, 130, 135

Abbreviations (Boreal): B.-Belemnella, Bt.-Belemnitella, G.-Goniotheutis, I.-Inoceramus, A.-Actinocamax

Abbreviations (Planktonic Foraminifera): Ga.-Globotruncanella, B.-Biticinella, G.-Globotruncana, P.-Planomalina, Gl.-Globigerinelloides, H.-Hedbergella, R.-Rotalipora, T.-Ticinella

Geochronologic scale	Stage	Sub.	Pelagic macrofossil zones: Tethyan	Pelagic macrofossil zones: Boreal	Planktonic Foraminifera zones		
TERTIARY							
Cretaceous, Late, Senonian, Late	MAASTRICHTIAN	L	Pachydiscus neubergicus	B. casimirovensis	UC17	Ga. mayaroensis	1
				Bt. junior	UC16	G. contusa	2
					UC15	G. stuarti	2
		E	Acanthoscaphites tridens	B. occidentalis	UC14	G. gansseri	2
				B. lanceolata	UC13	G. scutilla	2
	CAMPANIAN	L	Bostrychoceras polyplocum	Bt. langei / Bt. minor	UC12	G. calcarata	1
				Bt. minor	UC11	G. subspinosa	2
			Hoplitoplacenticeras vari	Bt. mucronata s. s.	UC10	G. stuartiformis	2
				Bt. mucronata s.s./Bt. m. senior			
		E	Delawarella delawarensis	Bt. m. senior / G. q. gracilis	UC9	G. elevata	2
				G. quadrata gracilis			
			Placenticeras bidorsatum	G. quadrata s. s.			
				G. granulata / G. quadrata s. s.			
Senonian, Early	SANTONIAN	L	Placenticeras syrtale (w. Eupachydiscus isculensis)	G. granulata			
				G. westfalica / G. granulata			
		E	Texanites texanus	G. westfalica; I. undulatus plicatus	UC8	G. concavata - G. elevata	3
	CONIACIAN	L	Parabevalites emscheri (Protexanites, Paratexanites, T. pseudotexanus)	I. subquadratus	UC7	G. sigali - G. concavata	3
				I. involutus			
		E	Barroisiceras haberfellneri	I. koeneni	UC6	G. renzi - G. sigali	3
"Middle Cretaceous"	TURONIAN	L	Romaniceras deveriai	I. deformis			
		M	Romaniceras ornatissinum; Romaniceras bizeti	I. vancouverensis; I. lamarcki	UC5	"G." helvetica	2
		E	Mammites nodosoides (Kanabiceras septemseriatum, Metoicoceras whitei, Inoceramus labiatus)	I. labiatus (M. nodosoides)	UC4	H. lehmanni ("zone à grandes globigérines")	2
	CENOMANIAN	L	Calycoceras naviculare	A. plenus; Calycoceras cf. naviculare; I. pictus	UC3	R. cushmani	2
		M	Acanthoceras rhotomagense	Acanthoceras rhotomagense	UC2	R. gandolfii - R. reicheli	3
		E	Mantelliceras mantelli	Schloenbachia varians	UC1	R. gandolfii - R. greenhornensis	4
				Neohibolites ultimus			
	ALBIAN	L (Vrac.)	Stoliczkaia dispar	Stoliczkaia dispar	LC19	P. buxtorfi - R. apenninica	3
					LC18	R. ticinensis - P. buxtorfi	3
		L	Mortoniceras inflatum	Mortoniceras inflatum	LC17	T. (B.) breggiensis	1
			Diploceras cristatum	Diploceras cristatum			
		M	Hoplites lautus / H. nitidus	Hoplites lautus; Hoplites loricatus	LC16	T. praeticinensis	2
			Hoplites dentatus	Hoplites dentatus	LC15	T. bejaouaensis - Gl. gyr. - T. primula	3
		E	Douvilleiceras mammilatum	Douvilleiceras mammilatum	LC14	T. bejaouaensis - Gl. gyroidinaeformis	3
			Leymerella tardefurcata	Leymerella regularis; L. tardefurcata; P. schrammeni			
Early	APTIAN	L (Cl.)	Diodochoceras nodosocostatum	Hypacanthoplites jacobi	LC13	Gl. ferreolensis - T. bejaouaensis	3
		L (Gargasian)	Cheloniceras subnodosocostatum	Parahoplites nutfieldensis; P. melchioris	LC12	H. trocoidea - Gl. ferreolensis	3
			Aconoceras nisus	Cheloniceras martinioides	LC11	Gl. algerianus	1
				Tropaeum bowerbanki	LC10	Schackoina cabri	1
		E (Bedoulan)	Deshayeites deshayesi (Puzosia matheroni)	D. deshayesi	LC9	Gl. blowi	2
				D. forbesi			
				Praed. fissicostatus	LC8	H. sigali	2
	BARREMIAN	L	Silesites seranonis: Pulchella provincialis	Parancyloceras bidentatum / P. scalare			
				Simancyloceras stolleyi / C. sparsicostata			
				Ancyloceras innexum / Simancyloceras pingue			
				Paracrioceras denckmanni			
		E	"Nicklesia" pulchella: Pulchella caicedi	Paracrioceras elegans	LC7	H. aff. H. simplex	1
			Pulchella didayi	Hoplocrioceras fissicostatum			
			Pulchella pulchella	Hoplocrioceras rarocinctum			
Neocomian	HAUTERIVIAN	L	Pseudothurammina angulicostate	Simbirskites discofalcatus	LC6	"H". hoterivica	1
			Subsaynella sayni	S. gottschei; S. staffi			
		E	Crioceras duvali	S. inversum; Endomoceras regale	LC5		
			Acanthodiscus radiatus	E. noricum			
			Lyticoceras l. s. sp.	E. amblygonium			
	VALANGINIAN	L	Saynoceras verrucosum: callidiscus	Arnoldia - Astieria	LC4	Calpionellites	
			Himantoceras trinodosum	Dichotomites spp.			
			S. verrucosum	Prodichotomites polytomus			
		E	Kilianella roubaudi: campylotoxus	Polyptychites middendorfi / P. clarkei			
			K. roubaudi	P. bancoi / P. euomphalus			
			pertransiens	Platylenticeras involutum	LC3	Calpionellopsis	
				P. heteroplearum			
				P. robustum			
	BERRIASIAN	L	Berriasella boissieri	"Wealden"	LC2	Calpionella elliptica	
		E	Berriasella grandis		LC1	Calpionella alpina	
JURASSIC					JU		

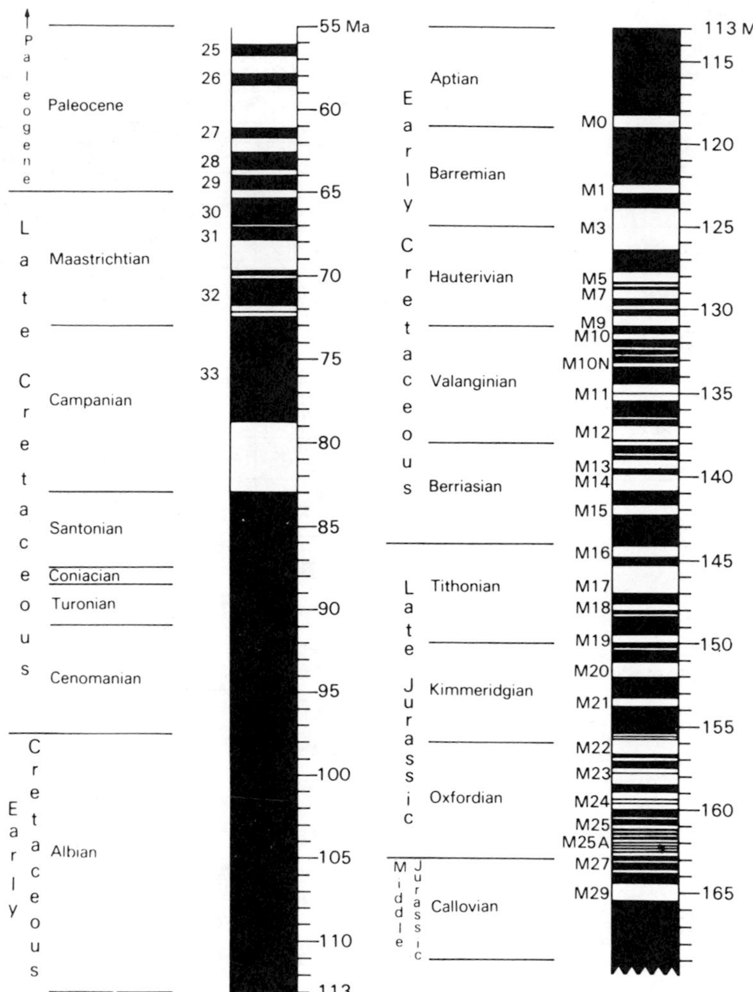

Fig. 3.38. Middle Jurassic to Paleocene magnetostratigraphy (Harland et al., 1982).

1. The biostratigraphic basis for chronostratigraphic subdivision is inadequate except, perhaps, for about the last 100 Ma of the Precambrian.
2. Radiometric determinations must be made mainly on igneous rocks or metamorphic events, which commonly provide only minimum or maximum ages for sedimentary successions. The determinations themselves are subject to possible errors as high as ±100 Ma.
3. Most chronologies that have evolved are based on igneous, metamorphic or tectonic events for specific, local Shield areas. One of the lessons of plate tectonics is that these are extremely unlikely to have any global significance (Chap. 9).

One of the largest and best exposed areas of Precambrian rocks is the Canadian Shield. Here a Precambrian time scale has been erected using an approach comparable (though with less accurate control) to that usually used for the Phanerozoic. Stockwell (1973) redefined the old terms Archean and Proterozoic, and subdivided the latter into three eras, as shown in Figure 3.42. The boundaries for these eons and eras are drawn at the end of major orogenic events, as defined by the cooling phase of late orogenic granites or pegmatites. Each of these orogenies is defined in a type area within the Canadian Shield. The ages of these events are based on numerous U–Pb and Rb–Sr whole-rock determinations. These do not give precisely the same results, and because of uncertainty in the ^{87}Rb decay constant at the time

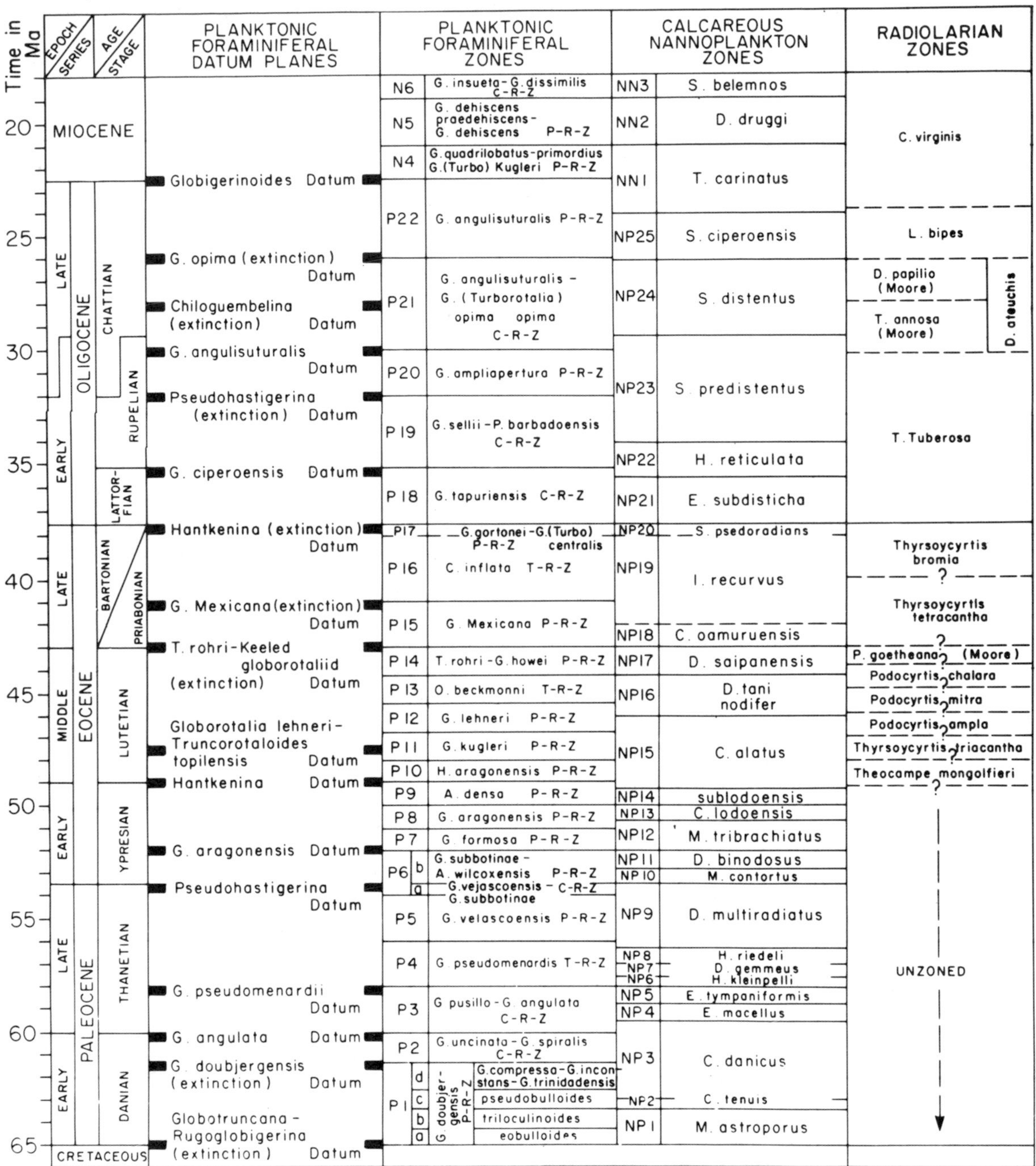

Fig. 3.39. Chronostratigraphy and biostratigraphy of the Paleogene (after Berggren, 1972). This diagram is included to illustrate Berggren's use of multiple biostratigraphic criteria, but revisions are continuing (e.g. see Harland et al., 1982).

the scale was developed two columns for Rb–Sr ages are given in Figure 3.42. This scale has been found to be satisfactory for use throughout the Canadian Shield and is the official scale used in all maps and reports of the Geological Survey of Canada.

James (1972) published the deliberations of a committee of the U.S. Geological Survey which reached quite different conclusions. This committee recommended that the subdivision of the Precambrian should be purely temporal (geochronometric), setting aside the principle of subdivision

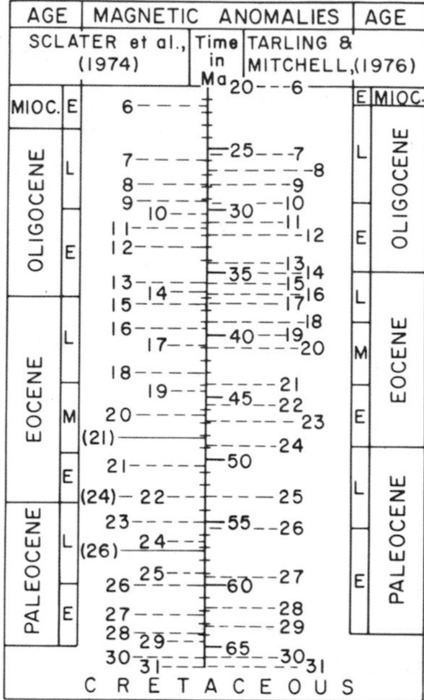

Fig. 3.40

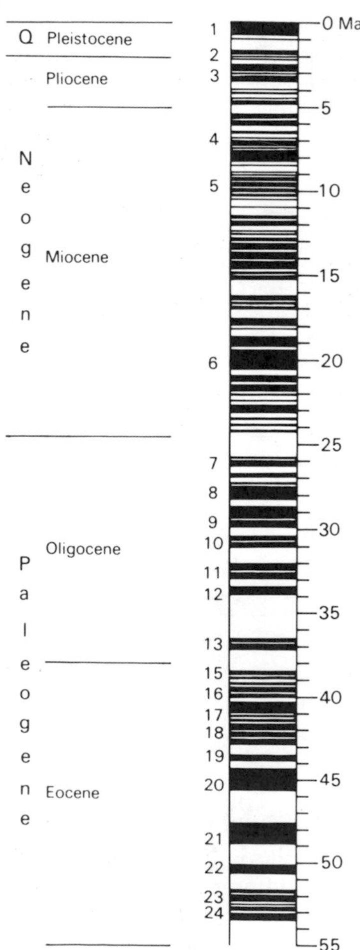

Fig. 3.41

EON	ERA	SUB-ERA	EVENT	AGE OF BOUNDARY (Ma) U-Pb scale	Rb-Sr scale (constant 1.47)	Rb-Sr scale (constant 1.39)
PROTEROZOIC	HADRYNIAN					
	HELIKIAN	NEOHELIKIAN	Grenvillian Orogeny	Ca 1000	Ca 1010	Ca 1070
		PALEOHELIKIAN	Elsonian Event	? 1400	?	?
	APHEBIAN		Hudsonian Orogeny	Ca 1800	? 1750	? 1850
ARCHEAN			Kenoran Orogeny	Ca 2560	? 2540	? 2690

Fig. 3.42

Fig. 3.40. Comparison between magnetic polarity time scales of Sclater et al. (1974), Tarling and Mitchell (1976) and, in brackets, Berggren et al. (1978). (Reproduced, with permission, from Berggren et al., 1978)

Fig. 3.41. Eocene to Recent magnetostratigraphy (Harland et al., 1982).

Fig. 3.42. The Precambrian time scale adopted by the Geological Survey of Canada (reproduced, with permission, from Stockwell, 1973).

based on stratotypes. It also rejected the idea of using major orogenies as time markers, pointing out that these may last for tens of millions of years. The committee rejected terms such as Archean and Proterozoic, on the basis that there was no agreement on their definition. Indeed there is much confusion on this point (a subject we return to below). Similar conclusions had earlier been reached by Goldich (1968), who proposed a scale consisting of equal time slices independent of any specific rock units or areas. However, James's (1972) committee chose not to go quite this far, dividing the Precambrian into four units based on groupings of sedimentary episodes, orogenies, and times of plutonism (Fig. 3.43).

This same approach was adopted by the IUGS Working Group on the Precambrian for the United States and Mexico (Harrison and Peterman, 1980, 1982). They returned to the terms Archean and Proterozoic, and subdivided each into three eras, as shown in Figure 3.43. Note that most of the proposed boundaries differ from others shown in this figure (Australian and Russian scales are included for comparison). This points up the difficulty—perhaps futility—of attempting to reach a tectonically based consensus on the chronostratigraphy of what are, in effect, randomly accreted terranes having widely varying tectonic histories.

These problems can be exemplified by the definition of the Archean–Proterozoic boundary. The IUGS Commission on Precambrian Stratigraphy recommended that this be fixed at the conveniently rounded-off value of 2.5 Ga (James, 1978). At best, this is a convenient convention, but it has raised difficulties. It is widely recognized that crustal processes were quite different in the Archean and Proterozoic eons (section 9.5.1), but the transition from one crustal style to the other occurred at different times in different parts of the globe. The 2.5-Ga boundary seems appropriate for North America, but one of the key indicators of Proterozoic-style crust, the development of recognizable, stable cratons with shallow-water cover sediments, can be documented in rocks as old as 3.0 Ga, possibly 3.5 Ga, in southern Africa (Swaziland Supergroup: see section 9.5.2). As an example of the kind of problem that can occur, we need only to turn to Windley's (1977) comprehensive summary of the Earth's crustal evolution. Geochronology and chronostratigraphy, as such, are not discussed at all in this book. Windley refers to Archean rocks as pre-2.5 Ga old (op. cit., p. 1), yet classifies the Stillwater Complex, Montana, which is 2.75 Ga old, as Early Proterozoic (op. cit., p. 65).

Turning now to the Proterozoic, there seems to be considerable promise that a practical basis for a broad biostratigraphic subdivision will be developed within a few years. Russian geologists have long used acritarchs (Timofeev, 1973) and stromatolites (Semikhatov, 1974) as biostratigraphic indicators. Their subdivision of the Precambrian, based in part on stromatolite morphology for the post-1600 Ma period, is given in Figure 3.43. However, stromatolite assemblages in Proterozoic rocks elsewhere do not permit very precise correlation with the Russian standard (Preiss,

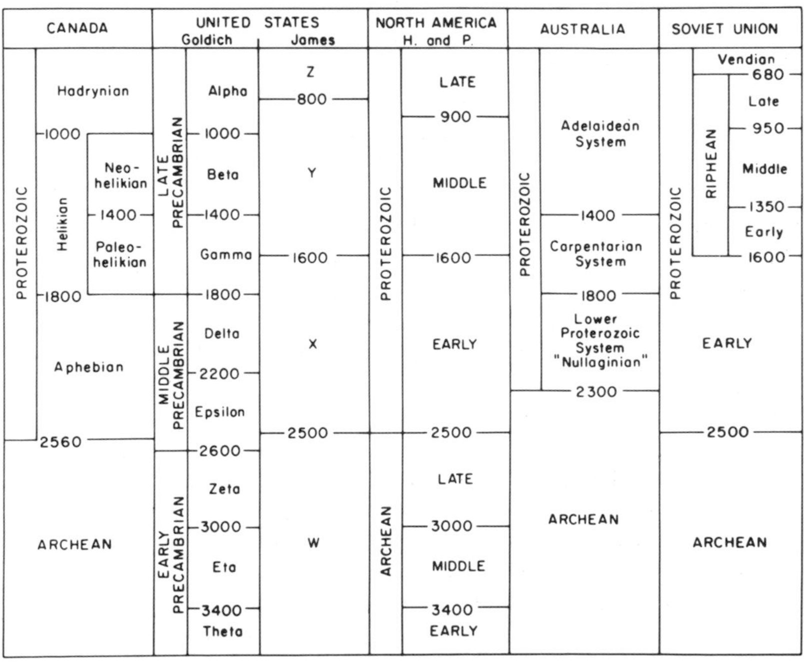

Fig. 3.43. Comparison of various Precambrian time scales (compiled from Dunn et al., 1966; Goldich, 1968; James, 1972; Stockwell, 1973; Preiss, 1977; Harrison and Peterman, 1980, 1982).

1977). Hofmann (1977) pointed out that the nomenclature of stromatolite study is very confused, and needs to be properly quantified before this subject can be properly tested. Preiss (1977) stated, "since the classification of stromatolites is based on morphological features not of microorganisms, but of the organo-sedimentary structures built by them, it would be inappropriate to use a biological binomial nomenclature", a practice which is still common. Sedimentary environment exerted a major control on stromatolite morphology, and similar morphologies were built by different microbiotas at different times during the Proterozoic (Schopf, 1977).

A more promising approach may be the study of the microbial communities that built the stromatolites. Schopf (1977) reported the recognition of potential index fossils and crude range biozones for the Proterozoic, and further refinements are anticipated. These organisms occur fairly frequently in Proterozoic sediments and therefore the method has the potential to be of considerable practical use.

Between about 670 Ma and the beginning of the Cambrian the globe was characterized by a distinctive assemblage of soft-bodied Metozoan organisms, termed the Ediacarian fauna, after the Ediacara Hills in South Australia. Elements of the fauna have been found in many parts of the world. Cloud and Glaessner (1982) proposed the erection of a new geochronologic–chronostratigraphic unit, the Ediacarian Period/System, and they further proposed that, because the fauna shows clear evolutionary links to those in the Phanerozoic and represents the first known multicellular (Metozoan) animal life, the base of the Phanerozoic should be redefined downward to include the Ediacarian. They argued that because the Ediacarian cannot be both Phanerozoic and Precambrian, the name of the latter should be changed to the old term Cryptozoic. The definition of an Ediacarian Period and System is an excellent way to recognize advances in chronostratigraphy, but the accompanying suggestions to revise the Phanerozoic and rename the Precambrian are bound to be controversial. The term Phanerozoic is derived from the Greek words for "visible" and "life", and is an excellent term for the Cambrian to Recent. Logic would certainly suggest that the widespread Ediacaran fauna be included. However, it could be argued that the word Proterozoic ("early life"), the name of one of the Precambrian eras, is just as satisfactory. The strength of tradition and the desire for consistency are also factors to consider. As noted in section 3.6.3, the most important characteristic of a chronostratigraphic boundary is that there is general agreement amongst the scientific community about its position. The actual location of the boundary is not important, as long as that location permits a precise definition which all can accept. This is the only way to ensure proper communication. The word Precambrian is widely understood, thoroughly entrenched in the literature, and not likely to be abandoned because of this single new development.

References

AGER, D.V., 1964: The British Mesozoic Committee; Nature, v. 203, p. 1059.

AGTERBERG, F.P., and GRADSTEIN, F.M., 1981: Workshop on quantitative stratigraphic correlation techniques: Ottawa, February 1980; J. Internat. Assoc. Math. Geol., v. 13, p. 81–91.

ALLEN, J.R.L., and WILLIAMS, B.P.J., 1978: Sedimentology and stratigraphy of the Townsend Tuff Bed in South Wales and the Welsh Borders; J. Geol. Soc. London, v. 138, p. 15–29.

ARMSTRONG, R.L., 1978: Pre–Cenozoic Phanerozoic time scale—computer file of critical dates and consequences of new and in-progress decay-constant revisions, *in* G.V. Cohee, M.F. Glaessner and H.D. Hedberg, eds., Contributions to the geologic time scale; Am. Assoc. Petrol. Geol. Studies in Geol. 6, p. 73–91.

BARNDT, J., JOHNSON, N.M., JOHNSON, G.D., OPDYKE, N.D., LINDSAY, E.H., PILBEAM, D., and TAHIRKHELI, R.A.H., 1978: The magnetic polarity stratigraphy and age of the Siwalik Group near Dhok Pathan Village, Potwar Plateau, Pakistan; Earth Plan. Sci. Letters, v. 41, p. 355–364.

BARNES, C.R., JACKSON, D.E., and NORFORD, B.S., 1976: Correlation between Canadian Ordovician zonations based on graptolites, conodonts and benthic macrofossils from key successions, *in* M.G. Bassett, ed., The Ordovician System: proceedings of a Palaeontological Association symposium, Birmingham, September 1974; Univ. Wales and Nat. Mus. Wales, Cardiff, p. 209–225.

BATHURST, R.G.C., 1976: Carbonate sediments and their diagenesis; 2nd ed., Develop. in Sedimentology, Elsevier, Amsterdam, 658 p.

BENNETTS, K.R.W., and PILKEY, O.H., 1976: Characteristics of three turbidites, Hispaniola–Caicos Basin; Geol. Soc. Am. Bull., v. 87, p. 1291–1300.

BERGGREN, W.A., 1972: A Cenozoic time-scale—some implications for regional geology and paleobiology; Lethaia, v. 5, p. 195–215.

BERGGREN, W.A., McKENNA, M.C., HAR-

DENBOL, J., and OBRADOVICH, J.D., 1978: Revised Paleogene polarity time scale; J. Geol., v. 86, p. 67–81.

BERGGREN, W.A. and VAN COUVERING, J.A., 1978: biochronology, *in* G.V. Cohee, M.F. Glaessner and H.D. Hedberg, eds., Contributions to the geologic time scale; Am. Assoc. Petrol. Geol. Studies in Geol. 6, p. 39–55.

BERRY, W.B.N., 1968: Growth of prehistoric time scale, based on organic evolution; W.H. Freeman and Co., San Francisco, 158 p.

BERRY, W.B.N., 1977: Graptolite biostratigraphy: a wedding of classical principles and current concepts, *in* E.G. Kauffman and J.E. Hazel, eds., Concepts and methods of biostratigraphy; Dowden, Hutchinson and Ross Inc., Stroudsburg, Pennsylvania, p. 321–338.

BOUCOT, A.J., 1953: Life and death assemblages among fossils; Am. J. Sci. v. 251, p. 25–40.

BURNABY, T.P., 1965: Reversed coiling trends in *Gryphaea arcuata*; Geol. J., v. 4, p. 257–278.

CARRIGY, M.A., 1971: Deltaic sedimentation in Athabasca Tar Sands; Am. Assoc. Petrol. Geol., v. 55, p. 1155–1169.

CARTER, R.M., 1974: A New Zealand case–study of the need for local time-scales; Lethaia, v. 7, p. 181–202.

CHLUPÁČ, I., 1972: The Silurian–Devonian boundary in the Barrandian; Bull. Can. Petrol. Geol., v. 20, p. 104–174.

CLOUD, P. and GLAESSNER, M.F., 1982: The Ediacarian Period and System: Metazoa inherit the earth; Science, v. 217, p. 783–792.

COHEE, G.V., GLAESSNER, M.F., and HEDBERG, H.D., eds., 1978: Contributions to the geologic time scale; Am. Assoc. Petrol. Geol., Studies in Geol. No. 6.

COMPTON, R.R., 1962, Manual of field geology; John Wiley and Sons, New York, 378 p.

CORDIANI, U.G., KAWASHITA, K., and FILHO, A.T., 1978: Applicability of the rubidium–strontium method to shales and related rocks, in G.V. Cohee, M.F. Glaessner and H.D. Hedberg, eds., Contributions to the geologic time scale; Am. Assoc. Petrol. Geol. Studies in Geol., 6, p. 93–117.

COX, A., 1969: Geomagnetic reversals; Science, v. 163, p. 237–245.

DALRYMPLE, G.G., and LANPHERE, M.A., 1969: Potassium–argon dating; W.H. Freeman and Co., San Francisco, 258 p.

DAVIES, G.R. and LUDLUM, S.D., 1973: Origin of laminated and graded sediments, Middle Devonian of Western Canada; Geol. Soc. Am. Bull., v. 84, p. 3527–3546.

DAWES, P.R., 1982: Note on the magnetic polarity time scales used in the Nares Strait volume, *in* P.R. Dawes and J.W. Kerr, eds., Nares Strait and the drift of Greenland: a conflict in plate tectonics; Meddelelser ǿm Gronland, Geoscience 8, p. 389–392.

DICKINSON, W.R. and RICH, E.I., 1972: Petrologic intervals and petrofacies in the Great Valley sequence, Sacramento Valley, California; Geol. Soc. Am. Bull., v. 83, p. 3007–3024.

DIXON, O.A., and JONES, B., 1978: Upper Silurian Leopold Formation in the Somerset–Prince Leopold Islands type area, Arctic Canada; Bull. Can. Petrol. Geol., v. 26, p. 411–423.

DIXON, O.A., NARBONNE, G.M., and JONES, B., 1981: Event correlation in Upper Silurian rocks of Somerset Island, Canadian Arctic; Am. Assoc. Petrol. Geol. Bull., v. 65, p. 303–311.

DOYLE, J.A., 1977: Spores and pollen: The Potomac Group (Cretaceous) Angiosperm sequence, *in* E.G. Kauffman and J.E. Hazel, eds., Concepts and methods of biostratigraphy; Dowden, Hutchinson and Ross Inc., Stroudsburg, Pennsylvania, p. 339–364.

DUNN, P.R., PLUMB, K.A., and ROBERTS, H.G., 1966: A proposal for time–stratigraphic subdivision of the Australian Precambrian; Geol. Soc. Australia J., v. 13, p. 593–608.

EDWARDS, L.E., 1982: Numerical and semi–objective biostratigraphy: review and predictions; Third North American Paleont. Conv., Proc., v. 1, p. I47–I52.

ELDREDGE, N., and GOULD, S.J., 1977: Evolutionary models and biostratigraphic strategies, *in* E.G. Kauffman and J.E. Hazel, eds., Concepts and methods of biostratigraphy; Dowden, Hutchinson and Ross Inc., Stroudsburg, Pennsylvania, p. 25–40.

ELLIOTT, T., 1978: Deltas, *in* H.G. Reading, ed., Sedimentary environments and facies; Blackwell, Oxford, p. 97–142.

GEORGE, T.N., HARLAND, W.B., AGER, D.V., BALL, H.W., BLOW, W.H., CASEY, R., HOLLAND, C.H., HUGHES, N.F., KELLAWAY, G.A., KENT, P.E., RAMSBOTTOM, W.H.C., STUBBLEFIELD, J., and WOODLAND, A.W., 1969: Recommendations on stratigraphical usage; Proc. Geol. Soc. London, no. 1656, p. 139–166.

GOLDICH, S.S., 1968: Geochronology in the Lake Superior region; Can. J. Earth Sci., v. 5, p. 715–724.

GOULD, S.J., 1972: Allometric fallacies and the evolution of *Gryphaea*: a new interpretation based on White's criterion of geometric similarity, *in* Th. Dobzhansky et al., eds., Evolutionary Biology; Appleton-Century-Crofts, New York, v. 6, p. 91–118.

GRADSTEIN, F., 1981: Oldest oceanic sediments and basement recovered by deep sea drilling; Geolog, v. 10, pt. 2, p. 40–43.

GRAY, J., and BOUCOT, A.J., eds., 1979: Historical biogeography, plate tectonics, and the changing environment; Oregon State Univ. Press, Corvallis, Oregon, 500 p.

HALLAM, A., 1959: On the supposed evolution of *Gryphaea* in the Lias; Geol. Mag., v. 96, p. 99–108.

HALLAM, A., ed., 1973: Atlas of paleobiogeography; Elsevier, Amsterdam, 531 p.

HALLAM, A., and BRADSHAW, M.J., 1979: Bituminous shales and oolitic ironstones as indicators of transgressions and regressions; J. Geol. Soc. London, v. 136, p. 157–164.

HANCOCK, J.M., 1977: The historic development of biostratigraphic correlation, *in* E.G. Kauffman and J.E. Hazel, eds., Concepts and methods of biostratigraphy; Dowden, Hutchinson and Ross Inc., Stroudsburg, Pennsylvania, p. 3–22.

HARLAND, W.B., 1978: Geochronologic scales, *in* G.V. Cohee, M.F. Glaessner and H.D. Hedberg, eds., Contributions to the Geologic time scale; Am. Assoc. Petrol. Geol. Studies in Geology 6, p. 9–32.

HARLAND, W.B., AGER, D.V., BALL, H.W., BISHOP, W.W., BLOW, W.H., CURRY, D., DEER, W.A., GEORGE, T.N., HOLLAND, C.H., HOLMES, S.C.A., HUGHES, N.F., KENT, P.E., PITCHER, W.S., RAMSBOTTOM, W.H.C., STUBBLEFIELD, C.J., WALLACE, P., and WOODLAND, A.W., 1972: A concise guide to stratigraphical procedure; J. Geol.Soc. London, v. 128, p. 295–305.

HARLAND, W.B., COX, A.V., LLEWELLYN, P.G., PICKTON, C.A.G., SMITH, A.G. and WALTERS, R., 1982: A geologic time scale; Cambridge Earth Science Series, Cambridge University Press, Cambridge, 131 p.

HARLAND, W.B., SMITH, A.G., and WILCOX, B., eds., 1964: The Phanerozoic time-scale; Quart. J. Geol. Soc. London, Supplement v. 120S.

HARPER, C.W., Jr., 1981: Inferring succession of fossils in time: the need for a quantitative and statistical approach; J. Paleont., v. 55, p. 442–452.

HARRISON, C.G.A., and FUNNELL, B.M., 1964: Relationship of palaeomagnetic reversals and micropalaeontology in two Late Cenozoic cores from the Pacific Ocean; Nature, v. 204, p. 566.

HARRISON, J.E., and PETERMAN, Z.E., 1980: North American Commission on Stratigraphic Nomenclature, Note 52—A preliminary proposal for a chronometric time scale for the Precambrian of the United States and Mexico; Geol. Soc. Am. Bull., v. 91, p. 377–380.

HARRISON, J.E. and PETERMAN, Z.E., 1982: North American Commission on Stratigraphic Nomenclature, Report 9—Adoption of geochronometric units for divisions of Precambrian time; Am. Assoc. Petrol. Geol. Bull., v. 66, p. 801–802.

HAY, W.W., and SOUTHAM, J.R., 1978: Quantifying biostratigraphic correlation; Ann. Rev. Earth Plan. Sci., v. 6, p. 353–375.

HAZEL, J.E., 1977: Use of certain multivariate and other techniques in assemblage zonal biostratigraphy: examples utilizing Cambrian, Cretaceous, and Tertiary benthic invertebrates, *in* E.G. Kauffman and J.E. Hazel, eds., Concepts and methods of biostratigraphy; Dowden, Hutchinson and Ross Inc., Stroudsburg, Pennsylvania, p. 187–212.

HEDBERG, H.D., ed., 1976: International Stratigraphic Guide; John Wiley and Sons, New York, 200 p.

HEIRTZLER, J.R., DICKSON, G.O., HERRON, E.M., PITMAN, W.C., and LE PICHON, X., 1968: Marine magnetic anomalies, geomagnetic field reversals and motions of the ocean floor and continents; J. Geophys. Res., v. 73, p. 2119–2136.

HOFMANN, H.J., 1977: On Aphebian stromatolites, and Riphean stromatolite stratigraphy; Precamb. Research, v. 5, p. 175–206.

HORNE, J.C., and FERM, J.C., 1978: Carboniferous depositional environments: eastern Kentucky and southern West Virginia, a field guide; Dept. Geol., Univ. South Carolina.

HORNE, J.C., FERM, J.C., CARUCCIO, F.T., and BAGANZ, B.P., 1978: Depositional models in coal exploration and mine planning in Appalachian region; Am. Assoc. Petrol. Geol. Bull., v. 62, p. 2379–2411.

HOWARTH, M.K., 1973: The stratigraphy and ammonite fauna of the Upper Liassic Grey Shales of the Yorkshire coast; Bull. Brit. Mus. Nat. Hist. Geol., 24, p. 237–277.

HSÜ, K.J., KELTS, K., and VALENTINE, J.W., 1980: Resedimented facies in Ventura Basin, California, and model of longitudinal transport of turbidity currents; Am. Assoc. Petrol. Geol. v. 64, p. 1034–1051.

HUGHES, N.F., ed., 1973: Organisms and continents through time; Spec. Papers in Paleontology 12 and Syst. Assoc. Pub. 9.

IRVING, E., 1966: Paleomagnetism of some Carboniferous rocks from New South Wales and its relation to geological events; J. Geophys. Res., v. 71, p. 6025–6051.

JAMES, H.L., 1972: Stratigraphic Commission Note 40—Subdivision of Precambrian—an interim scheme to be used by U.S. Geological Survey; Am. Assoc. Petrol. Geol. Bull., v. 56, p. 1128–1133.

JAMES, H.L., 1978: Subdivision of the Precambrian—A brief review and a report on recent decisions by the Subcommission on Precambrian Stratigraphy; Precamb. Research, v. 7, p. 193–204.

JELETZKY, J.A., 1978: Causes of Cretaceous oscillations of sea level in Western and Arctic Canada and some general geotectonic implications; Geol. Surv. Canada, Paper 77–18.

JOHNSON, G.D., JOHNSON, N.M., OPDYKE, N.D., and TAHIRKHELI, R.A.K., 1979: Magnetic reversal stratigraphy and sedimentary tectonic history of the Upper Siwalik Group, eastern Salt Range and southwestern Kashmir, *in* A. Farah and K. DeJong, eds., Geodynamics of Pakistan; Geol. Surv. Pakistan, p. 149–166.

JONES, B., and DIXON, O.A., 1975: The Leopold Formation: an Upper Silurian intertidal/supratidal carbonate succession on northeastern Somerset Island, Arctic Canada; Can. J. Earth Sci., v. 12, p. 395–411.

KAUFFMAN, E.G., 1977: Evolutionary rates and biostratigraphy, *in* E.G. Kauffman and J.E. Hazel, eds., Concepts and methods of biostratigraphy; Dowden, Hutchinson and Ross Inc., Stroudsburg, Pennsylvania, p. 109–142.

KAUFFMAN, E.G., and HAZEL, J.E., eds., 1977: Concepts and methods of biostratigraphy; Dowden, Hutchinson and Ross Inc., Stroudsburg, Pennsylvania, 658 p.

KELLER, H.M., TAHIRKHELI, R.A.K., MIRZA, M.A., JOHNSON, G.D., and JOHNSON, N.M., 1977: Magnetic polarity stratigraphy of the Upper Siwalik deposits, Pabbi Hills, Pakistan; Earth Plan. Sci. Letters, v. 36, p. 187–201.

KENNEDY, W.J., and COBBAN, W.A., 1977: The role of ammonites in biostratigraphy, *in* E.G. Kauffman and J.E. Hazel, eds., Concepts and methods of biostratigraphy; Dowden, Hutchinson and Ross Inc., Stroudsburg, Pennsylvania, p. 309–320.

KENNETT, J.P., ed., 1980: Magnetic stratigraphy of sediments; Dowden, Hutchinson and Ross Inc., Stroudsburg, Pennsylvania, Benchmark Papers in Geology 54, 438 p.

KERMACK, K.A., 1954: A biometrical study of *Micraster coranguinum* and *M. (Isomicraster) senonensis*; Phil. Trans. Roy. Soc. B., v. 237, p. 375–428.

LAMBERT, R. St.J., 1964: The relationship between radiometric ages obtained from plutonic complexes and stratigraphical time, *in* W.B. Harland, A.G. Smith and B. Wilcock, eds., The Phanerozoic time scale; Quart. J. Geol. Soc. London, Supplement v. 120S, p. 43–54.

LAMBERT, R. St.J., 1971: The pre-Pleistocene Phanerozoic time-scale: a review, *in* The Phanerozoic time scale—a supplement; Geol. Soc. London Spec. Publ. 5, p. 9–31.

LARSON, R.L., and PITMAN, W.C. III, 1972: World-wide correlation of Mesozoic magnetic anomalies, and its implications; Geol. Soc. Am. Bull., v. 83, p. 3645–3662.

LOWRIE, W., and ALVAREZ, W., 1981: One hundred million years of geomagnetic polarity history; Geology, v. 9, p. 392–397.

McKERROW, W.S., 1971: Palaeontological prospects—the use of fossils in stratigraphy; J. Geol. Soc. London, v. 127, p. 455–464.

McDOUGALL, I., 1978: Potassium-argon isotopic dating method and its application to physical time scale studies; in G.V. Cohee, M.F. Glaessner and H.D. Hedberg, eds. Contributions to the geologic time scale; Am. Assoc. Petrol. Geol. Studies in Geol. 6, p. 119–126.

McLAREN, D.J., 1970: Presidential address: Time, life and boundaries; J. Paleont, v. 44, p. 801–815.

McLAREN, D.J., 1973: The Silurian–Devonian boundary; Geol. Mag., v. 110, p. 302–303.

MIALL, A.D., 1979: Mesozoic and Tertiary geology of Banks Island, Arctic Canada: the history of an unstable craton margin; Geol. Surv. Canada Mem. 387.

MIDDLEMISS, F.A., RAWSON, P.F., and NEWALL, G., eds., 1971: Faunal provinces in space and time; Geol. J. Spec. Issue 4.

MILLER, F.X., 1977: The graphic correlation method in biostratigraphy, *in* E.G. Kauffman and J.E. Hazel, eds., Concepts and methods of biostratigraphy; Dowden, Hutchinson and Ross, Inc., Stroudsburg, Pennsylvania, p. 165–186.

MOORBATH, S., 1964: The rubidium–strontium method, *in* W.B. Harland, A.G. Smith and B. Wilcock, eds., The Phanerozoic time scale, Quart. J. Geol. Soc. London, Supplement v. 120S, p. 87–100.

MOORBATH, S., 1967: Recent advances in the application and interpretation of radiometric age data; Earth Sci. Rev., v. 3, p. 111–133.

MORLEY, L.W., and LAROCHELLE, A., 1964: Paleomagnetism as a means of dating geological events, *in* F.F. Osborne, ed., Geochronology in Canada, Roy. Soc. Can. Spec. Publ. 8, p. 39–51.

MYHR, D.W., and MEIJER-DREES, N.C., 1976: Geology of the southeastern Alberta Milk River gas pool, *in* M.M. Lerand, ed., The sedimentology of selected clastic oil and gas reservoirs in Alberta; Can. Soc. Petrol. Geol., p. 96–117.

NELSON, A.R., 1981: Quaternary glacial and marine stratigraphy of the Qivitu Peninsula, northern Cumberland Peninsula, Baffin Island, Canada; Geol. Soc. Am. Bull., Pt. II, v. 92, p. 1143–1261.

ODIN, G.S., 1975: Les glauconies: Constitution, formation, age; Thèse de doctorat d'état des science naturelles; Paris, Univ. P. et M. Curie, 250 p.

ODIN, G.S., 1978: Results of dating Cretaceous Paleogene sediments, Europe, *in* G.V. Cohee, M.F. Glaessner and H.D. Hedberg, eds., Contributions to the geologic time scale; Am. Assoc. Petrol. Geol. Studies in Geol. 6, p. 127–141.

OPDYKE, N.D., 1972: Paleomagnetism of deep-sea cores; Rev. Geophys. Space Phys., v. 10, p. 213.

OPDYKE, N.D., GLASS, B., HAYS, J.D., and FOSTER, J., 1966: Paleomagnetic study of Antarctic deep-sea cores; Science, v. 154, p. 349–357.

PICARD, N.D., 1964: Paleomagnetic correlation of units within the Chugwater (Triassic) Formation, west-central Wyoming; Am. Assoc. Petrol. Geol. Bull., v. 48, p. 269–291.

PREISS, W.V., 1977: The biostratigraphic potential of Precambrian stromatolites; Precamb. Research, v. 5, p. 207–219.

PRESTON, F.W., and HENDERSON, J.H., 1964: Fourier series characterization of cyclic sediments for stratigraphic correlation; State Geol. Surv. Kansas Bull. 169, p. 415–425.

PUTNAM, P.E. and OLIVER, T.A., 1980: Stratigraphic traps in channel sandstones in the Upper Mannville (Albian) of east-central Alberta; Bull. Can. Petrol. Geol., v. 28, p. 489–508.

RICCI-LUCCHI, F., and VALMORI, E., 1980: Basin-wide turbidites in a Miocene, over-supplied deep-sea plain: a geometrical analysis; Sedimentology, v. 27, p. 241–270.

ROWE, A.W., 1899: An analysis of the genus *Micraster*, as determined by rigid zonal collecting from the zone of *Rhynchonella cuvieri* to that of *Micraster coranguinum*; Quart. J. Geol. Soc. London, v. 55, p. 494–547.

RYAN, W.B.F., CITA, M.B., RAWSON, M.D., BURCKLE, L.H., and SAITO, T., 1974: A paleomagnetic assignment of Neogene stage boundaries and the development of isochronous datum planes between the Mediterranean, the Pacific and Indian Oceans in order to investigate the response of the World Ocean to the Mediterranean "salinity crisis"; Riv. Italiana Paleontologia e Stratigrafia, v. 80, p. 631–688.

SCHOPF, J.W., 1977: Biostratigraphic usefulness of stromatolitic Precambrian microbiotas: a preliminary analysis; Precamb. Research, v. 5, p. 143–174.

SCHULTZ, E.H., 1982: The chronosome and supersome: terms proposed for low-rank chronostratigraphic units; Bull. Can Petrol. Geol., v. 30, p. 29–33.

SCLATER, J.G., JARRARD, R.D., McGOWRAN, B., and GARTNER, S., 1974: Comparison of the magnetic and biostratigraphic time scales since the Late Cretaceous, *in* Initial Repts. Deep Sea Drilling Proj, v. XXII, p. 381–386.

SEMIKHATOV, M.A., 1974: Stratigrafiya i geokhronologiya proterozoya [Proterozoic stratigraphy and geochronology]; Trans. Geol. Inst. Acad. Sci. USSR. 256, Publishing House Nauka, Moscow, 302 p. (in Russian).

SHACKLETON, N.J., and OPDYKE, N.D., 1976: Oxygen-isotope and paleomagnetic stratigraphy of Pacific Core V28–239, Late Pliocene to latest Pleistocene; Geol. Soc. Am. Mem. 145, p. 449–464.

SHAW, A.B., 1964: Time in stratigraphy; McGraw-Hill, New York, 365 p.

SHERIFF, R.E., 1976: Inferring stratigraphy from seismic data; Am. Assoc. Petrol. Geol. Bull., v. 60, p. 528–542.

SILVER, L.T. and SCHULTZ, P.H., eds., 1982: Geological implications of impacts of large asteroids and comets on the earth; Geol. Soc. Am. Spec. Paper 190.

SMITH, A.G., BRIDEN, J.C., and DREWRY, G.E., 1973: Phanerozoic world maps, *in* N.F. Hughes, ed., Organisms and continents through time; Spec. Papers in Palaeont. 12 and Syst. Assoc. Pub. 9, p. 1–42.

SMITH, T.F., and WATERMAN, M.S., 1980: New stratigraphic correlation techniques; J. Geol., v. 88, p. 451–457.

STAPLIN, F.L., 1969: Sedimentary organic matter, organic metamorphism and oil and gas occurrence; Bull. Can. Petrol. Geol., v. 17, p. 47–66.

STAPLIN, F.L., 1977: Interpretation of thermal history from colour of particulate organic matter—a review; Palynology, v. 1, p. 9–18.

STEIGER, R.H. and JAGER, E., 1978: Subcommission on geochronology: convention on the use of decay constants in geochronology, *in* G.V. Cohee, M.F. Glaessner and H.D. Hedberg, eds., Contributions to the geologic time scale; Am. Assoc. Petrol. Geol. Studies in Geol. 6, p. 67–71.

STOAKES, F.A., 1980: Nature and control of shale basin fill and its effect on reef growth and termination: Upper Devonian Duvernay and Ireton Formations of Alberta, Canada; Bull. Can. Petrol. Geol., v. 28, p. 345–410.

STOCKWELL, C.H., 1973: Revised Precambrian time scale for the Canadian Shield; Geol. Surv. Canada Paper 72–52.

SUBCOMMISSION ON A MAGNETIC POLARITY TIME SCALE, 1973: Magnetic polarity time scale; Geotimes, v. 18, p. 21–22.

SURLYK, F., and BIRKELUND, T., 1977: An integrated stratigraphical study of fossil assemblages from the Maastrichtian White Chalk of Northwestern Europe, *in* E.G. Kauffman and J.E. Hazel, ed., Concepts and methods of biostratigrpahy; Dowden, Hutchinson and Ross, Inc., Stroudsburg, Pennsylvania, p. 257–282.

SYLVESTER-BRADLEY, P.C., 1977: Biostratigraphical tests of evolutionary theory, *in* E.G. Kauffman and J.E. Hazel, eds., Concepts and methods of biostratigraphy; Dowden, Hutchinson and Ross Inc., Stroudsburg, Pennsylvania, p. 41–64.

TARLING, D.H., and MITCHELL, J.G., 1976: Revised Cenozoic polarity time scale; Geology, v. 4, p. 133–136.

TIMOFEEV, B.V., 1973: Mikrofitofossilii proterozoya i rannego paleozoya [Plant microfossils from the Proterozoic and lower Paleozoic], *in* T.F. Vozzhennikova and B.V. Timofeev, eds., Mikrofossili Drevneishikh Otlozhenii [Microfossils of the oldest deposits]; Proc. 3rd. Internat. Palynological Conf., Publishing House Nauka, Moscow, p. 7–12 (in Russian).

TRUEMAN, A.E., 1922: The use of *Gryphaea* in the correlation of the Lower Lias; Geol. Mag., v. 59, p. 256–268.

TURNER, P., 1980: Continental red beds; Develop. Sedimentology 29, Elsevier, Amsterdam, 562 p.

VAIL, P.R., TODD, R.G., and SANGREE, J.B., 1977: Seismic stratigraphy and global changes of sea level, Part five: Chronostratigraphic significance of seismic reflections, *in* C.E. Payton, ed., Seismic stratigraphy—applications to hydrocarbon exploration; Am. Assoc. Petrol. Geol. Mem. 26, p. 99–116.

VALENTINE, J.W., 1973: Evolutionary paleoecology of the marine biosphere; Prentice Hall Inc., Englewood Cliffs, New Jersey, 511 p.

VAN COUVERING, J.A., and BERGGREN, W.A., 1977: biostratigraphical basis of the Neogene time scale, *in* E.G. Kauffman and J.E. Hazel, eds., Concepts and methods of biostratigraphy; Dowden, Hutchinson and Ross Inc., Stroudsburg, Pennsylvania, p. 283–306.

VAN EYSINGA, F.W.B., 1975: Geological time table; 3rd ed., Elsevier, Amsterdam.

VAN HINTE, J.E., 1969: The nature of biostratigraphic zones; First Internat. Conf. Planktonic Microfossils, Geneva, Proc.v. 2, p. 267–272.

VAN HINTE, J.E., 1976a: A Jurassic time scale; Am. Assoc. Petrol.Geol. Bull., v. 60, p. 489–497.

VAN HINTE, J.E., 1976b: A Cretaceous time scale; Am. Assoc. Petrol. Geol. Bull., v. 60, p. 498–516.

VINE, F.J., and MATTHEWS, D.H., 1963: Magnetic anomalies over ocean ridges; Nature, v. 199, p. 947–949.

VISSER, C.F., and JOHNSON, G.D., 1978: Tectonic control of Late Pliocene molasse sedimentation in a portion of the Jhelum re-entrant, Pakistan; Geol. Rundschau., v. 67, p. 15–37.

WARDLAW, N.C., and REINSON, G.E., 1971: Carbonate and evaporite deposition and diagenesis, Middle Devonian Winnipegosis and Prairie Evaporite formations of south-central Saskatchewan; Am. Assoc. Petrol. Geol. Bull., v. 55, p. 1759–1786.

WATKINS, N.D., 1976: Polarity subcommission sets up some guidelines; Geotimes, v. 21, p. 18–20.

WATSON, R.A., 1983: A critique chronostratigraphy; Am. J. Sci., v. 283, p. 173–177.

WENDT, I., 1980: Project 89: calibration of stratigraphic methods; Geological Correlation no. 8, Rept.

Int. Geol. Correlation Pgm., IUGS–UNESCO, Paris, p. 164–165.

WESTGATE, J.A., 1980: Dating methods of Pleistocene deposits and their problems: V. Tephrochronology and fission–track dating; Geosci. Canada, v. 7, p. 3–10.

WINDLEY, B.F., 1977: The evolving continents; John Wiley and Sons, London, 385 p.

YOUNG, G.M., 1979: Correlation of Middle and Upper Proterozoic strata of the northern rim of the North Atlantic craton; Trans. Roy. Soc. Edin., v. 70, p. 323–336.

ZIEGLER, A.M., COCKS, L.R.M., and McKERROW, W.S., 1968: The Llandovery transgression of the Welsh borderland; Paleontology, v. 11, p. 736–782.

CHAPTER 4

Facies analysis

4.1 Introduction

The study and interpretation of the textures, sedimentary structures, fossils and lithologic associations of sedimentary rocks on the scale of an outcrop, well section or small segment of a basin comprise the subject of facies analysis. In writing this chapter the author has been strongly aware of the several excellent texts now available that deal exclusively with facies analysis, for example, the introductory text by Walker (1979a) and the more advanced treatments by J.L. Wilson (1975) and Reading (1978). Several other books contain a more condensed treatment of the same material (e.g. Blatt et. al., 1980; Leeder, 1982). Therefore little would be gained by presenting a complete discussion of facies models. Accordingly this chapter focuses on a discussion of analytical methods, and reviews the kinds of information that can be obtained from sediments, based on the observations described in Chapter 2. It is hoped that this material will provide students with an introduction to modern facies modeling methods. Its practice on specific ancient examples can then be carried out with one of the advanced texts (mentioned above) in hand, and the final section of this chapter should serve to complement these, by providing a discussion and updating of some of the most recent advances in our ideas.

To extend a facies analysis to an entire basin it is necessary to come to grips with problems of stratigraphic correlation and to apply various basin mapping techniques. Stratigraphic methods are described in Chapter 3, and basin mapping, including the use of isopach and facies ratio maps, paleocurrent analysis, and so on, is dealt with in Chapter 5. The end product of this work is a paleogeographic synthesis, depicting an interpretation of the stratigraphic and geographic evolution of the basin through time. Certain large-scale basin fill patterns have emerged over the years from work of this kind. These are referred to as depositional systems, and are described in Chapter 6. The difference between facies analysis and depositional systems analysis is essentially one of scale. However, the demarcation is not clear-cut, and some overlap between Chapters 4 and 6 is unavoidable. This point is returned to in section 4.7.

4.2 The meaning of facies

The meaning of the word **facies** has been much debated in geology (e.g. Moore, 1949; Teichert, 1958; Krumbein and Sloss, 1963). It is widely used in sedimentary geology, but also has a somewhat different meaning in the area of metamorphic petrology (Fawcett, 1982). Reading (1978, Chap. 2) provided an excellent discussion of the modern sedimentological uses of the term, and the methods of interpreting individual facies and facies relationships.

Nowadays the word facies is used in both a descriptive and an interpretive sense, and the word itself may have either singular or plural meaning. Descriptive facies include **lithofacies** and **biofacies**, both of which terms are used to refer to certain observable attributes of sedimentary rock bodies, which can be interpreted in terms of depositional or biological process. (When used without a prefix in this book the word facies is intended to mean either lithofacies or biofacies.) An individual lithofacies is a rock unit defined on the basis of its distinctive lithologic features, including composition, grain size, bedding characteristics and sedimentary structures. Each lithofacies represents an individual depositional event. Lithofacies may be grouped into lithofacies associations or assemblages, which are characteristic of particular depositional environments. These assemblages form the basis for defining lithofacies models; they commonly are cyclic. A biofacies is defined on the basis of fossil components, including either body fossils or trace fossils. The term biofacies is normally used in the sense of an assemblage of such components. For the purpose of sedimentological study a deposit may be divided into a series of facies units, each of which displays a distinctive assemblage of lithologic or biologic features. These units may be single beds a few millimeters thick or a succession of beds tens to hundreds of meters thick. For example a river deposit may consist of decimeter-thick beds of a "conglomerate lithofacies" interbedded with a "crossbedded sandstone facies". Contrast this with the biofacies terms used to describe the fill of many major early Paleozoic basins. Commonly this may be divided into units hundreds of meters thick comprising a "shelly biofacies", containing such fossils as brachiopods and trilobites, and a "graptolitic biofacies". At the other extreme J.L. Wilson (1975) recommended the use of microfacies in studying thin sections of carbonate rocks, and defined 24 standard types.

The scale of an individual lithofacies or biofacies unit depends on the level of detail incorporated in its definition. It is determined by the variability of the succession or by the nature of the research undertaken (basin-wide reconnaissance versus a detailed local study), or by the availability of rock material for examination. Facies units defined on the basis of outcrop, core, well-cutting or geophysical criteria tend to refer to quite different scales and levels of detail. Geophysicists in the petroleum industry refer to "seismic facies", but this is not comparable to the small-scale type of facies discussed in this chapter. Seismic facies are referred to in Chapters 5 and 6. To a large extent the scales at which facies units are defined reflect criteria of convenience. The term is thus a very flexible and convenient one for descriptive purposes.

The term facies can also be used, usually for lithofacies assemblages, in an interpretive sense, for groups of rocks that are thought to have been formed under similar conditions. This usage may emphasize specific depositional processes, such as "till facies" or "turbidite facies". Alternatively, it may refer to a particular depositional environment, such as "shelf carbonate facies" or "fluvial facies", encompassing a wide range of depositional processes.

Widespread use is still made of two nineteenth century Swiss stratigraphic terms that have acquired a generalized facies meaning encompassing lithologic characteristics, depositional environment and tectonic setting. The first of these is the "flysch facies", comprising marine sediments, typically turbidites and other sediment gravity flow deposits, formed on continental margins. The "molasse facies" consists of nonmarine and shallow marine sediments, mainly sandstones and conglomerates, formed within and flanking fold belts during and following their elevation into mountain ranges. Both of these facies types may make up major stratigraphic units hundreds or thousands of meters thick and extending for hundreds or thousands of kilometers. Continued use of the terms flysch and molasse is not recommended because of ambiguities about their implications for tectonic setting. The vague association with orogenic belts is inadequate now that we have the theory of plate tectonics to assist us in interpreting the relationship between sedimentation and tectonics (Chap. 9).

Lithostratigraphy and lithofacies analysis are two contrasting approaches to the study of sedimentary rocks. The first is the traditional descriptive approach. The second is based on detailed facies descriptions, which provide the basis for the genetic study of sediments using facies models. Lithofacies and biofacies analysis can, and should, be used to assist in stratigraphic studies, because by understanding the depositional environments and paleogeography existing at the time a rock unit was formed we are much better placed to make predictions and extrapolations about lateral changes in thickness and composition. Obviously this will be invaluable for correlation purposes, and can make for a much more logical definition of formal lithostratigraphic units (see section 6.8). Biofacies analysis is crucial in the definition and comprehension of biostratigraphic units.

4.3 Recognition and definition of facies types

4.3.1 Philosophy and methods

In order to make sense of the lithologic variability present in most sedimentary basins it is necessary to generalize, categorize and simplify what we see in well sections and outcrops. It is necessary because the human brain can only absorb a limited amount of detail and process it at any one time. In sedimentology we find that much of the variability disguises a limited range of basic lithofacies and biofacies types, and that variations between these types represent minor random environmental fluctuations or are even the result of the accident of exposure or the position of a thin section cut. The existence of this natural pattern is what makes facies studies, facies modeling and paleogeographic reconstruction possible. For example, for many years sedimentologists categorized most deep-sea deposits in terms of only five basic lithofacies. These were the A, B, C, D and E divisions of the classic Bouma turbidite sequence (Bouma, 1962). (Certain thick-bedded and coarse-grained beds did not fit the pattern, and little attention was paid to them until the advent of the submarine fan model, which showed that Bouma sequences tend to occur mainly on the outer, non-channelized part of a fan. Nowadays the lithofacies spectrum for submarine fans is rather more complex, as discussed in section 4.6). Recently Miall (1977, 1978) showed that most deposits of braided rivers could be described using about 20 lithofacies types. J.L. Wilson (1975) in a major review of Phanerozoic carbonate rocks concluded that most could be described satisfactorily by drawing from a list of only 24 standard microfacies. Eyles et al. (1983) proposed a simple lithofacies classification scheme for the description of glacial-marine and glacial-lacustrine deposits.

Facies modelers have yet to categorize all sediment types in every variety of depositional environment; for example, no one has yet attempted to erect a universal lithofacies scheme for clastic rocks comparable to that of Wilson for carbonates. Similarly many fossil groups await detailed biofacies analysis. For this reason every basin analyst should study each basin with a fresh eye and, at least in the preliminary stages of the work, erect a local facies scheme without too much dependence on previously published work. Slavish adherence to such readily available research keys may result in minor but critical lithofacies types and lithofacies relationships being missed, or forced to fit an inappropriate mold. New facies models or a better definition of old ones will never come about if geologists are content merely with such replication studies. "Hummocky cross-stratification" is an excellent example of a structure which sedimentologists had been looking at without seeing for many years, until focused upon and given a name by Harms et al. (1975). Suddenly it was realized that this distinctive style of crossbedding is characteristic of many storm-generated ancient deposits. Its recognition has led to the reappraisal of numerous shallow marine sequences, to the extent that the term has become something of a cliche in recent years (e.g. see Byers and Dott, 1981).

How, then, does the basin analyst perform a facies analysis of an undescribed rock sequence or succession of fossils? The method for measuring and describing stratigraphic sections is discussed in detail in Chapter 2. For the purpose of lithofacies analysis the focus must be on recognizing associations of attributes that are repeated through the section (or parts of the entire basin). Lithofacies may be distinguished by the presence of bedding units with a characteristic sedimentary structure or structures, a limited grain size range, a certain bed thickness, perhaps a distinctive texture or color (color is subject to diagenetic change and should not be used as a primary criterion in definition; see section 2.3.2). Biofacies represent associations within the same stratigraphic interval of a limited suite of genera or species. Biofacies studies may be carried out on single taxonomic groups, because many of these, if examined by specialists, can yield highly detailed paleoecological (and hence depositional) information. The definition of biofacies is an exercise in defining assemblages and will be discussed in the next section.

In order to recognize these associations of lithologic attributes or fossil types it may be useful to set up a checklist and tabulate their occurrence. Field data recording forms, such as were discussed in Chapter 2, may be useful for this purpose. Statistical analysis is commonly used by paleoecologists to establish biofacies from such tabulated data. It is nowadays less used for erecting individual lithofacies because many of the common associations of lithologic attributes are well enough known to be recognized without the aid of statistics.

4.3.2 Field examples of facies schemes

Two examples of lithofacies schemes that have been erected from field data are given in this sec-

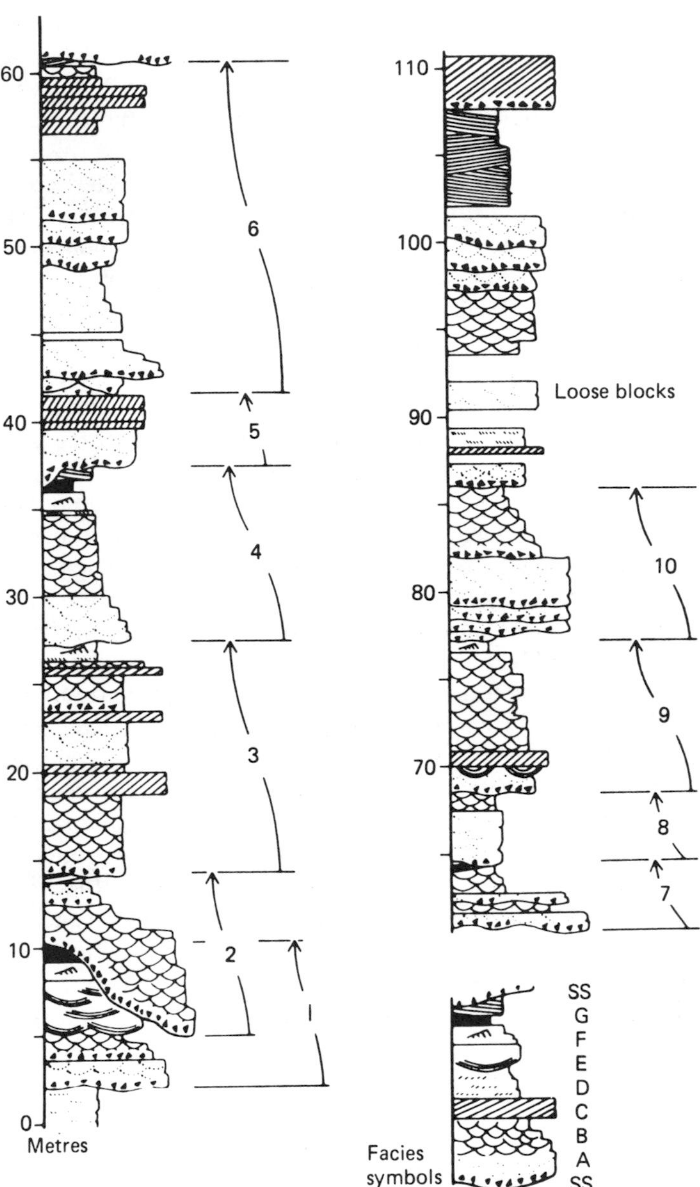

Fig. 4.1. A stratigraphic section plotted using a standardized facies scheme and the variable-width column technique. Fluvial cycles of the Battery Point Formation (Devonian), Quebec (Cant and Walker, 1976).

tion to illustrate the methods. A simple biofacies association is also described.

Cant and Walker (1976) described a Devonian fluvial section in eastern Quebec, and subdivided it into eight lithofacies. The section is shown in Figure 4.1, in which lithologic symbols are keyed to the lithofacies scheme. Note the use in this illustration of the variable-width column plotting technique for drawing stratigraphic sections that is discussed in section 2.3.4. Here are some examples of their lithofacies descriptions, edited here in order to focus on salient features.

Well Defined Trough Cross-Bedded Facies (B). This facies is composed of well-defined sets of trough crossbedding . . . , with trough depths averaging 15 to 20 cm (range 10 to 45 cm). The troughs are regularly stacked on top of each other, but in some individual occurrences of the facies, trough depths decrease upward . . . The sets are composed of well sorted medium sand . . . A few of the coarser sets have granules and pebbles concentrated at their bases.

Asymmetrical Scour Facies (E). This facies consists of large, asymmetrical scours and scour fillings, up to 45 cm deep and 3 m wide . . . The scours cut into each other and into underlying troughed facies (A and B), and occurrences of the asymmetrical scour facies have a flat, erosionally truncated top . . . The main difference between the scour fillings and the two troughed facies (A and B) lies in the geometry of the infilling strata. In the asymmetrical scours, the layers are not at the angle of repose, but are parallel to the lower bounding surface.

Rippled Sandstone and Mudstone Facies (F). This facies includes cross-laminated sandstones . . . , and alternating cross-laminated sandstones and mudstones [an example of the latter] is 1.5 m thick and consists of three coarsening-upward sequences, which grade from basal mudstones into trough cross-laminated fine sandstone, and finally into granule sandstone. The sandstones capping each coarsening-upward sequence have sharp, bioturbated tops.

Note the use of varied criteria in defining these lithofacies, and the absence, at this stage, of any environmental interpretation.

A carbonate example was described by R.C.L. Wilson (1975), who studied the Upper Jurassic Oolite Series (actually a formation, but an incorrect, older nomenclature is still in use) of Dorset, England. He recognized four lithofacies, as shown in Figure 4.2. Note in this illustration the relationship between lithostratigraphic and lithofacies units. The four lithofacies are as follows (described using the carbonate classification of Folk, 1962):

1. Coarsening-upward units shown by two beds . . . (1) *Chlamys qualicosta* bed: intramicrite-oomicrite-oosparite-poorly washed biosparite. (2) Pisolite: quartz sands and phyllosilicate clay-intramicrite-oomicrite-oosparite-oncolites . . .
2. Cross-bedded sets of oosparite showing 20° to 25° dips and sharp contacts either with phyllosilicate clays with nodular micrites . . . or bioturbated oolite . . . Some minor flaser bedding and clay drapes over current ripples also occur . . .
3. Association of *Rhaxella* biomicrites . . .
4. Sheet deposits (5 to 10 cm) and large accretion sets (30 cm) of oomicrite and biomicrite with subsidiary oosparite and biosparite. Shell debris often shows imbricate structure, and the oomicrites are texturally inverted sediments, being a mixture of extremely well-sorted oolites in a micrite matrix. Some sets showing alternating current directions occur . . .

The careful reader will note two points about these descriptions. Firstly, a mixture of lithofacies and biofacies criteria is used. This commonly is desirable in fossiliferous successions where the fossils are distinctive rock-forming components. Secondly, several of the lithofacies consist of several different carbonate rock types interbedded on a small scale. Each of these could be described as a lithofacies in its own right, in which case Wilson's four facies become facies assemblages. It is a question of the scale of description that is the most suitable for the purpose at hand. Those engaged in analyzing entire basins or major stratigraphic intervals in a basin cannot afford to spend too much time on fine detail. Decisions must be made with the core or outcrop in front of you about how thick the descriptive units are going to be, and how to combine thin beds for the purpose of lithofacies description. Generally lithofacies thicknesses in the order of a few decimenters to a few meters have been found to be the most useful. Smaller scale subdivisions may be erected for selected examples of well-exposed sections or continuous core if desired, to illustrate particular points in a facies description.

Wilson suspected that his Oolite Series beds were tidal in origin, and he tabulated various features of each of the four facies using a range of criteria suggested by Klein (1971) for recognizing tidal deposits (Fig. 4.3). Many of these make specialized use of paleocurrent evidence, which is discussed in section 4.5.4. Figure 4.3 illustrates the use of a checklist for lithofacies description. This can be a useful technique for field observation if the geologist suspects in advance what it is he or she is about to study. It is then possible to incorporate all the special observational detail and environmental criteria available from studies of the appropriate modern environment or other ancient rock units. There is, of course, the danger in doing so that other features might be missed and interpretation becomes a self-fulfilling prophecy. Tables of this kind are not recommended for illustration of written or oral presentation of a lithofacies study except, perhaps, as an appendix. They are visually difficult to absorb, and the geologist does better to provide summary descriptions, such as those given above, plus photographs and graphic logs.

Howard (1972) gave an excellent example of the use of trace fossils in erecting simple biofacies for the purpose of environmental interpretation. He studied the Upper Cretaceous, shallow marine rocks exposed at Book Cliffs, Utah, and erected five loosely defined trace fossil biofacies. In conjunction with lithofacies criteria this enabled Howard to suggest environmental interpretations for each assemblage, as shown in Figure 4.4. For example the stratigraphically lowest facies consists of thoroughly bioturbated silt with the

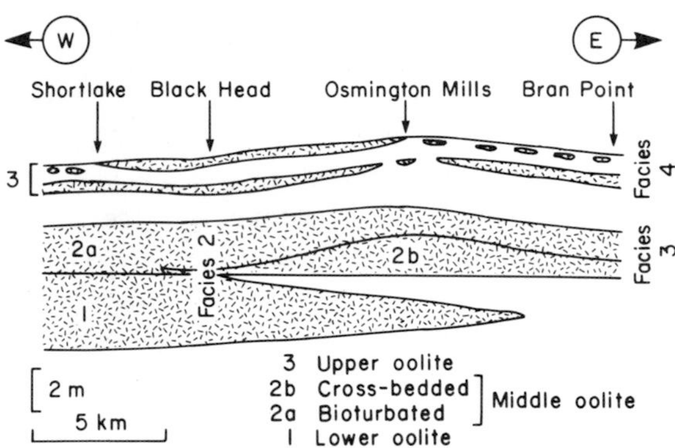

Fig. 4.2. Carbonate facies and their distribution in the Osmington Oolite (Jurassic), southern England (R.C.L. Wilson, 1975).

	Facies of Osmington Oolites			
	1	2	3	4
A. 1. Cross stratification with sharp set boundaries		■		■
2. Herringbone cross-stratification			■	■
4. Parallel laminae	■			
5. Complex internal organization of dunes, etc.		■	■	■
B. 6. Reactivation surfaces		■		■
7. Bimodal distribution of set thicknesses		■		
9. Unimodal distribution of dip direction of cross strata		■	■	
10. Orientation of cross-strata parallels sand-body trend			?	
C. 13. Small current ripples to larger current ripples			■	
19. Symmetrical ripples			■	
D. Flaser bedding			■	
F. 35 & 36 channel log deposits			■	
H. Burrowing, etc.	■	■	■	

Fig. 4.3. Distribution of selected lithofacies attributes in the Osmington Oolite (R.C.L. Wilson, 1975).

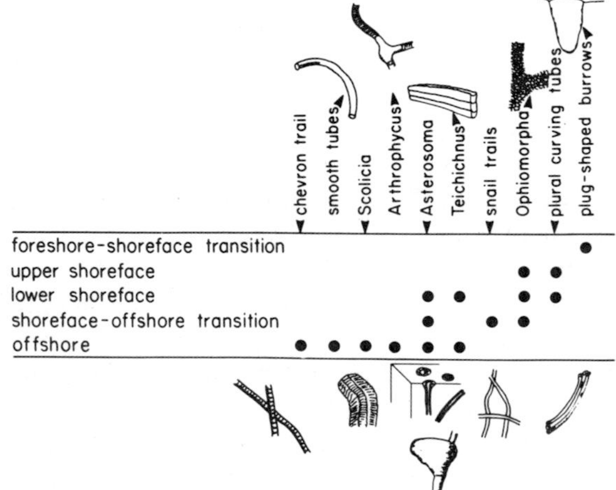

Fig. 4.4. Distribution of trace fossils in a Cretaceous marginal marine sandstone (Howard, 1972).

variety of trace fossils listed. Cleaner sands are associated with abundant *Ophiomorpha*, which is a shrimp burrow. In the shallowest waters the turbulent conditions were unsuitable for most invertebrates. Note the occurrence of only one trace fossil type in the foreshore–shoreface transition. In the deposits of the foreshore itself no trace fossils were recorded.

The use of statistical techniques in erecting more complex biofacies assemblages is discussed in the next section.

4.3.3 Establishing a facies scheme

The examples in the previous section serve to illustrate the methods used in three different kinds of basin problem to define facies types. The objective of the basin analyst should be to erect a facies scheme that can encompass all the rock types present in his or her project area. If the project encompasses several or many stratigraphic units deposited under widely varying environments it may make more sense to erect a separate scheme for each lithostratigraphic unit, because the lithofacies and, most likely, the biofacies assemblages will be different for each such unit.

Facies schemes should be kept as simple as possible, otherwise they defeat the whole purpose of carrying out a facies analysis. Some authors have subdivided their rocks into twenty or thirty lithofacies and erected subclasses of some of these. This gives the appearance of meticulous research and great analytical precision, but such schemes are difficult to understand or absorb. Remember that the purpose of a facies analysis is to aid in interpretation and, in basin analysis, this is often best accomplished by judicious simplification. It is rare that many more than half a dozen distinct lithofacies occur together in intimate stratigraphic association. Commonly two to four lithofacies occur in repeated, monotonous succession through tens or hundreds of meters of section. In defining only a few facies to cover thick and varied successions the problem of what to group together and what to separate is always present. What to do about thin beds, or whether and how to split up units that show gradational contacts? The only answer is to do what works. One approach is to carry out a very detailed description of the first well or section of a new project, and to use the resulting notes to erect a preliminary facies scheme. This is then employed for describing each new section. It has the advantage that instead of writing down basic descriptions for every bedding unit, it may be adequate simply to assign it to the appropriate facies and record in one's notes any additional observations that seem necessary, such as grain size differences or additional sedimentary structures relative to the original descriptive scheme. The scheme itself may thus be modified as one goes along. This approach has worked well for the author for many years of studying clastic sedimentary sequences in outcrops and core, in the Canadian Arctic and elsewhere. It is more difficult to apply this method to carbonate rocks, because of their finer-scale variability and the difficulty of seeing all the necessary features in weathered outcrop. One approach is to take into the field a rock saw and other necessary equipment for constructing etched slabs or peels. These can then be examined and described as field work proceeds. (J.L. Wilson, 1975, pp. 56–60).

Friend et al. (1976) avoided these problems entirely by recording only attributes of a sequence, not lithofacies as such.They then used statistical techniques to group these attributes into lithofacies assemblages, as discussed in the next section. Their field recording methods are described in section 2.3.1.

With continual study of the same kind of rocks a standard set of facies classes will gradually emerge and can be used for rapid field or laboratory description. For the writer this led to the erection of a generalized lithofacies scheme for describing the deposits of braided rivers, based on field studies of various ancient deposits and a review of literature on both ancient deposits and modern rivers. The initial scheme contained ten lithofacies types (Miall, 1977). Further research by the writer and use by other workers led to the addition of about ten more lithofacies types (Miall, 1978), but the original ten include most of the common ones. It seems likely that this scheme could be applied to most kinds of fluvial deposit and to the fluvial component of deltaic successions. Table 4.1 lists these lithofacies, showing the codes used for note taking, and a sedimentological interpretation of each. Greek letters refer to Allen's (1963a) classification of cross-bedding. The lithofacies codes consist of two parts, a capital letter of modal grain size (G = gravel, S = sand, F = fines) and a lower case letter or letters chosen as a mnemonic of a distinctive texture or structure of each lithofacies. The three lithofacies B, E and F of Cant and Walker (1976), discussed in the previous section, are St, Ss and Fl in this scheme. Le Blanc Smith (1980) has developed this fluvial facies scheme still further by incorporating additional structures and information on grain size.

J.L. Wilson's microfacies scheme for carbonates contains 24 types. Figure 4.5 illustrates his standard legend, facies numbers and abbreviated description. Figures 4.6 and 4.7 illustrate the use of these two schemes in the drawing up of stratigraphic sections.

4.4 Facies associations and models

The term **facies association** was defined by Potter (1959) as "a collection of commonly associated sedimentary attributes", including "gross geometry (thickness and areal extent); continuity and shape of lithologic units; rock types. . . . sedimentary structures, and fauna (types and abundances)." A facies association (or assemblage) is therefore based on observation, perhaps with some simplification. It is expressed in the form of a table, a statistical summary, or a diagram of typical stratigraphic occurrences (e.g. a vertical profile). A **facies model** is an interpretive device, which is erected by a geologist to explain the observed facies association. A facies model may be developed at first to explain only a single stratigraphic unit, and similar units may then be studied in order to derive generalized models.

Two main methods have evolved for defining and describing lithofacies associations. The first is the use of multivariate statistics such as cluster analysis and factor analysis to determine natural groupings of sedimentary attributes (e.g. Imbrie and Purdy, 1962; Purdy, 1963; Harbaugh and Demirmen, 1964; Friend et al., 1976). General statistical methods have been described by Harbaugh and Merriam (1968). The second method focuses on the order of occurrence of lithofacies in a sedimentary section, building from the obvious deduction that lithofacies that commonly rest upon one another must have shown some environmental association in nature. Various semiquantitative techniques have been used to study lithofacies sequences (e.g. Selley, 1970), but the most useful method is Markov

Table 4.1. A lithofacies scheme for fluvial deposits*

Facies code	Lithofacies	Sedimentary structures	Interpretation
Gms	massive, matrix supported gravel	none	debris flow deposits
Gm	massive or crudely bedded gravel	horizontal bedding, imbrication	longitudinal bars, lag deposits, sieve deposits
Gt	gravel, stratified	trough crossbeds	minor channel fills
Gp	gravel, stratified	planar crossbeds	linguoid bars or deltaic growths from older bar remnants
St	sand, medium to v. coarse, may be pebbly	solitary (theta) or grouped (pi) trough crossbeds	dunes (lower flow regime)
Sp	sand, medium to v. coarse, may be pebbly	solitary (alpha) or grouped (omikron) planar crossbeds	linguoid, transverse bars, sand waves (lower flow regime)
Sr	sand, very fine to coarse	ripple marks of all types	ripples (lower flow regime)
Sh	sand, very fine to very coarse, may be pebbly	horizontal lamination, parting or streaming lineation	planar bed flow (l. and u. flow regime)
Sl	sand, fine	low ange (<10°) crossbeds	scour fills, crevasse splays, antidunes
Se	erosional scours with intraclasts	crude crossbedding	scour fills
Ss	sand, fine to coarse, may be pebbly	broad, shallow scours including eta cross-stratification	scour fills
Sse, She, Spe	sand	analogous to Ss, Sh, Sp	eolian deposits
Fl	sand, silt, mud	fine lamination, very small ripples	overbank or waning flood deposits
Fsc	silt, mud	laminated to massive	backswamp deposits
Fcf	mud	massive, with freshwater molluscs	backswamp pond deposits
Fm	mud, silt	massive, desiccation cracks	overbank or drape deposits
Fr	silt, mud	rootlets	seatearth
C	coal, carbonaceous mud	plants, mud films	swamp deposits
P	carbonate	pedogenic features	soil

*From Miall, 1978.

chain analysis, which has been much used in recent years, particularly to study clastic sequences (e.g. Doveton, 1971; Miall, 1973; Cant and Walker, 1976). Krumbein and Dacey (1969), Dacey and Krumbein (1970) and Davis (1973) have discussed the general theory and methods.

Students of biofacies have used multivariate statistics widely for erecting biofacies associations (e.g. Mello and Buzas, 1968; Macdonald, 1975; Ludvigsen, 1978). Sequence studies of the Markov type have not yet been attempted.

4.4.1 Use of multivariate statistics

Imbrie and Purdy (1962) and Purdy (1963) carried out a landmark study of carbonate facies which involved a statistical analysis of data from modern sediments of the Great Bahama Bank. Purdy performed point count analyses under the microscope of 218 samples from the northern part of this bank. He recognized twelve major organic and inorganic components, as follows: coralline algae, *Halimeda* (calcareous alga), Peneroplidae

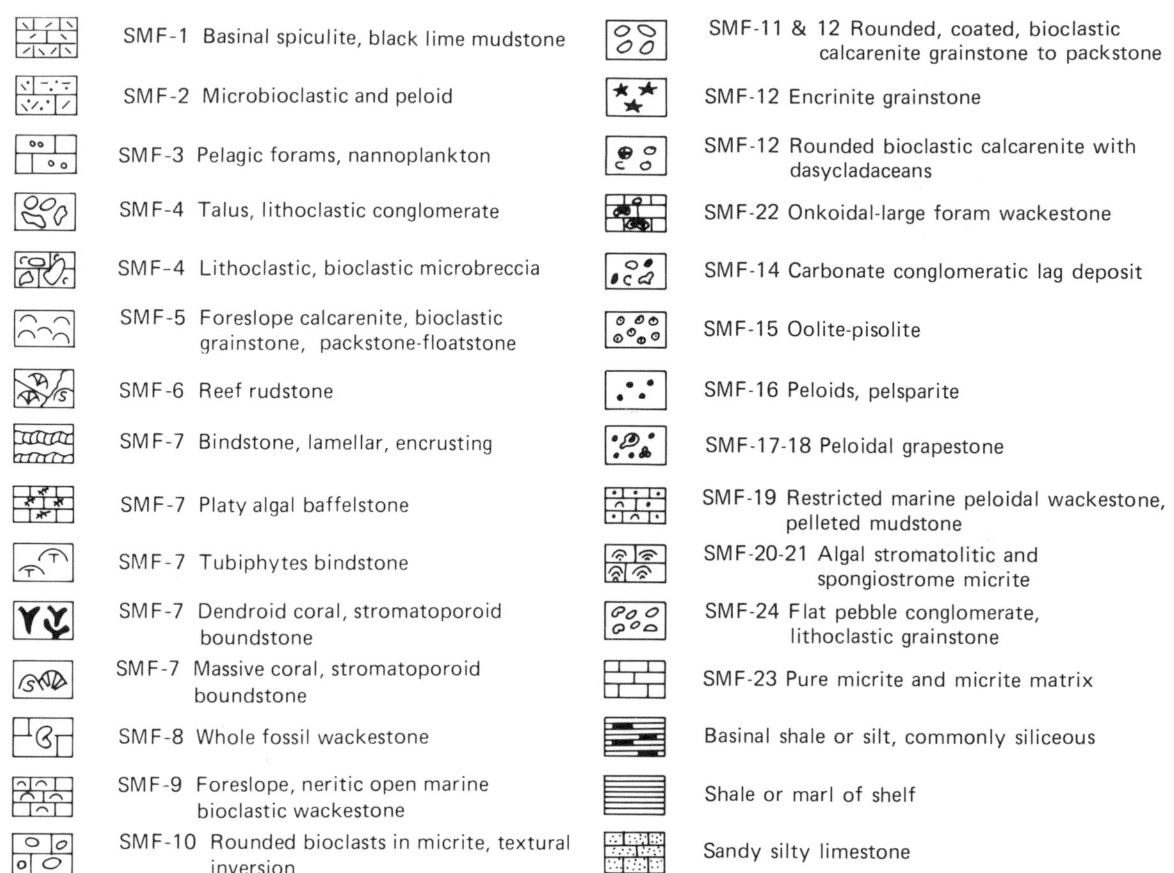

Fig. 4.5. The standard carbonate microfacies scheme of J.L. Wilson (1975).

(a family of foraminifera), other foraminifera, corals, molluscs, fecal pellets, mud aggregates, grapestone, oolite, cryptocrystalline grains and <1/8 mm (silt) fraction. Cluster analysis was then used to determine which components tended to occur together; the results are shown in Figure 4.8. From this information five major lithofacies were defined, and by referring back to the sample data these could be mapped as shown in Figure 4.9. Biofacies have also been mapped in the same area (Newell et al., 1959; Coogan, in Multer, 1971) and their distribution follows a very similar pattern (not shown). Purdy (1963) interpreted the lithofacies data in terms of physical and chemical processes operating over the bank. The work has been widely used and quoted as an example of a lithofacies study in carbonate rocks.

Walker (1979a, p.1) stated that such methods are unsuitable for clastic rocks, "where most of the important information (sedimentary and biological structures) cannot readily be quantified." However, given adequate outcrop (or core) data this is not necessarily true. Friend et al. (1976) used a complex set of statistical manipulations to analyze the nonmarine Devonian clastic sediments of East Greenland. Their field data were collected on computer processible forms, as discussed in section 2.3.1, and this greatly speeded up the process. Their method was, briefly, as follows. Each section was divided up into 10 m thick segments for the purpose of analysis. This thickness satisfactorily encompassed small-scale bed-by-bed variation, without obscuring major stratigraphic trends. Computer data processing techniques were then used to print out the total thickness in each 10 m segment of the follow 22 properties: red color; non-red color; fine, medium, coarse silt; very fine, fine, medium, coarse, and very coarse sandstone; conglomerate; limestone; flat bedding in sandstone; symmetrical ripples; asymmetrical ripples; planar cross-stratification; trough cross-stratification; deformation; con-

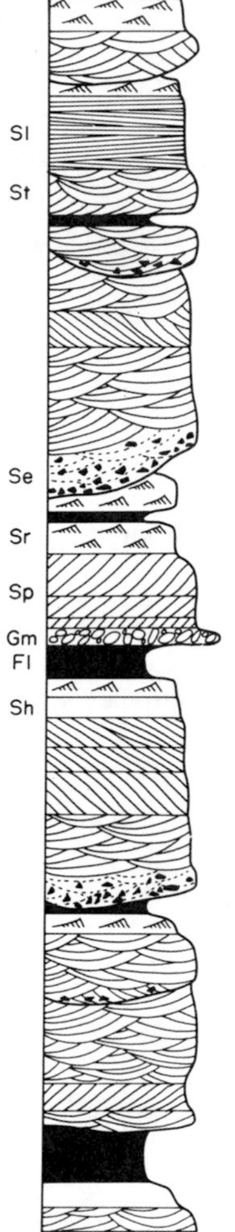

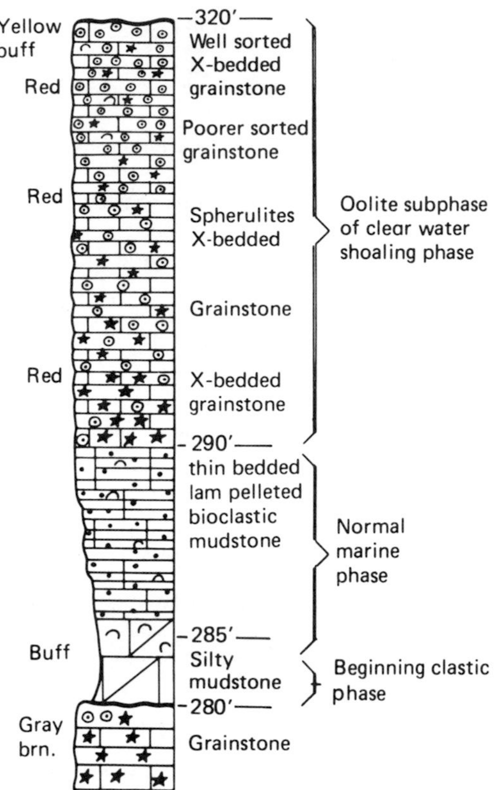

Fig. 4.7. Use of the carbonate microfacies scheme for plotting a limestone section (J.L. Wilson, 1975).

◁ **Fig. 4.6.** Use of the fluvial lithofacies scheme of Table 4.1 for plotting a section in braided fluvial deposits (Miall, 1978).

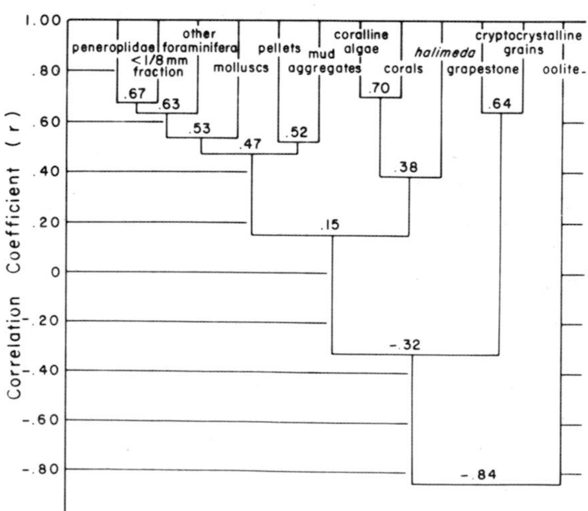

Fig. 4.8. Cluster analysis (R-mode) of facies components on the modern Bahama Platform (Purdy, 1963).

cretions; trace fossils; flat bedding in silt; lenticular bedding in silt. Some of these parameters are clearly interdependent, for example, planar and trough cross-bedding are found only in coarser grained sandstones. These data were therefore subjected to factor analyses to classify the parameters into principal components or "reaction groups" (Purdy's 1963 term). Although expressed statistically, these groups are equivalent to conventional lithofacies. The analysis resulted in the definition of seven of these groups. The 10-m samples were then compared by Q-mode cluster analysis using the distribution of the seven principal components therein as input data. The result was the definition of eight major sample types grouping together 32 subtypes. They are numbered 1A through 8; their characteristics are summarized in Figure 4.10, and four examples of typical 10-m segments are shown in Figure 4.11.

Fig. 4.9. Distribution of lithofacies on the Bahama Platform (Purdy, 1963).

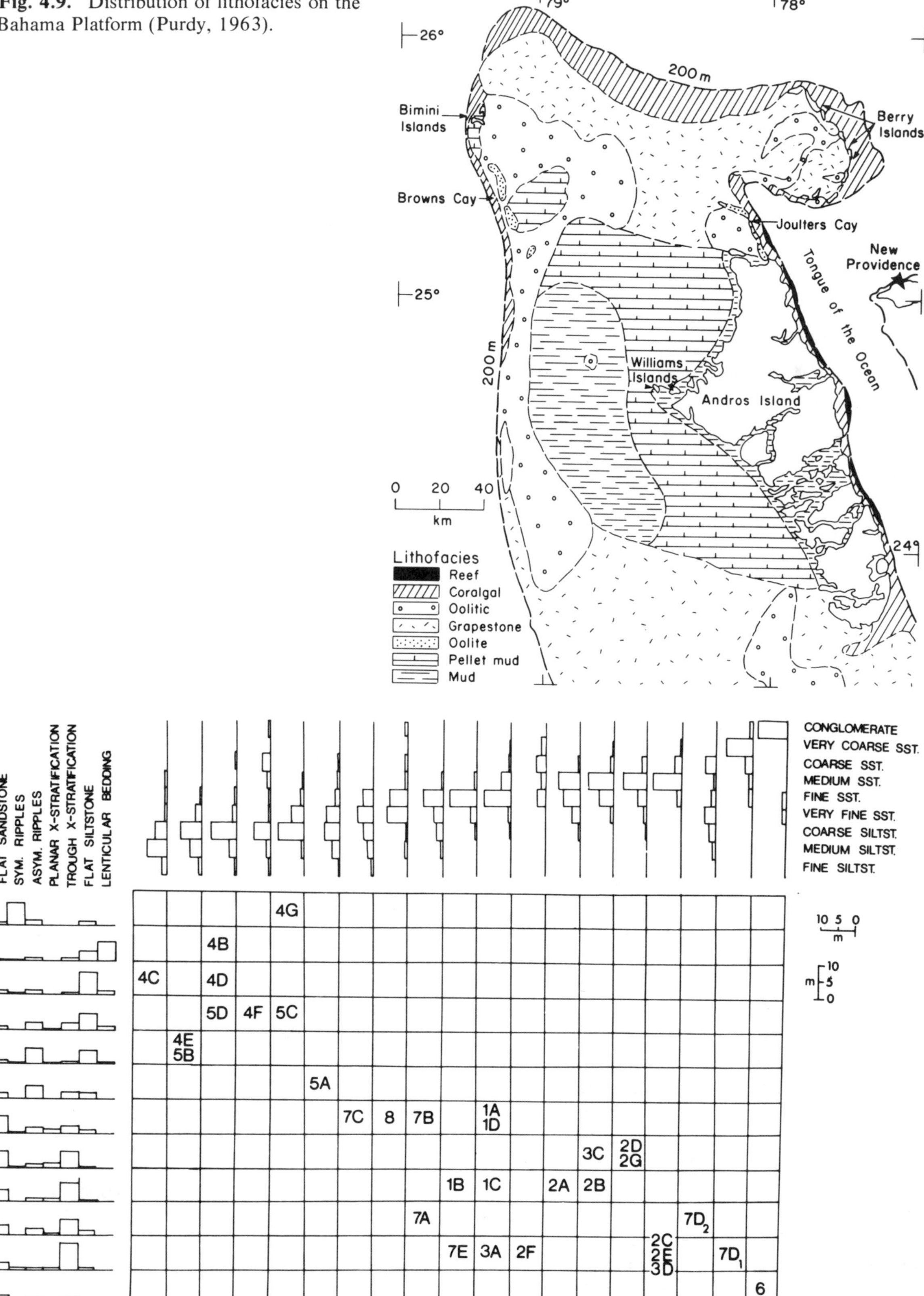

Fig. 4.10. Distribution of lithofacies attributes in the reaction groupings of Friend et al. (1976).

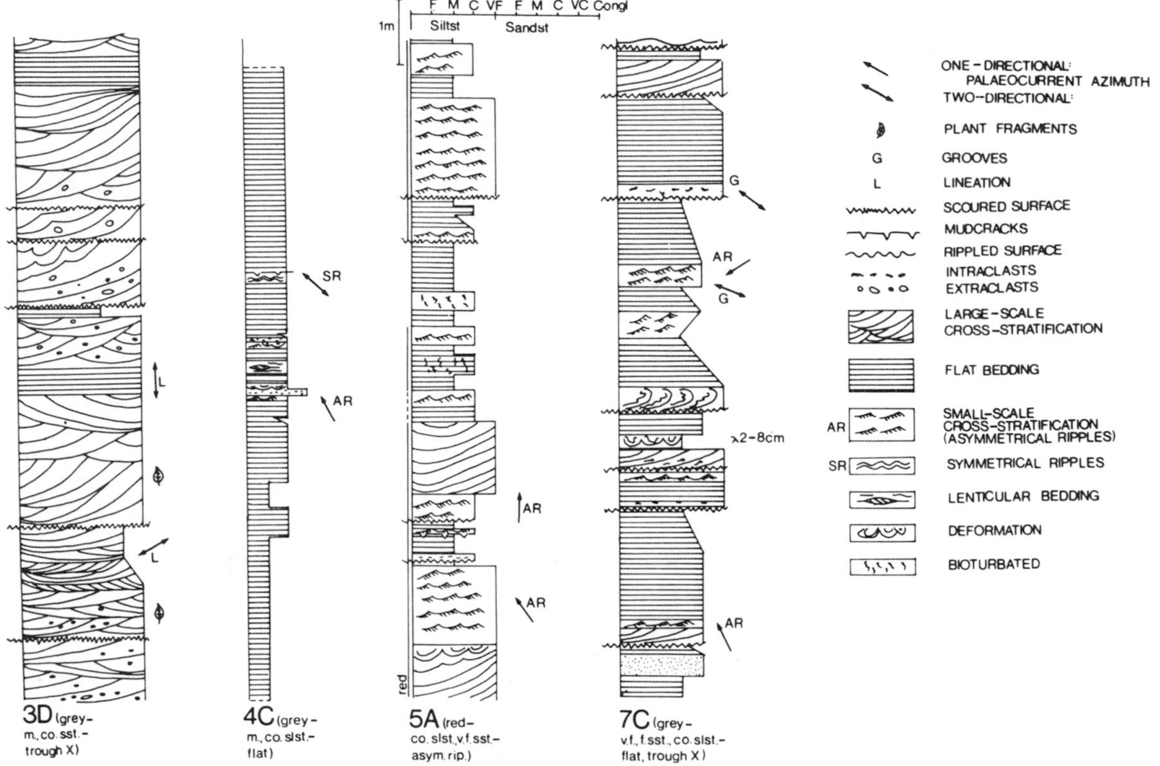

Fig. 4.11. Typical 10-m sections in the nonmarine Devonian sediments of east Greenland, illustrating four of the sample groups (Friend et al., 1976).

These 32 subtypes can be considered as statistically defined lithofacies assemblages.

Clearly, analyses of this type can only be conducted if field data have been rigorously collected from a very large total thickness of section. For major sedimentary basins containing essentially uniform lithofacies assemblages the methodology might be an excellent way to map subtle lithofacies variations as a basis for interpreting paleogeographic variations, tectonic control, etc. This was, in fact, the purpose of the exercise by Friend et al. (1976), and subsequent papers by them (numbered in sequence with the first, but with various author listings) investigated the East Greenland Devonian basin in great detail.

The study of lithofacies assemblages is rarely as detailed as this. Data sets frequently are inadequate, or they lend themselves to simple visual classification. It may not be necessary for a good basin analysis to construct such elaborate classifications if a few key parameters such as bed thickness, sandstone–mudstone ratio, etc., can be mapped, as described in Chapter 5. Nevertheless, the work by Friend et al. (1976) demonstrates how complex the actual lithologic world really is, and this is important to remember after one has read some of the elegant simplifications being produced by modern facies modelers. See, for example, many of the chapters in Walker (1979a), particularly the one on sandy fluvial systems, and contrast this with the work of Friend and his colleagues.

To complete the picture a single example is discussed here of the use of multivariate statistics in erecting biofacies. Cluster analysis appears to be widely accepted by paleoecologists as one of the best statistical techniques for biofacies analysis but, as noted by Ludvigsen (1978), there exists some controversy about what is defined by Q-mode versus R-mode analysis. Q-mode clustering compares samples, and therefore outlines areally and stratigraphically distinct groups (biotopes). These may have significance as biozones, as discussed in Chapter 3. Biofacies are normally considered as being defined by R-mode analysis, which compares attributes. But "the use of this terminology has been criticized by Mello and Buzas (1968) and Hazel (1975) who maintained

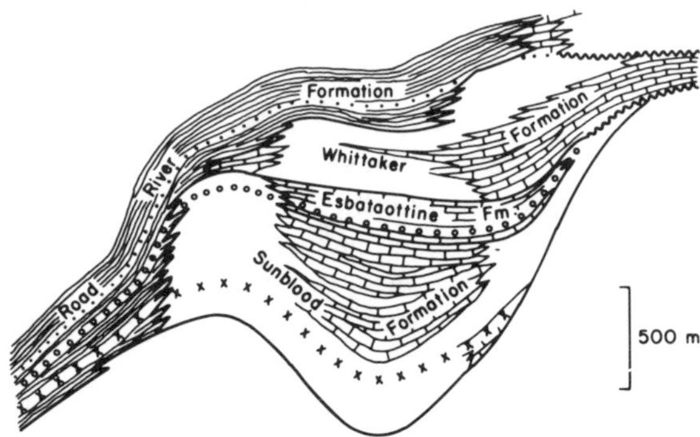

Fig. 4.12. Stratigraphic relations in the Ordovician shelf margin deposits of the Mackenzie Mountains, Northwest Territories (reproduced, with permission, from Ludvigsen, 1978).

that a biofacies should not be defined only in compositional terms, but like lithofacies and metamorphic facies, should have an areal component and should be mappable" (Ludvigsen, 1978, p. 11).

Ludvigsen (1978) used both Q- and R-mode methods to define biofacies of Middle Ordovician trilobites in the Mackenzie Mountains of northwest Canada. The rocks show a major lithofacies change from basinal and slope mudstones to platform carbonates (Fig. 4.12). The area deepened gradually, so that the slope facies encroached on the platform. A particularly rapid transgression is noted in the middle of the Whittaker Formation, at the beginning of the Trentonian stage. Ludvigsen made 55 collections totalling 36 trilobite genera through the platform rocks. They were divided according to age (stage) into Chazyan (C), Blackriverian (B) and Trentonian (T). Q- and R-mode cluster analyses were performed using unweighted pair-group clustering of Jaccard coefficients. The data were then listed in the order of clustering, and the data matrix replotted as shown in Figure 4.13. Clusters in the dendrograms (not shown) are indicated by brackets around this table. Where these bracketed rows and columns intersect they outline biofacies groupings. Some genera were ecologically tolerant and appear in more than one biofacies, and some range through two stages, so that successive biofacies have similar compositions. For example biofacies C_1 and B_1 both consist of the genera *Bathyurus* and *Ceraurus*. Together they define

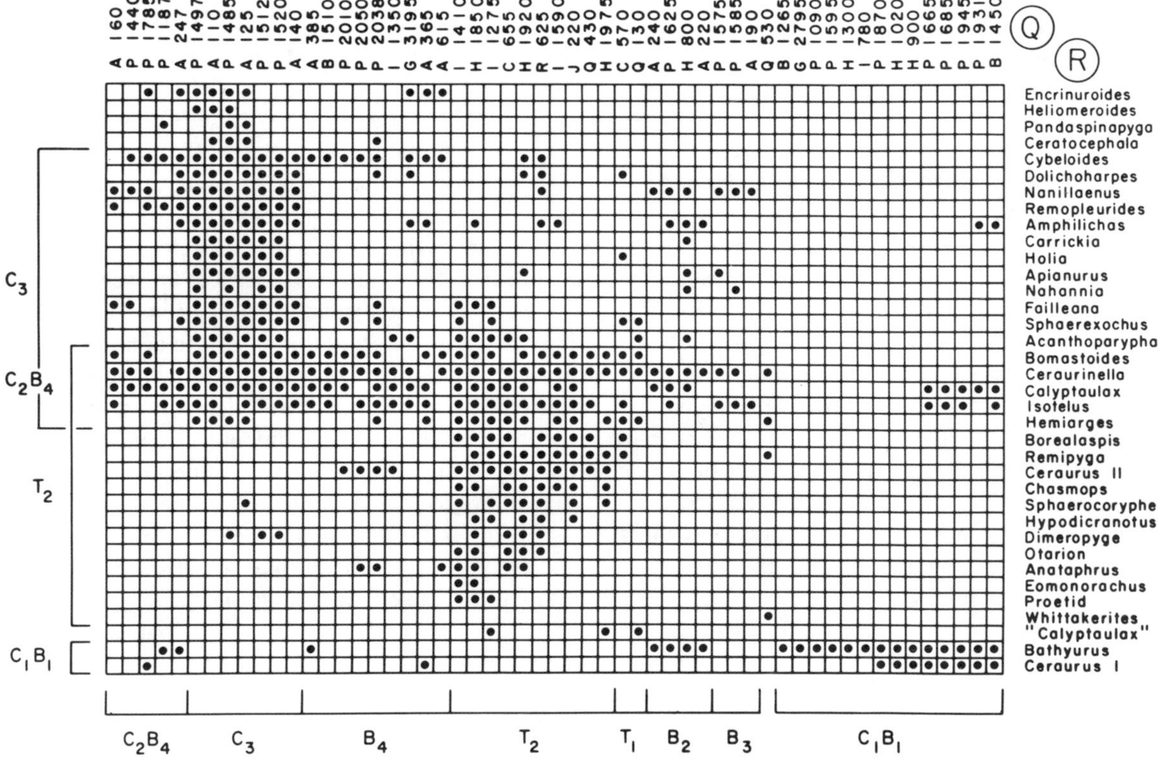

Fig. 4.13. Distribution of trilobite genera in field collections, plotted to show cluster groups generated using Q- and R-mode analysis (reproduced, with permission, from Ludvigsen, 1978).

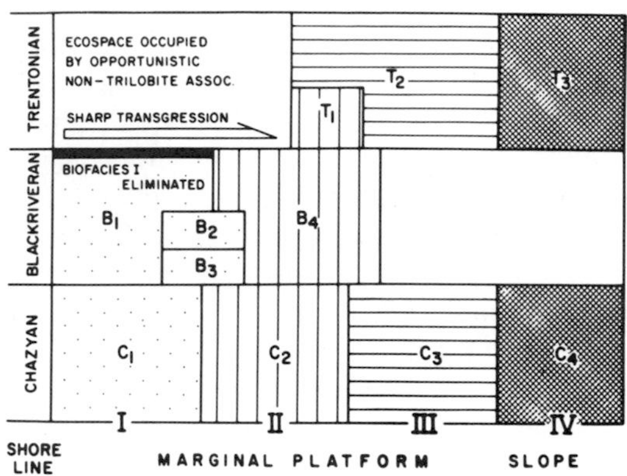

Fig. 4.14. Distribution of Ordovician biofacies in time and space (reproduced, with permission, from Ludvigsen, 1978).

what Ludvigsen named Biofacies I, which persisted through Chazyan and Blackriverian time. Sample location, stratigraphy and trilobite ecology indicate that this biofacies occurred in shallow platform environments. Other interpretations from these data permitted Ludvigsen to draw up the summary diagram shown in Figure 4.14. The slope biofacies C_4 and T_3 were defined using qualitative data not shown in Figure 4.13. Many interesting interpretations of facies shifts and sea level change were made by Ludvigsen using these data.

4.4.2 Cyclicity and the Markov chain method

Lithofacies group together into assemblages because they represent various types of depositional events which frequently occur together in the same overall depositional environment. For example, a submarine fan environment typically contains canyon, channel, levee, overbank and proximal slope subenvironments, each of which produces a distinctive lithofacies. These lithofacies become stacked into stratigraphic units because the environments shift through time, permitting different lithofacies to accumulate along any given vertical axis. The nature of these environmental shifts often is predictable, which means that the resulting lithofacies successions are equally predictable. For example on the submarine fan, channels may shift in position and subenvironments within a turbidity current move (very rapidly) downslope. Deltas and tidal flats prograde seaward, river channels and tidal inlets migrate, and so on. These processes may repeat themselves many times. This is the basis for the principle of cyclic sedimentation. It was clearly understood by one of the founders of modern sedimentary geology, Johannes Walther. He enunciated what has come to be known as Walther's law, the "Rule of Succession of Facies", which states:

> The various deposits of the same facies area and, similarly, the sum of the rocks of different facies areas were formed beside each other in space, but in a crustal profile we see them lying on top of each other . . . it is a basic statement of far-reaching significance that only those facies and facies-areas can be superimposed, primarily, that can be observed beside each other at the present time (Walther, 1894, as translated from the German by Middleton, 1973).

The value of this rule to basin analysts is that it means lateral facies relationships can be predicted by studying them in vertical succession. Stratigraphic sections such as outcrops measured up cliffs or drill hole sections comprise most of our data base. Lateral facies relationships are much more difficult to study because they tend to be gradual and stretched out over considerable distances. Few outcrops are long enough to reveal such features in their entirety. Paleogeographic reconstruction therefore relies heavily on the examination of vertical profiles.

It must be stressed that Walther's Law only applies to continuous sections without significant stratigraphic breaks. An erosion surface may mark the removal of the evidence of several or many subenvironments. It is therefore important to observe the nature of contacts between lithofacies units when describing sections. Contacts may be sharp or gradational. Sharp contacts may or may not imply significant sedimentary breaks. Periods of erosion or non-deposition may be revealed by truncated sedimentary structures, extensive boring, penecontemporaneous deformation, hardgrounds or submarine diagenesis.

Given this constraint, the study of cyclic sedimentation has been one of the most popular and fruitful in sedimentology because as Reading (1978, p. 5) stated: "it enabled geologists to bring order out of apparent chaos, and to describe concisely a thick pile of complexly interbedded sedimentary rocks". Interpretations of cyclic sedimentation form the basis of most facies models.

There has been considerable controversy over the methods used to define sedimentary cycles. Early attempts were subjective and qualitative. The culmination of the analysis commonly was the erection of a single "model" or "ideal" cycle (e.g. Visher, 1965). This has been criticized by Duff and Walton (1962) on the grounds that in fact this ideal cycle may rarely be represented in the section. Random environmental changes, local reversals of the sequence or the effect of some external event such as a flash flood or tectonic pulse commonly intercede to modify the succession. The many analyses of the Carboniferous Coal Measure cyclothems prior to about 1960 illustrate this difficulty well, but cannot be entered into here for reasons of space (modern ideas about these cycles are discussed in Chapter 8).

In recent years the trend has been to use the technique of Markov chain analysis to analyze for cyclicity. This has the virtue of a certain statistical rigor. Like the multivariate techniques described in the preceding section it results in a grouping of lithofacies into an assemblage, but it also has the additional property of revealing the order in which the lithofacies tend to occur. It can reveal not only a single model cycle, but also the several or many statistically most probable variations on this theme. A Markov processes is one "in which the probability of the process being in a given state at a particular time may be deduced from knowledge of the immediately preceding state" (Harbaugh and Bonham-Carter, 1970, p. 98). There are several varieties of Markov chain analysis, which test for dependency on two or more preceding states (as though it required two or more facies to come together to give birth to the next in line), but the most useful method for basin analysis purposes is the single dependency model (Miall, 1973).

The analysis starts with a transition count matrix. This is a two-dimensional array which tabulates the number of times that all possible vertical lithofacies transitions occur in a given stratigraphic succession. The lower bed of each transition couplet is given by the row number of the matrix, and the upper bed by the column number. Lithofacies are assigned numbers or letter codes for the purpose of analysis, as described in section 4.3. Input for the analysis may consist of manually tabulated data from field notes or data retrieved from a computer-processible file, such as the one developed by Friend et al. (1976). There are two ways to erect a transition count matrix. Transitions may be recorded whenever lithofacies change in character, regardless of bed thickness, or transitions may be constructed by sampling the section at fixed vertical intervals through the section. The first method is called "embedded Markov chain analysis" (Krumbein and Dacey, 1969). It is the one preferred by most workers, because it emphasizes change, and therefore focuses on depositional processes. Transitions from a lithofacies into itself are not measured, and therefore a row of zeros runs from top-left to bottom-right in the transition count matrix. An example is illustrated in Figure 4.15A (from Miall, 1979a). The second method gives a more accurate measure of lithofacies thicknesses, but at the expense of accuracy in measuring step-by-step change. Too large a sampling interval bypasses thin beds, whereas too small an interval will overload the matrix with transitions from each lithofacies into itself.

A series of simple mathematical operations, easily performed on a hand-held electronic calculator, produces a difference matrix, as shown in Figure 4.15. Positive entries in this matrix are those transitions which occur with greater than random frequency. By following through these

Transition count matrix, grouped Eocene data from the Cyclic Member, Eureka Sound Formation

Lithofacies and code	A	B	C	D	E	Row sums
Shale (A)	0	34	38	0	5	77
Shale and sand (B)	15	0	26	2	3	46
Sand (C)	41	11	0	12	11	75
Soil (D)	4	1	2	0	7	14
Lignite (E)	19	3	3	0	0	25
Total						237

A

Difference probability matrix, grouped Eocene data from the Cyclic Member, Eureka Sound Formation

Lithofacies and code	A	B	C	D	E
Shale (A)	.00	.15	.02	−.09	−.09
Shale and sand (B)	−.08	.00	.17	−.03	−.07
Sand (C)	.07	−.14	.00	.07	−.01
Soil (D)	−.06	−.13	−.19	.00	.39
Lignite (E)	.40	−.10	−.23	−.07	.00

B

Fig. 4.15. Markov analysis of a deltaic succession (Eocene, Banks Island, Arctic Canada), showing **A** transition count matrix and **B** difference matrix (Miall, 1979a).

positive values it is then possible to build up a path diagram showing lithofacies relationships, as shown in Figure 4.16. Alternatively the diagram may be redrawn as a composite model cyclic sequence, as illustrated in Figure 4.17. The transition paths shown in Figure 4.16 were interpreted as those of a coarsening-upward, prograding deltaic cycle (Miall, 1979a), and those in Figure 4.17 as a fining-upward fluvial cycle (Rust, 1978b). In both cases alternative paths are indicated. Both these analyses illustrate simple facies relationships with few lithofacies states. Much more complex patterns may emerge if more than about six states are used, and much more transition count data will be needed to produce reliable path diagrams. Nevertheless, even where cyclic tendencies are weak the method can still bring out genetic relationships between lithofacies that might otherwise have been missed. Cant and Walker (1976) discussed an example of the use of such data in carrying out detailed environmental interpretations.

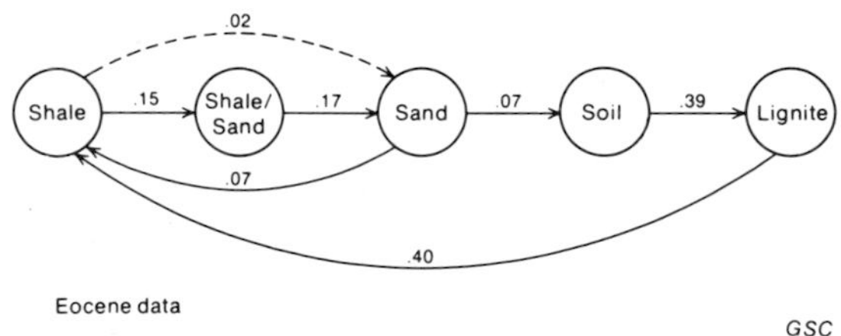

Fig. 4.16. Path diagram derived from Figure 4.15B, showing deltaic coarsening-upward cycles (Miall, 1979a).

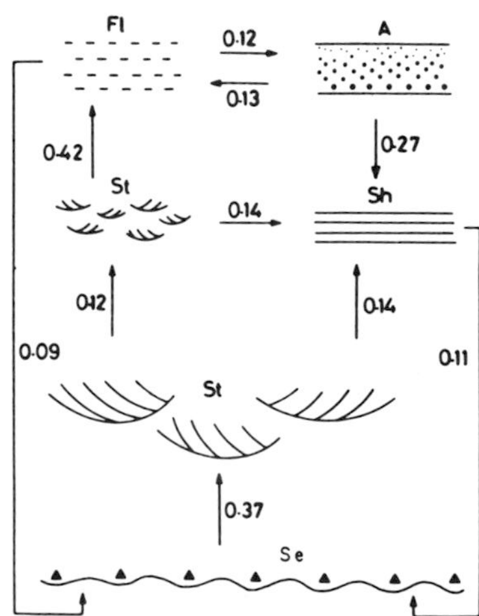

Fig. 4.17. Path diagram for a Devonian fluvial unit in the Canada Arctic (M.R. Gibling, in Rust, 1978b).

Statistical tests of randomness may be performed on these lithofacies relationships, as discussed by Hobday et al. (1975) and Hiscott (1981a).

As it stands Markov analysis takes no account of the nature of the contacts between the various lithofacies states. Transitions that cross erosional contacts may prove to have a high probability of occurrence, and this is obviously important if the cyclic processes are to be interpreted correctly. Miall and Gibling (1978) designed a method of documenting the nature of lithofacies contacts, termed the "contact matrix". Transitions between lithofacies states are coded in the field or in core as erosional, gradational or unknown (poor exposure). From these data two preliminary matrices are derived in the same way that transition count matrices are constructed. A "gradation matrix" tabulates the number of contacts between each lithofacies pair that are gradational. Elements in this matrix are termed G_{ij}. An "erosion matrix" consisting of elements E_{ij} is derived in the same way. The contact matrix consists of elements C_{ij} and is calculated as follows:

$$C_{ij} = \frac{G_{ij} - E_{ij}}{G_{ij} + E_{ij}}.$$

Values in the contact matrix range from +1 to −1. Negative scores indicate a preponderance of scouring or sediment breaks, positive scores indicate a dominance of gradational contacts. Zero scores indicate equal numbers of contact types—or no data. An example of a contact matrix is given in Table 4.2. This is for a sequence of braided fluvial cycles, and uses Miall's (1977, 1978) lithofacies code scheme (see Table 4.1). The facies are arranged in order of decreasing grain size and sedimentary structure scale from right to left and bottom to top. Cells in the lower left are therefore fining-upward transitions and in Table 4.2 these all have positive scores. Coarsening-upward transitions (top, right) all have negative scores. This clearly confirms that gradational relationships are virtually all of fining-upward type and shows that each cycle has an erosional base.

There is a temptation to use the Markov chain method to reduce complex stratigraphic data to single, simple cyclic models, such as Cant and

Table 4.2. Example of a contact matrix*

	Fm	Fl	Sr	Sh	Sp	St
Fm	—	—	−1.00	−0.33	—	−1.00
Fl	1.00	—	−1.00	−0.33	−1.00	−1.00
Sr	0.40	0.40	—	−1.00	—	−1.00
Sh	1.00	0.56	1.00	—	—	−1.00
Sp	—	—	1.00	—	—	—
St	1.00	0.78	1.00	0.33	—	—

*From Miall and Gibling, 1978.

Walker's (1976) "summary sequence". This can be very misleading, because it repeats the old mistake noted by Duff and Walton (1962) of concentrating on "ideal" or "model" cycles. This is not the object of the exercise. The strength of the method is its ability to show quantitatively how lithofacies relationships vary from such models.

4.4.3 The theory of facies models

The concept of a facies model as a summary of a depositional environment and its products is a recent one in sedimentology. The term was first discussed at a conference reported by Potter (1959), but was used in the sense we now imply by the term "facies assemblage". The difference is critical. A facies assemblage is essentially descriptive, whereas a facies model attempts to provide an interpretation of a particular type of facies assemblage in terms of depositional environments. Much use is usually made of comparisons with actual modern environments, and so it is now common practice to refer to these mental constructs as "actualistic facies models" (although Shea, 1982, has criticized the use of the word actualistic, partly on etymological grounds).

Walker (1979a) has described the process of constructing a facies model, using turbidites as an example. Data from various sources (modern sediments, ancient deposits) are collected, and the wealth of information sifted to determine the important features that they have in common ("local variability distilled away" in Fig. 4.18). Various statistical procedures may be used to achieve this, such as those discussed in the previous section. The objective is to produce a general summary of turbidites.

> If we distill enough individual turbidites, we can end up with a perfect 'essence of turbidite'—now called the Bouma model. But what is the essence of any local example and what is its 'noise'? Which aspects do we dismiss and which do we extract and consider important? Answering these questions involves experience, judgement, knowledge and argument among sedimentologists. (Walker, 1979a, p. 5).

Walker stated that a facies model should fulfill four functions:

1. It must act as a norm, for purposes of comparison.
2. It must act as a framework and guide for future observations.
3. It must act as a predictor in new geological situations.
4. It must act as a basis for environmental interpretation.

This is illustrated in Figure 4.18. Facies models are a powerful tool for the interpretation of poorly exposed sediments, because they suggest certain critical observations or clues for interpreting the sedimentary record. As with all such summaries and models, however, uncritical use may lead to loss of information or misinterpretation, because it is tempting to observe strata in terms of a preconceived model. If a facies model is being used properly in new field situations, each use may generate a refinement of the model or it may lead to the recognition of situations where the particular model is inappropriate. The development of a new model could then follow.

What does a facies model look like? The answer depends on how the model was constructed. Paleogeographic sketches, block diagrams and vertical profile logs are typical components of

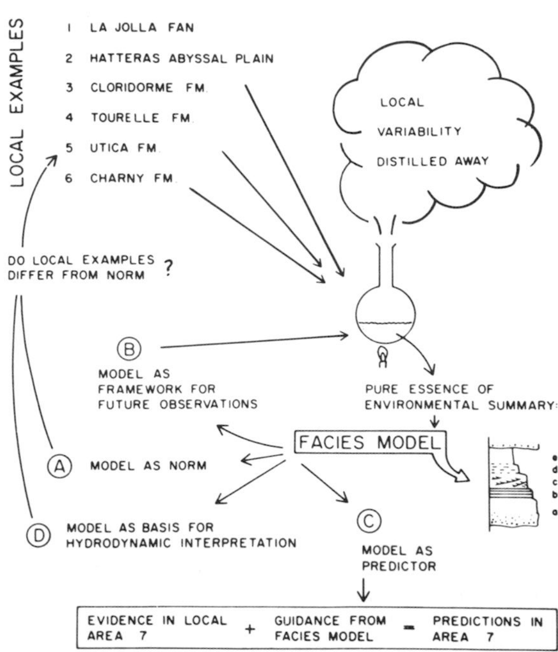

Fig. 4.18. The development of a facies model, using turbidity current deposits as an example (Walker, 1979a).

a published facies model. Some models are based primarily on geomorphology (e.g. deltas, alluvial fans), others on transport processes (e.g. turbidite, glacial till and eolian dune models), and others on organic processes (e.g. reefs). Many are governed by climatic and tectonic variables, which have not, in all cases, been exhaustively explored.

Many models are based primarily on the study of modern environments; for example the threefold subdivision of deltas erected by Galloway (1975) used maps of typical deltas to convey information on lithofacies geometry. For continental environments there is no problem seeing and describing a wealth of detail in modern settings, and this has led to the clogging of the facies models literature with at least a dozen models for fluvial deposits alone (Miall, 1980). Strangely, the same stage has not been reached for eolian environments although this is changing rapidly, as noted in section 4.6.4.

Facies models for subaqueous environments are at present only loosely based on studies of modern settings, because the techniques of marine geology (geophysical profiling, shallow coring) have not, until recently, permitted very precise descriptions of lithofacies geometry and relationships. The first comprehensive submarine fan model of Mutti and Ricci-Lucchi (1972) was based mainly on well-exposed Cenozoic fan deposits in Italy, and interpretations of other ancient deposits figure prominently in recent discussions (e.g. Walker, 1978). Emerging models for storm-dominated shelf deposits incorporate hummocky cross-stratification (e.g. Hamblin and Walker, 1979), a structure not yet even reliably documented from modern environments. Howard (in Byers and Dott, 1981, p. 342) explained how it may be obscured in cores of recent sediments by our methods of sampling the sea bottom, and has offered what may be the first observation from a modern offshore region (Howard and Reineck, 1981).

Many of these points are reviewed at greater length in section 4.6, but this summary should serve to illustrate the varied state of the art of facies modeling at the present day. It is very much a case of work in progress, and basin analysts potentially can add much to our fund of knowledge by careful studies of newly explored stratigraphic units.

4.4.4 Strengths and weaknesses of facies modeling

We have all heard that cliche "the present is the key to the past". It was one of the great generalizations to emerge from James Hutton's enunciation of the principle of uniformitarianism toward the end of the eighteenth century. Charles Lyell's work a half century later seemed to nail it down forever as a cornerstone of the still unborn science of sedimentology. Yet it is only true in a limited sense. Undoubtedly the study of modern depositional environments provides the essential basis for modern facies studies, but there are at least a dozen major problems which emerge to confuse this work. It turns out that the past is a very important key to the present, that many geomorphological processes can best be understood by adopting a geological perspective and looking at the ancient record. The art of facies modeling is therefore a two-way process.

The greatest advantage in studying modern environments is that we can observe and measure sedimentary processes in action. We can measure current strength in rivers and the oceans; we can observe at least the smaller bedforms moving and evolving; we can measure temperatures and salinities, and study the physics and chemistry of carbonate sedimentation, and the critical effects of organic activity in the photic zone (shallow-depth zone in the sea affected by light penetration). We can sample evaporite brines in inland lakes and deep oceans such as the Red Sea, and deduce the processes of concentration and precipitation. We can put down shallow core holes and study the evolution of the sedimentary environment through the Recent to the present day. Many physical, chemical, thermodynamic and hydrodynamic sedimentary models have resulted from this type of work, particularly during about the last two dozen years. As noted in the previous section, for some sedimentary environments work of this type has constituted the major advance in our understanding. Examples of facies models that depend largely on studies of modern environments include those for deltas (Coleman and Wright, 1975; Galloway, 1975), barrier islands (Bernard et al., 1959; Dickinson et al., 1972), tidal inlets (Kumar and Sanders, 1974), clastic tidal flats (Evans, 1965; Van Straaten, 1951, 1954; Reineck and Singh, 1973), tidal deltas (Hayes, 1976) and sabkhas (Shearman, 1966; Kinsman, 1966, 1969).

Some of the most difficult problems to resolve when using actualistic models stem from the inadequate time scale available to us for observation purposes. A very persistent geologist may be stubborn enough to pursue the same field project for ten, perhaps even twenty years. Aerial photographs might push observations of surface form back as far as about 1920 at the earliest. Old maps may go back for another hundred years or more, but become increasingly unreliable. Weather records may be available back into the nineteenth century, stream gauge data has been collected for only a few decades. It is difficult to assess the relevance of perhaps a 10^2 a record to a geological unit that may have taken 10^6 a or more to accumulate. Have the last 100 or so years been typical? Were the same sorts of processes occurring at the same rates in the distant geological past? As discussed below the answer is a qualified maybe.

The most important aspects of this question are the difficulty of assessing the importance of ephemeral events and judging the preservation potential of deposits we can see forming at the present day. Reading (1978, p. 10) and Dott (1983) have called attention to the need to distinguish between "normal" and "catastrophic" sedimentation. Normal processes "persist for the greater proportion of time. Net sedimentation is usually slow. It may be nil or even negative if erosion predominates. Normal processes include pelagic settling, organic growth, diagenesis, tidal and fluvial currents" (Reading, 1978). Reading distinguished catastrophic processes as those which "occur almost instantaneously. They frequently involve 'energy' levels several orders of magnitude greater than normal sedimentation. They may deposit a small proportion of the total rock and give rise to only an occasional bed, or they may deposit a large proportion of the total rock and so become the dominant process of deposition." Examples of catastrophic processes include flash floods in rivers, hurricanes and sediment gravity flows. Although geologists have studied many modern flash flood deposits and the effects of several recent hurricanes (e.g. Hayes, 1967), we cannot be sure that their magnitude and frequency at the present day are the same as that in some past period of interest, without attempting to obtain some geological perspective from studying the ancient record. The most violent and geologically important event may be one which only occurs every 200 or 500 years and has not yet been seen. Sediment gravity flows are thought to be the chief agent of erosion of submarine canyons, but in spite of years of oceanographic observation no major flows have actually been observed there (Shepard et al., 1979).

Dott (1983) and Ager (1981) argued that many stratigraphic sequences contain more gaps than record, and that significant proportions of the sedimentary record are deposited in a very short time by particularly violent dynamic events. Such events (hurricanes, sediment gravity flows, flash floods, etc.) are rare and difficult to study in action. Most of our energy as sedimentologists is expended in studying the less violent processes that occupy most of geological time but may contribute volumetrically far less sediment to the total record.

Studies of modern environments also suffer from the fact that many of the deposits we see forming at the present day are lost to erosion and never preserved. Thus the geological record may be biased. The bias may be in favor of more deeply buried sediments which the geologist, scratching the surface with trenches and box cores, never sees. For example Picard and High (1973) published a detailed study of the sedimentary structures of modern ephemeral streams, but many of the structures were themselves sufficiently ephemeral to be rarely, if ever, found in the ancient record. Cant (1976) argued that fluvial flash flood deposits are preferentially preserved because they infill deep scours below the normal level of fluvial erosion. Many of our fluvial facies models are based on studies of modern rivers in upland regions undergoing net degradation. How relevant are they to research in some of the great ancient alluvial basins which, by the thickness of preserved deposits, demonstrate a long history of aggradation? Facies models for barrier and shoreface environments are subject to similar constraints. For years the classic Galveston Island (Texas) model of coarsening-upward beach accretion cycles dominated geological thinking (Bernard et al., 1959), but more recently it has been realized that the barrier sediments may be removed by lateral migration of deep tidal inlets, and that many barriers consist of superimposed inlet deposits with a superficial skin of wave-formed shoreline sediments (Kumar and Sanders, 1974). Hunter et al. (1979) argued that many shallow subtidal deposits are systematically removed by rip currents and are never preserved in the geological record.

Some of the best studies of modern environments are those which use shallow drill cores to penetrate into pre-Recent deposits. Such sediments can be said to have "made it" into the geological record, and yet they can be placed in the context of a still extant, and presumably little modified, modern environment. Some work on modern turbidites (Bennetts and Pilkey, 1976), anastomosed rivers (Smith and Smith, 1980) and reefs (Adey, 1975; Adey et al., 1977; Shinn et al., 1979) is of this type.

In two important ways the present is quite unlike the past, and is therefore an unreliable laboratory for reconstructing sedimentary environments. The Pleistocene ice age generated several geologically rapid changes of sea level culminating in a major rise and transgression during the last 10,000–20,000 a. Secondly, the present configuration of continents and oceans is a unique pattern, different from any in the past because of the long history of sea-floor spreading, rifting, subduction and suturing. This plate movement has had an important effect on some of the broader aspects of facies models.

Because of the recent sea level rise modern continental shelf, shallow marine and coastal plain deposits around the world have been formed under transgressive conditions. Sea level was approximately 150 m lower during the Plio-Pleistocene glacial phases, rivers graded their profile to mouths located near the edge of present continental shelves, carbonate platforms such as the Bahamas were exposed to subaerial erosion and may have developed extensive karst systems. Submarine canyons were deeply entrenched by active subaerial and submarine erosion (Shepard, 1981; McGregor, 1981). During the rapid transgressions that followed in interglacial phases shoreline sands were continually reworked. On the Atlantic shelf off North America, extensive barrier islands were formed and receded into their present position (Swift, 1975). River valleys became drowned and filled with estuarine and deltaic deposits. Submarine canyons commonly were deprived of their abundant supply of river-borne detritus as this was now deposited on the landward side of a widening and deepening continental shelf. This had a drastic effect on the rate of growth of some submarine fans. Carbonate sedimentation began afresh in warm, detritus-free waters, over resubmerged platforms, but the local water depths, circulation patterns, and facies distribution may have been partly controlled by erosional topography (Purdy, 1974; although this is disputed by Adey, 1978). We therefore have excellent modern analogues for studies of rapid transgressions in the geological record, but few for rapid regressions or for periods of still-stand. The Mississippi and other large deltas are good examples of regressive deposits built since the last ice age, but they represent an environment that may be characterized by unusually rapid progradation. Some have maintained that most modern continental shelves are covered by relict sediments, implying that their study may not be of much geological relevance (Emery, 1968), but more recent work has shown that the dynamic effects of tidal currents and storms do in fact result in continual change (Swift et al., 1971).

Rapid transgressions may have occurred commonly during other ice ages in the geological past, as discussed in Chapter 8, but most other changes in the geological past were somewhat slower. For two lengthy periods in the Phanerozoic our suite of modern analogues is particularly inadequate. These were the times of high global sea level stand during the Ordovician–Silurian and again in the Cretaceous, when vast areas of the world's continents were covered by shelf seas. We simply do not have modern equivalents of these huge inland seas, and many of the large shelf seas that do exist, such as the Bering Sea and Yellow Sea, have yet to be studied in detail. Tidal range may have been amplified over these shelves, and wave energy reduced by friction over the shallow bottom. R.N. Ginsburg (in Byers and Dott, 1981) discussed the problem of interpreting Cambro–Ordovician carbonate banks of the North American craton. These are up to hundreds of kilometers long and thousands of kilometers wide. "The dilemma is how the vast extent of the banks, most of which suggest carbonate production and deposition in but a few meters or less of water, could all be bathed in normal marine or slightly restricted water. Would not such banks form major circulation barriers?" (Ginsburg, as reported by L.C. Pray, in Byers and Dott, 1981). Perhaps the banks are actually diachronous, and developed by seaward progradation, or they may have been crossed by "irregular to channelized deeper water areas facilitating water circulation". Tidal currents certainly would have been able to assist with the latter, as would waves and currents generated during storms.

Climatic patterns and the network of oceanic currents are controlled by global plate configurations. In many respects our present geography is unique. Therefore we have modern analogues for situations that may not have existed in the past and, conversely, we cannot replicate certain conditions that did exist. For example the Mesozoic Tethyan Ocean and the Pangea supercontinent had profound effects on climate and water circulation, and hence on sedimentation patterns, and we have only generalized models for interpreting them. Some of these broader effects on sediment facies are considered in Chapter 8 and 9. The study of paleoclimates is beyond the scope of this book, but is considered in an excellent review by Frakes (1979).

The last group of problems with the actualistic modeling method arises from the important effects organisms have on sedimentary processes. Plants and animals have, of course, both evolved profoundly since Archean time. Therefore in many environments sedimentation is controlled or modified by sediment–organism relations that did not exist, or were different, in the geological past. Many authors, beginning with Schumm (1968), have speculated on the implications of the evolution of land plants on fluvial patterns. Vegetation has a crucial effect on stabilizing channel banks, colonizing bars and islands, controlling chemical weathering, sediment yield and discharge fluctuations. Our typical braided and meandering fluvial facies models reflect these effects as they have been studied in temperate, humid, and hot and cold arid climates. Fluvial facies models seem likely to be quite different for thickly vegetated humid tropical climates (Baker, 1978; Miall, 1980), but have yet to be studied in detail. Going back in time the evolution of grasses in Mesozoic time must have changed geomorphic patterns profoundly. Abundant land vegetation is thought to have first appeared in the Devonian, and prior to that time the majority of river channels may have been unconfined, ephemeral and braided, as in modern arid regions. Modern deserts are commonly used as analogues for pre-Devonian rivers, but there is no reason why they had to be arid. However, we have no modern analogue for a humid, vegetation-free environment with which to study the pre-Devonian except, possibly, the south coast of Iceland.

Turning to shallow marine environments the same kinds of difficulties apply. H.E. Clifton (in Byers and Dott, 1981) pointed out that salt marsh vegetation, so important on modern tidal flats, did not exist prior to the Cenozoic. Similarly foraminifera, which provide sensitive bathymetric indicators for younger sediments, particularly those of Cenozoic age, did not exist in the Paleozoic. James (1979a) discussed the implications of these various evolutionary developments on changes in carbonate depositional environments.

The ecology of forms which are now extinct may be difficult to interpret, which makes them less useful in facies studies (section 4.5.7). Instead of providing independent, unambiguous environmental information, as do still-living forms, it may be necessary to interpret them with reference to their sedimentary context, which itself may be of uncertain origin. Functional morphology is studied to determine probable habits, but many uncertainties may remain. In the

Precambrian most sedimentary units are entirely devoid of fossils, and here problems of lithofacies interpretation without supporting fossil evidence may become acute. Many Precambrian units have been reinterpreted several times for this reason, as different environmental criteria are brought to bear on particular problems. Long (1978) discussed many of these difficulties with reference to the recognition of fluvial deposits in the Proterozoic.

4.5 Review of environmental criteria

In Chapter 2 a summary of major environmental indicators is given under the heading of what to look for in outcrop and subsurface sections. This list is by no means exhaustive; for example it concentrates almost exclusively on what can be seen with the naked eye. However, it includes many of the sedimentary features vital to a generalized environmental interpretation and hence to an effective basin analysis. In this section some indications are given of how to use this information. Beginners at the art of basin analysis, or those who wish to understand what their specialist colleagues are up to, may find this section both enlightening and confusing. The numerical precision or well-defined statistical error of laboratory-oriented geological studies is not to be found in the area of facies interpretation. Statistical methods may be used to aid in analyzing the composition of facies assemblages, but the business of interpreting their meaning depends heavily on qualitative study. The sedimentologist must be aware of the meaning and limitations of all the facies criteria visible, and must be able to weigh the evidence of all these against each other. A good knowledge of published facies models and modern analogues is, of course, essential.

A common basin analysis problem is that the geologist is faced with a new outcrop or core showing certain assemblages of lithologies, textures, structures and, perhaps, fossils, and may have no idea which facies model to turn to for assistance in making an environmental interpretation. The intent of this section is to review very briefly the types of interpretation that can be made from the principal kinds of sedimentological observation, in order to provide some environmental clues and an entry into some of the crucial literature. Discussions of the physical and chemical conditions of formation of most sediment types are beyond the scope of this book and are discussed in more detail by Blatt et al. (1980) and Leeder (1982).

In some of the notes which follow siliciclastics and carbonates are treated separately because they require a different approach. This is particularly the case with interpretations of grain size and texture. For others, such as hydrodynamic sedimentary structures, there is no reason not to consider all rock types under the same heading. Structures such as crossbedding and ripple marks are commonly regarded as the domain of the sandstone specialists, but they also occur in carbonates and evaporites. Chemical sediments are often studied by geologists whose first interest is in the chemistry of their formation and diagenesis, but could usually benefit from the approach taken to sedimentology by clastic sedimentologists.

The emphasis throughout is on features that are most useful for constructing depositional environments and paleogeography. Diagenesis and geochemistry are not dealt with in this book except those aspects of them which contribute to an understanding of basin maturity (Chap. 7).

It cannot be overemphasized that very few sedimentological criteria have an unambiguous environmental interpretation. Many years ago it was thought that ripple marks only occurred in shallow water, until oceanographers started taking photographs of the bottom of the oceans. More recently dish structures were thought to be indicators of submarine grain flows, but have now been found in fluvial and other deposits (Nilsen et al., 1977). The common trace fossil *Ophiomorpha* is a good indicator of shallow marine environments but J. Coleman (personal communication, 1976) reported finding it many miles inland in the deposits of the modern Mekong River. Marine it certainly is, but the Mekong River has an extensive marine "salt wedge" which flows far upstream during high tide. *Ophiomorpha* has also been found in outer shelf environments (Weimer and Hoyt, 1964). Reliance should never be placed on a single structure or feature of the rocks, or on a single method of analysis, for making environmental interpretations. The geologist must review and assess all the evidence available.

Detailed descriptions of actual facies models are not given here, but in section 4.6 a review of some of the most recent developments in the area is offered for more advanced students.

4.5.1 Grain size and texture

In a general sense the grain size of a clastic sediment indicates the relative amount of energy required to emplace the grains in their final resting place. This energy might have been derived from the force transmitted by air or water movement, or it may represent downward movement under gravity. Most clastic sediments represent a combination of both these processes.

Grain size interpretation in siliciclastic rocks tends to be much simpler than it does in carbonates and evaporites because diagenesis in these chemical sediments frequently obscures original grain relationships. Grain size may be related entirely to in-place primary or diagenetic crystal growth, and not at all to transport processes. However, some chemical sediments show evidence of having behaved as clastic detritus at some stage in their formation, so that considerations of grain dynamics might provide useful environmental information.

There are two ways in which general grain size data yield environmental information in siliciclastic sediments. Local vertical variations in mean or maximum grain size frequently are cyclic or rhythmic, and these, coupled with variations in sedimentary structures, are diagnostically powerful. Analysis of sequences for cyclic patterns has been discussed earlier in this chapter (section 4.4.2) and is dealt with in greater detail in section 4.5.8. The second aspect of grain size information is the environmental information contained in the size distribution of individual samples. This subject has received an enormous amount of attention from sedimentologists, who have proposed a wide variety of statistical techniques for distinguishing the deposits of various depositional environments on the basis of some supposed environmental signature retained in the sample. Most of this attention has been directed toward the study of sandstones. It has long been realized that the hydraulic sorting effects of waves, wind and unidirectionally flowing water result in the movement of different populations of grains. Most sandstones are mixtures of several populations, but it has long been the hope that the right analytical methods would infallibly recognize these subpopulations based on some size or sorting criteria, enabling the depositional environment to be recognized. All such techiques depend on careful laboratory size analyses, preferably carried out using sieves or a settling tube or, failing this in the event of lithified samples, counts of grains in thin section. Various methods were discussed by Friedman and Sanders (1978), Solohub and Klovan (1970), Visher (1969), and Glaister and Nelson (1974). None of these methods is fully reliable because of the problem of inherited size distributions in the case of second- or multicycle sands, and the effects of diagenesis (cementation, secondary porosity development) on lithified sandstones. Extensive laboratory work and data analysis are required to complete a rigorous grain size analysis. Even then the results are usually ambiguous and, in this writer's opinion, do not justify the great effort expended. Accordingly the use of this tool is not normally recommended. The method does, perhaps, have some uses in the analysis of well cuttings or side-wall core, where these are the only samples available, because the technique can be applied to small samples, whereas most of the observational techniques discussed here cannot. Even here, visual inspection of geophysical logs may provide equally reliable environmental information in a fraction of the time.

A useful technique in the study of conglomerates is the measurement of maximum clast size. Normally this is determined by averaging the intermediate diameter of the ten largest clasts present at the sample level. Such measurements have been used in the study of grading and cyclic changes in subaerial alluvial fans (Gloppen and Steel, 1981; Steel, 1974) and in subaqueous resedimented conglomerates (Nemec et al., 1980). It is found that there is a direct relationship between maximum particle size and bed thickness for deposits formed by subaerial or subaqueous debris flows, indicating that the beds were formed by single depositional events without subsequent reworking (Bluck, 1967; Nemec et al., 1980).

Grain size and textural studies in conglomerates have been used by Walker (1975) to erect models of deposition on submarine fans (Figs. 4.19, 4.65). The disorganized-bed model is thought to represent rapid, clastic deposition on steep slopes, perhaps in the feeder submarine canyon at the head of a fan. Debris flows passing on to the inner fan are thought to exhibit first an inverse-to-normally graded texture, passing downcurrent into a graded-stratified type. Inverse grading develops as a result of dispersive pressure, possibly including a "kinetic sieve" mechanism whereby smaller clasts fall down between the larger ones. Hein (1982) has extended this work to pebbly and massive sandstones.

For sediment gravity flows in general, grain size, fabric and texture are, together with sedimentary structures and vertical profiles, important criteria for distinguishing flow type (section 4.6.9).

In the conglomerate deposits of alluvial fans and other gravelly rivers grain size variations commonly reveal crude stratification and large scale crossbedding (section 4.5.4). The stratification is the product of longitudinal bar growth or superimposition of debris flows. Individual bar deposits may show an upward decrease in grain size (Miall, 1977). True grading is rare in fluvial gravels formed by traction transport.

Many resedimented conglomerates and pebbly sandstones contain a clast fabric with the long (a) axis dipping upstream, indicating that they were deposited from a dispersed sediment mass without bedload rolling (Walker, 1975; Hein, 1982). This contrasts with the common imbrication of flat clasts in conglomerates deposited by traction transport, such as in gravelly rivers (Rust, 1972), in which the a axis may be transverse to flow. The fabric is therefore a useful indicator of transport mode.

For most carbonate sediments a different approach must be taken to grain size and texture. Most grains, both fine and coarse, are locally produced by organic activity. The mean grain size or size of largest grains may mean little in terms of local hydraulics. Dunham (1962) has pointed out that it is more useful to focus on the fine material in the rock, because this is an accurate measure of the strength of winnowing currents. Dunham (1962) devised a system for classifying the texture of carbonate rocks based on whether the sediment is a self-supporting grain framework or whether the grains are enclosed in a micrite matrix. The spectrum of packing categories erected by Dunham is shown in Table 4.3. The terms mudstone, wackstone, etc., are widely used for describing the texture of carbonates. In general the amount of current winnowing energy implied by the rock type name increases from left to right.

A similar carbonate textural spectrum was devised by Folk (1962), who used a two-part terminological system (Fig. 4.20). The suffixes -micrite and -sparite are used for predominant matrix type. Sparry rocks are packstones and grainstones from which micrite matrix has been removed by winnowing and replaced by coarse calcite cement. Documenting textural relations in this way is an important first step in subdividing the various subenvironments of carbonate platforms. Wilson (1975, Chap. 1) discussed the subject at greater length.

Schreiber et al. (1976) and Schreiber (1981) have shown that many evaporite deposits, particularly those formed in marginal marine environments, can be treated as clastic deposits in that they show similar bedding and sedimentary structures. The same is true for some carbonate rocks, particularly oolite sand shoals (Ball, 1967) and carbonate turbidites (Mountjoy et al., 1972), which are formed by clastic sorting and redistribution processes although consisting of carbonate particles. Considerations of grain size (cyclic or lateral changes, grading) and texture (packing) might provide useful environmental information in these cases.

4.5.2 Petrology

The composition of the major rock-forming constituents of siliciclastic sediments (including mudrocks) is not directly diagnostic of environment. Krynine (1942) suggested that many sandstone types were characteristic of particular tectonic–sedimentary environments, e.g. arkoses supposedly represent nonmarine sediments derived from granitic orogenic complexes, greywackes represent "early geosynclinal" sedimentation. These interpretations are no longer followed, although Dickinson and Suczek (1979) and Schwab (1981) have shown that sandstone composition may closely reflect plate tectonic setting (section 9.4). However, this does not necessarily translate into depositional environment. For example volcaniclastic forearc sediments may be deposited in fluvial, lacustrine, marginal marine, shelf or submarine fan settings, depending on continental margin configuration (Chap. 9).

Certain minor components of sandstones may be strongly suggestive of depositional environment. For example glauconite pellets form only in

GRADED-STRATIFIED
GRADED-BED
INVERSE-TO-NORMALLY GRADED
DISORGANIZED-BED
NO INVERSE GRADING
STRAT., CROSS-STRAT.
IMBRICATED
NO INVERSE GRADING
NO STRAT.
IMBRICATED
NO STRAT
IMBRICATED
NO GRADING
NO INVERSE GRADING
NO STRAT.
IMBRIC. RARE
THESE THREE MODELS SHOWN IN SUGGESTED RELATIVE POSITIONS DOWNCURRENT

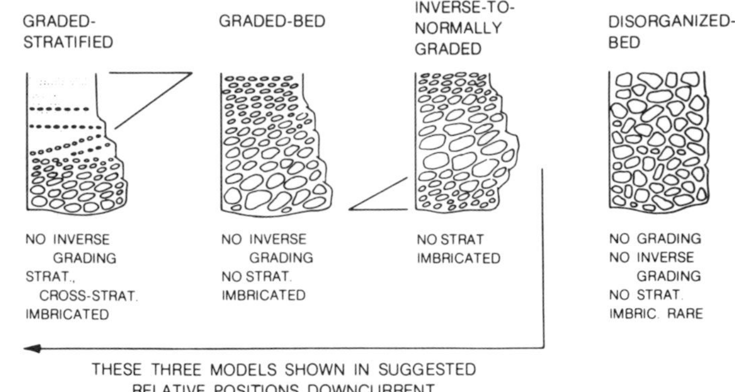

Fig. 4.19. Stratigraphic-textural-fabric models for resedimented conglomerates (Walker, 1975).

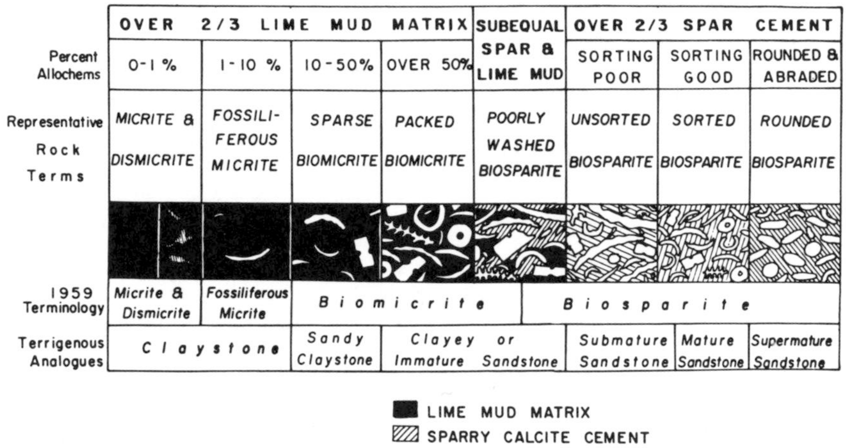

OVER 2/3 LIME MUD MATRIX
SUBEQUAL SPAR & LIME MUD
OVER 2/3 SPAR CEMENT
Percent Allochems
0-1 %
1-10 %
10-50%
OVER 50%
SORTING POOR
SORTING GOOD
ROUNDED & ABRADED
Representative Rock Terms
MICRITE & DISMICRITE
FOSSILI-FEROUS MICRITE
SPARSE BIOMICRITE
PACKED BIOMICRITE
POORLY WASHED BIOSPARITE
UNSORTED BIOSPARITE
SORTED BIOSPARITE
ROUNDED BIOSPARITE
1959 Terminology
Micrite & Dismicrite
Fossiliferous Micrite
Biomicrite
Biosparite
Terrigenous Analogues
Claystone
Sandy Claystone
Clayey or Immature Sandstone
Submature Sandstone
Mature Sandstone
Supermature Sandstone
LIME MUD MATRIX
SPARRY CALCITE CEMENT

Fig. 4.20. The textural spectrum in limestones (Folk, 1962).

shallow marine environments (Odin and Matter, 1981) and are rare as detrital or resedimented grains. Carbonaceous debris from plants is typical of nonmarine environments. Paleosol development is indicated by lenses of calcium carbonate (caliche, calcrete) and other minerals. Abundant red iron staining indicates oxygenated environments, typically either preservation of oxidized states in detrital particles (Van Houten, 1973), or the production of oxidized colors during early diagenesis (Walker, 1967). Red beds therefore are mostly indicative of nonmarine or high intertidal environments (Turner, 1980), although there are exceptions (e.g. Franke and Paul, 1980).

Within a given basin detrital composition may reflect variations in source area or depositional

Table 4.3. Classification of carbonate rocks according to depositional texture (Dunham, 1962).

<table>
<tr><td colspan="5">Depositional Texture recognizable</td><td rowspan="5">Depositional texture not recognizable

Crystalline carbonate

(Subdivide according to classifications designed to bear on physical texture or diagenesis.)</td></tr>
<tr><td colspan="4">Original components not bound together during depositions</td><td rowspan="4">Original components were bound together during deposition... as shown by intergrown skeletal matter, lamination contrary to gravity, or sediment-floored cavities that are roofed over by organic or questionably organic matter and are too large to be interstices.</td></tr>
<tr><td colspan="3">Contains mud (particles of clay and fine silt size)</td><td rowspan="3">Lacks mud and is grain-supported</td></tr>
<tr><td colspan="2">Mud-supported</td><td rowspan="2">Grain-supported</td></tr>
<tr><td>Less than 10% grains</td><td>More than 10% grains</td></tr>
<tr><td>Mudstone</td><td>Wackstone</td><td>Packstone</td><td>Grainstone</td><td>Boundstone</td><td></td></tr>
</table>

Table 4.4. Differences between siliciclastic and carbonate sediments*

Carbonate sediments	Siliciclastic sediments
The majority of sediments occur in shallow tropical environments	Climate is no constraint. Sediments occur worldwide and at all depths
The majority of sediments are marine	Sediments are both terrestrial and marine
The grain size of sediments generally reflects the size of organism skeletons and calcified hard parts	The grain size of sediments reflects the hydraulic energy in the environment
The presence of lime mud often indicates the prolific growth of organisms whose calcified portions are mud size crystallites	The presence of mud indicates settling out from suspension
Shallow water lime sand bodies result primarily from localized physicochemical or biological fixation of carbonate	Shallow-water sand bodies result from the interaction of currents and waves
Localized buildups of sediments without accompanying change in hydraulic regimen alter the character of surrounding sedimentary environments	Changes in the sedimentary environments are generally brought about by widespread changes in the hydraulic regime
Sediments are commonly cemented on the seafloor	Sediments remain unconsolidated in the environment of deposition and on the sea-floor
Periodic exposure of sediments during deposition results in intensive diagenesis, especially cementation and recrystallization	Periodic exposure of sediments during deposition leaves deposits relatively unaffected
The signature of different sedimentary facies is obliterated during low grade metamorphism	The signature of sedimentary facies survives low-grade metamorphism

*Reproduced with permission, from James, 1979a.

environment. These data may therefore be used as a paleocurrent indicator (section 5.5) and may assist in stratigraphic correlation of units formed under the same hydraulic conditions. Davies and Ethridge (1975) showed that detrital composition varied between fluvial, deltaic, beach and shallow marine environments on the Gulf Coast and in various ancient rock units as a result of hydraulic sorting and winnowing processes and chemical destruction. Thus, although composition is not environmentally diagnostic, it may be useful in extending interpretations from areas of good outcrop or core control into areas where only well cuttings are available.

For carbonate sediments, in contrast to siliciclastics, petrographic composition is one of the most powerful environmental indicators. Most carbonate grains are organic in origin, including micrite mud derived from organic decay or mechanical attrition, sand-sized and larger particles consisting of organic fragments, fecal pellets, grapestone and ooliths, all of which are produced in part by organic cementation processes, and boundstones or biolithites, formed by framework-building organisms. Most carbonate particles are autochthonous; therefore an examination of the composition of a carbonate sediment is of crucial importance.

These and other differences between carbonate and siliciclastic sediments are summarized in Table 4.4 (from James, 1979a). Interpretation of ancient carbonate sediments is complicated by the fact that the organisms which generate carbonate particles have changed with time. However, there are many similarities in form and behavior between modern and extinct groups, so that actualistic modeling can usually be carried out with caution (see also section 4.5.7). Another problem is that diagenetic change is almost ubiquitous in carbonate rocks and may obscure primary petrographic features. Ginsburg and Schroeder (1973) showed that some carbonates are converted contemporaneously from reef boundstones or grainstones to wackstones by continual boring, followed by infill of fine grained sediment and cement. Mountjoy (1980) suggested that many carbonate mud mounds may owe their texture to this process.

Most environmental interpretation of carbonates is based on thin-section examination using a microfacies description system such as that erected by Wilson (1975; Fig. 4.5 of this book). Wilson was able to define a set of standard facies belts for the subenvironments of a carbonate platform and slope, each characterized by a limited suite of microfacies reflecting the variations in water depth, water movement, oxygenation and light penetration (section 4.6.8). A classic example of a carbonate petrology study, the analysis of modern sediments of the Bahama Platform is described briefly in section 4.4.1 (see Figs. 4.8, 4.9). Each facies assemblage can readily be related to environmental variables, such as the quiet water pelletoidal facies and the high energy skeletal sand and oolite lithofacies.

Many carbonates consist of dolomite rather than limestone. In most cases the dolomite is clearly a replacement, as indicated by the presence of dolomite rhombs penetrating allochemical particles such as shell fragments or ooliths. In other cases there may be evidence for a primary or penecontemporaneous origin, which may be environmentally useful information. Primary dolomite crusts form in association with evaporites in areas where seawater is evaporated at high rates, such as on supratidal flats and in shallow tidal lagoons. Modern examples include parts of Bonaire, sabkhas on the Persian Gulf, Deep Springs Lake, California and Coorong Lagoon, Australia (Blatt et al., 1980, pp. 512–522; Friedman, 1980). The dolomite–evaporite association, together with the evidence of certain evaporite or desiccation textures (section 4.5.6) is strongly environmentally diagnostic. The dolomitization processes depend on "seepage refluxion" or "evaporative pumping" of seawater to the surface under hot, arid conditions (Adams and Rhodes, 1960). However, this is only one mechanism for the production of dolomite. Most probably it forms at depth by a diagenetic mixing of seawater and freshwater, a process known as the Dorag model (Badiozamani, 1973; Land, 1973). In rare cases dolomite can also occur as detrital grains.

The composition of evaporite minerals is not a good guide to the depositional environment of evaporites. A sample of normal seawater, if evaporated to dryness, yields a sequence of precipitates in the following order: calcite, gypsum, halite, epsomite, sylvite and bischofite. However, the composition of the final deposit in nature may vary considerably because of the effects of temperature, the availability of earlier-formed components for later reaction, and the rate and nature of replenishment of the water supply. Evaporites may form in a variety of marine and nonmarine environments (Schreiber et al., 1976; Schreiber, 1981), and it is not their composition as much as their internal structure and lithologic associations that are the best clues to depositional environment (e.g. the association with penecontemporaneous dolomite mentioned above).

Certain other chemical deposits contain useful environmental information. Chert is common as a replacement mineral in carbonate sediments, where it forms nodules and bedded layers, commonly containing replacement casts of fossils, ooliths, etc. Knauth (1979) suggested that such chert was formed by the mixing of fresh and marine waters in a shallow subsurface, marginal marine setting. Chert also occurs in abyssal oceanic sediments in association with mafic and ultramafic igneous rocks. Radiolarians, sponge spicules and diatoms are common. These are some of the typical components of ophiolites, which are now regarded as remnants of oceanic crust and indicate a former deep water environment (Grunau, 1965; Barrett, 1982).

Iron-rich rocks occur in a variety of settings. Their chemistry is controlled by Eh and pH conditions, as shown in Figure 4.21, and this diagram may be used as a rough guide to depositional environment. Pyrite and siderite are common as early diagenetic crystals and nodules in the reduced environment of organic-rich muds, particularly in fluvial or coastal (deltaic, lagoonal) swamps. Occasionally such deposits may be present in rock-forming abundance, as in the pisolitic bog iron ores. Plant remains, impressions and replaced (petrified) wood are commonly associated with these forms of iron. Pyrite is also associated with unoxidized, disseminated organic particles in the black muds of anoxic lake and ocean basins (section 8.3.3).

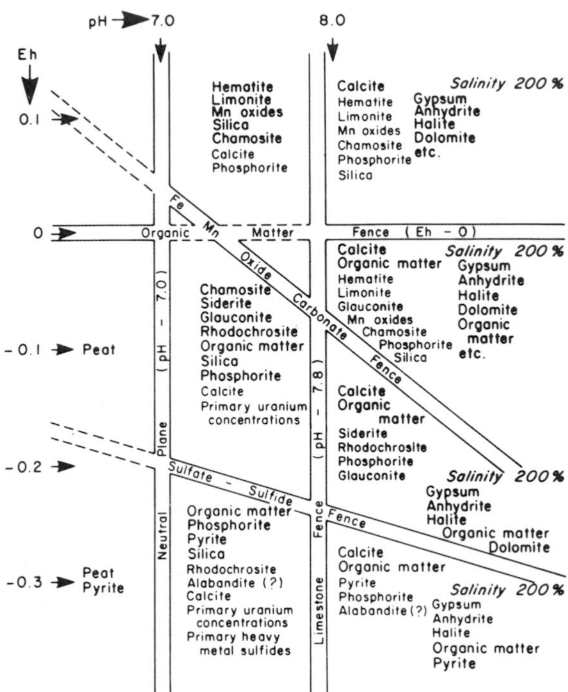

Fig. 4.21. Eh–pH stability fields of some common chemical sediments (Krumbein and Garrels, 1952, Journal of Geology v.60).

Hematite and chamosite iron ores are locally important in the Phanerozoic. Most are oolitic and display typical shallow water sedimentary structures. The iron probably is an early diagenetic replacement. Precambrian iron formations are widespread in rocks between about 1.8 and 2.6 Ga. Their mineralogy is varied and unusual (Eichler, 1976); it may reflect formation in an anoxic environment (Cloud, 1973), although this has been disputed (section 9.5.2).

Manganese- and phosphate-rich rocks are locally important. Both commonly occur as crusts and replacements on disconformity and hardground surfaces, where they are taken as indicators of non-deposition or very slow sedimentation. Phosphates are particularly common on continental margins in regions of upwelling oceanic currents. More detailed discussions of these sediments are given by Blatt et al., (1980).

Coal always indicates subaerial swamp conditions, usually on a delta plain or river floodplain. However, lacustrine coals and coal formed in barrier–lagoon settings have also been described. Calcrete (or caliche) occurs in alluvial and coastal environments and is an excellent indicator of subaerial exposure (e.g. Bown and Kraus, 1981; James, 1972).

4.5.3 Bedding

Are the various rock types present in a stratigraphic unit interbedded in major packages several or many meters thick, or are they interlaminated on a scale of a few centimeters or millimeters? Is the bedding fine or coarse? Is it flat, disturbed, undulatory, distinct or indistinct (gradational)? These questions, while rarely providing answers that are uniquely diagnostic of depositional environment, may provide important supplementary information. In a general way bed thickness is proportional to depositional energy level.

Finely laminated sediments are mostly formed in quiet water environments. These may include laminated pelagic mudrocks, prodelta deposits, deep water evaporites, thin-bedded basin-plain turbidites, delta plain lagoonal and fluvial overbank muds. Laminated fluvial sheetflood sandstones are an exception to this pattern (see next section). Thicker beds form in a variety of high-energy wave or current dominated environments. Reef rocks, formed in extremely high-energy conditions, may lack bedding entirely.

An interbedding of contrasting lithofacies may be environmentally indicative. For example wavy, flaser and lenticular bedding (Reineck and Wünderlich, 1968) record the alternation of quiet water mud sedimentation and higher-energy flow conditions under which rippled sand is deposited. This can occur during tidal reversals on exposed mudflats, on fluvial floodplains, or below normal wave base on the shelf, at depths affected by infrequent storm waves. Another common bedding association is that produced by the alternation of storm and fair weather processes on the shoreface. A sequence: basal gravel → laminated sand → bioturbated or rippled sand, indicates storm suspension followed by decreasing energy levels, and then a return to low-energy wave activity and bioturbation during periods of fair weather (Kumar and Sanders, 1976).

As noted in section 4.5.1, maximum grain size and bed thickness commonly are correlated in deposits formed by individual sediment gravity flow events. This can provide invaluable interpretive data.

4.5.4 Sedimentary structures produced by hydrodynamic molding of the bed

The interpretation of hydrodynamic sedimentary structures is one of the most important components of facies analysis, particularly in siliciclastic sediments. Certain carbonate and evaporite deposits contain similar structures and can be studied in the same way.

The basis for interpreting structures formed in aqueous environments is the flow regime concept. This fundamental theory states that flow of a given depth and velocity over a given bed of noncohesive grains will always produce the same type of bed configuration and therefore the same internal stratification. If such structures are predictable their presence can be used to interpret flow conditions. These fundamental ideas were first enunciated following an extensive series of flume experiments by Simons and Richardson (1961), and were developed further for use by geologists by Simons et al. (1965), Harms and Fahnestock (1965), Southard (1971) and Harms et al. (1975). It is now realized that bedforms are controlled mainly by three parameters, sediment grain size, flow depth and flow velocity, and a series of experiments has demonstrated the sequences of bedforms produced as these parameters are varied. For example Figure 4.22 shows the stability fields of ripples, dunes and other bedforms for flow depths of about 20 cm. Figure 4.23 illustrates the velocity–depth relations for these bedforms for sand of 0.45 to 0.54 mm diameter. The conditions of water turbulence under which these various forms are produced are shown in Figure 4.24. Small-scale ripples, dunes and sand waves are forms out of phase with surface water movement; indeed, their form may bear no relation to surface water patterns at all. These bedforms have traditionally been classified as lower flow regime forms. The upper flow regime is characterized by antidunes and standing waves, which are in phase with surface water waves. The transition from lower to upper flow regime is marked by a streaming out of transverse turbulent eddies into longitudinal eddies. An intermediate "upper flat bed condition" is marked by streaming flow, which aligns the sand grains and produces primary current lineation (parting lineation) (Fig. 2.20B).

How can these flume data be used to interpret ancient sediments? Firstly, Allen (1968) and Harms et al. (1975) demonstrated the relationships between bedforms and sedimentary structures. For example planar tabular crossbedding is produced by the migration of straight-crested megaripples such as sand waves, whereas trough crossbedding develops from the migration of dunes (Fig. 4.25). Allen (1968) demonstrated the dependence of dune and ripple shape on water depth (Fig. 4.26). Secondly, the flow regime concept may be used to interpret ordered sequences of sedimentary structures in terms of gradations in flow conditions. Examples are discussed below of applications to fluvial point bar deposits, Bouma turbidite sequences, and wave-formed sedimentary structures. This is by no means an exhaustive listing. For example the concepts have been adapted by Dott and Bourgeois (1982) to the interpretation of hummocky crossbedding, a product of storm wave activity (see below and section 4.6.7).

Allen (1970) discussed flow patterns around a meander bend, in particular the effect on bedforms of the helical flow which develops in each meander, resulting in the movement of grains obliquely up the point bar slope (Fig. 4.27). Decreasing depth and velocity up this slope result in decreasing grain size and scale of sedimentary structures. Using flow regime data such as that shown in Figures 4.22 and 4.23 Allen (1970) was able to predict the types of sedimentary structures from depth–velocity–grain size conditions. His series of hypothetical profiles is shown in Figure 4.28. These can be matched to real examples of fining-upward cycles in the Devonian of Wales and the Appalachian region, demonstrating that Allen's model was of considerable value in reconstructing paleohydraulic conditions. Bridge (1978) has subsequently elaborated the model.

The Bouma sequence of thin-bedded, outer submarine fan turbidite deposits also contains a succession of structures that can be interpreted in terms of flow regime. The basal A member (Fig. 4.29) is formed by grains settling from suspension. Flow velocities decrease upwards, so that the plane bedded unit B, which commonly contains parting lineation, is formed under upper flow regime flat bed conditions, and the rippled unit C represents the lower flow regime (Harms and Fahnestock, 1965). Silty unit D is deposited from the dilute tail of the turbidity current as flow ceases altogether. This interpretation has been of

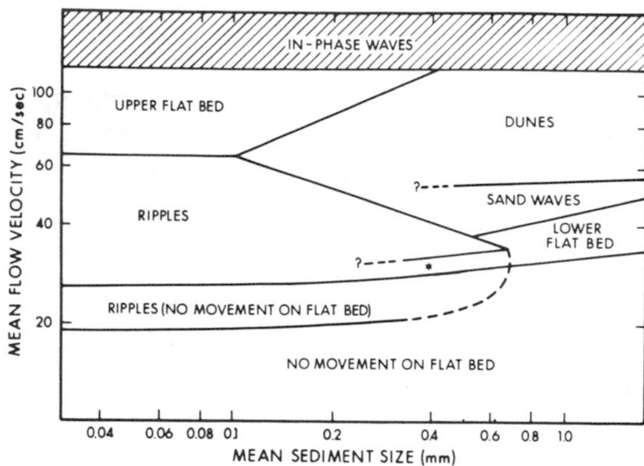

Fig. 4.22. Stability fields of bedforms in sand and silt in a flow depth of 20 cm (Harms et al., 1975).

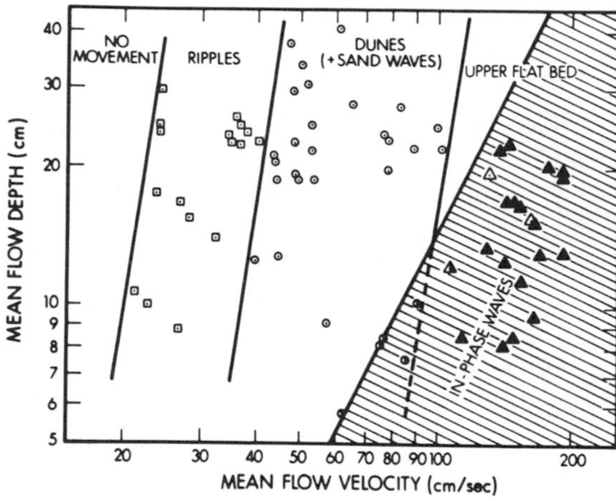

Fig. 4.23. Stability fields of bedforms in sand of about 1ϕ (medium to coarse) (Harms et al., 1975).

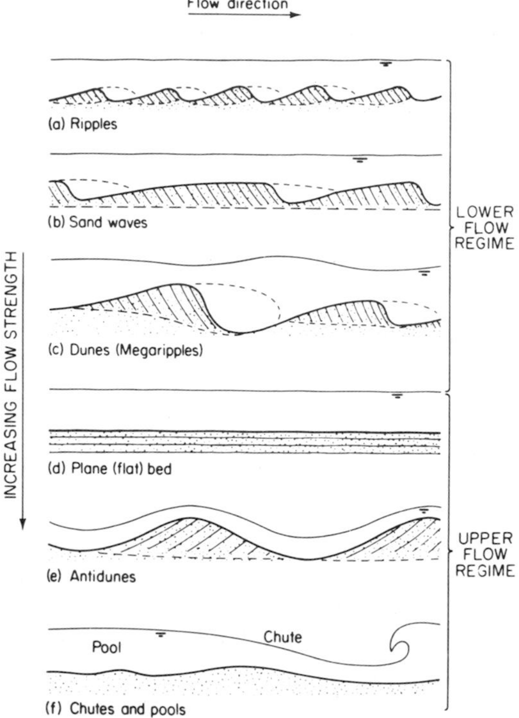

Fig. 4.24. The flow regime concept, illustrating the general succession of bedforms which develops with increasing flow velocity. Dashed lines indicate areas of flow separation. Note internal stratification (after Simons et al., 1965; Blatt et al., 1980). From H. Blatt, G.V. Middleton and R. Murray, Origin of Sedimentary Rocks, 2nd edition, © 1980, p. 137. Reprinted by permission of Prentice-Hall, Inc., Englewood Cliffs, New Jersey.

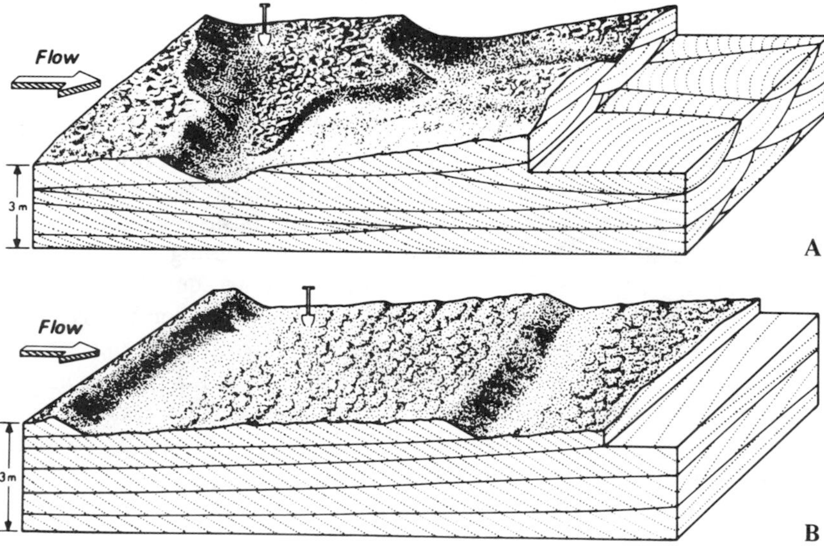

Fig. 4.25. Relationships between bedforms and sedimentary structures. **A** linguoid dunes and trough crossbedding; **B** sand waves and planar crossbedding (Harms et al., 1975).

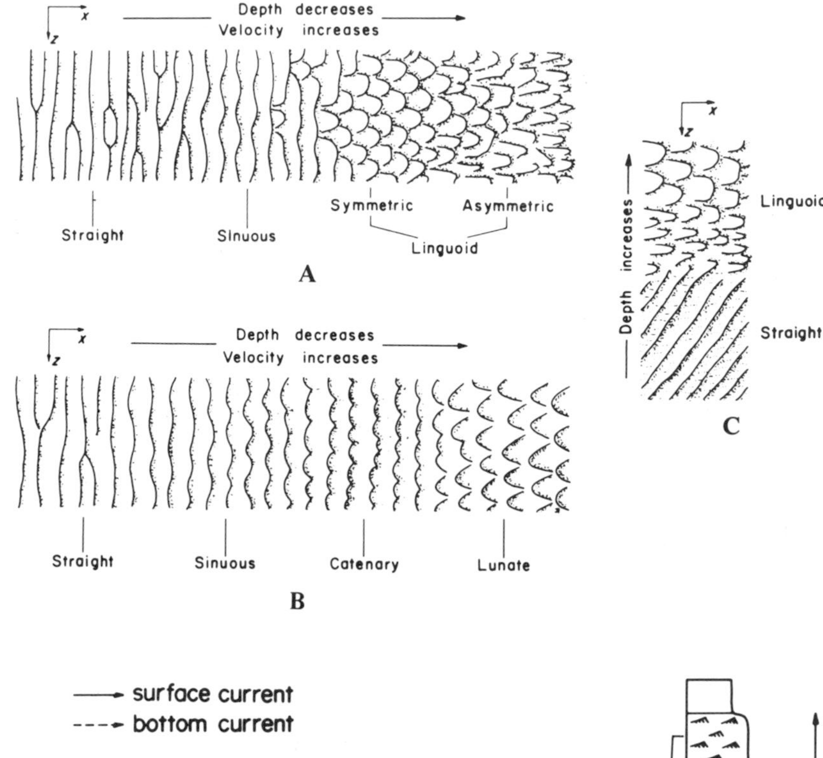

Fig. 4.26. Variations in bedform morphology with depth and velocity. **A** large-scale ripples (dunes or megaripples); **B** small-scale ripples; **C** where depth varies transverse to flow, large scale ripples (Allen, 1968).

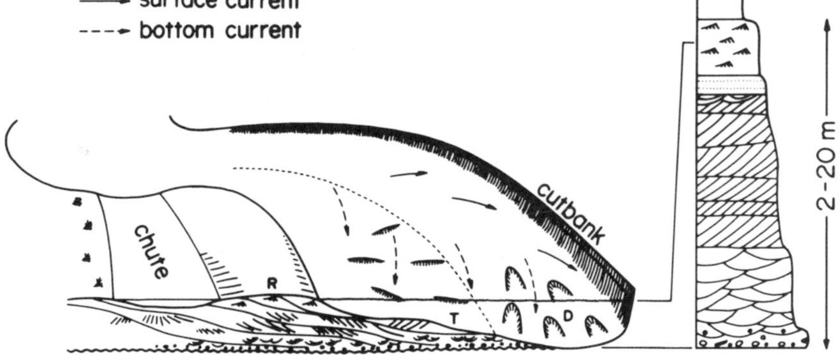

Fig. 4.27. Facies model for sedimentation on a point bar by lateral accretion inside a migrating meander. D = dunes, T = transverse bars or sand waves, R = ripples.

considerable use in understanding the mechanics of turbidity currents.

Clifton et al. (1971) carried out one of the first detailed studies of the sedimentary structures that form on coastlines under breaking waves. They recognized a direct relationship between wave type, resulting water motion and structure type (Fig. 4.30). The gradation from asymmetric ripple to outer planar facies represents a shoreward increase in orbital velocity and a transition from lower flow regime ripples through a dune facies to an upper flow regime plane bed condition. These structures all dip landward. The inner rough facies is characterized by seaward-dipping ripples and dunes of the lower flow regime, and the inner planar facies by plane beds, antidunes and standing waves formed under high-energy, upper flow regime, shallow swash conditions. These facies all move up and down the shore with the rise and fall of the tide, producing a complex but distinctive series of structural assemblages that Clifton et al. (1971) showed could be recognized in the ancient record.

Many environmental deductions can be made from the details of internal structure of hydrodynamic sedimentary structures, and from orientation (paleocurrent) information. Three general groups of structures can be distinguished:

1. structures formed by unimodal water currents in rivers, deltas, parts of ebb and flood tidal deltas in inlets, submarine fans and continental slopes (contour currents)
2. structures formed by reversing (bimodal) water currents such as tides and wave oscilla-

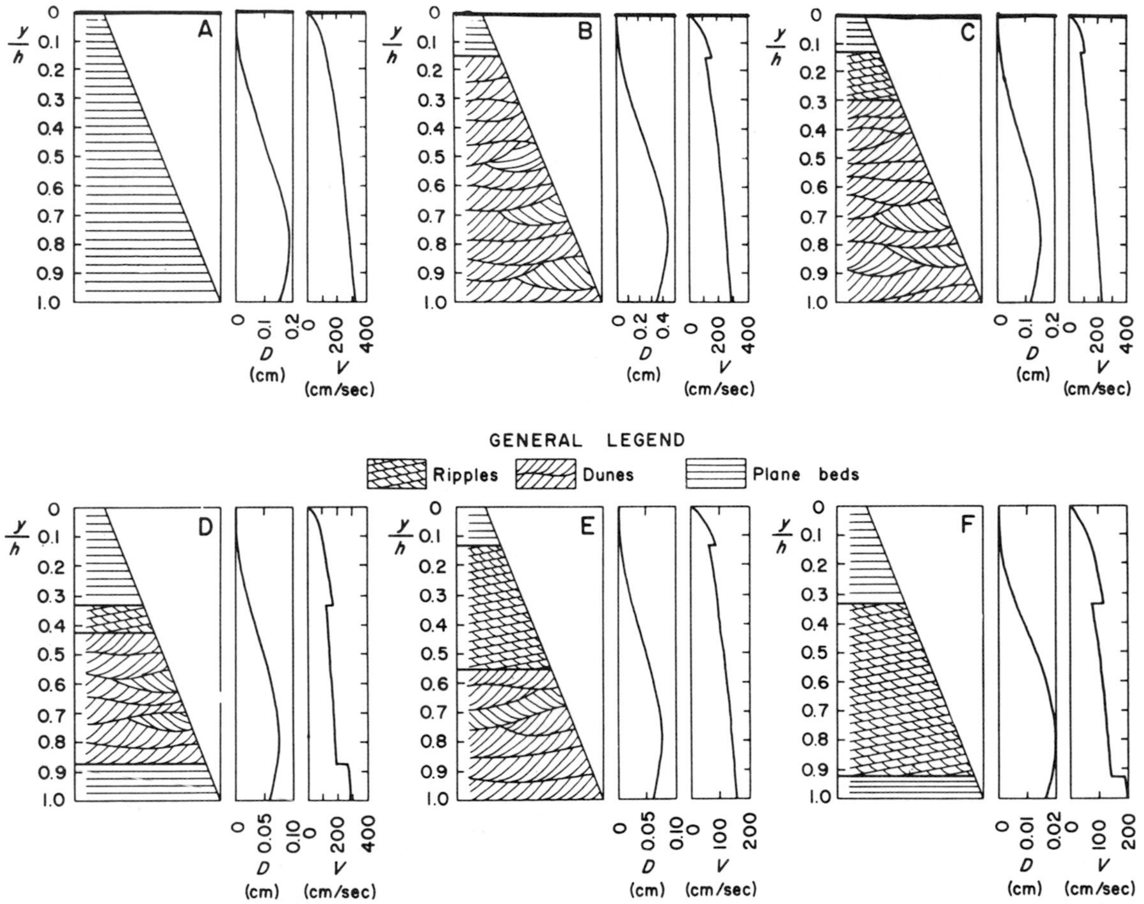

Fig. 4.28. Hydraulic model for point bar sedimentation, showing variations in vertical profile reflecting variations in grain size, $D =$ and flow velocity. $V = y/h$ indicates position on point bar with respect to total depth (Allen, 1970).

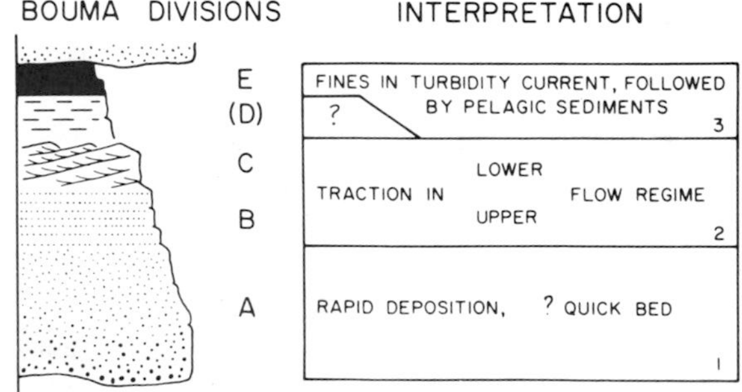

Fig. 4.29. Hydraulic interpretation of a Bouma turbidite sequence. Unit *D* probably represents deposition from the dilute tail of a turbidity current (Walker, 1978).

tion in shelf and marginal marine environments and in lakes

3. structures formed by eolian currents in coastal dune complexes, inland sand seas and some alluvial–lacustrine environments.

Unimodal currents are readily recognized from unimodal foreset orientations (e.g. Fig. 2.18A, C, D, F, G), but such patterns are not necessarily environmentally diagnostic. For example, as discussed in section 4.6, it has been found that in such areas of strongly reversing currents as tidal inlets and their associated deltas, ebb and flood currents are segregated into different parts of the system. Structures in a single outcrop of a tidal delta may therefore be misinterpreted as fluvial in origin, based on structure type and paleocurrent patterns. Simple paleocurrent models such as those of Selley (1968) should therefore be used

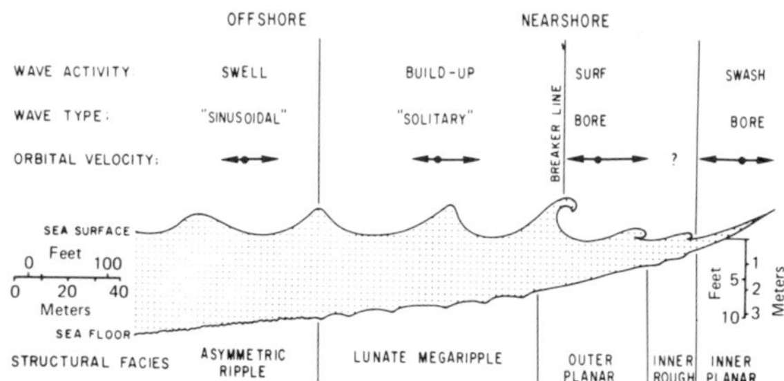

Fig. 4.30. Distribution of sedimentary structures (structural zones or facies) beneath shoaling waves (Clifton et al., 1971).

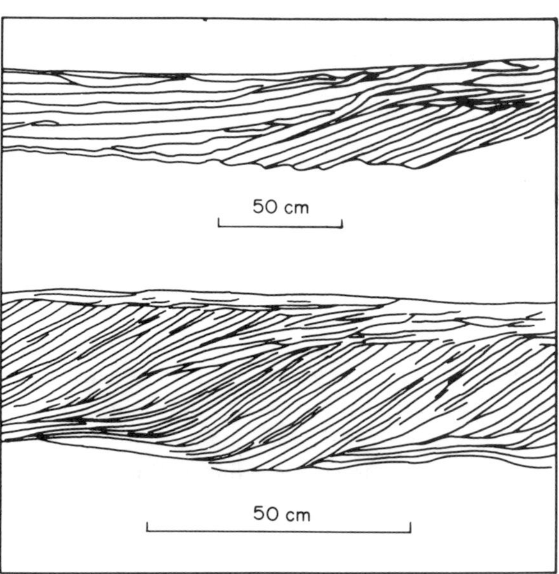

Fig. 4.31. Planar crossbed sets showing reactivation surfaces (Miall, 1977; after Collinson, 1970).

with caution. Other evidence, such as fauna, might yield clues as to the correct interpretation. Crossbedding structures may contain evidence of stage fluctuation, in the form of reactivation surfaces (Collinson, 1970), as shown in Figure 4.31. These are erosion surfaces formed during a fall in water level but, again, they are not environmentally diagnostic as water levels rise and fall in rivers, deltas and tidal environments.

Reversing currents can be recognized from such structures as herringbone crossbedding (Fig. 2.18I) or wave–ripple cross-lamination (Fig. 4.32), in which foreset dip directions are at angles of up to 180° to each other. Herringbone crossbedding is a classic indicator of reversing tidal currents, but it can also form under oscillatory wave-generated flow conditions (Clifton et al., 1971), and even in fluvial environments, where bars migrate toward each other across a channel. Because of the segregation of ebb and flood currents in estuaries and inlets herringbone crossbedding is, in fact, not common in many marginal marine deposits. The reversing ripples, chevron ripples, lenticular foresets and variable symmetry and orientation of wave-formed ripple cross-lamination (Figs. 4.32, 2.18B) are strongly diag-

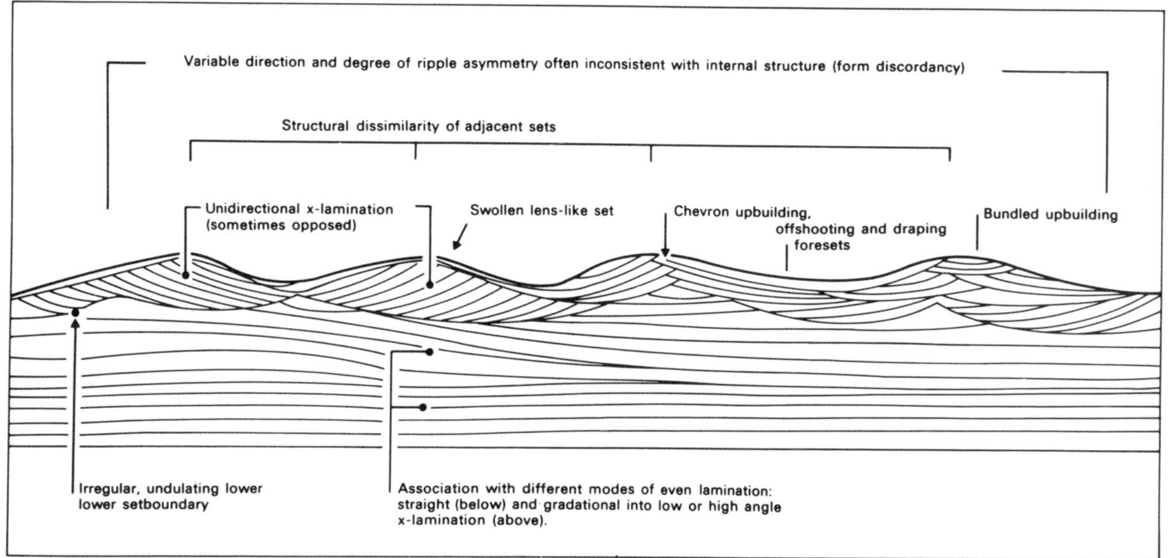

Fig. 4.32. Distinctive characteristics of wave formed-ripples (after Boersma (unpublished Ph.D., 1970) in de Raaf et al., 1977; Reading, 1978).

nostic of a low-energy wave environment, such as a gently shelving marine beach or a lake margin (de Raaf et al., 1977; Allen, 1981). Similar structures could also form in abandoned meanders or floodplain ponds in an alluvial environment, but would comprise a less conspicuous part of the overall succession. In many marginal marine environments crossbedding will be formed by both waves and tides, resulting in very complex paleocurrent patterns (e.g. Klein, 1970). Careful documentation of structure types and their orientations may be necessary to distinguish the precise environment and mode of origin, but such work may also yield invaluable information on sand body geometry, shoreline orientation, beach and barrier configuration, etc., as discussed in section 5.9.

Recognition of eolian crossbedding was, until recently, one of the major problems in clastic sedimentology. Much reliance was placed on the idea that eolian dunes are large (tens of meters high) resulting in very large-scale crossbedding. However, giant crossbed sets have now been recognized in fluvial (Coleman, 1969; Jones and McCabe, 1980) and shelf (Jerzykiewicz, 1968; Nio, 1976; Flemming, 1978) environments, so that this argument is no longer valid. Brookfield (1977) and Gradzinski et al. (1979) have recently examined the mechanics of dune construction and migration and have presented some useful ideas on the nature of large-scale crossbed bounding surfaces. Hunter (1977a, 1977b, 1980, 1981) and Kocurek and Dott (1981) have studied the details of sand movement by wind on modern and ancient dunes and have shown that several distinctive crossbedding and lamination patterns are invariably produced. Walker and Harms (1972) and Steidtmann (1974) carried out useful detailed studies of ancient eolian units. All this work has brought us to the point where eolian crossbedding should now be relatively simple to recognize, even in small outcrops.

Gradzinski et al. (1979) illustrated the gross structure of dune deposits in the Tumlin Sandstone (Triassic of Poland), showing the presence of three types of bounding surface (Figure 4.33): Main bounding surface formed by migration of transverse dunes and truncation of underlying sets; second order surfaces which are commonly shallow scoop shapes, bounding cosets of cross-strata; and third order surfaces, analogous to reactivation surfaces, developed by changes in turbulent eddy currents flowing along the dune slip face. These currents commonly generate wind ripples with crests oriented parallel to the dip of the slip face.

Hunter's contribution has been to demonstrate the variety of mechanisms by which sand is deposited on eolian dunes. Climbing ripple migration, grainfall and sandflow are the principal processes, all of which can be readily recognized from the details of internal structure. For example sandflow crossbedding is formed by avalanching of noncohesive sand on slip faces. It forms units that are distinctly lens-shaped in cross section and wedge-shaped down dip.

The last structure which should be mentioned in this brief review is hummocky cross stratification (HCS: Harms et al., 1975) which has been recognized in many ancient shelf deposits (Byers and Dott, 1981; Dott and Bourgeois, 1982). The structure is readily recognizable in small outcrops but would be hard to distinguish in core (Fig. 4.34). Facies associations and stratigraphic position in ancient rocks suggest it is the product of storm wave activity in the inner shelf, and a few tentative identifications of this structure have been made in comparable modern settings (Howard and Reineck, 1981). It has become a widely used environmental indicator (see section 4.6.7).

4.5.5 Sedimentary structures produced by hydrodynamic erosion of the bed

Few of these are environmentally diagnostic, although their presence may add weight to an interpretation made from other evidence.

As discussed in Chapter 2 there are two main classes of erosional structure at the outcrop scale, macroscopic and mesoscopic. Macroscopic structures comprise channels, scours and low relief erosion surfaces. Examples are illustrated in Figure 2.21A–D. They can occur in practically any environment as a result of current activity. More diagnostic than the channel itself is the channel fill, which can be analyzed in terms of vertical profile, lithology, sedimentary structures, etc. Subordinate features on an erosion surface may also provide some environmental information. Lag concentrates of pebbles, bioclastic debris or phosphatized fossils may be interpreted in terms of such processes as condensed sedimenta-

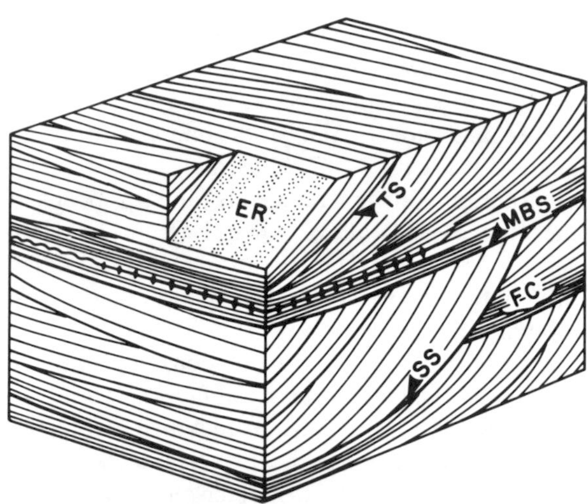

Fig. 4.33. Structure of typical ancient eolian dune deposit. *MBS* = main bounding surface, *SS* = second order bounding surface, *TS* = third order surface, *ER* = eolian ripples, *FC* = fluvial channel or sheet-flood deposits (after Gradzinski et al., 1979).

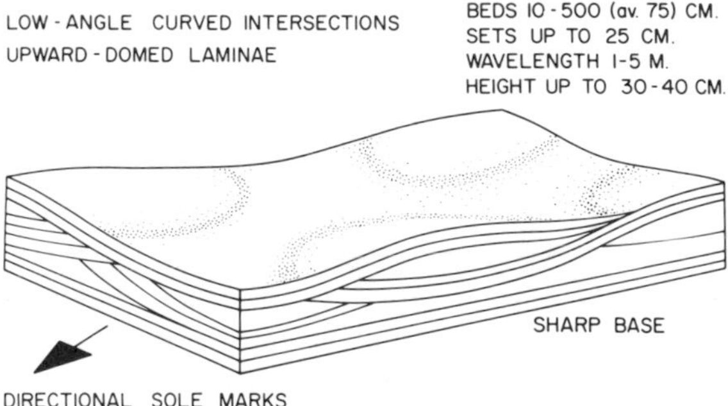

Fig. 4.34. Hummocky cross-stratification (Hamblin and Walker, 1979; reproduced, with permission, from Walker, 1979c).

tion or wind deflation, storm erosion and deposition, or submarine non-deposition (hardgrounds). Desiccated surfaces may break up, yielding breccias of partly lithified mud, silt or carbonate clasts, which can then be interpreted as the result of subaerial exposure (see section 2.3.2).

Mesoscopic erosional features include the variety of sole markings described and illustrated in Chapter 2. These are abundant in submarine fan deposits and submarine basin plain sediments, particularly in sandy turbidites. Flute marks, in particular, indicate a style of vortex turbulence that is common at the base of turbidity currents. However, they have been observed in a wide range of other settings, including fluvial traction current deposits. Turbidity currents themselves are not restricted to the deep oceans but are common in lakes and subaqueous ice margin environments.

4.5.6 Liquefaction, load and fluid loss structures

A few of these are environmentally diagnostic, others are practically ubiquitous in occurrence. Load and slump structures, convolute bedding, and other products of thixotropy, can occur wherever sediments are saturated, which is to say practically anywhere.

Water escape features such as dish and pillar structures are also found in the deposits of many environments (Nilsen et al., 1977), although they are particularly abundant in fluidized sediment gravity flows.

As noted in Chapter 2, desiccation and synaeresis produce subtly different sedimentary structures. Desiccation cracks are an indicator of subaerial exposure, synaeresis of salinity changes in tidal, lagoonal or lacustrine settings. Desiccation may be accompanied by breakup of the uppermost beds and resedimentation as an intraformational breccia.

Displacive evaporite growth in sabkhas produces a variety of distinctive structures which are useful diagnostic indicators (section 2.3.2).

4.5.7 Paleoecology

In Chapter 2 we discuss briefly the three main types of body fossil preservation—those in growth positions, soft-bodied organisms that would not survive significant transportation and the far more common death assemblage. In section 3.5.1 we touch on the problems of biostratigraphic interpretation of death assemblages and reworked fossils, and the difficulties of correlating faunas and floras limited in their distribution by factors of ecology or biogeography (not necessarily the same thing). In section 5.9 we discuss the use of oriented fossils as paleocurrent indicators. All these points are relevant when we focus on the use of body fossils as indicators of depositional environment.

Paleoecology is the study of the relationships between fossil organisms and their environment.

In most paleoecological research and in the resulting books and papers the primary focus is on the animals, with the use of such criteria as species interrelationships and fossil–sediment characteristics in the analysis of the ecology of some fossil species or community. Such research typically is carried out by specialists whose background usually is paleontology or biology, and these individuals may have little or no interest in regional stratigraphy or basin history. Here we wish to reverse the process, to use fossils as paleogeographic indicators. In the Phanerozoic record fossils are amongst the most powerful environmental indicators available, but of course their distribution and usefulness in the Precambrian is very limited.

Even the simplest paleoecological observations can be invaluable. The presence of abundant bioturbation or (for example) a brachiopod fauna may be the key evidence in distinguishing the deposits of tidal channels and their point bars, tidal inlets and deltas from certain fluvial and delta plain facies which can be sedimentologically very similar (Barwis, 1978). At the other extreme are sophisticated statistical studies of particular communities or fossil groups, such as the trilobite population analysis of Ludvigsen (1978), discussed in section 4.4.1. Most paleoecological study by stratigraphers and basin analysts falls somewhere between these extremes.

Fossil data are subject to two biases, which may distort our observations. Firstly there is the bias of unskilled or incomplete collection, which is particularly likely to arise when paleoecological analysis is attempted by the nonspecialist or when it is rushed through, as in a reconnaissance survey by a prospecting team. Ager (1963) demonstrated this with data assembled by B.W. Sparks, who used two methods to collect molluscs from Quaternary deposits in southern Britain (Figure 4.35). A bulk sample was analyzed by sieving, and the distribution compared with that of a collection made by hand-picking from the outcrop surface. The bias in the second collection in favor of larger or more brightly colored species is startling.

The second bias is that introduced by the geological obstacle course organisms are put through before they end up under the geologist's hammer. The various hazards and possible paths through them are summarized in Figure 4.36. The study of this flow diagram is itself the subject of a specialized science called **taphonomy**. Many of the processes of transportation, removal or breakup of a fossil depend on sedimentological processes and are therefore of interest to basin analysts. Figure 4.37 illustrates the fate of dinosaur remains in Cretaceous fluvial deposits of the Red Deer Valley, Alberta (from Dodson, 1971). The state of preservation may yield much information on local transport energy and hence on depositional environment.

One of the first and most obvious observations that must be made is to relate biofacies to lithofacies. The relationships may be obvious, including marked contrasts in facies such as that shown in Figure 4.38 (from Hecker, 1965). Here there is a red bed lithofacies characterized by fish, a brackish lithofacies with *Lingula*, and marine deposits with corals, stromatoporoids and abundant bioturbation. This pattern of interbedded sediments and fossil types is characteristic of Early Paleozoic continental–marine transitions around the world.

Many excellent paleoecological–sedimentological studies have been made of carbonate reefs, such as the Permian reef complex of the Guadalupe Mountains in the southwestern United States (Newell et al., 1953), Silurian reefs of the Great Lakes (Lowenstam, 1950, 1957) and Devonian reefs of Alberta (Andrichuck, 1958; Klovan, 1964). Figure 4.39 illustrates the lithofacies and biofacies subdivisions of some Silurian carbonate banks (Wilson, 1975; Sellwood, 1978). More subtle fossil–sediment relationships may only be apparent after careful statistical studies, such as the brachiopod analysis carried out by Jones (1977). In some cases, biofacies are not obviously related to lithofacies, as in the Welsh Paleozoic brachiopod assemblages discussed in section 3.5.1.

How are these observations on biofacies actually used to indicate depositional environment? In four main ways: 1. by comparison with the ecology of living relatives; 2. by deductions about functional morphology; 3. by deductions from position and preservation; 4. by deductions from assemblages and associations.These lines of reasoning are not intended to be mutually exclusive and, indeed, in many ancient deposits there is an overlap in their application.

To start with the first point, comparisons with modern relatives become more difficult the further back in time we proceed, as we trace

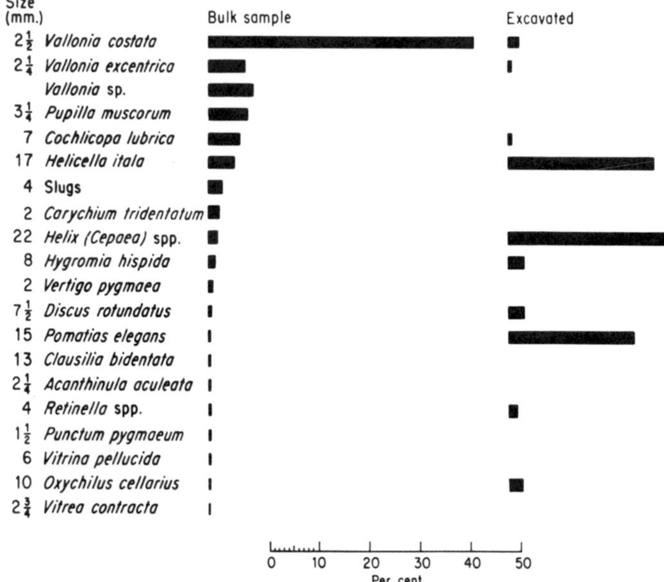

Fig. 4.35. Distribution of species in two collections made from the same outcrop by two different methods. From D.V. Ager, Principles of Paleoecology, © 1963, McGraw-Hill, New York. Reproduced with permission of the publisher.

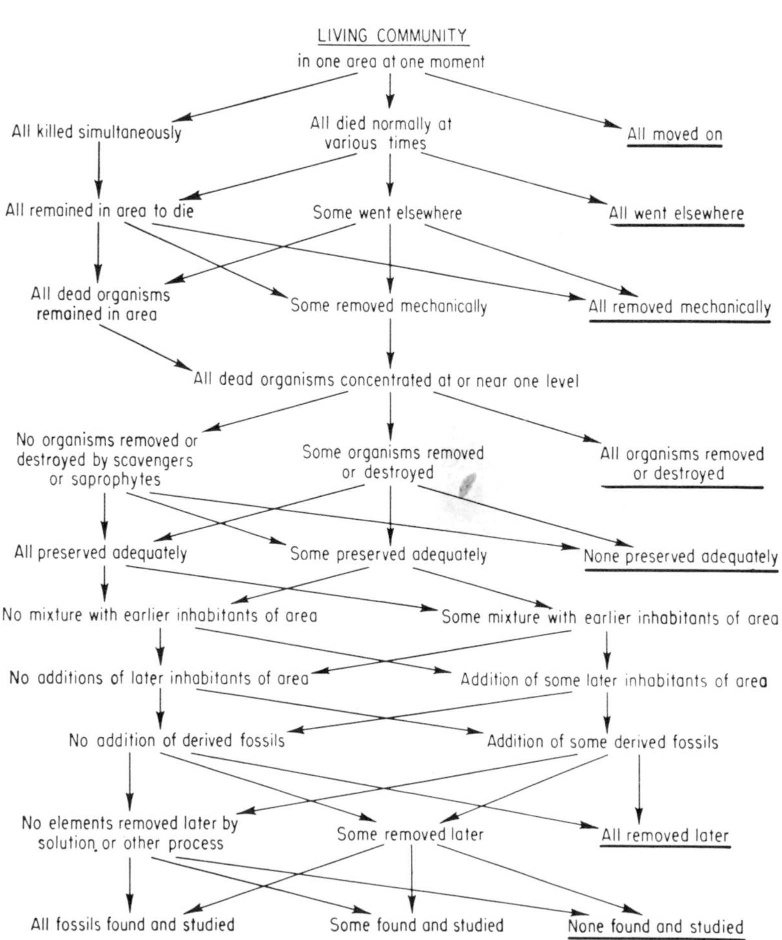

Fig. 4.36. The various possibilities for the preservation or destruction of a living animal community and its eventual collection as a fossil assemblage. From D.V. Ager, Principles of Paleoecology, © 1963, McGraw-Hill, New York. Reproduced with permission of the publisher.

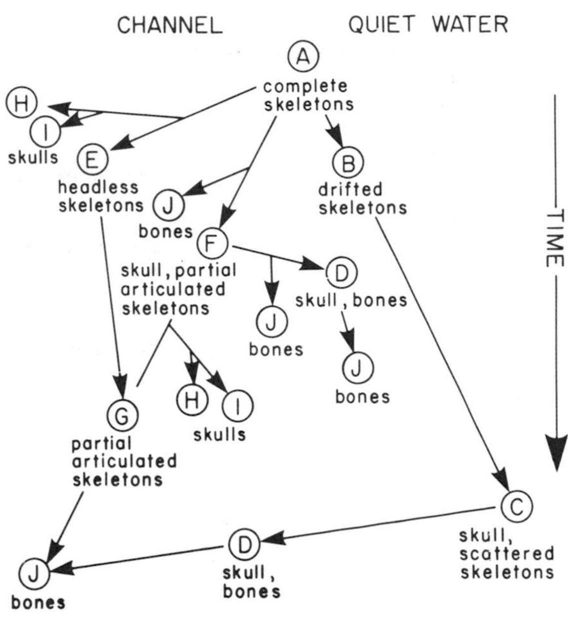

Fig. 4.37. The taphonomy of dinosaur skeletons, based on studies of the Cretaceous rocks of Red Deer Valley, Alberta (Dodson, 1971).

evolutionary lines back to more and more distant ancestors, for whom there is no guarantee of unchanged habits. This is where examination of functional morphology, position and associations might provide invaluable supportive evidence. The corals are a classic example of a group of organisms that has undergone marked evolutionary change during the Phanerozoic, yet seems to have continued to inhabit rather similar ecological niches throughout this time. Modern hermatypic (colonial, reef forming) corals inhabit depths of water down to about 90 m, but are only abundant above about 20 m. They require a minimum water temperature of about 22°C, with an optimum of 25 to 29°C. Strong sunlight, continuous nutrient supply, minimal salinity variations and clean, sediment-free water are also necessary. These requirements restrict colonial corals to a present day latitudinal range of 30°S to 30°N, with variations depending on the extent of cold or warm oceanic currents, sediment input from large rivers, etc. (Ager, 1963). For

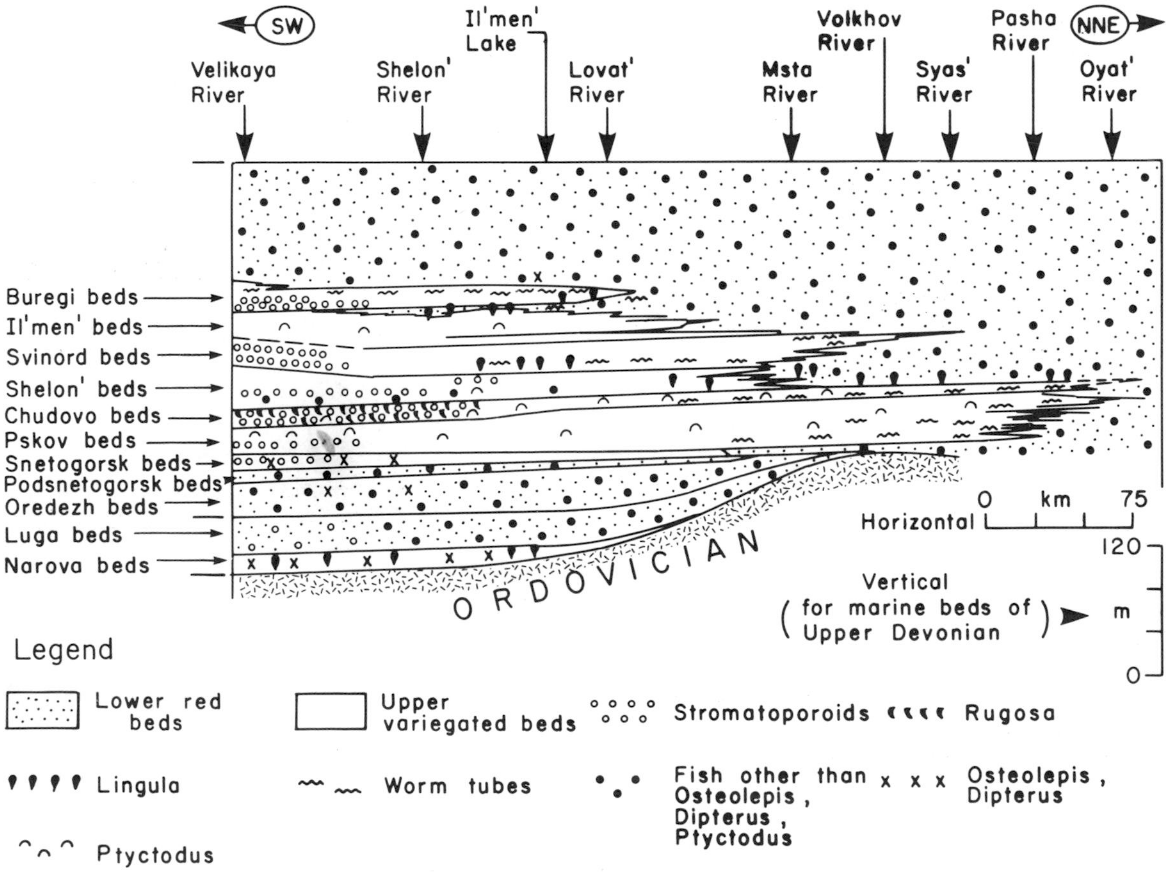

Fig. 4.38. Lithofacies–biofacies relationships for some Upper Devonian rocks in Russia (Hecker, 1965).

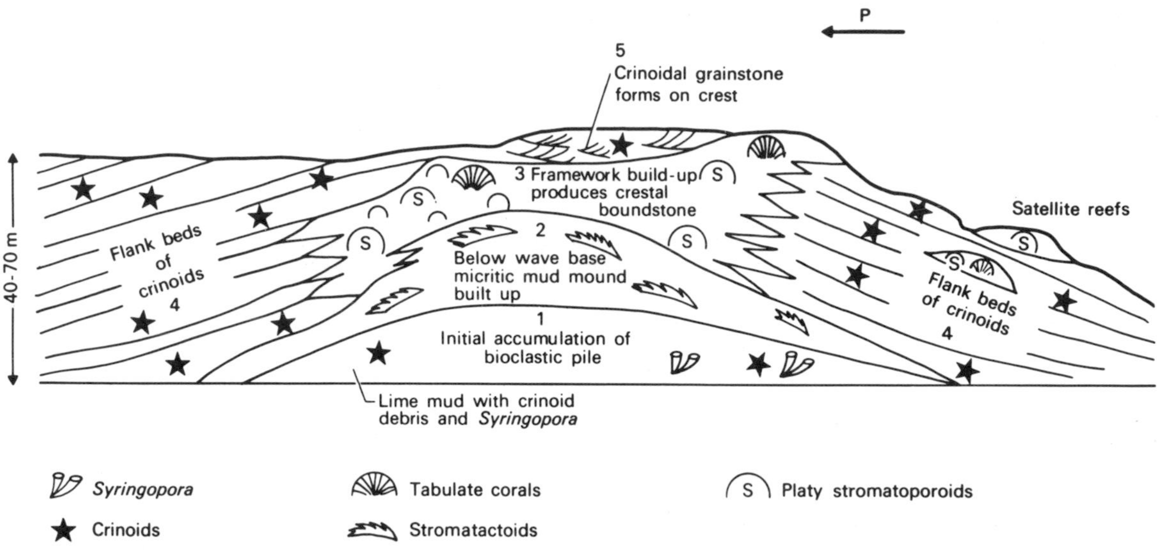

Fig. 4.39. Lithofacies–biofacies composition of carbonate buildups in the mid-Silurian shelf of the American Midwest (Sellwood, 1978; after J.L. Wilson, 1975). (Reproduced, with permission, from J.L. Wilson, 1975).

Mesozoic and Cenozoic corals, which are closely related to modern forms (orders Scleractinia and Alcyonaria), environmental deductions based on modern ecology seem safe. Paleozoic corals belong to two different orders, Rugosa and Tabulata, yet their gross appearance, lithofacies and biofacies associations (particularly the symbiotic association with algae) remain similar, as is evident from the various Paleozoic reef studies quoted above. Heckel (1972) suggested, however, that they may have occupied less agitated environments, with the now extinct stromatoporoids filling part of this niche. Table 4.5 summarizes some other modern–ancient comparisons for carbonate-producing organisms.

A group of organisms that has become extremely important in recent years for unraveling the history of continental margins is the Foraminifera. Comparisons of ancient and modern assemblages yield invaluable paleobathymetric data for Cenozoic and later Mesozoic time, as discussed below.

Many ancient groups, such as the trilobites, graptolites and many vertebrates, including early fish, amphibians and reptiles, are now extinct, as are numerous genera and species of still thriving phyla. Therefore comparisons with the ecology of modern relatives becomes tenuous or impossible. Ecological and environmental interpretations may then depend partly or wholly on deductions made from functional morphology. For example in section 3.5.1 we discuss the evolution of the Cretaceous echinoid genus *Micraster*, which developed divergent stocks demonstrating habits of deeper and deeper burrowing into the Chalk substrate. This genus is now extinct, but Nichols (1959) was able to demonstrate the change in burrowing habit by careful examination of the functional morphology and comparison with similar features in modern echinoids. *Micraster* tests show an increase with time in the number of pores required for respiratory tube feet, an obvious adaption to deeper burial.

There has been much debate about the habitats of trilobites, which show considerable variation in morphology. Some have large eyes, suggesting dim light (deep water?), others are blind and may have lived below the photic zone. Long horizontal spines may have been a device to inhibit sinking in soft substrate, and the same effect may have been achieved by the broad, flat exoskeletons of some species. Others were burrowers, for example *Bathyrus*, which has a concave-up, wedge shape adapted to wriggling into the substrate. Ludvigsen (1978) discussed these and other morphological features and related them to trilobite biofacies zones and their position on the Ordovician platform of Western Canada (see section 4.4.1).

A widely used climatic indicator is the shape of tree leaves. It is known from present-day distributions that in tropical regions the leaves of

Table 4.5. Modern and ancient carbonate-producing organisms and their sedimentary products*

Modern organism	Ancient counterpart	Sedimentary aspect
Corals	Archaeocyathids, Corals, Stromatoporoids, Bryozoa, Rudistid bivalves, Hydrozoans	The large components (often in place) of reefs and mounds
Bivalves	Bivalves, Brachiopods, Cephalopods, Trilobites and other arthropods	Remain whole or break apart into several pieces to form sand and gravel-size particles
Gastropods, Benthic Foraminifers	Gastropods, Tintinids, Tentaculitids, Salterellids, Benthic Foraminifers, Brachiopods	Whole skeletons that form sand and gravel-size particles
Codiacean algae—*Halimeda*, sponges	Crinoids and other pelmatozoans, Sponges	Spontaneously disintegrate upon death into many sand-size particles
Planktonic foraminifers	Planktonic foraminifers, Coccoliths (post-Jurassic)	Medium sand-size and smaller particles in basinal deposits
Encrusting foraminifers and coralline algae	Coralline algae, Phylloid algae, Renalcids, Encrusting Foraminifers	Encrust on or inside hard substrates, build up thick deposits or fall off upon death to form lime sand particles
Codiacean algae—*Penicillus*	Codiacean algae, *Penicillus*-like forms	Spontaneously disintegrate upon death to form lime mud
Blue–green algae	Blue-green algae (especially in Pre-Ordovician)	Trap and bind fine-grained sediments to form mats and stromatolites

*Reproduced, with permission, from James, 1979a.

dicotyledonous trees are entire (smooth), whereas in temperate regions they are more commonly dentate. Edwards (1936) showed that this characteristic could be used to interpret fossil floras, even for extinct or unidentifiable forms. Examples are illustrated in Figure 4.40.

Factors of abundance, state of preservation and position may yield important environmental information. Fossil forms that normally grew attached to a substrate may occur as disoriented, broken and abraded individuals or as hydraulically accumulated shell beds, suggesting storm wave activity. Isolated individuals may be found out of context; for example I have found rare abraded orthocone cephalopods and compound corals in laminated stromatolitic dolomites formed on a Silurian tidal flat in the Canadian Arctic. Ballance et al. (1981) reported finding coconuts in Miocene turbidites, and suggested that they may have been transported offshore by tsunamis, the same events that probably triggered the turbidity currents. Fossils found in place, such as coral colonies, pelecypods in their burrows and upright trees attached to their roots are particularly diagnostic of depositional environment (so long as their habits are known) because such occurrences usually are devoid of any ambiguity about possible transportation into an unrelated environment. However, Jeletzky (1975) and Cameron (1975) have engaged in heated controversy over the significance of shallow-water invertebrate fossils found in nodules in a turbidite succession on Vancouver Island. Are the nodules still in place where they were formed around the fossils or is there evidence of abrasion indicating transportation to deeper water? The geologist should be prepared for the most unlikely event. For example upright trees, weighed down by boulders in their roots, have been found in the flood deposits of Mt. St. Helens, and offshore

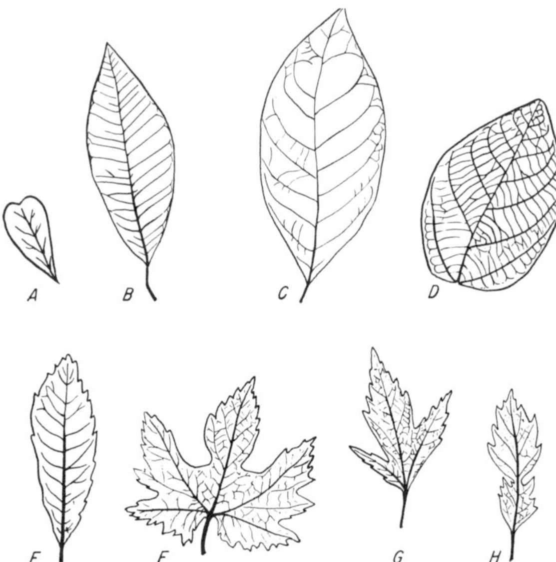

Fig. 4.40. Leaf shape as a reflection of climate. **A–D**: tropical plants typically have "entire" margins; assemblage from Lower Eocene, southern England. **E–H**: temperate plants typically have "dentate" margins; assemblage from Miocene, southern France (Edwards, 1936; Ager, 1963). From D.V. Ager, Principles of Paleoecology, © 1963, McGraw-Hill, New York. Reproduced with permission of the publisher.

near islands in the South Pacific, where they were emplaced by storms.

Preferred orientations of attached or transported fossils may have paleocurrent significance. Corals, crinoids, stromatolites and other immobile organisms may grow in particular directions in response to current conditions, and accumulations of elongated fossils such as graptolites, gastropods, logs or belemnites may be aligned hydraulically. This is discussed further in Chapter 5. Deductions based on preservation or position require particularly careful field observation and normally cannot be made on loose talus.

Perhaps the most reliable paleoecological interpretations are those based on assemblages and associations in an entire fossil community. Deductions made from extinct groups can be checked against information yielded by those with living relatives, and a picture of ecological relationships within the community can be attempted. Some forms may yield very generalized environmental information, including many pelagic groups such as the graptolites, whose occurrence is governed more by energy conditions at the site of postmortem deposition, than by habitat during life. Interpretations based on tolerant groups may be finely tuned if other, more selective species can be found. Particularly accurate interpretations may be possible if species with different but slightly overlapping habitat ranges can be found together. For example in the Lower Cretaceous of northern Texas arenaceous foraminifera are associated with the oyster *Ostrea carinata* in deposits assumed to be close to the shoreline, but the association only occurs in limestones and marls, not in mudstones. Modern arenaceous foraminifera are known to be tolerant of brackish water. *O. carinata* therefore appears to be tolerant of reduced salinities but not of turbid water. By contrast the oyster *Gryphaea washitaensis* occurs in all rock types but never in association with arenaceous foraminifera. It therefore seems to have preferred open marine conditions but had no preference for either turbid or nonturbid water (Laughbaum, 1960; Ager, 1963). Another example is the association of ostracodes, gastropods and the brachiopod *Lingula* in some Silurian dolomite beds in the Canadian Arctic. This is interpreted as a brackish-water tidal-flat assemblage, and contrasts strongly with the rich coral, brachiopod, crinoid and trilobite assemblage of interbedded open marine limestones (Miall and Gibling, 1978). Heckel (1972) provided tables relating the modern distribution of fossilizable invertebrate groups to water salinity, turbidity and depth, e.g. Figure 4.41, and within each group individual genera or species may have much more restricted tolerance ranges, as in the various examples given above.

Some of the most detailed assemblage studies carried out for basin interpretation purposes are those on ancient reefs of all ages, and on Cenozoic foraminifera. For example Figure 4.42 shows the vertical changes in assemblage through Niagaran reefs of the Great Lakes (Lowenstam et al., 1956). The reefs were built on a soft substratum in quiet water and gradually extended up into the zone of breaking waves.

Studies of foraminifera paleoecology began in the Cenozoic of the Gulf Coast and California in the nineteen thirties (Natland, 1933). The initial stimulus for the work was the need to correlate petroleum-bearing sediments, and this remains one of its most important applications. However, the same data are now finding increasing application in the interpretation of the subsidence history

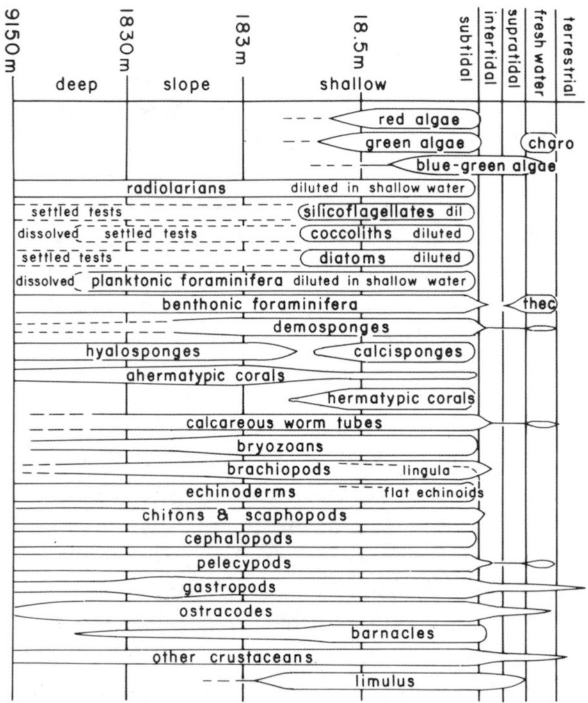

Fig. 4.41. Depth distribution of common types of fossil invertebrates (reproduced, with permission, from Heckel, 1972).

of continental margins, a subject of considerable relevance to our attempts at understanding the mechanisms of plate tectonics. An excellent example of this is the work on subsurface Cenozoic strata of the continental margin off Labrador and Baffin Island by Gradstein and Srivastava (1980). They distinguished four biofacies and interpreted their environments based on associated faunas and comparisons with other modern and ancient foraminifera assemblages. Considerable use was made of DSDP data. The biofacies are as follows:

1. Nonmarine: spores and pollen, no foraminifera.
2. Shallow neritic: marginal marine to inner shelf (<100 m deep). Diagnostic foraminiferal assemblages are of very low generic and specific diversity, with rare or no planktonics. In the Late Neogene sections *Cibicidiodes, Elphidium, Cassidulina, Bulimina, Melonis,* and gastropods and bryozoans occur in low numbers.
3. Deep neritic: 100–200 m water depths. Foraminiferal generic and specific diversity varies, planktonics occur locally. Cenozoic benthonic genera include *Uvigerina, Pullenia, Gyroidina, Alabamina*. Paleogene assemblages are impoverished.
4. Bathyal: upper slope, water depths 200–1000 m. Foraminiferal generic and specific diversity is high and the assemblages are dominated by coarse, often large-sized agglutinated taxa. Over 50 species, including rich *Cyclammina* spp. Locally planktonics are abundant (mostly turborotaliid and globigerinid forms). In a few wells Paleogene calcareous benthonics occur in low numbers, including *Pleurostomella, Osangularia, Stilostomella* and *Nuttalides*.

Gradstein and Srivastava (1980) constructed subsidence history plots for five wells off the Newfoundland–Labrador coast, one of which is shown in Figure 4.43. The line defining the upper boundary of the shaded area represents changes in water depth with time, as determined from biofacies data. The lower line, defining the locus

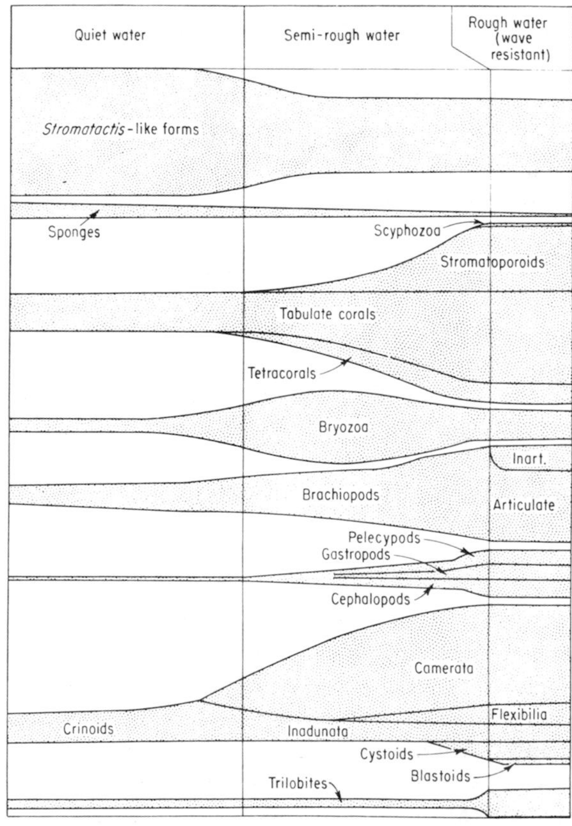

Fig. 4.42. Vertical variations in fossil assemblages of Niagaran reefs, Great Lakes (Lowenstam et al., 1956).

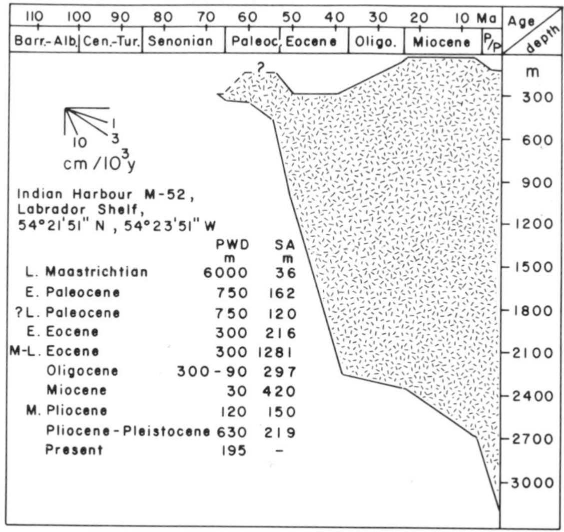

Fig. 4.43. Subsidence curve for an offshore well, Labrador shelf. See text for explanation (Gradstein and Srivastava, 1980).

of the pre-Cretaceous "basement" unconformity, was constructed by subtracting cumulative sediment thickness from the water depth curve. It can be seen that subsidence rates have varied markedly since the Cretaceous, the fastest subsidence occurring in the Eocene, which geophysical evidence indicates was a time of active sea-floor spreading in the immediately adjacent oceanic crust.

Benedict and Walker (1978) attempted a similar type of paleobathymetric analysis of a Paleozoic sequence, using chemical, sedimentary and paleoecological criteria.

Trace fossils are widely used environmental indicators in marine sediments, where they are most abundant. They have one advantage over body fossils, in that they are almost invariably in place; derived and transported trace fossils are very rare. Trace fossils may be used in two ways: 1. to indicate local sedimentation and erosion patterns, 2. to indicate depositonal environment.

Howard (in Frey, 1975; in Basan, 1978) discussed the ways in which trace fossils may be used to evaluate local erosion and sedimentation. The density of bioturbation in a bed varies inversely with the rate of sedimentation, so that where sediment supply is low and invertebrate life abundant primary structures such as bedding may be completely destroyed. Intervals of rapid sedimentation, such as the passage of a turbidity current or a storm deposit, may contain little or no penetration by organisms, except perhaps for escape burrows made vertically through the bed. Types of bioturbation may vary from one lithology to the next, because of different behavior patterns of the organisms in response to different sediment types. Some examples of specific responses of selected burrow types to varying sedimentation patterns are shown in Figure 4.44, based on observations by Goldring (1964).

Important and widely quoted work by Seilacher (1967) showed how trace fossils could be used to interpret paleobathymetry. A set of five biofacies were erected, as shown in Figure 4.45. In the shallowest waters waves and currents keep nutrients in suspension. Animals are subject to violent conditions and build deep vertical burrows such as *Skolithus*. In less turbulent waters faunal diversity increases, sediment feeding becomes more important and a variety of crawling and grazing feeding trails appear (*Cruziana* biofacies). Below wave base sedimentation is slower, the oxygen content of the sediments is lower and nutrient supply more sparse. Sediment mining organisms of the *Zoophycos* and *Nereites* biofacies are characteristic. The latter, with its complex but highly systematic grazing patterns, is particularly typical of abyssal submarine fan deposits. The details of this zonation vary from basin to basin and depend on sediment type, water temperature, salinity and circulation patterns. Also, as Howard (in Basan, 1978) pointed out, different organisms may make very similar structures and conversely the same species may make different structures when engaged in different activities or when interacting with substrates of different composition. Interpretations must therefore always be made in conjunction with other facies studies, and the geologist would be wise to consult published work on trace fossil assemblages in other basins of a similar age.

In carbonate environments algal mats and stromatolites are excellent indicators of intertidal deposition. Their "upper limit is controlled by climate; in arid areas they cannot grow above the high intertidal into the supratidal zone, whereas in areas of high rainfall where the supratidal zone is moist or flooded for days at a time, mats are prolific. The lower limit is more variable and appears to be controlled by the presence of gastropods that eat algae" (James 1979b). In hypersaline zones gastropods cannot survive and mats grow

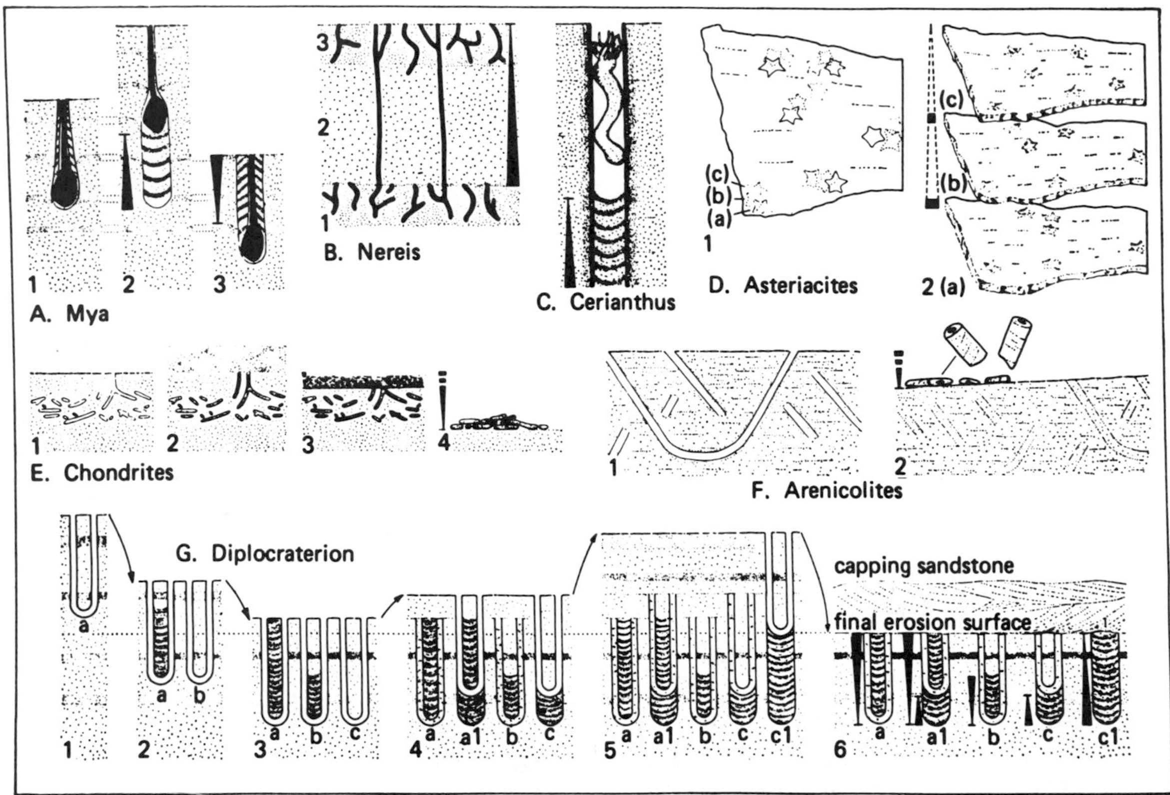

Fig. 4.44. Amount of sedimentation or erosion as indicated by adjustments to depth and modes of preservation of various lebensspuren. Heights of solid arrows show amount of deposition or erosion. **A** movement pattern of pelecypod *Mya*, which has a single siphon. With stationary sedimentary surface (1), growing organism gradually burrows deeper; bottom of structure wider than top. With rapid sedimentation (2), organism migrates toward surface, leaving infilled burrow the width of the shell. With degradation of surface (3), organism migrates downward, leaving burrow of same width but having different internal structure. **B** movement pattern of polychaete worm *Nereis*. Older colonized surface (1) is rapidly covered by sediment (2) and, during deposition, paths of escape are directed upward. With stabilization, new colonization surface (3), has irregular "normal" burrows. Structures in (1) and (3) are generally mucus lined; in (2) they are unlined. **C** movement pattern of anemone *Cerianthus*, an organism dwelling in a single tube. With sedimentation, animal moves upward, leaving an unfilled or passively filled burrow. A similar pattern might be expected in traces such as *Skolithus* and *Monocraterion*. **D** movement represented by trace fossil *Asteriacites lumbricalis*, resting place of a stelleroid. With sedimentation, animal migrated upward, in stages a-c; combined (1) and separate (2) plan of all impressions. **E** preservation patterns of trace fossil *Chondrites*. Tunnel system (1) is infilled (2), following a change in type of sediment being deposited (bed junction preservation). Slight degradation of surface (3) removes the proximal shafts before further sediment, of a different type, accumulates (concealed bed junction preservation). Renewed degradation of surface winnows away sediment, leaving mucus-lined infilled tunnels as burial preservations (4). **F** preservation pattern of trace fossil *Arenicolites curvatus*. Sedimentary surface containing open U-tubes (1) has been degraded (2), the mucus-cemented tube fragments accumulating in an intraformational conglomerate; sediment has filled the tubes. **G** movement pattern represented by trace fossil *Diplocraterion yoyo*. In Upper Devonian Baggy Beds of North Devon, this trace occurs in various configurations shown in (6); all have been truncated to common erosion surface. Repeated phases of erosion and sedimentation (1-5) evidently led to development of the various types. (1), development of burrow (a). With degradation of surface, this tube migrates downward, and at intervals, new tubes (b,c) are constructed (2,3). Sedimentation follows (4,5) but some tubes are abandoned. (6), all tubes are abandoned, and erosion reduces them to a common base. (From Goldring, 1964).

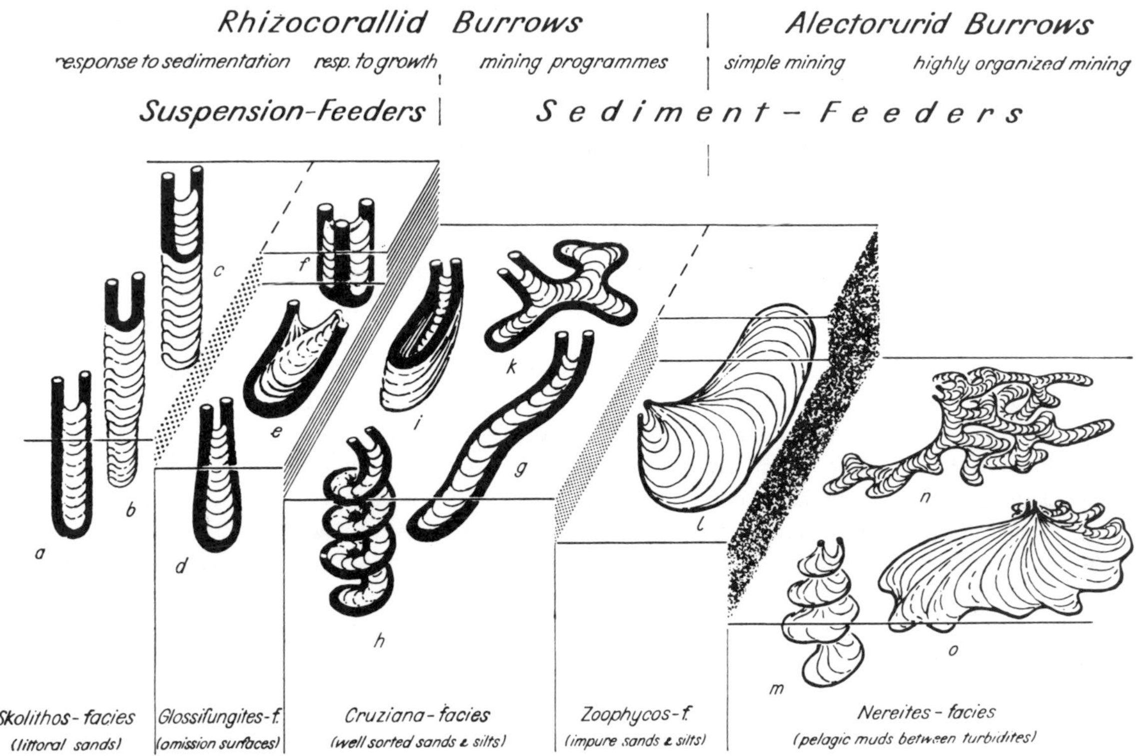

Fig. 4.45. Bathymetric zonation of trace fossils (Seilacher, 1967).

into the subtidal zone. After deposition and burial mats commonly rot away, but they leave voids as a result of disappearance of organic materials, entrapped gas or shrinkage. The resulting distinctive type of porosity has been referred to as laminoid fenestrate, loferite or birdseye. Stromatolites are present well back into the Precambrian and may have lived in similar environments throughout this time.

4.5.8 Vertical profiles

The importance of Walther's Law and of the vertical profile in facies analysis is discussed in section 4.4.2. The recognition of cyclic sequences has become one of the most widely used tools for reconstructing depositional environments in the subsurface. One of the reasons for this is that cyclic changes commonly are readily recognizable in geophysical logs and are often interpreted without access to core or well cuttings. Whether such interpretations are always correct is another question. The sedimentological literature is full of shorthand references to fining-upward or coarsening-upward cycles, or to fining-and-thinning or coarsening-and-thickening upward. Grain size, bed thickness and scale of sedimentary structures commonly are correlated in clastic rocks, so that the cyclicity may be apparent from several types of observation. French sedimentologists tend to use the terms positive and negative cycles; dipmeter analysts recommend coding cyclic changes diagrammatically in red and blue, but I am embarrassed to have to admit that I am not sure if fining-upward cycles are positive and red or negative and blue or the opposite, and (or) vice versa. Such terms are obviously not helpful if one cannot remember which way they are used, and a simple descriptive terminology seems preferable.

Vertical profiles formed a crucial component of the first facies models, including the point bar model of Allen (1963b) and Bernard and Major (1963), the barrier island model of Bernard et al. (1959) and the Bouma (1962) turbidite sequence. They were the focus of a classic paper on facies models by Visher (1965), and their recognition in the subsurface has been covered in many articles and textbooks (e.g. Pirson, 1977; Fisher et al., 1969; Sneider et al., 1978; Selley, 1979). They still provide an essential basis of many modern facies models (Reading, 1978; Walker, 1979a), because they provide a simple way of synthesizing many different types of observation, including all the environmental criteria described in this chapter. Methods of statistical analysis of cyclic sedimentation are described in section 4.4.2.

Clastic cycles are of two basic types, those indicating an increase in transport energy upward, and those demonstrating a decrease. Both types can be caused by a variety of sedimentary, climatic and tectonic mechanisms. Beerbower (1964) divided these into *autocyclic* and *allocyclic* controls. Autocyclic mechanisms are those which result in the natural redistribution of energy within a depositional system. Examples include the meandering of a channel in a river, tidal creek or submarine fan, subaerial flood events, subaqueous sediment gravity flows, channel switching on a subaerial or submarine fan or a delta (avulsion), storms and tidal ebb and flood. All of these potentially can produce cyclic sequences. Allocyclic mechanisms are those in which change in the sedimentary system is generated by some external cause. Tectonic control of basin subsidence, sediment supply and paleoslope tilt, eustatic sea level change and climate change are the principal types of allocyclic mechanism. These are large scale basinal sedimentary controls and are dealt with mainly in Chapters 6 and 8. A sedimentary basin may be affected by several of these processes at the same time, so that it is not uncommon to find that there are two or three scales of cyclicity nested in a vertical profile. Allocyclic cycles tend to be thicker and more widespread in their distribution than autocyclic cycles. The latter generally are formed only within the confines of the subenvironment affected by the particular autocyclic process. This assists the geologist in distinguishing and interpreting sedimentary cycles, but such interpretations may still be far from easy, and it is recommended that environmental interpretation not end with a discussion of cyclic sedimentation as is so often the case in both surface and subsurface basin research. Miall (1980) discussed these problems with reference to fluvial deposits.

Coarsening- and thickening-upward cycles are the most varied in their origins. Several distinct types are produced by coastal regression and progradation (lateral accretion), where there is a gradation from low energy environments offshore to higher energy in the shoaling wave and intertidal zones. Examples are illustrated in Figure 4.46. Other types are formed where there is a steep slope and abundant sediment supply, and the flow system which develops over it attempts to grade itself to a balance by "filling in" the basin margin with a wedge of sediment. Coarsening-upward cycles are formed under these circumstances by prograding submarine fans and alluvial fans, particularly where the relief is maintained or even increased by active tectonic uplift (Fig 4.46). Other examples of coarsening-upward cycles (not illustrated) are those produced by crevasse splays in fluvial and deltaic settings, washover fans building from a barrier landward into a lagoon and fluvioglacial sequences formed in front of an advancing continental ice sheet.

Fining- and thinning-upward cycles commonly occur in fluvial environments as a result of lateral channel migration (point bar sequence) or vertical channel aggradation. Alluvial fans may also show fining-upward cycles where they form under conditions of tectonic stability. These three types are shown in Figure 4.47. Other illustrated examples are the tidal creek point bar and intertidal beach progradation sequences. Sediments deposited by catastrophic runoff events, including fluvial flash floods and debris flows, and many types of subaqueous sediment gravity flow also show a fining-upward character. Elliott (1978) discussed fining-upward sequences generated on a transgressing coastline.

Many of these cycles are superficially similar, and it may require careful facies and paleocurrent studies to distinguish them. Information on lateral variability may be crucial but, of course, this usually is unobtainable in the case of subsurface studies.

For carbonate environments less emphasis has traditionally been placed on sequence or profile, and more on the grain type, fauna and structures of individual beds. Assemblages of such attributes commonly are environmentally diagnostic, whereas in the case of siliciclastic sediments much ambiguity may be attached to their interpretation, and such additional features as vertical profile and lateral facies relationships assume a greater importance. The range of environments in which carbonates are formed is much narrower than that of siliciclastics; they are confined mainly to shallow continental shelves, platforms or banks and adjacent shorelines and continental

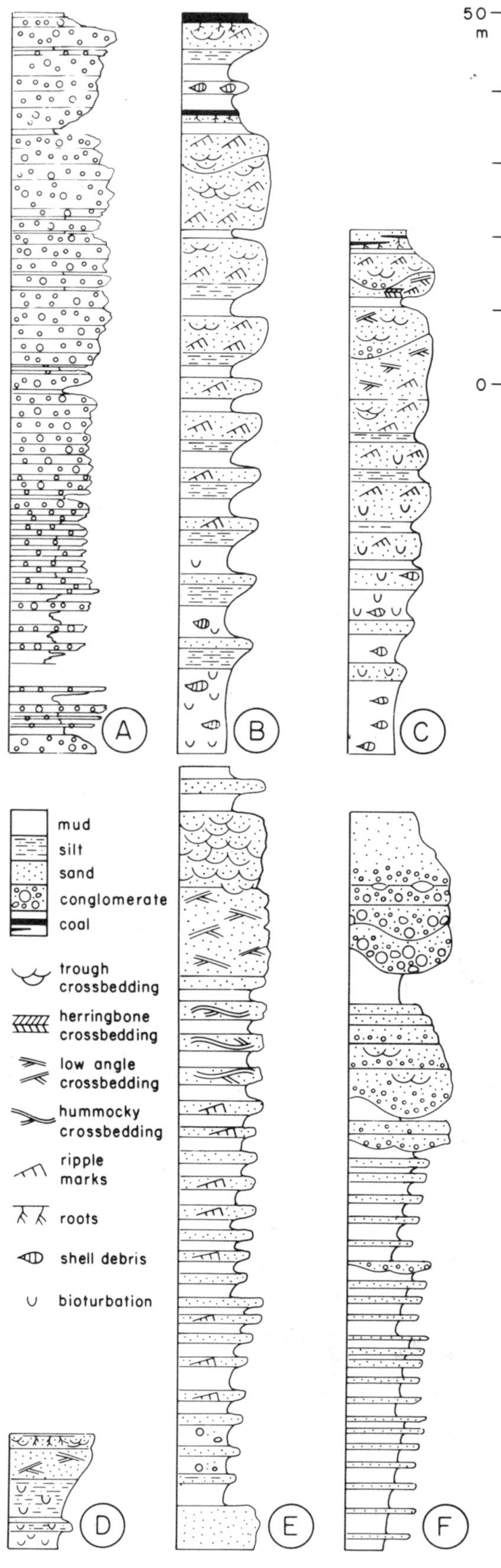

margin environments. Yet enormous variability is apparent in these various settings, particularly in shallow-water and coastal regions, and this is another reason why standard vertical profile models have not become as popular as they have with clastic sedimentologists.

James (1979b) and Ginsburg (1975) discussed shoaling-upward sequences formed in shallow subtidal to supratidal settings. These are common in the ancient record, reflecting the fact that carbonate sedimentation is generally much greater than the rate of subsidence. Shallowing-up sequences therefore repeatedly build up to sea level and prograde seaward. Lateral shifts in the various subenvironments are common. James (1979b) offered four generalized sequences as models of vertical profiles which could develop under different climatic and energy conditions (Fig. 4.48). Ginsburg and Hardie (1975) and Ginsburg et al. (1977) developed an exposure index, representing the percent of the year an environmental zone is exposed by low tides. By studying tide gauges and careful surveying of part of the modern Andros Island tidal flat they were able to demonstrate that a variety of physical and organic sedimentary structures is each present over a surprisingly narrow tidal exposure zone. This idea has considerable potential for interpreting shoaling-upward sequences (Fig. 4.49), as demonstrated by Smosna and Warshauer (1981). Lofer cycles of the Alpine Triassic developed under conditions of fluctuating water level. Most sedimentation occurs during progradation, and these are therefore unusual in being deepening-upward cycles (Fischer, 1964; Wilson, 1975). The vertical profile is illustrated in Figure 4.50. Smith (1977) documented a shallow–deep–shallow cycle in the Mississippian of Montana, but it is not known how general the model is.

Carbonate buildups or reefs may contain an internal cyclicity that is the result of upward reef growth. James (1979b) suggested that the vertical profile may show an upward transition from an ini-

◁ **Fig. 4.46.** Typical examples of thickening- and coarsening-upward profiles. **A** prograding alluvial fan (Steel et al., 1977); **B** river-dominated delta (Miall, 1979b); **C** wave-dominated delta (Miall, 1979b); **D** Barrier island—the Galveston Island model (Davies et al., 1971); **E** prograding, storm-dominated shoreline (Hamblin and Walker, 1979); **F** submarine fan (Walker, 1979b).

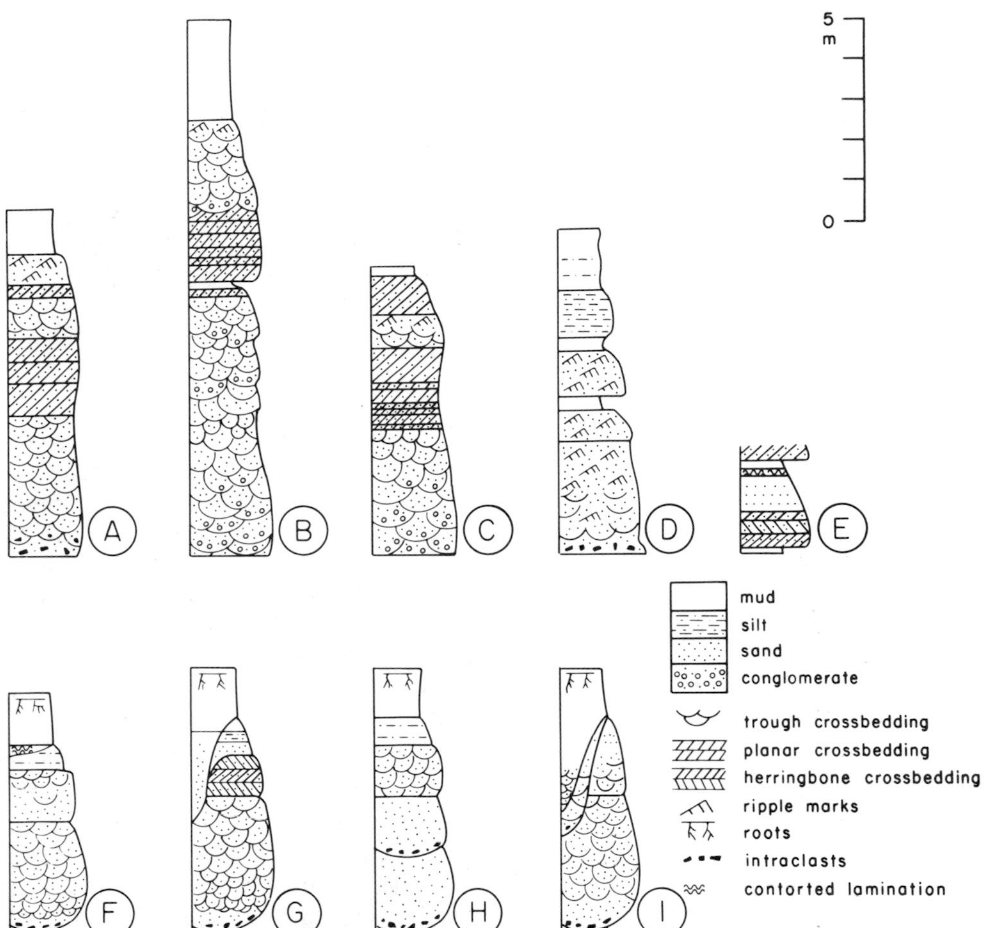

Fig. 4.47. Typical examples of thinning- and fining-upward profiles. **A** sandy braided river; **B**, **C** point bars in high sinuosity rivers; **D** degrading alluvial fan; **E** sandy tidal flat; **F-I** tidal creek point bars (**A-D** from Miall, 1980; **E** from Klein, 1971; **F-I** from Barwis, 1978).

tial "pioneer" or "stabilization" phase to "colonization", "diversification" and "domination" phases, characterized by distinctive textures and faunas. In practice, most ancient reefs are the product of numerous sedimentation episodes separated by diastems or disconformites, attesting to fluctuating water levels (e.g. Upper Devonian reefs of Alberta: Mountjoy, 1980). Analysis of vertical profiles of repeated cyclic patterns therefore may not be very helpful for basin analysis purposes, although such work may be useful for documenting small-scale patterns of reef growth (e.g. Wong and Oldershaw, 1980).

Deep-water carbonates comprise a variety of allochthonous, shelf-derived breccias and graded calcarenites, contourite calcarenites and hemipelagic mudstones, cut by numerous intraformational truncation (slide) surfaces (Cook and Enos, 1977; McIlreath and James, 1979). The lithofacies assemblages are distinctive, but variations in slope topography and the random occurrence of sediment gravity flows seem to preclude the development of any typical vertical profile. Organic stabilization and submarine cementation of carbonate particles probably prevent the development of carbonate submarine fans comparable to those formed by siliciclastic sediments, with their distinctive channel and lobe morphology and characteristic vertical profile.

Cyclic sequences are common in evaporite-bearing sediments, reflecting a sensitive response of evaporite environments to climate change, brine level or water chemistry. Vertical profile models are therefore of considerable use in en-

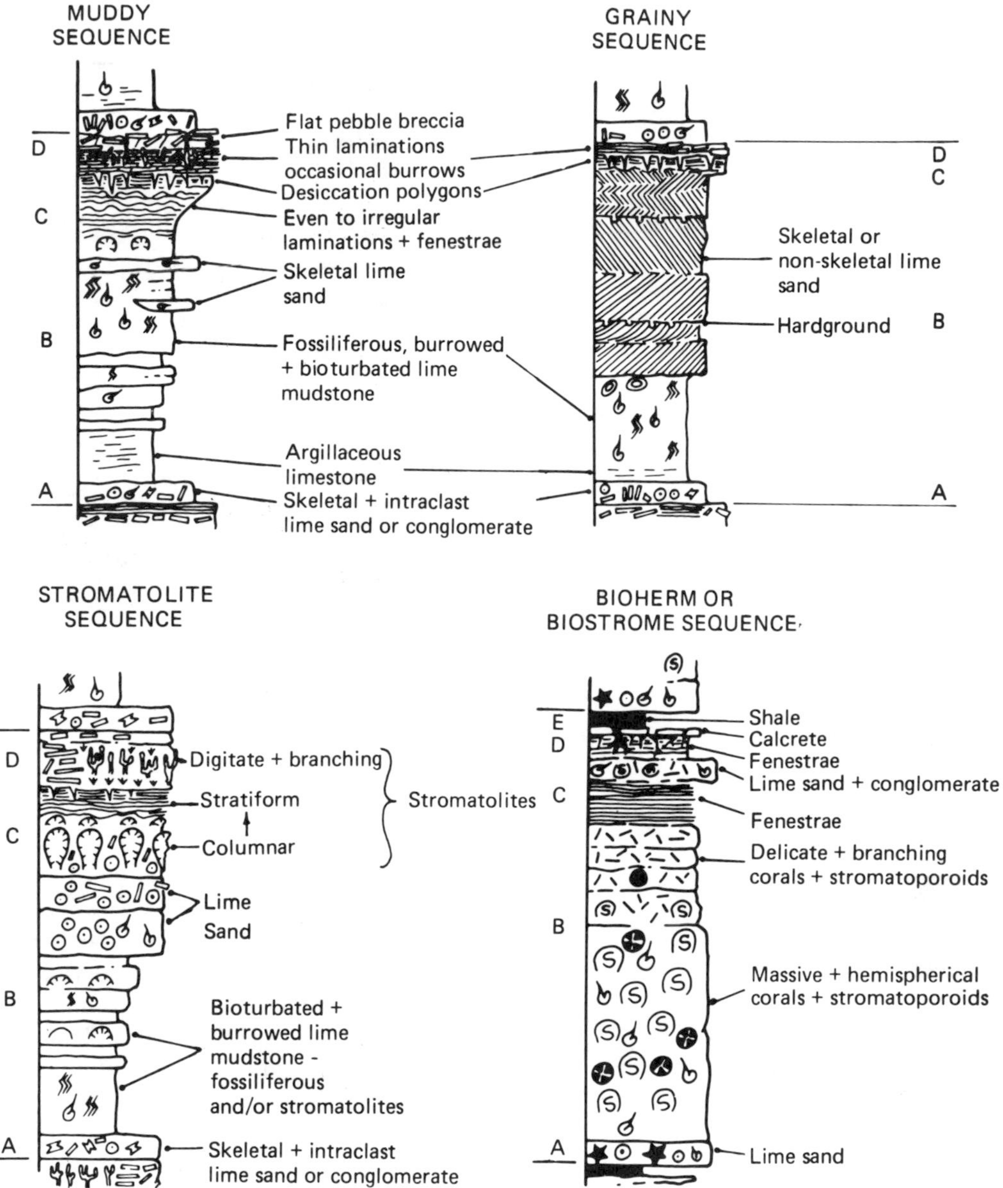

Fig. 4.48. Examples of shoaling-upward carbonate sequences formed in shallow subtidal to intertidal environments (James, 1979b).

vironmental interpretation. One of the most well-known of these is the coastal sabkha, based on studies of modern arid intertidal to supratidal flats on the south coast of the Persian Gulf (Shearman, 1966; Kinsman, 1969; Kendall, 1979a). Coastal progradation and growth of displacive nodular anhydrite results in a distinctive vertical profile that has been widely applied (indeed, over-applied) to ancient evaporite-bearing rocks (Fig. 4.51). Kendall (1979a) discussed variations in this profile model, reflecting differences in climate and water chemistry that arise in other coastal and playa lake margin settings.

As noted elsewhere, evaporites can occur in a variety of other lacustrine and hypersaline marine settings. They mimic many kinds of shallow to deep marine carbonate and siliciclastic facies, and a range of sedimentary criteria is required to demonstrate origin. Vertical profile is only one of these, but may be useful particularly when ex-

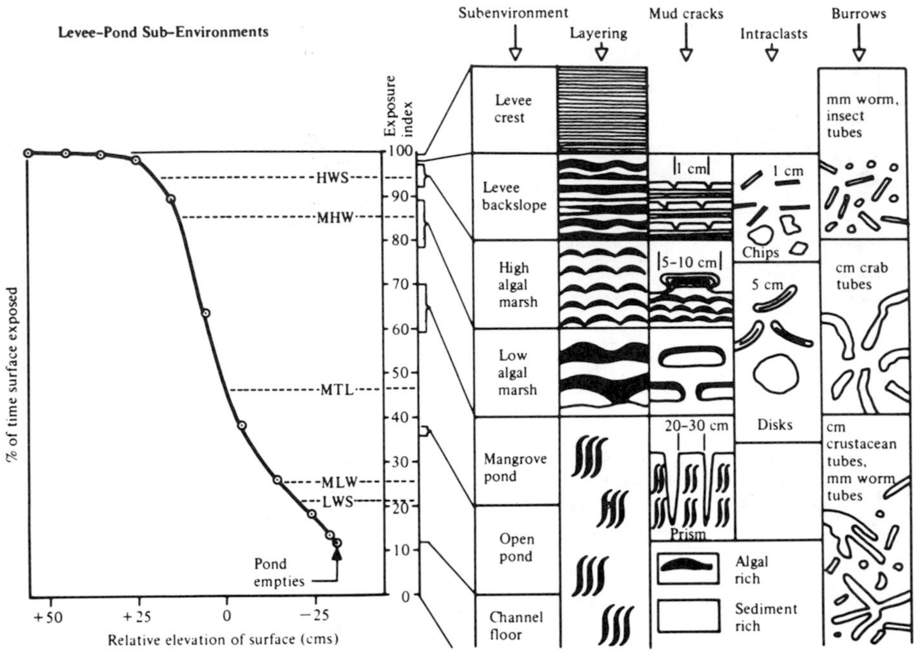

Fig. 4.49. Zonal distribution of sedimentary structures on a tidal flat and their relationship to the exposure index. Symbols are self-explanatory except that sigmoidal ornament at base of layering column indicates bioturbation (Ginsburg and Hardie, 1975).

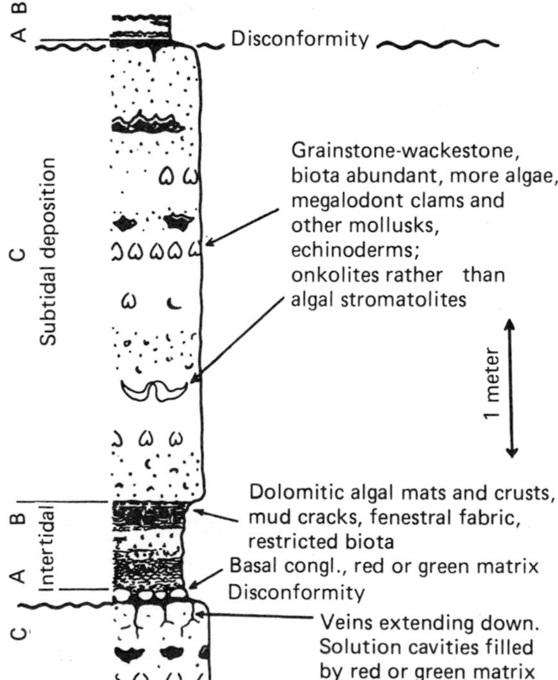

Fig. 4.50. A typical Lofer cycle. C member typically contains cavities produced by desiccation and solution during succeeding drop in sea level (Fischer, 1964; J.L. Wilson, 1975).

amining subsurface deposits in core. For example sulfates which accumulate below wave base commonly display a millimeter scale lamination interbedded with carbonate and/or organic matter, and possibly including evaporitic sediment gravity flow deposits (Kendall, 1979b). The latter may even display Bouma sequences (Schreiber et al., 1976). Shoaling-upward intertidal to supratidal cycles have been described by Schreiber et al. (1976) and Vai and Ricci-Lucchi (1977) in Messinian (Upper Miocene) evaporites of the Mediterranean basin (Fig. 4.51). Caution is necessary in interpreting these cycles because they may not indicate a build-up or progradation under stable water levels, but may be the product of brine evaporation and falling water levels. Many cycles of recharge and evaporation have been proposed for major evaporite basins such as the Mediterranean (Hsü et al., 1973).

Lacustrine environments in general are characterized by a wide variety of vertical profiles, reflecting many cyclic processes involving changes in water level and water chemistry. Many of these contain a chemical sediment component. Van Houten (1964) described a shoaling-upward,

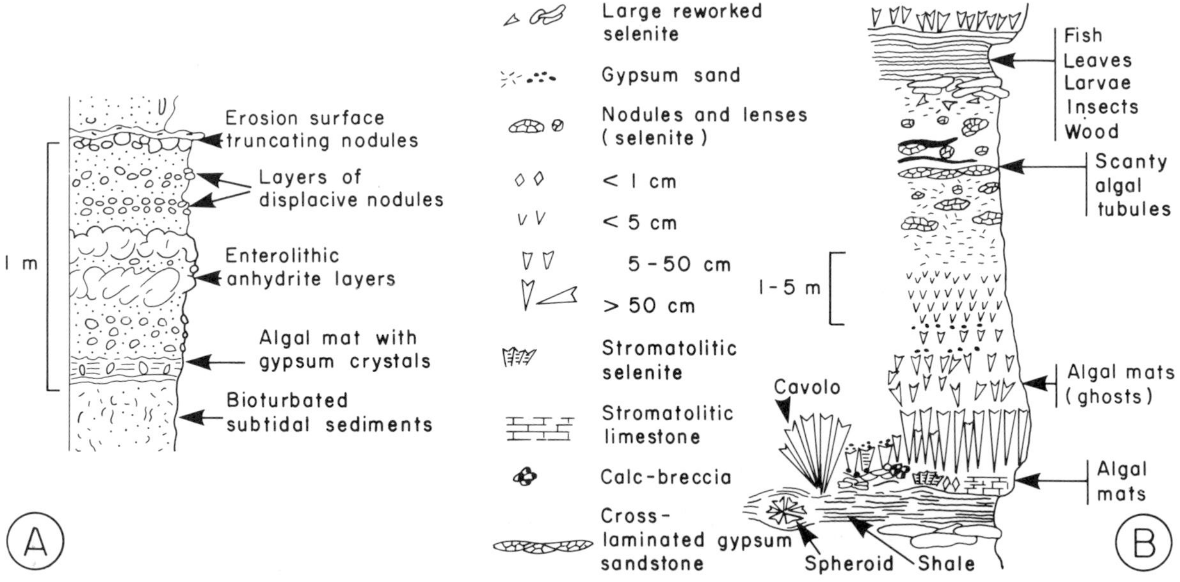

Fig. 4.51. Two typical evaporite cycles. **A** coastal sabkha (Shearman, 1966; Kendall, 1979a); **B** Subtidal to intertidal cycle formed by desiccation (Vai and Ricci-Lucchi, 1977).

coarsening-upward type of cycle in the Lockatong Formation (Triassic) of New Jersey. The cycles are about 5 m thick and consist, in upward order, of black, pyritic mudstones, laminated dolomitic mudstone and massive dolomitic mudstone with desiccation cracks and bioturbation. Chemical cycles are also present, which have an upper member of analcime-rich mudstone. The cycles are interpreted as the product of short-term climate change, with differences between the two types of cycle reflecting a greater tendency toward humidity or aridity, respectively. Eugster and Hardie (1975) described transgressive–regressive playa margin cycles in the oil shale-bearing Green River Formation of the Rocky Mountains. Donovan (1975) erected five profile models for cycles which occur as a result of changes in lake level and fluvial–deltaic lake margin progradation in the Devonian–Orcadian Basin of northern Scotland. Numerous other examples could be quoted.

Much has been written on the recognition of cyclic sedimentation from geophysical logs (Fisher et al., 1969; Skipper, 1976; Pirson, 1977). At present the technique is best suited to the study of clastic cycles. As explained in section 2.4.5 the gamma ray, spontaneous potential and resistivity logs are sensitive indicators of sand–mud variations, and so are ideally suited to the identification of fining- and coarsening-upward cycles. These appear as "bell-shaped" and "funnel-shaped" log curves, respectively, and various subtleties of environmental change may be detected by observing the convexity or concavity of the curves, smooth versus serrated curves, presence of nested cycles of different thicknesses, and so on. Fisher et al. (1969) published a series of typical profiles for coastal and marginal marine clastic environments based on examples from the Gulf Coast (Fig. 4.52). Other examples from the Beaufort–Mackenzie Basin (from Young et al., 1976) are illustrated in Figure 4.53. These curves are commonly interpreted in the absence of core or cuttings. As should be apparent from the preceding pages similar cycles may be produced in different environments, and so this is a risky procedure. However, by paying close attention to appropriate facies models and scale considerations (cycle thickness, well spacing) good paleogeographic reconstructions can be attempted. The availability of core in a few crucial holes may make all the difference. Figure 4.54 illustrates a fluvial fining-upward cycle with the typical bell-shaped log profile. The log has been tied to a core which confirms the expected vertical changes in grain size and sedimentary structures. Other examples of typical core and log profiles are given by Shawa (1974).

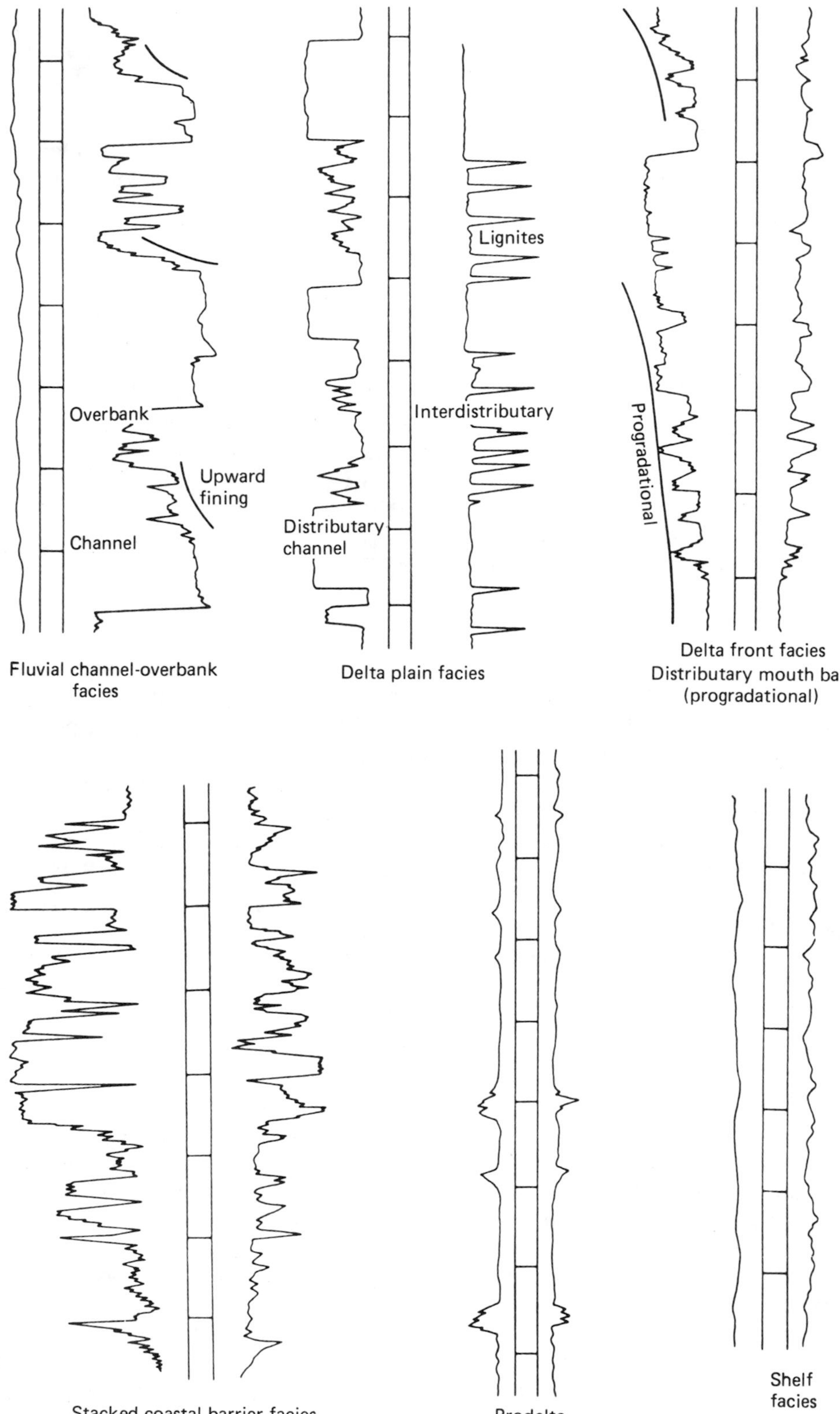

Fig. 4.52. Examples of typical geophysical log profiles through coastal plain and shelf clastic sequences, based on examples from the Gulf Coast. Left log, gamma ray or S.P., right log, resistivity. Center bar shows scale subdivisions of 30 m (100 ft) (Fisher et al., 1969).

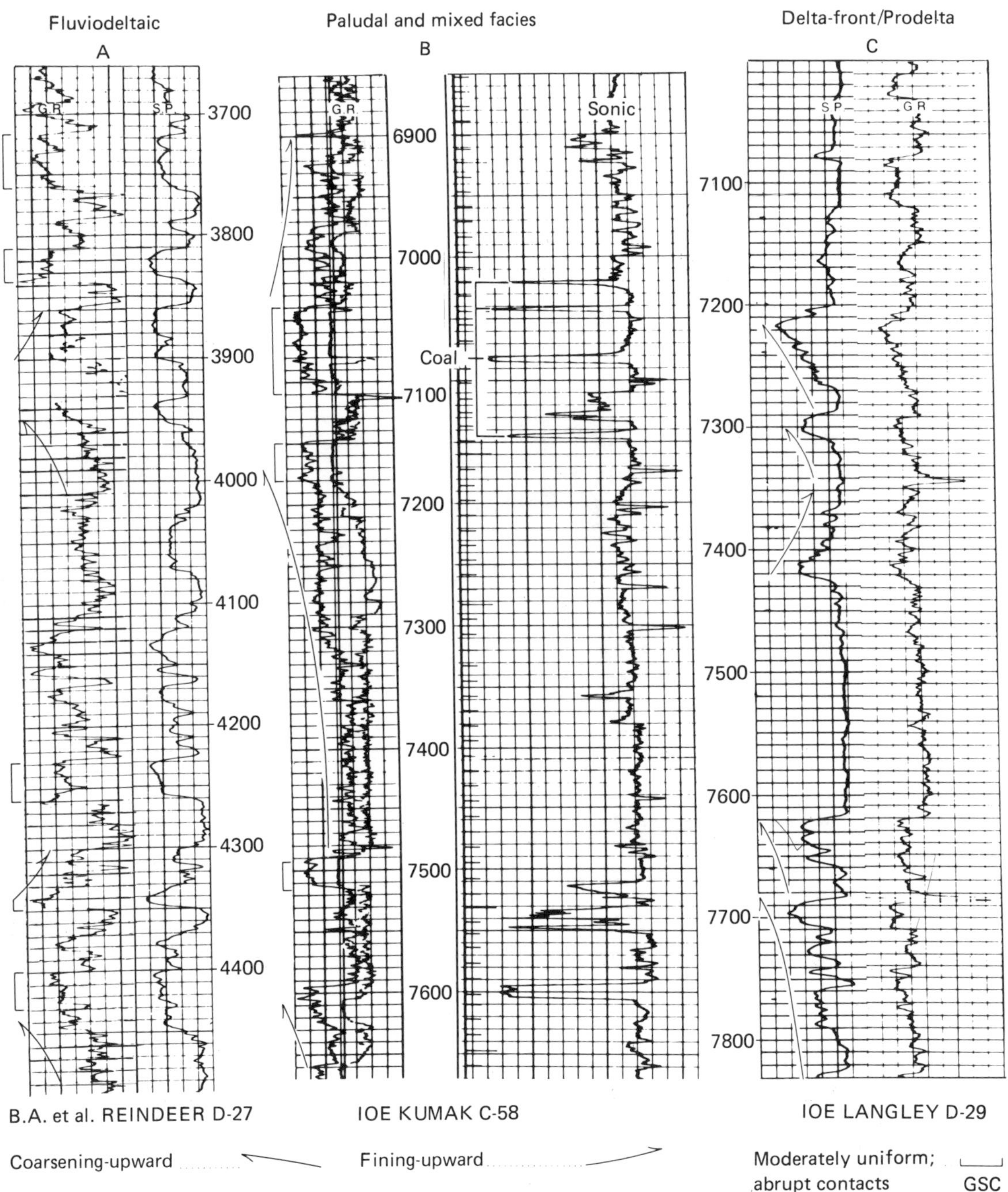

Fig. 4.53. Some examples of actual log profiles from the Beaufort–Mackenzie Basin (Young et al., 1976).

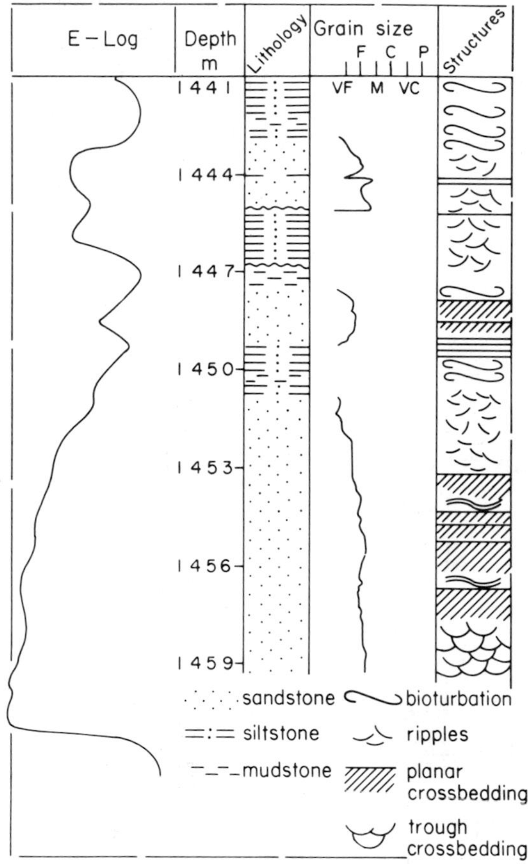

Fig. 4.54. Core and log character of a fluvial fining-upward cycle (Lerand and Oliver, in Shawa, 1974).

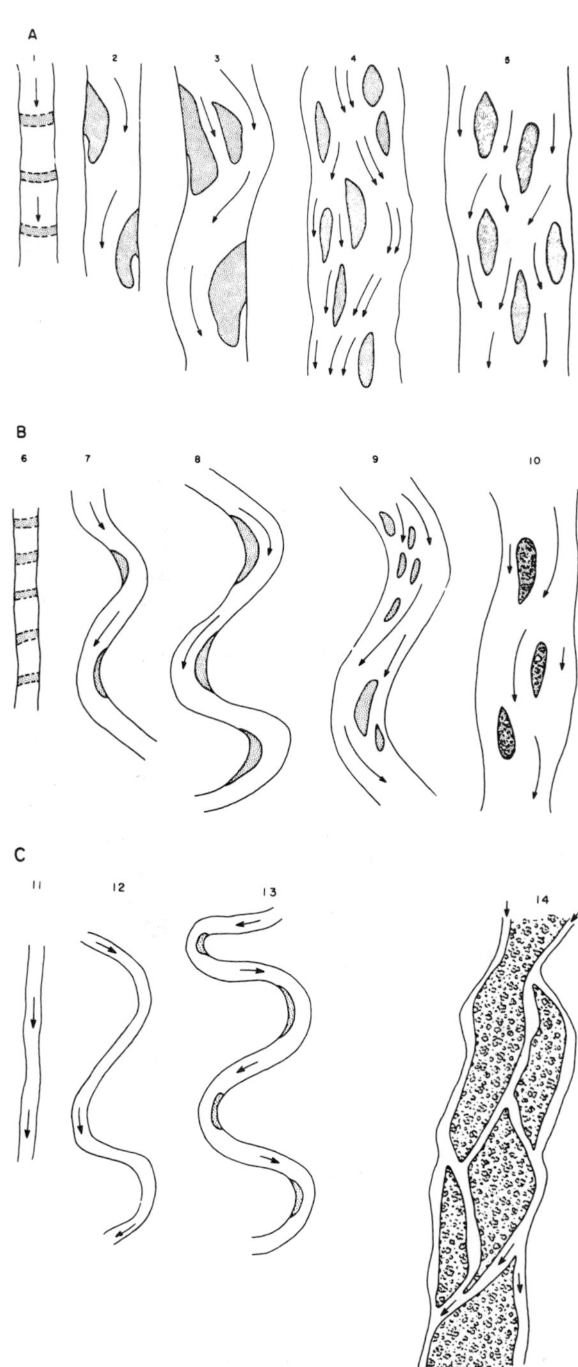

Fig. 4.55. River channel patterns. **A** bedload rivers; **B** mixed-load rivers; **C** suspended-load rivers (Schumm, 1981).

4.6 Facies models in the nineteen eighties

Many excellent textbooks and review articles deal with modern facies models, including Reading (1978), Walker (1979a), and chapters in Blatt et al. (1980), Reineck and Singh (1980), Friedman and Sanders (1978) and Leeder (1982). It would be superfluous to attempt to improve on this wealth of source material, nor is there space for it here. However, it would seem useful to review some of the current developments as a guide to research in the area for advanced students. As noted in sections 4.4.3 and 4.4.4 the state of the art is a very varied one; our ideas on some environments are undergoing active development, and there is a danger in offering a review such as this that it will be out-of-date before it appears in print.

One of the problems with current facies model research is the tendency for workers to erect a new facies model for every new study of an ancient rock unit. Such models may provide excellent tools for local paleogeographic interpretation but their general applicability commonly is questionable. There is a great need for review and synthesis at this time, rather than a proliferation of new models.

The notes below are not intended as a primer for the beginner. They assume a familiarity with facies models at least to the level of Blatt et al. (1980), Walker (1979a), or Leeder (1982).

4.6.1 Fluvial environments

Models of fluvial sedimentation represent an attempt to combine geomorphological observations of modern rivers with interpretations of ancient sequences.

Our understanding of modern fluvial processes has improved greatly in recent years. Schumm's (1963) classification of rivers into suspended-load, mixed-load and bedload rivers with corresponding variations in channel type, together with Leopold and Wolman's (1957) description of braided, meandering and straight channels were the foundations of this work. Recently Rust (1978a) attempted a quantitative classification of alluvial channels which recognizes the overlap in some of the descriptive categories. Rust proposed to use channel sinuosity and a braiding parameter to define four principal river types, braided, meandering, straight and anastomosed. Galay et al. (1973), Mollard (1973) and Schumm (1981) illustrated the gradational variations between these end members. Schumm (1981) illustrated fourteen channel types (Fig. 4.55) and discussed the delicate geomorphic balances that could permit rivers to metamorphose from one type to another.

However, the geomorphic picture is even more complicated than this. Baker (1978) showed that many of our concepts of river behavior are derived from temperate or cold regions and do not apply to tropical humid regions covered by dense rain forests, such as the Amazon Basin. Smith and Smith (1980) discussed how base level control can cause vertical aggradation, channel stability and the development of anastomosed systems. Long (1978) and Cotter (1978) demonstrated the validity of an earlier suggestion by Schumm (1968) to the effect that vegetation has such an important effect on stabilizing channel patterns that, before the evolution of extensive land vegetation in the Devonian, fluvial styles may have been entirely different. Miall (1977), Friend (1978) and Tunbridge (1981) emphasized the importance and distinctiveness of the deposits of unchannelized flash floods. Friend (1978) and Parkash et al. (1983) discussed terminal fans, which are inland, ephemeral distributary systems that deposit their sediment load as the runoff infiltrates into the channel bed. Schumm (1981) discussed the importance of geomorphic thresholds in controlling discontinuous erosional and depositional behavior.

The translation of this geomorphic complexity into models for interpreting the ancient record has not yet met with complete success. Until recently sedimentologists have focused on the diagnostic potential of lithofacies assemblages and vertical profiles (Cant and Walker, 1976; Miall, 1977, 1978; Jackson, 1978; Rust, 1978b). However, similar profiles can arise from different geomorphic and tectonic processes, resulting in the superimposition (nesting) of several cycle types within each other (Miall, 1980; section 4.5.8). Undoubtedly lateral control is the way to tackle this problem. Careful tracing of bedding units and entire channel fill deposits can reveal within-channel depositional patterns, channel geometry and shifting behavior. A growing number of workers now construct and publish detailed drawings or interpreted photomosaics of long cliff or roadside exposures. Friend (1983) offered a classification of lithosome geometry that recognizes the variables of channel behaviour and de-emphasizes the importance of vertical profile. There are two important components of such a classification, external shape and internal form. Many deposits grow by either lateral accretion, as on a point bar, or by vertical channel aggradation and progressive abandonment, or by a process combining both these styles.

In practice reliance on interpretation of vertical profiles will continue because of their importance in subsurface work. Their usefulness would be enhanced by a close drilling network (less than about 200–300 m) and the availability of continuous core. Improvements in stratigraphic correlation by using magnetic reversals, graphic–biostratigraphic correlation or some of the other methods described in Chapter 3 would also be of considerable use in facies modeling (Miall, 1983b).

4.6.2 Deltaic environments

Until a few years ago delta studies were dominated by interpretations based on comparisons with the modern Mississippi delta. However, extensive comparative studies of deltas on a worldwide basis led Coleman and Wright (1975) to recognize a much wider range of variability in deltaic form and process. The Mississippi is a particular type of delta in which the receiving basin (Gulf of Mexico) is characterized by a low tide range and low wave energy. Where either of these marine energy sources becomes dominant delta morphology and sedimentology are very different. A classification of deltas which is now widely accepted is that by Galloway (1975) into wave-, tide-, and river-dominated types (Fig. 4.56).

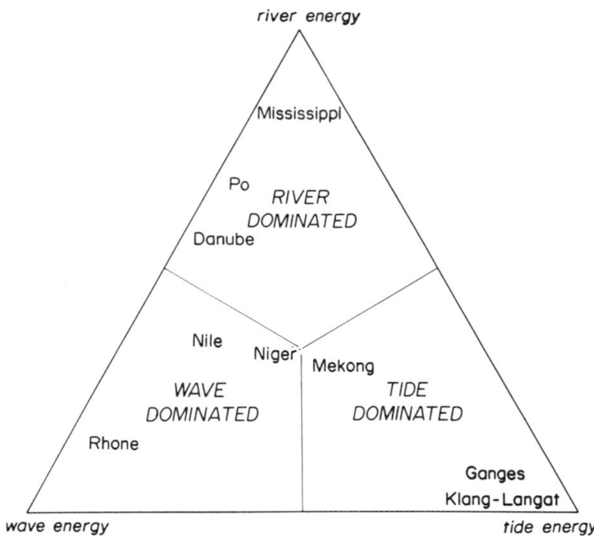

Fig. 4.56. Delta classification (Galloway, 1975).

Little important new work has been done on deltas since the appearance of these two papers, although considerable attention has been paid to various components of deltas that can occur in non-deltaic settings: river channels and flood plains, and wave-dominated environments. Bedforms in a large ancient delta distributary have been described by Jones and McCabe (1980).

Recent papers by Coleman et al. (1983) and Winker and Edwards (1983) have focused attention on the importance of subaqueous deformation structures on delta front and prodelta depositional surfaces. A variety of slump, slide and extensional fault structures are formed, particularly on river-dominated deltas, where rapid seaward progradation commonly leads to oversteepening and instability. Growth faulting may lead to significant local thickening of stratigraphic units and add considerably to the difficulties of subsurface correlation and mapping.

The classifications and processes described by Coleman and Wright (1975) and Galloway (1975) have been widely used in studies of ancient deltas, particularly studies of coal sedimentology in river-dominated deltas (e.g. Miall, 1979a; Horne et al., 1978; Flores, 1981). Publications describing ancient wave- and tide-dominated deltas are sparse. In fact only one or two ancient deposits interpreted as the product of a tidal delta are known to the writer (Eriksson, 1979; Hobday and Von Brunn, 1979). Heward (1981) suggested that many ancient wave-dominated deltas have been misinterpreted as barrier–lagoon or beach deposits through an over-application of the barrier island model. He argued that longshore or shelf sources of sand would be inadequate to form the thick, regressive marginal marine sands known in the ancient record and that the influence of rivers seems to be indicated. There is clearly scope for considerable research in the area of wave- and tide-dominated delta types.

4.6.3 Lacustrine environments

Lakes are amongst the most varied of all depositional environments, even though they occupy a relatively small percentage of the earth's surface at the present day (about 1%, according to Collinson, 1978). There are no universal facies models for the lacustrine environment, and recent books and review articles on this subject are not so much syntheses of ideas as catalogues of case studies, each example varying markedly from the next in terms of lithofacies, geochemistry, thickness and extent (Matter and Tucker, 1978; Collinson, 1978; Ryder, 1980; Picard and High, 1981). The documentation of this variability is perhaps the most important achievement of recent facies studies of lake sediments. This is by no means an academic subject. Vast petroleum reserves occur as oil shales in the Green River Formation of the Uinta Basin, and as liquid petroleum in China (e.g. Taching Oil Field), the western United States, West Africa and elsewhere (Ryder, 1980).

Most major lake deposits owe their origin to tectonic isolation of a sedimentary basin from the sea. The advent of plate tectonics has helped explain the origins of these basins, many of which are surprisingly broad and deep. Typical tectonic settings for lakes include rifts (Lake Baikal, East African Rift System), transform plate margins (Dead Sea, Cenozoic basins of California), remnant ocean basins (Black Sea?), some foreland basins (Cenozoic Uinta Basin) and intracratonic depressions (Lake Eyre). Tectonic control of sedimentation is discussed at length in Chapter 9.

Another major sediment control is water chemistry. This can vary widely, reflecting variations in the inflow/evaporation balance, nature and quantity of riverine dissolved sediment load, temperature, water level, and water body structure (presence of stratification, seasonal overturning, etc). Collinson (1978) provided a useful review of this topic. Hardie et al. (1978) discussed subenvironments and subfacies that exist in saline lakes and showed how lacustrine brines evolve during evaporative concentration (Fig. 4.57).

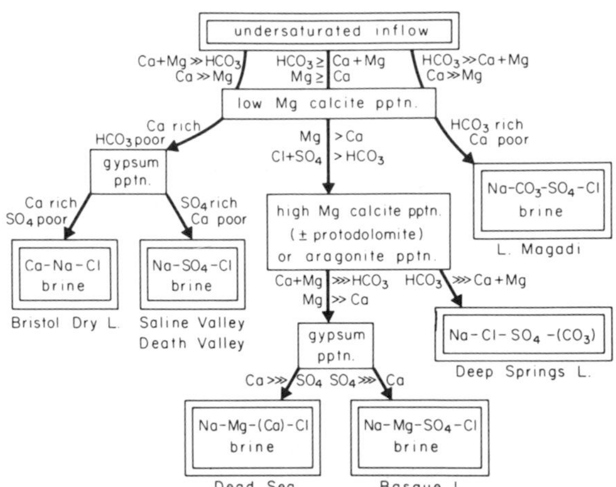

Fig. 4.57. Evolution of brines by concentration in saline lakes (Hardie et al., 1978).

Analysis of lake sediments may involve an examination of sediments and processes more commonly associated with marine or alluvial environments, including turbidites, deltas, evaporitic carbonate mudflats and fluvial sheet floods. These facies may be complexly interbedded within small stratigraphic thicknesses because of marked variations in water level resulting from tectonic or climatic fluctuations (e.g. Donovan, 1973, 1975; Link and Osborne, 1978; Smoot, 1978). Analysis of such deposits requires considerable sedimentological versatility.

4.6.4 Eolian environments

Eolian facies models were, until recently, one of the major areas of weakness in clastic sedimentology. This is perhaps because many studies of modern deposits (e.g. Bigarella, 1972; McKee, 1966) examined small dunes, which do not appear to have been typical of the ancient record. Much recent work has focused on documenting the external form of dunes including the use of satellite imagery of some of the large and innaccessible deserts of Africa and Asia (Cooke and Warren, 1973; McKee, 1979). This work has demonstrated a complexity of dune types that has so far not been recognized in the ancient record. Fossil dune deposits demonstrate a surprising similarity and relative simplicity of form, as demonstrated by the useful summary papers in McKee (1979). The reason is perhaps that stratigraphically significant desert sand accumulations only occur in major intracratonic sand seas such as the Sahara, where preservation depends on active subsidence and burial of the larger, more deeply based dune types (draas). These have not yet been intensively examined by sedimentologists.

A more successful trend in current research has been to re-examine the internal structure of modern and ancient dunes and to focus on the processes of sand accumulation. This research by Hunter, Brookfield, Kocurek and Gradzinski has now provided us with a wide range of diagnostic environmental criteria with which to recognize and interpret ancient dune deposits. Because it relates entirely to the study of crossbedding this work is discussed (briefly) in section 4.5.4.

4.6.5 Clastic shorelines

One of the first modern facies models was that for the Galveston Island barrier–lagoon system of the Texas coast (Bernard et al., 1959). This established a trend in sedimentological thought, in that the regressive barrier–lagoon model dominated ideas about sandy coastal environments for a number of years. It was soon realized that Galveston Island represents only one example of a spectrum of barrier types that can be formed under a wide range of oceanic conditions, in this case low tide range and low wave energy. Other papers followed (Davies et al., 1971; Dickinson et al., 1972) which amplified the basic model, that of a coarsening-upward profile with varying assemblages of fossil remains and sedimentary structures. Low-angle crossbedding, representing beach accretion surfaces, gained acceptance as one of the most useful diagnostic criteria.

Research into muddy tidal flat sediments has formed a separate theme, based on much classic work in Holland (Van Straaten, 1954; Reineck, 1972; Reineck and Wünderlich, 1968) and eastern England (Evans, 1965).

Recent advances in coastal studies have proceeded along five main lines: 1. studies of wave processes and wave-formed sedimentary structures using scuba, box coring and hydraulic monitoring; 2. recognition of the importance of tidal range in controlling coastal morphology and many important sedimentary processes; 3. recognition of the distinctiveness and relatively high preservation potential of tidal inlets and their associated deltas; 4. recognition of the importance of storms in affecting coastal and shallow subtidal sedimentation; 5. clarifying the similarities and differences between muddy fluvial systems and meandering tidal creeks. Practically all these achievements are based on work in modern environments, and much research remains to be done to apply the results to ancient sediments.

Wave-formed sediments on barred and nonbarred coasts have been examined by Clifton et al. (1971) and Hunter et al. (1979) on the Oregon coast and by Davidson-Arnott and Greenwood (1974, 1976) on the New Brunswick coast. In each case sedimentary structures were examined by box coring and related to the various subenvironments beneath shoaling and breaking waves. As described briefly in section 4.5.4, Clifton et al. (1971) were able to determine flow regime conditions for each zone of sedimentary structures, and all these papers examined lithofacies generation and preservation as a contribution toward the development of a facies model for wave-dominated shorelines. The distinctive internal structures of wave-formed ripples are described in section 4.5.4.

One of the most useful developments in the attempt to understand coastal processes was the discovery that the behavior of barrier islands, tidal flats, tidal inlets and tidal deltas is related in a predictable way to tide range. Hayes (1976) demonstrated that in microtidal regimes (tide range <2 m) waves are normally the most important sediment moving force, and long, unbroken barrier islands like Galveston Island are common. Mesotidal coasts (2–4 m tides) are characterized by shorter barrier islands cut by numerous tidal inlets, with ebb and flood deltas formed by the strong reversing currents passing in and out of the lagoon. On macrotidal coasts (>4 m) tidal energy inhibits the development of barrier islands and, instead, broad tidal flats and salt marshes are formed, cut by active tidal channels. Offshore linear sand ridges may be developed, oriented by the pattern of strong reversing flow and possibly oblique to shoreline trends. The assemblage of lithosomes and their internal structures and orientations will obviously be markedly different in each case. Blatt et al. (1980) offered a hypothetical vertical profile for a macrotidal tidal flat (Fig. 4.58), which should be compared with the microtidal barrier model illustrated in Figure 4.46D.

Tidal inlets may produce deposits with a greater preservation potential than the barrier which they cut through. This was the conclusion of Hoyt and Henry (1967) and Kumar and Sanders (1974) who showed that inlets migrate laterally where there are longshore currents, scouring away barrier sediments and replacing them with a laterally-accreted inlet–fill sequence containing high energy bimodal–bipolar crossbedding (Fig. 4.59). The composition of the barrier sequence that is finally preserved will depend on the balance between the rates of subsidence, regression and inlet migration. Tidal deltas at the lagoonal and seaward ends of the inlets also consist of distinctive sedimentary assemblages (Hayes, 1976). Flood deltas, formed in the lagoon, are dominated by flood-oriented crossbedding because ebb currents passing back over them are too weak to significantly modify the deposits. Conversely, ebb deltas at the seaward end of the lagoon are dominated by ebb-oriented structures formed during low tidal stages, and are

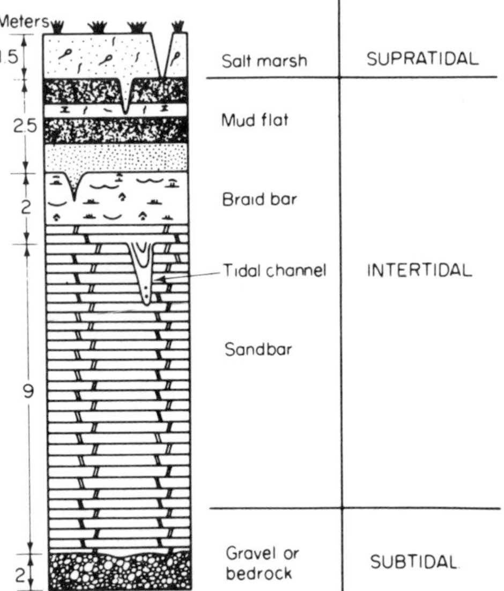

Fig. 4.58. Hypothetical profile of a prograding macrotidal coast. From H. Blatt, G.V. Middleton and R. Murray, Origin of Sedimentary Rocks, 2nd edition, © 1980, p. 658. Reprinted by permission of Prentice-Hall, Inc., Englewood Cliffs, New Jersey.

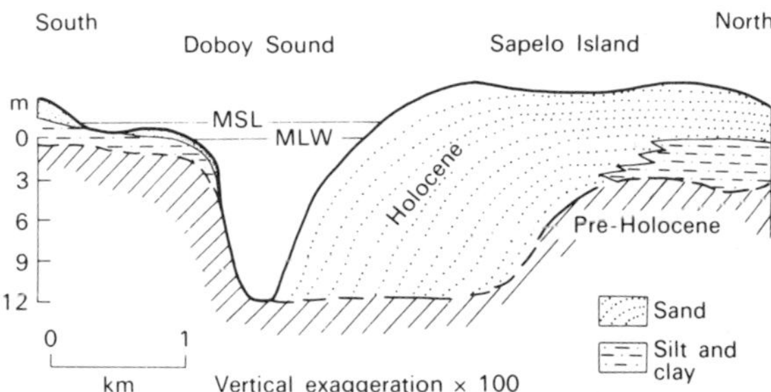

Fig. 4.59. Section parallel to shore through a migrating tidal inlet, showing accumulation of lateral accretion deposits (Hoyt and Henry, 1967; Reading, 1978).

covered by several meters of water when flood currents are at their peak, so that they too are little modified. Flood and ebb currents also tend to be segregated into different parts of the inlet–delta system. In each case the resulting deposits may yield strongly unimodal paleocurrent patterns on a local scale, which is unexpected for a marginal marine environment.

The influence of storms on marginal marine environments has been recognized for what in sedimentology is quite a long time (Hayes, 1967; Hobday and Reading, 1972; Kumar and Sanders, 1976; McGowen and Scott, 1976). Deep scour, development of washover fans, distinctive shoreface sequences and reopening of tidal inlets are amongst the more significant effects. Recent work by Harms et al. (1975), Walker (1979c) and Hamblin and Walker (1979) on shallow subtidal environments has added a new component to this pattern (see section 4.6.7).

Lastly there is the work in tidal creeks by Bridges and Leeder (1976) and Barwis (1978). These are usually high sinuosity channels which develop point bars quite similar to those in rivers. Reversing flow directions (though ebb dominant), abundant bioturbation, marine fauna and association with thick tidal flat muds are the main differences.

4.6.6 Arid shorelines and evaporites

One of the major breakthroughs in facies model research in the late sixties and early seventies was the recognition and acceptance of the sabkha model for the progradation of low-relief coastlines and tidal flats under hot, arid conditions, based on studies in the Persian Gulf (Kinsman, 1966, 1969; Shearman, 1966; Purser, 1973). The characteristic vertical profile is discussed briefly in section 4.5.8. Many ancient evaporites were reinterpreted as sabkha deposits but, as so often happens when geologists get a new idea into their heads, the model was applied with more enthusiasm than precision and many reappraisals have been necessary.

The most important recent development in the study of evaporites has been the recognition and resolution of the Mediterranean Messinian problem by Hsü et al. (1973) and Schreiber et al. (1976). The Messinian (Upper Miocene) evaporites contain abundant evidence of shallow water clastic deposition in the form of crossbedding, clastic textures and the presence of sabkha sequences (Fig. 4.60), yet Ryan (1976) was able to demonstrate that the evaporites were formed in a basin the floor of which lay at about 2.5 km below present-day sea level throughout Messinian time. The inescapable conclusion is that the Mediterranean was subjected to periods of extreme desiccation, the result of which may have been to raise sea levels throughout the rest of the world (see section 8.4.4). Kendall (1979b) concluded that

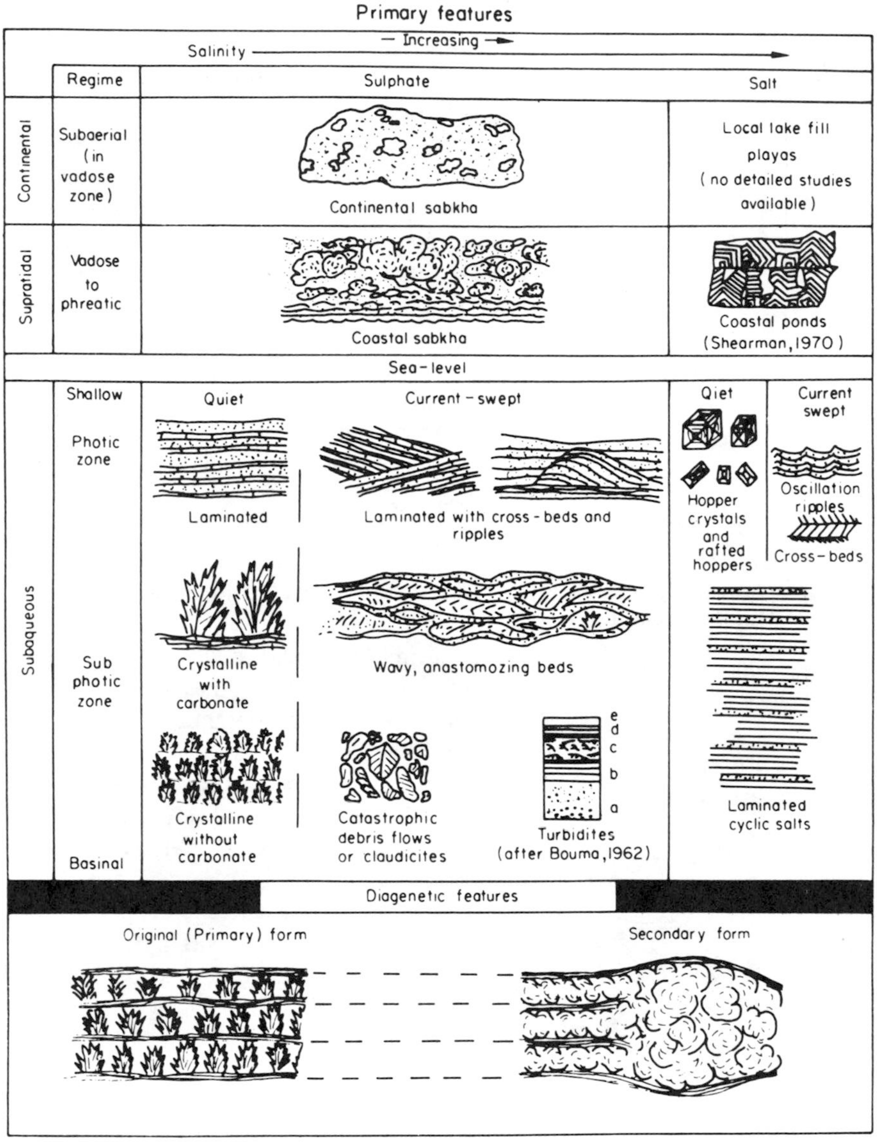

Fig. 4.60. Textures and structures of Messinian evaporites (Schreiber et al., 1976).

there are three principal ways in which evaporites can form, ranging from the shallow environments of the Messinian and the Middle Devonian Elk Point evaporites of western Canada to the deep basin (>1 km) of the Zechstein (Permian) of Germany (Fig. 4.61).

4.6.7 Clastic Continental shelves

Unlike most other clastic environments, modern continental shelves have not been well studied from a sedimentological perspective. A considerable literature deals with sediment movement patterns and surface bedforms in such areas as the Atlantic shelf off the United States and the North Sea, but there are difficulties in using this information for interpreting the ancient record, because of uncertainties as to general mechanisms and the details of internal structure and facies. Brenner (1980) subdivided shelves into 1. a nearshore zone extending to depths of about 20 m; 2. inner shelf, depths up to between 50 and 200 m; 3. outer shelf, a transition zone to the open ocean, with depths up to 500 m. The nearshore zone is characterized by intense wave activity, the influence of which can be recognized by a variety of distinctive sedimentary structures (Johnson, 1978; Raaf et al., 1977; Fig. 4.32 of this book).

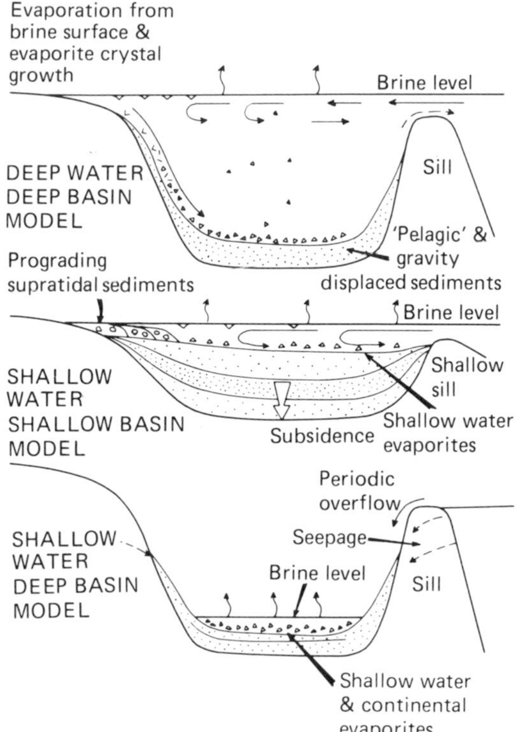

Fig. 4.61. Three models for the development of "basinal" evaporites (Kendall, 1979b).

The inner shelf may be dominated by tidal activity or storms, both of which generate a variety of large-scale bedforms, up to and including giant sand ridges that may be tens of kilometers long, several kilometers wide and several tens of meters high (Fig. 6.15). These are too big to map from facies analysis of individual sections, although they have been recognized by depositional systems analysis of some ancient shelf sea deposits (Figs. 6.16, 6.17). Brenner (1980) reported that they commonly show coarsening-upward cycles similar to those of barrier islands, but without any signs of emergence. As Walker and Middleton (1979) have noted, large tidal ridges are incorrect analogues for the large, shallow marine crossbed sets described by Johnson (1978, Figs. 9.30, 9.36), Nio (1976) and Jerzykiewicz (1968). The only modern shallow marine bedforms that appear to be good analogues are those described by Flemming (1978), which are formed by the Agulhas Current on the shelf of southeast South Africa (see section 6.5 for some additional discussion of shelf sand ridges).

The relative influence of tides and storms has been much debated in the literature. Klein (1977) and Klein and Ryer (1978) argued that tidal influences increase with shelf width, and that the large epeiric seas that covered the continents during times of high sea level (e.g. Ordovician–Silurian, Jurassic–Cretaceous) generated tide-dominated sedimentary deposits. They pointed out that the lack of really large shelf seas at the present day makes it difficult to model the sedimentology of these rocks, but provided tables of lithofacies types and sedimentary structures to facilitate generalized tidal interpretations.

The influence of storms in scouring and reworking shelf and marginal marine deposits has long been recognized, based on studies of hurricane activity on the Gulf and Atlantic coasts (McGowen and Scott, 1976; Hayes, 1967). Hayes (1967) demonstrated that in 1961 Hurricane Carla scoured the barrier island off Laguna Madre, and that sediment was redeposited as an extensive graded bed locally exceeding 9 cm in thickness. Both turbidity currents and wave-generated rip currents may be activated during storms. Walker (1979c) and Hamblin and Walker (1979) provided one example where turbidity currents seem to have been important, but Kreisa (1981) showed that entirely different transport processes may occur in different basins depending on bottom topography, storm frequency and storm track.

Harms et al. (1975), Walker (1979c), Hamblin and Walker (1979), Bourgeois (1980) and Dott and Bourgeois (1982) described hummocky cross-stratification (HCS), a distinctive structure (Fig. 4.34) that appears to be formed only by storm activity in shallow subtidal environments. Hamblin and Walker (1979) erected a facies model for Jurassic shelf deposits of western Alberta in which it was interpreted that storms also generated turbidity currents which resedimented shallow marine sands in the outer shelf. Their vertical profile model is illustrated in Figure 4.46. Dott and Bourgeois (1982) synthesized data on the occurrence of HCS in several shelf units of various ages. They showed that it commonly forms part of a sequence a few decimeters thick developed during waning storm activity. The main HCS zone may be followed by flat lamination, cross lamination, and then by the mudstones of fair weather sedimentation. Variations on this cycle reflect variations in water depth, wave height and sediment supply. Amalgamated partial sequences are common. Mount (1982) developed

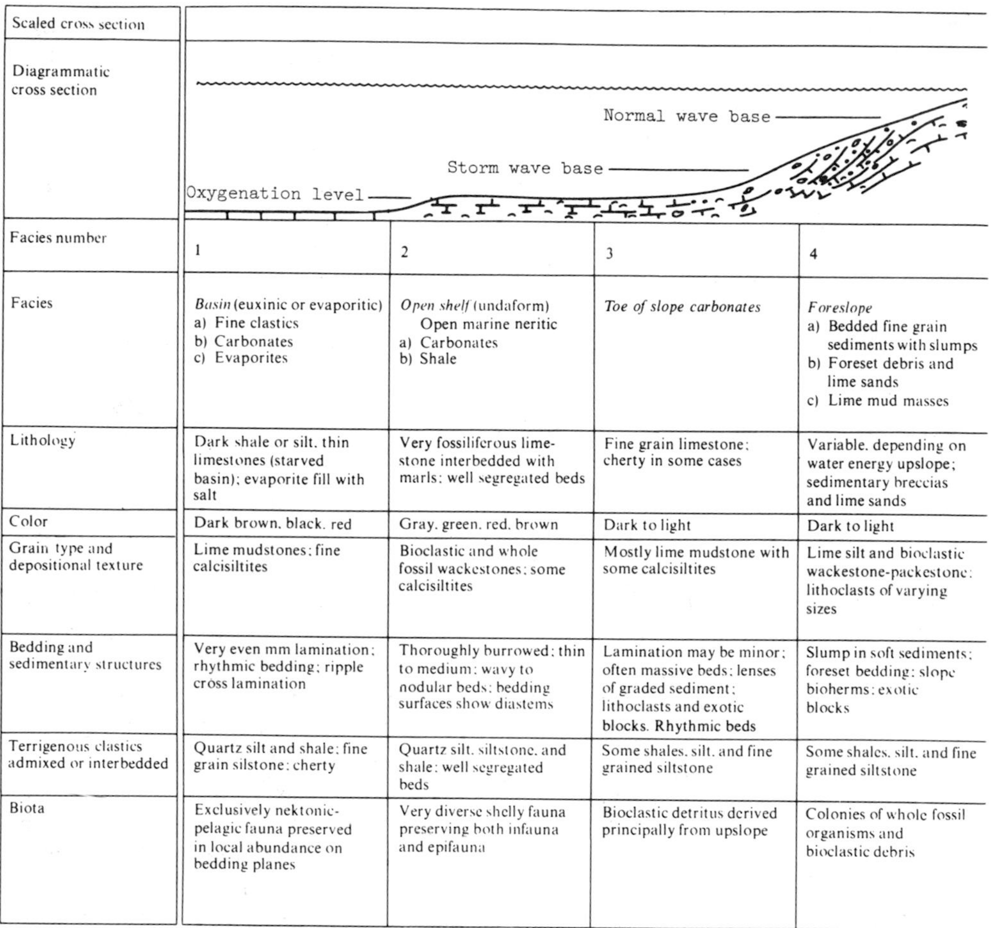

Scaled cross section				
Diagrammatic cross section				
Facies number	1	2	3	4
Facies	*Basin* (euxinic or evaporitic) a) Fine clastics b) Carbonates c) Evaporites	*Open shelf* (undaform) Open marine neritic a) Carbonates b) Shale	*Toe of slope carbonates*	*Foreslope* a) Bedded fine grain sediments with slumps b) Foreset debris and lime sands c) Lime mud masses
Lithology	Dark shale or silt, thin limestones (starved basin); evaporite fill with salt	Very fossiliferous limestone interbedded with marls; well segregated beds	Fine grain limestone; cherty in some cases	Variable, depending on water energy upslope; sedimentary breccias and lime sands
Color	Dark brown, black, red	Gray, green, red, brown	Dark to light	Dark to light
Grain type and depositional texture	Lime mudstones; fine calcisiltites	Bioclastic and whole fossil wackestones; some calcisiltites	Mostly lime mudstone with some calcisiltites	Lime silt and bioclastic wackestone-packestone; lithoclasts of varying sizes
Bedding and sedimentary structures	Very even mm lamination; rhythmic bedding; ripple cross lamination	Thoroughly burrowed; thin to medium; wavy to nodular beds; bedding surfaces show diastems	Lamination may be minor; often massive beds; lenses of graded sediment; lithoclasts and exotic blocks. Rhythmic beds	Slump in soft sediments; foreset bedding; slope bioherms; exotic blocks
Terrigenous clastics admixed or interbedded	Quartz silt and shale; fine grain silstone; cherty	Quartz silt, siltstone, and shale; well segregated beds	Some shales, silt, and fine grained siltstone	Some shales, silt, and fine grained siltstone
Biota	Exclusively nektonic-pelagic fauna preserved in local abundance on bedding planes	Very diverse shelly fauna preserving both infauna and epifauna	Bioclastic detritus derived principally from upslope	Colonies of whole fossil organisms and bioclastic debris

Fig. 4.62. The standard carbonate shelf-slope facies belts of J.L. Wilson (1975).

a similar model based on his independent work.

Other useful discussions of storm-generated sedimentary structures and sequences have been given by Bourgeois (1980) and Kreisa (1981); the latter paper includes a table of diagnostic criteria that have been used to interpret storm generated deposits in the ancient record.

4.6.8 Carbonate environments

In many respects carbonate facies studies can be pursued in a similar manner to siliciclastic studies, in that many carbonate deposits are accumulated by clastic processes controlled by waves, tides and gravity. However, the ubiquity of organic processes in carbonate seas introduces a significant complication, and many carbonate deposits are modified penecontemporaneously by submarine cementation and other types of early diagenesis. For example, reef and carbonate slope build-ups commonly have steeply-dipping sediment–water interfaces, much steeper than the angle of repose of loose talus, sometimes even vertical or overhanging. This is due to organic binding and early cementation.

The most important contribution to the study of carbonate facies models in recent years is the book by J.L. Wilson (1975) on Phanerozoic carbonates. Wilson demonstrated that carbonate sediments could be described using a standard set of twenty-four microfacies types (section 4.3.3). He showed that lithofacies, structures and biota defined nine standard facies assemblages, which could be used to describe practically all carbonate rocks in the ancient record (Fig. 4.62). Not all nine assemblages are present in every rock unit; for example the Mississippian of the Williston Basin consists of assemblages 1, 2, 6, 8 and 9 in lateral relationship from basin to platform

5	6	7	8	9
Organic (ecologic) reef a) Boundstone mass b) Crust on accumulation of organic debris and lime mud; bindstone c) Bafflestone	*Sands on edge of platform* a) Shoal lime sands b) Islands with dune sands	*Open platform* (normal marine, limited fauna) a) Lime sand bodies b) Wackestone-mudstone areas, bioherms c) Areas of clastics	*Restricted platforms* a) Bioclastic wackestone, lagoons and bays b) Litho-bioclastic sands in tidal channels c) Lime mud-tide flats d) Fine clastic units	*Platform evaporites* a) Nodular anhydrite and dolomite on salt flats b) Laminated evaporite in ponds
Massive limestone-dolomite	Calcarenitic-oolitic lime sand or dolomite	Variable carbonates and clastics	Generally dolomite and dolomitic limestone	Irregularly laminated dolomite and anhydrite, may grade to red beds
Light	Light	Dark to light	Light	Red, yellow, brown
Boundstones and pockets of grainstone; packstone	Grainstones well sorted; rounded	Great variety of textures; grainstone to mudstone	Clotted, pelleted mudstone and grainstone; laminated mudstone; coarse litho-clastic wackestone in channels	
Massive organic structure or open framework with roofed cavities; Lamination contrary to gravity	Medium to large scale crossbedding; festoons common	Burrowing traces very prominent	Birdseye, stromatolites, mm lamination, graded bedding, dolomite crusts on flats. Cross-bedded sand in channels	Anhydrite after gypsum; nodular, rosettes, chickenwire, and blades; irregular lamination; carbonate caliche
None	Only some quartz sand admixed	Clastics and carbonates in well segregated beds	Clastics and carbonates in well segregated beds	Windblown, land derived admixtures; clastics may be important units
Major frame building colonies with ramose forms in pockets; in situ communities dwelling in certain niches	Worn and abraided coquinas of forms living at or on slope; few indigenous organisms	Open marine fauna lacking (e. g. echinoderms, cephalopods, brachiopods); mollusca, sponges, forams, algae abundant; patch reefs present	Very limited fauna, mainly gastropods, algae, certain foraminifera (e. g. miliolids) and ostracods	Almost no indigenous fauna, except for stromatolitic algae

(Wilson, 1975, p. 352). Wilson's models are widely used in interpreting the ancient record.

Shoaling-upward cycles are extremely common in platform carbonate deposits (J.L. Wilson, 1975; James, 1979b); much research has been carried out to document their vertical profile (section 4.5.8) and interpret their textures and structures in terms of shallow subtidal to intertidal environments and processes. At present there is active discussion about the controls on this cyclicity, which revolves around the rates of carbonate sedimentation and the importance of sea level change (Wilkinson, 1982). The recognition of the frequency of eustatic sea level change during at least the Phanerozoic (Chap. 8) has added considerable fuel to the debate. Shoaling-upward cycles may be caused by tidal flat progradation and lateral shifting of subenvironments under conditions of slow but steady subsidence. New cycles may be initiated by deep storm scour or by transgression during a rise in sea level. Wilkinson (1982) suggested that there is an optimum depth for carbonate production (shallow but not exposed) and that a slight increase in the local rate of relative sea level rise may result in outward and upward tidal flat building losing pace with submergence, so that transgression occurs and a new cycle of lateral growth is intiated.

Reefs continue to receive considerable attention from carbonate geologists, both because of their intrinsic ecological interest and because of their importance as petroleum reservoirs. Detailed studies of modern reefs (e.g., Adey et al., 1977; James and Ginsburg, 1980; Shinn et al., 1979) provide much useful data for interpreting the ancient record (e.g. J.L. Wilson, 1975; James, 1979c; Mountjoy, 1980). Coring of Holocene reef build-ups and scuba diving examination of reef surfaces and blasted-open reef interiors have provided particularly useful insights.

Until recently it was thought that the morphology of modern reefs is controlled by the Pleistocene karst surfaces on which they have grown (Purdy, 1974; Sellwood, 1978). For example Purdy suggested that atolls formed because, during low sea level stands, more dissolution occurred in the center of an exposed platform than at the margins. The latter therefore became an elevated rim and subsequent reef growth maintained this relief. Adey (1978) disputed this interpretation. He showed that vertical reef accretion is far more rapid than formerly thought, up to 15 m/1000 a, but that actual rates are dependent on nutrient availability, wave energy and water clarity. These factors control the composition of the biota, some components of which grow faster than others. Reef shape throughout the later Cenozoic therefore reflects environmental variables, not erosional topography. This has important implications for interpreting the ancient record.

Recent studies of the Upper Devonian reefs of Alberta (Mountjoy, 1980; Stoakes, 1980) showed that their development was controlled by a series of eustatic sea level changes, such as are described in Chapter 8. The reefs consist of several or many partially developed or eroded shoaling-upward cycles, commonly separated by diastems. Basin fill sediments between the buildups consist either of fine detrital carbonates plus mudstone, as in Alberta, or evaporites as in, for example, the Silurian "pinnacle" reefs of the Michigan Basin (Mesolella et al., 1974). The reefs had a considerable relief above the basin floor, and the basin fill sediments banking against them therefore are considerably younger. The contact is that of an onlap rather than interfingering. This may be difficult to demonstrate from scattered wells, but is clearly shown on high-quality seismic records (Brown, 1981). The recognition of the cyclic nature of deposition has considerable consequences for the interpretation of major reef bodies. They probably were deposited during short intervals of rapid build-up, alternating with longer intervals of still-stand, submarine cementation and hardground development, or even exposure, karstification and vadose diagenesis (see also section 8.3.2). Walther's Law can only be used with extreme caution in interpreting facies relationships in such complex deposits.

Some of the most spectacular recent advances in carbonate facies studies have been made in deep water deposits, as a result of the DSDP project, development of submersibles and seismic methods, and the stimulus of finding major oil fields in deep water deposits, such as the Tamabra Limestone (Cretaceous) of the Poza Rica Trend, Mexico (Enos, 1977; Fig. 5.10 of this book). These studies demonstrate "that all deep carbonate bank margins are not simply areas of accumulation of 'fore-reef talus' and pelagic sediments" (Mullins and Neumann, 1979). The work of Mullins and Neumann (1979), Schlager and Chermak (1979), the many excellent papers in Cook and Enos (1977) and the summary article by McIlreath and James (1979) described a variety of distinctive facies types, topographic patterns and facies assemblages, reflecting tectonic control of the bank edge and the effects of waves and oceanic currents in dispersing the sediment (Fig. 6.31).

Bank margins vary from steep, possibly fault-controlled cliffs to gentle slopes extending for up to hundreds of kilometers. Spectacular talus slope breccias occur on the steeper slopes and are flanked by a variety of sediment gravity flow deposits, including debris flows, liquefied or fluidized flows and turbidites. These are interbedded with calcareous pelagic and hemipelagic sediments composed of planktonic organic detritus or reworked fine platform sediment. Deep-water bioherms may occur on some slopes. These deposits commonly are cut by spectacular intraformational truncation surfaces (beautifully illustrated by Davies, 1977), which record the development of major slides.

The facies assemblage on any particular margin may vary markedly along strike, depending on the bank topography, strength of waves and oceanic currents. McIlreath and James (1979) proposed four slope models, and a similar but more detailed study by Mullins and Neumann (1979), based on the Bahama Bank margin, revealed seven slope types (Fig. 6.31). Peri-slope talus breccias are coarser and thicker adjacent to cliff margins. Detritus may be carried into the basin by submarine canyons, but commonly there are a series of small gullies cutting the slope, fed from numerous sources on the bank rather than a single point source. These gullies channel the sand into a series of ephemeral, overlapping and coalescing calcarenite lobes rather than single large submarine fans. On windward margins sediment accumulation is more rapid than on leeward margins, where it may be swept away. Ocean currents may rework the debris into contourites or remove it altogether, producing an erosion surface.

Pelagic facies are first recognized in carbonate rocks of Upper Silurian age, but did not become volumetrically important until the evolution of planktonic foraminifera and coccoliths in the Jurassic. It has been estimated that in the last 100 Ma pelagic sediments have constituted 67% of world carbonate deposition (Hay et al., 1976). They were particularly important during the Late Cretaceous high sea level stand (Chap. 8), when vast areas of Europe and North America were covered by a detritus-free epeiric sea in which the distinctive chalk facies accumulated (Hancock and Scholle, 1975; Scholle, 1980).

4.6.9 Clastic continental slope and deep oceanic environments

The four most important advances in facies studies in this area are the improved understanding of the history and processes of submarine canyons, the development of ideas about sediment gravity flows, the evolution of the submarine fan model, and the recognition of atypical or "non-fan" settings for many sediment gravity flow deposits.

Shepard (1981) reviewed the origin of submarine canyons and demonstrated that they have multiple causes. Erosion by sediment gravity flows is a major process, but subaerial erosion is now known to have been important during periods of low sea level. Shepard (1981), McGregor (1981) and Brown (1981) demonstrated that canyons may be repeatedly cut and filled during oscillations in sea level. Such cycles may last for tens of millions of years and develop a local erosional relief up to 2000 m. Considerable stratigraphic complexities may result that could be difficult to resolve from outcrop or well data, but may be obvious on seismic records (section 6.7).

Giant submarine slides up to 450 m thick and several hundred kilometers in length have been identified by seismic profiling on some continental slopes (Dingle, 1980), and smaller features occur off major river mouths (Coleman and Garrison, 1977; Coleman et al., 1983). These are triggered by earthquake shock or deltaic oversteepening. Few comparable structures have been identified in the ancient record, except for smaller features, which are common on carbonate continental slopes.

Middleton and Hampton (1976) provided a classification of sediment gravity flows that did much to systematize observations on their deposits (Fig. 4.63). They achieved this by focusing on type of flow (laminar, turbulent) and grain support mechanisms, and showed that there is a broad spectrum of flow types from dense, viscous debris flows to low concentration turbidity currents. Lowe (1979) discussed some problems with this classification and proposed a revised nomenclature (Fig. 4.64). He pointed out that a flow may have been maintained by multiple support and transport mechanisms and that these could evolve downstream in each flow event.

The submarine fan model was first developed for ancient deep-water deposits by Mutti and Ricci-Lucchi (1972), an important paper based on well-exposed Cenozoic deposits in northern Italy. These authors attempted to incorporate available information on modern fans into the model, but at the time seismic resolution and core penetration were inadequate to provide very tight controls on interpreting the ancient record. To some extent this is still true, although instrumentation has improved considerably with the introduction of deep-towed geophysical packages. Most subsequent developments in our ideas about ancient submarine fans have been made by the Italian school (e.g. Mutti, 1974, 1977; Ghibaudo, 1980; Ricci-Lucchi, 1975), although useful syntheses and reviews have been written by Walker (e.g. 1978). His development of a fan facies model is an excellent example of the synthesizing and distillation process that is necessary to generate a facies model (Walker, 1979a, 1979b); the method is discussed in section 4.4.3 and his fan model is illustrated in Figure 4.65.

Much work has been carried out on modern submarine fans by marine geologists (e.g. Normark, 1970, 1978; Normark et al., 1979; Curray and Moore, 1971; Damuth and Kumar, 1975; Nelson, 1976). This work provides important constraints on submarine fan facies models but, because of the observational problems mentioned above, most of the detailed depositional information must be derived from ancient deposits.

Submarine fan models are still in an active state of development and it is apparent that much more

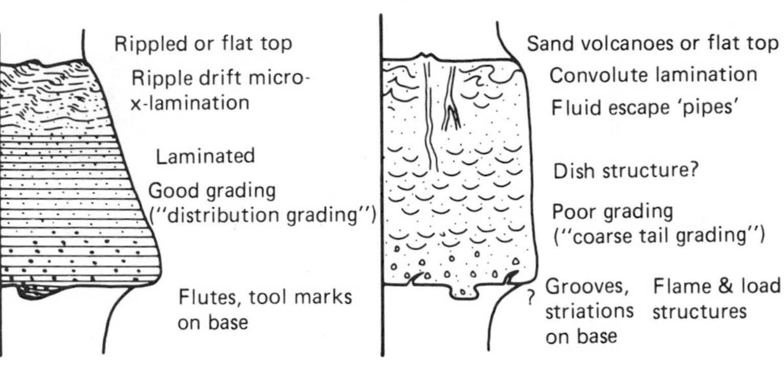

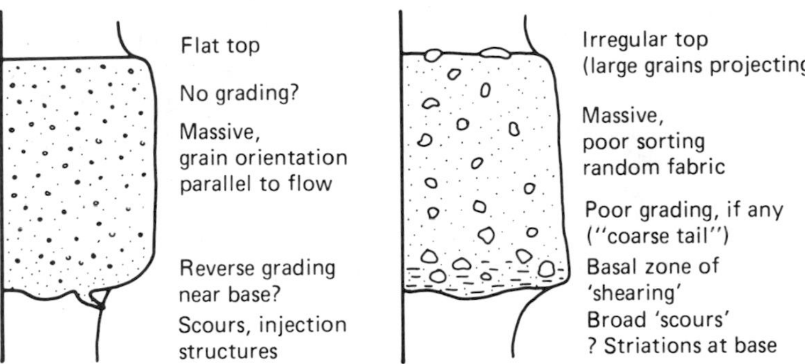

Fig. 4.63. Structures and textures of sediment gravity flow deposits (Middleton and Hampton, 1976). Reproduced, with permission, from Marine Sediment Transport and Environmental Management, edited by D.J. Stanley and D.J.P. Swift, © 1976, John Wiley and Sons, Inc., New York.

field data are required before it will be safe to claim that we completely understand this environment. Several recent discussions in the literature (Nilsen, 1980; Hiscott 1981b; and replies by Walker, Normark and Ghibaudo) provide a useful focus on the current problems, many of which stem from the fact that our ideas commonly are based on examinations of only small segments of ancient fans. Very few complete basin analyses, of the type proposed in this book, have been undertaken. There is considerable argument about the terminology of fan subenvironments (inner, middle, outer fan), and about the presence and interpretation of suprafan lobes. Some of this may be due to the fact, only now becoming generally recognized, that fan morphology and facies depend strongly on the style of sediment input and on sediment grain size. Some fans are fed by submarine canyons that channel coarse, relatively unsorted debris from the shelf. These are characterized by coarse, channeled, inner and mid fan deposits, referred to as "suprafans" by Normark (1978) and Nilsen (1980). Other fans are developed in front of deltas, which trap much of the coarse debris. These fans have one or more active inner fan channels, beyond the mouths of which high-energy turbidity currents form unchannelized blanket deposits. Most of the coarse sediment remains in the main channel or bypasses the

FLOW BEHAVIOR	FLOW TYPE		SEDIMENT SUPPORT MECHANISM
FLUID	FLUIDAL FLOW	TURBIDITY CURRENT	FLUID TURBULENCE
		FLUIDIZED FLOW	ESCAPING PORE FLUID (FULL SUPPORT)
		LIQUEFIED FLOW	ESCAPING PORE FLUID (PARTIAL SUPPORT)
PLASTIC (BINGHAM)	DEBRIS FLOW	GRAIN FLOW	DISPERSIVE PRESSURE
		MUDFLOW OR COHESIVE DEBRIS FLOW	MATRIX STRENGTH MATRIX DENSITY

Fig. 4.64. Classification of sediment gravity flows and their sediment support mechanisms (Lowe, 1979).

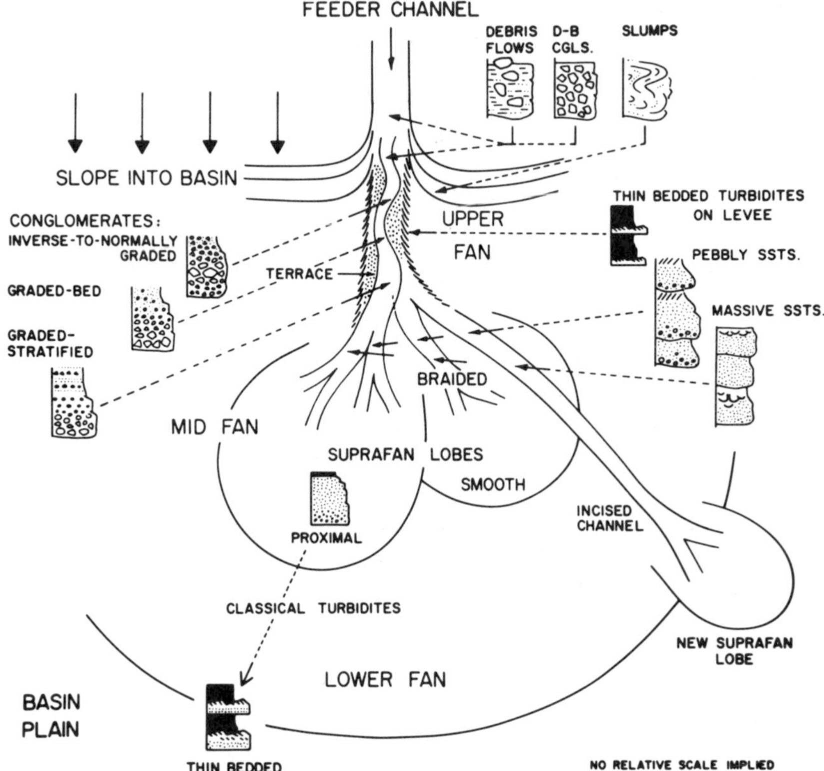

Fig. 4.05. The submarine fan model of Walker (1979b).

inner fan and is deposited on the mid or outer fan. The unchanneled "classical turbidites" formed there comprise the "suprafan lobes" of Walker (1978, 1979b). This use of the term suprafan for two different situations is an obvious source of confusion. The situation has not been helped by the introduction by the Italian workers (e.g. Ghibaudo, 1980) of the terms "poorly efficient" (or inefficient) and "highly efficient" for the two types of fan. What the human concept of efficiency has to do with sediment dispersal and fan morphology is unclear.

Detailed work on well-exposed ancient deposits and sophisticated new geophysical surveys of modern fans are providing some particularly intriguing data about fan channels. Hein and Walker (1982) showed from an outcrop study of a Cambro–Ordovician sequence that channel fills could consist of graded, stratified and crossbedded conglomerates revealing internal scouring and channeling, terraces and bar deposits. The similarity to subaerial braided channels is striking. Damuth et al. (1983) used side-scan sonar techniques to map channels on the modern Amazon fan. Most of the fans are highly sinuous, and are associated with levees, cutoffs and abandoned meanders. The resolving power of the sonar technique was only adequate to identify features in the order of hundreds of meters across. It is therefore unclear what is the internal morphology of these channels. The Hein and Walker (1982) and Damuth et al. (1983) studies show that the scales of work on ancient and modern fans are now approaching each other, but in this particular case it is unclear whether or not the two studies relate to different aspects of one standard type of fan channel. The scale of braiding observed in the outcrop study could just about be contained within the larger meanders seen on the Amazon fan. Is this what all fan channels are like? Or are we seeing preliminary evidence of a broad array of channel types comparable in the range of geometries, scales and sediment types to those of fluvial environments? The latter seems more likely, in which case a whole new era of fan studies may be about to begin.

Other problems with submarine fans include the possible misidentification of large scours or giant flutes as channels in the ancient record (Normark et al., 1979), and the problem of rigorously defining what constitutes a cyclic sequence (Hiscott, 1981b). The nature of the cyclicity is used as a major environmental criterion in most submarine fan models, and so this point may be crucial.

The use of the submarine fan model itself may be inappropriate for many sediment gravity flow deposits, as these can also accumulate in canyon fills, base-of-slope aprons and basin sheets and lobes. Active tectonics may completely obscure autocyclic sedimentary patterns, particularly in the forearc and accretionary basins of convergent plate margins (section 9.3.2). Recognition of these various basin settings requires the application of basin-wide mapping techniques (Chap. 5) before the detailed procedures of facies analysis can be applied, and the larger-scale features of the overall depositional system recognized.

Some examples of submarine fan depositional systems that do not fit simple facies models are discussed in section 6.7.

Low-viscosity, high-velocity turbidity currents may flow far out into the deep basin-plain, where they form thin-bedded deposits in which individual units may be traceable for many kilometers (e.g. Bennetts and Pilkey, 1976; Ricci-Lucchi and Valmori, 1980). Paleocurrent patterns commonly reveal flow directions along the axis of the sedimentary basin, particularly in arc-related basins, rather than perpendicular to shore, as on many mature divergent-margin continental slopes (Chap. 9).

These basin-plain turbidites may be interbedded with deposits formed by ocean-bottom currents, as first recognized by Heezen and Hollister (1964). The strongest and therefore most important of these are the western boundary undercurrents, consisting of cold water masses derived from the poles and deflected to the western margins of the Atlantic, Pacific and Indian Oceans by the Coriolis effect. Their deposits are termed contourites, because the currents tend to flow parallel to continental margin topography. Stow and Lovell (1979) distinguished two types of contourite, a sandy type, formed by strong undercurrents flowing at velocities of 10–20 cm/sec, rarely up to more than 40 cm/sec, and a muddy type, deposited from the nepheloid layer—a layer of turbid water up to 1 km thick associated with the oceanic currents or shelf circulation.

Distinction of contourites from basin-plain turbidites may be difficult in the ancient record, but Stow and Lovell (1979) suggested a range of criteria that may be indicative if used in combination. Sedimentary structures may be distinctive—contourites commonly contain abundant ripples. Their paleocurrent patterns should be distinctive, particularly if they can be related to regional basin configuration. Petrological composition, perhaps reflecting different sources, may also be useful. For example carbonate-rich turbidites may be deposited and preserved below the level of carbonate undersaturation and dissolution (carbonate compensation depth; 4000–5000 m below sea level) as a result of rapid sedimentation and burial. Interbedded contourites may be carbonate-free because of slower sedimentation and reworking.

Pelagic sediments in the deep oceans consist of a wide variety of fine clastic and chemical sediments. Our understanding of this environment has, needless to say, increased dramatically since the start of the Deep Sea Drilling Project in 1968 (Hsü and Jenkyns, 1974; Jenkyns, 1978). For the deposits of present-day oceans sophisticated sedimentary models such as that of Berger and Winterer (1974) are available. These authors proposed a theory of oceanic plate stratigraphy integrating the effects on the carbonate compensation depth of plate movement through different climatic zones and sea-floor subsidence with time, as a plate expanded from a spreading center. However, most ancient deposits of deep oceans are obducted remnants associated with ophiolites along tectonically complex suture zones. It may no longer be possible to reconstruct basin configuration or regional paleogeography. Jenkyns (1978) discussed a variety of case studies. Close association with submarine volcanics and their hydrothermal products is a common theme.

4.6.10 Glacial environments

Subaerial glacial environments are comparatively well understood as a result of the enormous amount of work carried out on Quaternary deposits (e.g. Jopling and McDonald, 1975; Lawson, 1982). New work on facies analysis of tills is showing them to be complex deposits with varying composition and texture caused by frequently changing ice flow directions. Interbedded glacio–fluvial deposits may represent sediments formed in subglacial tunnels (Eyles et al., 1982).

Submarine glaciation is much less well under-

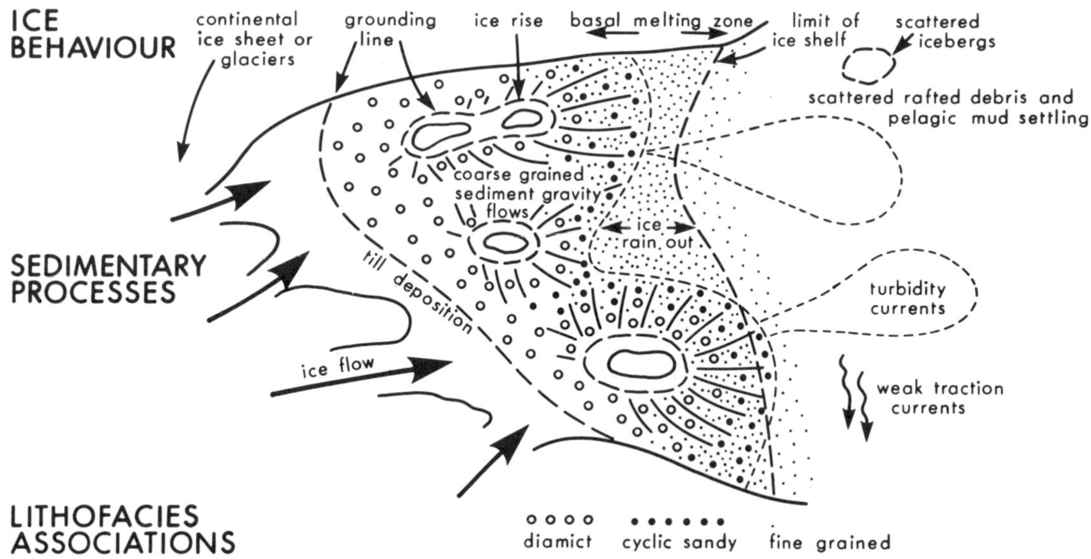

Fig. 4.66. Hypothetical model of glaciomarine sedimentation beneath an ice shelf. The zone of sediment gravity flow deposits could extend for tens to hundreds of kilometers. (Reproduced from Miall, 1983a)

stood because of the inaccessibility of the deposits in front of modern tidewater glaciers or beneath ice shelves. A useful review was provided by Edwards (1978), who discussed the various major ancient deposits known from the geological record and erected some generalized facies models.

Much useful data are starting to accumulate from studes of Recent glaciomarine sediments on the Antarctic shelf (e.g. Kurtz and Anderson, 1979; Wright and Anderson, 1982) and off Alaska (Powell, 1981), although the Antarctic data are limited to short cores in which structures have been disturbed by the drilling process. Their usefulness is therefore limited.

Quaternary geologists have erected a complex terminology for till types, but these are based mainly on observations of processes in modern glacial environments and not on observed facies criteria. Recently, application of facies analysis methods to some Quaternary and ancient glacial deposits has shown that many tills are resedimented deposits laid down by slumping or sediment gravity flows (Eyles et al., 1983; Miall, 1983a). The genetic connection to glaciation may be rather distant, consisting of reworking of original ice-laid deposits by ice-push and slumping, which then initiated mass movement. The use of the word "till" for these deposits seems inappropriate, and the non-genetic term diamict is preferred. This is not a new idea, but advances in our understanding of sediment gravity flows (reviewed in the preceding section) have given basin analysts powerful new methods for reinterpreting ancient glacial sequences.

Downslope the diamicts may be interbedded with sandy fluidized flow or grain flow deposits derived from meltwater tunnels, and all these facies pass downslope into thick-bedded turbidites interbedded with pelagic laminites containing ice-rafted debris. Figure 4.66 is a speculative ice shelf sedimentary model developed for the Gowganda (Huronian) glaciation of northern Ontario. It carries implications for a complex stratigraphic evolution of the sedimentary basin, with numerous local facies changes.

Similar methods need to be applied to many other Quaternary and older glacial sequences. Many of these have been extensively studied using classical stratigraphic methods, but these have provided little insight into depositional environments or basin evolution.

4.7 Conclusions and scale considerations

The focus of this chapter has been on analytical methods. Details of all the major facies models

Table 4.6. Scales of sedimentological analysis

1.	Scales	Facies Analysis	Depositional Systems Analysis
2.	General definitions	Outcrop or local-scale facies assemblages and environmental interpretations, within-basin (autocyclic) cyclic mechanisms, facies models	Basin-wide paleogeography and stratigraphic architecture, effects of contemporaneous tectonics, sea level change, climate change
3.	Selected analytical methods	Facies Analysis scale	Depositional Systems scale
	Sedimentary structures and paleocurrents	Bar and bedform geometry and relative arrangement, local hydraulics, identification of river, wave and tide influence	Basin dispersal patterns and relationship to tectonic elements
	Vertical profile studies	Environmentally diagnostic, autocyclic mechanisms	Allocyclic mechanisms
	Petrology (siliciclastic)	Authigenic components may be environmentally diagnostic	Basin dispersal patterns, plate-tectonics setting of source rocks
	Petrology (carbonate)	Basis of most lithofacies identifications and paleogeographic reconstructions at all scales.	
	Paleoecology	Local environmental interpretations	Regional paleobathymetry, gross salinity and temperature changes
4.	Selected environments	Facies Analysis scale	Depositional systems scale
	Fluvial	Bar type, channel geometry, local fluvial style	Areal and stratigraphic variations of fluvial style, relation of dispersal patterns to contemporaneous tectonic elements
	Eolian	Dune type and geometry, interdune environments	Erg paleogeography, relation to playa and fluvial systems
	Deltaic	Discrimination of river, wave and tide influence is main objective at all levels of analysis as these are main controls on all scales of cyclicity and facies geometry.	
	Shelf	Types of HCS and storm sequence, tidal sand wave evolution	Relative dominance of tides versus storms
	Continental Rise	Fan channel geometry and facies, channel migration and lobe progradation	Canyon-fan-basin plain relations, paleogeography of fan complexes
	Carbonate Platform	Local environments, tidal exposure, reef paleoecology	Discrimination of windward and leeward shelf margins, reef distribution
	Evaporite Basins	Recognition of deep or shallow water origin	Large scale cyclicity and relation to water level fluctuations

have been omitted because this material is covered in several excellent textbooks, to which reference is made elsewhere. Readers may wish to mentally insert one of these books between section 4.5 and 4.6, using the latter as a review and update of recent developments.

At this point it would be useful to draw together a number of ideas appearing in different parts of this book that touch on stratigraphic scale.

Figure 1.1 is an attempt to illustrate a global stratigraphic hierarchy and the kinds of analytical methods that are used to investigate each level of this hierarchy. In section 2.2 we discuss different types and scales of basin analysis project, the kinds of data collection typically undertaken, and the problems and opportunities each offers. Elsewhere in this chapter a distinction has been drawn between facies analysis and facies models on the one hand and depositional systems analysis on the other hand. These represent different levels of the stratigraphic hierarchy and corresponding analytical complexity. This point is illustrated in

Table 4.6. Selected analytical methods are listed with an indication of the kinds of information obtained at the smaller, facies analysis scale and the larger, depositional systems scale. Similarly, each depositional environment can be analyzed at the two different scales, and some examples are given to demonstrate this idea. The table is not exhaustive, but is offered as an illustration of these scale considerations. To tackle the larger scale many different basin mapping techniques are required, which is why the discussion of depositional systems is preceded by Chapter 5. In this chapter, also, the question of sedimentary structure scale is returned to, and its implications for paleocurrent analysis are discussed (section 5.9.4).

To some extent the difference between facies models and depositional systems is artificial; the boundary between the two is vague and, for some environments, such as deltas, the distinction is all but impossible to define. Yet it is a useful approach to take, because it helps to distinguish and clarify the purpose of a number of different procedures we perform more or less simultaneously as a basin analysis is carried out.

References

ADAMS, J.E. and RHODES, M.L., 1960: Dolomitization by seepage refluction; Am. Assoc. Petrol. Geol. Bull., v. 44, p. 1912–1920.

ADEY, W., 1975: The algal ridges and coral reefs of St. Croix; their structure and Holocene development; Atoll Res. Bull., No. 187, 67 p.

ADEY, W.H., 1978: Coral reef morphologenesis: a multidimensional model; Science, v. 202, p. 831–837.

ADEY, W., MACINTYRE, R., and STUCKENRATH, R.D., 1977: Relict barrier reef system off St. Croix: its implications with respect to late Cenozoic coral reef development in the Western Atlantic; Proc. 3rd Internat. Coral Reef Symp., v. 2, p. 15–21.

AGER, D.V., 1963: Principles of paleoecology; McGraw-Hill, New York, 371 p.

AGER, D.V., 1981: The nature of the stratigraphical record; 2nd ed., Halsted Press, John Wiley and Sons Inc., Somerset, New Jersey, 122p.

ALLEN, J.R.L., 1963a: The classification of cross-stratified units, with notes on their origin; Sedimentology, v. 2, p. 93–114.

ALLEN, J.R.L., 1963b: Henry Clifton Sorby and the sedimentary structures of sands and sandstones in relation to flow conditions; Geol. Mijnb., v. 42, p. 223–228.

ALLEN, J.R.L., 1968: Current ripples; North Holland Pub. Co., Amsterdam, 433 p.

ALLEN, J.R.L., 1970: Studies in fluviatile sedimentation: a comparison of fining-upward cyclothems with special reference to coarse-member composition and interpretation; J. Sediment. Petrol., v. 40, p. 298–323.

ALLEN, P., 1981: Wave-generated structures in the Devonian lacustrine sediments of south-east Shetland and ancient wave conditions; Sedimentology, v. 28, p. 369–380.

ANDRICHUK, J.M., 1958: Stratigraphy and facies analysis of Upper Devonian reefs in Leduc, Stettler and Redwater area, Alberta; Am. Assoc. Petrol. Geol. Bull., v. 42, p. 1–93.

BADIOZAMANI, K., 1973: The Dorag dolomitization model—application to the Middle Ordovician of Wisconsin; J. Sediment. Petrol., v. 43, p. 965–984.

BAKER, V.R, 1978: Adjustment of fluvial systems to climate and source terrain in tropical and subtropical environments, *in* A.D. Miall, ed., Fluvial sedimentology; Can. Soc. Petrol. Geol. Mem. 5, p. 211–230.

BALL, M.M., 1967: Carbonate sand bodies of Florida and the Bahamas; J. Sediment. Petrol., v. 37, p. 556–591.

BALLANCE, P.F., GREGORY, M.R., and GIBSON, G.W., 1981: Coconuts in Miocene turbidites in New Zealand: possible evidence for tsunami origin of some turbidity currents; Geology, v. 9, p. 592–595.

BARRETT, T.J., 1982: Stratigraphy and sedimentology of Jurassic bedded chert overlying ophiolites in the North Apennines, Italy; Sedimentology, v. 29, p. 353–373.

BARWIS, J.H., 1978: Sedimentology of some South Carolina tidal–creek point bars, and a comparison with their fluvial counterparts, *in* A.D. Miall, ed., Fluvial sedimentology; Can. Soc. Petrol. Geol. Mem. 5, p. 129–160.

BASAN, P.B., ed., 1978: Trace fossil concepts; Soc. Econ. Paleont. Mineral. Short Course No. 5, 201 p.

BEERBOWER, J.R., 1964: Cyclothems and cyclic depositional mechanisms in alluvial plain sedimentation, *in* D.F. Merriam, ed., Symposium on cyclic sedimentation; Kansas Geol. Surv. Bull. 169, v. 1, p. 31–42.

BENEDICT, G.L., III, and WALKER, K.R., 1978: Paleobathymetric analysis in Paleozoic sequences and its geodynamic significance; Am. J. Sci., v. 278, p. 579–607.

BENNETTS, K.R.W., and PILKEY, O.H., Jr., 1976: Characteristics of three turbidites, Hispaniola–Caicos Basin; Geol. Soc. Am. Bull., v. 87, p. 1291–1300.

BERGER, W.H. and WINTERER, E.L., 1974: Plate stratigraphy and the fluctuating carbonate line, *in* K.J. Hsü and H.C. Jenkyns, eds., Pelagic sediments: on land and under the sea; Internat. Assoc. Sedimentologists Spec. Publ. 1, p. 11–48.

BERNARD, H.A. and MAJOR, C.F., Jr., 1963: Recent meander belt deposits of the Brazos River: an alluvial "sand" model (Abstract); Am. Assoc. Petrol. Geol. Bull., v. 47, p. 350–351.

BERNARD, H.A., MAJOR, C.F. Jr., and PARROTT, B.S., 1959: The Galveston Barrier Island

and environs—a model for predicting reservoir occurrence and trend; Trans. Gulf Coast Assoc. Geol. Soc., v. 9, p. 221–224.

BIGARELLA, J.J., 1972: Eolian environments: their characteristics, recognition and importance, *in* J.K. Rigby and W.K. Hamblin, eds., Recognition of ancient sedimentary environments; Soc. Econ. Paleont. Mineral. Spec. Publ. 16, p. 12–62.

BLATT, H., MIDDLETON, G.V. and MURRAY, R., 1980: Origin of sedimentary rocks; 2nd ed., Prentice-Hall Inc., Englewood Cliffs, New Jersey, 782 p.

BLUCK, B.J., 1967: Deposition of some Upper Old Red Sandstone conglomerates in the Clyde area: a study in the significance of bedding; Scott. J. Geol., v. 3, p. 139–167.

BOUMA, A.H., 1962: Sedimentology of some flysch deposits; Elsevier, Amsterdam, 168 p.

BOURGEOIS, J., 1980: A transgressive shelf sequence exhibiting hummocky stratification: the Cape Sebastian Sandstone (Upper Cretaceous), southwestern Oregon; J. Sediment. Petrol., v. 50, p. 681–702.

BOWN, T.M. and KRAUS, M.J., 1981: Lower Eocene alluvial paleosols (Willwood Formation, northwest Wyoming, U.S.A.) and their significance for paleoecology, paleoclimatology, and basin analysis; Palaeogeog. Palaeoclimatol., Palaeoec., v. 34, p. 1–30.

BRENNER, R.L., 1980: Construction of process-response models for ancient epicontinental seaway depositional systems using partial analogs; Am. Assoc. Petrol. Geol. Bull., v. 64, p. 1223–1244.

BRIDGE, J.S., 1978: Palaeohydraulic interpretation using mathematical models of contemporary flow and sedimentation in meandering channels, *in* A.D. Miall, ed., Fluvial sedimentology; Can. Soc. Petrol. Geol. Mem. 5, p. 723–742.

BRIDGES, P.H. and LEEDER, M.R., 1976: Sedimentary model for intertidal mudflat channels with examples from the Solway Firth, Scotland; Sedimentology, v. 23, p. 533–552.

BROOKFIELD, M.E., 1977: The origin of bounding surfaces in ancient aeolian sandstones; Sedimentology, v. 24, p. 303–332.

BROWN, L.F. Jr., 1981: Seismic stratigraphy: reconfirming basic principles but mandating modern lithogenetic concepts; Geol. Soc. Am. Abs. with Prog., v. 13, p. 418.

BYERS, C.W. and DOTT, R.J. Jr., 1981: SEPM Research Conference on modern shelf and ancient cratonic sedimentation—the orthoquartzite-carbonate suite revisited; J. Sediment. Petrol., v. 51, p. 329–347.

CAMERON, B.E.B., 1975: Geology of the Tertiary rocks north of Latitude 49°, west coast of Vancouver Island; in Geol. Surv. Can. Paper 75–1A, p. 17–19.

CANT, D.J., 1976: Selective preservation of flood stage deposits in a braided fluvial environment; Geol. Assoc. Can. Progr. Abs., v. 1, p. 77.

CANT, D.J. and WALKER, R.G., 1976: Development of a braided-fluvial facies model for the Devonian Battery Point Formation; Can. J. Earth Sci., v. 13, p. 102–119.

CLIFTON, H.E., HUNTER., R.E. and PHILLIPS, R.L., 1971: Depositional structures and processes in the non-barred, high energy nearshore; J. Sediment. Petrol., v. 41, p. 651–670.

CLOUD, P., 1973: Paleoecological significance of banded iron-formation; Econ. Geol., v. 68, p. 1135–1143.

COLEMAN, J.M., 1969: Brahmaputra River: Channel processes and sedimentation; Sediment. Geol., v. 3, p. 129–239.

COLEMAN, J.M. and GARRISON, L.E., 1977: Geological aspects of marine slope stability, northwestern Gulf of Mexico; Mar. Geotechnol., v. 2, p. 9–44.

COLEMAN, J.M., PRIOR, D.B. and LINDSAY, J.F., 1983: Deltaic influences on Shelfedge instability processes, *in* D.J. Stanley and G.T. Moore, eds., The Shelfbreak: critical interface on continental margins; Soc. Econ. Paleont. Mineral Spec. Publ. 33, p. 121–138.

COLEMAN, J.M. and WRIGHT, L.D., 1975: Modern river deltas: variability of processes and sand bodies, *in* M.L. Broussard, ed., Deltas, models for exploration; Houston Geol. Soc., Houston, p. 99–149.

COLLINSON, J.D., 1970: Bedforms of the Tana River, Norway; Geogr. Annaler, v. 52A, p. 31–55.

COLLINSON, J.D., 1978: Lakes, *in* H.G. Reading, ed., Sedimentary environments and facies; Blackwell, Oxford, p. 61–79.

COOK, H.E. and ENOS, P., eds., 1977: Deep-water carbonate environments; Soc. Econ. Paleont. Mineral. Spec. Publ. 25.

COOKE, R.U. and WARREN, A., 1973: Geomorphology in deserts; Batsford, London, 374 p.

COTTER, E., 1978: The evolution of fluvial style, with special reference to the central Appalachian Paleozoic, *in* A.D. Miall, ed., Fluvial sedimentology; Can. Soc. Petrol. Geol. Mem. 5, p. 361–384.

CURRAY, J.R. and MOORE, D.G., 1971: Growth of the Bengal deep-sea fan and denudation in the Himalayas; Geol. Soc. Am. Bull., v. 82, p. 563-572.

DACEY, M.F. and KRUMBEIN, W.C., 1970: Markovian models in stratigraphic analysis; J. Internat. Assoc. Math. Geol., v. 2, p. 175–191.

DAMUTH, J.E., KOLLA, K., FLOOD, R.D., KOWSMANN, R.O., MONTEIRO, M.C., GORINI, M.A., PALMA, J.J.C., and BELDERSON, R.H., 1983: Distributary channel meandering and bifurcation patterns on the Amazon deep-sea fan as revealed by long-range side-scan sonar (GLORIA); Geology, v. 11, p. 94–98.

DAMUTH, J.E. and KUMAR, N., 1975: Amazon Cone: morphology, sediments, age, and growth pattern; Geol. Soc. Am. Bull., v. 86, p. 863–878.

DAVIDSON-ARNOTT, R.G.D. and GREENWOOD, B., 1974: Bedforms and structures associated with bar topography in the shallow water wave environments, Kouchibouguac Bay, New Brunswick, Canada; J. Sediment. Petrol., v. 44, p. 698–704.

DAVIDSON-ARNOTT, R.G.D. and GREENWOOD, B., 1976: Facies relationships on a barred coast, Kouchibouguac Bay, New Brunswick,

Canada, *in* R.A. Davis Jr., and R.L. Ethington, eds., Beach and nearshore sedimentation; Soc. Econ. Paleont. Mineral. Spec. Publ. 24, p. 149–168.

DAVIES, D.K. and ETHRIDGE, F.G., 1975: Sandstone composition and depositional environments; Am. Assoc. Petrol. Geol. Bull., v. 59, p. 239–264.

DAVIES, D.K., ETHRIDGE, F.G. and BERG, R.R., 1971: Recognition of barrier environments; Am. Assoc. Petrol. Geol. Bull., v. 55, p. 550–565.

DAVIES, G.R., 1977: Turbidites, debris sheets, and truncation structures in Upper Paleozoic deep-water carbonates of the Sverdrup Basin, Arctic Archipelago, *in* H.E. Cook and P. Enos, eds., Deep-water carbonate environments; Soc. Econ. Paleont. Mineral. Spec. Publ. 25, p. 221–248.

DAVIS, J.C., 1973: Statistics and data analysis in geology; John Wiley and Sons, New York, 550 p.

DeRAAF, J.F.M., BOERSMA, J.R. AND VAN GELDER, A., 1977: Wave generated structures and sequences from a shallow marine succession, Lower Carboniferous, County Cork, Ireland; Sedimentology, v. 24, p. 451–483.

DICKINSON, K.A., BERRYHILL, H.L., Jr. and HOLMES, C.W., 1972: Criteria for recognising ancient barrier coastlines, *in* J.K. Rigby and W.K. Hamblin, eds., Recognition of ancient sedimentary environments, Soc. Econ. Paleont. Mineral. Spec. Publ. 16, p. 192–214.

DICKINSON, W.R. and SUCZEK, C.A., 1979: Plate tectonics and sandstone composition; Am. Assoc. Petrol. Geol. Bull., v. 63, p. 2164–2182.

DINGLE, R.V., 1980: Large allochthonous sediment masses and their role in the construction of the continental slope and rise off southwestern Africa; Mar. Geol., v. 37, p. 333–354.

DODSON, P., 1971: Sedimentology and taphonomy of the Oldman Formation (Campanian), Dinosaur Provincial Park, Alberta (Canada); Palaeogeog., Palaeoclim., Palaeoec., v. 10, p. 21–74.

DONOVAN, R.N., 1973: Basin margin deposits of the Middle Old Red Sandstone at Dirlot, Caithness; Scott. J. Geol., v. 9, p. 203–212.

DONOVAN, R.M., 1975: Devonian lacustrine limestones at the margin of the Orcadian Basin, Scotland; J. Geol. Soc. London, v. 131, p. 489–510.

DOTT, R.H., Jr., 1983: 1982 SEPM Presidential Address: Episodic sedimentation—How normal is average? How rare is rare? Does it Matter?; J. Sediment. Petrol., v. 53, p. 5–23.

DOTT, R.H., Jr., and BOURGEOIS, J., 1982: Hummocky stratification: significance of its variable bedding sequences; Geol. Soc. Am. Bull., v. 93, p. 663–680.

DOVETON, J.H., 1971: An application of Markov chain analysis to the Ayrshire Coal Measures succession; Scott. J. Geol., v. 7, p. 11–27.

DUFF, P. McL. D. and WALTON, E.K., 1962: Statistical basis for cyclothems: a quantitative study of the sedimentary succession in the east Pennine Coalfield; Sedimentology, v. 1, p. 235–255.

DUNHAM, R.J., 1962: Classification of carbonate rocks according to depositional texture, *in* W.E. Ham, ed., Classification of carbonate rocks; Am. Assoc. Petrol. Geol. Mem. 1, p. 108–121.

EDWARDS, M.B., 1978: Glacial environments, *in* H.G. Reading, ed., Sedimentary environments and facies; Blackwell, Oxford, p. 416–438.

EDWARDS, W.N., 1936: The flora of the London Clay; Proc. Geol. Assoc. London, v. 47, p. 22–31.

EICHLER, J., 1976: Origin of the Precambrian banded iron formations, *in* K.H. Wolf, ed., Handbook of strata-bound and strataform ore deposits; Elsevier, v. 7, p. 157–201.

ELLIOTT, T., 1978: Deltas, *in* H.G. Reading, ed., Sedimentary environments and facies; Blackwell, Oxford, p. 97–142.

EMERY, K.O., 1968: Relict sediments on continental shelves of the world; Am. Assoc. Petrol. Geol. Bull., v. 52, p. 445–464.

ENOS, P., 1977: Tamabra Limestone of the Pozo Rico Trend, Cretaceous, Mexico, *in* H.E. Cook and P. Enos, eds., Deep water carbonate environments; Soc. Econ. Paleont. Mineral. Spec. Publ. 25, p. 221–248.

ERIKSSON, K.A., 1979: Marginal marine depositional processes from the Archaean Moodies Group, Barberton Mountain Land, South Africa: evidence and significance; Precamb. Res., v. 8, p. 153–182.

EUGSTER, H.P. and HARDIE, L.A., 1975: Sedimentation in an ancient playa–lake complex: The Wilkins Peak Member of the Green River Formation of Wyoming; Geol. Soc. Am. Bull., v. 86, p. 319–334.

EVANS, G., 1965: Intertidal flat sediments and their environments of deposition in the Wash; Quart. J. Geol. Soc. London, v. 121, p. 209–245.

EYLES, N., EYLES, C.H., and MIALL, A.D., 1983: Lithofacies types and vertical profile models; an alternative approach to the description and environmental interpretation of glacial diamict sequences; Sedimentology, v. 30, p. 393–410.

EYLES, N., SLADEN, J.A., and GILROY, S., 1982: A depositional model for stratigraphic complexes and facies superimposition in lodgement tills; Boreas, v. 11, p. 317–333.

FAWCETT, J.J. 1982: Facies (metamorphic facies); McGraw-Hill Encyclopedia of Science and Technology, 5th ed., McGraw-Hill, New York, p. 303.

FISCHER, A.G., 1964: The Lofer cyclothems of the Alpine Triassic, *in* D.F. Merriam, ed., Symposium on cyclic sedimentation; Bull. Geol. Surv. Kansas 169, p. 107–149.

FISHER, W.L., BROWN, L.F., SCOTT, A.J. and McGOWEN, J.H., 1969: Delta systems in the exploration for oil and gas; Texas Bur. Econ. Geol., 78 p.

FLEMMING, B.W., 1978: Underwater sand dunes along the southeast African continental margin—observations and implications; Mar. Geol., v. 26, p. 177–198.

FLORES, R., 1981: Coal deposition in fluvial paleoenvironments of the Paleocene Tongue River Member of the Fort Union Formation, Powder River area, Powder River Basin, Wyoming and Montana, *in* F.G. Ethridge and R. Flores, eds., Recent and ancient nonmarine depositional environments: models for exploration; Soc. Econ. Paleont. Mineral. Spec. Publ. 31, p. 169–190.

FOLK, R.L., 1962: Spectral subdivision of limestone types, *in* W.E. Ham, ed., Classification of carbonate rocks; Am. Assoc. Petrol. Geol. Mem. 1, p. 62–84.

FRAKES, L.A., 1979: Climates throughout geologic time; Elsevier, Amsterdam, 310 p.

FRANKE, W. and PAUL, J., 1980: Pelagic redbeds in the Devonian of Germany—deposition and diagenesis; Sediment. Geol., v. 25, p. 231–256.

FREY, R.W., ed., 1975: The study of trace fossils; Springer-Verlag, New York, 562 p.

FRIEDMAN, G.M., 1980: Dolomite is an evaporite mineral: evidence from the rock record and from sea-marginal ponds of the Red Sea; Soc. Econ. Paleont. Mineral. Spec. Publ. 28, p. 69–80.

FRIEDMAN, G.M. and SANDERS, J.E., 1978: Principles of sedimentology; John Wiley and Sons, New York, 792 p.

FRIEND, P.F., 1978: Distinctive features of some ancient river systems, *in* A.D. Miall, ed., Fluvial sedimentology, Can. Soc. Petrol. Geol. Mem. 5, p. 531–542.

FRIEND, P.F., 1983: Towards the field classification of alluvial architecture or sequence, *in* J.D. Collinson and J. Lewin, eds., Modern and ancient fluvial systems; Internat. Assoc. Sedimentologists Spec. Publ. 6, p. 345–354.

FRIEND, P.F., ALEXANDER-MARRACK, P.D., NICHOLSON, J. and YEATS, A.K., 1976: Devonian sediments of east Greenland I; Medd. øm Gronland, Bd. 206, No.1.

GALAY, V.J., KELLERHALS, R. and BRAY, D.I., 1973: Diversity of river types in Canada, *in* Fluvial Processes and sedimentation, Proc. Hydrology Symp., Edmonton., National Research Council, Canada, p. 217–250.

GALLOWAY, W.E., 1975: Process framework for describing the morphologic and stratigraphic evolution of the deltaic depositional systems, *in* M.L. Broussard, ed., Deltas, models for exploration; Houston Geol. Soc., Houston, p. 87–98.

GHIBAUDO, G., 1980: Deep-sea fan deposits in the Macigno Formation (Middle–Upper Oligocene) of the Gordana Valley, Northern Apennines, Italy; J. Sediment. Petrol. v. 50, p. 723–742.

GINSBURG, R.N., ed., 1975: Tidal deposits: a casebook of recent examples and fossil counterparts; Springer-Verlag, Berlin, 428 p.

GINSBURG, R.M. and SCHROEDER, J.H., 1973: Growth and submarine fossilization of algal cup reefs, Bermuda; Sedimentology, v. 20, p. 575–614.

GINSBURG, R.N. and HARDIE, L.A., 1975: Tidal and storm deposits, northeastern Andros Island, Bahamas, *in* R.N. Ginsburg, ed., Tidal deposits: a casebook of recent examples and fossil counterparts; Springer-Verlag, Berlin, p. 201–208.

GINSBURG, R.N., HARDIE, L.A., BRICKER, O.P., GARRETT, P. and WANLESS, H.R., 1977: Exposure index: a quantitative approach to defining position within the tidal zone, *in* L.A. Hardie, ed., Sedimentation on the modern carbonate tidal flats of northwestern Andros Island, Bahamas; Johns Hopkins Univ. Studies in Geol., v. 22, p. 7–11.

GLAISTER, R.P. and NELSON, H.W., 1974: Grain-size distributions, an aid in facies identification; Bull. Can. Petrol. Geol., v. 22, p. 203–240.

GLOPPEN, T.G. and STEEL, R.J., 1981: The deposits, internal structure and geometry in six alluvial fan–fan delta bodies (Devonian–Norway)—A study in the significance of bedding sequences in conglomerates, *in* F.G. Ethridge and R.M. Flores, eds., Recent and ancient nonmarine depositional environments: models for exploration; Soc. Econ. Paleont. Mineral Spec. Publ. 31, p. 49–70.

GOLDRING, R., 1964: Trace fossils and the sedimentary surface in shallow-water marine sediments, *in* L.M.J.U. van Straaten, ed., Deltaic and shallow marine deposits; Developments in Sedimentology, Elsevier, no. 1, p. 136–143.

GRADSTEIN, F.M. and SRIVASTAVA, S.P., 1980: Aspects of Cenozoic stratigraphy and paleoceanography of the Labrador Sea and Baffin Bay; Palaeogeog., Palaeoclim., Palaeoec., v. 30, p. 261–295.

GRADZINSKI, R., GAGOL, J., and SLACZKA, A., 1979: The Tumlin Sandstone (Holy Cross Mts., central Poland): Lower Triassic deposits of aeolian dunes and interdune areas; Acta Geologica Polonica, v. 29, p. 151–175.

GRUNAU, H.R., 1965: Radiolarian cherts and associated rocks in space and time; Eclogae Geol. Helv., v. 58, p. 157–208.

HAMBLIN, A.P. and WALKER, R.G., 1979: Storm-dominated shallow marine deposits; the Fernie-Kootenay (Jurassic) transition, southern Rocky Mountains; Can. J. Earth Sci., v. 16, p. 1673–1690.

HANCOCK, J.M. and SCHOLLE, P.A., 1975: Chalk of the North Sea, *in* A.W. Woodland, ed., Petroleum and the continental shelf of Europe; John Wiley and Sons, New York, v. 1, p. 413–425.

HARBAUGH, J.W. and BONHAM-CARTER, G., 1970: Computer simulation in geology; Wiley-Interscience, New York, 98 p.

HARBAUGH, J.W. and DEMIRMEN, F., 1964: Application of factor analysis to petrologic variations of Americus Limestone (Lower Permian), Kansas and Oklahoma; Spec. Distrib. Publ. Geol. Surv. Kansas 15, p. 1–40.

HARBAUGH, J.W. and MERRIAM, D.F., 1968: Computer applications in stratigraphic analysis; Wiley, New York, 282 p.

HARDIE, L.A., SMOOT, J.P. and EUGSTER, H.P., 1978: Saline lakes and their deposits: a sedimentological approach, *in* A. Matter and M.E. Tucker, eds., Modern and ancient lake sediments; Internat. Assoc. Sedimentologists Spec. Publ. 2, p. 7–41.

HARMS, J.C. and FAHNESTOCK, R.K., 1965: Stratification, bed forms, and flow phenomena (with an example from the Rio Grande), *in* G.V. Middleton, ed., Primary sedimentary structures and their hydrodynamic interpretation; Soc. Econ. Paleont. Mineral. Spec. Publ. 12, p. 84–115.

HARMS, J.C., SOUTHARD, J.B., SPEARING, D.R. and WALKER, R.G., 1975: Depositional environments as interpreted from primary sedimentary structures and stratification sequences; Soc. Econ. Paleont. Mineral Short Course 2, 161 p.

HAY, W.W., SOUTHAM, J.R. and NOEL, M.R., 1976: Carbonate mass balance—cycling and deposition on shelves and in deep sea (Abstract); Am. Assoc. Petrol. Geol. Bull., v. 60, p. 678.

HAYES, M.O., 1967: Hurricanes as geological agents: Case studies of Hurricanes Carla, 1961, and Cindy, 1963; Rept. Invest. Bureau Econ. Geol., Austin, Texas, No. 61.

HAYES, M.O., 1976: Morphology of sand accumulation in estuaries; an introduction to the symposium, *in* L.E. Cronin, ed., Estuarine Research, v. 2, Geology and Engineering; Academic Press, London, p. 3–22.

HAZEL, J.E., 1975; Ostracode biofacies in the Cape Hatteras, North Carolina area; Bull. Am. Paleont., v. 65, p. 463–487.

HECKEL, P.H., 1972: Recognition of ancient shallow marine environments, *in* J.K. Rigby and W.K. Hamblin, eds., Recognition of ancient sedimentary environments, Soc. Econ. Paleont. Mine al. Spec. Publ. 16, p. 226–296.

HECKER, R.F., 1965: Introduction to paleoecology (translated from Russian); American Elsevier Pub. Co., New York, 166 p.

HEEZEN, B.C. and HOLLISTER, C.D., 1964: Deep-sea current evidence from abyssal sediments; Mar. Geol., v. 1, p. 141–174.

HEIN, F.J., 1982: Depositional mechanisms of deep-sea coarse clastic sediments, Cap Enragé Formation, Quebec; Can. J. Earth Sci., v. 19, p. 267–287.

HEIN, F.J., and WALKER, R.G., 1982: The Cambro–Ordovician Cap Enragé Formation, Quebec, Canada: Conglomeratic deposits of a braided submarine channel with terraces; Sedimentology, v. 29, p. 309–330.

HEWARD, A.P., 1981: A review of wave-dominated clastic shoreline deposits; Earth Sci. Rev., v. 17, p. 223–276.

HISCOTT, R.M., 1981a: Chi–square tests for Markov chain analysis; J. Int. Assoc. Math. Geol., v. 13, p. 53–68.

HISCOTT, R.M., 1981b: Deep-sea fan deposits in the Macigno Formation (Middle–Upper Oligocene) of the Gordana Valey, northern Apennines, Italy: Discussion; J. Sediment. Petrol., v. 51, p. 1015–1021.

HOBDAY, D.K. and READING, H.G., 1972: Fair weather versus storm processes in shallow marine sand bar sequences in the late Pre-Cambrian of Finnmark, North Norway; J. Sediment. Petrol., v. 42, p. 318–324.

HOBDAY, D.K., TAVENER-SMITH, R. and MATTHEW, D., 1975: Markov analysis and the recognition of palaeoenvironments in the Ecca Group near Vryheid, Natal; Trans. Geol. Soc. S. Africa, v. 78, p. 75–82.

HOBDAY, D.K. and VON BRUNN, V., 1979: Fluvial sedimentation and paleogeography of an early Paleozoic failed rift, southeastern margin of Africa; Palaeogeog., Palaeoclim., Palaeoec., v. 28, p. 169–184.

HORNE, J.C., FERM, J.C. CARUCCIO, F.T. and BAGANZ, B.P., 1978: Depositional models in coal exploration and mine planning in Appalachian region; Am. Assoc. Petrol. Geol. Bull., v. 62, p. 2379–2411.

HOWARD, J.D., 1972: Trace fossils as criteria for recognizing shorelines in the stratigraphic record, *in* J.K. Rigby and W.K. Hamblin, eds., Recognition of ancient sedimentary environments, Soc. Econ. Paleont. Mineral Spec. Publ. 16, p. 215–225.

HOWARD, J.D. and REINECK, H.-E., 1981: Depositional facies of high-energy beach-to-offshore sequence: comparison with low energy sequence; Am. Assoc. Petrol. Geol. Bull., v. 65, p. 807–830.

HOYT, J.H. and HENRY, V.J., 1967: Influence of island migration on barrier island sedimentation; Geol. Soc. Am. Bull., v. 78, p. 77–86.

HSÜ, K.J., CITA, M.B. and RYAN, W.B.F., 1973: The origin of the Mediterranean evaporites, *in* W.B.F. Ryan, K.J. Hsü et al., Initial Reports of the Deep Sea Drilling Project, v. 13, p. 1203–1231.

HSÜ, K.J. and JENKYNS, H.C., eds., 1974: Pelagic sediments: on land and under the sea; Internat. Assoc. Sedimentologists Spec. Publ. 1.

HUNTER, R.E., 1977a: Basic types of stratification in small eolian dunes; Sedimentology, v. 24, p. 361–388.

HUNTER, R.E., 1977b: Terminology of cross-stratified sedimentary layers and climbing-ripple structures; J. Sediment. Petrol., v. 47, p. 697–706.

HUNTER, R.E., 1980: Depositional environments of some Pleistocene coastal terrace deposits, southwestern Oregon—case history of a progradational beach and dune sequence; Sediment. Geol., v. 27, p. 241–262.

HUNTER, R.E., 1981: Stratification styles in eolian sandstones: some Pennsylvanian to Jurassic examples from the Western Interior U.S.A., *in* F.G. Ethridge and R. Flores, eds , Recent and ancient nonmarine depositional environments: models for exploration; Soc. Econ. Paleont. Mineral. Spec. Publ. 31, p. 315–329.

HUNTER, R.E., CLIFTON, H.E. and PHILLIPS, R.L., 1979: Depositional processes, sedimentary structures and predicted vertical sequences in barred nearshore systems, southern Oregon Coast; J. Sediment. Petrol., v. 49, p. 711–726.

IMBRIE, J. and PURDY, E.G., 1962: Classification of modern Bahamian carbonate sediments, *in* W.E. Ham, ed., Classification of carbonate rocks; Am. Assoc. Petrol. Geol. Mem. 1, p. 253–272.

JACKSON, R.G., II, 1978: Preliminary evaluation of lithofacies models for meandering alluvial streams, *in* A.D. Miall, ed., Fluvial sedimentology, Can. Soc. Petrol. Geol. Mem. 5, p. 543–576.

JAMES, N.P., 1972: Holocene and Pleistocene calcareous crust (caliche) profiles: criteria for subaerial exposure; J. Sediment. Petrol., v. 42, p. 817–836.

JAMES, N.P., 1979a: Facies models 9. Introduction to carbonate facies models, *in* R.G. Walker, ed., Facies Models, Geosci. Can. Reprint Series 1, p. 105–107.

JAMES, N.P., 1979b: Facies models 10. Shallowing-upward sequences in carbonates, *in* R.G. Walker, ed.,

Facies Models, Geosci. Can. Reprint Series 1, p. 109–120.

JAMES, N.P., 1979c: Facies models 11. Reefs, *in* R.G. Walker, ed., Facies Models; Geosci. Can. Reprint Series 1, p. 121–132.

JAMES, N.P. and GINSBURG, R.N., 1980: The seaward margin off Belize barrier and atoll reefs; Internat. Assoc. Sedimentologists Spec. Publ. 3.

JELETZKY, J.A., 1975: Hesquiat Formation (new): a neritic channel and interchannel deposit of Oligocene age, western Vancouver Island, British Columbia (92E); Geol. Surv. Can. Paper 75–32.

JENKYNS, H.C., 1978: Pelagic environments, *in* H.G. Reading, ed., Sedimentary environments and facies; Blackwell, Oxford, p. 314–371.

JERZYKIEWICZ, T., 1968: Sedimentation of the youngest sandstones of the Intrasudetic Cretaceous Basin; Geologia Sudetica, Warsaw, v. 4, p. 409–462.

JOHNSON, H.D., 1978: Shallow siliciclastic seas, *in* H.G. Reading, ed., Sedimentary environments and facies; Blackwell, Oxford, p. 207–258.

JONES, B., 1977: Variations in the Upper Silurian brachiopod *Atrypella phoca* (Salter) from Somerset and Prince of Wales Islands, Arctic Canada; J. Paleont., v. 51. p. 459–479.

JONES, C.M. and McCABE, P.J., 1980: Erosion surfaces within giant fluvial cross-beds of the Carboniferous in northern England; J. Sediment. Petrol., v. 50, p. 613–620.

JOPLING, A.V., and McDONALD, B.C., eds., 1975: Glaciofluvial and glaciolacustrine sedimentation; Soc. Econ. Paleont. Mineral.Spec. Publ. 23, 320 p.

KENDALL, A.C., 1979a: Facies models 13. Continental and supratidal (sabkha) evaporites, *in* R.G. Walker, ed., Facies Models; Geosci. Can. Reprint Series 1, p. 145–157.

KENDALL, A.C., 1979b: Facies models 14. Subaqueous evaporites, *in* R.G. Walker, ed., Facies Models, Geosci. Can. Reprint Series 1, p. 159–174.

KINSMAN, D.J.J., 1966: Gypsum and anhydrite of recent age, Trucial Coast, Persian Gulf, *in* J.L. Rau, ed., Second Symposium on Salt, Northern Ohio Geol. Soc., Cleveland, Ohio, v. 1, p. 302–326.

KINSMAN, D.J.J., 1969: Modes of formation, sedimentary associations, and diagenetic features of shallow-water supratidal evaporites; Am. Assoc. Petrol. Geol. Bull., v. 53, p. 830–840.

KLEIN, G. deV., 1970: Depositional and dispersal dynamics of intertidal sand bars; J. Sediment. Petrol., v. 40, p. 1095–1127.

KLEIN, G. deV., 1971: A sedimentary model for determining paleotidal range; Geol. Soc. Am. Bull., v. 82, p. 2585–2592.

KLEIN, G. deV., 1977: Tidal circulation model for deposition of clastic sediment in epeiric and mioclinal shelf seas; Sediment. Geol., v. 18, p. 1–12.

KLEIN, G. deV., and RYER, T.A., 1978: Tidal circulation patterns in Precambrian, Paleozoic, and Cretaceous epeiric and mioclinal shelf seas; Geol. Soc. Am. Bull., v. 89, p. 1050–1058.

KLOVAN, J.E., 1964: Facies analysis of the Redwater Reef complex, Alberta, Canada; Bull. Can. Petrol. Geol., v. 12, p. 1–100.

KNAUTH, L.P., 1979: A model for the origin of chert in limestone; Geology, v. 7, p. 274–277.

KOCUREK, G. and DOTT, R.H., Jr., 1981: Distinctions and uses of stratification types in the interpretation of eolian sand; J. Sediment. Petrol., v. 51, p. 579–595.

KREISA, R.D., 1981: Storm-generated sedimentary structures in subtidal marine facies with examples from the Middle and Upper Ordovician of southwestern Virginia; J. Sediment. Petrol., v. 51, p. 823–848.

KRUMBEIN, W.C. and DACEY, M.F., 1969: Markov chains and embedded chains in geology; J. Internat. Assoc. Math. Geol., v. 1, p. 79–96.

KRUMBEIN, W.C. and GARRELS, R.M. 1952: Origin and classification of chemical sediments in terms of pH and oxidation-reduction potentials; J. Geol., v. 60, p. 1–33.

KRUMBEIN, W.C. and SLOSS, L.L., 1963: Stratigraphy and Sedimentation; Freeman, San Francisco, 2nd ed., 660 p.

KRYNINE, P.D., 1942: Differential sedimentation and its products during one complete geosynclinal cycle; Proc. 1st Pan Amer. Congr. Mining Eng. Geol., Pt. 1, v. 2, p. 537–560.

KUMAR, N. and SANDERS, J.E., 1974: Inlet sequence: a vertical succession of sedimentary structures and textures created by the lateral migration of tidal inlets; Sedimentology, v. 21, p. 491–532.

KUMAR, N., and SANDERS, J.E., 1976: Characteristics of shoreface deposits; modern and ancient; J. Sediment. Petrol., v. 46, p. 145–162.

KURTZ, D.D., and ANDERSON, J.B., 1979: Recognition and sedimentologic description of recent debris flow deposits from the Ross and Weddell Seas, Antarctica; J. Sediment. Petrol., v. 49, p. 1159–1170.

LAND, L.S., 1973: Holocene meteoric dolomitization of Pleistocene limestones, North Jamaica; Sedimentology, v. 20, p. 411–424.

LAUGHBAUM, L.R., 1960: A Paleoecologic study of the Upper Denton Formation, Tarrant, Denton, and Cooke Counties, Texas; J. Paleont., v. 34, p. 1183–1197.

LAWSON, D.E., 1982: Mobilization, movement and deposition of active subaerial sediment flows, Matanuska Glacier, Alaska; J. Geol., v. 90, p. 279–300.

LE BLANC SMITH, G., 1980: Logical-letter coding system for facies nomenclature; Witbank Coalfield; Trans. Geol. Soc. S. Africa, v. 83, p. 301–312.

LEEDER, M.R., 1982: Sedimentology: Process and product; George Allen and Unwin, London, 344 p.

LEOPOLD, L.B. and WOLMAN, M.G., 1957: River channel patterns; braided, meandering and straight; U.S. Geol. Survey Prof. Paper 282–B.

LINK, M.H. and OSBORNE, R.H., 1978: Lacustrine facies in the Pliocene Ridge Basin Group: Ridge Basin, California, *in* A. Matter and M.E. Tucker, eds., Modern and ancient lake sediments; Internat. Assoc. Sedimentologists Spec. Publ. 2., p. 169–187.

LONG, D.G.F., 1978: Proterozoic stream deposits: some problems of recognition and interpretation of ancient sandy fluvial systems, *in* A.D. Miall, ed., Fluvial sedimentology; Can. Soc. Petrol. Geol. Mem. 5, p. 313–341.

LOWE, D.R., 1979: Sediment gravity flows: their classification and some problems of application to natural flows, *in* L.J. Doyle and O.H. Pilkey Jr., eds., Geology of Continental Slopes; Soc. Econ. Paleont. Mineral. Spec. Publ. 27, p. 75–82.

LOWENSTAM, H.A., 1950: Niagaran reefs in the Great Lakes area; J. Geol., v. 58, p. 430–487.

LOWENSTAM, H.A., 1957: Niagaran reefs in the Great Lakes area; Geol. Soc. Am. Mem. 67, v. 2, p. 215–248.

LOWENSTAM, H.A., WILLMAN, H.B. and SWANN, D.H., 1956: The Niagaran Reef at Thornton, Illinois; Am. Assoc. Petrol. Geol. and Soc. Econ. Paleont. Mineral. Guidebook, p. 1–19.

LUDVIGSEN, R., 1978: Middle Ordovician trilobite biofacies, southern Mackenzie Mountains, *in* C.R. Stelck and B.D.E. Chatterton, eds., Western and Arctic Canadian biostratigraphy; Geol. Assoc. Can. Spec. Paper 18, p. 1–37.

MACDONALD, K.B., 1975: Quantitative community analysis: recurrent group and cluster techniques applied to the fauna of the Upper Devonian Sonyea Group, New York; J. Geol., v. 83, p. 473–499.

MATTER, A. and TUCKER, M., eds., 1978: Modern and ancient lake sediments; Internat. Assoc. Sedimentologists Spec. Publ. 2.

McGOWEN, J.H. and SCOTT, A.J., 1976: Hurricanes as geologic agents on the Texas Coast, *in* L.E. Cronin, ed., Estuarine Research, v. 2, Geology and Engineering; Academic Press, London, p. 23–46.

McGREGOR, B.A., 1981: Ancestral head of Wilmington Canyon; Geology, v. 9, p. 254–257.

McILREATH, I.A. and JAMES, N.P., 1979: Facies models 12. Carbonate slopes, *in* R.G. Walker, ed., Facies Models; Geosci. Can. Reprint Series 1, p. 133–143.

McKEE, E.D., 1966: Structures of dunes at White Sands National Monument, New Mexico (and comparison with structures of dunes from other selected areas); Sedimentology, v. 7, p. 1–69.

McKEE, E.D., ed., 1979: A study of global sand seas; U.S. Geol. Survey Prof. Paper 1052.

MELLO, J.F. and BUZAS, M.A., 1968: An application of cluster analysis as a method of determining biofacies; J. Paleont., v. 42, p. 747–758.

MESOLELLA, K.J., ROBINSON, J.D., McCORMICK, L.M. and ORMISTON, A.R., 1974: Cyclic deposition of Silurian carbonates and evaporites in the Michigan Basin; Am. Assoc. Petrol. Geol. Bull., v. 58, p. 34–62.

MIALL, A.D., 1973: Markov chain analysis applied to an ancient alluvial plain succession; Sedimentology, v. 20, p. 347–364.

MIALL, A.D., 1977: A review of the braided river depositional environment; Earth Sci. Rev., v. 13, p. 1–62.

MIALL, A.D., 1978: Lithofacies types and vertical profile models in braided river deposits: a summary, *in* A.D. Miall, ed., Fluvial sedimentology; Can. Soc. Petrol. Geol. Mem. 5, p. 597–604.

MIALL, A.D., 1979a: Mesozoic and Tertiary geology of Banks Island, Arctic Canada: the history of an unstable craton margin; Geol. Surv. Can. Mem. 387.

MIALL, A.D., 1979b: Facies models 5: Deltas, *in* R.G. Walker, ed., Facies Models; Geosci. Can. Reprint Series 1, p. 43–56.

MIALL, A.D., 1980: Cyclicity and the facies model concept in fluvial deposits; Bull. Can. Petrol. Geol., v. 28, p. 59–80.

MIALL, A.D., 1983a: Glaciomarine sedimentation in the Gowganda Formation (Huronian), northern Ontario; J. Sediment. Petrol., v. 53, p. 477–491.

MIALL, A.D., 1983b: Basin analysis of fluvial sediments, *in* J.D. Collinson and J. Lewin, eds., Modern and ancient fluvial systems; Internat. Assoc. Sedimentologists Spec. Publ. 6, p. 279–286.

MIALL, A.D. and GIBLING, M.R., 1978: The Siluro–Devonian clastic wedge of Somerset Island, Arctic Canada, and some regional paleogeographic implications; Sediment. Geol., v. 21, p. 85–127.

MIDDLETON, G.V., 1973: "Johannes Walther's Law of the correlation of facies"; Geol. Soc. Am. Bull., v. 84, p. 979–988.

MIDDLETON, G.V. and HAMPTON, M.A., 1976: Subaqueous sediment transport and deposition by sediment gravity flows, *in* D.J. Stanley and D.J.P. Swift, eds., Marine sediment transport and environmental management; John Wiley, New York, p. 197–218.

MOLLARD, J.D., 1973: Airphoto interpretation of fluvial features, *in* Fluvial Processes and Sedimentation, Proc. Hydrology Symp., Edmonton, National Research Council, Canada, p. 341–380.

MOORE, R.C., 1949: Meaning of facies; Geol. Soc. Am. Mem. 39, p. 1–34.

MOUNT, J.E., 1982: Storm-surge-ebb origin of hummocky cross-stratified units of the Andrews Mountain Member, Campito Formation (Lower Cambrian), White-Inyo Mountains, eastern California; J. Sediment. Petrol., v. 52, p. 941–958.

MOUNTJOY, E., 1980: Some questions about the development of Upper Devonian carbonate buildups (reefs), Western Canada; Bull. Can. Petrol. Geol., v. 28, p. 315–344.

MOUNTJOY, E.W., COOK, H.E., PRAY, L.C. and McDANIEL, P.N., 1972: Allochthonous carbonate debris flows—worldwide indicators of reef complexes, banks or shelf margins; 24th. Internat. Geol. Congr., Montreal., Sect. 6, p. 172–189.

MULLINS, H.T. and NEUMANN, A.C., 1979: Deep carbonate bank margin structure and sedimentation in the northern Bahamas, *in* L.J.Doyle and O.H. Pilkey Jr., eds., Geology of continental slopes; Soc. Econ. Paleont. Mineral. Spec. Publ. 27, p. 165–192.

MULTER, H.G., 1971: Field guide to some carbonate rock environments, Florida Keys and western Bahamas; Fairleigh Dickinson Univ., Madison, New Jersey, 158 p.

MUTTI, E., 1974: Examples of ancient deep-sea fan deposits from circum-Mediterranean geosynclines, *in*

R.H. Dott Jr. and R.H.Shaver, eds., Modern and ancient geosynclinal sedimentation; Soc. Econ. Paleont. Mineral Spec. Publ. 19, p. 92–105.

MUTTI, E., 1977: Distinctive thin-bedded turbidite facies and related depositional environments in the Eocene Hecho Group (south-central Pyrenees, Spain); Sedimentology, v. 24, p. 107–132.

MUTTI, E. and RICCI-LUCCHI, F., 1972: Le turbiditi dell'Appennino settentrionale: Introduzione all' analisi di facies; Soc. Geol. Ital. Mem. 11, p. 161–199.

NATLAND, M.L., 1933: The temperature- and depth-distribution of some Recent and fossil foraminifera in the southern California region; Bull. Scripps Inst. Oceanog., v. 3, p. 225–230.

NELSON, C.H., 1976: Late Pleistocene and Holocene depositional trends, processes, and history of Astoria deep-sea fan, northeast Pacific; Mar. Geol., v. 20, p. 129–173.

NEMEC, W., PORĘBSKI, S.J. and STEEL, R.J., 1980: Texture and structure of resedimented conglomerates: examples from Ksiąz Formation (Fammenian-Tournaisian), southwestern Poland; Sedimentology, v. 27, p. 519–538.

NEWELL, N.D., RIGBY, J.K., FISCHER, A.G., WHITEMAN, A.J., HICKOX, J.E. AND BRADLEY, J.S., 1953: The Permian reef complex of the Guadalupe Mountains Region, Texas and New Mexico; W.H. Freeman, San Francisco, 236 p.

NEWELL, N.D., IMBRIE, J., PURDY, E.G. and THURBER, D.L., 1959: Organism communities and bottom facies, Great Bahama Bank; Bull. Am. Mus. Nat. Hist. 117, p. 177–228.

NICHOLS, D., 1959: Changes in the Chalk heart-urchin *Micraster* interpreted in relation to living forms; Phil. Trans. Roy. Soc. London, Ser. B., v. 242, p. 347–437.

NILSEN, T., 1980: Modern and ancient submarine fans: discussion of papers by R.G. Walker and W.R. Normark; Am. Assoc. Petrol. Geol. Bull., v. 64, p. 1094–1101.

NILSEN, T., BARTOW, J.A., STUMP, E. and LINK, M.H., 1977: New occurrences of dish structure in the stratigraphic record; J. Sediment. Petrol., v. 47, p. 1299–1304.

NIO, S-D., 1976: Marine transgression as a factor in the formation of sand wave complexes; Geol. Mijnb., v. 55, p. 18–40.

NORMARK, W.R., 1970: Growth patterns of deep-sea fans; Am. Assoc. Petrol. Geol. Bull., v. 54, p. 2170–2195.

NORMARK, W.R., 1978: Fan valleys, channels, and depositional lobes on modern submarine fans: characteristics for recognition of sandy turbidite environments; Am. Assoc. Petrol. Geol. Bull., v. 62, p. 912–931.

NORMARK, W.R., PIPER, D.J.W. and HESS, G.R., 1979: Distributary channels, sand lobes, and mesotopography of Navy Submarine Fan, California borderland, with applications to ancient fan sediments; Sedimentology, v. 26, p. 749–774.

ODIN, G.S and MATTER, A., 1981: De glauconiarum origine; Sedimentology, v. 28, p. 611–642.

PARKASH, B., AWASTHI, A.K. and GOHAIN, K., 1983: Lithofacies of the Markanda terminal fan, Kurukshetra District, India, *in* J.D. Collinson and J. Lewin, eds., Modern and ancient fluvial systems; Internat. Assoc. Sedimentologists Spec. Publ. 6, p. 337–344.

PICARD, M.D. and HIGH, L.R., Jr., 1973: Sedimentary structures of ephemeral streams; Dev. in Sediment. 17, Elsevier, Amsterdam, 223 p.

PICARD, M.D. and HIGH, L.R., Jr.,1981: Physical stratigraphy of ancient lacustrine deposits, *in* F.G. Ethridge and R.M. Flores, eds., Recent and ancient nonmarine depositional environments: models for exploration, Soc. Econ. Paleont. Mineral. Spec. Publ. 31, p. 233–259.

PIRSON, S.J., 1977: Geologic well log analysis, 2nd ed.; Gulf Pub. Co., Houston, 377 p.

POTTER, P.E., 1959: Facies models conference; Science, v. 129, p. 1292–1294.

POWELL, R.D., 1981: A model for sedimentation by tidewater glaciers; Ann. Glaciology, v. 2, p. 129–134.

PURDY, E.G., 1963: Recent calcium carbonate facies of the Great Bahama Bank; J. Geol., v. 71, p. 334–355, p. 472–497.

PURDY, E.G., 1974: Karst-determined facies patterns in British Honduras: Holocene carbonate sedimentation model; Am. Assoc. Petrol. Geol. Bull., v. 58, p. 825–855.

PURSER, B.H., ed., 1973: The Persian Gulf: Holocene carbonate sedimentation and diagenesis in a shallow epicontinental sea; Springer-Verlag, Berlin, 471 p.

READING, H.G., ed., 1978: Sedimentary environments and facies; Blackwell, Oxford, 557 p.

REINECK, H.E., 1972: Tidal flats, *in* J.K. Rigby and W.K. Hamblin, eds., Recognition of ancient sedimentary environments, Soc. Econ. Paleont. Mineral. Spec. Publ. 16, p. 146–159.

REINECK, H.E. and SINGH, I.B., 1980: Depositional sedimentary environments—with reference to terrigenous clastics; 2nd ed., Springer-Verlag, New York, 549 p.

REINECK, H.E. and WÜNDERLICH, R., 1968: Classification and origin of flaser and lenticular bedding; Sedimentology, v. 11, p. 99–104.

RICCI-LUCCHI, F., 1975: Depositional cycles in two turbidite formations of the northern Apennines (Italy); J. Sediment. Petrol., v. 45, p. 3–43.

RICCI-LUCCHI, F. and VALMORI, E., 1980: Basin-wide turbidites in a Miocene, over-supplied deep-sea plain: a geometrical analysis; Sedimentology, v. 27, p. 241–270.

RUST, B.R., 1972: Pebble orientation in fluviatile sediments; J. Sediment. Petrol., v. 42, p. 384–388.

RUST, B.R., 1978a: A classification of alluvial channel systems, *in* A.D. Miall, ed., Fluvial Sedimentology; Can. Soc. Petrol.Geol. Mem. 5, p. 187–198.

RUST, B.R., 1978b: Depositional models for braided alluvium, *in* A.D. Miall, ed., Fluvial Sedimentology, Can. Soc. Petrol. Geol. Mem. 5, p. 605–626.

RYAN, W.B.F., 1976: Quantitative evaluation of the depth of the western Mediterranean before, during and after the Late Miocene salinity crisis; Sedimentology, v. 23, p. 791–814.

RYDER, R.T., 1980: Lacustrine sedimentation and hydrocarbon occurrences: a review, *in* Am. Assoc. Petrol. Geol. Fall Ed. Conf. 103 p.

SCHLAGER, W. and CHERMAK, A., 1979: Sediment facies of platform-basin transition, Tongue of the Ocean, Bahamas, *in* L.J. Doyle and O.H. Pilkey, Jr., eds., Geology of continental slopes; Soc. Econ. Paleont. Mineral. Spec. Publ. 27, p. 193–208.

SCHOLLE, P.A., 1980: Petroleum potential of chalks with examples from Texas, Colorado–Kansas, and the North Sea; Am. Assoc. Petrol. Geol. Fall Ed. Conf., 27 p.

SCHREIBER, B.C., 1981: Marine evaporites: Facies development and relation to hydrocarbons and mineral genesis; Am. Assoc. Petrol. Geol. Fall Ed. Conf., 44 p.

SCHREIBER, B.C., FRIEDMAN, G.M., DECIMA, A. and SCHREIBER, E., 1976: Depositional environments of the Upper Miocene (Messinian) evaporite deposits of the Sicilian Basin; Sedimentology, v. 23, p. 729–760.

SCHUMM, S.A., 1963: A tentative classifiction of alluvial river channels; U.S. Geol. Survey Circ. 477.

SCHUMM, S.A., 1968: Speculations concerning paleohydrologic control of terrestrial sedimentation; Geol. Soc. Am. Bull., v. 79, p. 1573–1588.

SCHUMM, S.A., 1981: Evolution and response of the fluvial system; sedimentological implications, *in* F.G. Ethridge and R. Flores eds., Recent and ancient nonmarine depositional environments: models for exploration; Soc. Econ. Paleont. Mineral. Spec. Publ. 31, p. 19–29.

SCHWAB, F.L., 1981: Evolution of the western continental margin, French–Italian Alps: sandstone mineralogy as an index of plate tectonic setting; J. Geol., v. 89, p. 349–368.

SEILACHER, A., 1967: Bathymetry of trace fossils; Mar. Geol., v. 5, p. 413–428.

SELLEY, R.C., 1968: A classification of paleocurrent models; J. Geol., v. 76, p. 99–110.

SELLEY, R.C., 1970: Studies of sequence in sediments using a simple mathematical device; Quart. J. Geol. Soc. London, v. 125, p. 557–581.

SELLEY, R.C., 1979: Dipmeter and log motifs in North Sea submarine-fan sands; Am. Assoc. Petrol. Geol. Bull., v. 63, p. 905–917.

SELLWOOD, B.W., 1978: Shallow-water carbonate environments, *in* H.G. Reading, ed., Sedimentary environments and facies; Blackwell, Oxford, p. 259–313.

SHAWA, M.S., ed., 1974: Use of sedimentary structures for recognition of clastic environments; Can. Soc. Petrol. Geol., 66 p.

SHEA, J.H., 1982: Twelve fallacies of uniformitarianism; Geology, v. 10, p. 449–496.

SHEARMAN, D.J., 1966: Origin of marine evaporites by diagenesis; Trans. Inst. Min. Metall. Bull., v. 75, p. 208–215.

SHEPARD, F.P., 1981: Submarine canyons: multiple causes and long-time persistence; Am. Assoc. Petrol. Geol. Bull., v. 65, p. 1062–1077.

SHEPARD, F.P., MARSHALL, N.F., McLOUGHLIN, P.A. and SULLIVAN, G.G., 1979: Currents in submarine canyons and other sea valleys; Am. Assoc. Petrol. Geol. Studies in Geol. 8.

SHINN, E.A., HALLEY, R.B., HUDSON, J.H., LIDZ, B. and ROBBIN, D.M., 1979: Three dimensional aspects of Belize patch reefs; Am. Assoc. Petrol. Geol. Bull., v. 63, p. 528.

SIMONS, D.B. and RICHARDSON, E.V., 1961: Forms of bed roughness in alluvial channels; Am. Soc. Civil. Eng. Proc., v. 87, No. HY3, p. 87–105.

SIMONS, D.B., RICHARDSON, E.V., and NORDIN, C.F., 1965: Sedimentary structures generated by flow in alluvial channels, *in* G.V. Middleton, ed., Primary sedimentary structures and their hydrodynamic interpretation; Soc. Econ. Paleont. Mineral. Spec. Publ. 12, p. 34–52.

SKIPPER, K., 1976: Use of geophysical wireline logs for interpreting depositional processes; Geosci. Can., v. 3, p. 279–280.

SMITH, D.L., 1977: Transition from deep- to shallow-water carbonates, Paine Member, Lodgepole Formation; central Montana, *in* H.E. Cook and P. Enos, eds., Deep-water carbonate environments; Soc. Econ. Paleont. Mineral. Spec. Publ. 25, p. 187–202.

SMITH, D.G. and SMITH, N.D., 1980: Sedimentation in anastomosed river systems: examples from alluvial valleys near Banff, Alberta; J. Sediment. Petrol., v. 50, p. 157–164.

SMOOT, J.P., 1978: Origin of the carbonate sediments in the Wilkins Peak Member of the lacustrine Green River Formation (Eocene), Wyoming, U.S.A., *in* A. Matter and M.E. Tucker, eds., Modern and ancient lake sediments; Internat. Assoc. Sedimentologists Spec. Publ. 2, p. 107–126.

SMOSNA, R. and WARSHAUER, S.M., 1981: Rank exposure index on a Silurian carbonate tidal flat; Sedimentology, v. 28, p. 723–731.

SNEIDER, R.M., TINKER, C.N. and MECKEL, L.D., 1978: Deltaic environment reservoir types and their characteristics; J. Petrol. Tech., v. 20, p. 1538–1546.

SOLOHUB, J.T. and KLOVAN, J.E., 1970: Evaluation of grain-size parameters in lacustrine sediments; J. Sediment Petrol., v. 40, p. 81–101.

SOUTHARD, J.B., 1971: Representation of bed configurations in depth–velocity–size diagrams; J. Sediment. Petrol., v. 41, p. 903–915.

STEEL, R.J., 1974: New Red Sandstone floodplain and piedmont sedimentation in the Hebridean Province; J. Sediment. Petrol., v. 44, p. 336–357.

STEEL, R.J., MAEHLE, S., NILSEN, H., RØE, S.L. and SPINNANGR, Å., 1977: Coarsening-upward cycles in the alluvium of Hornelen Basin (Devonian), Norway: Sedimentary response to tectonic events; Geol. Soc. Am. Bull., v. 88, p. 1124–1134.

STEIDTMANN, J.R., 1974: Evidence for eolian origin of cross-stratification in sandstone of the Casper Formation, southernmost Laramie Basin, Wyoming; Geol. Soc. Am. Bull., v. 85, p. 1835–1842.

STOAKES, F.A., 1980: Nature and control of shale basin fill and its effect on reef growth and termination: Upper Devonian Duvernay and Ireton Formations of

Alberta, Canada; Bull. Can. Petrol. Geol., v. 28, p. 345–410.

STOW, D.A.V. and LOVELL, J.P.B., 1979: Contourites: their recognition in modern and ancient sediments; Earth Sci. Rev., v. 14, p. 251–291.

SWIFT, D.J.P., 1975: Barrier island genesis: evidence from the Middle Atlantic shelf of North America; Sediment. Geol., v. 14, p. 1–43.

SWIFT, D.J.P., STANLEY, D.J. and CURRAY, J.R., 1971: Relict sediments on continental shelves: a reconsideration; J. Geol., v. 79, p. 322–346.

TEICHERT, C., 1958: Concept of facies; Am. Assoc. Petrol. Geol. Bull., v. 42, p. 2718–2744.

TUNBRIDGE, I.P., 1981: Sandy high-energy flood sedimentation—some criteria for recognition, with an example from the Devonian of S.W. England; Sediment. Geol., v. 28, p. 79–96.

TURNER, P., 1980: Continental red beds; Dev. in Sediment., Elsevier, No. 29.

VAI, G.B. and RICCI-LUCCHI, F., 1977: Algal crusts, autochthonous and clastic gypsum in a cannibilistic evaporite basin: a case history from the Messinian of Northern Apennines; Sedimentology, v. 24, p. 211–244.

VAN HOUTEN, F.B., 1964: Cyclic lacustrine sedimentation, Upper Triassic Lockatong Formation, central New Jersey and adjacent Pennsylvania, *in* D.F. Merriam, ed., Symposium on cyclic sedimentation; Geol. Surv. Kansas Bull. 169, p. 495–531.

VAN HOUTEN, F.B., 1973: Origin of red beds: a review—1961–1972, *in* F.A. Donath ed., Ann. Rev. Earth Plan. Sci., v. 1, p. 39–62.

VAN STRAATEN, L.M.J.U., 1951: Texture and genesis of Dutch Wadden Sea sediments; Proc. 3rd. Internat. Congress Sedimentology, Netherlands, p. 225–255.

VAN STRAATEN, L.M.J.U., 1954: Composition and structure of Recent marine sediments in the Netherlands; Leidse Geol. Meded., v. 19, p. 1–110.

VISHER, G.S., 1965: Use of vertical profile in environmental reconstruction; Am. Assoc. Petrol. Geol. Bull., v. 49, p. 41–61.

VISHER, G.S., 1969: Grain size distributions and depositional processes; J. Sediment. Petrol., v. 39, p. 1074–1106.

WALKER, R.G., 1975: Generalized facies models for resedimented conglomerates of turbidite association; Geol. Soc. Am. Bull., v. 86, p. 737–748.

WALKER, R.G., 1978: Deep-water sandstone facies and ancient submarine fans: models for exploration for stratigraphic traps; Am. Assoc. Petrol. Geol. Bull., v. 62, p. 932–966.

WALKER, R.G., ed., 1979a: Facies models; Geosci. Can. Reprint Series 1, 211 p.

WALKER, R.G., 1979b: Facies models 8. Turbidites and associated coarse clastic deposits, *in* R.G. Walker, ed., Facies Models, Geosci. Can. Reprint Series 1, p. 91–103.

WALKER, R.G., 1979c: Facies models 7. Shallow marine sands, *in* R.G. Walker, ed., Facies Models; Geosci. Can. Reprint Series 1, p. 75–89.

WALKER, R.G. and MIDDLETON, G.V., 1979: Facies models 4. Eolian sands, *in* R.G. Walker, ed., Facies Models; Geosci. Canada Reprint Series 1, p. 33–41.

WALKER, T.R., 1967: Formation of red beds in modern and ancient deserts; Geol. Soc. Am. Bull., v. 78, p. 353–368.

WALKER, T.R. and HARMS, J.C., 1972: Eolian origin of Flagstone beds, Lyons Sandstone (Permian), Type area, Boulder County, Colorado; Mountain Geol., v. 9, p. 279–288.

WALTHER, J., 1893–1894: Einleitung in die Geologie als Historische Wissenschaft; Jena, Fischer Verlag, 3 vols.

WEIMER, R.J. and HOYT, H.J., 1964: Burrows of *Callianassa major* Say, geologic indicators of littoral and shallow neritic environments; J. Paleont., v. 38, p. 761–767.

WILKINSON, B.H., 1982: Cyclic cratonic carbonates and Phanerozoic calcite seas; J. Geol. Educ., v. 30, p. 189–203.

WILSON, J.L., 1975: Carbonate facies in geologic history; Springer-Verlag, New York, 471 p.

WILSON, R.C.L., 1975: Upper Jurassic Oolite shoals, Dorset coast, England, *in* R.N. Ginsburg, ed. Tidal Deposits; Springer-Verlag, New York, p. 355–362.

WINKER, C.D., and EDWARDS, M.B., 1983: Unstable progradational clastic shelf margins, *in* D.J. Stanley and G.T. Moore, eds., The Shelfbreak: critical interface on continental margins; Soc. Econ. Paleont. Mineral Spec. Publ. 33, p. 139–158.

WONG, P.K. and OLDERSHAW, A.E., 1980: Causes of cyclicity in reef interior sediments, Kaybob Reef, Alberta; Bull. Can. Petrol. Geol., v. 28, p. 411–424.

WRIGHT, R., and ANDERSON, J.B., 1982: The importance of sediment gravity flow to sediment transport and sorting in a glacial marine environment: Eastern Weddell Sea, Antarctica; Geol. Soc. Am. Bull., v. 93, p. 951–963.

YOUNG, F.G., MYHR, D.W. and YORATH, C.J., 1976: Geology of the Beaufort–Mackenzie Basin; Geol. Surv. Can. Paper 76–11.

CHAPTER 5

Basin mapping methods

5.1 Introduction

Facies analysis, as described in Chapter 4, begins with the art of interpreting single outcrops or, at most, small areas, such as a well-exposed cliff section. Paleogeographic analysis of an entire sedimentary basin requires the documentation of many such outcrops (or well sections), their correlation by stratigraphic methods (Chap. 3), and the use of various sedimentological mapping techniques. The latter are the subject of this chapter.

Stratigraphic units undergo thickness and facies changes across a basin, reflecting contemporaneous paleogeography, subsidence patterns and location of sediment sources. Most of these aspects of basin development are controlled by tectonics, by the development of bounding faulted or folded uplifts and by subsidence caused by thermal contraction of continental margins, basement failure or thrust-sheet loading (Chap. 9). Eustatic sea level changes are also of profound paleogeographic importance (Chap. 8). Such effects leave their imprint on sediments via their control on paleoslope, by the structuring of land masses which guide oceanic currents, wave advance or air flow, and by the development of depocenters and their control on water depths or rates of internal scour.

The art of basin mapping is one of reconstructing paleogeography and fill geometry from very limited evidence. Skills at data synthesis, interpretation and extrapolation are of prime importance. Maps drawn only from outcrop data are rarely sufficiently accurate or useful for subsurface prediction because they do not contain adequate three-dimensional distribution of data points.

Two broad classes of data are used most frequently in basin mapping. The first class deals with facies thickness and ratio data, and petrographic information. Such data reflect environment and paleogeography but suggest no pattern until they have been measured at many control points and the results contoured. This approach is discussed in sections 5.2, 5.3 and 5.5. Geophysical methods are discussed in section 5.4 and computer methods for manipulating these data are discussed in section 5.6. The use of stratigraphic cross sections is illustrated in section 5.8. The second class of data refers to oriented objects or structures within the rocks. Each of these has directional (paleogeographic) significance, although statistical methods are required to extract local or regional trends from the data sets. The study of this class of data is what is normally implied by the term paleocurrent analysis. It is discussed in section 5.9. Examples of the synthesis of these various types of information are described in section 5.10. Many of these mapping methods may be used simultaneously, and the order of presentation in this chapter is not intended to suggest a rigid sequence of mapping procedures. For example it is commonly advisable to start a detailed stratigraphic analysis by constructing cross sections in order to clarify correlations.

Many other specialized techniques are discussed by Potter and Pettijohn (1977) and King (1972), and in numerous research publications, but those dealt with here are the most widely used in regional basin studies for the purpose of petroleum, coal and ore prospecting and evaluation.

One new mapping technique that has generated some interest recently is the use of satellite imagery to locate subtle structures in a search for giant oil and gas fields. So far the technique has only been used retrospectively. Halbouty (1980) published Landsat images for 15 areas containing known giant fields and analyzed them to determine what, if anything, could have been predicted about the fields in the absence of any ground data. The results are singularly unconvincing. Vague tonal changes or disconnected lineaments are apparent, but in most cases bear little or no relationship to the fields. It seems unlikely that such images will ever be of more than passing interest for basin mapping purposes, even for the remotest frontier regions, where the real reconnaissance work is based on conventional air photography and aerial geophysics. (These are not available for many areas of the world for political reasons, and satellite images are commonly used as a substitute. However, their scale is small and they provide no stereoscopic coverage.) Satellite photos are, of course, no use at all for offshore basins.

5.2 Structure and isopach contouring

Structure contour and isopach maps are as essential as a stratigraphic correlation table for interpreting basin history. They may or may not contain much direct paleogeographic information but they are required in order to illustrate basin shape and orientation and basin fill geometry.

Structure contours are drawn to illustrate the attitude of selected stratigraphic horizons within the basin. They are always drawn relative to sea level. Control points are derived from whatever surface and subsurface information is available. Depth data in petroleum exploration wells are converted to "depth subsea" by subtracting the elevation of the kelly bushing (K.B.) of the drill rig. It may be useful to illustrate fill geometry by drawing several structure contour maps at carefully selected horizons. Commonly these are chosen at major unconformities, which demarcate discrete stratigraphic packages within the basin fill. Most important of all is that drawn on the "basement" unconformity. The term basement is a useful one in basin studies, but it should be used with caution as it may mean different things to different people. Normally the term refers to the base of the structurally conformable sequence which the geologist is engaged in mapping. This is usually an unconformity, and may be a profound one with, for example, Mesozoic or Cenozoic rocks resting on Precambrian. Elsewhere the basement surface may be within the Phanerozoic, for example the basement of the Sverdrup Basin in the Canadian Arctic is of Cambrian to Devonian age. The basin fill itself is Mississippian to Tertiary, and rests on a profound unconformity developed during a major middle to Late Devonian orogeny. Petroleum geologists commonly refer to "economic basement", which means the depth of the section beyond which all hope for oil and gas has been abandoned. It is drawn at an unconformity across which major diagenetic structural or metamorphic changes are apparent.

Structure contours are, of course, of prime importance in the exploration for structural petroleum traps. In the initial stages of exploration they may be drawn entirely from seismic data, using depths of prominent reflectors plotted as "two-way travel time" (the time taken for the sound wave to travel down to the reflector and back). As well information becomes available these times are converted to depths and the maps redrawn.

On a broad scale structure contours, particularly at deeper levels within the basin, reveal the location of sub-basins, depocenters and axes of uplift. They may outline subtle syndepositional topography that had important effects on local paleogeography and facies patterns. Detailed local maps may reveal drape structures over channel fills, or reefs, or drainage nets eroded into unconformity surfaces. Detailed well control may yield this information directly, or it may be reconstructed and interpreted from magnetic, gravity or seismic data. Several examples of structure maps showing some of these features are illustrated in section 5.4.

In some areas it may be useful to emphasize local structure (fault blocks, drapes, etc.) by subtracting regional dips from the local data. Computer methods are particularly useful for this purpose, as discussed in section 5.6.

Paleogeological maps or **subcrop maps** are maps showing the units outcropping at an unconformity surface. They can only be reconstructed from extensive outcrop or subsurface data. Such maps reveal the extent of tectonic deformation and erosion that occurred before deposition of the post-unconformity strata. They may show buried folds or faults, and therefore are useful for reconstructing ancient stress patterns. Busch (1974) illustrated examples where detailed well control permitted the construction of paleogeological maps that showed drainage patterns, just as do geological maps of present-day land sur-

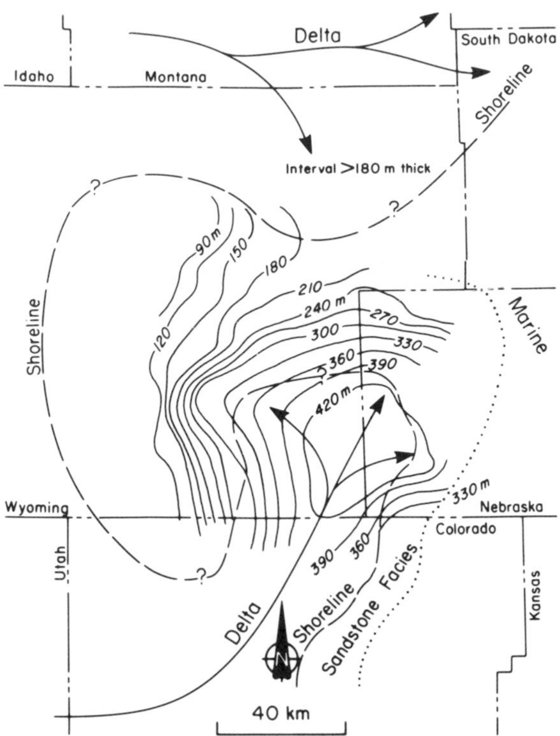

Fig. 5.1. An isopach map outlining an Upper Cretaceous deltaic depocenter, Rocky Mountains area, Wyoming–Colorado, (Weimer, 1970).

faces underlain by gently dipping strata and cut by a network of stream valleys. **Worm's-eye-view maps** show the distribution of units which onlap an unconformity. Their main use is in illustrating the pattern of basin fill, shifting shorelines or gradual burial of a pre-existing erosional topography.

An isopach is a line of equal thickness. Isopach maps should be drawn for an entire basin fill and for selected stratigraphic intervals of interest in order to illustrate basin fill geometry. In areas of structural complexity it may be impossible to draw meaningful or sedimentologically useful structure contour maps, whereas isopach maps reveal the basin fill in its original, undeformed form and reveal something about contemporaneous structure. Isopachs are particularly useful for illustrating the location and outline of major depocenters, such as large deltas (Fig. 5.1), and for delineating syndepositional intrabasinal upwarps or "schwelle" over which the section is thinned (Fig. 5.2). Isopachs are usually constructed only for units bounded by conformable contacts. If they are drawn for unconformity-bounded units interpretations must take into account the effects of post-depositional structural deformation and erosion. Isopachs will also be distorted if the unit to be plotted cannot be precisely correlated.

5.3 Lithofacies maps

A variety of methods has been devised to map lithofacies variations. Most have been in use for many years, but the techniques for constructing and interpreting them have become more refined as more has been learned about depositional environments.

When constructing facies maps the geologist should always work with as thin a stratigraphic unit as is possible. The objective of the exercise is to reconstruct paleogeography, but in every basin this was constantly changing. Delta and submarine fan distributaries switched, shorelines transgressed or regressed, carbonate platforms prograded or were deeply scoured by hurricanes, and so on. By working with a thin unit we can hope to reconstruct the paleogeography at some more or less frozen instant in time. Too thick an interval and all the subtleties are lost in a sea of generalizations. This is why the refinements of stratigraphic correlation discussed in Chapter 3 are so important.

5.3.1 Multicomponent maps

Lithofacies maps may be plotted either as ratios or as isopachs of selected lithofacies components. Ratio maps depict the variations between two or three lithofacies end members which are selected to emphasize variations across the basin. Potter and Pettijohn (1977) defined a clastic ratio which can easily be calculated from the records of surface or subsurface stratigraphic sections. The clastic ratio is defined as the ratio of total cumulative thickness of clastic deposits to thickness of nonclastics:

$$\frac{\text{conglomerate} + \text{sandstone} + \text{mudstone}}{\text{limestone} + \text{dolomite} + \text{evaporite} + \text{coal}}$$

Values are calculated and plotted for as many control points as are available, and are then contoured by hand or with the use of a computer contouring program. The main use of the clastic ratio is in mapping the extent of a marginal clastic belt

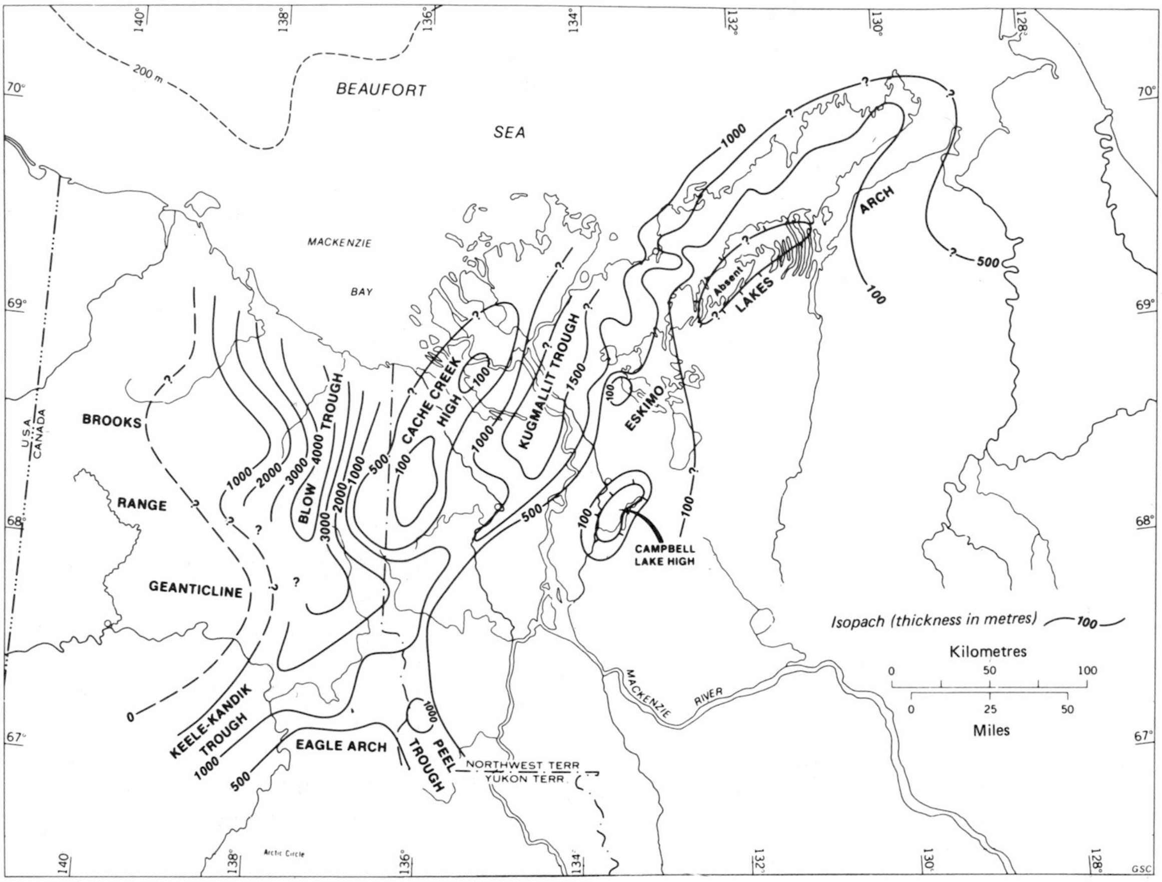

Fig. 5.2. Isopach map of Barremian–Albian strata, Mackenzie Delta region, Canada, which delineates depocenters and local structural highs (Young et al. 1976).

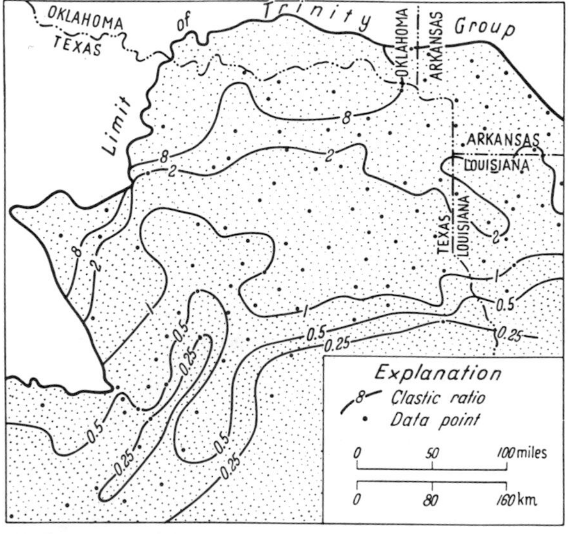

Fig. 5.3. A clastic ratio map for part of the Cretaceous section of the Gulf Coast region. Note higher concentration of clastics near northern edge of basin, which is the direction of the sediment source (Potter and Pettijohn, 1977).

flanking a carbonate platform or evaporite basin. An example is illustrated in Figure 5.3.

Another technique is to recognize three principal lithofacies components of a stratigraphic unit (e.g. sandstone–siltstone–mud or limestone–sandstone–mud), determine the cumulative thickness of each in the control point sections, and recalculate the values (if necessary) to sum to 100%. A ternary diagram is drawn up using these three components as end members, and it is subdivided into fields as shown in Figure 5.4. Each field in the triangle is colored or ornamented, and the control point values examined to determine in which field they fall. The map can then be plotted and contoured with the same colors or patterns. The subdivision of the triangle into fields is arbitrary and may be chosen to group or subdivide component mixtures in whatever way is most useful.

The problem with the clastic ratio and ternary plot methods is that they may be very crude

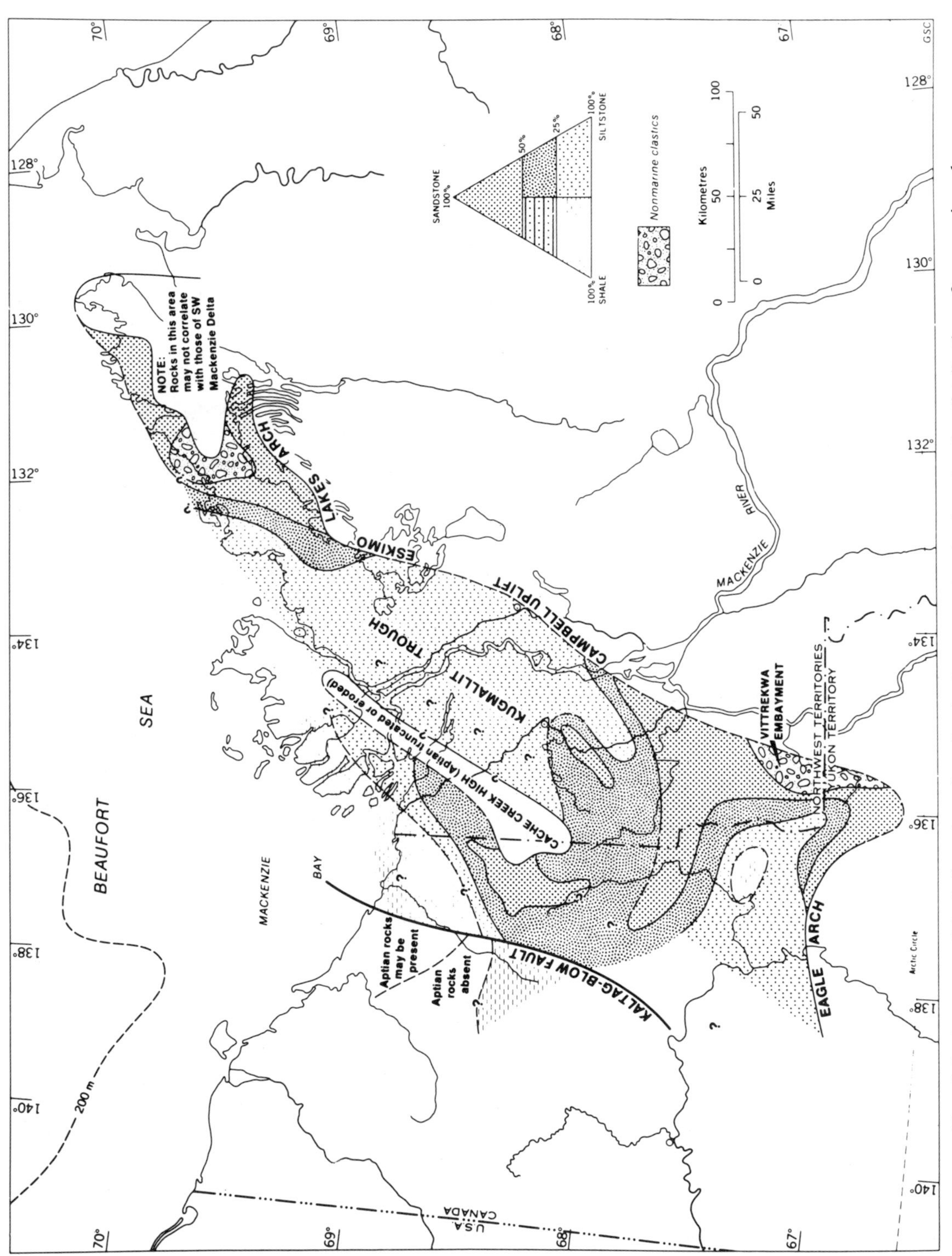

Fig. 5.4. A ternary lithofacies plot of early Aptian sediments, Mackenzie Delta region, Canada. Maps of this type give a general idea of transport directions and depositional energy, but in this case the clearest information is given by the distribution of nonmarine clastics, an additional lithofacies subdivision that is not part of the ternary plot (Young et al., 1976).

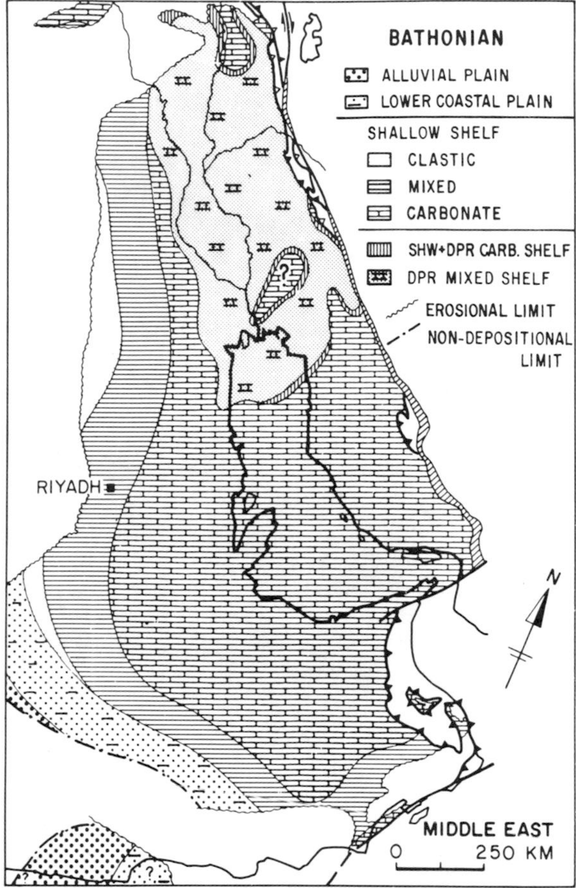

Fig. 5.5. A lithofacies assemblage map; Bathonian sediments of the Persian Gulf region (Murris, 1980).

reflections of trends in depositional evironments. This is particularly true of ternary sand–silt–mud or other clastic combinations, because these attempt to force what in nature is a simple two end-member system into a three-component mold. In clastic environments what produces lithofacies variations is the hydrodynamic sorting of coarse to fine grains—how coarse and how fine depends on current strength and the grain size range of available detritus. The best maps are therefore those which exploit this two-dimensional process. The fields in the ternary diagram of Figure 5.4 cannot readily be interpreted in terms of environmental variables. To make the paleogeography clearer Young et al. (1976) have added an additional nonsystematic category for nonmarine clastics, and its occurrence on the map shows more clearly than the ternary data where the principal sediment sources are located. The clastic ratio device may be useful where there is a clear facies change from clastic to non-clastic across the map area. Even here subtle variations in clastic grain size or carbonate–evaporite relationships will be obscured.

5.3.2 Lithofacies assemblage maps

There are two solutions to the problems outlined above, which make for better lithofacies mapping. They represent opposite extremes of complexity, but both are extremely useful in illustrating the full range of paleogeographic variability in a basin.The first method is to plot the distribution of lithofacies assemblages. The assemblages may be defined in any of the ways discussed in Chapter 4, using descriptive criteria or statistical methods. Once the assemblages are defined, however arbitrary or mathematical the process, each control point is assigned to one of the assemblages and the results plotted and contoured. The classic Bahama Platform lithofacies map referred to in Chapter 4 (Fig. 4.9) is a good example of this type of map. Another example, from an ancient shelf, is illustrated in Figure 5.5 (from Murris, 1980), and shows a northeastward gradation across the Arabian craton from a marginal coastal plain clastic belt to shelf and deeper-water carbonate areas. Murris has defined transitional zones between his lithofacies assemblages such as the mixed clastic–carbonate shelf. If this makes for a more meaningful map there is nothing wrong with the approach, but it may require arbitrary decisions about cutoffs between the lithofacies assemblage classes.

5.3.3 Single component maps

Probably the best and most subtle way to illustrate paleogeographic patterns is to focus on one lithofacies or lithofacies assemblage at a time. Data for each are treated separately and interpreted in terms of its own depositional controls. Each map then reflects a specific process or set of related processes. There are various ways to plot the data. Total thickness is the simplest, but these values will be distorted if the stratigraphic unit has an eroded top. This problem can be circumvented by calculating the thickness of the lithofacies as a percent of total unit thickness, and then only at the feather edge of the unit are values likely to be distorted, where the thin basal portion remaining may not be representative. In many clastic depositional systems the only important lithofacies components are sandstone and mudstone. A commonly used technique is to plot sandstone/mudstone ratio, which is a variation of the percentage map. Conglomerate, if present, should be added to the sandstone total. Many thousands of "total sand isopach" or "sand isolith" maps have been drawn up by petroleum geologists, particularly to illustrate sandstone reservoirs in stratigraphic traps, where paleogeographic control of reservoir geometry is of paramount importance. The technique is also widely used for carbonate rocks.

Commonly, petroleum geologists plot porous section (e.g. "porosity feet isopleths") rather than total section. These are the values they are really interested in, and they commonly are just as useful in determining or illustrating paleogeography. Porosity is controlled by lithofacies or grain size, and even diagenetic modifications of porosity tend to be guided by original composition because of its control on porosity paths for migrating fluids. The same approach would be useful in the exploration for stratabound ores, but has been less widely exploited by mining geologists.

Some examples of these one-component maps are given in Figures 5.6 to 5.11. Figure 5.6 illustrates the use of "total porous section" data in mapping delta systems on a regional scale (Miall, 1976a). These Triassic deltas were derived from the craton overlying the Canadian Shield to the east. Downdip to the west they "shale-out" into a monotonous mud–silt sequence containing thin turbidites. This map groups together all the sandy subfacies such as channel fills, crevasse splays, distributary mouth bars, etc. These can be identified and separated by working on a larger scale, providing the well control is available. Examples are shown in Figures 5.7 and 5.8. The first of these shows total-sand isopachs which define a valley-fill sandstone of Mississippian age (Lyons and Dobrin, 1972). The large scale and the U-shaped plan geometry of the sandstone body show that it is not a meandering channel in the conventional sense, as meandering generates sheet sandstones by lateral accretion. Figure 5.8 illustrates the fill of a major fluvial channel and a digitate crevasse splay sandstone extending out from it (Galloway, 1981). Note that well spacing here is between about 20 and 200 m. This example is from the Cenozoic of the Gulf Coast. Figure 5.9 illustrates a different type of environment, a linear barrier island sand trend in a Cretaceous unit (San Juan Basin, New Mexico; Sabins, 1972). The barrier sand has been separated from related facies for plotting purposes. Superimposition of sand isopachs and structural contours shows that the barrier is aligned parallel to structural strike.

Some carbonate examples are illustrated in the next two maps. Figure 5.10 shows the distribution of middle Cretaceous limestones in the Golden Lane area of Mexico. The Golden Lane Atoll is clearly demarcated, with a thinner sheet of deep-water talus deposits (Tamabra Limestone) banked against it (Enos, 1977). Figure 5.11 shows porosity thickness isopachs of a flanking unit in a Paleocene bioherm, Libya (Brady et al., 1980).The outline of the entire bioherm is shown by a dashed line. It developed the flanking biohermal unit during a time of emergence, when depositional slopes were thought to have been too great on the west side of the reef for sedimentation to occur.

Other facies mapping techniques that might prove useful in particular cases include plotting bed thickness or the number of sandstone beds in the section or the number of sandstone units greater than (say) 5 m in thickness, and so on. Anything that reflects paleogeographic changes across the map area should be tried, and it is up to the ingenuity of the geologist to explore these. Padgett and Ehrlich (1978) used a special technique to illustrate fluvial drainage patterns in a Carboniferous coal basin. They mapped the area of mined-out coal for a single seam. Areas not mined correspond to dendritic channels that scoured into the original peat swamp (Fig. 5.12). At a different stratigraphic level the same technique revealed the opposite effect: the coal seam itself outlined a channel, which became filled with peat following abandonment. Friend and Moody-Stuart (1972) used the mean percentage of sand in fluvial fining-upward cycles as a lithofacies mapping tool (some of their other methods are described in section 5.5). Rees (1972) discussed other quantitative techniques such as "distance–function" or "facies–departure" maps. He also provided examples of multicomponent facies maps similar to those described in section 5.3.1. Walker (1967) defined a number of measurements to be used to calibrate proximal–distal relations in turbidite beds, including the ABC index, which reflects the fact that first the A and then the B and C units in a

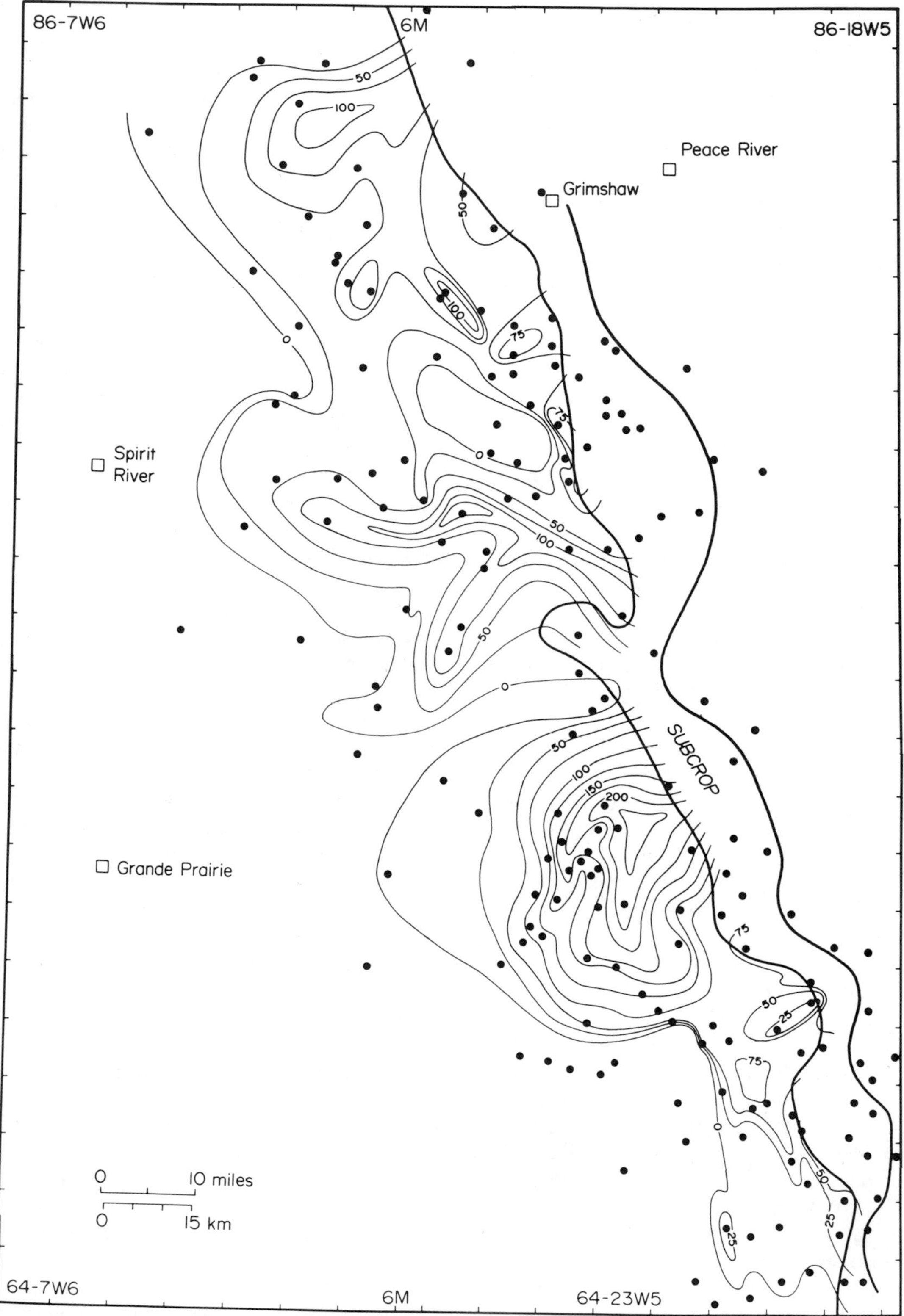

Fig. 5.6. Distribution of total porous section in one of four informal members of the Triassic Toad Grayling Formation, northwest Alberta. Lobate and birdsfoot river-dominated deltas and possible barrier sands parallel to the basin margin are delineated (Miall, 1976a).

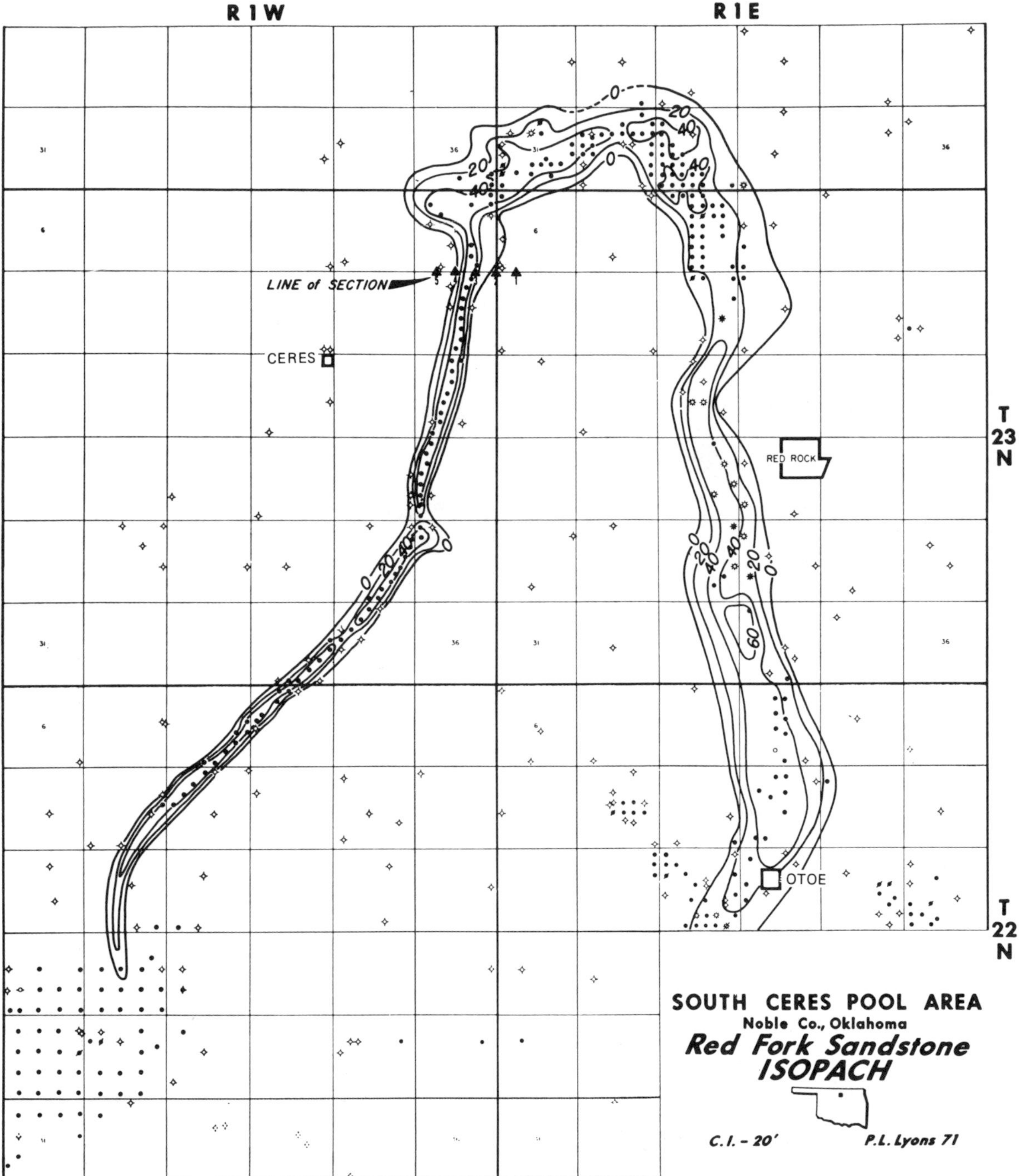

Fig. 5.7. Sand isopach of a Mississippian valley-fill sand in Oklahoma, which forms a stratigraphic petroleum trap (Lyons and Dobrin, 1972).

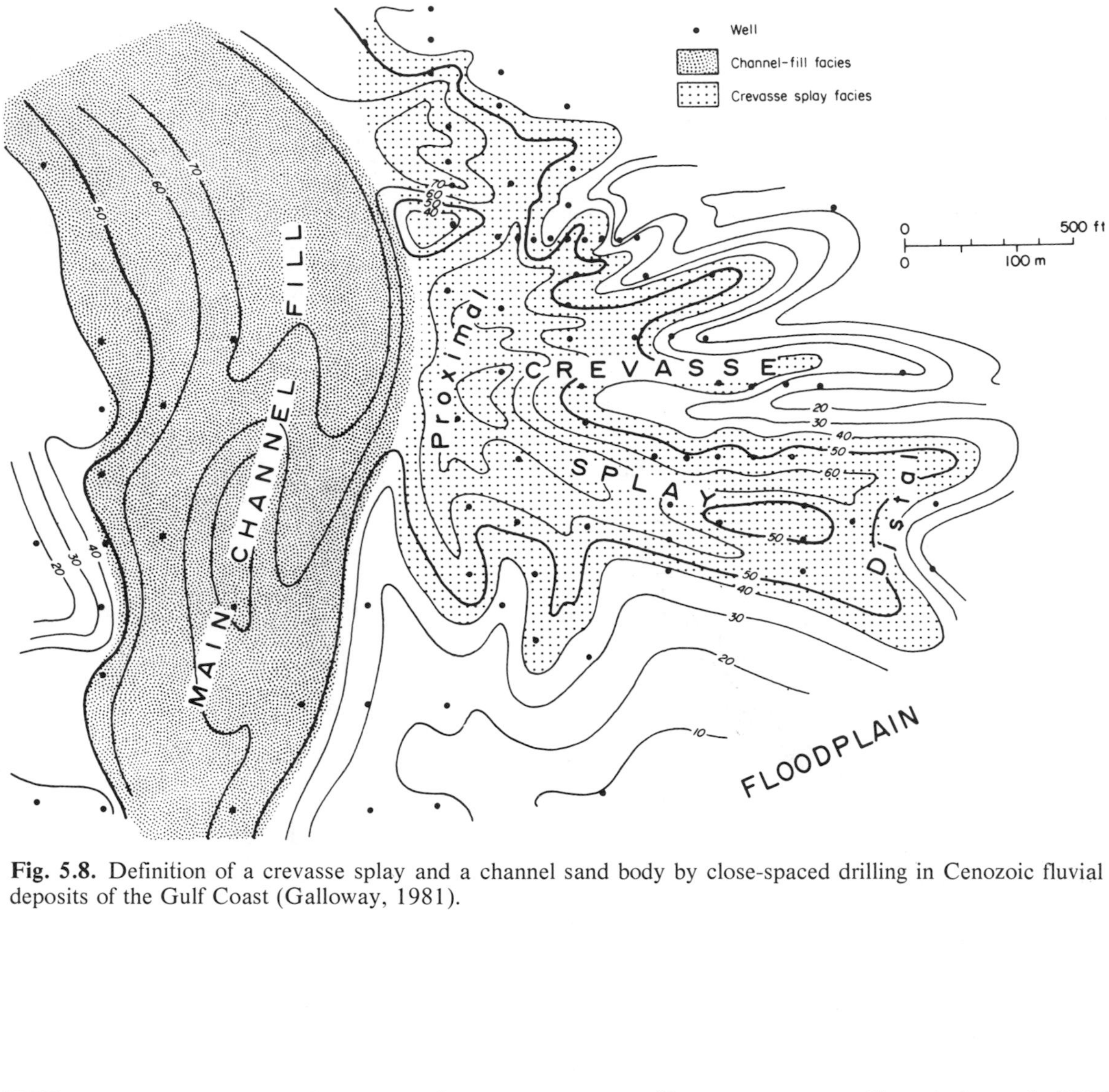

Well
Channel-fill facies
Crevasse splay facies
0 500 ft
0 100 m
MAIN CHANNEL FILL
Proximal
CREVASSE
SPLAY
Distal
FLOODPLAIN

Fig. 5.8. Definition of a crevasse splay and a channel sand body by close-spaced drilling in Cenozoic fluvial deposits of the Gulf Coast (Galloway, 1981).

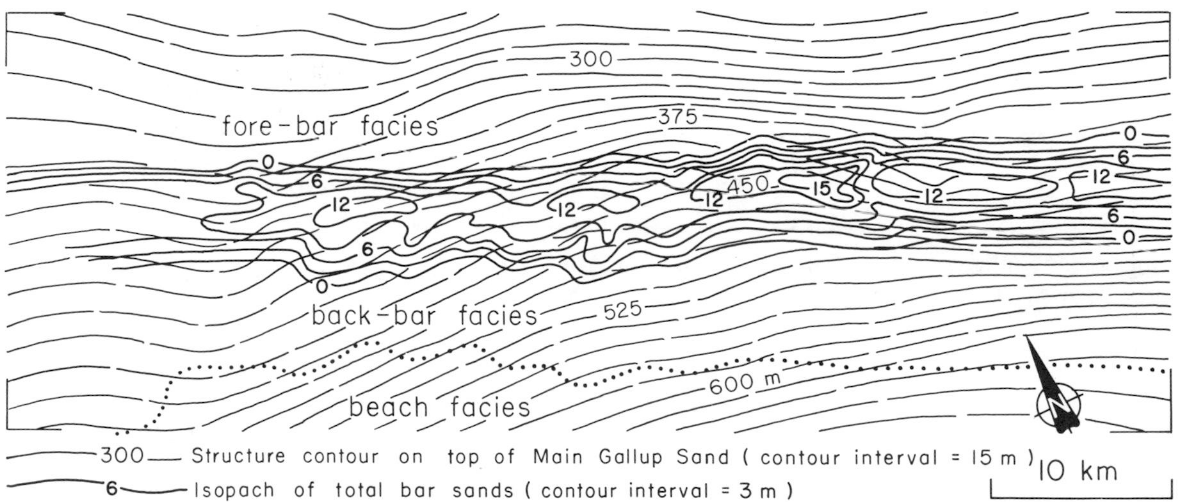

Fig. 5.9. Isopach of a Cretaceous barrier bar sandstone in New Mexico (Sabins, 1972).

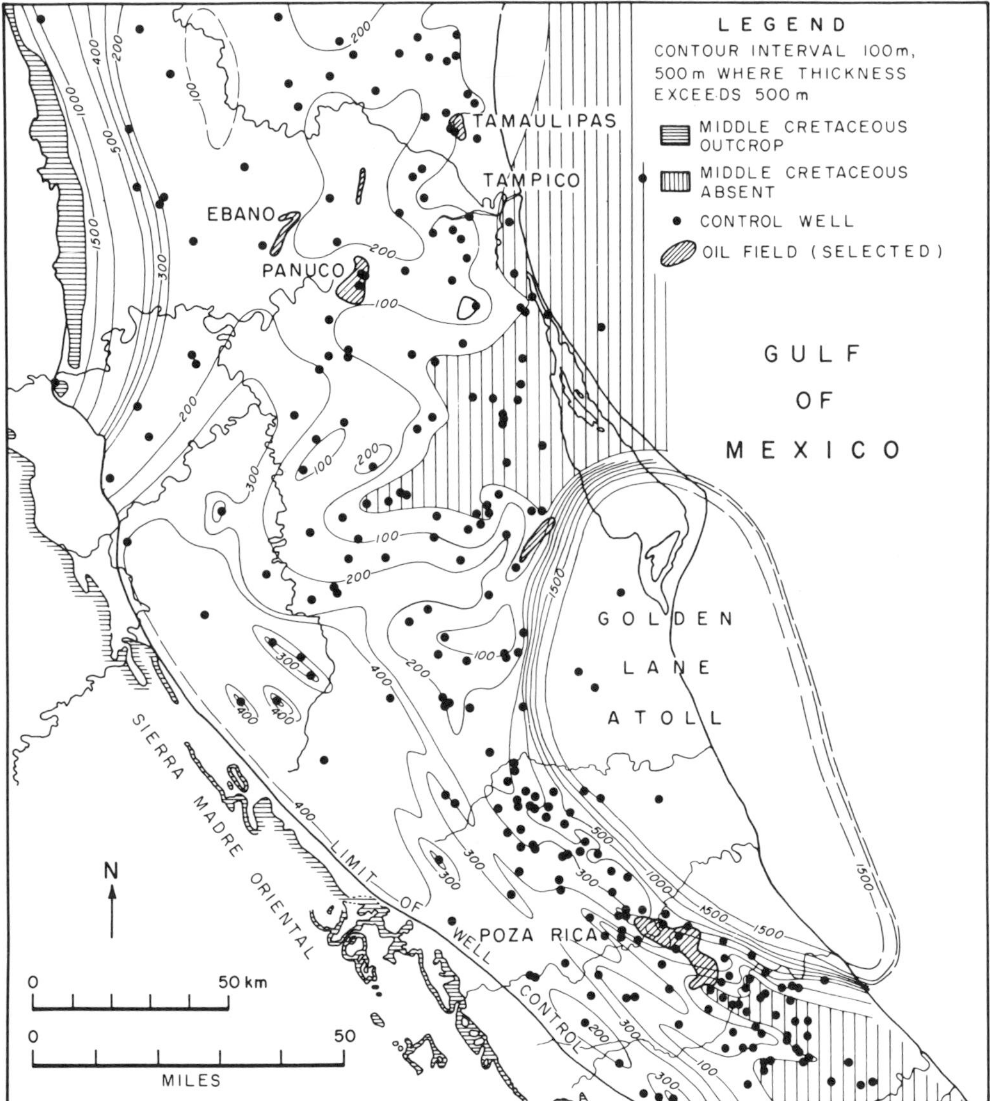

Fig. 5.10. Isopach of middle Cretaceous limestones in the Golden Lane area, Mexico. The facies change from reef to basin is clearly shown by thickness changes, the thin limestone corresponding to the deep water talus deposits of the Tamabra Limestone (Enos, 1977).

Bouma sequence wedge out downslope as turbidity current flow strength decreases. However, this mapping approach is now rarely used because it is recognized that submarine fan sequences may have several sources, and that a wide variance in flow directions is possible on any one fan, reflecting the complex topography of the depositional lobes.

5.4 Geophysical techniques

Geophysical methods are used at all stages in basin exploration. Regional gravity, aeromagnetic or refraction seismic techniques provide information on deep crustal structure and broad basin configuration before the first well has been drilled. High-resolution reflection seismic is employed to determine internal structure, and is now being exploited to yield data on stratigraphic sequences (Chap. 8), and even depositional facies and some fine details of internal basin architecture (section 5.4.1). Geophysical well logs are discussed elsewhere in this book (Chap. 2), but the dipmeter is a special kind of tool, and is treated in two parts of this chapter (sections 5.4.2 and 5.9.6). As exploration of a basin proceeds geophysical and lithologic data derived from wells or diamond drill

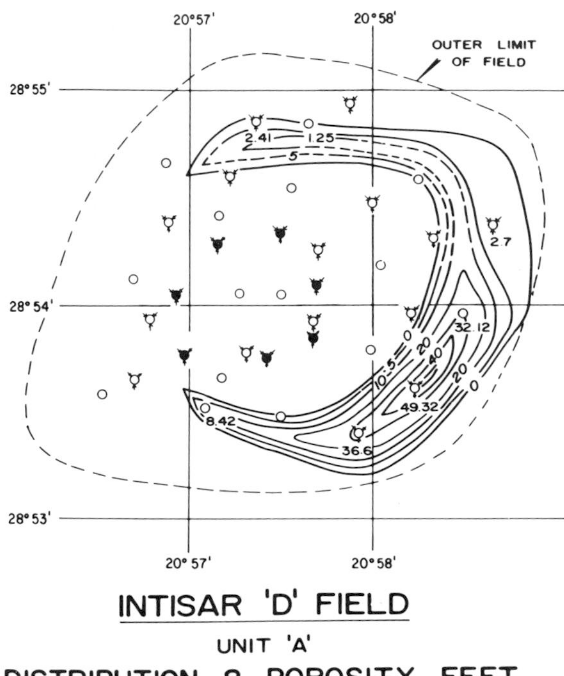

Fig. 5.11. Isopach of porous section in a flanking unit, Intisar bioherm (Paleocene), Libya (Brady et al., 1980).

holes can be used to improve interpretations of the gravity or seismic data. In fact such information is essential if seismic records are to be calibrated and their processing finely tuned to reveal the maximum amount of information. Even in well-explored basins seismic interpretation of the areas between wells, or deeper parts of the section, may be of considerable value, so that a basin analyst may be using many different mapping techniques simultaneously and interactively.

5.4.1 Seismic maps and sections

Introduction. For many years seismic data collection and interpretation have absorbed a significant percentage of the exploration budget of petroleum companies, because of the central role seismic maps and sections have played in delineating potential structural traps. In recent years data acquisition and processing methods have improved to the extent that we can now "see" many details of internal stratigraphic geometry, and these can be interpreted in terms of depositional environment and paleogeography. Seismic modeling is also an important field of development. Seismic has therefore become an even more vital tool for basin analysts. The terms **seismic stratigraphy** and **seismic facies** are much in use amongst geologists; enthusiasts like to suggest

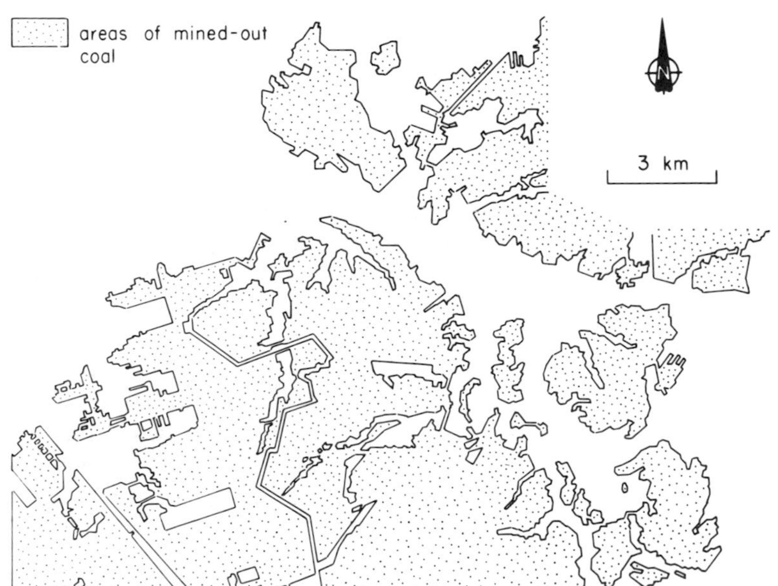

Fig. 5.12. Mined-out areas of a Carboniferous coal seam, West Virginia.Unshaded areas delineate a sandstone-filled channel system (Padgett and Ehrlich, 1978).

that they are in possession of the ultimate weapon for exploring our piece of inner space. Enough is known about facies geometries and reflection patterns that, indeed, a lot can be said about basin fill composition and style based only on skilled interpretation of good seismic records. To that extent seismic is a tool for studying lithofacies and should, perhaps, be dealt with in Chapter 4. However, seismic facies studies are largely an exercise in making interpretations from stratigraphic geometry. Because this is a spatial parameter, and because seismic is most powerful when used in combination with actual lithologic data from well records, seismic stratigraphy and seismic facies are best considered as basin-mapping tools.

Many useful papers have been published on the use of seismic in stratigraphic studies (e.g. Sheriff, 1976; Lyons and Dobrin, 1972), but the subject has undergone a profound change with the publication in 1977 of some of the recent developments in the science by the geophysical research groups in Exxon and its affiliated companies. P.R. Vail, R.M. Mitchum and their co-workers were responsible for the bulk of what has become the seismic stratigraphic bible, AAPG Memoir 26 (Payton, 1977). This is required reading for all basin analysts. Application of this material specifically to sandstone-dominated sequences is covered in a useful book by Anstey (1978). In the paragraphs below is presented a brief (nontechnical, nonmathematical) description of what seismic reflections actually mean lithologically, followed by a summary of some of the more interesting features of basin architecture that can be revealed by seismic.

The nature of the seismic record. A seismic reflection is generated at the interface between two materials with different acoustic impedance. The latter is given by the product of the density of the material and the velocity of the seismic wave. The larger the difference in impedance between two lithologies, the stronger the reflection. Where a soft lithology lies on a harder one the deflection is positive and is normally represented by a waveform deflection to the right, colored in to render it more visible. Where the reverse situation occurs the polarity of the deflection is to the left (left blank). Colors are now commonly used to exaggerate amplitude differences, and these may emphasize major facies variations. Published seismic sections show a series of vertical wave traces, each representing a shot point. These are usually located about 25 to 50 m apart. Nowadays each trace is obtained by combining (**stacking**) the records of many adjacent geophones. The vertical scale of the section is **two-way travel time**, not depth. For each time, horizon depth varies according to velocity, so that the structure revealed on a section may be spurious. For example a mass of high-velocity rock such as a dense carbonate reef or channel fill may overlie what looks like a gentle anticline. This is a case of **velocity pull-up**, where the more rapid acoustic transmission through the dense material returns reflections from the underlying layers sooner than occurs on either side. Computer processing can produce a depth-corrected section if accurate velocity information is available.

It is important to remember that it is contacts that produce reflections, not rock bodies as such. For example, tight, cemented sandstone or dense, nonporous carbonate have higher impedencies than most mudrocks, so that a contact of one of these with an overlying mudstone will yield a strong positive reflection. Conversely a gas-filled sandstone has a very low impedance. Where it is overlain by a mudstone there will be a strong negative deflection. This appears as a blank area in the seismic record, and is the basis of the **bright-spot technique** for locating gas reservoirs. However, there are many conditions where interbedded units may have contrasting lithology but very similar impedencies, so that they yield little or no seismic reflections. Examples would include a contact between a tight sandstone and a porous carbonate, a very porous oil or water-saturated sandstone and a mudstone. Tables 5.1 and 5.2 (from Anstey, 1978) demonstrate a range of such conditions.

Imagine an angular unconformity with lateral facies changes in the overlying unit and a gently dipping, truncated sequence of different units below. The seismic reflection from the unconformity could vary from strongly positive to strongly negative, with some areas completely invisible. This would be a difficult reflection to follow accurately on seismic records (Fig. 5.13).

Seismic records are also complicated by **multiples** and **diffractions** (Fig. 5.13). Multiples are reflections from a shallow layer that are reflected downward again from a near-surface level such as the base of the weathering zone. This internal reflection is then redirected toward the surface

Table 5.1. Seismic reflection strength and polarity for shale resting on sandstone*

Depth	Porosity of sand	Saturant in sand	Reflection strength	Reflection polarity
shallow	poor	liquid	strong	positive
		gas	medium	positive
	medium	liquid	medium	positive
		gas	weak	positive
	good	liquid	weak	pos. or neg.
		gas	very strong†	negative
deep	poor	liquid	strong	positive
		gas	medium	positive
	medium	liquid	medium	positive
		gas	weak	positive or negative
	good	liquid	weak	positive or negative
		gas	weak to medium	negative

*From Anstey, 1978. Sandstone-to-shale reflections, for the same situations, have the same strength but opposite polarity. †Bright-spot conditions.

Table 5.2. Seismic reflection strength and polarity for sandstone resting on carbonate*

Porosity of sand	Saturant in sand	Reflection strength	Reflection polarity
poor	liquid	weak-to-medium	positive
	gas	medium	positive
medium	liquid	medium-to-strong	positive
	gas	strong	positive
good	liquid	strong	positive
	gas	very strong†	positive

*From Anstey, 1978. Carbonate-to-sandstone reflections, for the same situations, have the same strength but opposite polarity. No distinction of depth is made in this table, since depth has less effect than in the Shale-to-sandstone case. The table assumes negligible porosity in the carbonate. The likely effect of porosity in the carbonate is to weaken all reflections, but not generally to change the polarity. It is conceivable that the interface between a tight sand and a gas-filled porous carbonate could be negative. †Also bright–spot conditions, but geologically uncommon.

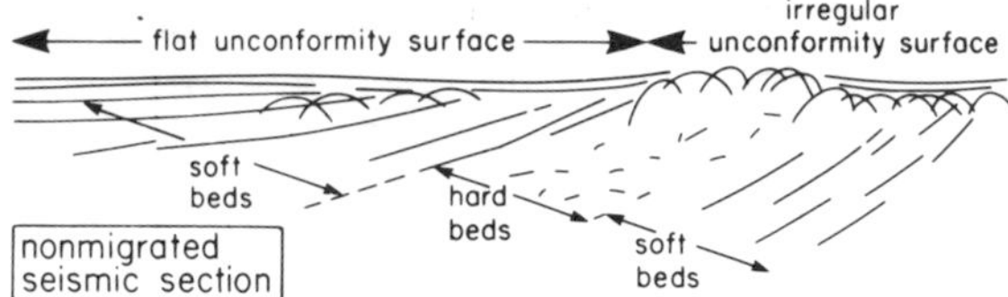

Fig. 5.13. Seismic diffractions, and reflections developed by an unconformity surface with variable impedencies.

from the original point and appears in the seismic record as a parallel layer several tenths of a second lower down in the section. Multiples may be strong enough to obscure deep reflections, but nowadays they are relatively easy to remove by processing. Diffractions are reflections from steeply dipping surfaces such as faults, channel margins, erosional relief on an unconformity, and so on. Their appearance in the record occurs because seismic waves are not linear, laser-like energy beams but spherical wave fronts. The pulse from each shot point therefore "sees" reflection events over a circular region, the diameter of which can be shown by calculation to range from about 100 to 1200 m, depending on depth and wavelength. An event on the periphery of the wave front will be "seen" later than one at the same depth directly below the shot point and, if at a high angle from the horizontal, may produce complex internal reflections. The resulting diffraction patterns may be very pronounced. They are a useful indicator of a steep reflecting event but may confuse other reflections and can be removed by a processing routine called **migration**, which repositions the reflections in their correct spatial position. The circular reflection area (called the Fresnel zone) also accounts for the fact that diffractions can be produced by events off the line of section, for example if the seismic line skims past a reef body or a buried fault line scarp. An example of this is discussed later (Chap. 6).

Structural mapping. Our understanding of the internal complexities of major deformed belts, such as the Appalachian or Cordilleran fold-thrust belt, is almost entirely based on seismic reflection data (e.g. Bally et al., 1966). Recently, deep seismic penetration using the most sophisticated of industry techniques has produced evidence of dramatic crustal shortening in areas such as the Cordilleran foreland region of Wyoming and the southern Appalachians. This has profound implications for the plate tectonic interpretation of major suture zones, as discussed in Chapter 9.

In the area of stratigraphic–sedimentologic basin analysis seismic data may be of considerable assistance in delineating the configuration of important internal sufaces, bearing in mind the limitations on the seismic record noted above. Some examples of the kinds of data useful to the basin analyst are discussed in section 5.2.

The best reflections on a seismic record may be traceable for hundreds or thousands of kilometers, through the available grid of seismic lines. It is a common procedure to plot and contour the two-way travel times of such reflections, producing an analogue structural map. Two examples illustrating features of sedimentological interest are shown in Figures 5.14 and 5.15. The first of these shows drape over Upper Devonian reefs in Alberta (from Evans, 1972). Reefs are composed of carbonate accumulations which resist compaction under burial pressure. They usually are encased in less resistant off-reef lithologies such as mudrocks or evaporites. Successive units therefore drape over the reefs and the drape may extend for hundreds of meters upsection. In Western Canada Devonian reefs commonly may be "seen" in structure contours drawn on Cretaceous horizons, and it seems likely that the subtle variations in subsidence rates even had slight effects on paleogeography for several hundred million years.

Structure contours drawn on unconformities may reveal erosional relief, buried fault line scarps, etc. Figure 5.15 illustrates a dendritic drainage pattern eroded into uplifted Cretaceous rocks in Mexico. Conglomerates derived from the upland to the southwest were deposited as a fanglomerate wedge downslope (Hélu et al., 1977).

Seismic facies. Seismic data are used in stratigraphic applications primarily to define depositional systems (Chap. 6) and regional or global sequences (Chap. 8). However, the internal configuration or pattern of reflections within such seismically defined units may contain much essential depositional information. This is the subject of seismic facies analysis. It should be repeated at this point that the smallest features

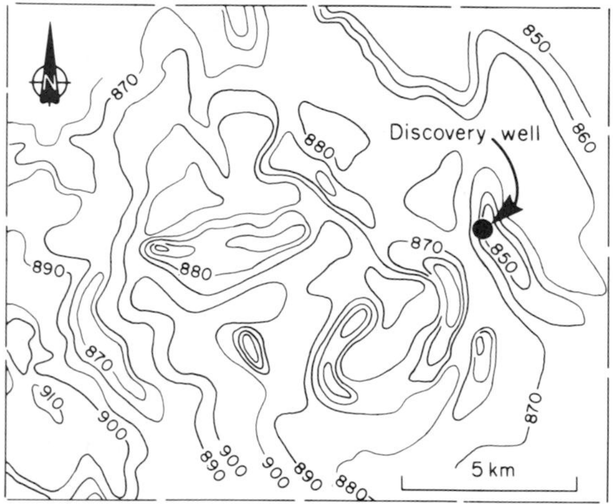

Fig. 5.14. Seismic structure map, contoured in two-way travel time, showing drape over Middle Devonian (Zama) reefs, Alberta. The reflecting horizon is the top Slave Point Formation, a shale–limestone contact about 200 m above the reefs (Evans, 1972).

resolvable on conventional industry seismic are several meters or tens of meters in vertical amplitude (see Fig. 8.11), so that the term facies has a quite different meaning from that used in Chapter 4.

Figure 5.16 illustrates the main styles of seismic facies reflection patterns (Mitchum et al., 1977). Most of these are best seen in sections parallel to depositional dip. Parallel or subparallel reflections indicate uniform rates of deposition; divergent reflections result from differential subsidence rates such as in a half-graben or across a shelf-margin hinge zone. Prograding reflections comprise an important class of seismic facies patterns. They are particularly common on continental margins, where they represent deltaic or continental slope outgrowth. The dipping surfaces, like large scale crossbed foresets, are termed **clinoforms**. As noted in section 6.2 they demon-

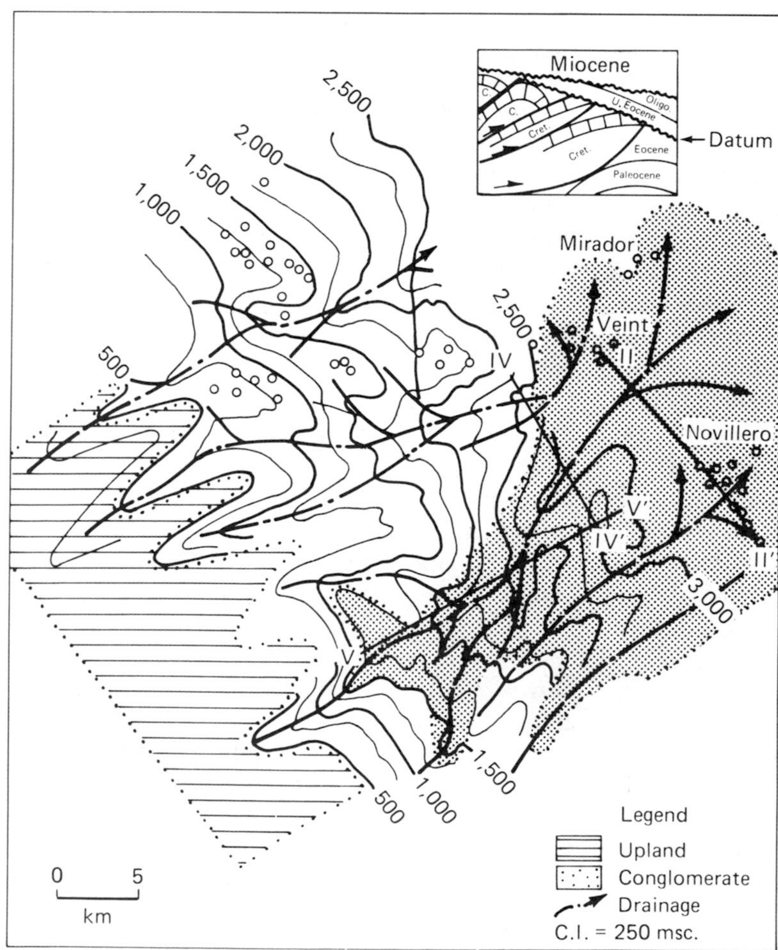

Fig. 5.15. Seismic structure map defining dendritic drainage channels on an unconformity surface, Cretaceous–Tertiary contact, Veracruz Basin, Mexico (Hélu et al., 1977).

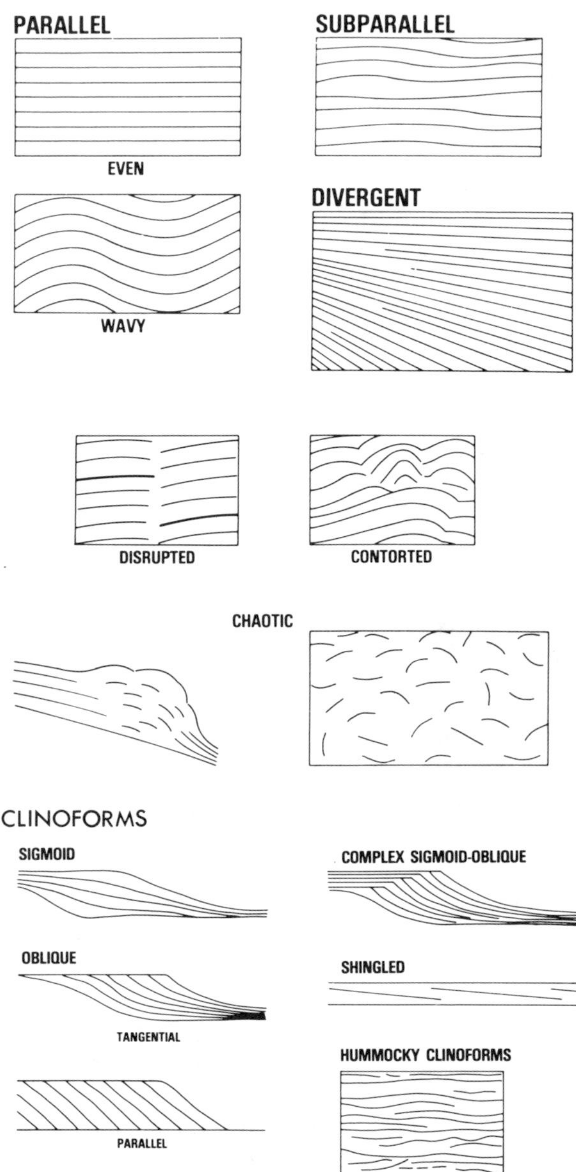

Fig. 5.16. Typical seismic reflection patterns, illustrating the concept of seismic facies (Mitchum et al., 1977).

strate that in many basins lateral accretion is as important a part of basin filling as vertical aggradation. Variations in patterns of progradation reflect different combinations of depositional energy, subsidence rates, water depth and sea level position. Sigmoid clinoforms tend to have low depositional dips, typically less than 1°, whereas oblique clinoforms may show depositional dips up to 10°. Parallel–oblique patterns show no topsets. This usually implies shallow water depths with wave or current scour and sediment bypass to deeper water, perhaps down a submarine canyon that may be revealed on an adjacent seismic cross section. Many seismic sequences show a very complex prograding stratigraphy, of which the "complex sigmoid-oblique" pattern in Figure 5.16 is a simple example. This diagram illustrates periods of sea level still-stand, with the development of truncated topsets (termed toplap, see Chap. 6), alternating with periods of relative sea level rise (or more rapid basin subsidence) allowing the lip of the prograding sequence to build upward as well as outward. Mitchum et al. (1977) described the "hummocky clinoform" pattern as consisting of "irregular discontinuous subparallel reflection segments forming a practically random hummocky pattern marked by nonsystematic reflection terminations and splits. Relief on the hummocks is low, approaching the limits of seismic resolution . . . The reflection pattern is generally interpreted as strata forming small, interfingering clinoform lobes building into shallow water," such as the upbuilding or offlapping lobes of a delta undergoing distributary switching. Submarine fans may show the same hummocky reflections. Shingled patterns typically reflect offlapping sediment bodies on a continental shelf.

Chaotic reflections may reflect slumped or contorted sediment masses, or those with abundant channels or cut-and-fill structures. Disrupted reflections are usually caused by faults. Lenticular patterns are likely to appear most frequently in sections oriented perpendicular to depositional dip. They represent the depositional lobes of deltas or submarine fans.

Some examples of these patterns are illustrated in Figure 5.17. All these features are several hundreds of meters thick and several kilometers or even tens of kilometers across, which explains why they were not properly documented until the advent of high-quality seismic. Few individual outcrops are big enough to reveal them, and on stratigraphic cross sections constructed from outcrop or well data such large features become lost in a welter of detail, particularly in the absence of good stratigraphic marker beds. Progradational sequences are now seen to be particularly common on divergent continental margins (Chap. 9); their discovery based on conventional well data was regarded as a major breakthrough (e.g. Oliver and Cowper, 1963). Documentation of these major architectural features is one of the basic

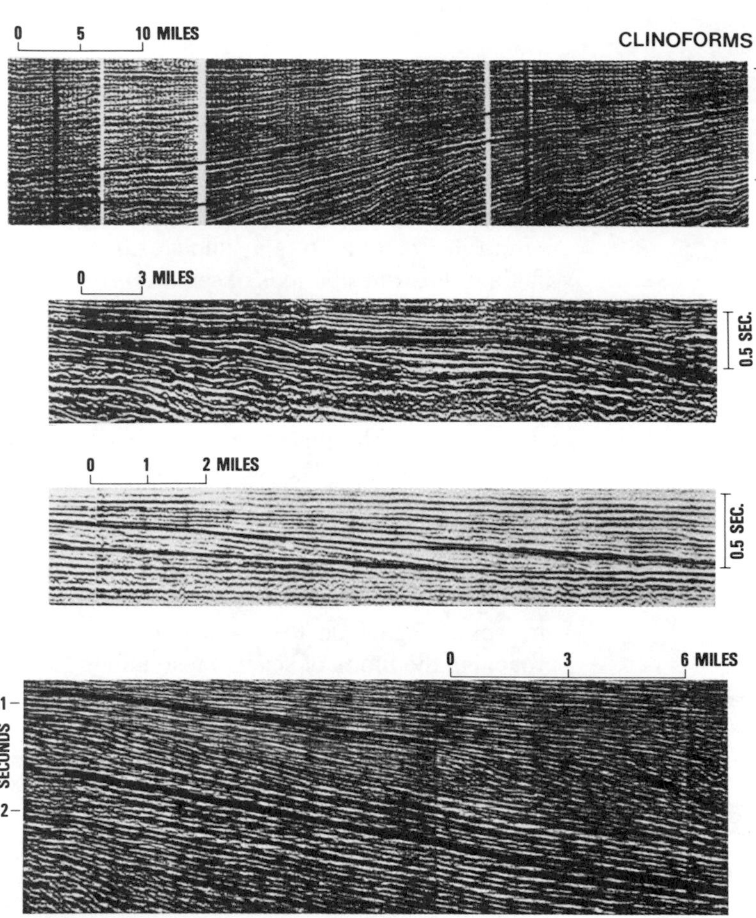

Fig. 5.17. Examples of seismic lines showing typical seismic facies patterns.

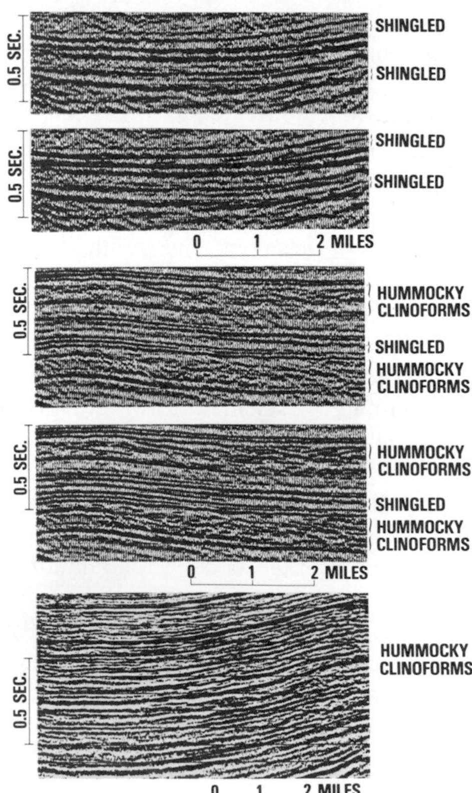

building blocks of the depositional systems approach to basin analysis, as discussed in Chapter 6.

Some clues as to actual lithology can be gained from reflection continuity, amplitude and spacing. Continuity depends on the areal extent of individual lithofacies units. For example it is likely to be excellent in outer submarine fan or basin plain environments characterized by thin-bedded turbidites. It will also be excellent in basin center lacustrine and low-energy shelf settings. It will be at its poorest in alluvial fan, braided fluvial, fan–delta and coarse inner submarine fan environments, where facies changes are rapid and the beds are cut by numerous channels. These may give rise to chaotic or even reflection-free patterns. Reefs will also tend to be chaotic or reflection free, because of their seismically massive character. In some environments reflection continuity may vary considerably throughout the section. For example prograding deltas may be characterized by discontinuous or hummocky clinoform reflections, capped by a persistent horizontal or gently dipping reflection corresponding to the thin, fine-grained, transgressive deposits of the abandonment phase.

Amplitude and spacing are controlled by vertical lithofacies variability. They will be at their highest where the section is characterized by cyclic sequences consisting of contrasting lithologies, and where the cycle thickness is greater than the limit of seismic resolution (namely a few meters or tens of meters). Fluvial fining-upward cycles (thick meandering river and valley fill examples), prograding deltaic or barrier cycles and the coarsening-upward cycles of submarine fan lobes are good examples.

An exotic new tool is the use of three-dimensional seismic data. A tight grid of seismic lines is run across an area at line spacings of a few tens of meters. This permits the plotting of horizontal seismic sections. Where the structure is simple or can be removed by suitable processing routines, the resulting sections display areal amplitude variations for a single stratigraphic horizon. These variations depend on lithology, so that the sections may reveal detailed subsurface facies variations that can readily be interpreted in term of depositional environment and paleogeography. Brown et al. (1980) described an example of this technique applied to interpreting Neogene fluvial and deltaic deposits in the Gulf of Thailand. Channels and sand bars are clearly delineated. Three-dimensional seismic is expensive, and it may never become a standard mapping tool.

5.4.2 The dipmeter

Modern geophysical well logs (section 2.4.5) show a sensitive response to lithologic variation. One of the most sensitive is the microresistivity device, which can record lithologic units as thin as a few centimeters. The dipmeter tool consists of four microresistivity recorders arranged at 90° to each other around the hole. Comparison of the resulting curves will yield many points at which correlations of the same bed (or other hole intersection event) can be made between two, three or all four curves (Fig. 5.18). Any slight differences in depth of the correlated event reflect dip, and it is simply a matter of applying the principle of the three-point problem, familiar to all first-year undergraduates, in order to solve for dip azimuth and angle.The dipmeter will respond to any kind of dipping surface in the rocks, and therefore has considerable potential in structural, facies and paleocurrent analysis, as shown below.

The dipmeter tool has been used as an aid in structural mapping for many years (e.g. Stratton and Hamilton, 1947), but the method has become much more sophisticated with the advent of digitized logs and the use of the computer for correlating the curves and plotting the results. For example, modern pattern recognition techniques are described by Vincent et al. (1977). Data processing is a difficult problem because of the possibility of spurious curve correlations (producing noise), aggravated by variations in lithology (and hence correlatable event spacing) with depth. As is the case with seismic, processing and interpretation are most effective when adapted to specific local conditions. Very few dipmeter patterns have a unique interpretation, and so they cannot be correctly interpreted in the absence of lithologic data.

Dipmeter logs are routinely run by some petroleum companies in order to assist with prospect evaluation. The data are normally presented as a "tadpole plot", examples of which are shown in the following figures. The head of each tadpole is positioned on a grid to show dip angle, and the tail is oriented to indicate azimuth, north always being toward the top of the plot. The azimuth and dip of hole deviation usually are also shown, and must be taken into account for detailed interpretations. Logging companies can also provide other kinds of data displays, such as dip histograms, stereonet plots, and orientation rose diagrams for selected intervals.

At first sight a typical dipmeter plot may look like a confusing welter of detail. Figure 5.18 is a typical example. The first task is to separate the various sources of dip using simple graphical techniques or more complex statistical routines. Rider (1978) illustrated the use of a histogram method, which shows the range of dips recorded over a selected vertical interval (Fig. 5.19). In this case the readings fell into three clearly defined classes—low structural dips, sedimentary dips, and those higher than 35°, which is about the maximum angle of repose for loose sediment. This third class may include spurious noise, and computer processing routines can be designed to eliminate it. In order to render sedimentary dips more comprehensible it may be desirable to remove structural dip (and hole deviation) and replot the data. Again, the computer can perform

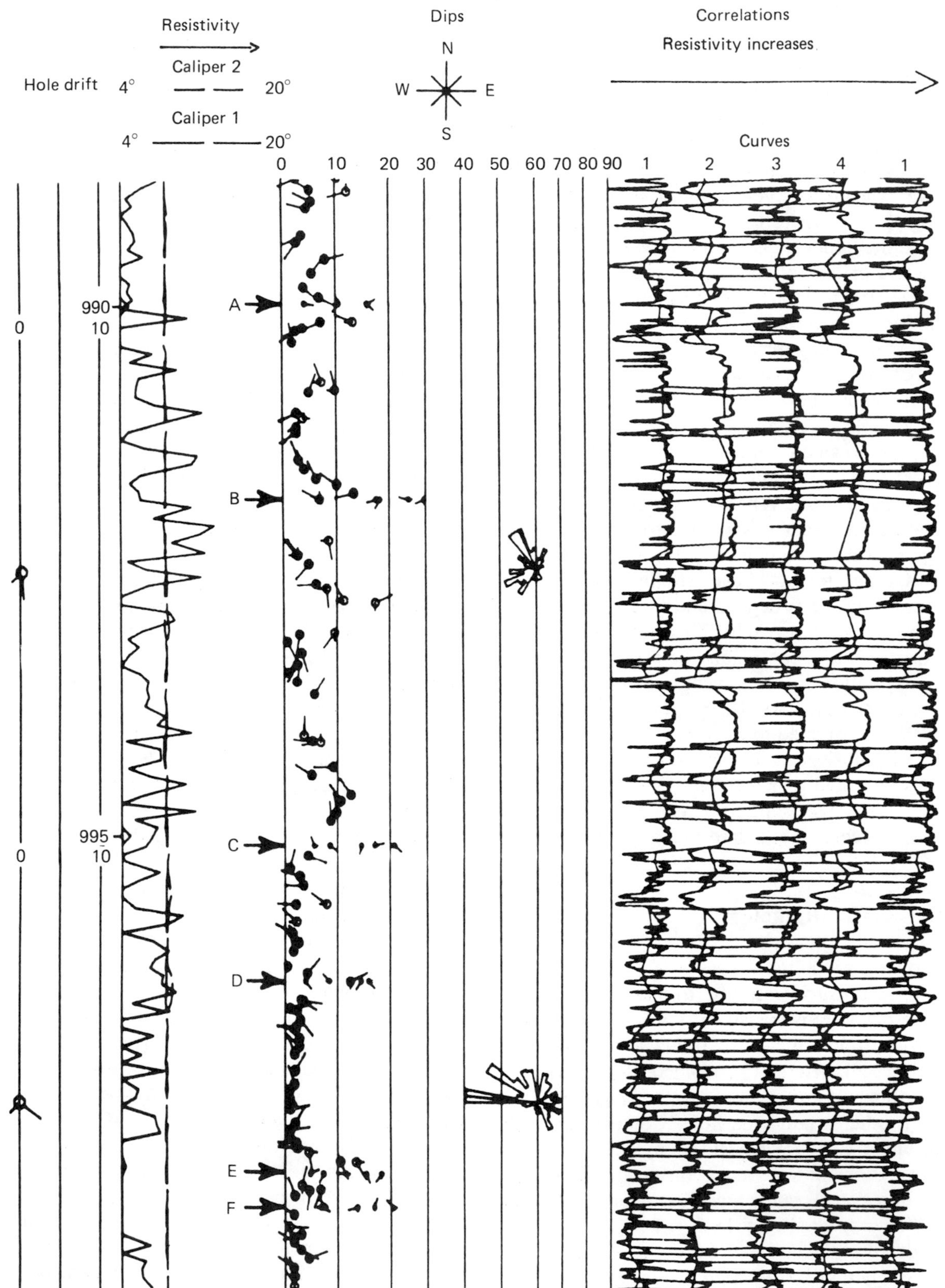

Fig. 5.18. Typical computer plot of dipmeter microresistivity recording (right) and calculated dips (tadpole plots). Hole deviation plot and summary rose diagrams are also shown (From: Vincent, Ph., Gartner, J.E., and Arrali, G. "Geodip: An approach to detailed dip determination using correlation by pattern recognition", SPE 6823, February, Journ. of Pet. Tech. (1979) pp. 232–240. © SPE-AIME).

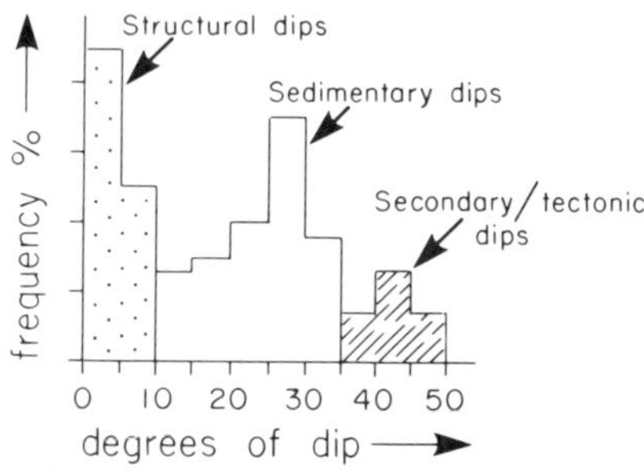

Fig. 5.19. Histogram of calculated dips for a selected stratigraphic interval, showing three classes of dip (Rider, 1978).

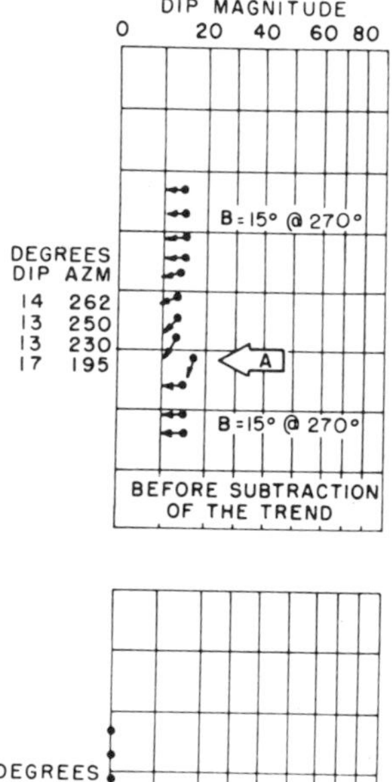

Fig. 5.20. Correction of dipmeter readings for regional structure, which revealed a typical sedimentary pattern of upward decrease in dip (Jageler and Matuszak, 1972).

all the necessary computations and plots. It is necessary only for the geologist to recognize what the structural dip is for input into the computer. This is usually clear, as it will normally be indicated by numerous consistent readings over a considerable vertical interval, a platoon of well-disciplined tadpoles for each unit such as a mudstone or fine-grained limestone lacking any sedimentary, tectonic or drape influences. Jageler and Matuszak (1972) provided an example of structural correction, which reveals a clear sedimentary dip pattern (Fig. 5.20). This is discussed below.

What do dipmeter data mean? Schlumberger (1970) illustrated thirty-four distinctive dipmeter patterns, which can be interpreted in terms of faults, folds, unconformities, channels, depositional dips or drape over massive sedimentary bodies. Gilreath (1977) discussed some of the more common examples. Rider (1978) and Shields (1974) discussed the use of the dipmeter for interpreting depositional environments and crossbedding. The difficulty is that few of the possible patterns have a unique interpretation, and without detailed local lithofacies and structural information from core, outcrop or seismic data it may not be possible to make effective use of the information. Many of the references listed above suggest using the dipmeter to recognize crossbedding patterns, but until the development of microresistivity devices this was strictly hypothetical, because log sensitivity was adequate only to "see" the largest crossbed structures such as eolian draas. Rider (1978) pointed out that crossbedding may contain several kinds of dipping surfaces (foresets, set bounding planes, reactivation surfaces, scours) and that dipmeter sampling through a typical crossbedded interval with set thicknesses in the order of a few decimeters may produce very scattered and unintelligible patterns. However, where the dipmeter can be correlated with a core, a powerful paleocurrent analysis tool may be available, as discussed in section 5.8. Very little published work of this type is available, possibly because of its confidential nature.

Fault and fold interpretations are not discussed here, for reasons of space.

Jageler and Matuszak (1972) gave an excellent example of the use of the dipmeter in delineating a reef, based on the depositional dips of the reef itself and drape over the resistant reef mass. Figure 5.21 illustrates two structural contour maps, one drawn on well information alone and one using dipmeter information to control local contour orientation and spacing. The reef was first discovered when well #1 was drilled, revealing a 35° dip, probably representing the windward talus slope.

Dipmeter readings commonly show a regular

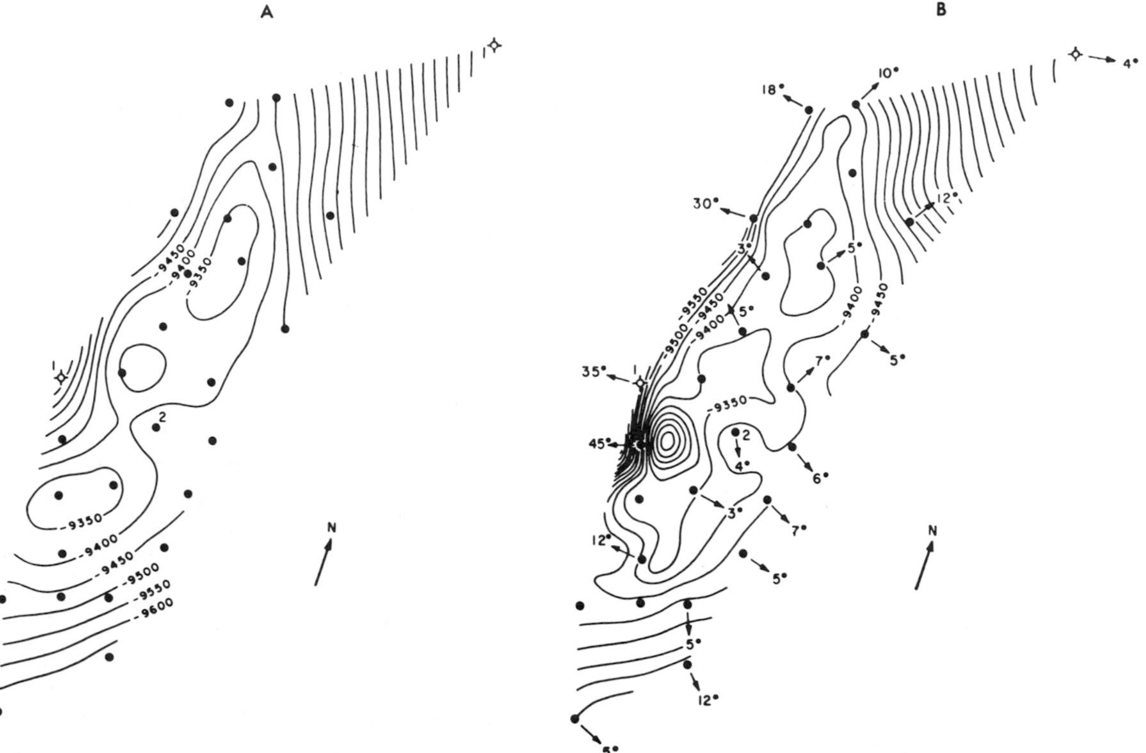

Fig. 5.21. Use of a dipmeter to correct and refine a structure contour map of a reef body, West Texas. **A** map drawn using well data only; **B** dipmeter readings have been added and the map replotted (Jageler and Matuszak, 1972).

upward or downward increase in angle, while maintaining constant azimuth. Examples are illustrated in Figure 5.22. An upward increase in dip angle over a thickness of a few tens of meters may indicate a fault or surface creep below an unconformity surface (Fig. 5.22D). Upward decrease in dip has a greater variety of causes. Sediments onlapping an irregular erosion surface, draped over a reef (Fig. 5.22C) or sandstone pod (Fig. 5.22B), or infilling a channel or valley (Fig. 5.22A) may show similar patterns. Associated data, such as the random readings within the reef (Fig. 5.22C), or crossbedding (Fig. 5.22B), plus geophysical log character and lithology should all be used to discriminate these patterns.

The use of the dipmeter in paleocurrent analysis is described in section 5.9.6.

5.4.3 Other geophysical methods

A wide variety of other geophysical techniques is used in basin stratigraphic and facies mapping. A few are illustrated here, but the subject is a large one and cannot be treated fully in this book.

Gravity and magnetic data commonly are used to delineate broad basin configuration. For example Figure. 2.3 shows how regional gravity data and limited well control were used to reconstruct the broad geometry of the Jurassic to Eocene basin fill underlying Banks Island in the western Canadian Arctic. The well information indicated that the configuration of Bouguer anomaly contours is a good analogue of the gross structure of the basal unconformity of this section. Facies studies showed that the same structures were in existence during the Cretaceous–Paleogene, and had an important effect on the paleogeographic evolution of the area (Miall, 1979).

Facies patterns on carbonate shelves are similarly controlled by structure, as may be revealed by regional gravity studies. A classic example of this is the lithofacies distribution over the Florida–Bahamas platform. Gravity and refraction seismic data show that the configuration of the platform is controlled by basement faults

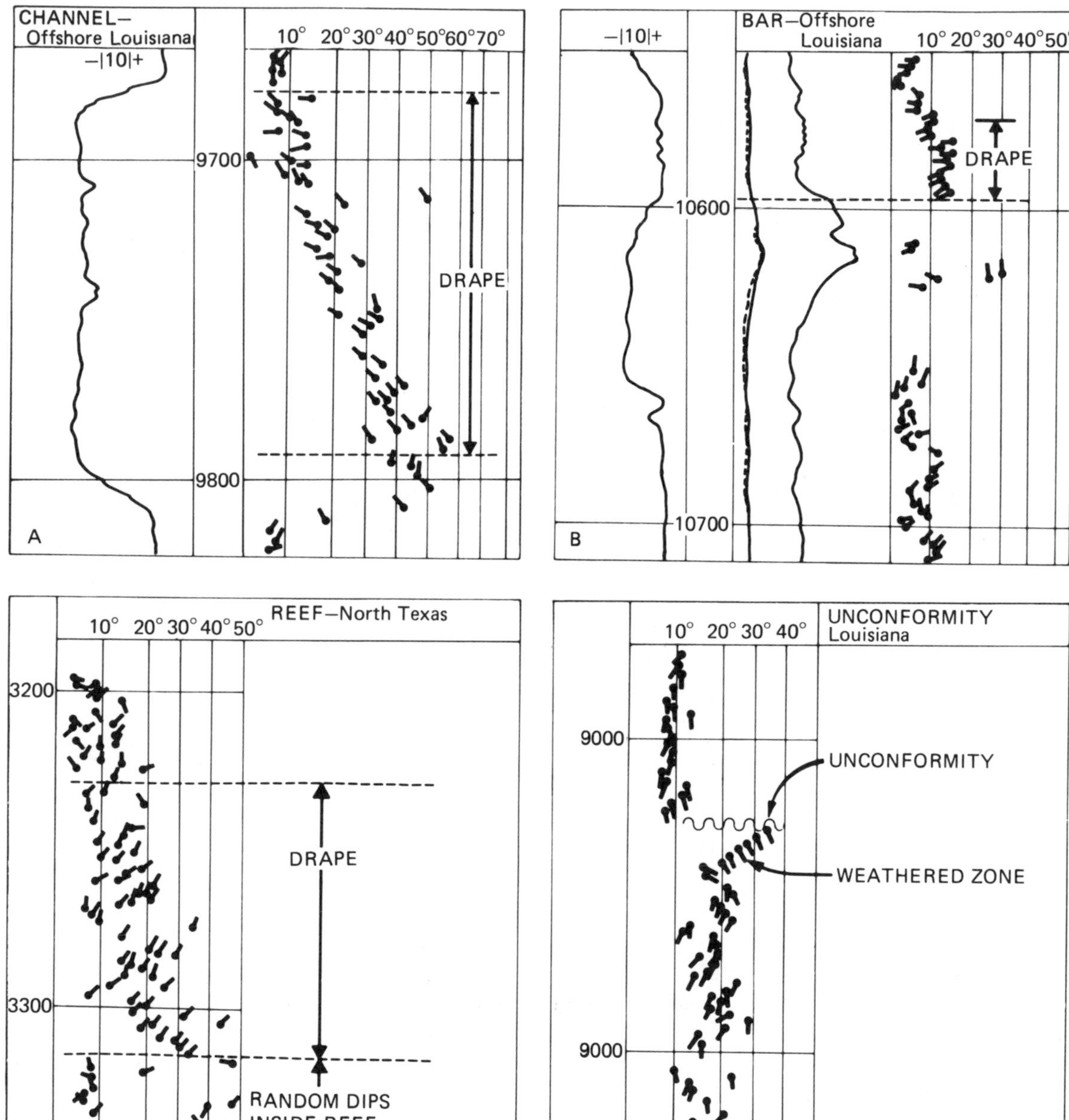

Fig. 5.22. Examples of typical sedimentary dipmeter patterns. **A** drape in a channel fill; **B** drape over a bar deposit; **C** drape over a reef body and random dips within the reef; **D** dip distortion by surface creep in a weathering zone (Gilreath, 1977).

(Talwani, 1960; Ball, 1972). A similar relationship has been determined for the Upper Devonian reefs of Alberta. A north–northeast trending grain is apparent in gravity contours and is reflected in reef trends (Fig. 5.23). Both examples have been related to rejuvenated Precambrian structures, providing the elevated foundation on which reefs grow (Martin, 1967; Ball, 1972).

Basement control generally is of profound importance in determining broad basin fill patterns. Regional geophysical data pertaining to deep crustal structure, such as gravity, refraction seismic and deep-reflection seismic, are therefore of crucial importance in basin analysis. Additional examples are given in Chapter 9 where this subject is discussed at some length.

Well log data are used for stratigraphic correla-

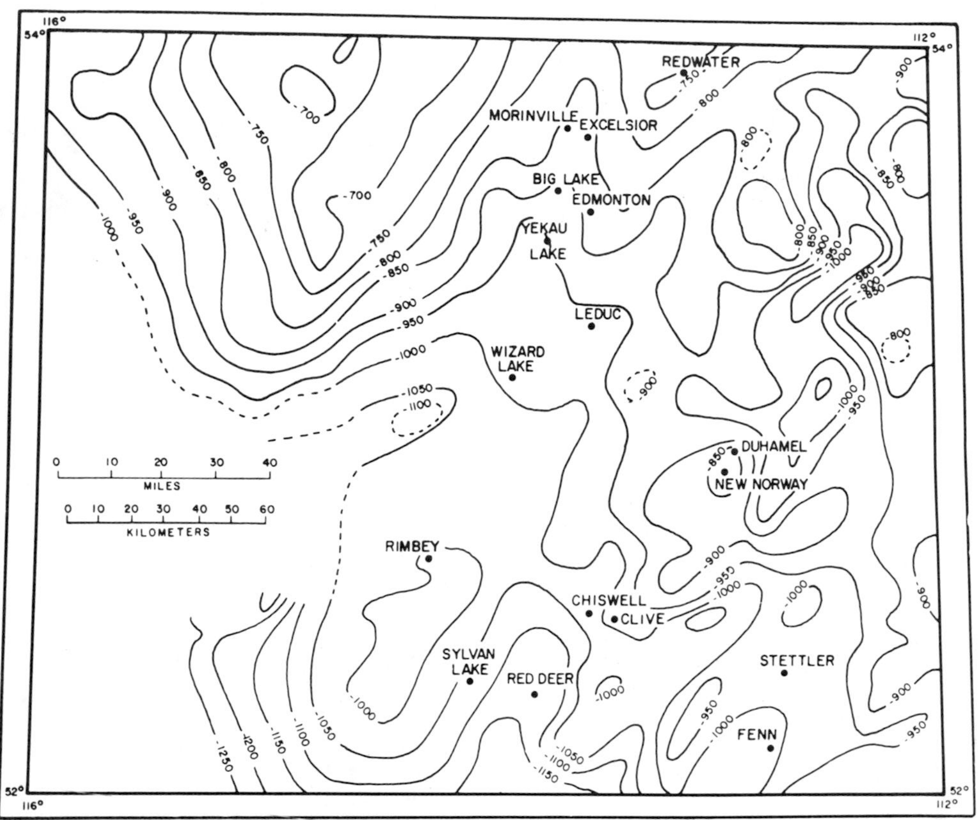

Fig. 5.23. Bouguer gravity anomaly map of central Alberta, showing a north–northeasterly grain, reflecting structure in the Precambrian basement. Most Devonian reefs are located on or near gravity highs (e.g. Redwater, Leduc, Duhamel, New Norway, Clive, Sylvan Lake, Wizard Lake) (Ball, 1972).

tion and for lithologic interpretation. They may also be used directly for mapping purposes, although this is a technique that has not been widely employed. McCrossan (1961) constructed a resistivity map for the lower and middle Ireton Formation of central Alberta which can be used to predict proximity to reefs (Fig. 5.24). The Ireton is an Upper Devonian basinal shale which drapes against the reefs. Calcareous content, and thus resistivity, increase close to the reefs due to the presence of reef talus. The resistivity data therefore provide a useful analogue paleogeographic map.

5.5 Clastic petrographic data

As clastic sediments are dispersed by currents or sediment gravity flows they are subjected to size sorting, abrasion, breakage and rounding. They may retain a distinctive detrital composition, although easily weathered minerals such as ferromagnesian grains will gradually be destroyed. As a result of these processes a particular lithosome may become fingerprinted with a distinctive size range and composition, and downcurrent changes in the various petrographic parameters may be used as paleocurrent indicators. The use of petrographic data in facies studies and correlation is discussed briefly in section 4.5.2. This section focuses on the use of the information as another form of lithofacies mapping, in order to assist in paleogeographic reconstruction. The same data may also be useful in making plate tectonic interpretations, as discussed in section 9.4.

Petrographic data are simplest to use where the sediments have been deposited by unidirectional currents, as in fluvial, deltaic and submarine fan environments. Size and sorting parameters do not convey much, if any, directional information when the sediments have been formed by multidirectional currents, as in most shallow marine

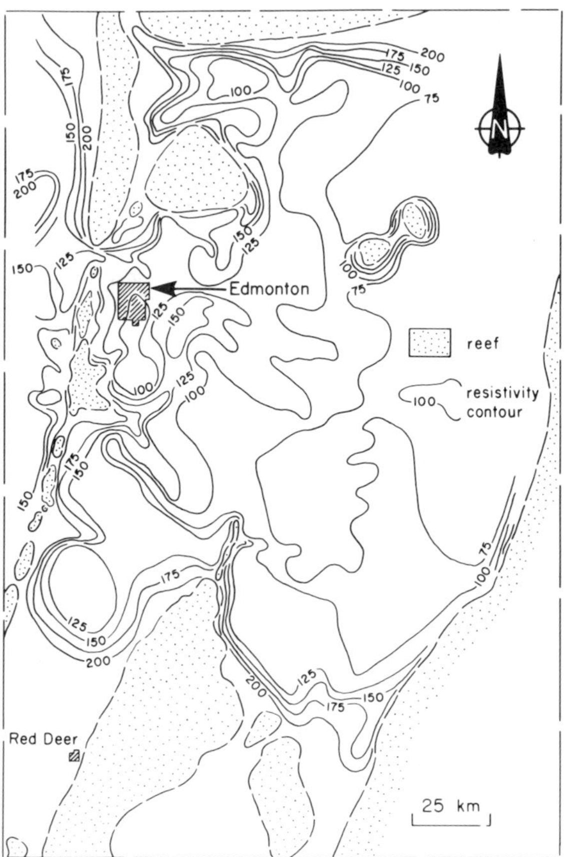

Fig. 5.24. Resistivity map of the Ireton Formation (Devonian), central Alberta. High values indicate an increase in carbonate content adjacent to reef bodies (McCrossan, 1961).

and shelf environments. Few petrographic studies of this type have been carried out in submarine fans, in part because most such deposits contain abundant paleocurrent indicators such as sole markings (section 5.8), and in part because most exposed fans are in complex structural belts where areal trends are difficult to reconstruct. Therefore most of the remaining discussion in this section relates to fluvial and fluviodeltaic deposits, and may have uses in surface or subsurface exploration for petroleum, coal, uranium or other economic deposits.

Grain size changes due to abrasion and breakage during downstream transport are rapid in conglomeratic rocks, but the rate of change with distance of transport drops off asymptotically, becoming slow in coarse sand and negligible in fine sand to silt. Much experimental work has been carried out in an attempt to quantify the rate of change, so that grain size in the field could be interpreted directly in terms of distance of transport. However, each detrital lithology responds differently to current transport, and rivers themselves vary markedly in competency at any one point, so that research of this type is unlikely to meet with any success. Nevertheless, mapping of relative grain size changes in conglomeratic and coarse sandy sediments has considerable use as a paleocurrent indicator.

Detrital petrographic composition reflects the geology of the sediment source area. For nonmarine rocks this means exposed upland regions. Downcutting by rivers reveals progressively older rocks, so that the derived sediments may display an "inverted stratigraphy" effect, older sediments containing detritus from the young cover rocks, and the upper strata containing material derived from local basement. Documentation of this unroofing history may provide much useful information on timing of orogeny, volcanism, plutonism, etc. The petrography of the detritus may not necessarily bear a simple relationship to that of the source area because of the destruction of unstable grains in transit, particularly in humid environments where chemical weathering may be rapid (e.g. Franzinelli and Potter, 1983). Nevertheless, studies of petrographic variation within the basin itself may provide much useful information. Interpretation of source area for marine rocks may be a more complex procedure. For example, submarine fans may contain deltaic detritus plus shelf carbonate or siliciclastic material swept down on to the continental rise via submarine canyons.

What measurements should be made? For grain size studies in conglomeratic rocks the simplest and most useful mapping parameter is maximum clast size, which is usually given as the average of the intermediate clast diameter of the ten largest clasts present at a sample point. For sand-sized grains mean or maximum grain size may be used. Both can be measured in outcrop or thin section by visual estimation against a grain size chart (Fig. 2.11). This permits measurements to the nearest phi grade, which is quite accurate enough for our purposes (sieve analyses and other laboratory studies are time-consuming, and usually add little). Detrital composition of conglomerates can be studied by counting and identifying one hundred or more clasts at each sample point. Sandstone composition should be studied in thin-

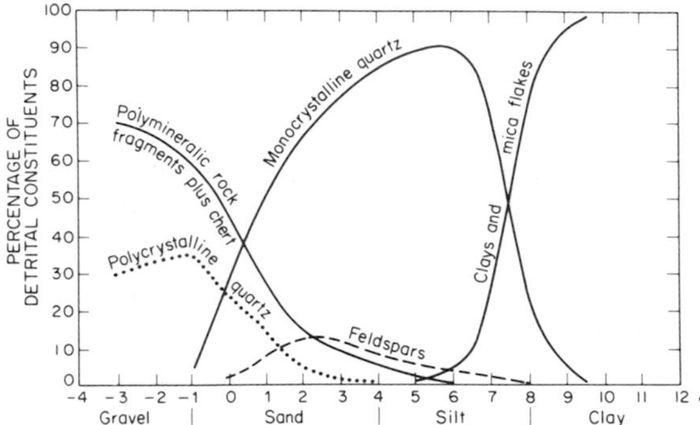

Fig. 5.25. Relationship between grain size and composition of the detrital fraction in siliciclastic sediments. From H. Blatt, G.V. Middleton and R.C. Murray, Origin of Sedimentary Rocks, 2nd edition, © 1980, p. 321. Reprinted by permission of Prentice-Hall, Inc., Englewood Cliffs, New Jersey.

section using a movable microscope stage and a point counter. It is recommended that at least 300 grains be identified and counted per sample point. The stage should be set to advance at each count by an amount equal to or slightly greater than mean grain diameter. The number of points counted can be adjusted according to the compositional variability of the sample. Pure quartz arenites do not need such intensive study but immature, lithic arenites with many varieties of rock fragments may need more detailed examination. Mineral types have different densities and so they vary in their hydrodynamic behavior. They also have different fragmentation patterns. The detrital composition of a clastic rock may therefore vary markedly with grain size (Fig. 5.25) and so, for mapping purposes, it is advisable to control measurements according to a standard grain size range, say, 1 phi class. These methods will usually result in a measurement repeatability of two or three percentage points or better, and this, again, is accurate enough for our purposes. Analytical procedures are discussed in detail by Dickinson (1970) and Pettijohn et al. (1973).

Much use has been made of heavy minerals (density greater than about 2.8) in provenance studies, even though most sands typically contain less than 1% by weight. They are extracted from loose or disaggregated sand by separation in dense liquids such as bromoform, and examined in grain mounts. Considerable skill is needed to identify grain types correctly, and the technique is in general more difficult to use than thin-section analysis of the "light" fraction.

Grain size and compositional measurements may be averaged over a vertical section at each mapping location, or restricted to individual correlatable stratigraphic units. Plotting of areal and vertical changes may both provide useful paleogeographic information, and it is up to the basin analyst to exploit the data in the most effective way. Petrographic data may be analyzed manually, graphically or statistically to define petrographic assemblages or provinces, typifying a particular source area geology (Suttner, 1974). Alternatively particular mineral species may be treated separately. Some workers have concentrated on identifying and mapping varieties of quartz, feldspar, zircon or other grain types, requiring a great deal of highly skilled research, but possibly providing paleogeographic (dispersal and source) information unobtainable in any other way. German workers seem to be particularly interested in this approach (e.g. Heim, 1974; Schnitzer, 1977) and Allen (1972) has carried out some noteworthy studies.

Resistant grains such as quartz and chert, plus the heavy minerals zircon, tourmaline, rutile and garnet are commonly recycled through several episodes of erosional derivation, sedimentation, lithification, uplift and re-erosion. Usually they show abundant evidence of this in the form of a high degree of rounding and the presence of authigenic overgrowths. It is usually difficult or impossible to determine the source of such grains because, in most cases, they could have come from several different stratigraphic units in the source area. Variations in color, crystal habit or inclusions may help, but require more detailed research.

Extensive treatments of sandstone petrology and its application in sedimentological studies are given by Pettijohn et al. (1973) and Blatt et al. (1980). In the remainder of this section some examples of field studies are discussed to illustrate the kind of contribution petrography can make to regional basin analysis.

Friend and Moody-Stuart (1972) studied the Devonian fluvial sediments of Spitsbergen and drew a variety of lithofacies and petrographic maps. Figures 5.26 and 5.27 illustrate two of these. Mean grain size was calculated for each locality by averaging the grain size values obtained at 25-cm intervals up each section. The rocks were divided into three stratigraphic units and a separate map drawn for each (Fig. 5.26).

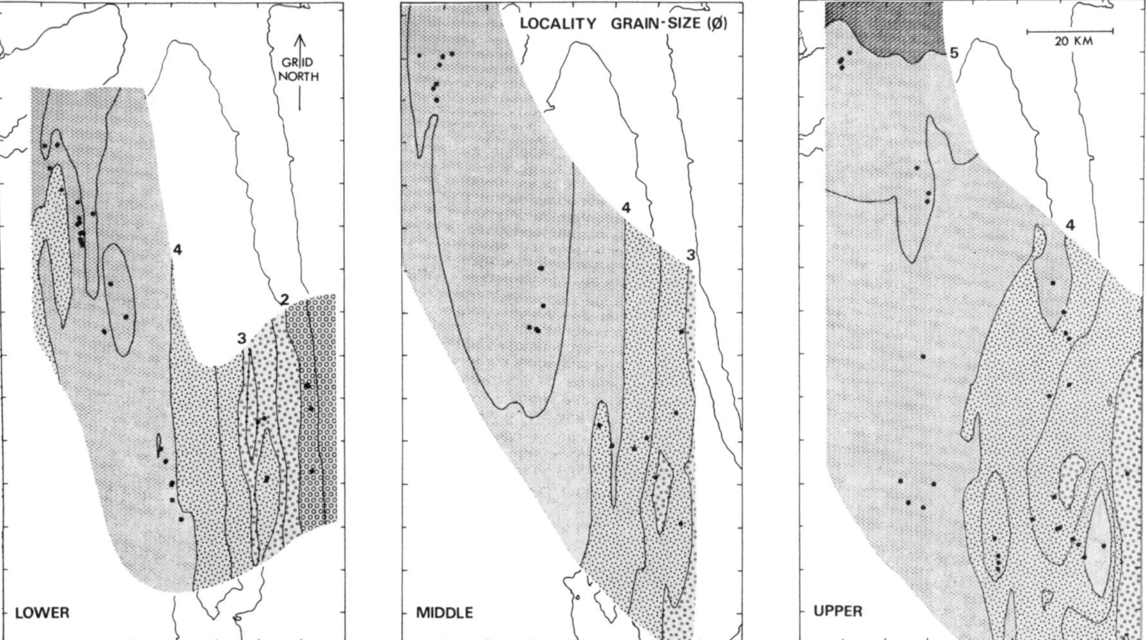

Fig. 5.26. Locality mean grain size for each of three stratigraphic subdivisions of the fluvial Wood Bay Formation (Devonian), Spitsbergen, contoured by trend-surface analysis (Friend and Moody-Stuart, 1972).

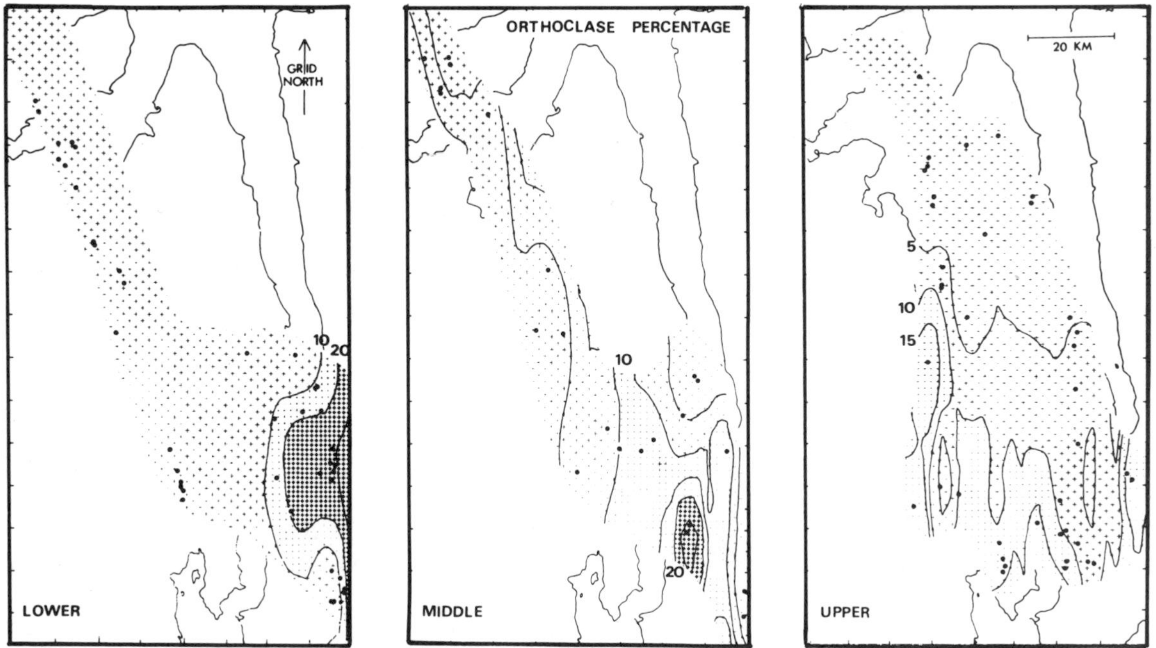

Fig. 5.27. Orthoclase content, plotted and contoured as in Fig. 5.26 (Friend and Moody-Stuart, 1972).

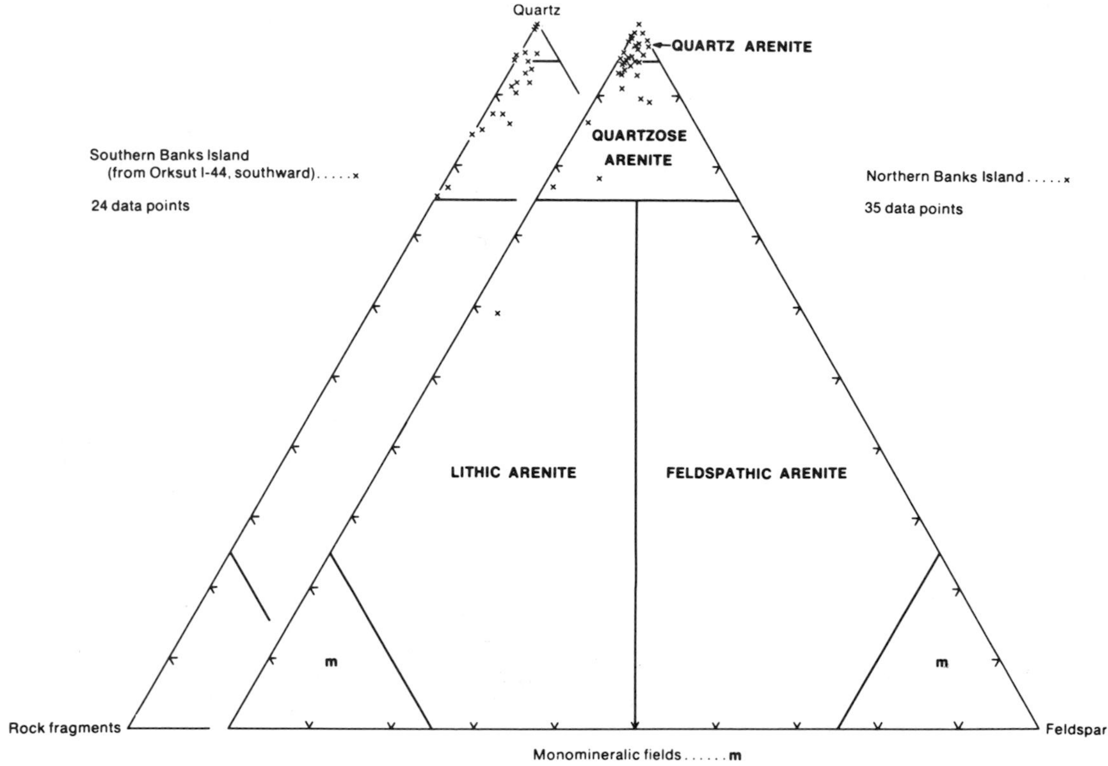

Fig. 5.28. Detrital petrography of the fluvial Isachsen Formation (Lower Cretaceous) of Banks Island, Arctic Canada. The two plots are for the north and south halves of the island. The sandstone classification is that of Okada (1971) (Miall, 1979).

The data points were contoured using a trend-surface analysis method (section 5.6). Figure 5.27 shows detrital orthoclase content in the sandstones, based on measurements in over 200 thin-sections. Statistical studies by the authors showed that part, but not all, of the variation in orthoclase content could be explained as being due to differences in grain size—the feldspar tends to be concentrated in the coarser sandstones. What all these maps show is that sediments were fed into the basin from the west, south and southeast, with river systems flowing generally northward.

Miall (1979) analyzed the petrography of the Isachsen Formation, a Lower Cretaceous braided river deposit in Banks Basin, Arctic Canada. The bulk of the formation consists of texturally and mineralogically mature medium to coarse sandstone, showing abundant evidence of a recycled origin. Paleocurrent evidence indicates northward-flowing trunk rivers with tributaries entering from the east (the west side of the basin is covered by younger strata but may also have been a source area). Initial attempts to recognize petrographic provinces in the sandstone were unsuccessful. Ternary plots show that the samples from the north and south end of the basin contain essentially the same mineral composition (Fig. 5.28). Cluster analysis revealed a weak trend for feldspar to be more common in the north, and sedimentary rock fragments are slightly more abundant in the south. This was then emphasized by replotting the thin-section data in terms of three minor components (Fig. 5.29), which shows that chert is also slightly more abundant in the north. It was found that a few rare components, such as detrital quartz–feldspar intergrowths, quartz with mica, tourmaline or zircon inclusions, and zoned quartz, are all present almost exclusively in the north end of the basin. Analyses of these data in terms of potential sediment sources resulted in the map shown in Figure 5.30. The bulk of the quartz sand and the sedimentary rock fragments were derived from mature Proterozoic

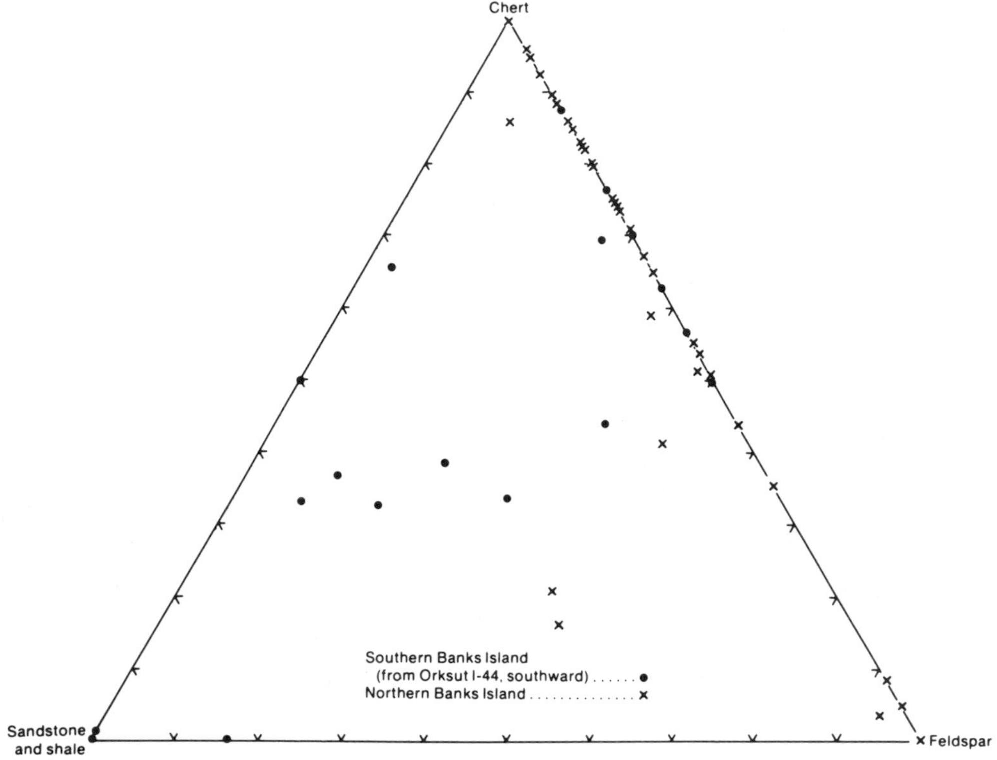

Fig. 5.29. Distribution of minor components in the Isachsen Formation, Banks Island (Miall, 1979).

sediments, chert was fed into the north end of the basin from various Lower Paleozoic sources, and the feldspar plus rare quartz and feldspar components appear to have been derived from a small Archean granodiorite pluton. The Melville Island Group is a Middle to Upper Devonian sandstone unit which outcropped over wide areas adjacent to the north end of the basin, but appears to have contributed little detritus. It is mainly fine grained and has the wrong texture and mineralogy.

Rahmani and Lerbekmo (1975) carried out a detailed heavy mineral analysis of the Cretaceous–Paleocene "molasse" sandstones derived from the Canadian Rocky Mountains. Factor analysis was used to determine mineral associations, and these resulted in the definition of a series of petrographic provinces (Fig. 5.31). It was found, unexpectedly, that these provinces were distributed in belts subparallel to structural strike. It was concluded that the dominant fluvial systems flowed longitudinally down the depositional basin from source areas to the west and northwest, not transversely, as would have initially been expected. The mineral provinces could be distinguished because of variations in source area geology.

Additional discussion of this topic, with examples and many references, is contained in Potter and Pettijohn (1977, Chap. 8).

Grain roundness, sphericity and shape have been much studied as potential paleocurrent indicators, but the results do not seem to justify the effort expended except, perhaps, where no other mapping methods can be found to work, or very detailed local studies are required. Roundness and sphericity are hard to measure and are very dependent on grain size and lithology. Grain shape is simpler to measure, depending only on determinations of the lengths of the three principal diameters of each clast. Sneed and Folk (1958) recognized various classes of compact, platy and elongated shapes using these data. It seems likely that only in conglomerates could the data be collected accurately enough and only in conglomerates are downstream changes likely to be significant. For example Bluck (1965) found

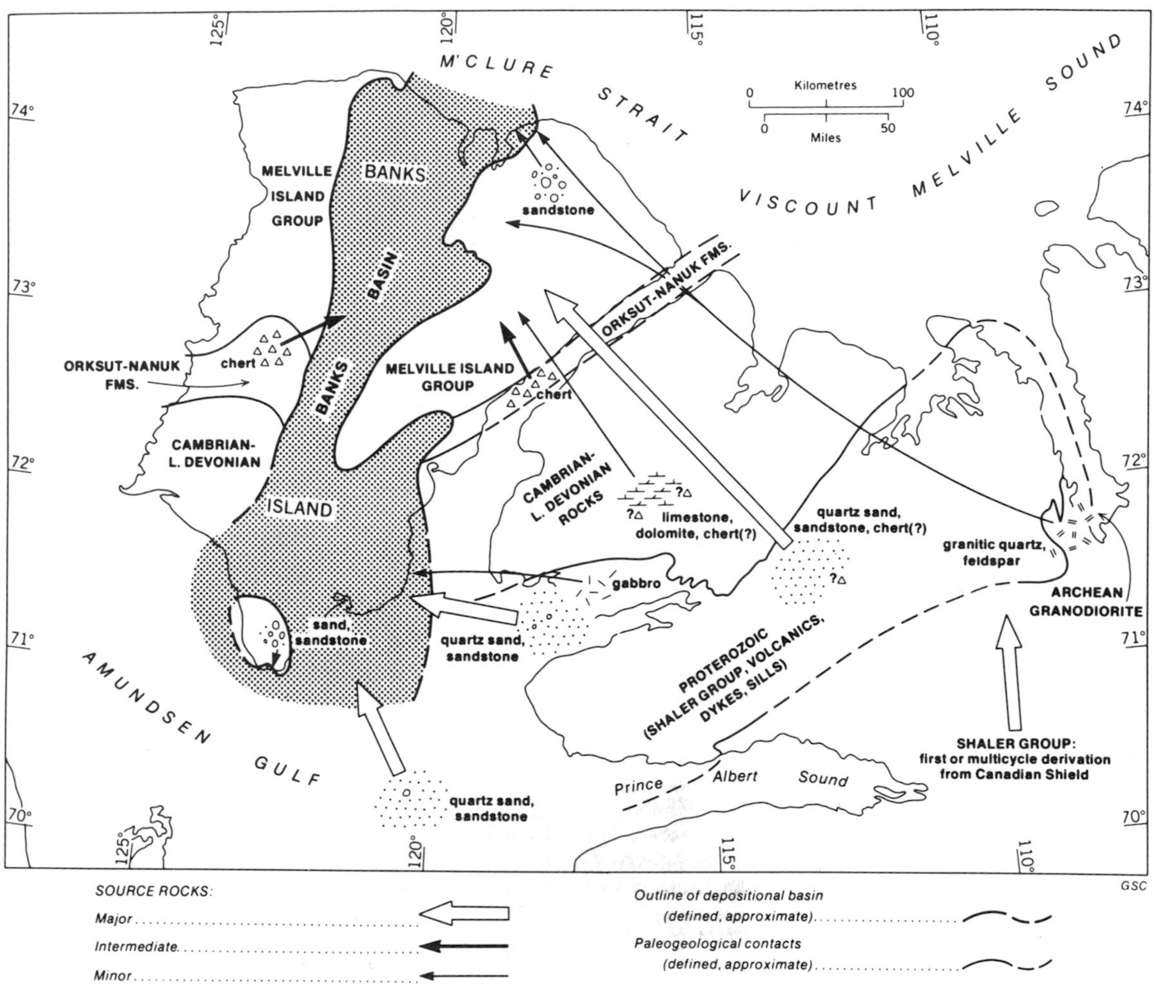

Fig. 5.30. Paleogeology of the Banks Island area during the Early Cretaceous, and the depositional basin of the Isachsen Formation, showing interpreted source rocks of the Isachsen sandstones (Miall, 1979).

that in the deposits of a small Triassic alluvial fan there was a downstream increase in the proportion of platy clasts because these offered flatter surfaces for more effective sediment transport. Ehrlich and Weinberg (1970) proposed the use of automated Fourier analyses methods for grain shape characterization, with promising results.

5.6 Computer mapping methods

The enthusiasts of the computer revolution tell us that in a few years many of us will be able to perform our jobs without ever leaving home, with the aid of portable "intelligent" terminals. While many petroleum geologists pass much of their basin analysis career without looking at very many real rocks, the thought that they could now carry on, perhaps, without even actually getting out of bed, seems a little disquieting. Whatever will happen to the old cliche that geologists are hard outdoor types?

Computers and computer processible data banks are used in basin analysis for two main purposes.

1. For storage and housekeeping of large data files, such as well records.
2. For retrieval of selected data sets and perfor mance of various statistical or mapping routines.

Over about the last twenty years files have been established by government agencies and commercial houses for the well records of the major sedimentary basins. These files, covering areas such as the Western Canada Sedimentary Basin, contain millions of items of information including, for each well, its name, location, completion date, name of operator, geological formation tops, drill stem test results, depths of cored intervals, and possibly much more (e.g. Farmer, 1981; Clark, 1981).

The American Stratigraphic Company and Canadian Stratigraphic Service routinely log most wildcat and many other exploration wells in North America, totaling about 70,000 wells. For many years they have offered these logs to industry and other users on a commercial basis. These logs are now digitized and stored in computer processible data files. In addition to the basic information listed above, the files also contain the following lithologic information:

rock type and percentage
grain, crystal or fragment size
framework percentage
rock accessories percentage
rock builders percentage
sandstone components percentage
diagenesis type and percentage
sandstone sorting and rounding
rock color
porosity grade and type
oil shows
minor accessories
fossils
minerals and their frequency of occurrence

Each of these items is recorded for all descriptive units or "beds", which range from 30 cm to several tens of metres in thickness. The files are regularly updated, and therefore provide an efficient storage system of basic stratigraphic data.

Retrieval programs are available to extract and list, plot in map or section form, and contour, if required, most of these data items. The retrieval interval can be specified by formation, or by specified slices of a formation ("slice maps"), or by defining a fixed thickness interval ending at the top or base of the formation.

Many sophisticated software packages are available to process these data. Basic contouring can be performed using repetitive complex mathematical manipulations. An example is illustrated in Figure 5.32. Some programs can allow for and plot out faults. The same programs can be used to contour structural data from wells or seismic

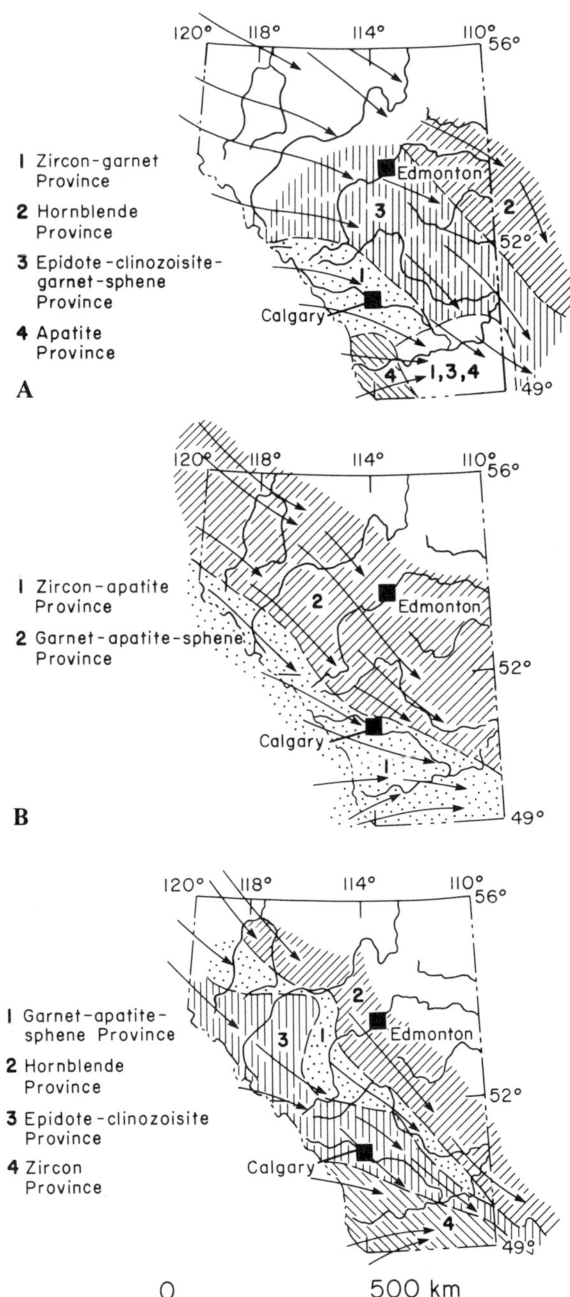

Fig. 5.31. Heavy mineral provinces, as determined by factor analysis, in three formations within the upper molasse wedge of Alberta. Arrows show interpreted dispersal directions. **A** Belly River Formation (Upper Cretaceous); **B** Edmonton Formation (Cretaceous–Tertiary); **C** Paskapoo Formation (Tertiary) (Rahmani and Lerbekmo, 1975).

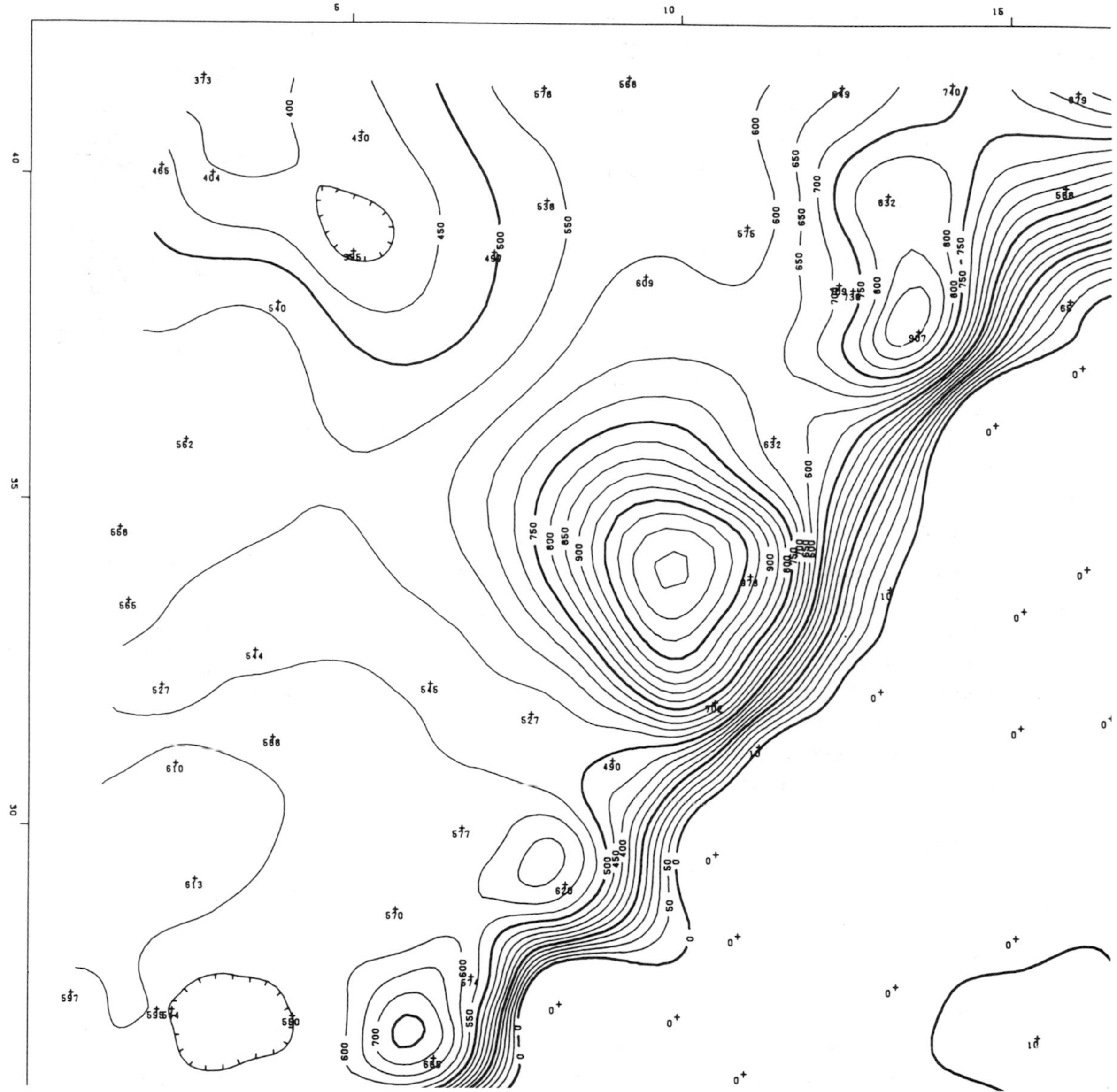

Fig. 5.32. A typical example of an isopach map plotted and contoured by computer. Courtesy Canadian General Electric Corp. and Scientific Computer Applications, Inc.

records, isopach maps, and lithofacies maps of all kinds (in fact virtually all the maps illustrated in sections 5.2 to 5.5). They can also draw three-dimensional projections of map surfaces and can plot cross sections on any selected datum plane. Walters (1969), Davis (1973) and Waters (1981) discuss some of the principles of automated contouring and provide information on software packages.

Similar computer processible data management systems have been developed for mining exploration and development, for example the Geolog System of International Geosystems Corporation, Vancouver. These are suitable for field logging of DDH core and outcrops, and processing routines can plot maps and cross sections, contoured with grain size, fracture density, assay results, etc. Corrections for hole deviation are routinely performed.

One of the principal advantages of all these computer based data management systems is that they serve to standardize field and core observations, with the provision of checklists and spaces on input forms to fill out each item. Standard codes and estimation procedures are used, so that data from different areas or collected by different operators, can readily be combined if required for basin mapping purposes.

The main virtue of computer-produced maps and sections is the ease and rapidity with which they can be drawn. A typical exploration problem might be concerned with one or two formations penetrated by anywhere from several dozen to several thousand wells. To plot and contour all these maps by hand is a laborious procedure. The data can be retrieved and plotted by computer in minutes.

Bole (1981) discussed the use of an interactive cathode ray tube (CRT) display system for retrieving, manipulating, editing and plotting data from a well file, permitting very rapid testing of different ideas in a project area. Lee (1981) discussed a statistical procedure for quantifying lithofacies by correlating selected descriptive attributes from those stored in digitized well files. The method results in excellent computer drawn facies maps that would take a great deal of time to produce by hand.

Whether computer contoured maps and sections are "better" than hand-drawn ones is another question. For maps with numerous control points they are probably equally meaningful but, as I have argued elsewhere (Miall, 1975), where the control is sparse there probably is little to be gained from using the computer other than its speed and neatness. Some computer experts would not agree. They point out that there may be numerous ways of contouring a given data set, and suggest that only a clinically unbiased computer algorithm can produce a "correct" map. My answer to this is that no computer has ever attended my facies models course. Data points on an isopach or lithofacies map are not arbitrary digits, but are reflections of subtle trends which only a skilled basin analyst can interpret. From regional information or paleocurrent data or some piece of core data it might be possible to predict shoreline orientation or channel or reef trends, and the contours can be drawn accordingly. As Robinson (1981) pointed out, computers cannot yet build pre-conceived trends into their contouring output, whereas hand-drawn maps can. Such maps demonstrate the analyst's ideas about an area, and in the exploration business the generation of ideas which go beyond established, routine facts is the most important contribution a geologist can make. The ideas may reflect several or many different interpretations, but although many companies may explore the same piece of territory using essentially the same data, it is usually the one with the best geological ideas that finds the mine or brings in the oil producer.

A more basic problem with commercial data files is the accuracy of their stratigraphic information. Most retrieval programs for petroleum exploration data are keyed to formation tops but, as we discuss in several places throughout this book, formations may be unreliable guides to paleogeography. They may be difficult to correlate and, particularly in some of the better explored sedimentary basins, they may be based on a local stratigraphic section and are of little use regionally (Chap. 3). Application of the depositional systems approach to stratigraphy may demonstrate that many formations have a quite arbitrary meaning (Chap. 6). The basin analyst may therefore be required to recheck all the relevant formation tops used in the data bank, or input his or her own, based on an up-to-date interpretation of the stratigraphy. The computer's advantage of speed and efficiency is thereby lost.

Setting aside these problems, there is no question that the computer is eminently suited to perform any kind of statistical data reduction technique, many of which have broad geological applications. The use of multivariate statistics in facies analysis is discussed in section 4.4; an example of the use of factor analysis in studying heavy mineral dispersal trends is discussed in section 5.5. This type of work plus all the various multicomponent and ratio maps discussed in section 5.3 are readily produced by statistical manipulation of a data bank. A particular favorite with users of well files is **trend surface analysis**. The technique is described by Davis (1973).

The purpose of trend-surface analysis is to separate map data into two components, regional trends and local fluctuations.The regional component can then be subtracted mathematically, leaving "residuals", which correspond to the local variation. The definition of regional and local is subjective, and can be adapted to the particular mapping exercise at hand. For example a local trend might be defined as channel and crevasse sand accumulations within a delta lobe, or individual lobes within a large delta complex. The

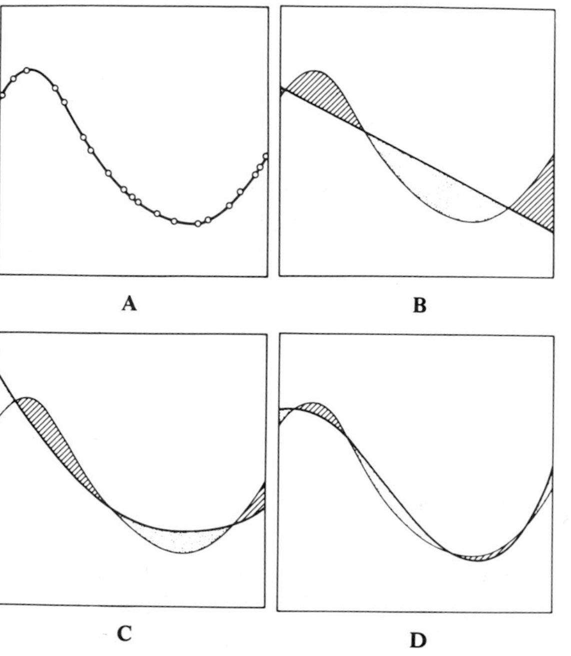

Fig. 5.33. Fitting a curve to data points—a two-dimensional representation of the trend-surface analysis process. Residuals are shaded. **A** curve drawn through the original data; **B** straight line (plane) fit; **C** parabolic fit; **D** cubic fit. Note the different position of the residuals in each case. Reproduced, with permission, from J.C. Davis, Statistics and Data Analysis in Geology, © 1973, John Wiley and Sons, Inc., New York.

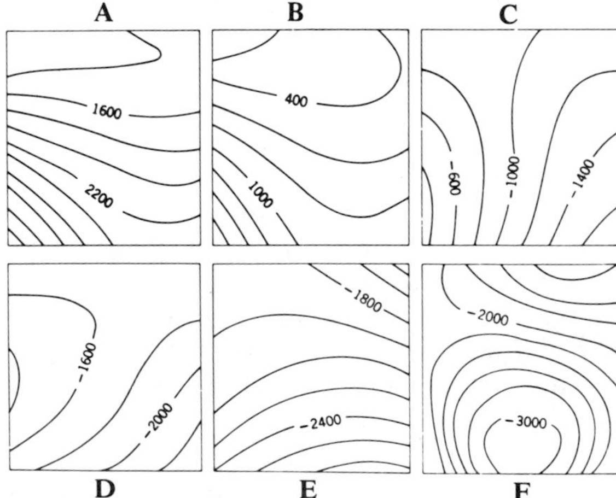

Fig. 5.34. Third degree trend-surfaces determined for a series of stratigraphic tops in northwestern Kansas (Merriam and Harbaugh, 1964). **A** Cretaceous, **B** Permian, **C** Pennsylvanian, **D** Mississippian, **E** Ordovician, **F** Precambrian.

most important decision the geologist has to make before proceeding with the analysis is to define the nature of the regional trend. Is it a plane (first degree trend), a cylindrical or parabolic curve (second degree), or some more complex three-dimensional surface (third and higher degrees)? There are no geological or statistical rules to guide this decision, only experience and judgment about what makes most sense. Examples of how this affects the definition of residuals are shown in two dimensions in Figure 5.33. Residuals will always be generated in a trend-surface analysis, but they will have no geological meaning unless the trend that has been extracted also corresponds to some geological reality. For example on many continental margins regional structure can be approximated over limited areas by a dipping plane or gentle curve. Application of a first or second degree trend-surface analysis may then produce a pattern of residuals corresponding to drape over basement fault blocks or buried reefs. This might be extremely useful in the exploration for hydrocarbon traps. Higher order trend-surface analysis will also result in complex patterns of residuals, but these may have no geological meaning whatever.

A geologist may be more interested in the regional trend than in the residuals, in which case derivation of ever more complex polynomial surfaces might be required. Statistical tests are available to determine the goodness of fit of these surfaces to the original data. Again, the decision must be based on geological grounds as to which surface is most appropriate. The highest degree surfaces may fit the data very accurately, but only by accomodating all the local variability. Such a result is not likely to be very useful. Davis (1973, p. 336) pointed out that realistic-looking trends can be produced by random data, and he discussed statistical tests for determining where to stop.

Two examples of trend-surface analysis are given in Figures 5.34 and 5.35. The first of these shows a series of third degree trend-surfaces of stratigraphic tops in northwestern Kansas (from Merriam and Harbaugh, 1964). The second example shows a fourth degree trend and residuals for sand distribution in a Pennsylvanian delta (from Wermund and Jenkins, 1970). The regional trend corresponds to the major lobe or depocenter, and individual channel fills and bar fingers are clearly shown. Moving averages and other more

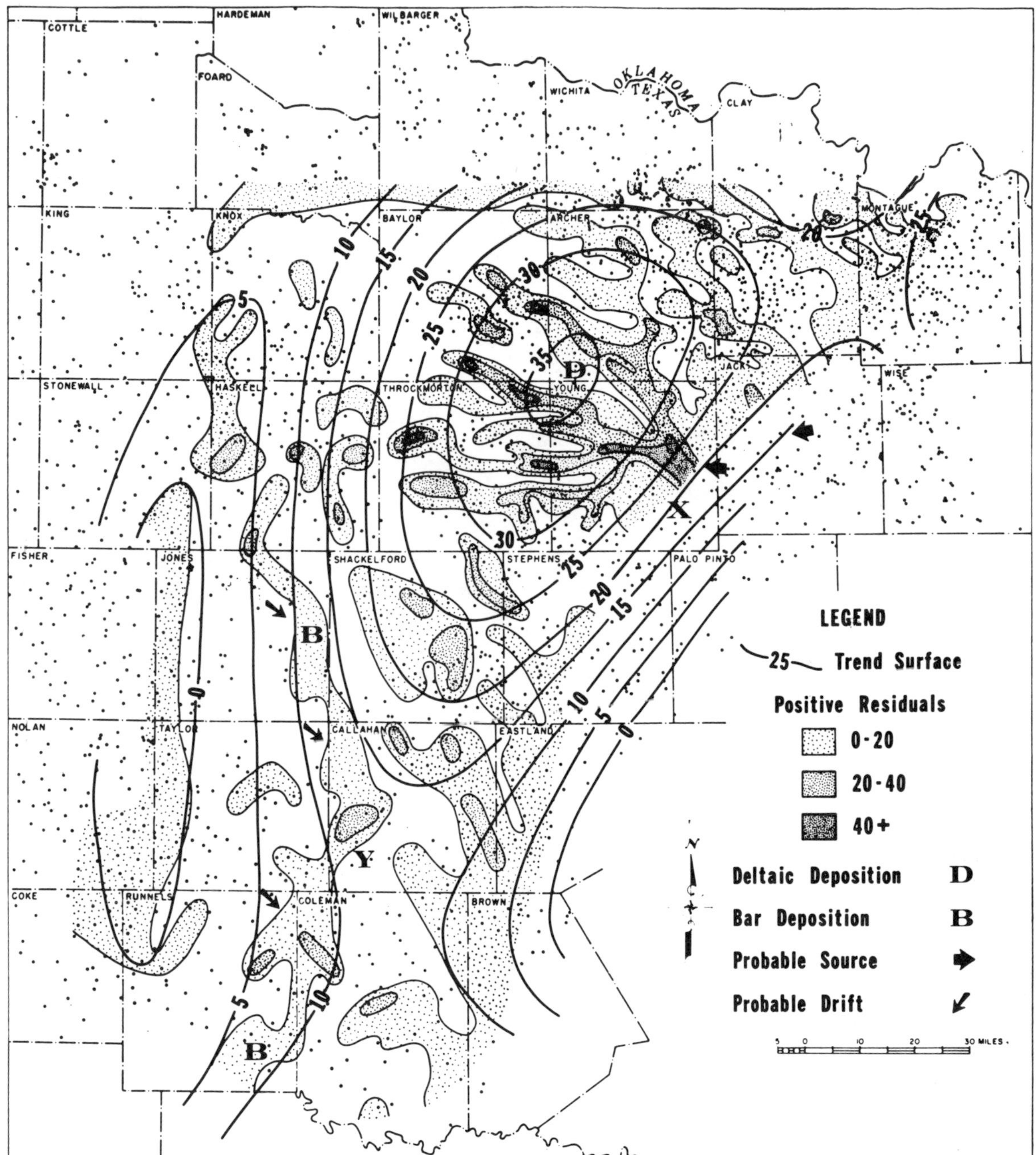

Fig. 5.35. Trend-surface analysis of a Pennsylvanian delta in north-central Texas. The deltaic depocenter is outlined by a fourth degree surface drawn on sand thicknesses. Channel fills and other individual sandstone bodies are outlined by the residuals (Wermund and Jenkins, 1970).

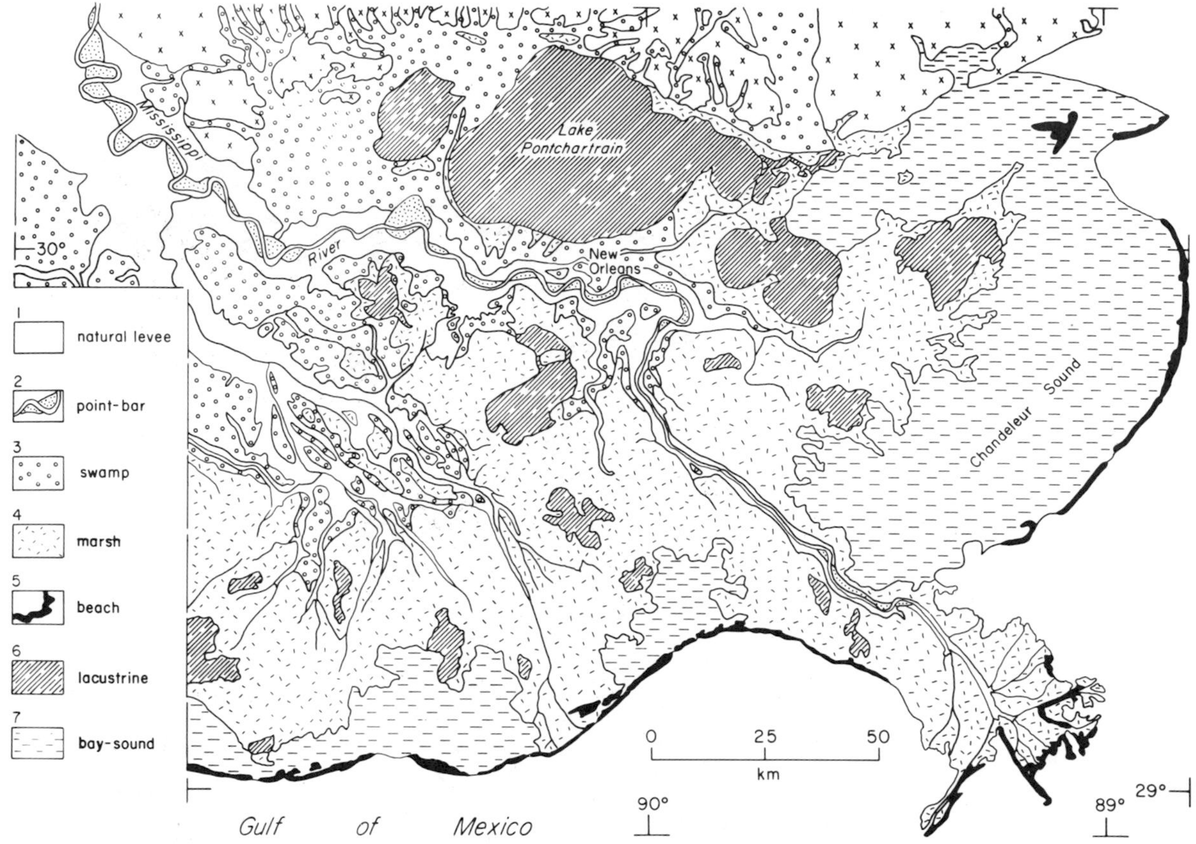

Fig. 5.36. Depositional subenvironments of the modern Mississippi delta (McCammon, 1975).

advanced computer map analysis techniques are discussed by Davis (1973) and Miller and Kahn (1962).

5.7 Sampling density and map reliability

Most lithostratigraphic maps are plotted from inadequate data. In any kind of exploration program it is necessary to draw conclusions and make predictions based on a limited number of outcrops or exploratory holes. The subsurface geologist will commonly be required to make decisions about where to place the next hole in order to pursue underground a particular facies type or a stratigraphic hydrocarbon or mineral prospect.

There are two sorts of questions to be considered here:

1. What sort of sample density, spacing or distribution pattern is adequate to ensure that our lithofacies, petrographic or seismic maps are an accurate reflection of reality?
2. Given all the mapped distribution and trend information available, what is the likely value of one, or a few, more sample points (read exploration holes)?

These questions are the delight of statisticians, but the answers require consideration of some geological realities as well as the ramifications of statistical probability.

McCammon (1975) described an interesting experiment which explored the problem of how many sample points were necessary to define the facies patterns in a typical sedimentary environment, using as his sample the fluvial and deltaic subfacies of the modern Mississippi delta (Fig. 5.36). He divided the map into a grid of 4,034 sample points, which is about one per 10 km^2, and redrew the facies map using these data (Figure 5.37A). Assuming that this was a buried system and that a geologist could sample only a few of the total points by drilling, how should these be distributed to provide the best reproduction of the original grid-derived facies pattern? In the first maps McCammon distributed the limited sample over a widely spaced grid and filled in the remaining points of the original 10 km^2 grid by assuming that these were the same facies as that of the sampled point nearest to them. Figure 5.37B shows the map produced when 167 points were used to map the original 4,034. It resulted in 65% of the original points being classified correctly. Other methods were tried in which 100-odd sample points were distributed *knowing* the distribution of the original 4,034. Where should they be placed to provide to provide the most accurate simulation? The best result was obtained by placing the points in the center of each facies area, and dividing larger areas into equidimensional subareas (Fig. 5.37C). However, this only increased the accuracy to 70.8%, which suggests that even where some information is available on facies patterns, a grid sampling plan is about as efficient as any other method in producing accurate facies maps. Rarely is this achieved in practice, because most exploratory holes are drilled for a variety of purposes, giving rise to rather patchy distributions.

Robinson (1981) discussed a similar exercise of mapping subsurface patterns using limited data, focusing on the problem of delineating paleodrainage nets. Only where sample spacing was equal to average channel width was the computer-drawn map a satisfactory reproduction of the original.

This is not quite the whole story, because even where sample points can be chosen with prior knowledge, as in Figure 5.37C, automated contouring routines cannot incorporate this same information. Usually other geological information, such as outcrop or dipmeter paleocurrent data, or outcrop facies data that can be extrapolated into the subsurface, can be used to define facies trends. This is where facies analysis techniques can be most useful. Known basement fault control or ancient shoreline orientation may help improve the map, and geologists also have the advantage that they can use Walther's Law and other facies model principles to improve the interpretation.

The second of our questions is an important one in the case of prospect exploration. That one additional hole which locates an economic prospect has achieved the objective of the facies mapping exercise, whether or not the facies patterns have been mapped correctly. Skilled use of geology and geophysics in petroleum exploration results in a ratio of successful to dry holes ranging between about 1:4 and 1:10, depending on the exploration maturity of the basin and the subtlety of the traps.

The geologist must constantly bear in mind the scale of the facies units being mapped versus the scale of the sampling pattern. For example, grid sampling every 20 km is not likely to be very successful if the objective is to map fluvial point bars and crevasse splays. Or a map showing lobate sand isolith contours may suggest a simple delta pattern but is unlikely to be this if the lobes are hundreds of kilometers across (more likely a coastal plain complex). Constant pattern and scale comparisons with appropriate modern analogues are necessary to keep the interpretations consistent with reality. In fact the relationship between the size of fairly regular, detectable features, such as pinnacle reefs, and the spacing of control points can be calculated statistically for sample grids, in order to indicate the probability of interception (Singer and Wickman, 1969; Davis, 1973). This is not possible for irregularly-shaped geological features or random sample distributions.

5.8 Stratigraphic cross-sections

Reference is made at numerous points in this book to structural and stratigraphic cross-sections, as illustrations of stratigraphic correlations, facies interpretations, basin models, etc. Such sections are a standard mapping tool, and it may seem superfluous to include a section on them in this chapter. However, there are several points about stratigraphic sections that are not well enough understood, and so it would seem to be useful to comment upon them at this point.

Cross sections are, of course, vital tools in stratigraphic correlation and in structural interpretation. Constructing stratigraphic sections from outcrop logs or the first exploratory wells in a basin provides the geologist with a kind of workbench on which to refine basin interpretations. Typically the logs are pinned to the office wall or loosely taped to a large card, and correlation lines drawn in using soft pencil or pins and coloured thread. Days may be spent contem-

A

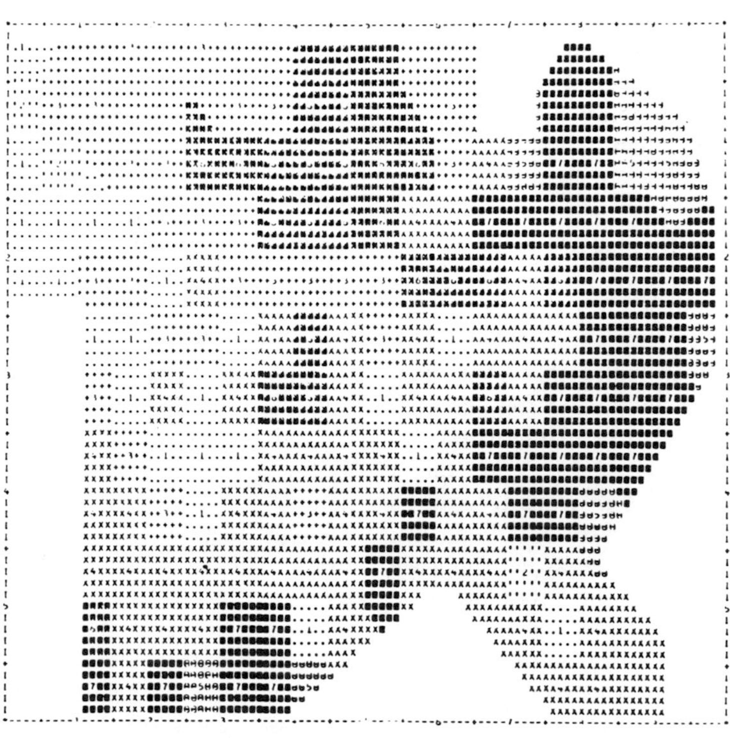

B

Fig. 5.37. Reproduction of the map pattern of Fig. 5.36 using a limited number of sample points. See text for explanation of the three methods used (McCammon, 1975).

Fig. 5.37 C.

plating this pastiche, rearranging the correlation lines and raising or lowering the logs to try out different stratigraphic relationships. Many thoughtful employers now provide their offices with large cork boards or magnetic walls and small magnets to facilitate this process (and to save on redecoration bills). The results of this work commonly end up as large folded charts tucked in at the end of company reports or published monographs.

Initially the focus will be on stratigraphic correlation, but as analysis proceeds cross sections can be used to study detailed facies relationships, perhaps tracing individual lithosomes from one well section to the next. Ideas gained in this work can be used as a basis for drawing lithofacies maps and, in fact, maps and sections should be constructed interactively and simultaneously, as far as possible, so that the geologist can build up a three-dimensional image of the basin.

There are essentially three types of cross section that are useful in basin analysis work:

1. detailed lithosome sections showing small scale basin architecture
2. generalized facies correlations, used to illustrate depositional systems or systems tracts
3. Basin cross sections drawn to show major stratigraphic sequences, gross facies patterns and some elements of regional structure, possibly in simplified cartoon form.

In all cases the act of drawing up the section from raw data may provide the geologist with unforeseen insights. These three types, with some examples, are discussed below.

Detailed lithosome sections are typically drawn with the following range of scales: horizontal, 1 cm = 40 m to 5 km; vertical, 1 cm = 5 to 300 m. Vertical exaggeration may vary for convenience, but is always useful in order to clarify facies relationships. Horizontal scale may be arbitrary, with well sections spaced at equal intervals, but there is no particular advantage to this approach and it may introduce distortions. An example of a type 1 section is shown in Figure 5.38, and illustrates the complex interfingering of Mississippian shelf limestones with sabkha evaporites in the Williston Basin, North Dakota (from Wilson, 1980). A different kind of example is shown in Figure 5.39, from Busch (1974), showing two deltaic channels cut into delta plain coals and floodplain fines in the Maracaibo Basin,

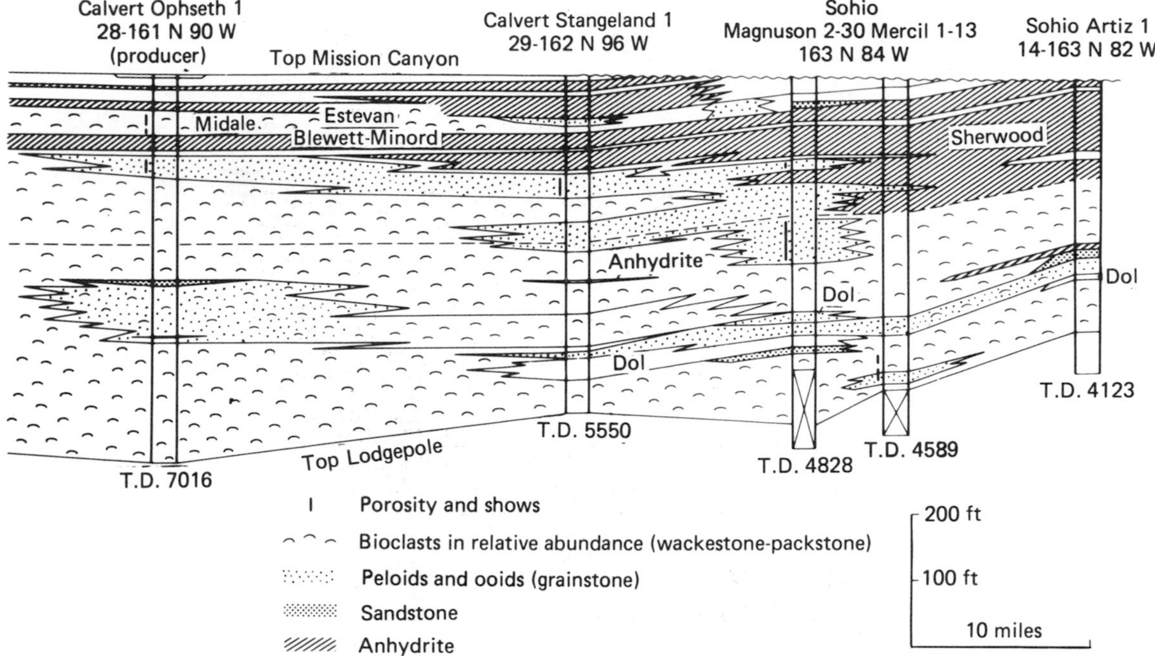

Fig. 5.38. Example of a detailed stratigraphic cross section showing relationship of major lithosomes, Mississippian of North Dakota (Wilson, 1980).

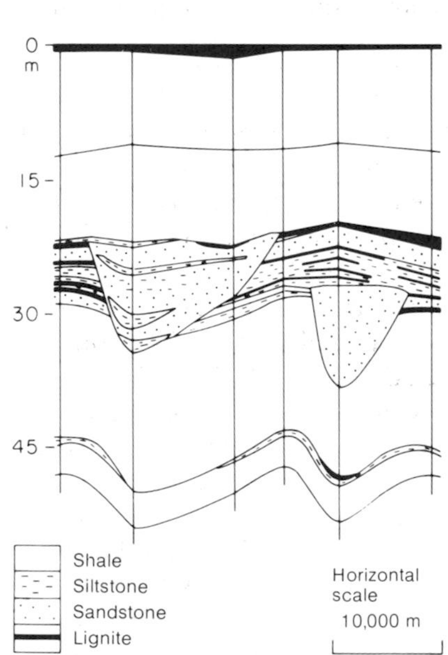

Fig. 5.39. A stratigraphic cross section revealing deltaic channels in the Maracaibo Basin, Venezuela (Busch, 1974).

Venezuela. The folds below the channels were caused by compaction of the fine sediments and loading of the denser channel fill. Other examples of detailed lithosome sections illustrated in this book are Figures 3.2 to 3.7, 3.26, 4.2, 4.39, 8.8, 8.22 and 8.25. It is an instructive exercise to compare these, examining scales, choice of datum, documentation of section control, style of illustrating facies changes, and so on.

Sections illustrating depositional systems are illustrated in Chapter 6. Typical scales are: horizontal, 1 cm = 3 to 30 km; vertical, 1 cm = 50 to 500 m. Seismic sections are often illustrated at these scales. Examples are given in Figures 5.40 and 5.41 showing clastic wedges on the Gulf Coast and southern Africa. Both these examples show how successful illustrations can be constructed by judicious simplification. Such schematic diagrams convey a great deal of information to the reader in the briefest of examinations. Figure 5.41 is an interesting example of a section which contrives to show three types of stratigraphy at once: depositional systems or facies relationships, stratigraphic nomenclature, and geochronologic correlations.

Small-scale basin cross sections are constructed at scales of: horizontal, 1 cm = 5 to 100 km;

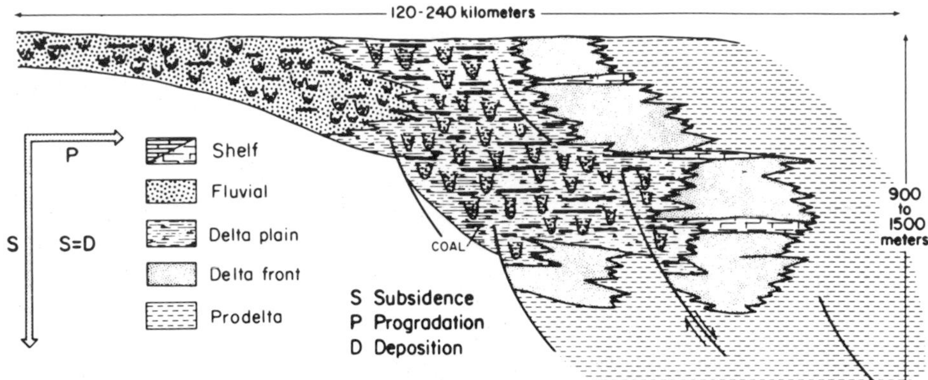

Fig. 5.40. Generalized stratigraphic cross section through the Eocene sediments of the Gulf Coast, showing interrelationships of the various depositional systems (Brown and Fisher, 1977).

vertical, 1 cm = 1 to 5 km. These are almost invariably structural cross sections, but in addition to illustrating structural style they can be used to delineate major stratigraphic sequences and gross facies relationships. They usually are extended to basement, and may contain geophysical data pertaining to deep crustal structure. Figure 5.42 shows the structure and stratigraphy of the continental margin off Newfoundland, based on seismic and limited well data (from McWhae, 1981). This region is still in an early stage of exploration, yet the stratigraphic relationships are already well understood on a regional scale. Figure 5.43 illustrates the reconstructed stratigraphy of the mainly Precambrian Hecla Hoek "geosynclinal trough", Svalbard (from Birkenmajer, 1981). This is an example where structure has been omitted because of its complexity. Stratigraphic relations

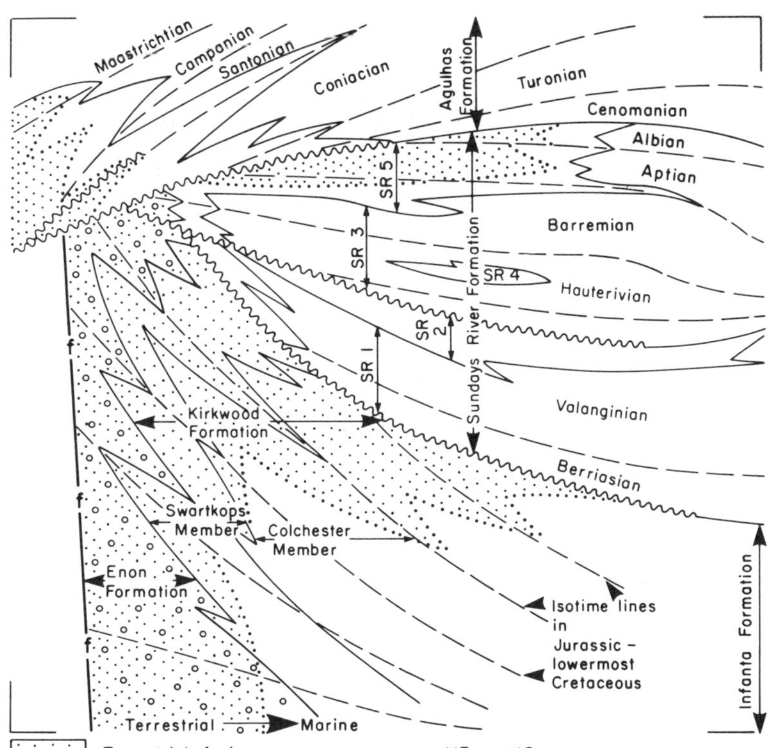

Fig. 5.41. Generalized stratigraphic relationships of Jurassic–Cretaceous strata, southern Africa, showing depositional systems, stratigraphic nomenclature and geochronology (Winter, 1979).

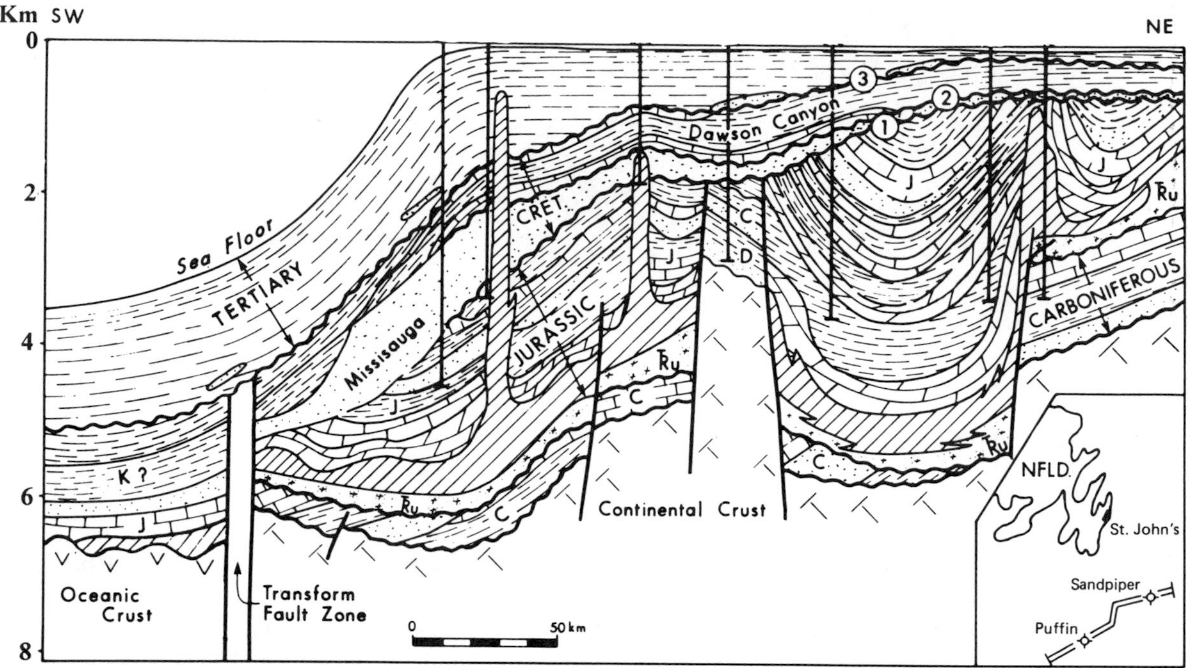

Fig. 5.42. Generalized basin cross section, showing schematic structure and facies, continental margin off Newfoundland (McWhae, 1981).

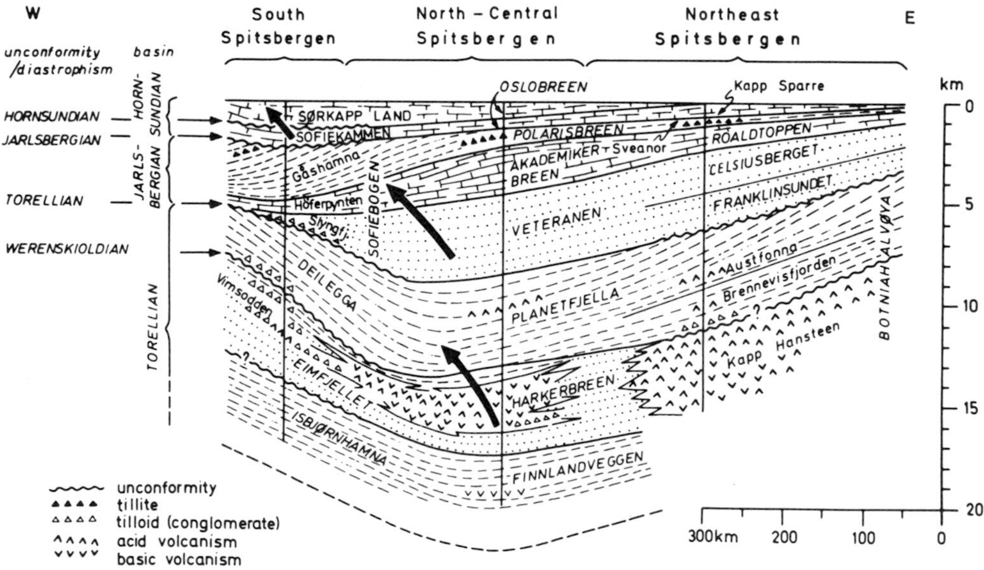

Fig. 5.43. Reconstructed stratigraphic relationships in the Hecla Hoek "geosynclinal trough", Svalbard (Birkenmajer, 1981).

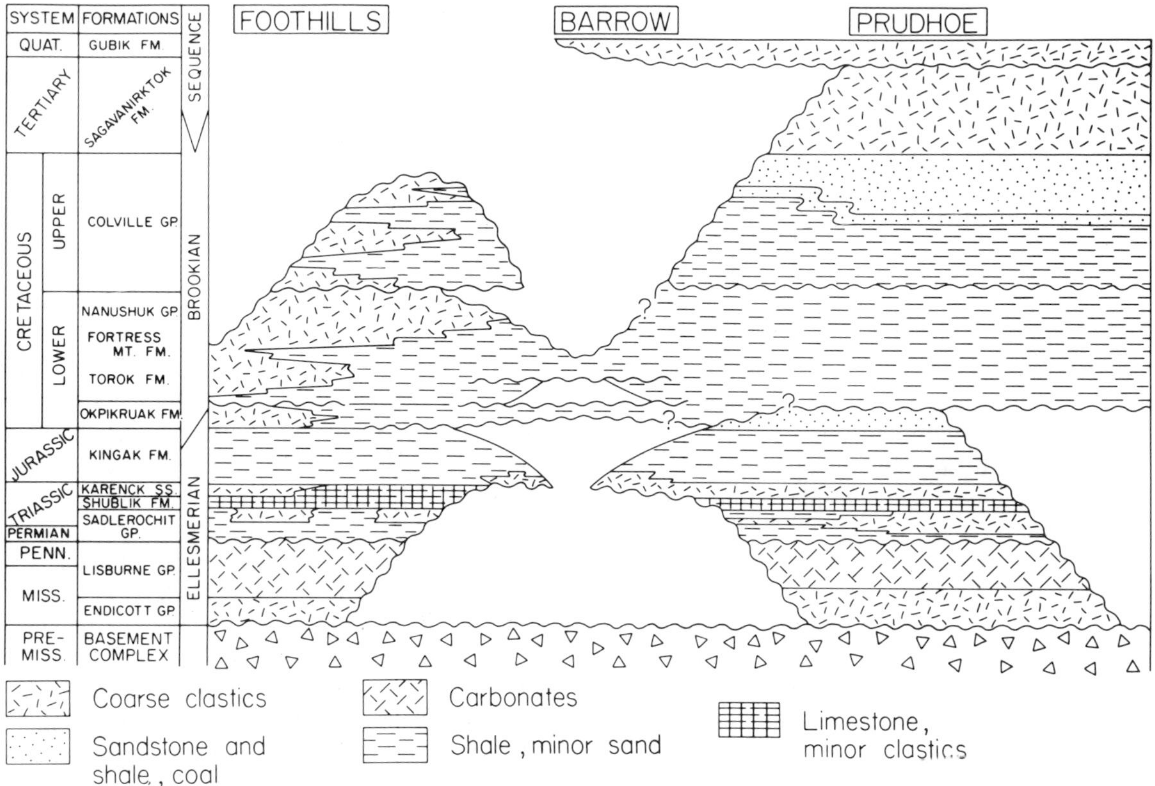

Fig. 5.44. Example of a schematic stratigraphic cross section or generalized correlation diagram, showing lithofacics assemblages, North Slope of Alaska (Jamison et al., 1980).

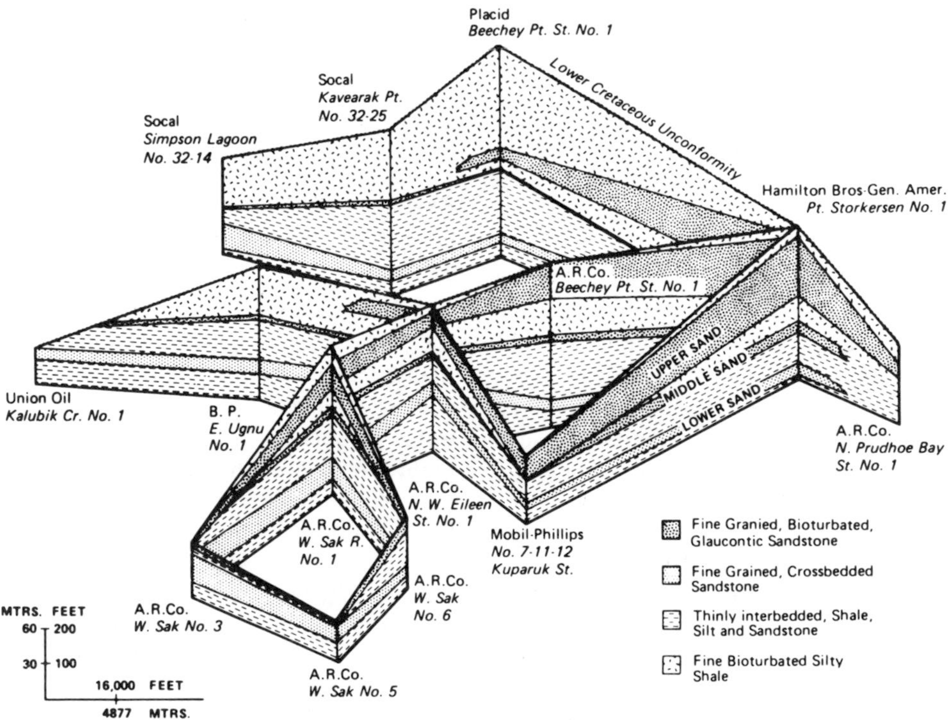

Fig. 5.45. Example of a fence diagram, Kuparuk River Formation (Lower Cretaceous); Prudhoe Bay area (Jamison et al., 1980).

are clearly seen, yet the reader is left with an uncomfortable feeling of not knowing how close to the presently preserved reality this cross section really is.

The Svalbard illustration is one step closer to a variation of the regional reconstruction known as a **schematic stratigraphic cross section**. This is a combination of correlation table and facies diagram. Vertical and horizontal scales usually are variable and arbitrary, although the vertical scale may be drawn using radiometrically based time or chronostratigraphic increments (e.g. stages) rather than thickness. Thus, these diagrams may contain gross spatial distortions, but they can be extremely useful in conveying correlation and facies information. An example is illustrated in Figure 5.44.

An entirely different type of cross section is the **fence diagram**. An example illustrating the Kuparuk Formation (one of the Prudhoe Bay area units shown in Figure 5.44) is given in Figure 5.45. Fence diagrams are an attempt to improve on conventional two-dimensional diagrams by creating a three-dimensional, perspective effect. Vertical sections are placed in their correct relative position as far as is possible. These diagrams are hard to read because parts of the section are hidden by fences in front, and the same regional thickness and facies variations may be better displayed in map form.

5.9 Paleocurrent analysis

5.9.1 Introduction

The art of paleocurrent analysis was invented by the quintessential nineteenth-century English amateur, Henry Clifton Sorby, who published his first paper on the subject in 1852. He understood the formation of "ripple drifted" and "drift bedded" structures (climbing ripples and large scale crossbedding, respectively) and, in seven years, made about 20,000 recorded observations, a total that has probably never been equaled. Most of Sorby's work remained unpublished. He was far ahead of his time, as paleocurrent work did not become a routine analytical procedure until the nineteen fifties. His work remained unappreciated and largely unknown until exhumed by Pettijohn (1962) and reinterpreted by Allen (1963b) (see also Miall, 1978).

The technique could not become a routine component of basin analysis until proper attention was paid to the description and classification of sedimentary structures (McKee and Weir, 1953; Pettijohn, 1957; Allen, 1963a), and a start was made on the investigation of bedform hydraulics. The major breakthrough here was the development of the flow regime concept by Simons et al. (1965). These two topics are discussed at some length in sections 2.3.2 and 4.5.4 to 4.5.5, respectively.

Paleocurrent analysis is primarily an outcrop study. Oriented core can be used, but is rarely available. In petroleum exploration wells unoriented core may be oriented using a dipmeter, but correlation between the two commonly is difficult. The dipmeter itself may provide some paleocurrent data under ideal circumstances, as discussed in section 5.9.6. Because of these limitations paleocurrent analysis may be of little use in petroleum exploration. It can assist in interpreting the geology of basin margins but, as discussed in section 2.2.1, this may not be particularly helpful. For coal, metals and other types of mining which deal with surface or large underground outcrops paleocurrent analysis may, in contrast, be extremely useful.

The technique can provide information on four main aspects of basin development:

1. Direction of local or regional paleoslope
2. Depositional environment
3. Direction of sediment supply
4. Geometry and trend of lithologic units.

Examples of these applications are discussed in the ensuing paragraphs.

5.9.2 Types of paleocurrent indicator

Sedimentary structures and fabrics used in facies and paleocurrent studies are described and illustrated in section 2.3.2. The following notes explain briefly how they yield current directions.

1. Ripple marks and crossbedding (Fig. 2.18): the inclination of foreset directions is generally downcurrent, because of the grain avalanching mechanism. Needless to say it is not always this simple. Smith (1972) demonstrated that planar crossbed sets in rivers commonly advance obliquely to flow direction. Trough crossbeds do not yield accurate flow directions unless the analyst can observe the orientation of the trough axis.

Hunter (1981) discussed the flow of air around large eolian dunes and suggested that these too commonly advance obliquely to wind direction. The use of statistical procedures and appropriate facies data usually circumvents these problems.

2. Channels and scours (Fig. 2.21) occur in many environments and may indicate the orientation of major erosive currents such as those generating river or tidal channels, delta or submarine fan distributaries, and so on. However, the larger channels, which are those most likely to be of regional significance, are usually too large to be preserved in outcrops. They may be readily apparent on seismic records, as discussed in Chapter 6.

3. Parting lineation or primary current lineation, the product of plane bed flow conditions (Fig. 2.20). This structure is only visible on bedding plane exposures. It usually yields directional readings of low variance because it forms during high energy flow in river, delta or tidal channels, when bars or other obstructions to flow are under water and flow sinuosity is low. The structure indicates orientation but not direction of flow, because of the ambiguity between two equally possible readings at 180° to each other. Usually this can be resolved with reference to other structures nearby.

4. Gravel fabric. Clast transport by traction or in sediment gravity flows commonly produces a measurable fabric. Imbrication in traction current deposits occurs where platy clasts are stacked up in a shingled pattern, with their flattest surface dipping upstream and resting on the next clast downstream (Fig. 2.19). Because gravel is only moved under high energy flow conditions it tends to show low directional variance, like parting lineation. Rust (1975) found that variance decreased with increasing clast size, suggesting reduced flow sinuosity with increased stream power.

5. Sole markings (Fig. 2.22). These are typically associated with the deposits of turbidity currents and fluidized or liquefied flow, although they also occur less commonly in other clastic environments where vortex erosion may occur at the base of a flow, and "tools" can be swept down a bedding surface. Their greatest use, however, is in the investigation of submarine canyon and fan deposits. They are best seen on the undersides of bedding surfaces, where sandstone has formed a cast of the erosional feature in the underlying bed. Tool markings yield information on orientation but not direction, like parting lineation; flute marks are longitudinally asymmetric, with their deepest end lying upstream.

6. Oriented plants, bones, shells, etc. These do not respond systematically to the aligning effects of currents unless they are elongated. There may be ambiguity as to whether they are oriented transverse or perpendicular to current patterns, and there are other difficulties, for example, the tendency of fossils such as high-spired gastropods to roll in an arc. Some of these details are discussed in section 5.9.5. Fossils are usually only useful for local, specialized paleocurrent studies.

7. Slump structures generated on depositional slopes contain overfolds which may be aligned parallel to strike. They are therefore a potential paleoslope indicator, particularly on the relatively steep slopes of prodeltas and actively prograding continental margins. Friction at the margin of the slump may cause it to rotate out of strike alignment, so that a statistical approach to measurement is desirable. Potter and Pettijohn (1977, Chap. 6) Rupke (1978), and Woodcock (1979) reviewed their use as paleoslope indicators.

Several other paleocurrent indicators are reviewed by Potter and Pettijohn (1977), including some, such as sand grain orientation and magnetic anisotropy, requiring detailed laboratory analysis of oriented samples. They are mainly of academic interest, where they may help solve a particular local problem. The seven types of paleocurrent indicator listed above can be quickly and routinely measured in the field, and have been widely used in basin studies.

5.9.3 Data collection and processing

Paleocurrent data should be carefully documented in the field. In the past paleocurrent trends have sometimes been reported on the basis of a geologist's mental estimates of the range of indicated directions but, even if this is correct, it may result in a significant loss of useful information. For every paleocurrent observation recorded in field notes the following information should be included:

1. location and (if relevant) precise position in a stratigraphic section
2. structure type
3. indicated current direction

4. scale of structure (thickness of crossbed, depth and width of channel, mean or maximum clast size in imbricate gravels)
5. local structural dip.

Current directions should be measured to the nearest 5° with a magnetic or sun-compass and corrected to true north wherever necessary. In the case of parting lineation and tool markings the correct orientation of two possibilities at 180° can usually be identified by referring to other types of current structure nearby. Measurement accuracy greater than ±5° is difficult to achieve and is, in any case, not necessary. Indicated current directions may need to be corrected for structural dip and fold plunge, otherwise significant directional distortions may result. This should be carried out as soon as possible, preferably after every day's work, in order that possible errors can be detected while there is still time to rectify them in the field. For linear structures such as parting lineation or sole markings, structural dips as high as 30° can be safely ignored, as they result in errors of less than 4°. However, foreset dip of crossbedding is significantly affected by structural dip, and should be corrected wherever structural dip exceeds 10°. This subject is discussed further by Potter and Pettijohn (1977, Chap. 10) who illustrated a correction technique using a stereonet. Ramsay (1961) provided graphical solutions. Parks (1970) presented computer routines for the necessary trigonometric calculations.

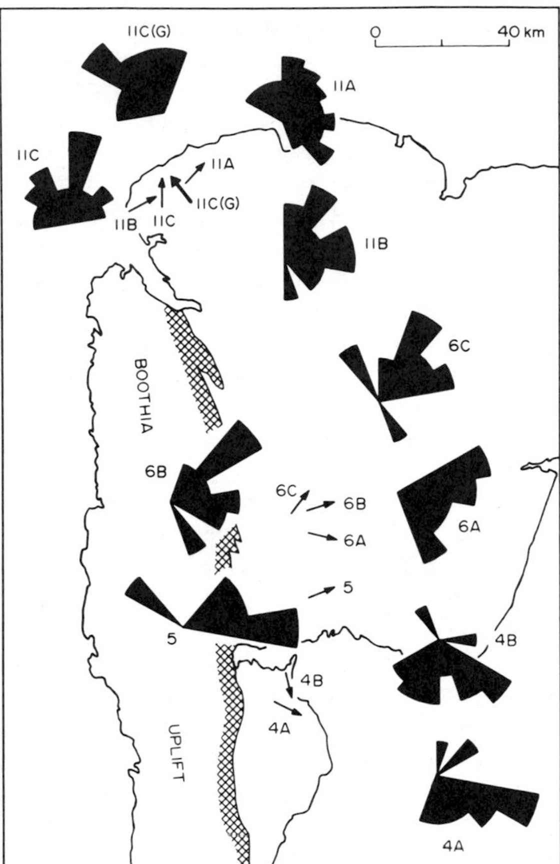

Fig. 5.46. Typical regional paleocurrent map, showing current rose diagrams and vector mean arrows for each of ten outcrop areas, with station numbers. The map represents 165 field readings. Devonian fluvial sandstones, Somerset Island, Arctic Canada (Miall and Gibling, 1978).

Sooner or later the question will arise as to how many readings should be made? There is no single or simple answer to this problem because it depends on how many measurable current indicators are available for observation and what are the objectives of the study. Olson and Potter (1954) discussed the use of grid sampling procedures and random selection of structures to measure, followed by calculation of reliability estimators to determine how many readings were necessary in order to be sure of determining correct directional trends. In this study they were concerned only with determining regional paleoslope. Nowadays we understand a great deal about air and water flow patterns in different depositional environments, and a case could be made for measuring and recording every visible sedimentary structure. Such detailed data can be immensely useful in amplifying environmental interpretations and clarifying local problems, as discussed in section 5.9.5. Some selection may have to be made in areas of particularly good exposure. A practical compromise is to record every available structure along measured stratigraphic sections and to fill in the gaps between sections with spot (gridded or random) samples. This procedure permits the elaboration of both local and regional paleocurrent trends.

If the trend itself is important, 25 readings per sample station is commonly regarded as the minimum necessary for statistically significant small samples. However, the same or fewer readings plotted in map or section form can yield a great deal of environmental detail, whether or not their mean direction turns out to be statistically significant. Several hundred or a few thousand readings may be necessary for a thorough analysis of a complete basin.

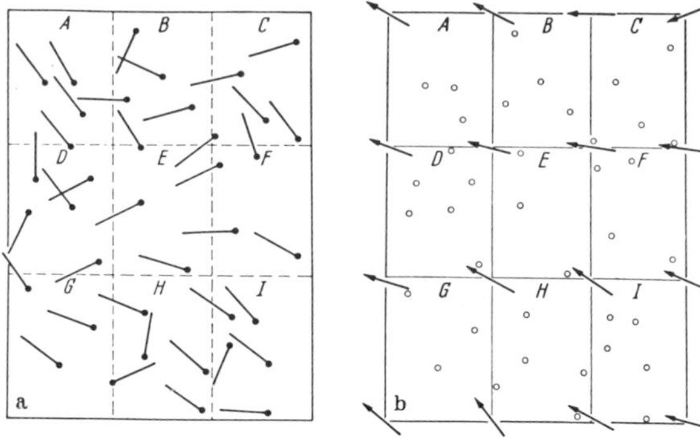

Fig. 5.47. Construction of a moving average paleocurrent map (right) from raw data (left) (Potter and Pettijohn, 1977).

A variety of statistical data reduction and data display techniques is available for paleocurrent work. The commonest approach is to group data into subsets according to stratigraphic or areal distribution criteria, display them visually in **current rose diagrams**, and calculate their mean and standard deviation (or variance). The method of grouping the data into subsets has an important bearing on the interpretations to be made from them, as discussed in the next section. A current rose diagram is simply a histogram converted to a circular distribution. The compass is divided into 20, 30, 40 or 45 degree segments and the rose drawn with segment radius proportional to number of readings or percent of total readings. A visually more correct procedure is to draw the radius proportional to square root of percent number of readings, so that segment area is proportional to percent. Examples are given in Figure 5.46.

Potter and Pettijohn (1977) and Curray (1956) discussed arithmetic and vector methods for calculating mean, variance, vector strength and statistical tests for randomness. Miall (1974) described a simple data weighting procedure for crossbedding which may bring out subtle relationships between orientation and bedform scale. Miall (1976b) and Cant and Walker (1976) demonstrated the display and interpretation of data collected from vertical stratigraphic sections (section 5.9.5).

Statistical procedures such as moving averages and trend analysis are available for smoothing local detail and determining regional trends. A moving average map is constructed from gridded data, as shown in Figure 5.47. An arbitrary grid is drawn on the map; the mean current direction for the data in each group of four squares is then calculated and shown by an arrow at the center of this area. Each data point is thus used four times (except at the edges of the map). Examples of this method applied to paleocurrent data have been published by Pelletier (1958), Sturm (1971), Andersen and Picard (1974), and Seeland (1978). Trend analysis requires rather more complex calculation. Agterberg et al. (1967) used this technique in the study of a Triassic delta.

The basin analyst should be wary of becoming too deeply enmeshed in the refinements of statistical methods. The use of probability tests is a useful curb on one's wilder flights of interpretive fancy, but there is a vast literature on the statistics of the circular distribution which seems to detract attention from the very simple questions: "Do the data make geological sense? Can they be correlated with trends derived from other methods such as lithofacies mapping or petrographic data?" Moving average and trend analysis are useful techniques for reducing masses of data to visually appealing maps, but they inevitably result in a loss of much interesting detail.

5.9.4 The bedform hierarchy

A geologist collects paleocurrent measurements over individual bedding planes, or through a local sequence in a quarry, or along a lengthy dipping section in a river cut, or several such sections throughout a map area. These readings differ from each other by varying amounts at different levels of this sampling hierarchy, but what do these differences mean? Statistically this is a classic example of a problem in analysis of variance (Olson and Potter, 1954; Kelling, 1969). But what do the differences mean geologically? The larger the sampling area or the thicker the sampled section the greater will be the number of depositional events that contributed to the data set, and normally it will mean an increase in directional variance. This generalization can be systematized by the concept of the bedform hierarchy, as first noted by Allen (1966, 1967). Flow fields and the sedimentary structures arising from them are of up to six orders of magnitude, ranging from individual ripples to entire depositional systems. Allen (1966) and Miall (1974) applied this idea to river systems (Fig. 5.48). In Table 5.3 an attempt is made to expand it to other environments. Jackson (1975) showed that these ranks depend

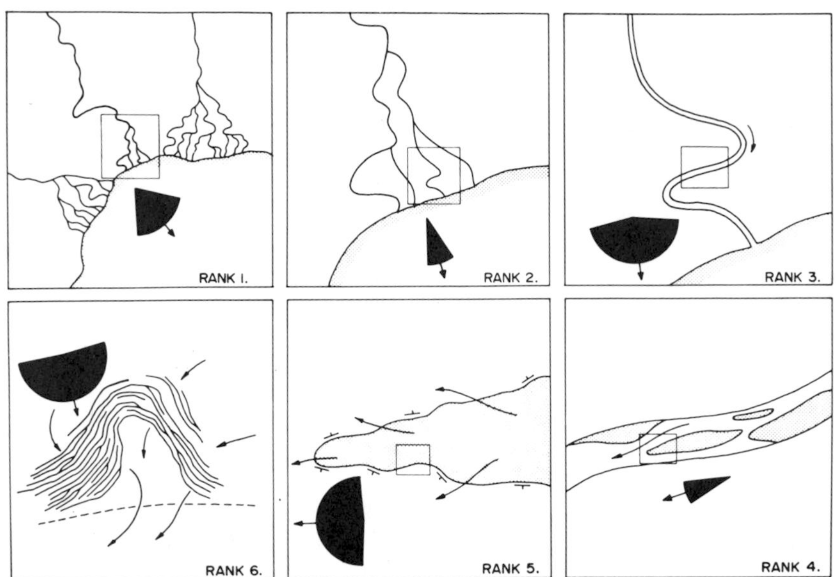

Fig. 5.48. The bedform hierarchy, illustrated for a fluvial–deltaic system. Current rose diagrams illustrate directional variability for the structures of each rank, which must be added to that of the next larger structures to determine total variability of the structures of each rank. Small squares show area enlarged for each lower rank. Structures are: rank 1. fluvial–deltaic system, 2. channels within a delta, 3. channel reaches, 4. lateral and mid-channel bars, 5. crossbedding flanking a bar, 6. ripple crests on a bar surface, oriented around a sheet run-off gulley (Miall, 1974).

on three orders of time scale and physical scale, and proposed that they be grouped into three dynamic types:

1. Microforms are structures of ranks 5 and 6, which are controlled by turbulent eddies in the fluid boundary layer. The time scale of variation is in the order of hours to seconds.
2. Mesoforms are structures of rank 4 and, to some extent, rank 3. They form in response to what Jackson (1975) termed dynamic events, such as hurricanes, seasonal floods or eolian sandstorms, when disproportionately large volumes of sediment are moved in geologically instantaneous time periods. The system may remain virtually unchanged between dynamic events.
3. Macroforms represent the long term accumulation of sediment in response to major tectonic, geomorphic and climatic controls.

It is important for mapping purposes to attempt to relate the sampling scale to this hierarchy, so that the sources of directional variance can be interpreted intelligently. For example a small quarry might be located entirely within the

Table 5.3. The bedform hierarchy and its application to three clastic environments

Rank	Dynamic type	Fluvial systems	Tidal inlets and deltas	Submarine fans
1	macroform	river system	inlet system	fan system
2	macroform	meander belt	inlet channel	individual fan
3	macroform	channel reach	ebb or flood delta	channel, suprafan lobe
4	mesoform	lateral, point bar	spillover lobe, ebb spit	sediment gravity flow
5	microform	large scale crossbedding	large scale crossbedding	flute
6	microform	ripple	ripple	vortex marks

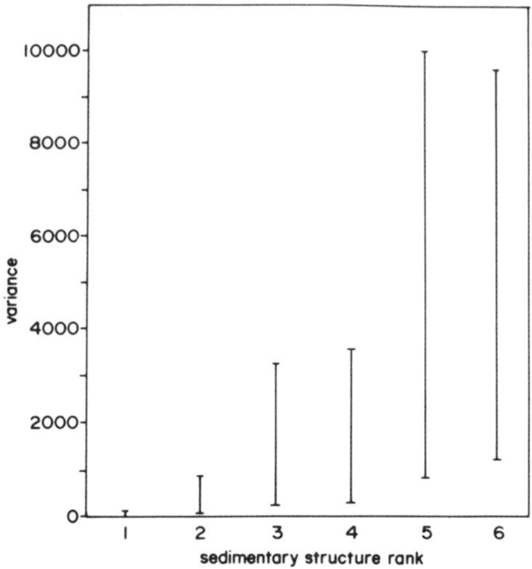

Fig. 5.49. The range of directional variance shown by sedimentary structures of different rank. Based on a compilation of data for fluvial systems by Miall (1974).

deposit of a single point bar or tidal delta, whereas a township (6 mi or 9.6 km square) could encompass an entire delta lobe or barrier–inlet system. When interpreting small-scale structures (ranks 5 and 6, the usual stuff of paleocurrent analysis) over large areas it is important to remember that directional variance within each rank is summed to that of the higher ranks as far as is locally appropriate. Miall (1974) demonstrated the implications of this for fluvial–deltaic systems such as those shown in Figure 5.48, by compiling data for the various ranks summed to that of the single river or meander belt (Fig. 5.49).

5.9.5 Environment and paleoslope interpretations

A recommended first level of analysis is to plot current rose diagrams and mean vectors for each outcrop or local outcrop group, separating the readings according to major facies variations. An example is illustrated in Figure 5.46. If contemporaneous basin shape and orientation can be deduced from these data or from other mapping techniques the paleocurrent data can be used interactively with lithofacies and biofacies criteria to interpret depositional environment, and to outline major depositional systems such as deltas or submarine fan complexes. The geologist should examine the relationships between mean current directions in different lithofacies and in different outcrops. The number of modes in the rose diagrams (modality) and their orientation with respect to assumed shoreline or lithofacies contours is also important information. Another useful approach is to plot individual readings or small groups of readings collected in measured sections at the correct position in graphic section logs (e.g. Figs. 5.50, 5.51, 5.52). Some examples of how to use these data are discussed below, and many regional case studies are considered in Chapters 6 and 9. Potter and Pettijohn (1977) provided extensive examples and an annotated bibliography.

Useful introductions to paleocurrent models have been published by Pettijohn (1962) and Selley (1968), but our increased understanding of clastic (siliciclastic, plus clastic carbonate and evaporite) depositional systems requires a fresh look at the problem, because some of the earlier models are now seen to be simplistic. Many ideas are to be found in Reading (1978), although this book does not deal explicitly with paleocurrent models. Harms et al. (1975) provided several useful case studies.

Paleocurrent distributions are commonly categorized as unimodal, bimodal, trimodal and polymodal. Each reflects a particular style of current dispersion. For example the rose diagrams in Figure 5.46 are unimodal to weakly bimodal, and vector mean directions are all oriented in easterly directions, roughly normal to the basin margin, with the exception of station 11C(G). The data were derived mainly from trough and planar crossbedding and parting lineation, and the environment of deposition is interpreted as braided fluvial (Miall and Gibling, 1978). In some areas, e.g. stations 6A, B and C, there is a suggestion of a fanning out of the current systems, suggesting possible deposition on large, sandy alluvial fans. The anomalous data set of station 11C(G) was derived from giant crossbed sets ranging from 1 to 6 m in thickness. They may be eolian dunes or large sand waves in a trunk river draining the basin east of Boothia Uplift. Field evidence is inconclusive, but the paleocurrent data can be reconciled with either interpretation.

In high-sinuosity rivers point bars dip at high angles to channel trends, whereas minor structures migrating down the point bars are oriented subparallel to channel direction and therefore parallel to the strike of point bar (epsilon)

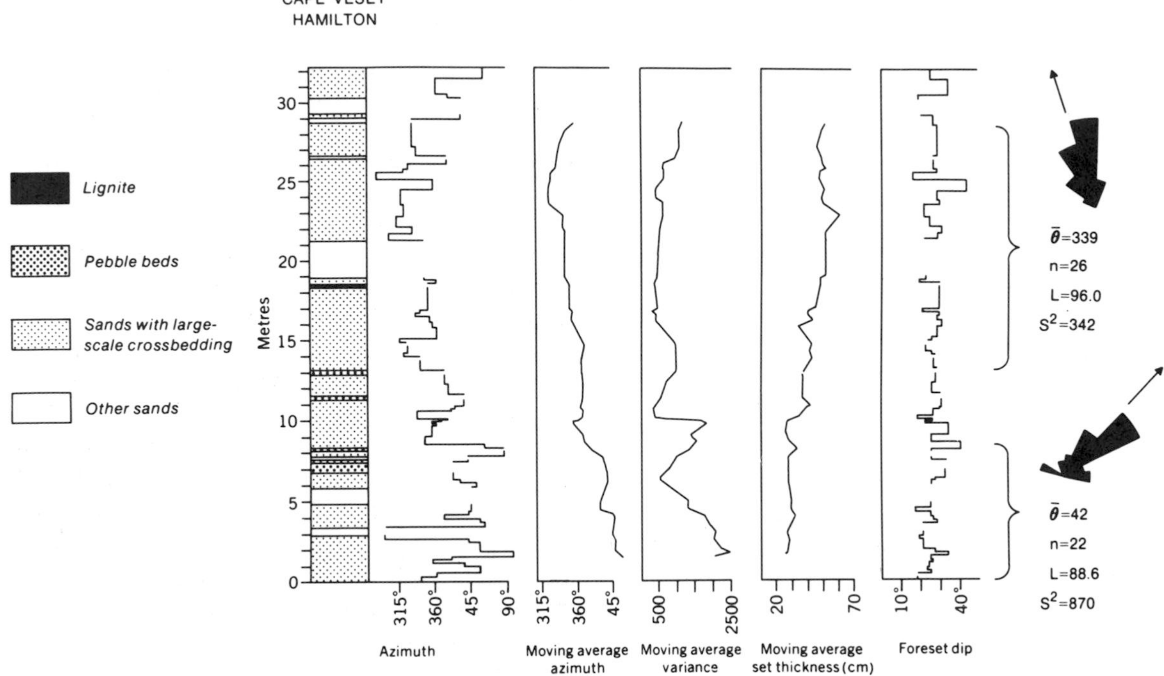

Fig. 5.50. Detailed vertical profile through the braided fluvial Isachsen Formation (Lower Cretaceous), northern Banks Island, Arctic Canada, showing crossbedding data. Moving averages are based on ten readings. See text for interpretation (Miall, 1976b, 1979).

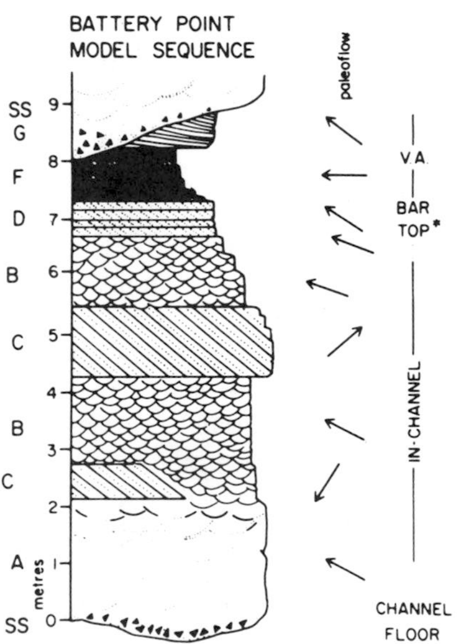

Fig. 5.51. Generalized vertical profile model for the fluvial Battery Point Formation (Devonian), Quebec, showing variation of crossbed orientations in different lithofacies types (north toward top of page) (Cant and Walker, 1976).

crossbedding (Fig. 4.27). This paleocurrent pattern is a useful diagnostic criterion for recognizing point bars and distinguishing them from other large, low-angle crossbed types, such as Gilbertian deltas (e.g. Puigdefabregas 1973; Mossop 1980).

Although fluvial deposits typically yield unimodal paleocurrent patterns on an outcrop scale, on a larger scale they may show much more complex patterns, for example centripetal patterns indicating internal drainage (e.g. Friend and Moody-Stuart, 1972) or bimodal patterns with the two modes at 90° and occurring in different lithofacies assemblages. Rust (1981) reported two examples of the latter case. In each case one mode, occurring in coarse conglomerates, was interpreted as the product of alluvial fans prograding transversely out from the basin margin. The other mode, in interbedded sandstones, represented a trunk river system draining longitudinally.

Miall (1976b, 1979) reported a similar case of interacting river systems, in which the respective lithofacies could only be separated by document-

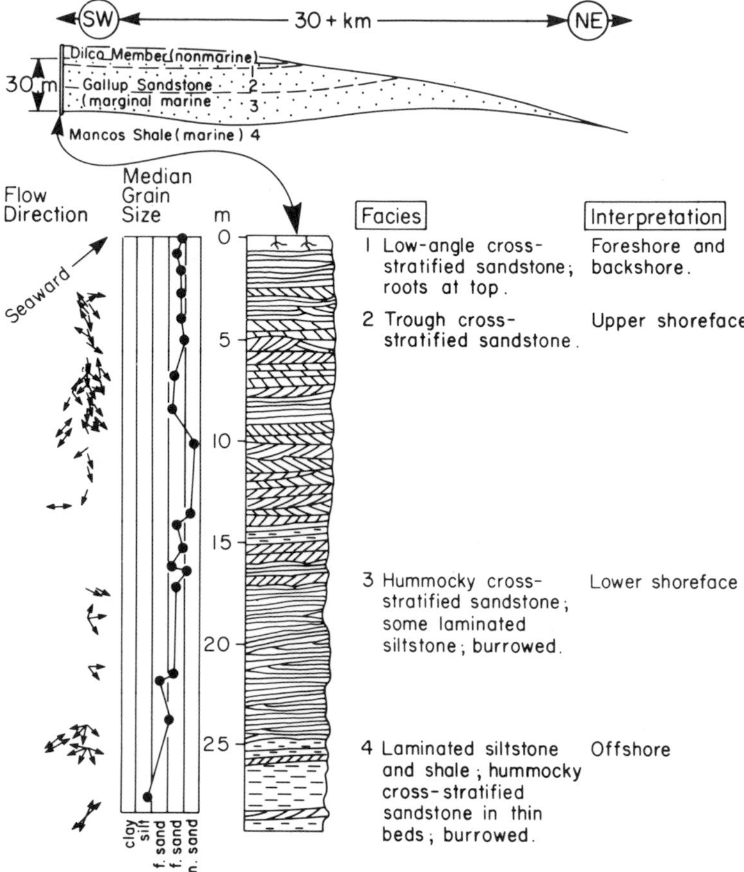

Fig. 5.52. Lithofacies profile, crossbed data and interpreted depositional environments of the Gallup Sandstone (Cretaceous), New Mexico (Harms et al., 1975).

ing subtle differences in crossbedding characteristics. Figure 5.50 displays data collected from a 32-m thick section through a Lower Cretaceous braided stream deposit at the margin of Banks Basin, Arctic Canada. The sediments consist of superimposed planar crossbed sets deposited from large sand waves or transverse bars. At around the 10-m level in the section they become thicker, shift in mean orientation by 63° and show less directional variance. These changes were interpreted as the result of rejuvenation of a tributary river draining into the basin from the southeast, possibly as a result of movement on one of the faults bounding the basin.

It is a common observation in fluvial deposits that planar crossbeds show a higher directional variance than trough crossbeds (Miall, 1977). Cant and Walker (1976) and Walker and Cant (1979) showed, in a very nice study comparing a modern river with an ancient deposit, why this is so. The deposits in both cases consisted of a series of fining-upward cycles, a model example of which is shown in Figure 5.51. Trough crossbeds in the Devonian unit show rather consistent current directions flowing toward the west–northwest, and were interpreted as the product of dunes migrating down the deeper channels. Planar crossbeds are oriented at high angles to this trend, as a result of deposition from large lobate bars migrating across the channel. This lithofacies and paleocurrent pattern is distinctive of certain types of low-sinuosity river, and can thus be used as an environmental model as well as a paleoslope indicator.

Coastal regions where rivers debouch into an area affected by waves and/or tides can give rise to very complex paleocurrent patterns. Bimodal, trimodal or polymodal distributions may result, for example in the tide-swept offshore bar described by Klein (1970). However, time–velocity asymmetry of tidal currents can result in local segregation of currents, so that they are locally

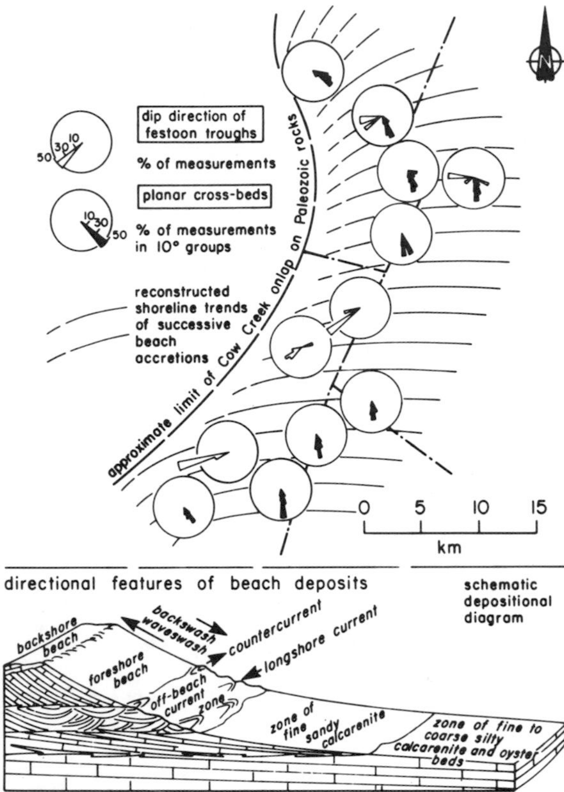

Fig. 5.53. Two types of crossbedding in the Cow Creek Limestone (Lower Cretaceous), central Texas, and their interpretation (Stricklin and Smith, 1973).

unimodally ebb- or flood-dominated. This could cause confusion at the outcrop level because the tidal deltas showing such paleocurrent patterns may consist of lithofacies that are very similar to some fluvial deposits. In wave-dominated environments current reversals generate distinctive internal ripple lamination patterns and herringbone crossbedding, as illustrated in Figure 4.32. Paleocurrent analyses of these bimodal crossbeds may yield much useful information on direction of wave attack and hence on shoreline orientation. The term paleoslope means little in this environment because there are a diversity of local slopes and because waves and tides are only marginally influenced by the presence and orientation of bottom slopes.

Several examples of sandy shoreline deposits and their paleocurrent patterns are given by Harms et al. (1975). One of the most instructive is the section through the Cretaceous Gallup Sandstone, New Mexico, illustrated in Figure 5.52. Individual trough and ripple orientations are shown at the left (north toward top). The shoreline is known from regional mapping to be oriented northwest—southeast. The top 2–3 m of section consists of low-angle cross-stratified beach accretion sets dipping at a few degrees toward the northeast (offshore). This structure is typical of intertidal wave swash zones (Fig. 4.46D). Facies 2 (about 3–14 m depth) consists of fine- to medium-grained sandstone with trough and some planar cross stratification. Foreset dip and trough axis orientations are mainly toward the northwest, southwest and south to southeast, with the latter predominating, suggesting onshore directed wave-generated currents, with local reversals, and longshore currents directed toward the southeast. The same longshore current pattern is apparent in the subtidal lower shoreface deposits, which are dominated by storm-generated hummocky cross-stratification. Wave ripples are present in the lowest facies, again showing predominantly onshore-directed currents.

Similar paleocurrent patterns in some ancient shallow marine calcarenites, oolites and coquinas were described by Knewtson and Hubert (1969) and Stricklin and Smith (1973), who compared them to the modern carbonate sand bodies in the Bahamas that were described by Ball (1967). Figure 5.53 shows the reconstructed depositional model and paleocurrent pattern of the Lower Cretaceous Cow Creek Limestone in central Texas. The upper member consists of 4.5 m of calcarenite and coquina forming large planar crossbed sets dipping toward the south at an average angle of 7°. The lower unit contains trough cosets oriented mainly in west–southwesterly directions, probably representing subtidal longshore currents.

Submarine fan and other deep marine deposits show three main paleocurrent patterns:

1. Individual submarine fans prograde out from the continental slope and therefore show radial paleocurrent patterns with vector mean directions oriented perpendicular to regional basin strike (e.g. Rupke, 1977). This pattern is typical of many continental margins, particularly divergent margins.
2. In narrow oceans and many arc-related basins deep water sedimentation takes place in a trough oriented parallel to tectonic strike. Sediment gravity flows, particularly low-viscosity turbidity currents, emerge from submarine fans and turn 90°, to flow longitudinally downslope, possibly

Fig. 5.54. Paleocurrent trends revealed by sole markings in the Lower Paleozoic turbidities of Hazen Trough, northern Canadian Arctic. Note the 90° swing in direction from transverse (southeasterly) to longitudinal (southwesterly) (Trettin, 1970).

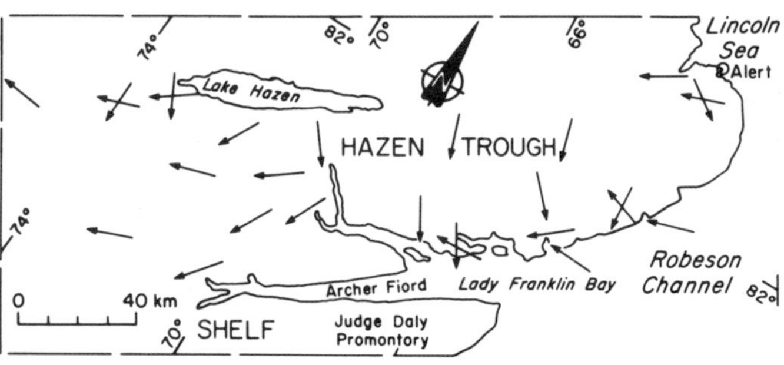

for hundreds of kilometers. Current directions may be reversed by tilting of the basin. Many examples of this pattern have been published (e.g. Trettin, 1970; Hesse, 1974; Ricci-Lucchi, 1975). Figure 5.54 illustrates a Lower Paleozoic example in the Canadian Arctic, in which the 90° turn in flow directions was meticulously documented. 3. Contour currents or boundary undercurrents flow parallel to continental margins and generate paleocurrent patterns oriented parallel to basin margins. Stow and Lovell (1979) discussed the use of facies criteria and paleocurrent data in distinguishing these deposits.

Fossils commonly yield good orientation patterns and many provide useful paleocurrent information, particularly in fine-grained sediments where other directional information may be lacking. Potter and Pettijohn (1977) and Jones and Dennison (1970) provided useful discussions and examples. Figure 5.55 summarizes data for several thousand orientation measurements made on various Ordovician to Devonian units in the Appalachians. Orientations may be generated either by rolling, as in *Tentaculites* and *Styliolina*, or by a "weather vane" effect, as in the case of the brachiopods. Stromatolite orientations commonly are impressively consistent, as demonstrated by Hoffman (1967) and Young and Long (1976).

The examples of paleocurrent studies discussed in this section show that depositional environment, bedform hierarchy and sampling scale all affect the consistency, variance and "modality" of the data. This fact should lay to rest the widely quoted remark by Potter and Pettijohn in the first (1963) edition of their book (repeated in 1977 edition, p. 111) that, with regard to crossbedding, "the most common variance of fluvial–deltaic deposits is in the range 4,000 to 6,000 ... whereas the smaller sample of marine deposits is is the range 6,000 to 8,000". This supposed difference has been used as a criterion for distinguishing depositional environments, although use of all the facies analysis and mapping methods described in Chapters 4 and 5 of this book should render the approach obsolete. Long and Young (1978) re-examined the idea and compiled more data, modifying the numbers to 1000

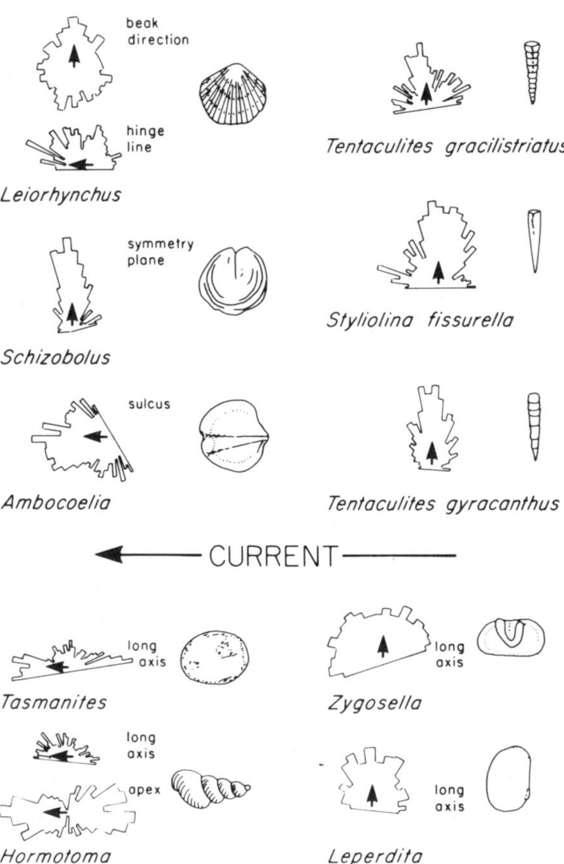

Fig. 5.55. Examples of orientation measurements made on fossil invertebrates in Ordovician to Devonian sediments of the Appalachians (Jones and Dennison, 1970).

to 4000 for fluvial rocks and 4000 to 8000 for marine rocks. They were concerned mainly with the use of the method for studying Precambrian rocks, where some analytical methods, such as biofacies analysis, are unavailable (except for stromatolites). However, the problem of the source of the variance remains, and anyone who has time to measure enough crossbeds for a variance study has time to carry out a proper basin analysis.

5.9.6 The dipmeter again

The principles of dipmeter data collection, processing and presentation, and uses of these data in facies studies are presented in section 5.4.2. In this section some brief comments and examples are provided showing the use of the tool in paleocurrent analysis.

Nurmi (1978) reported that the use of sophisticated computer processing routines such as those described by Vincent et al. (1977) may result in as many as fourteen dip readings over a 30-cm interval. Such detailed results require large scale graphical tadpole displays, and scales of 1:40 or even 1:24 are typical, which contrasts with the 1:1200 scale (1″ = 100′) commonly used for subsurface work in the North American petroleum industry. If readings can be obtained at nearly 2-cm intervals the dipmeter theoretically could "see" structures as small as ripples. However, these would not normally yield planar surfaces when cut by a normal-diameter petroleum exploration hole, and therefore the processing routines could not recognize them. The dipmeter can "see" medium to large-scale crossbed foresets, which offers considerable potential for paleocurrent analysis, but even here there are many pitfalls. Planar crossbed foresets should be particularly easy to recognize in dipmeter logs, particularly if they display the thickness range and orientation consistency of those illustrated in Figure 5.50. Trough crossbeds would be less easy to identify correctly, because a hole penetrating the edge of a set will yield an apparently spurious orientation, and the trough axis will be nearly horizontal. Set bounding surfaces, reactivation surfaces, channels, slumps, overturned crossbeds and the climb surface of climbing ripples may all yield apparently spurious readings. In practice, therefore, although practically every paper on dipmeter applications trumpets the possibilities for paleocurrent analysis, there are very few good published examples available (this may be an industry confidentiality problem). Only where the dipmeter log can be correlated with core—itself a tricky procedure—are reliable results likely to be obtained. Two examples are illustrated here.

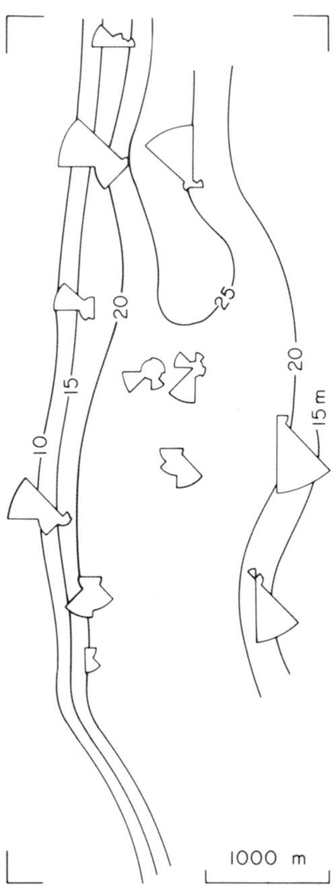

Fig. 5.56. Dipmeter and isopach data for a channel sandstone (Rider, 1978).

Rider (1978) illustrated a dipmeter analysis of a channel sandstone. Figure 5.56 shows sedimentary dips (the central mode in Fig. 5.19) within a channel sandstone. According to Rider crossbed set thickness is too small to have been picked up in his analysis. These dips, therefore, could represent set bounding surfaces, minor channels, drape over bars, etc. The southerly dips along the channel axis may be reactivation surfaces in large multi-set bar forms. Westerly dips oriented toward the channel margin could be of the same origin, as mid-channel bars in sandy rivers commonly migrate close to the bank, leaving a small slough (e.g. Cant and Walker, 1976).

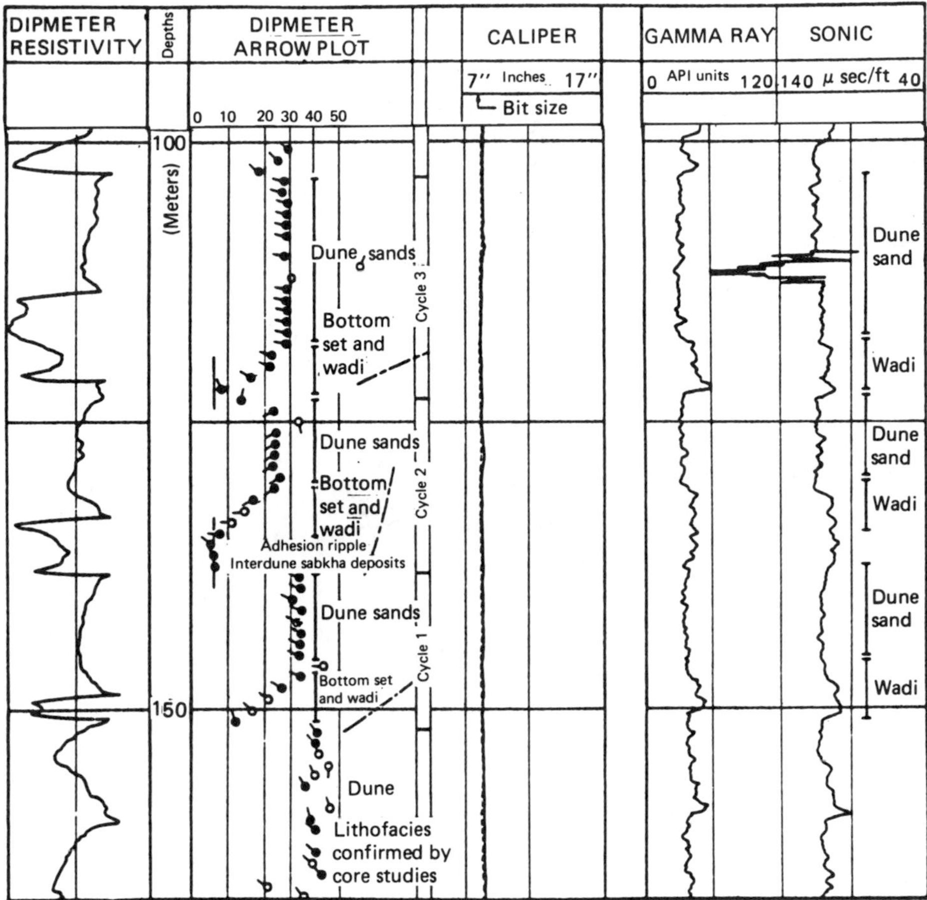

Fig. 5.57. Dipmeter data for a section through eolian crossbedding and associated facies in the Rotliegendes Sandstone (Permian), southern North Sea (Nurmi, 1978).

Some of the best dipmeter patterns are obtained from the large, consistently oriented crossbed sets of ancient eolian dune fields. Nurmi (1978) gave an example from the Rotliegendes Sandstone (Permian) of the southern North Sea Basin (Fig. 5.57). It is instructive to compare these dipmeter readings with the dune block diagram model illustrated in Figure 4.33. The long, curved bottom sets are particularly obvious in the dipmeter data, as is the overall westerly-directed paleocurrent pattern.

5.10 Paleogeographic synthesis

One of the most important skills in basin analysis is the art of synthesizing the various types of map data described in this chapter. One mapping approach may be used to supplement another or fill in data gaps or serve as a check on overall reliability. Where a vague paleogeographic trend is indicated by one method it is far more likely to be proved correct if it can be confirmed by another, independent approach.

The presentation of such a synthesis in a published report or oral presentation is as important a task as the groundwork on which it is based. Such presentations rely strongly on diagrams, so that the choice of the right kind of illustration is crucial. Several types of synthesis diagram may be used to illustrate a basin model, and experienced workers know that the work of designing and compiling such diagrams commonly provides them with valuable new insights. Judicious simplification may be very useful.

Interpretive maps and cross sections (commonly called cartoons) and block diagrams of

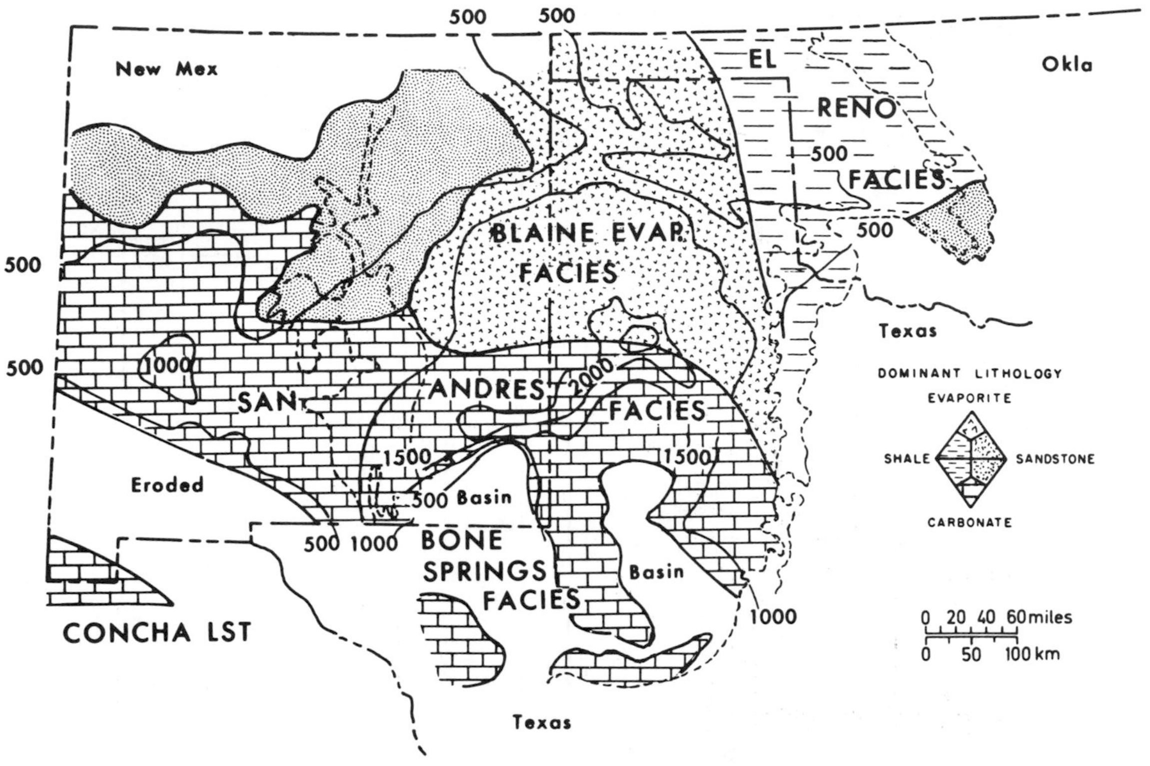

Fig. 5.58. Lithofacies and isopach map of Middle Permian sediments, Texas and New Mexico. The San Andres and Concha Formations are shelf limestones, and the Bone Springs is a black, laminated basinal limestone (Meissner, 1972).

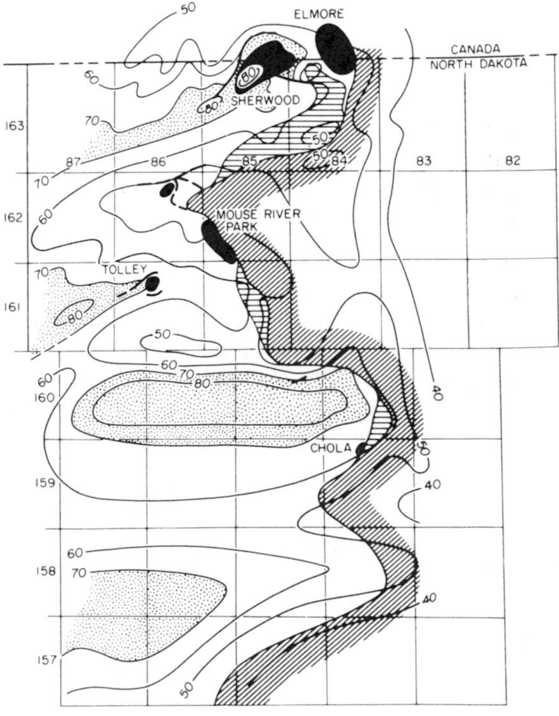

stratigraphic, structural or paleogeographic relationships are the most commonly used devices. Sequential illustrations may be used to demonstrate basin evolution through tectonic events or changes in sea level.

Paleogeographic maps are of three main types:

1. Those showing basic facies data, labeled and ornamented as to lithofacies type or stratigraphic name. Isopachs may be superimposed. This style of map is particularly widely used for carbonate sequences in which facies subdivisions contain

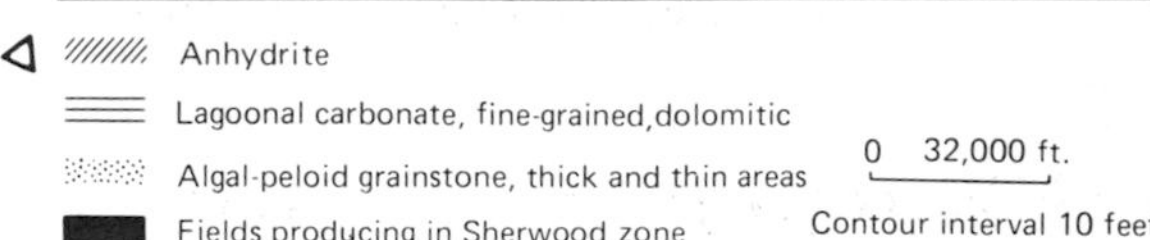

Fig. 5.59. Lithofacies and isopach map of Mississippian sediments, North Dakota. Linear shoals of grainstone pass eastward (updip) into evaporite, creating stratigraphic petroleum traps. Fig. 5.38 is an east–west cross section through the north part of this map area (Wilson, 1980).

Fig. 5.60. Isopach, lithofacies and paleocurrent data of Upper Cretaceous McNairy Formation, upper Mississippi Embayment. Paleocurrent data represent vector means for 35 sampling areas, based on 920 readings made on basin margin outcrops. A fluvial system is interpreted to have fed the basin from the northeast, and the area of sand–clay ratio >2 represents a delta plain, passing into a prodelta region in the south. This map is from one of the earliest integrated basin analysis studies in North America (Pryor, 1960). ▷

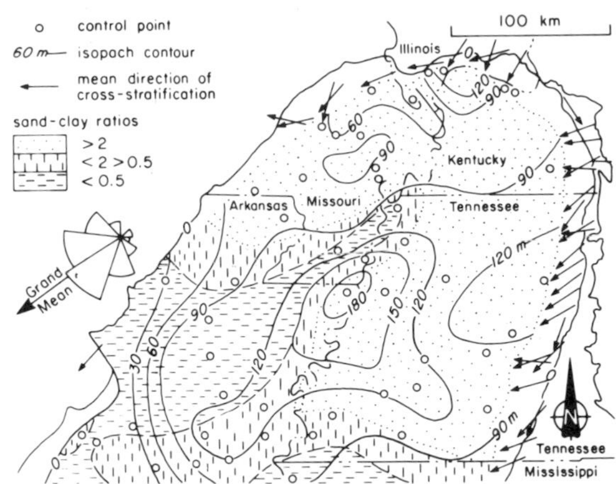

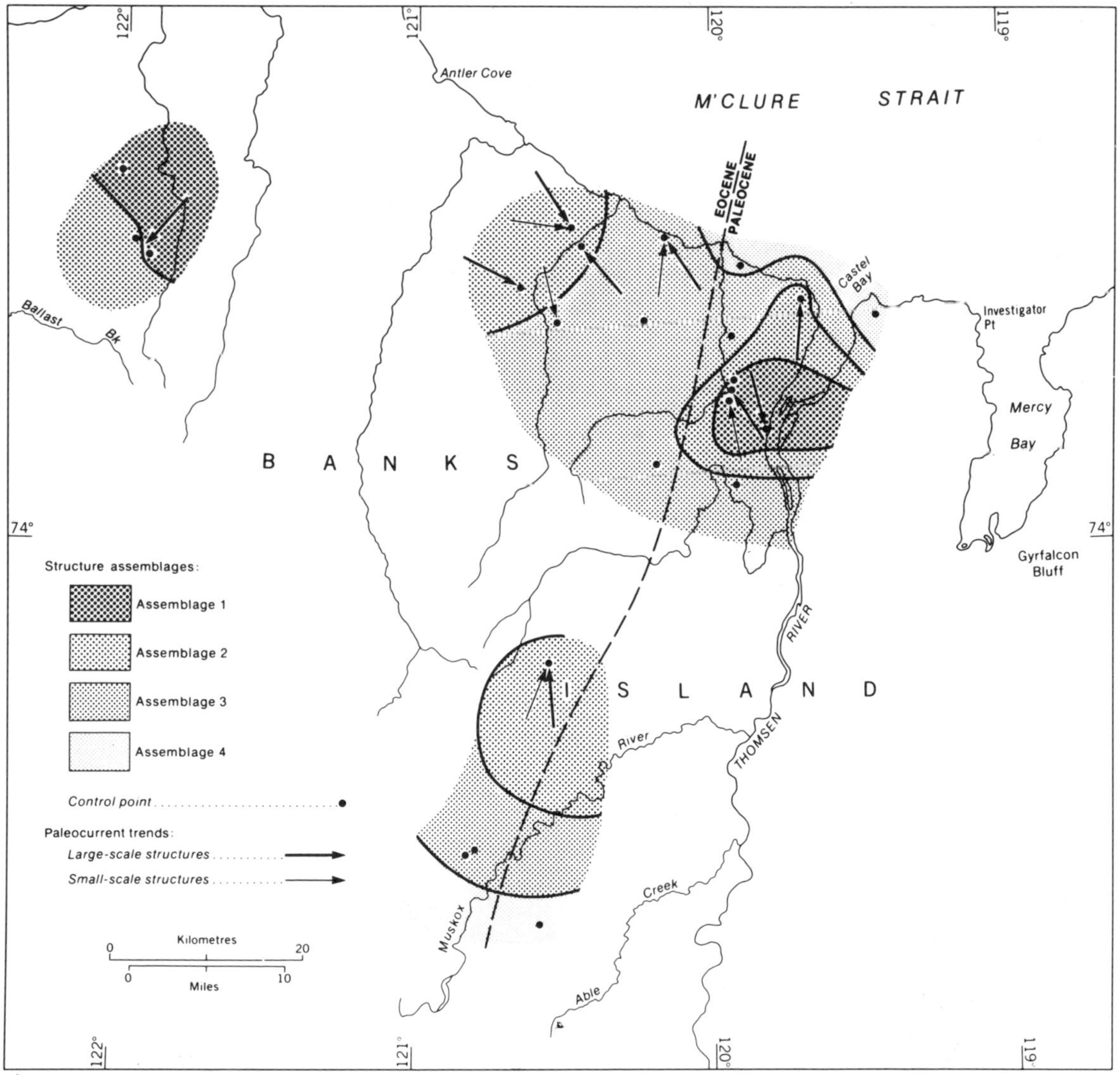

Fig. 5.61. Mapping birdsfoot deltas in the Eureka Sound Formation (Paleocene–Eocene), Banks Island, Arctic Canada. Structure assemblages are: 1. large-scale crossbedding, including epsilon sets (delta plain); 2. planar and trough cross-bedding (proximal delta front); 3. ripple marks (distal delta front); 4. few current structures (prodelta). Paleocurrent patterns appear random, but coincide well with mapped distribution of sedimentary structures (Miall, 1979).

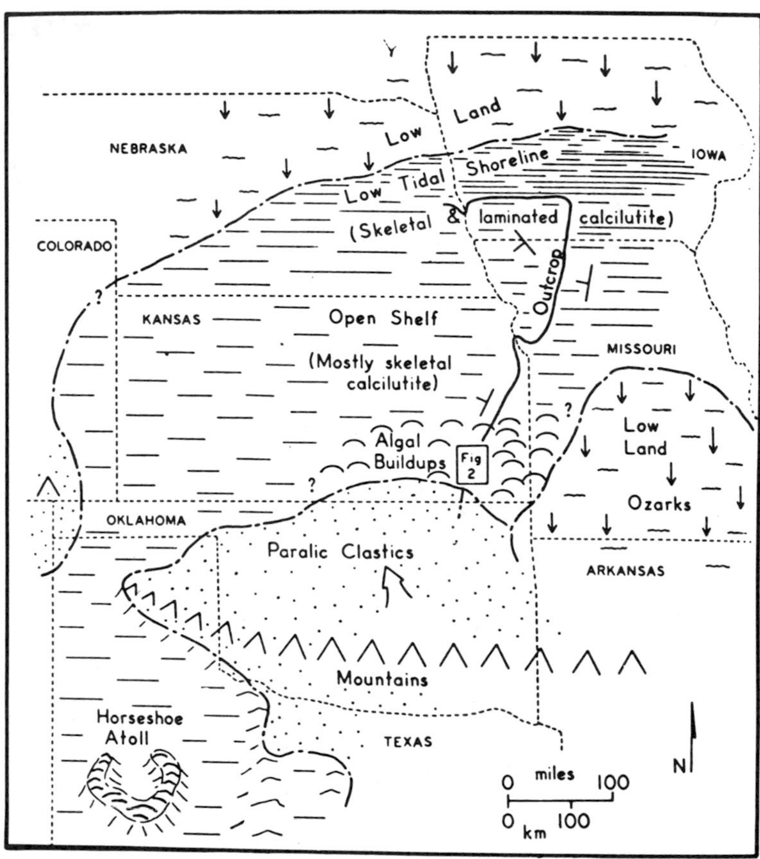

Fig. 5.62. A hybrid map showing generalized lithofacies data and interpreted depositional environments, Upper Pennsylvanian of U.S. Midcontinent (Heckel, 1972).

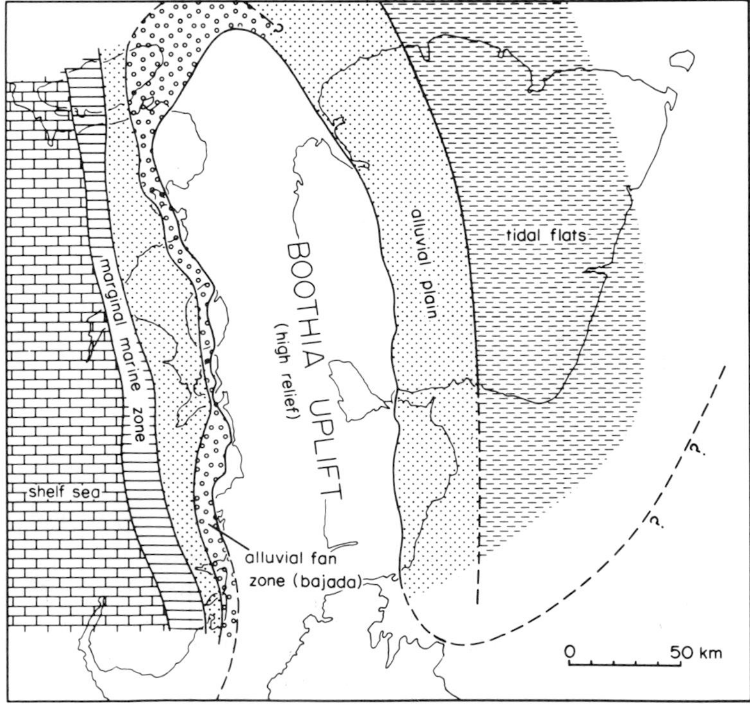

Fig. 5.63. Paleogeographic map for Somerset Island and eastern Prince of Wales Island during the earliest Devonian, showing the detrital apron flanking Boothia Uplift (Miall and Gibling, 1978).

most of the necessary data for paleogeographic interpretation. However, such maps are descriptive, not interpretive, although they can easily be read and interpreted by a geologist who has paid attention to the facies definitions. Examples are given in Figures 5.5, 5.58 and 5.59.

2. Maps combining two or more facies mapping techniques, such as facies subdivisons, clastic ratio or sand isolith contours, grain size trends and paleocurrent data. This style is much used for clastic sequences. These maps are also basically descriptive, and may need explanatory captions. Examples are illustrated in Figures 5.60 and 5.61.

3. Interpretive maps, in which descriptive data are very generalized, or may be omitted altogether. Two examples are provided in Figures 5.62 and 5.63.

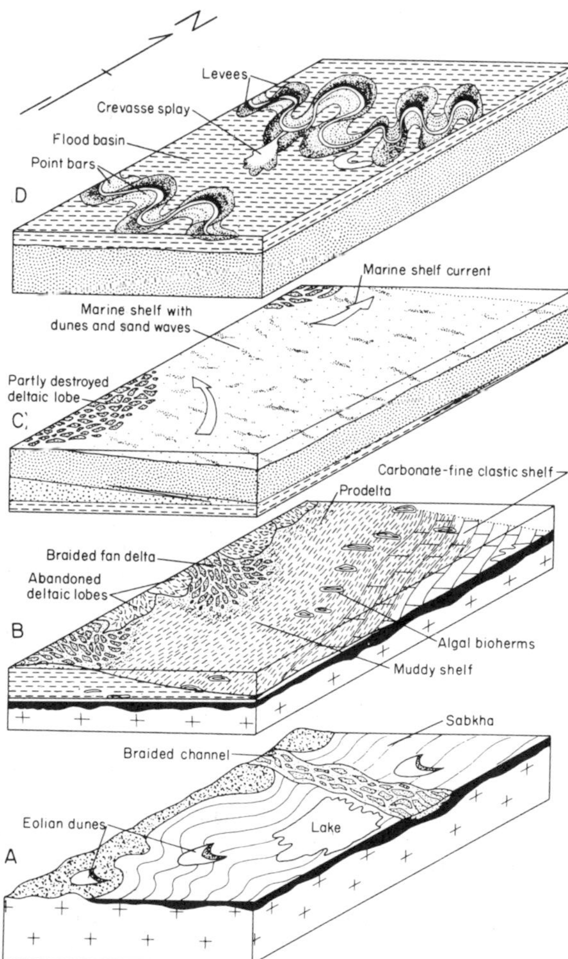

Fig. 5.64. Use of sequential block diagrams to indicate evolution of depositional environments through time, Umkondo Group (about 1800 Ma), southeast Zimbabwe (Tankard et al., 1982).

Block diagrams are very useful devices for summarizing paleogeographic interpretations. They are drawn to show schematic depositional environments, and an artist's touch is needed to make them look convincing. Some examples are shown in Figure 5.64, which illustrates another valuable technique, the use of sequential diagrams to demonstrate changing conditions during formation of a major stratigraphic unit. Care should be taken to indicate what particular piece of earth space is covered by the diagrams, using an index map.

References

AGTERBERG, F.P., HILLS, L.V. and TRETTIN, H.P., 1967: Paleocurrent trend analysis of a delta in the Bjorne Formation (Lower Triassic) of northwestern Melville Island, Arctic Archipelago; J. Sediment. Petrol., v. 37, p. 852–862.

ALLEN, J.R.L., 1963a: The classification of cross-stratified units, with notes on their origin; Sedimentology, v. 2, p. 93–114.

ALLEN, J.R.L., 1963b: Henry Clifton Sorby and the sedimentary structures of sands and sandstones in relation to flow conditions; Geol. Mijnb., v. 42, p. 223–228.

ALLEN, J.R.L., 1966: On bed forms and paleocurrents; Sedimentology, v. 6, p. 153–190.

ALLEN, J.R.L., 1967: Notes on some fundamentals of paleocurrent analysis, with reference to preservation potential and sources of variance; Sedimentology, v. 9, p. 75–88.

ALLEN, P., 1972: Wealden detrital tourmaline: implications for northwestern Europe; J. Geol. Soc. London, v. 128, p. 273–294.

ANDERSEN, D.W. and PICARD, M.D., 1974: Evolution of synorogenic clastic deposits in the Intermontane Uinta Basin of Utah, *in* W.R. Dickinson, ed., Tectonics and sedimentation; Soc. Econ. Paleont. Mineral. Spec. Publ. 22, p. 167–189.

ANSTEY, N.A., 1978: Seismic exploration for sandstone reservoirs; Internat. Human Resources Dev. Corp., Boston, 138 p.

BALL, M.M., 1967: Carbonate sand bodies of Florida and the Bahamas; J. Sediment. Petrol., v. 37, p. 556–591.

BALL, M.M., 1972: Exploration methods for stratigraphic traps in carbonate rocks, *in* R.E. King, ed., Stratigraphic oil and gas fields; Am. Assoc. Petrol. Geol. Mem. 16, p. 64–81.

BALLY, A.W., GORDY, P.L. and STEWART, G.A., 1966: Structure, seismic data and orogenic evolution of southern Canadian Rockies; Bull. Can. Petrol. Geol., v. 14, p. 337–381.

BIRKENMAJER, K., 1981: The geology of Svalbard, the western part of the Barents Sea, and the continental margin of Scandinavia, *in* A.E.M. Nairn, M. Churkin Jr., and F.G. Stehli, eds., The Ocean Basins and

Margins, v. 5, The Arctic Ocean, Plenum Press, New York, p. 265–321.
BLATT, H., MIDDLETON, G.V. and MURRAY, R.C., 1980: Origin of sedimentary rocks; 2nd ed., Prentice-Hall Inc., New Jersey, 782 p.
BLUCK, B.J., 1965: The sedimentary history of some Triassic conglomerates in the Vale of Glamorgan, South Wales; Sedimentology, v. 4, p. 225–245.
BOLE, G., 1981: Well data display system: an interactive exploration tool; Bull. Can. Petrol. Geol., v. 29, p. 215–223.
BRADY, T.J., CAMPBELL, N.D.J. and MAHER, C.E., 1980: Intisar "D" Oil Field, Libya, in M.T. Halbouty, ed., Giant oil and gas fields of the decade 1968–1978; Am. Assoc. Petrol. Geol. Mem. 30, p. 543–564.
BROWN, A.R., DAHM, C.G. and GRAEBNER, R.J., 1980: A stratigraphic case history using three-dimensional seismic data in the Gulf of Thailand; Geophysical Prospecting, v. 29, p. 327–349.
BROWN, L.F., Jr., and FISHER, W.L., 1977: Seismic–stratigraphic interpretation of depositional systems: examples from Brazilian rift and pull-apart basins, *in* C.E. Payton, ed., Seismic stratigraphy— applications to hydrocarbon exploration, Am. Assoc. Petrol. Geol. Mem. 26, p. 213–248.
BUSCH, D.A., 1974: Stratigraphic traps in sandstones—exploration techniques; Am. Assoc. Petrol. Geol. Mem. 21.
CANT, D.J. and WALKER, R.G., 1976: Development of a braided–fluvial facies model for the Devonian Battery Point Sandstone, Quebec; Can. J. Earth Sci., v. 13, p. 102–119.
CLARK, D.A., 1981: A system for regional lithofacies mapping; Bull. Can. Petrol. Geol., v. 29, p. 197–208.
CURRAY, J.R., 1956: The analysis of two-dimensional orientation data; J. Geol., v. 64, p. 117–131.
DAVIS, J.C., 1973: Statistics and data analysis in geology; John Wiley and Sons Inc., New York, 550 p.
DICKINSON, W.R., 1970: Interpreting detrital modes of graywacke and arkose; J. Sediment. Petrol., v. 40, p. 695–707.
EHRLICH, R., and WEINBERG, B., 1970: An exact method for characterization of grain shape; J. Sediment. Petrol., v. 40, p. 205–212.
ENOS, P., 1977: Tamabra Limestone of the Pozo Rico Trend, Cretaceous, Mexico, *in* H.E. Cook and P. Enos, eds., Deep-water carbonate environments; Soc. Econ. Paleont. Mineral. Spec. Publ. 25, p. 273–314.
EVANS, H., 1972: Zama—a geophysical case history, *in* R.E. King, ed., Stratigraphic oil and gas fields; Am. Assoc. Petrol. Geol. Mem. 16, p. 440–452.
FARMER, D.G., 1981: Computer files and the formal database in support of hydrocarbon exploration and development offshore; J. Geol. Soc. London, v. 138, p. 611–620.
FRANZINELLI, E., and POTTER, P.E., 1983: Petrology, chemistry and texture of modern river sands, Amazon river system; J. Geol., v. 91, p. 23–40.
FRIEND, P.F. and MOODY-STUART, M., 1972: Sedimentation of the Wood Bay Formation (Devonian) of Spitsbergen: regional analysis of a late orogenic basin; Norsk Polarinstitutt Skrifter Nr. 157.
GALLOWAY, W.E., 1981: Depositional architecture of Cenozoic Gulf Coastal Plain fluvial systems, *in* F.G. Ethridge and R.M. Flores, eds., Recent and ancient nonmarine depositional environments: models for exploration, Soc. Econ. Paleont. Mineral. Spec. Publ. 31, p. 127–156.
GILREATH, J.A., 1977: Dipmeter, *in* L.W. LeRoy, D.P. LeRoy and J.W. Raese, eds., Subsurface geology; Colorado School of Mines, Golden, Colorado, p. 389–396.
HALBOUTY, M.T., 1980: Geologic significance of Landsat data for 15 giant oil and gas fields, *in* M.T. Halbouty, ed., Giant oil and gas fields of the decade 1968–1978; Am. Assoc. Petrol. Geol. Mem. 30, p. 7–38.
HARMS, J.C., SOUTHARD, J.B., SPEARING, D.R. and WALKER, R.G., 1975: Depositional environments as interpreted from primary sedimentary structures and stratification sequences; Soc. Econ. Paleont. Mineral. Short Course 2.
HECKEL, P.H., 1972: Pennsylvanian stratigraphic reefs in Kansas, some modern comparisons and implications; Geol. Rundschau, v. 61, p. 584–598.
HEIM, D., 1974: Uber die feldspäte in Germanischen Buntsandstein, ihre Korngrössenabhängigkeit, verbreitung undpaleogeographische bedeutung; Geol. Rundschau, v. 63, p. 943–970.
HÉLU, P.C., VERDUGO, V.R., and BARCENAS, P.R., 1977: Origin and distribution of Tertiary conglomerates, Veracruz Basin, Mexico; Am. Assoc. Petrol. Geol., v. 61, p. 207–226.
HESSE, R., 1974: Long-distance continuity of turbidites: possible evidence for an Early Cretaceous trench–abyssal plain in the East Alps; Geol. Soc. Am. Bull., v. 85, p. 859–870.
HOFFMAN, P., 1967: Algal stromatolites: use in stratigraphic correlation and paleocurrent direction; Science, v. 157, p. 1043–1045.
HRISKEVICH, M.E., 1980: Exploration in Western Canada: a summary of the exploration development and future potential of Western Canada, *in* A.D. Miall, ed., Facts and principles of world petroleum occurrence; Can. Soc. Petrol. Geol. Mem. 6, p. 397–420.
HUNTER, R.E., 1981: Stratification styles in eolian sandstones: some Pennsylvanian to Jurassic examples from the Western Interior U.S.A., *in* F.G. Ethridge and R. Flores, eds., Recent and ancient nonmarine depositional environments: models for exploration; Soc. Econ. Paleont. Mineral. Spec. Publ. 31, p. 315–329.
JACKSON, R.G., II, 1975: Hierarchical attributes and a unifying model of bed forms composed of cohesionless material and produced by shearing flow; Geol. Soc. Am. Bull., v. 86, p. 1523–1533.
JAGELER, A.J. and MATUSZAK, D.R., 1972: Use of well logs and dipmeters in stratigraphic-trap exploration, *in* R.E. King, ed., Stratigraphic oil and gas fields; Am. Assoc. Petrol. Geol. Mem. 16, p. 107–135.
JAMISON, H.C., BROCKETT, L.D. and McINTOSH, R.A., 1980: Prudhoe Bay: a 10-year perspective, *in* M.T. Halbouty, ed., Giant oil and gas

fields of the decade 1968–1978; Am. Assoc. Petrol. Geol. Mem. 30, p. 289–314.

JONES, M.L. and DENNISON, J.M., 1970: Oriented fossils as paleocurrent indicators in Paleozoic lutites of southern Appalachians; J. Sediment. Petrol. v. 40, p., p. 642–649.

KELLING, G., 1969: The environmental significance of cross-stratification parameters in an Upper Carboniferous fluvial basin; J. Sediment. Petrol., v. 39, p. 857–875.

KING, R.E., ed., 1972: Stratigraphic oil and gas fields—classification, exploration methods, and case histories; Am. Assoc. Petrol. Geol., Mem. 16.

KLEIN, G. deV., 1970: Depositional and dispersal dynamics of intertidal sand bars; J. Sediment. Petrol., v. 40, p. 1095–1127.

KNEWTSON, S.L. and HUBERT, J.F., 1969: Dispersal patterns and diagenesis of oolitic calcarenites in the Saint Genevieve Limestone (Mississippian), Missouri; J. Sediment. Petrol., v. 39, p. 954–968.

LEE, P.J., 1981: The most predictable surface (MPS) mapping method in petroleum exploration; Bull. Can. Petrol. Geol., v. 29, p. 224–249.

LONG, D.G.F. and YOUNG, G.M., 1978: Dispersion of cross-stratification as a potential tool in the interpretation of Proterozoic arenites; J. Sediment Petrol., v. 48, p. 857–862.

LYONS, P.L. and DOBRIN, M.B., 1972: Seismic exploration for stratigraphic traps, *in* R.E. King, ed., Stratigraphic oil and gas fields; Am. Assoc. Petrol. Geol. Mem. 16, p. 225–243.

MARTIN, R., 1967: Morphology of some Devonian reefs in Alberta; a paleogeomorphological study, *in* International Symposium on the Devonian System; Alberta Soc. Petrol. Geol., v. 2, p. 365–385.

McCAMMON, R.B., 1975: On the efficiency of systematic point-sampling in mapping facies; J. Sediment. Petrol., v. 45, p. 217–229.

McCROSSAN, R.G., 1961: Resistivity mapping and petrophysical study of Upper Devonian inter-reef calcareous shales of central Alberta, Canada; Am. Assoc. Petrol. Geol. Bull., v. 45, p. 441–470.

McKEE, E.D. and WEIR, G.W., 1953: Terminology for stratification and cross-stratification in sedimentary rocks; Geol. Soc. Am. Bull., v. 64, p. 381–390.

McWHAE, J.R.H., 1981: Structure and spreading history of the northwestern Atlantic region from the Scotian Shelf to Baffin Bay, *in* J.W. Kerr and A.J. Fergusson, eds., Geology of the North Atlantic borderlands; Can. Soc. Petrol. Geol. Mem. 7, p. 299–332.

MEISSNER, F.F., 1972: Cyclic sedimentation in Middle Permian strata of the Permian basin, West Texas and New Mexico, *in* J.C. Elam and S. Chuber, eds., Cyclic sedimentation in the Permian Basin; 2nd ed., West Texas Geol. Soc., p. 203–232.

MERRIAM, D.F. and HARBAUGH, J.W., 1964: Trend-surface analysis of regional and residual components of geologic structure in Kansas; Kansas Geol. Surv. Spec. Dist. Publ. 11, 27 p.

MIALL, A.D., 1974: Paleocurrent analysis of alluvial sediments: a discussion of directional variance and vector magnitude; J. Sediment. Petrol., v. 44, p. 1174–1185.

MIALL, A.D., 1975: Computer applications in stratigraphic and sedimentary geology: notes from an iconoclast; Geoscience Canada, v. 2, p. 193–195.

MIALL, A.D., 1976a: The Triassic sediments of Sturgeon Lake South and surrounding areas, *in* M. Lerand, ed., The sedimentology of selected clastic oil and gas reservoirs in Alberta; Can. Soc. Petrol. Geol., p. 25–43.

MIALL, A.D., 1976b: Palaeocurrent and palaeohydrologic analysis of some vertical profiles through a Cretaceous braided stream deposit; Sedimentology, v. 23, p. 459–483.

MIALL, A.D., 1977: A review of the braided river depositional environment; Earth Sci. Revs., v. 13, p. 1–62.

MIALL, A.D., 1978: Fluvial sedimentology: an historical review, *in* A.D. Miall, ed., Fluvial sedimentology; Can. Soc. Petrol. Geol. Mem. 5, p. 1–47.

MIAL, A.D., 1979: Mesozoic and Tertiary geology of Banks Island, Arctic Canada: the history of an unstable craton margin; Geol. Surv. Can. Mem. 387.

MIALL, A.D. and GIBLING, M.R., 1978: The Siluro–Devonian clastic wedge of Somerset Island, Arctic Canada, and some regional paleogeographic implications; Sediment. Geol., v. 21, p. 85–127.

MILLER, R.L. and KAHN, J.S., 1962: Statistical analysis in the geological sciences; John Wiley and Sons, New York, 483 p.

MITCHUM, R.M. Jr., VAIL, P.R. and SANGREE, J.B., 1977: Seismic stratigraphy and global changes of sea level, Part 6: stratigraphic interpretation of seismic reflection patterns in depositional sequences, *in* C.E. Payton, ed., Seismic stratigraphy—applications to hydrocarbon exploration; Am. Assoc. Petrol. Geol. Mem. 26, p. 117–133.

MOSSOP, G.D., 1980: Facies control on bitumen saturation in the Athabasca Oil Sands, *in* A.D. Miall, ed., Facts and principles of world petroleum occurrence; Can. Soc. Petrol. Geol. Mem. 6, p. 609–632.

MURRIS, R.J., 1980: Hydrocarbon habitat of the Middle East, *in* A.D. Miall, ed., Facts and principles of world petroleum occurrence; Can. Soc. Petrol. Geol. Mem. 6, p. 765–800.

NURMI, R.D., 1978: Use of well logs in evaporite sequences, *in* W.E. Dean and B.C. Schreiber, eds., Marine evaporites; Soc. Econ. Paleont. Mineral. Short Course 4, p. 144–176.

OKADA, H., 1971: Classification of sandstone: analysis and proposal; J. Geol., v. 79, p. 509–525.

OLIVER, T.A. and COWPER, N.W., 1963: Depositional environments of the Ireton Formation, central Alberta; Bull. Can. Petrol. Geol., v. 11, p. 183–202.

OLSON, J.S. and POTTER, P.E., 1954: Variance components of crossbedding direction in some basal Pennsylvanian sandstones of the eastern Interior Basin: statistical methods; J Geol., v. 62, p. 26–49.

PADGETT, G. and EHRLICH, R., 1978: An analysis of two tectonically controlled integrated drainage nets of Mid-Carboniferous age in southern West Virginia, *in* A.D. Miall, ed., Fluvial Sedimentology; Can. Soc. Petrol. Geol. Mem. 5, p. 789–800.

PARKS, J.M., 1970: Computerized trigonometric

solution for rotation of structurally tilted sedimentary directional features; Geol. Soc. Am. Bull., v. 81, p. 537–540.

PAYTON, C.E., ed., 1977: Seismic stratigraphy—applications to hydrocarbon exploration; Am. Assoc. Petrol. Geol. Mem. 26.

PELLETIER, B.C., 1958: Pocono paleocurrents in Pennsylvania and Maryland; Geol. Soc. Am. Bull., v. 69, p. 1033–1064.

PETTIJOHN, F.J., 1957: Sedimentary rocks, 2nd ed., Harper, New York, 718 p.

PETTIJOHN, F.J., 1962: Paleocurrents and paleogeography; Am. Assoc. Petrol. Geol. Bull., v. 46, p. 1468–1493.

PETTIJOHN, F.J., POTTER, P.E. and SIEVER, R., 1973: Sand and sandstone; Springer-Verlag, New York, 618 p.

POTTER, P.E. and PETTIJOHN, F.J., 1977: Paleocurrents and basin analysis, 2nd ed.; Springer-Verlag, New York, 425 p.

PRYOR, W.A., 1960: Cretaceous sedimentation in upper Mississippi Embayment; Am. Assoc. Petrol. Geol. Bull., v. 44, p. 1473–1504.

PUIGDEFABREGAS, C., 1973: Miocene point-bar deposits in the Ebro Basin, northern Spain; Sedimentology, v. 20, p. 133–144.

RAHMANI, R.A. and LERBEKMO, J.F., 1975: Heavy mineral analysis of Upper Cretaceous and Paleocene sandstones in Alberta and adjacent areas of Saskatchewan, *in* W.G.E. Caldwell, ed., The Cretaceous system in the Western Interior of North America, Geol. Assoc. Can. Spec. Paper 13, p. 607–632.

RAMSAY, J.G., 1961: The effects of folding upon the orientation of sedimentation structures; J. Geol., v. 69, p. 84–100.

READING, H.G., ed., 1978: Sedimentary environments and facies; Blackwell, Oxford, 557 p.

REES, F.B., 1972: Methods of mapping and illustrating stratigraphic traps, *in* R.E. King, ed., Stratigraphic oil and gas fields; Am. Assoc. Petrol. Geol. Mem. 16, p. 168–221.

RICCI-LUCCHI, F., 1975: Sediment dispersal in turbidite basins: examples from the Miocene of northern Apennines; 9th. Internat. Sedimentol. Congr., Nice, Thème 5(2), p. 347–352.

RIDER, M.H., 1978: Dipmeter log analysis: an essay; 19th. Annual Logging Symp., Soc. Petrol. Well Log. Anal., Paper G.

ROBINSON, J.E., 1981: Well spacing and the identification of subsurface drainage systems; Bull. Can. Petrol. Geol., v. 29, p. 250–258.

ROBINSON, J.E. and MERRIAM, D.F., 1978: Recognition of subtle features in geological maps, *in* A.D. Miall, ed., Facts and principles of world petroleum occurrence; Can. Soc. Petrol. Geol. Mem. 6, p. 269–282.

RUPKE, N.A., 1977: Growth of an ancient deep-sea fan; J. Geol., v. 85, p. 725–744.

RUPKE, N.A., 1978: Deep clastic seas, *in* H.G. Reading, ed., Sedimentary environments and facies; Blackwell, Oxford, p. 372–415.

RUST, B.R., 1975: Fabric and structure in glaciofluvial gravels, *in* A.V. Jopling and B.C. McDonald, eds., Glaciofluvial and glaciolacustrine sedimentation; Soc. Econ. Paleont. Mineral.Spec. Publ. 23, p. 238–248.

RUST, B.R., 1981: Alluvial deposits and tectonic style: Devonian and Carboniferous successions in eastern Gaspe, *in* A.D. Miall, ed., Sedimentation and tectonics in alluvial basins, Geol. Assoc. Can. Spec. Paper 23, p. 49–76.

SABINS, F.F. Jr., 1972: Comparison of Bisti and Horseshoe Canyon stratigraphic traps, San Juan basin, New Mexico, *in* R.E. King, ed., Stratigraphic oil and gas fields; Am. Assoc. Petrol. Geol. Mem. 16, p. 610–622.

SCHLUMBERGER, LIMITED, 1970: Fundamentals of Dipmeter Interpretation, New York, 145 p.

SCHNITZER, W.A., 1977: Die quarzkornfarbenmethoden und ihre Bedeutung fur die stratigraphische und palaogeographische erforschung psammitischer sedimente; Erlanger Geol. Abh., Heft 103, 28 p.

SEELAND, D.A., 1978: Eocene fluvial drainage patterns and their implications for uranium and hydrocarbon exploration in the Wind River Basin, Wyoming; U.S. Geol. Surv. Bull. 1446.

SELLEY, R.C., 1968: A classification of paleocurrent models; J. Geol., v. 76, p. 99–110.

SHERIFF, R.E., 1976: Inferring stratigraphy from seismic data; Am. Assoc. Petrol. Geol. Bull., v. 60, p. 528–542.

SHIELDS, C., 1974: The dipmeter used to recognize and correlate depositional environment; 3rd. European Symp., Soc. Petrol. Well Log. Anal., Paper H.

SIMONS, D.B., RICHARDSON, E.V. and NORDIN, C.F., 1965: Sedimentary structures generated by flow in alluvial channels, *in* G.V. Middleton, ed., Primary sedimentary structures and their hydrodynamic interpretation; Soc. Econ. Paleont. Mineral. Spec. Publ. 12, p. 34–52.

SINGER, D.A. and WICKMAN, F.E., 1969: Probability tables for locating elliptical targets with square, rectangular, and hexagonal point-nets; Pennsylvania State Univ. Mineral. Sciences Experiment Station Spec. Publ. 1–69, 100 p.

SMITH, N.D., 1972: Some sedimentological aspects of planar cross-stratification in a sandy braided river; J. Sediment. Petrol., v. 42, p. 624–634.

SNEED, E.D. and FOLK, R.L., 1958: Pebbles in the Lower Colorado River, Texas, a study in particle morphogenesis; J. Geol., v. 66, p. 114–150.

STOW, D.A.V. and LOVELL, J.P.B., 1979: Contourites: their recognition in modern and ancient sediments; Earth Sci. Reviews, v. 14, p. 251–291.

STRATTON, E.F. and HAMILTON, R.G., 1947: Application of dipmeter surveys; Ann. Mtg. Soc. Petrol. Eng. of Am. Inst. Mech. Eng., Tulsa, Oklahoma.

STRICKLIN, F.L., Jr. and SMITH, C.I., 1973: Environmental reconstruction of a carbonate beach complex: Cow Creek (Lower Cretaceous) Formation of central Texas; Geol. Soc. Am. Bull., v. 84, p. 1349–1368.

STURM, E., 1971: High resolution paleocurrent analysis by moving vector averages; J. Geol., v. 79, p. 222–233.

SUTTNER, L.J., 1974: Sedimentary petrographic provinces: an evaluation, *in* C.A. Ross, ed., Paleogeographic provinces and provinciality; Soc. Econ. Paleont. Mineral. Spec. Publ. 21, p. 75–84.

TALWANI, M., 1960: Gravity anomalies in the Bahamas and their interpretation; unpublished Ph.D. thesis, Columbia Univ.

TANKARD, A.J., JACKSON, M.P.A., ERIKSSON, K.A., HOBDAY, D.K., HUNTER D.R. and MINTER, W.E.L., 1982: Crustal evolution of Southern Africa: 3.8 billion years of earth history; Springer-Verlag, New York, 523 p.

TRETTIN, H.P., 1970: Ordovician–Silurian flysch sedimentation in the axial trough of the Franklinian Geosyncline, northeastern Ellesmere Island, Arctic Canada, *in* J. Lajoie, ed., Flysch sedimentology in North America: Geol. Assoc. Can. Spec. Paper 7, p. 13–35.

VINCENT, Ph., GARTNER, J.E. and ATTALI, G., 1977: Geodip: an approach to detailed dip determination using correlation by pattern recognition; 6th Formation Evaluation Symp., Can. Well Log. Soc., Paper L.

WALKER, R.G., 1967: Turbidite sedimentary structures and their relationship to proximal and distal depositional environments; J. Sediment. Petrol., v. 37, p. 25–43.

WALKER, R.G. and CANT, D.J., 1979: Facies models 3. Sandy fluvial systems, *in* R.G. Walker, ed., Facies Models: Geosci. Can. Reprint Series 1, p. 23–31.

WALTERS, R.F., 1969: Contouring by machine: a user's guide: Am. Assoc. Petrol. Geol. Bull., v. 53, p. 2324–2340.

WATERS, N.M., 1981: Computer mapping: a review of what is available and what is useful for exploration purposes; Bull. Can. Petrol. Geol., v. 29, p. 182–196.

WEIMER, R.J., 1970: Rates of deltaic sedimentation and intrabasin deformation, Upper Cretaceous of Rocky Mountain region, *in* J.P. Morgan, ed., Deltaic sedimentation modern and ancient, Soc. Econ. Paleont. Mineral. Spec. Publ. 15, p. 270–292.

WERMUND, E.G. and JENKINS, W.A., Jr., 1970: Recognition of deltas by fitting trend surfaces to Upper Pennsylvanian sandstones in north-central Texas, *in* J.P. Morgan, ed., Deltaic sedimentation; Soc. Econ. Paleont. Mineral. Spec. Publ. 15, p. 256–269.

WILSON, J.L., 1980: A review of carbonate reservoirs, *in* A.D. Miall, ed., Facts and principles of world petroleum occurrence, Can. Soc. Petrol. Geol. Mem. 6, p. 95–117.

WINTER, H. de La R., 1979: Application of basic principles of stratigraphy to the Jurassic-Cretaceous interval in southern Africa; Geol. Soc. S. Africa Spec. Pub. 6, p. 183–196.

WOODCOCK, N.H., 1979: The use of slump structures as palaeoslope orientation estimators; Sedimentology, v. 26, p. 83–99.

YOUNG, F.G., MYHR, D.W. and YORATH, C.J., 1976: Geology of the Beaufort Mackenzie Basin; Geol. Surv. Can. Paper 76–11.

YOUNG, G.M. and LONG, D.G.F., 1976: Stromatolites and basin analysis: an example from the Upper Proterozoic of northwestern Canada; Palaeogeog., Palaeoclim., Palaeoec., v. 19, p. 303–318.

CHAPTER 6

Depositional systems

6.1. Introduction

The concept of depositional episodes and the depositional systems basin analysis method were developed largely in the Gulf Coast region as a means of analyzing and interpreting the immense thicknesses of sediment there that are so rich in oil and gas. The principles of the depositional system have never been formally stated, although they have been widely used, particularly by geologists at the Texas Bureau of Economic Geology (e.g. W.L. Fisher, L.F. Brown Jr., J.H. McGowen, W.E. Galloway, D.E. Frazier. See selected references quoted in section 1.2.2). Other geologists who made notable early contributions in this area include R.J. Weimer, D.A. Busch, J.C. Crowell, J.C. Ferm, H.A. Lowenstam, N.D. Newell, H.R. Wanless and many others.

The basis of the method is the application of Walther's Law and the facies model concept to large-scale depositional tracts, up to and including entire basins. The approach is essentially genetic stratigraphy, in which the focus of the analysis is on the interpretation of the interrelationships of large sediment bodies, based on an understanding of the depositional environments and syndepositional tectonics which controlled their formation. Sedimentological techniques, including facies analysis (Chap. 4) and basin mapping methods (Chap. 5), are of paramount importance, as is the application of refined biostratigraphic, radiometric or paleomagnetic correlation (Chap. 3). However, formal stratigraphic methods, such as the description of formations and members, erection of stratotypes, etc., (Chap. 3) are of secondary importance, and should only be attempted at the conclusion of the basin analysis exercise (section 6.8).

One of the purposes of this chapter is to describe some of the broad features of basin architecture and introduce a few more of the terms used by seismic stratigraphers (section 6.2). Although seismic stratigraphy was not of primary importance in the development of the depositional systems method, it has become increasingly important in application of the method to new basins, because of the ability of the tool to "see" large-scale stratigraphic geometry. Seismic data therefore form a prominent part of the remaining sections in the chapter (6.3 to 6.7) which describe some typical examples of depositional systems in a range of environmental settings. Included in this discussion are examples of how the depositional systems approach has been used to establish or modify existing stratigraphic terminology (section 6.8).

6.2 Stratigraphic architecture

The Law of Superposition of Strata is one of the oldest principles of sedimentary geology, and states that in any succession of strata the beds were deposited in a simple vertical order from base to top. This very obvious idea is the basis of the "layer cake" model of stratigraphy, which sees sediments as more or less uniform blankets or wedges, although perhaps showing thickness or facies changes from one part of the basin to another

Clearly, on the large scale, at the level of the system or series (Table 3.1) this is generally true. However, it overlooks an important point that has become increasingly clear as our knowledge of depositional processes and our ability to construct detailed stratigraphic cross sections have improved. This point is that in many environments lateral accretion of sediments is as important or more important than vertical aggradation. In a

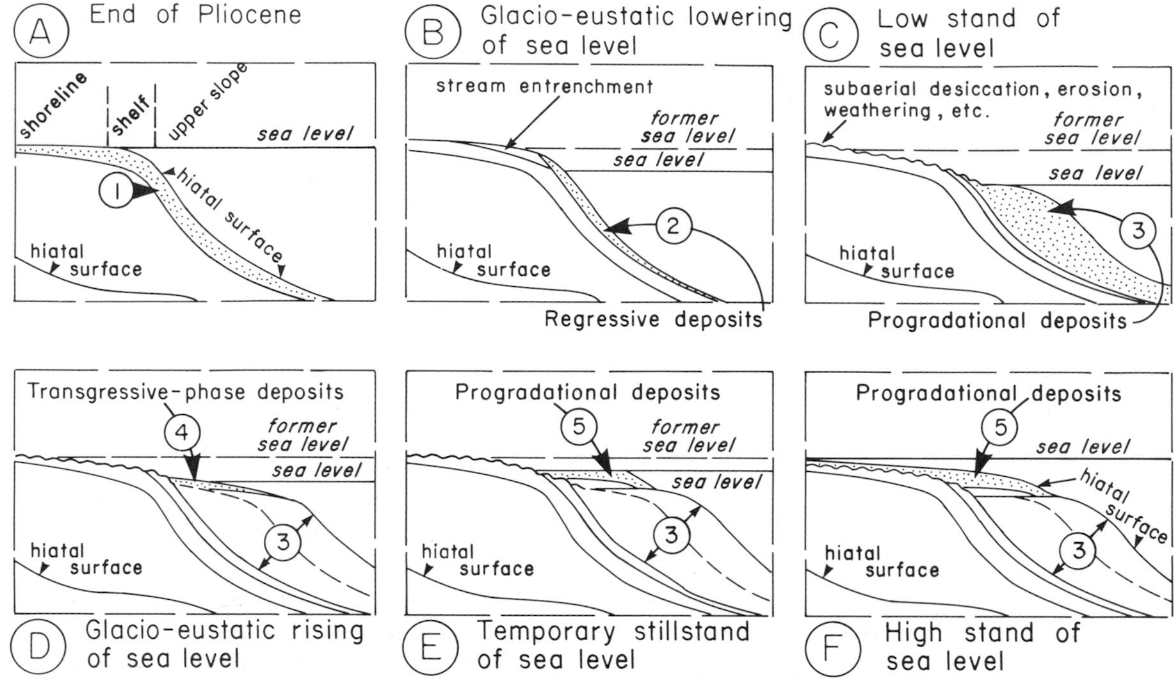

Fig. 6.1. A "depositional event" in the Mississippi Delta, as shown by Frazier (1974). Holocene sediments contain several or many such events as a result of the rapid changes in sea level during the ice age.

few environments, such as some cratonic carbonate shelves, evaporite basins and pelagic seas, sedimentation is uniform and develops stratigraphic units with an essentially tabular geometry. However, in most other environments distribution of the sediment is constrained by some factor such as river grade, base level, wave base or depth of marine light penetration, and sediment piles build "out" as well as, or instead of, "up". This is true of alluvial fans, submarine fans, prograding shorelines and many carbonate banks (as shown elsewhere in this chapter) but is nowhere so clearly illustrated as in prograding deltas or continental shelves, which show rapid lateral accretion on dipping clinoform surfaces (Fig. 5.16). Stratigraphic superposition is a complex process, and it results in a diversity of styles of stratigraphic architecture. For example, time increments represented by thick clinoform units on the prograding delta or shelf may be represented by a thin reworked zone or even a hiatus on the delta plain or the shelf platform, respectively. Also, the locus of progradation may switch laterally as a result of slope changes or channel diversion, so that a clinoform set may be followed laterally by a hiatus with the main stratigraphic time sequence being picked up elsewhere. Finally, sediment depocenters and stratigraphic relationships are markedly affected by subsidence and uplift and by changes in sea level, all of which generate offlap and onlap patterns. Sea level changes are extremely common in the geologic record (Chap. 8).

A simple hypothetical example of this complexity is shown in Figure 8.12, and a seismic section showing an actual example on a continental shelf is illustrated in Figure 8.10. Frazier (1974) discussed the application of these ideas to the Quaternary Mississippi Delta, affected as it has been by glacio-eustatic changes in sea level. Figure 6.1 illustrates a sequence of events repeated many times in the delta. He termed the development of a conformable stratigraphic package between hiatal surfaces a **depositional event**. A sequence of related events was termed a **depositional episode**.

It is obvious from diagrams such as these that unconformity surfaces may or may not be of regional significance. Many are temporary zones of sediment bypass or subaerial exposure during episodes of low relative sea level stand. They may or may not relate to tectonic movements or actual eustatic sea level changes.

Seismic stratigraphers have developed a range

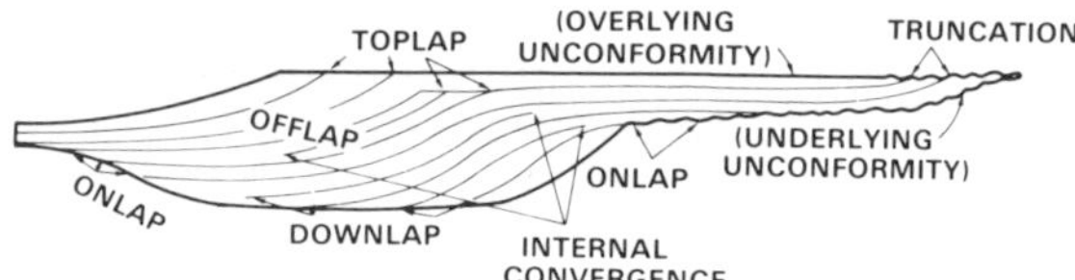

Fig. 6.2. Terminology for stratigraphic terminations, as developed by Mitchum et al. (1977) from seismic data.

of terms to describe bedding relationships within the stratigraphic packages that are formed during individual depositional episodes. These are illustrated in Figure 6.2 (from Mitchum et al., 1977). The seismic facies patterns shown in Figure 5.16 should be added to these.

Onlap and offlap are familiar terms but have been precisely defined by Mitchum (1977): **Onlap** is "a base discordant relation in which initially horizontal strata terminate progressively against an initially inclined surface, or in which initially inclined strata terminate progressively updip against a surface of greater initial inclination." **Offlap** is the progressive offshore shingling of sedimentary units within a conformable sequence, in which each successively younger unit leaves exposed a portion of the older unit on which it lies (Gary et al., 1972). These are both varieties of **lapout**, a general term for "lateral termination of strata at their depositional pinchout" (Mitchum, 1977). At the upper boundary of a depositional package this is termed **toplap**: "termination of strata against an overlying surface mainly as a result of nondeposition (sedimentary bypassing) with perhaps only minor erosion" (Mitchum, 1977). It should be remembered that seismic resolution is such that thin beds cannot be discriminated and apparent toplap may actually be a perfectly conformable but attenuated sequence.

At the base of a depositional package lapouts are termed **baselap**. Varieties of this pattern are termed onlap (defined above), downlap and uplap. **Downlap** is "a base-discordant relation in which initially inclined strata terminate downdip against an initially horizontal or inclined surface" (Mitchum, 1977). **Uplap** is a term used by Brown and Fisher (1977) to refer to depositional terminations against growth structures such as active diapirs, graben faults or growth faults.

The geometry or architecture of a depositional package or stratigraphic sequence is controlled by the interrelationship of three processes:

1. rate of sediment supply (clastic) or sediment generation (chemical)
2. rate of basin subsidence
3. rate of sea level change.

Facies changes, lapout relationships and hiatuses would provide a complete picture of the interrelationship of the first two processes were it not for the complications introduced by the third. Much recent work (that is discussed at length in Chapter 8) shows that long-term and short-term changes in global sea level have probably been occurring almost continuously, at least during the Phanerozoic. However, documenting the timing and amount of such changes is far from complete, and we are left with a great potential for confusion in unraveling the local stratigraphic record. For example, given adequate sediment supply and relatively slow subsidence it is possible for sedimentary progradation and regression to occur even during a rise in sea level. Conversely, rapid subsidence of a sediment-starved shelf could cause a transgression even during a time of falling sea level. It is therefore important to distinguish local events which can be interpreted directly from the sedimentary record, such as transgression and regression, from their ultimate cause, unless there is clear evidence for the latter. Figure 6.3 illustrates the relationship between rate of deposition and rate of subsidence in a deltaic complex, assuming stationary sea level (from Curtis, 1970). This should be compared with Figure 6.1, which shows a model based on sea level change. As discussed in Chapter 8, only where apparent changes in sea level can be precisely correlated regionally or globally is it safe to assume that sea level itself has actually changed.

6.3 Nonmarine depositional systems

Some of the world's most extensive clastic sequences have been formed in nonmarine and marginal marine environments. They may exceed 5 km in thickness and extend for several thousands of kilometers along strike. Depositional architecture is defined by the geometry of major coarse-grained (conglomerate, sandstone) sediment bodies, and is controlled by basin shape, paleoslope orientation and steepness, source area elevation and rate of relative basin subsidence. These factors are ultimately governed by tectonics.

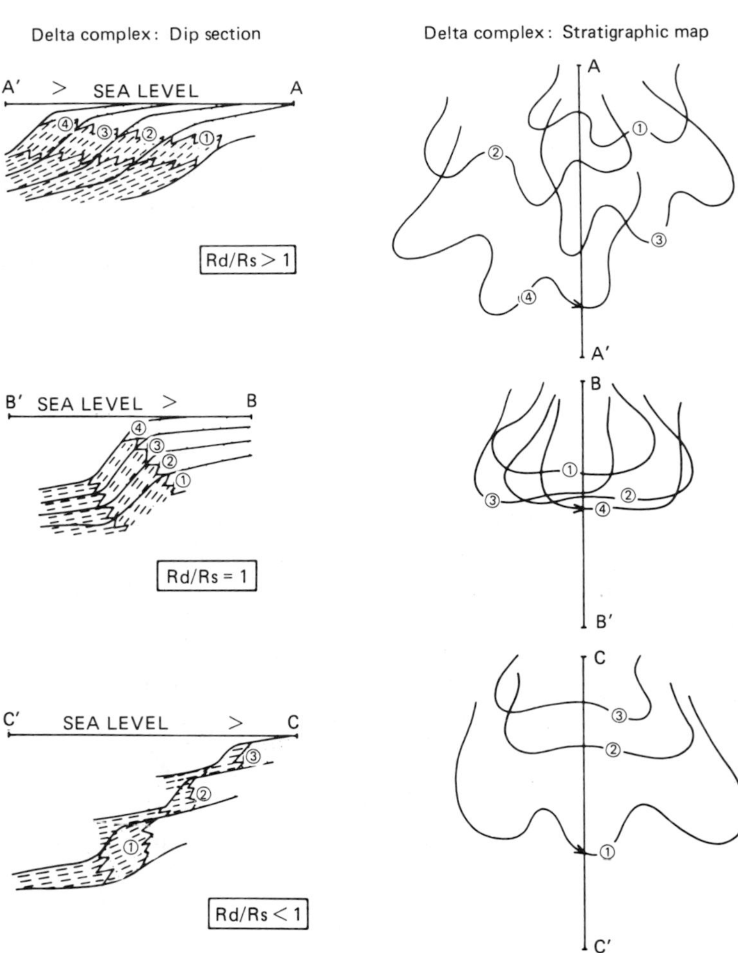

Fig. 6.3. Relationship between rate of deposition (Rd) and rate of subsidence (Rs) in a delta complex (Curtis, 1970).

Miall (1981) erected a classification of alluvial depositional systems which included alluvial fans, and fluvial and deltaic deposits. Although these environments include a wide variety of depositional patterns, as indicated by the proliferation of facies models (Chap. 4), they can be simplified into a few basic types for the purpose of studying large-scale variations of major depositional systems. In most such large fluvial basins only two or three basic facies patterns are present. The basin may be flanked by alluvial fans, and the major tributaries and trunk river commonly can be subdivided into low-sinuosity systems near the source and high-sinuosity systems near the mouth. The term alluvial fan may be used for any entirely nonmarine and nonlacustrine fluvial system whose channel network is distributary rather than contributary. Fans which prograde directly into a standing body of water are termed fan-deltas or coastal alluvial fans. The term delta should be restricted to environments where the distal end of the system is controlled by an interaction between marine (or lacustrine) and fluvial processes. This is not apparent with most fan-deltas which have sharply defined shorelines, and the term fan-delta may therefore be something of a misnomer (Rust, 1979). Alluvial fans are a sedimentary response to flow expansion at a basin margin, and so the identification of fan deposits in the ancient record is a useful indicator of sharp basin-margin relief, commonly fault-controlled.

Alluvial drainage systems can be divided into four parts:

1. Headwaters in upland source areas
2. Proximal environments, immediately below the fall line at the margins of the sedimentary basin
3. Medial environments

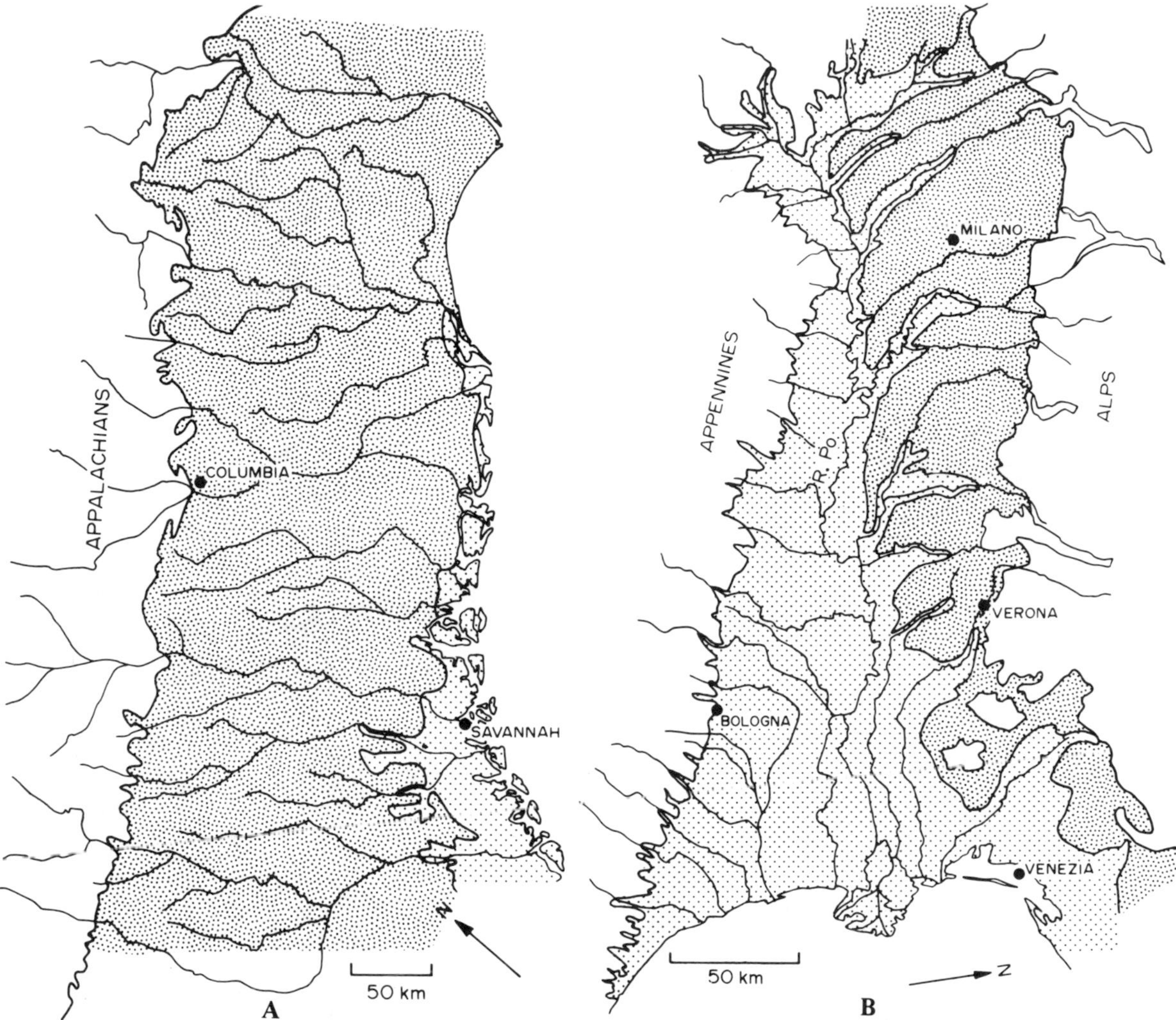

Fig. 6.4. A The Atlantic coastal plain of Georgia, North and South Carolina. Basin fill model 4 of Miall (1981). **B** The Po Basin, Italy. Basin fill model 7 of Miall (1981). Reproduced with permission from the Geological Association of Canada.

4. Distal environments, where the rivers interact with some terminating environment such as a lake, playa flat, eolian erg, tidal flat or delta.

The headwater zone is sedimentologically insignificant, although valley fill fluvial and lacustrine deposits may be preserved temporarily until removed by further valley incision.

Proximal, medial and distal parts of an alluvial system commonly have different paleoslopes. It is a common observation that in nonmarine basins within or adjacent to deformed belts, rivers tend to run either parallel or perpendicular to strike. This led Miall (1981) to make a simplifying assumption that permits a ninefold classification of alluvial basin types. If all tectonic environments are treated as linear and orthogonal rivers can be classified as either **transverse**, in which case they flow directly from the uplifted source area across structural grain, or **longitudinal**, flowing parallel to strike along the axis of the basin. These generalizations may hold only locally, for fold belts show syntaxes, island arcs are arcuate, and plate margins may be irregular. However, the distinction is an important one sedimentologically. Some systems, such as those on many modern coastal plains, consist entirely of transverse rivers (e.g. Atlantic coastal plain of Georgia and the Carolinas: Fig. 6.4A). In other basins the transverse rivers are tributaries to a longitudinal trunk river. The two may be charac-

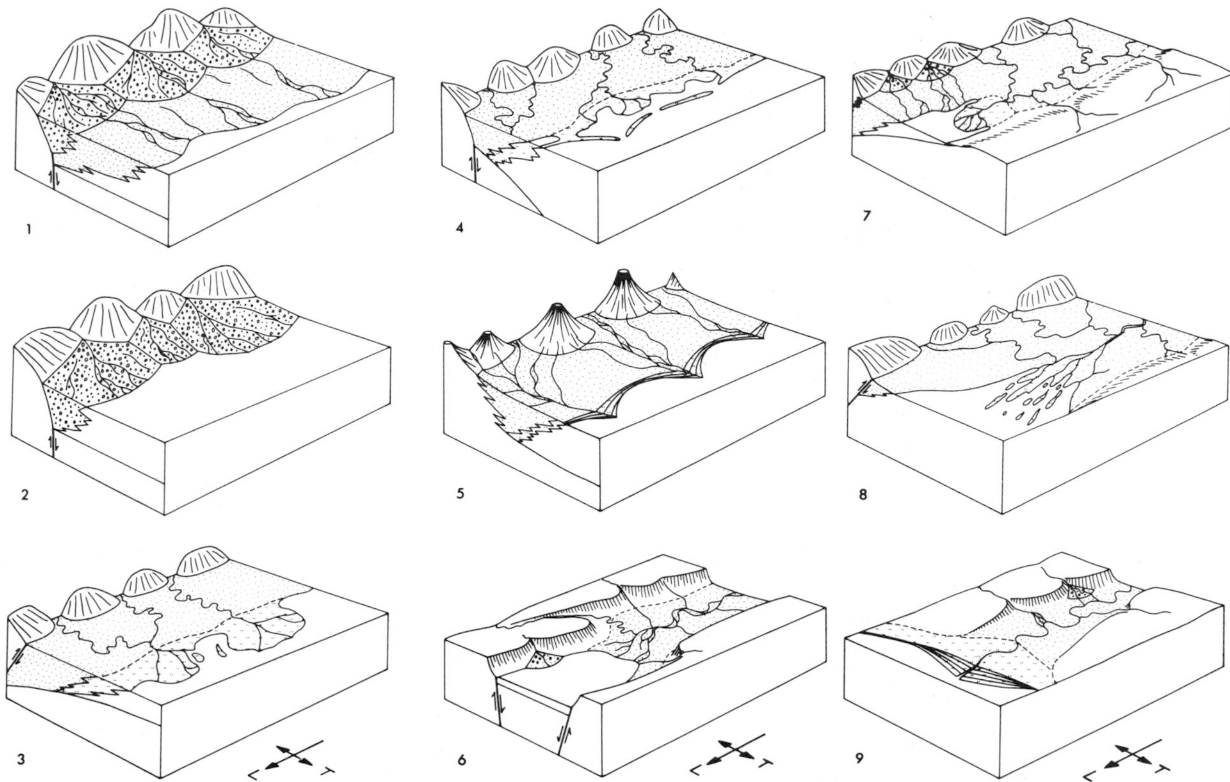

Fig. 6.5. Basin fill patterns, as defined by Miall (1981) Reproduced with permission from the Geological Association of Canada. T = transverse, L = longitudinal. See Table 6.1.

terized by quite different fluvial styles and, of course, paleocurrent patterns. The Po River, Italy, is a good example (Fig. 6.4B). Flanking active fold belts the transverse rivers may be alluvial fans such as the giant fans flanking the Himalayas, which drain into the longitudinal Ganga River. Some longitudinal rivers originate as major transverse rivers, emerging from structural re-entrants in the mountainous source area and curving downstream into a longitudinal orientation (e.g. Brahmaputra River).

Miall's (1981) nine alluvial basin fill models are listed in Table 6.1 and illustrated in Figure 6.5. The delta classification is discussed in sec-

Table 6.1. Alluvial basin fill patterns*

Model	Proximal	Medial	Distal
1	T fan	T braidplain	lake margin/non deltaic coast
2	T fan-delta	—	lake margin/sea coast
3	T fan/river	T river	river-dominated delta
4	T fan/river	T river	river-dom. delta with barrier-lagoon
5	T fan/river	T river	wave-dominated delta
6	T fan/river	L river	lake margin/estuary
7	T fan/river	L river	river-dominated delta
8	T fan/river	L river	tide-dominated delta
9	T fan/river	L river	wave-dominated delta

*From Miall, 1981 Reproduced with permission from the Geological Association of Canada (1981). T = transverse, L = longitudinal.

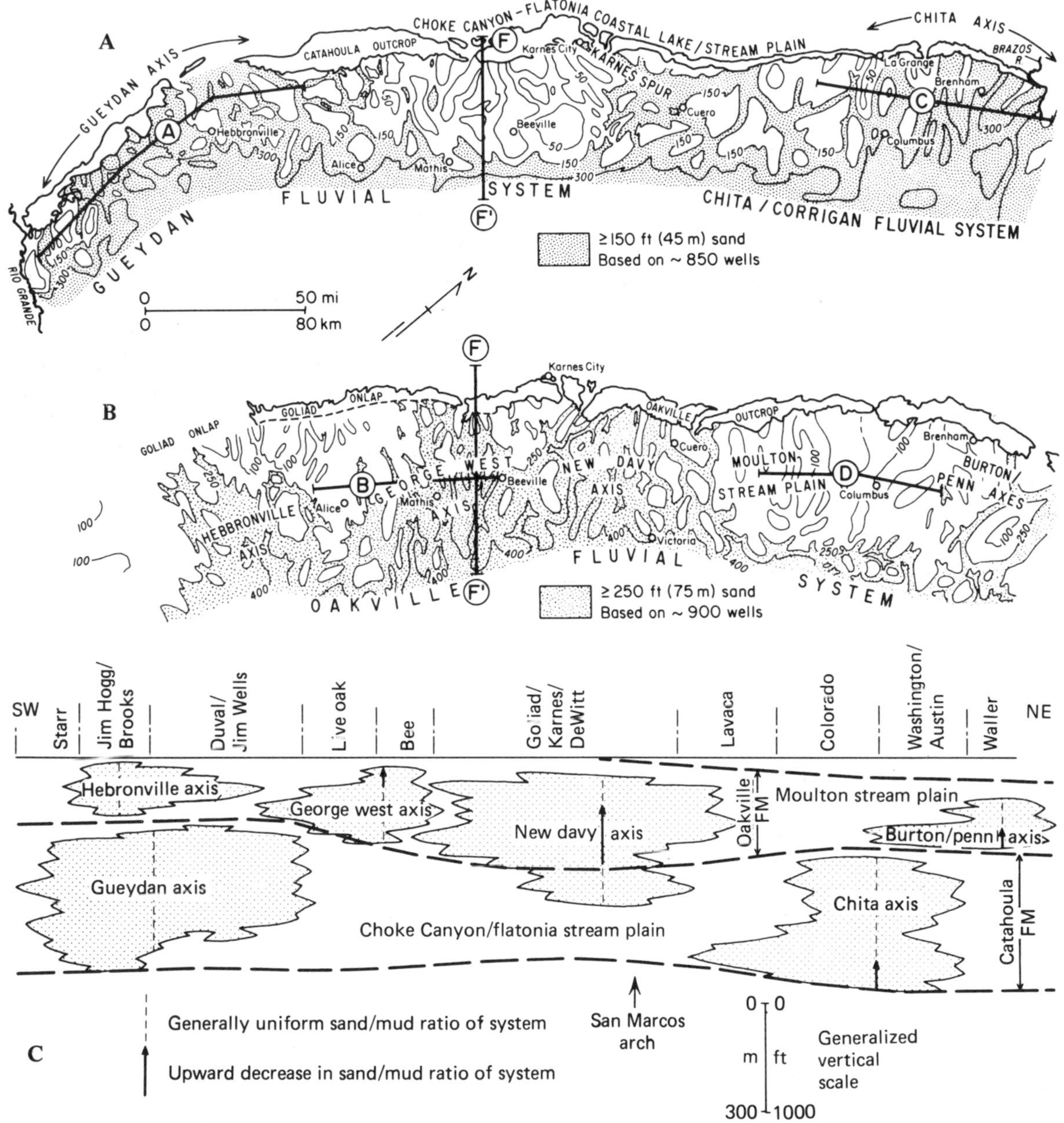

Fig. 6.6. Oligocene–Miocene Gulf Coast fluvial complexes. **A** simplified net sand isolith map, Catahoula Formation; **B** the same, Oakville Formation; **C** strike-parallel section showing position and thickness of major fluvial depositional axes (Galloway, 1981).

tion 6.4. The structural controls illustrated in the block diagrams are to some extent interchangeable. They are discussed at greater length in Chapter 9. No scale is implied in the diagrams. They range from small intermontane graben basins which typically show patterns 1 and 6, to the giant basins of the world's largest rivers, all of which are longitudinal (patterns 7, 8, 9). Miall (1981) discussed these models at some length and referred to many ancient examples. Only two ancient examples are illustrated here for reasons of space. The seismic character of alluvial systems is discussed as part of the delta-plain facies in the next section.

Galloway (1981) provided a synthesis of a depositional systems analysis of the Cenozoic fluvial deposits of the Gulf Coast (Fig. 6.6). These form part of an immense prograding clastic

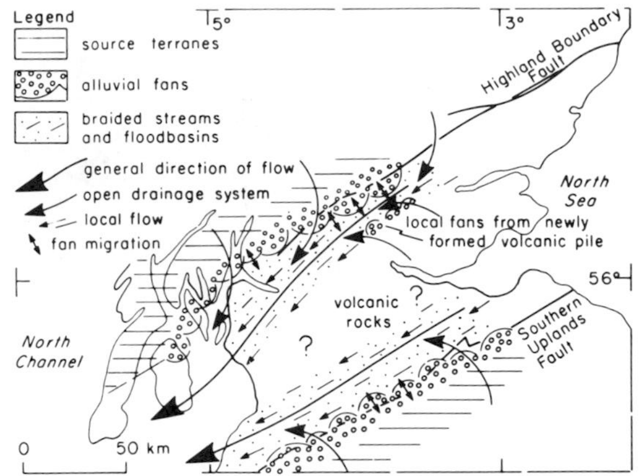

Fig. 6.7. Deposition of the Lower Old Red Sandstone in the Midland Valley of Scotland. This is an example of a transverse alluvial fan system feeding a longitudinal braidplain (Bluck, 1978, © John Wiley & Sons, Ltd. Reprinted with permission.)

system with environments ranging from fluvial to submarine fan. Local thicknesses of the Cenozoic section are in the order of 5 km. The river patterns are entirely transverse, consisting mainly of models 3 and 4. Fisher and McGowen (1967) showed that many of the rivers have stayed in essentially the same position from early Tertiary time to the present day, extending their mouths seaward by forming thick coastal sequences.

During the Devonian the Midland Valley of Scotland was occupied by a transverse–longitudinal river system (Fig. 6.7). This was one of many "Old Red Sandstone" basins flanking the suture of the proto-Atlantic (Iapetus) Ocean. Bluck (1978) showed that the Highland Boundary Fault, which defines the northern edge of the basin, was active at several times during the

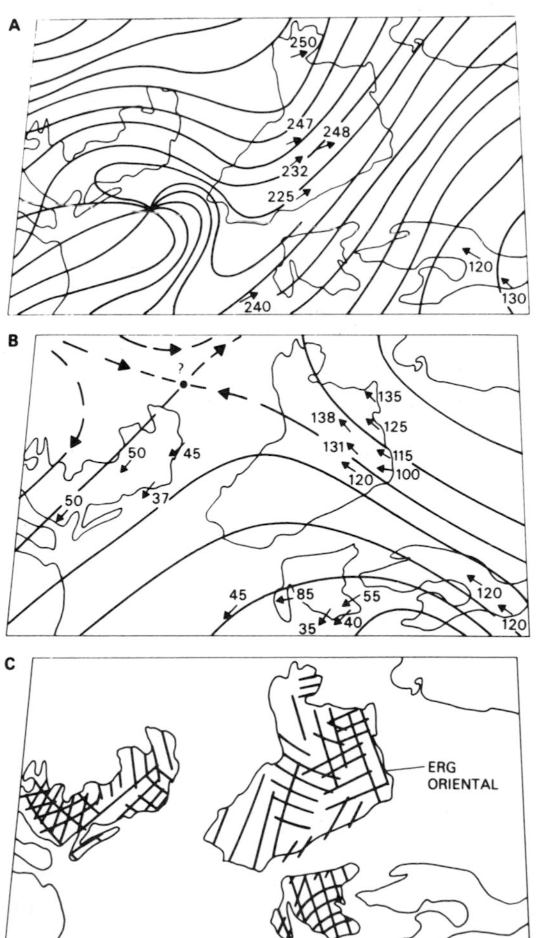

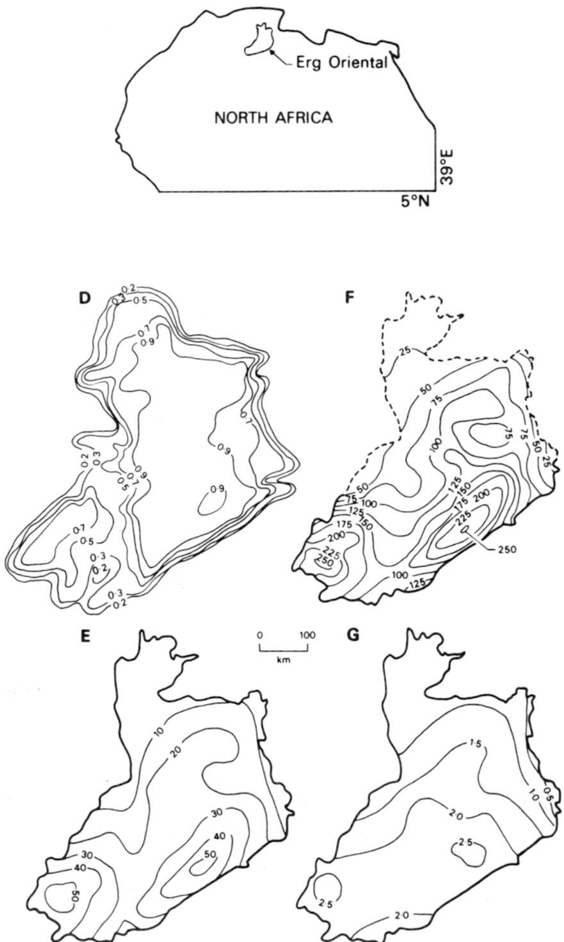

Fig. 6.8. The Erg Oriental, Algeria; **A** flow lines and bedform orientation directions, medium sand; **B** the same, fine sand; **C** prominent draa trends—different trends reflect different grain size ranges. **D** proportion of sand cover (versus bare bedrock); **E** mean sand thickness; **F** mean draa height (m); **G** mean draa wavelength (Wilson, 1973; Reading, 1978).

Devonian and was flanked by thick alluvial fan deposits. These built out southward and fed a longitudinal river system draining southwestward at first, although later it reversed and flowed northeastward toward the position of the present North Sea.

Eolian deposits occur in a variety of tectonic and climatic settings, but the major eolian depositional systems are the giant sand seas, termed **ergs** by Wilson (1973). At the present day ergs occur mainly in areas of less than 15 cm rainfall, and most occur within 30° north and south of the equator, except for some areas in central Asia. Major eolian deposits occur only within large continental masses, typically on cratons, and those in the ancient record have been widely used as paleoclimatic indicators.The largest erg at the present day is the Rub al Khali, Arabia (560,000 km^2), although closely related ergs in Russia and Australia have combined areas of 750,000 and 900,000 km^2, respectively (Wilson, 1973). Algerian ergs average 26 m in sand thickness, with a local maximum of 145 m (Fig. 6.8). Present-day ergs are confined to topographic basins. Internal elevated areas may be free of sand cover because of wind divergence around them or acceleration over them. This produces local undersaturation of sand in the air stream, and therefore increased erosive power, even during sandstorms. The topographic control of sand distribution leads to a high preservation potential for erg deposits, because basin areas are probably areas undergoing subsidence.

These data, and others given in an excellent paper by Wilson (1973) provide the basis for a powerful depositional systems model for interpreting the ancient record. Such spectacular and well-known eolian deposits as the Rotliegendes (Permian) of northwest Europe and the various Carboniferous to Jurassic units of the western United States, (Navajo, Lyons, Coconino, etc.; see Walker and Middleton, 1979, Table 1) can all be interpreted in terms of the erg model, for example the Navajo Sandstone (Lower Jurassic), which reaches a thickness of 700 m and extends over six states (Fig. 6.9). Kocurek (1981) gave an excellent summary of the Entrada sandstone (Jurassic) erg of northern Utah and Colorado.

Wind patterns in ergs are complex. This is illustrated by the orientation patterns of draas, the largest and most preservable of eolian bedforms (Fig. 6.8). Draa crests may form parallel, oblique or transverse to wind directions, and different grain sizes of sand have different sand flow trends because they are moved by different wind regimes (Fig. 6.8). This complexity is responsible for the enormous variability in bedform pattern that has now been thoroughly documented by satellite photography (McKee, 1979). The consistent paleocurrent orientations of major ancient dune deposits demonstrates, however, that much of this complexity is ephemeral. Crossbed thickness reaches about 30 m in ancient dune deposits, whereas modern draas are up to 430 m high, indicating that only the deeper parts of the bedform system are ever preserved. Interpretation of apparently consistent wind patterns in ancient deposits may tempt the basin analyst to make deductions about ancient global air circulation (e.g. trade wind patterns), and correlate this with paleolatitude, but as the discussion above suggests, this is probably a dubious procedure.

Eolian depositional systems interact at their margins with a variety of other environments, including ephemeral rivers and alluvial fans, playa lakes and arid shorelines (sabkhas). Glennie (1970, 1972) and Kocurek (1981) described some of the stratigraphic complexity that can result.

As noted in section 4.6.3, lacustrine environments are highly variable, and a discussion of lacustrine depositional systems would require a lengthy chapter of its own (selected references are given in that section). Many lakes are very broad and shallow, and their deposits consist of blanket-like bodies of highly variable lithofacies, reflecting climatically or tectonically induced fluctuations in water level and water chemistry. Basin center muds, evaporites or turbidites are interbedded with basin margin fluvial and deltaic clastics. Lake margins commonly show considerable stratigraphic complexity because of rapid variations in small scale depositional processes. An example is the Devonian Orcadian Basin of Scotland, as shown in Figure 6.10 (Donovan, 1975).

6.4 Coastal depositional systems

Two major types of depositional system are dealt with here, deltas and the barrier–lagoon–tidal inlet complex. Because of their importance to the petroleum industry these systems have been intensively studied by seismic stratigraphers and

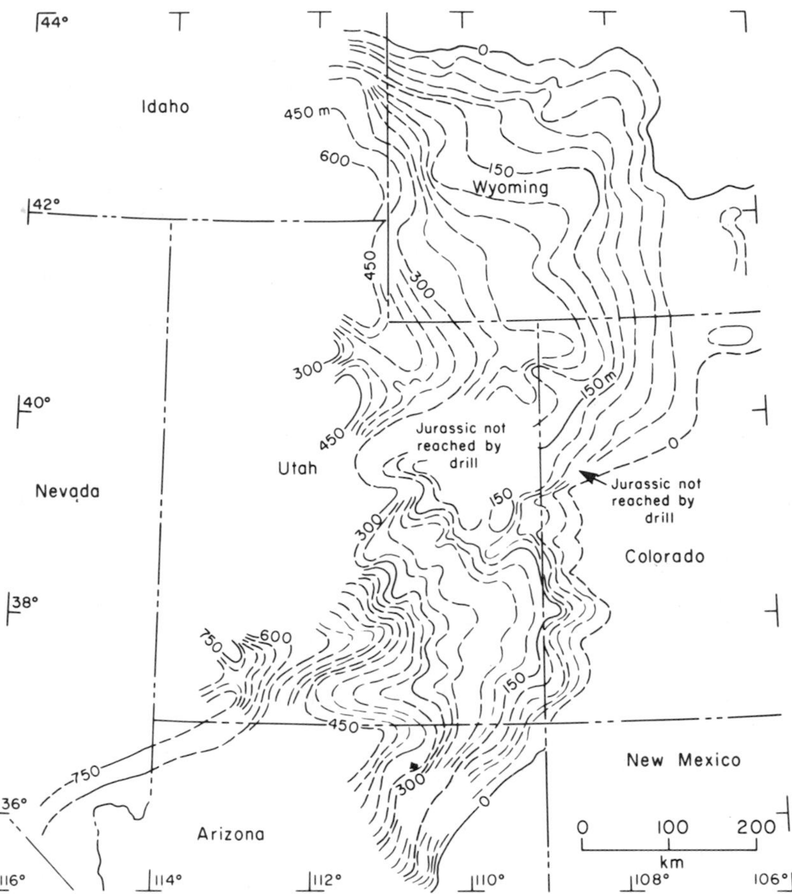

Fig. 6.9. Isopach map of Navajo and Kayenta Formations and correlative units (Triassic–Jurassic) (McKee, 1979).

their broad architectural characteristics are now well-known.

Deltaic systems are subdivided using the ternary river-, wave- and tide-dominated classification erected by Galloway (1975) (Fig. 4.56). Deltas and barrier complexes form the distal, terminating part of many alluvial drainage systems, and variations between them provide some of the criteria by which Miall (1981) defined the suite of nine alluvial basin fill models illustrated in Figure 6.5.

Deltaic deposits form some of the thickest and most extensive stratigraphic sequences to be found anywhere. This is because they act as traps for a significant percentage of the clastic detritus eroded from continents by the world's major rivers. Sedimentation rates are locally very high. The modern Mississippi delta has accumulated at rates of 6–12 m/1000 a during the last 6000 a (Curray, 1965; Frazier, 1967), which compares to typical average values for clastic continental margins of less than 0.5 m/1000 a (Miall, 1978). Rapid progradation of deltas and, to a lesser extent, regressive barriers, is one of the main mechanisms for the generation of clinoform patterns in seismic section (Fig. 5.16, 5.17). As noted in section 5.4.1 oblique clinoforms may show depositional dips approaching 10°. These are typical of coarse sandy deltas and fan deltas. Those with a minor sandy bedload develop sigmoidal clinoforms with dips less than 1° (Mitchum et al., 1977). Details of deltaic architecture depend on sediment grain size and the interaction of fluvial and marine sedimentary processes. Coleman and Wright (1975) developed a set of six delta models which incorporate the ternary classification of Galloway (1975) and display some of the intermediate variation between them. The models are depicted by sand thickness or gross isopach maps, and are of considerable use in categorizing a deltaic system from field or subsurface data (Fig. 6.11). As discussed by

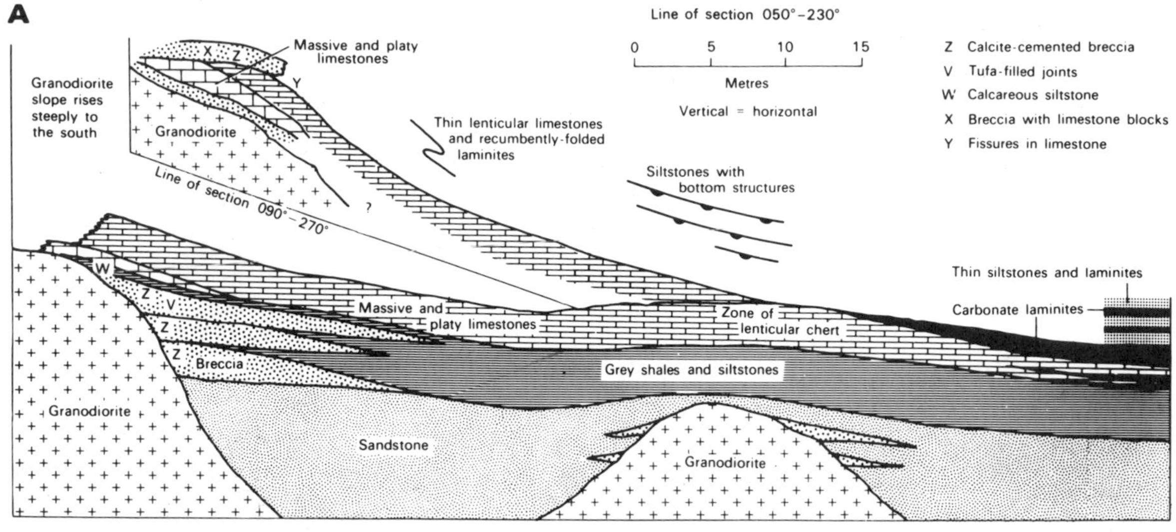

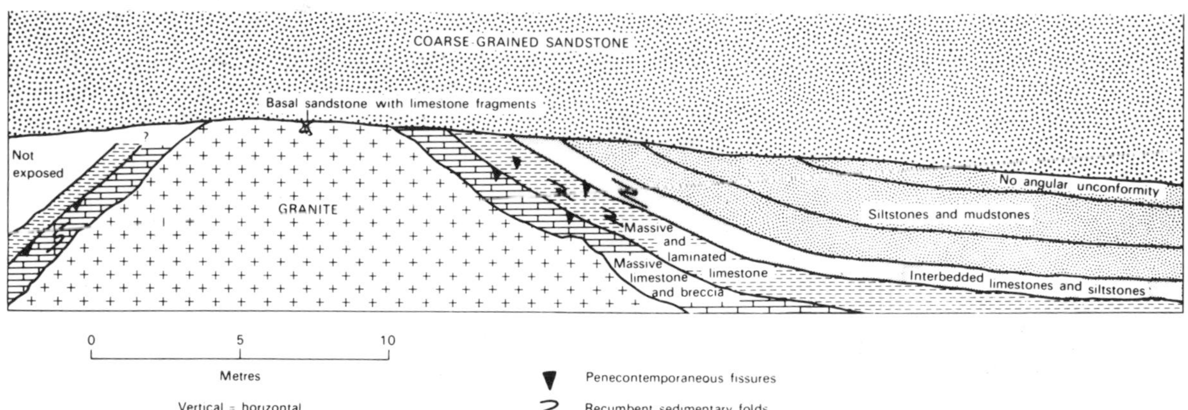

Fig. 6.10. Basin-margin facies relationships, Devonian Orcadian Basin, Scotland (Donovan, 1975; Reading, 1978).

Coleman and Wright (1975) and Miall (1981) the relative importance of fluvial and marine processes is largely controlled by basin size and shape and bottom slope. This, in turn, is governed mainly by tectonic setting, and is discussed in Chapter 9.

River-dominated deltas develop lobate- or birdsfoot-shaped depocenters, as revealed by isopach or sand isolith maps. Lobate sand trends are oriented at high angles to the shoreline (Fig. 6.11A). The sediment, once deposited, is not reworked except for the thin abandonment phase which occurs when individual lobes or major crevasse splays are abandoned by distributary switching. Wave-dominanted deltas show sand isoliths or gross isopachs that may be elongated parallel to shore, because of the effect of the waves in causing the delta front sand bodies (mouth bars) to coalesce laterally and retreat landward (Fig. 6.11F). Tide-dominated deltas show strongly linear trends of sand thicks and thins, reflecting tidal scour channels and sand ridges (Fig. 6.11B, C).

Growth faults and mud diapirs are commonly associated with thick deltaic successions. Growth faults are failure surfaces which develop as a result of sediment loading and basinward creep of the sediment pile. They are curved in plan view, concave seaward, and are **listric**, that is they flatten out to horizontal at depth (Figs. 6.12, 6.13). Fault movement commonly is greatest where sand lenses prograde onto the downdip block, and

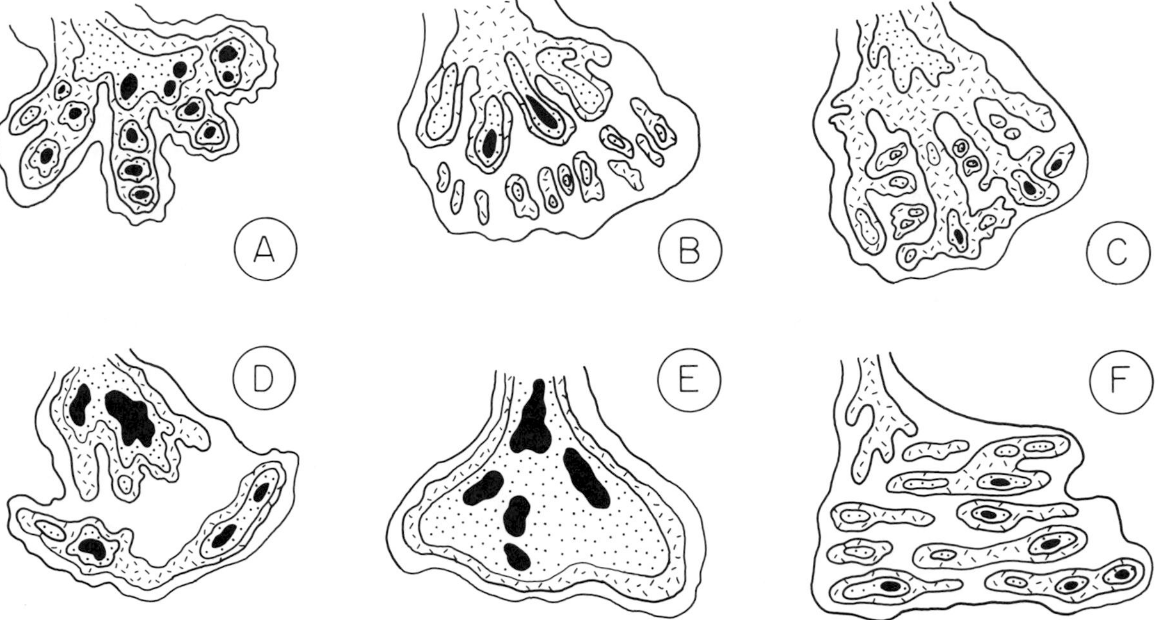

Fig. 6.11. Delta models of Coleman and Wright (1975). **A** Low wave and tide energy, low littoral drift; **B** Low wave energy, high tide range, low littoral drift; **C** Intermediate wave energy, high tide, low littoral drift; **D** Intermediate wave energy, low tide range; **E** high wave energy, low littoral drift; **F** high wave energy, strong littoral drift.

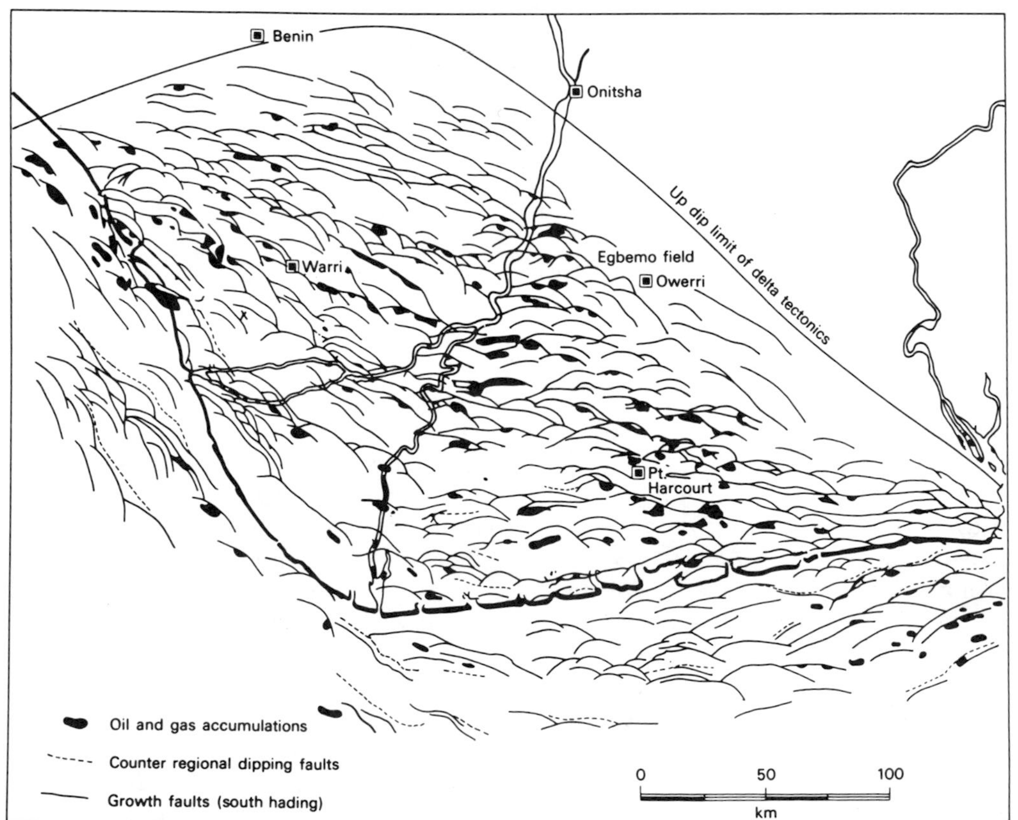

Fig. 6.12. Growth faults and petroleum accumulations in Niger Delta (Weber and Daukoru, 1975; Reading, 1978).

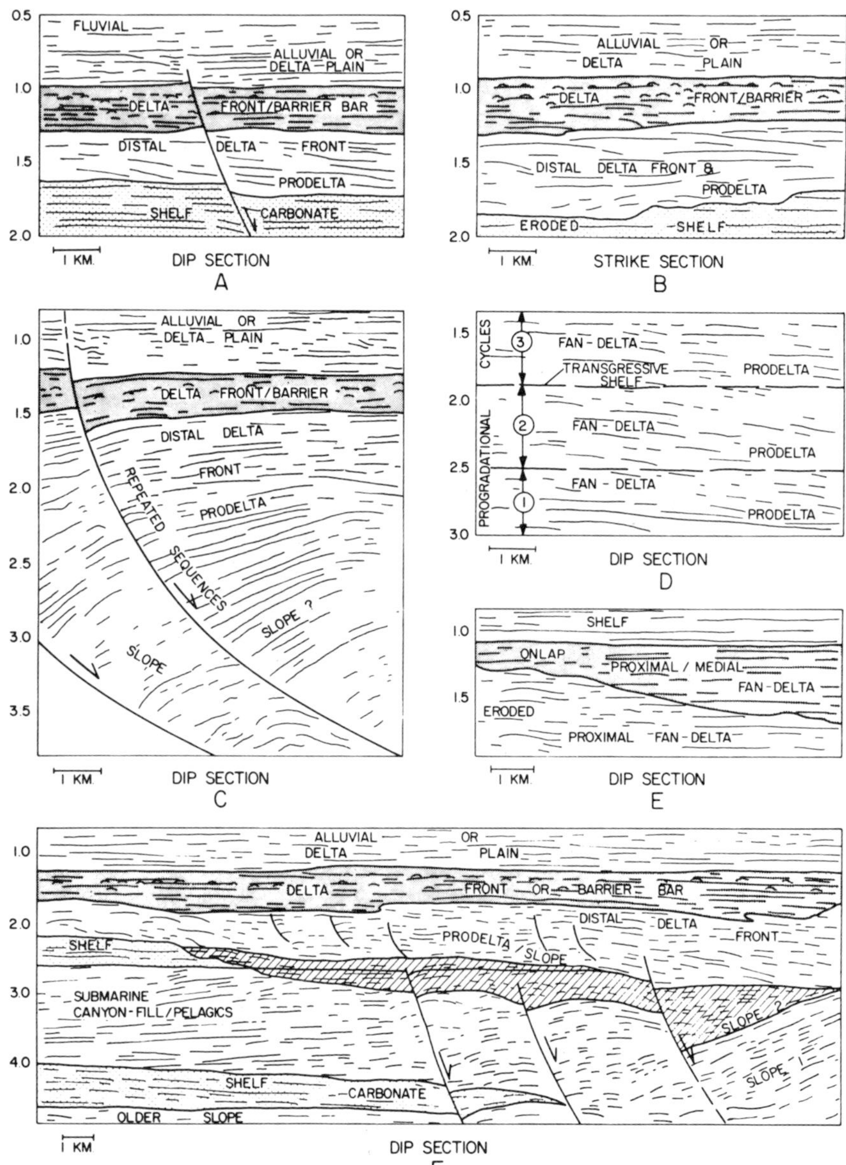

Fig. 6.13. Example of seismic reflection patterns from Brazil showing deltaic and associated depositional systems. Vertical time in seconds (Brown and Fisher, 1977).

slows or ceases when covered by less dense interdistributary muds (Edwards, 1976).

Mud diapirs are composed of prodelta sediments injected into overlying sediments when overlain by the delta front facies. The latter is composed of higher-density silts and sands and always overlies the prodelta facies wherever active progradation occurs.

A classic example of depositional systems analysis of an ancient deltaic tract is that by Fisher and McGowen (1967) of the Lower Wilcox Group (Eocene) of Texas. The unit locally reaches 1500 m in thickness, with sand isoliths exceeding 730 m. Fisher and McGowen analyzed the succession using outcrop sections and 2500 subsurface logs, relying extensively on methods of vertical profile analysis, such as those discussed in section 4.5.8.

Their summary map is shown in Figure 6.14. The area includes river-dominated deltas, fluvial, lagoonal, barrier bar and shelf depositional systems. The fluvial–deltaic drainage is essentially transverse, and this is an example of model 3 of Miall (1981). Flanking the deltas to the west is an extensive barrier–lagoon system (model 4 of Miall, 1981).

Brown and Fisher (1977) discussed the seismic reflection character of deltaic and barrier systems

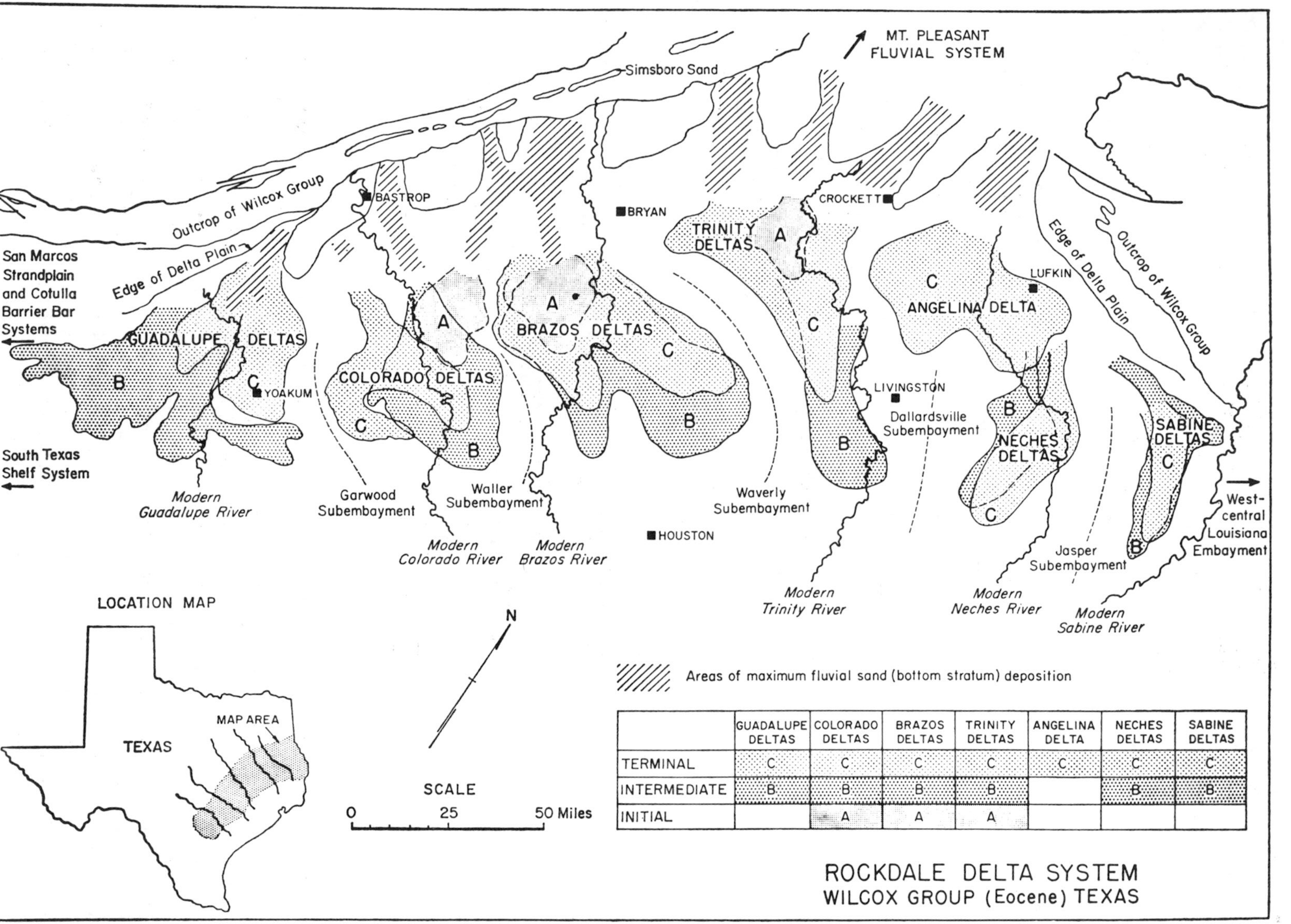

	GUADALUPE DELTAS	COLORADO DELTAS	BRAZOS DELTAS	TRINITY DELTAS	ANGELINA DELTA	NECHES DELTAS	SABINE DELTAS
TERMINAL	C	C	C	C	C	C	C
INTERMEDIATE	B	B	B	B		B	B
INITIAL		A	A	A			

Fig. 6.14. Deltaic depositional systems of Eocene age, Wilcox Group, Texas (Fisher and McGowen, 1967).

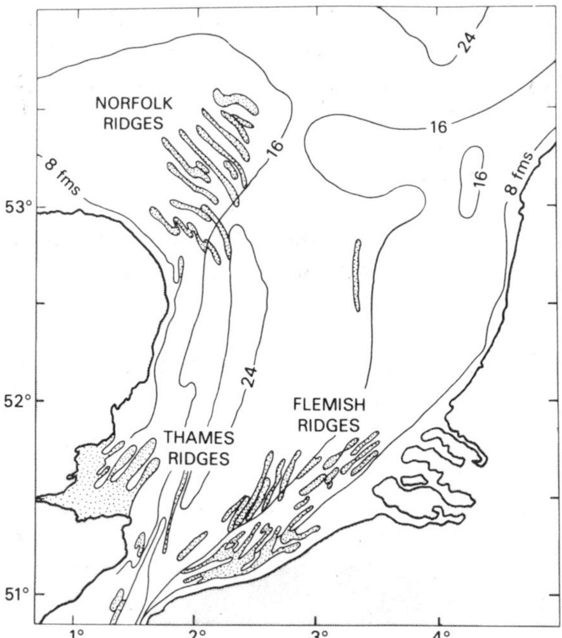

Fig. 6.15. Tidal sand ridges in the southern North Sea (Houbolt, 1968 Geol. Mijnbouw, V. 47, p. 245–273; Reading, 1978).

based on extensive work in Brazil, and illustrated it using selected, redrawn seismic cross sections (Fig. 6.13). The alluvial or delta plain facies typically shows horizontal, parallel to rarely divergent reflections. Sand-dominated (e.g. braided) systems may contain few velocity contrasts and thus yield weak reflections, whereas those containing well-defined channel sand and overbank mud sequences may yield good reflections. In strike section reflections may be mounded or chaotic, the latter where channel margins yield diffractions. Reflection continuity may be poor. Few channels will be deep enough to show up on the seismic records. Delta front reflections typically form the upper part of clinoform sets. In strike section they are commonly mounded or hummocky, reflecting offset delta lobes. Reflection continuity will be variable, the most continuous reflections being derived from thin delta abandonment facies, whereas channels and growth faults generate disrupted and diffracted reflections. The major part of a deltaic clinoform set is composed of the distal delta front to prodelta facies, where reflections are poorly layered to reflection free (mud-dominated). Chaotic reflections are caused by slump masses. Growth faults and rollover anticlines are common.

Barrier bar sands typically show shingled seismic facies patterns. In strike section they may be discontinuous or mounded, reflecting tidal channels and regressive accretion or superimposition of offlapping sand bodies.

6.5 Clastic shelves and associated depositional systems

Some general characteristics of shelf systems are discussed in section 4.6.7, in particular the poorly understood interaction of waves, tides and storms in moving sediment and generating bedforms. In the geological past shelf seas were of considerable importance during periods of high sea level stand, such as during the Ordovician–Silurian and Jurassic–Cretaceous. These seas were thousands of kilometers across—far larger than any shelf at the present day. Klein (1977) and Klein and Ryer (1978) pointed out that we need better information on the sedimentology of present day shelf seas such as the East China and Yellow Seas, North Sea and Hudson Bay, although even these cannot match in size the seas of the North American Western Interior. Such seas are very sensitive to eustatic sea level changes and the sediments which accumulated there are markedly cyclic, as discussed at length in Chapter 8.

The number of sedimentary units interpreted as shelf deposits is rapidly growing (see descriptions and references in Walker, 1979; Brenner, 1980). A common theme appears to be that they are essentially broad, tabular sediment bodies. They may be composed of large tidal or storm-generated sand ridges, such as those in the North Sea (Fig. 6.15) or the Atlantic Shelf off the United States (Houbolt, 1968; Swift et al., 1973). These are several kilometers in width and a few tens of meters high, and would therefore be impossible to identify from any but the largest outcrops. They could, however, be mapped by careful well correlation. As Off (1963) has shown, such ridges tend to run parallel to open shorelines or are parallel to the axes of gulfs or embayments, reflecting tidal current reversals. Orientation of sand isolith contours may therefore be a useful mapping tool. Brenner and Davies (1973, 1974) studied Upper Jurassic sand ridges in Wyoming which they interpreted as preservation of shelf ridge and swale topography. They probably were

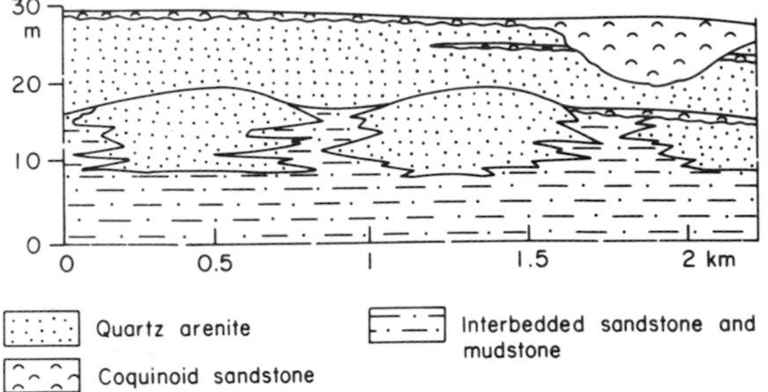

Fig. 6.16. Marine sand bar deposits, Upper Jurassic, Wyoming (Brenner and Davies, 1973, 1974; Brenner, 1980).

storm-generated, and were frequently breached by storm-surge channels (Fig. 6.16).

Several subsurface studies of shelf sands have been carried out, no doubt because they are excellent hydrocarbon reservoirs. The Cretaceous Viking and Cardium Formations of Alberta and the Sussex Sandstone of Wyoming are excellent examples (Evans, 1970; Tizzard and Lerbekmo, 1975; Brenner, 1978). Linear sand trends are clearly related to sand ridge morphology (Fig. 6.17). Both these units contain conglomerate lenses with pebbles several centimenters in diameter that have in the past been interpreted as beach deposits. However, it now seems likely that they were transported across the shelf from a shoreline source by storm activity, possibly in sediment gravity flows.

Many ancient shelf deposits are mixed carbonate–clastic successions, containing thin sand ridge or deltaic sand sheets, and carbonate banks. Galloway and Brown (1973) described an example from the Pennsylvanian of northern central Texas, in which a deltaic system prograded on to a very stable shelf. Distributary channels are incised into the underlying deposits. Widespread shelf limestones developed during periods of reduced fluvial sediment influx, and also occur in some interdeltaic embayments. Carbonate banks occur on the outer shelf edge, beyond which the sediments thicken dramatically into a clinoform slope system (Fig. 6.18). Handford and Dutton (1980) described a similar Pennsylvanian–Permian example from northernmost Texas.

Brown and Fisher (1977) discussed the seismic facies patterns of shelf systems. Examples are illustrated in Figure 6.19. Reflections are typically horizontal, parallel to slightly divergent in dip sections. Reflection continuity tends to be excellent and amplitude high, representing the tabular nature of sedimentary units. Shingled clinoforms may occur, indicating the presence of offlapping sand lenses. Figure 6.19 illustrates the lateral transition of shelf facies into reefs, fan deltas and submarine canyons.

Many shelves are cut by submarine canyons, which provide a sediment bypass to the deep ocean. These are discussed in section 6.7.

6.6 Carbonate and evaporite depositional systems

No simple classification of carbonate depositional systems can be offered, because of the dependence of carbonte sedimentation on such a wide variety of depositional controls. Such factors as temperature, carbonate saturation, salinity, water depth, nature of water currents, light penetration, water turbidity, nature of sediment substrate and rates of relative sea level change all determine sedimentation patterns, and these vary widely

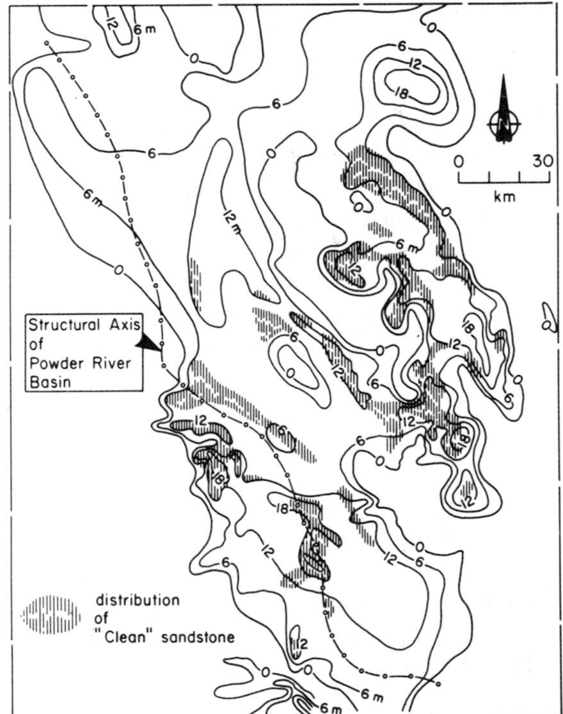

Fig. 6.17. Isolith sand map of Upper Sussex sandstone. Superimposed on isolith pattern is the distribution of "clean" sandstone, defining linear sand ridges (Brenner, 1978).

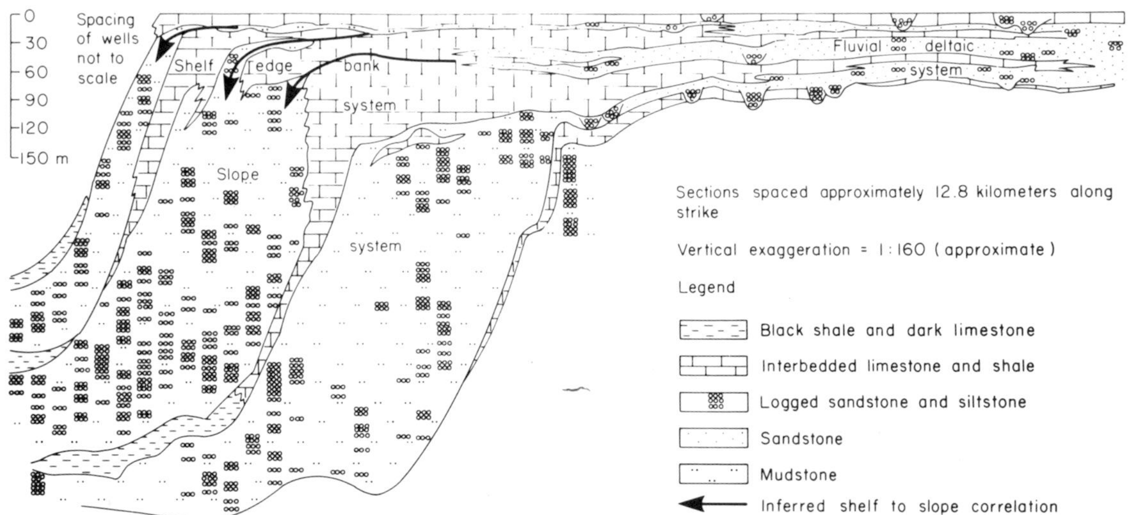

Fig. 6.18. A mixed carbonate-clastic shelf depositional system, Pennsylvanian, north-central Texas (Galloway and Brown, 1973).

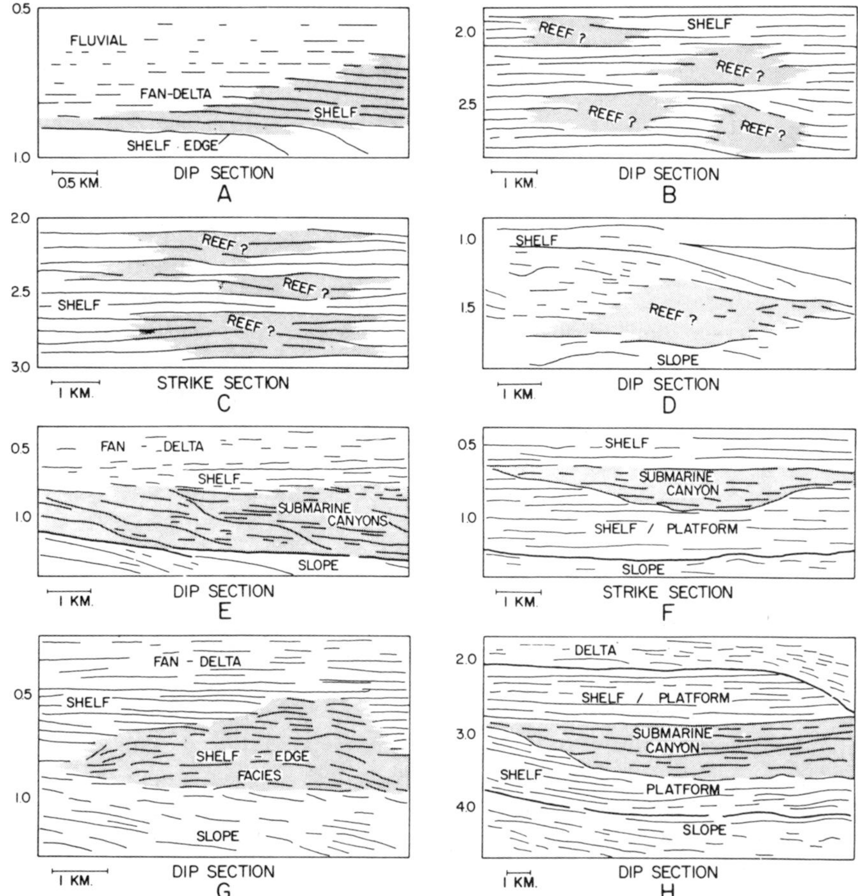

Fig. 6.19. Seismic facies patterns, showing examples of shelf and associated depositional systems from Brazil (Brown and Fisher, 1977).

and, in part, independently. Blatt et al. (1980) suggested a subdivision of carbonate platforms into two types: "The first is attached to such land masses as south Florida and the south coast of the Persian Gulf. The second is isolated shallow-water banks that rise from oceanic depths like the Bahama Platform or the coral atolls of the Pacific." The most important implication of such a simple classification relates to facies patterns. Attached platforms will show lateral gradations into clastic or evaporite facies on one side, whereas isolated platforms will be surrounded by deeper-water facies.

Krebs and Mountjoy (1972) offered a classification of Devonian carbonate buildups based on some generalizations about structural setting. They erected five groups.

1. buildups within a basin or geosyncline
2. buildups at a shelf or platform margin
3. buildups within a shelf or platform
4. buildups fringing a landmass
5. widespread biostromes forming shelves or platforms.

Subsidence rates and water circulation patterns depend strongly on structural setting and in turn control thickness, configuration and content of carbonate depositional systems. Wilson (1975) attempted to generalize the classification to all carbonate rocks, but this is difficult to do, in part because the terms geosyncline, shelf and platform are not adequately defined in the context of carbonate rocks.

In the discussion below some generalizations about carbonate depositional patterns are presented, based on the detailed analyses synthesized by Wilson (1975) and on seismic facies patterns. Some examples of carbonate systems are then given, following the Krebs and Mountjoy (1972) and Wilson (1975) classification scheme. These represent various combinations and configurations of Wilson's (1975) nine standard facies belts (Fig. 4.62).

Under optimum conditions calcium carbonate accumulation can be extremely rapid (exceeding 1 m/1000 a, according to Wilson and D'Argenio, 1982). Such conditions include warm, clear, shallow water of normal marine salinity. Unless subsidence is equally rapid and keeps pace with sedimentation a carbonate platform will rapidly become aggraded to sea level. However, because of organic binding and submarine cementation carbonates can build steep depositional slopes, and this allows them to prograde seaward rapidly. Figure 6.20 shows a common situation—a shelf undergoing differential subsidence, such as at a continental margin or on one side of an intracratonic basin. Figure 6.21 illustrates variations in these growth patterns depending on relative rates of subsidence and cementation. They are very reminiscent of the seismic clinoform patterns discussed in Chapter 5 (Fig. 5.16), although the figures are based on careful log and outcrop analysis (Meissner, 1972; Wilson, 1975), not on seismic data.

Whereas the analysis of clastic systems revolves around the analysis of detrital depocenters, the crucial component of many carbonate systems is the **carbonate buildup**. Wilson (1975) defined the latter as "a body of locally formed (laterally restricted) carbonte sediment which possesses topographic relief. This is a general and useful term because it carries no inference about internal composition." Buildups include offshore banks or barriers and shelf margin buildups, which may or may not be predominantly organic in origin. Also included are the various kinds of organically constructed wave-resistant features termed **reefs**. Figure 6.22 illustrates some of the carbonate buildups most easily recognized from seismic interpretation, and Figure 6.23 shows some typical seismic facies patterns that develop in and around buildups. Because of their usually strong velocity contrast with enclosing sediment, internally massive and reflection-free character, and presence of drape structures over the top of them, buildups are ideally suited to seismic mapping methods.

Two other types of carbonate mass are important. **Carbonate ramps** are tabular or wedge-shapped masses built on the flanks of positive areas down regional paleoslopes. Because of the occurrence of the zone of optimal carbonate production near sea level, sedimentation there is more rapid, and there is a tendency for a ramp to develop a buildup pattern such as that illustrated in Figure 6.20. Such a buildup may have a virtually flat top and is then termed a **carbonate platform** or **shelf**.

Wilson (1975) grouped shelf margins into three types, as illustrated in Figure 6.24. Variation between them probably depends on wave and current energy, nutrient supply and tectonics. Type III is favoured by initially steep (faulted) margins, rapid sedimentation and upwelling cur-

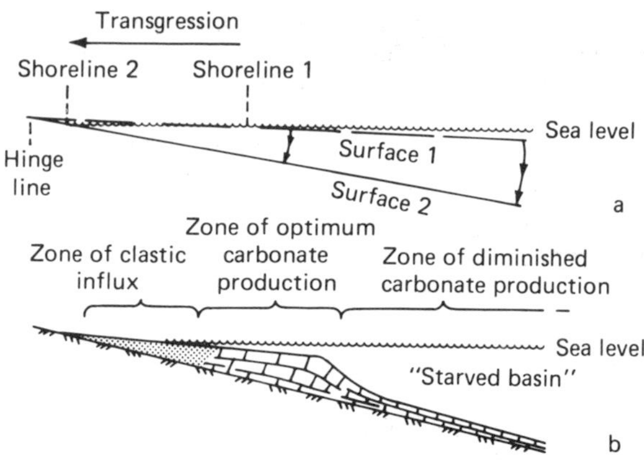

Fig. 6.20. Generation of shelf to basin topography by hinging shelf margin. Diagram below shows the generation of a thick shelf sequence in the zone of optimum carbonate production (Meissner, 1972).

Fig. 6.23. Typical seismic reflection patterns of carbonate buildups (Bubb and Hatlelid, 1977).

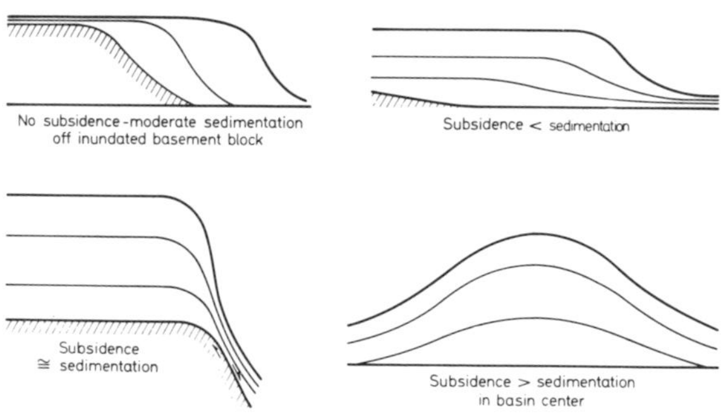

Fig. 6.21. Stratigraphic profiles in carbonate depositional systems reflecting variations in rates of subsidence and sedimentation (Wilson, 1975).

rents bringing in abundant nutrients for organic framework growth. Type II reflects lesser wave or current energy, an environment less favorable for framework building. Water movement is inadequate to remove much of the organic debris, which accumulates as lime sand. Type I represents the lowest energy environment, and includes significant accumulations of lime mud.

A few examples of the various types of carbonate platform and buildup are now described and illustrated. Firstly there are the vast platform carbonate blankets which developed during times of high global sea level, when large areas of the world's continents were covered by epeiric seas. Good examples are the Ordovician of North America (Figs. 6.25, 6.26; after Clark and

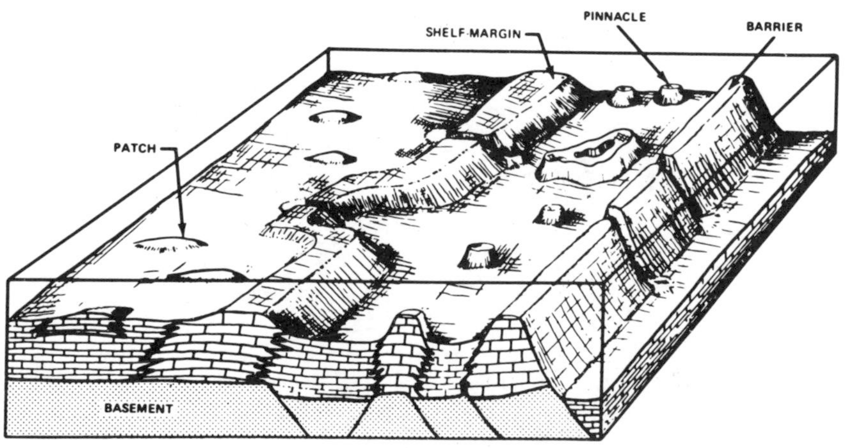

Fig. 6.22. Principal types of carbonate buildup recognizable from seismic data (Bubb and Hatlelid, 1977).

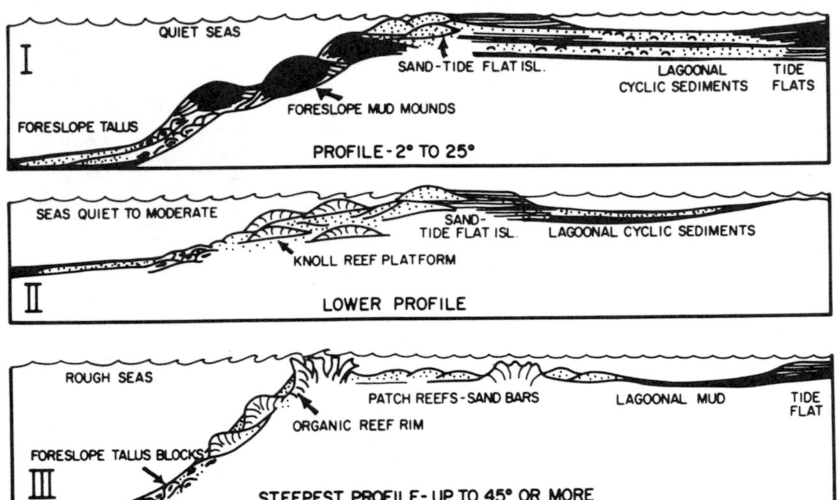

Fig. 6.24. Three types of carbonate shelf margin; I = lime-mud accumulations; II = knoll reef ramp or platform; III = organic reef rim (Wilson, 1975).

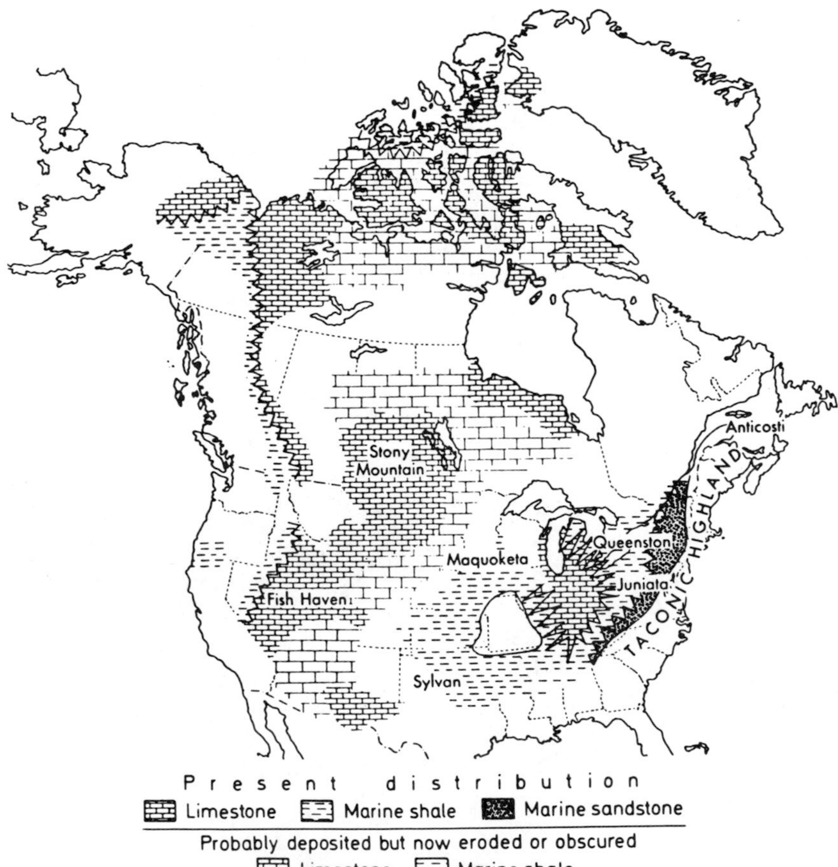

Fig. 6.25. Distribution and lithofacies of Upper Ordovician strata in North America (Clark and Stearn, 1968, © John Wiley & Sons, Ltd. Reprinted with permission).

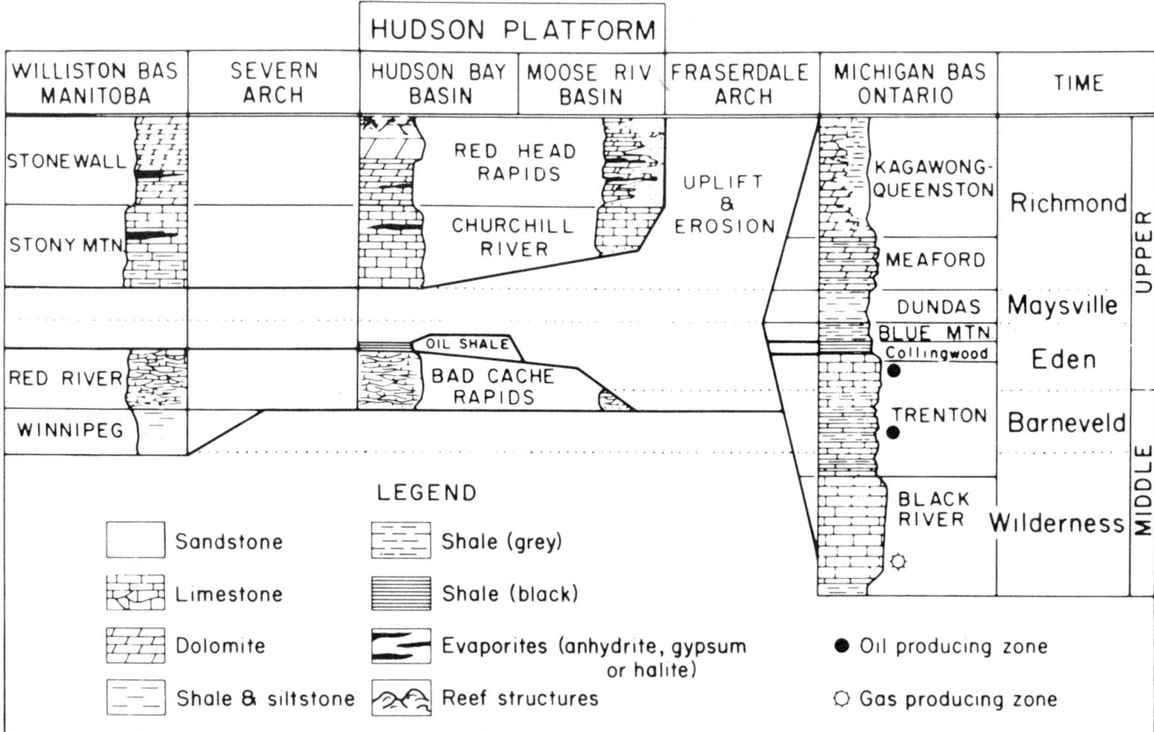

Fig. 6.26. Correlation of Ordovician strata across the northern interior of North America (Sanford and Norris, 1973).

Stearn, 1968; Sanford and Norris, 1973) and the Cretaceous Chalk of Western Europe. Lithostratigraphic correlations in some units can be traced for hundreds of kilometers (Figs. 6.26, 8.2) and formed the basis for the early recognition of regional stratigraphic sequences, as discussed in detail in Chapter 8.

Particularly thick platform and buildup deposits occur at the margins of cratons and in what were once termed miogeosynclines. Most of these would now be classified as divergent continental margins (Chap. 9). Modern examples include the Bahamas Bank and Yucatán; an excellent ancient example is the Upper Devonian platform and reef complex of Alberta (Figs. 6.27, 6.28; after Toomey et al., 1970; type 1 of Krebs and Mountjoy, 1972).

A common location for carbonate buildups is at the outer edges of major platforms or ramps, such as at the edges of cratonic blocks (type 2 of Krebs and Mountjoy, 1972). These include linear barrier systems, for example the Great Barrier Reef of Australia and Florida Keys. Ancient examples are the fringing reef around the Peace River cratonic uplift, Alberta (Fig. 6.27) and the famous Permian Reef Complex of the Delaware Basin, Texas (Fig. 6.29). The latter is a good example of carbonate clinoform development, with depositional dips up to 30°, depositional relief of up to 700 m, and seaward progradation of the shelf in the order of 20 km. The precise nature of the reef is still being debated, although this is a classic example of a carbonate buildup (Newell et al., 1953; Wilson, 1975).

The Portland Group (Upper Jurassic) of southern Britain is a good example of a carbonate ramp developed at the margin of a cratonic uplift or "swell". The succession is unusual in that it contains no significant buildups, only algal (stromatolitic) mounds or ridges in a lagoonal setting and oolite shoals (Fig. 6.30; after Townson, 1975).

Small carbonate buildups occur within platforms or cratonic basins in which moderate subsidence takes place (types 3 and 5 of Krebs and Mountjoy, 1972). There are very few modern examples of this setting, except perhaps parts of the Persian Gulf. However, ancient examples are numerous, and include Silurian and Devonian

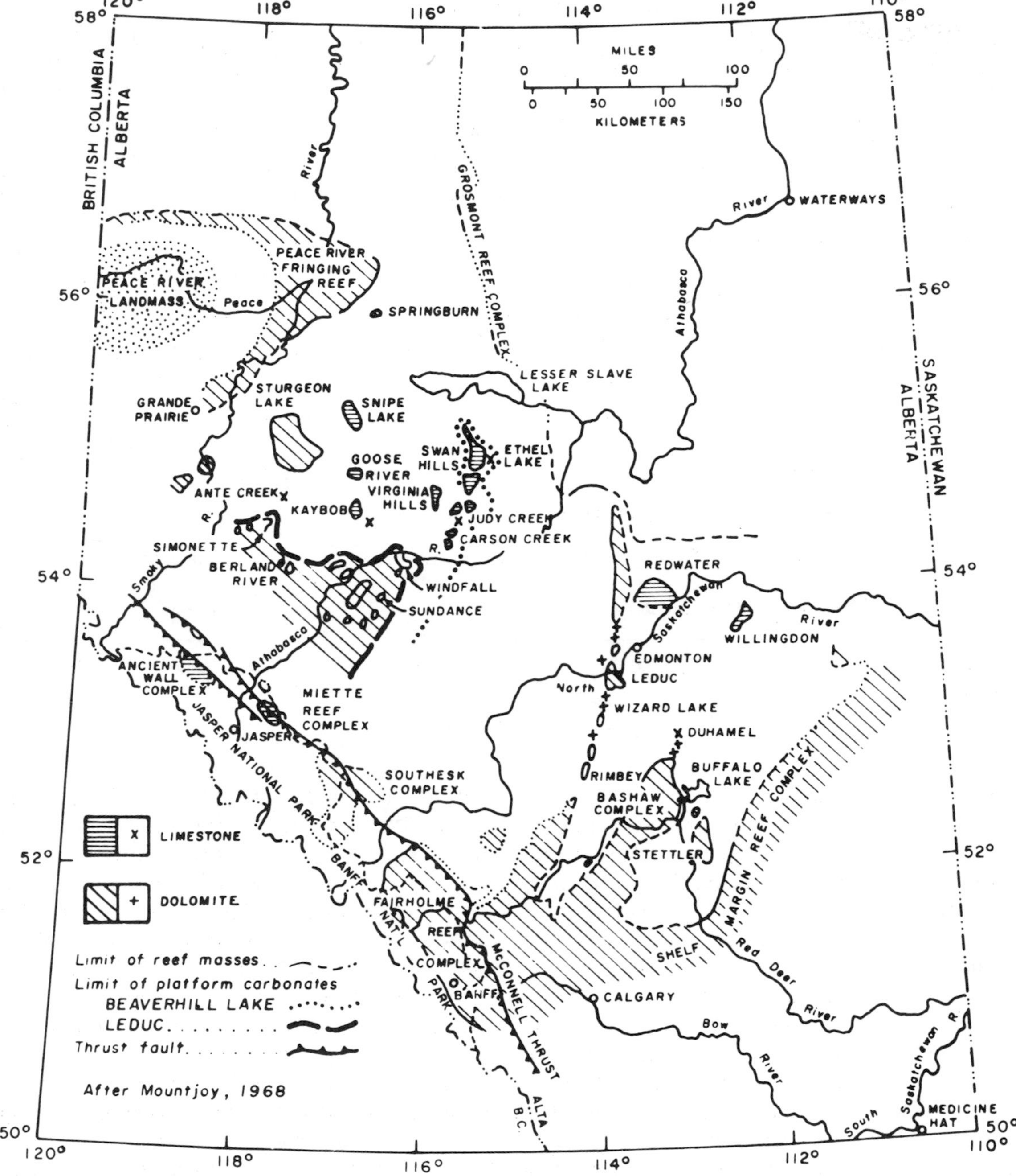

Fig. 6.27. Distribution of Upper Devonian carbonate buildups in Alberta (Toomey et al., 1970).

patch reefs and Mississippian Waulsortian mud mounds (Wilson, 1975).

It has become clear in recent years that major carbonate depositional systems hundreds of meters thick occur in deep water, on continental slopes and in oceanic basins. Some details of the depositional processes and products are described briefly in section 4.6.8. Most of the deposits are clastic, in the sense that they consist of shelf-derived carbonate debris moved downslope by slumps or sediment gravity flows, and commonly are winnowed by currents. Deep water bioherms (lithoherms) formed by in-place organic growth and cementation at depths of 600–700 m also

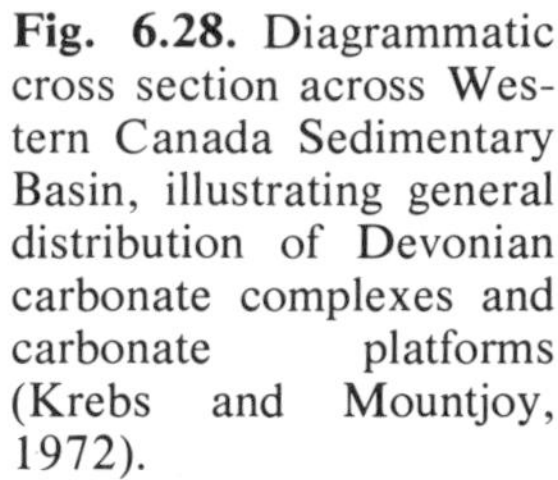

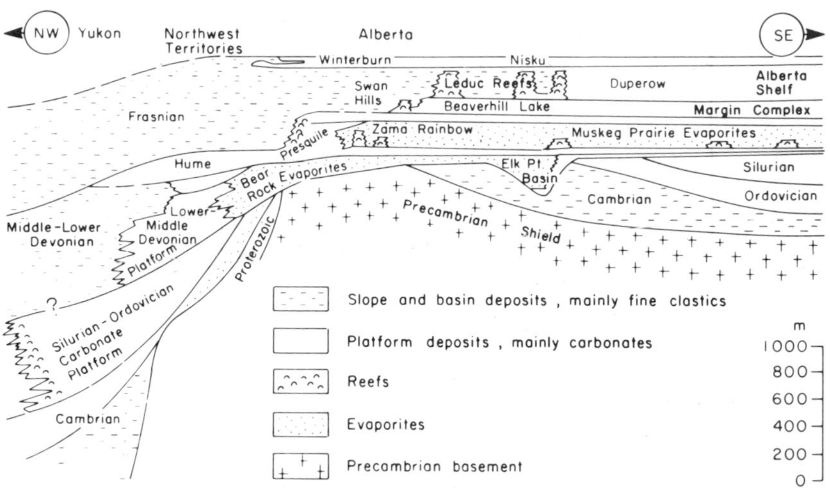

Fig. 6.28. Diagrammatic cross section across Western Canada Sedimentary Basin, illustrating general distribution of Devonian carbonate complexes and carbonate platforms (Krebs and Mountjoy, 1972).

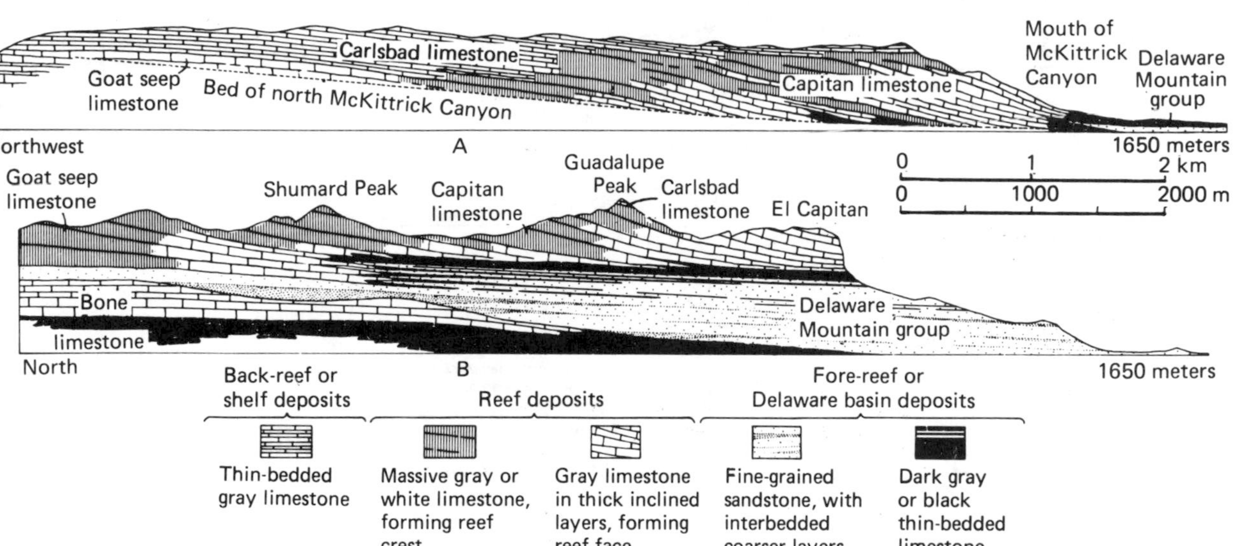

Fig. 6.29. Two sections through the Permian reef complex of Guadalupe Mountains, West Texas (King, 1948).

occur on slopes off the Bahamas. Mullins and Neumann (1979) erected seven depositional system models for carbonate slopes, based on shallow seismic and coring work off the Bahamas (Fig. 6.31). Ancient examples of several of these types are described in Cook and Enos (1977).

Evaporite deposits form in numerous small-scale, shallow water or continental environments such as playa lakes and coastal sabkhas (Kendall, 1979a). However, there are many examples of giant evaporite depositional systems which occur in large cratonic basins and small ocean basins. Examples are the Permian Zechstein of Poland, Germany, Denmark and the North Sea area (Fig. 6.32) and the Middle Devonian Elk Point Basin of western Canada (Fig. 6.33), both of which occur in cratonic settings; the Messinian (Miocene) of the Mediterranean, and Cretaceous evaporites of Brazil and West Africa, which developed during the early stages of Atlantic seafloor spreading.

There has been a great deal of debate about the origin of these saline giants, whether they represent sabkha evaporites on a large scale or deep-water deposits. Excellent discussions of this problem are given by Till (1978) and Kendall

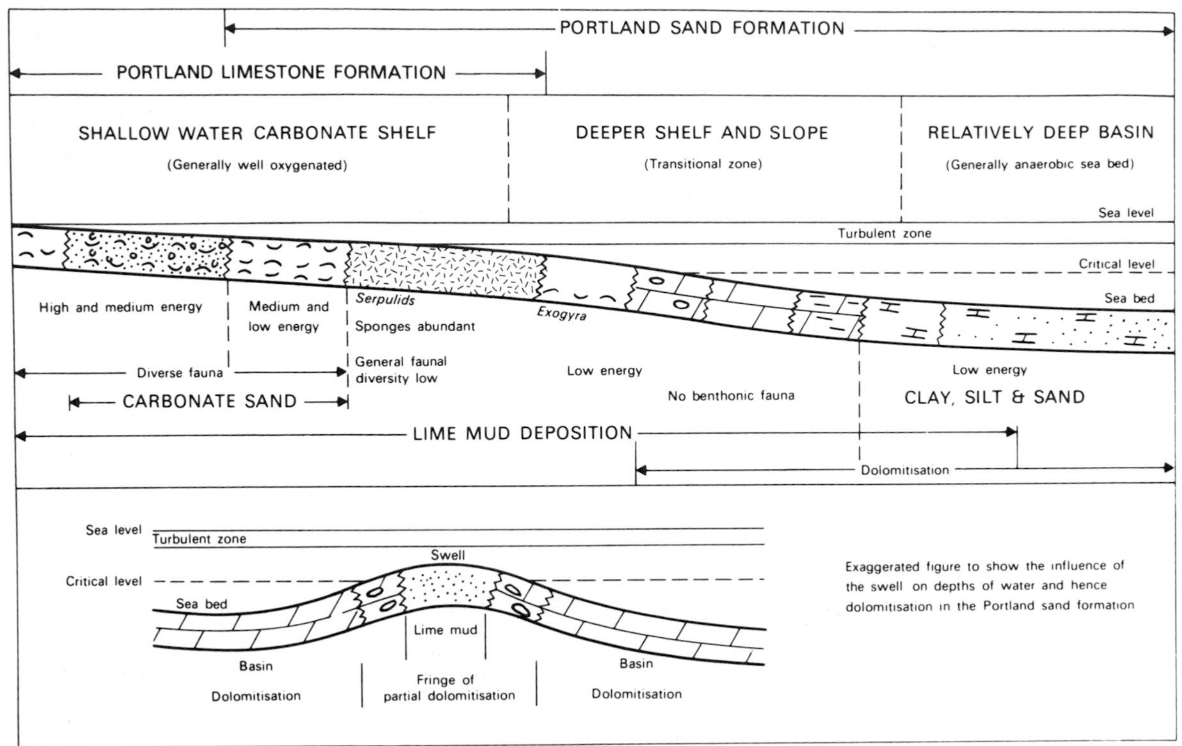

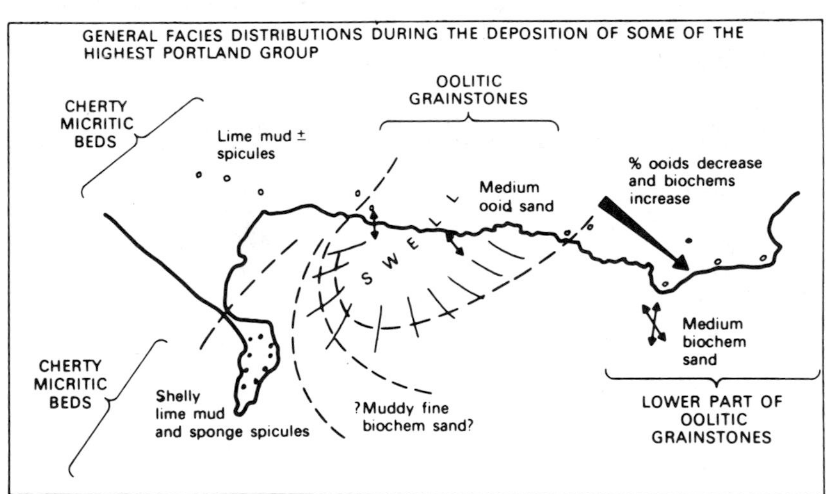

Fig. 6.30. Environments and facies distribution of the Portland Group, southern Britain (Townson, 1975; Reading, 1978).

(1979b). At present the evidence seems to point to a similar origin for most, if not all these deposits. Initially deep topographic basins contained deep-water masses of normal salinity. These were cut off from oceanic circulation by elevation of a tectonic land barrier, growth of a reef barrier, eustatic lowering of sea level, or a combination of these processes. Under appropriate climatic conditions evaporation would exceed riverine and reduced marine input, and the basin would become completely desiccated. This process explains the presence of both quiet (deep?) water laminated evaporites, some of which can be correlated over considerable distances (see Fig. 3.5), and a range of shallow-water and subaerial, sabkha-like deposits (Fig. 4.60) all in the same basin. The presence of reefs and evaporites in the same basin is readily explained by this hypothesis. Reefs grow during periods of deeper water and normal salinity, and are exposed to erosion and karstification during phases of evaporitive drawn-down of the water level. This process may

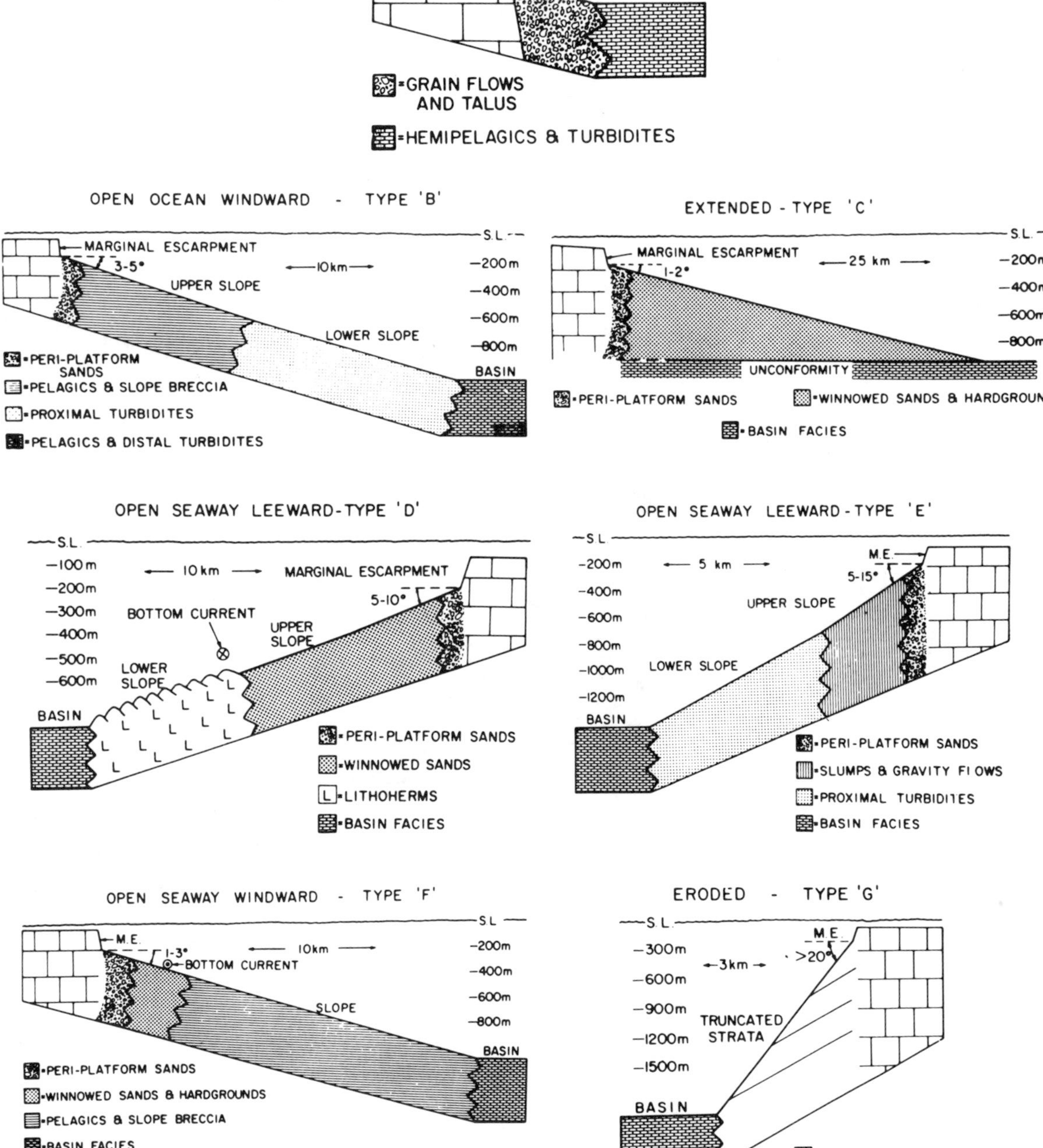

Fig. 6.31. Models of carbonate slope depositional systems, based on the northern Bahamas (Mullins and Neumann, 1979).

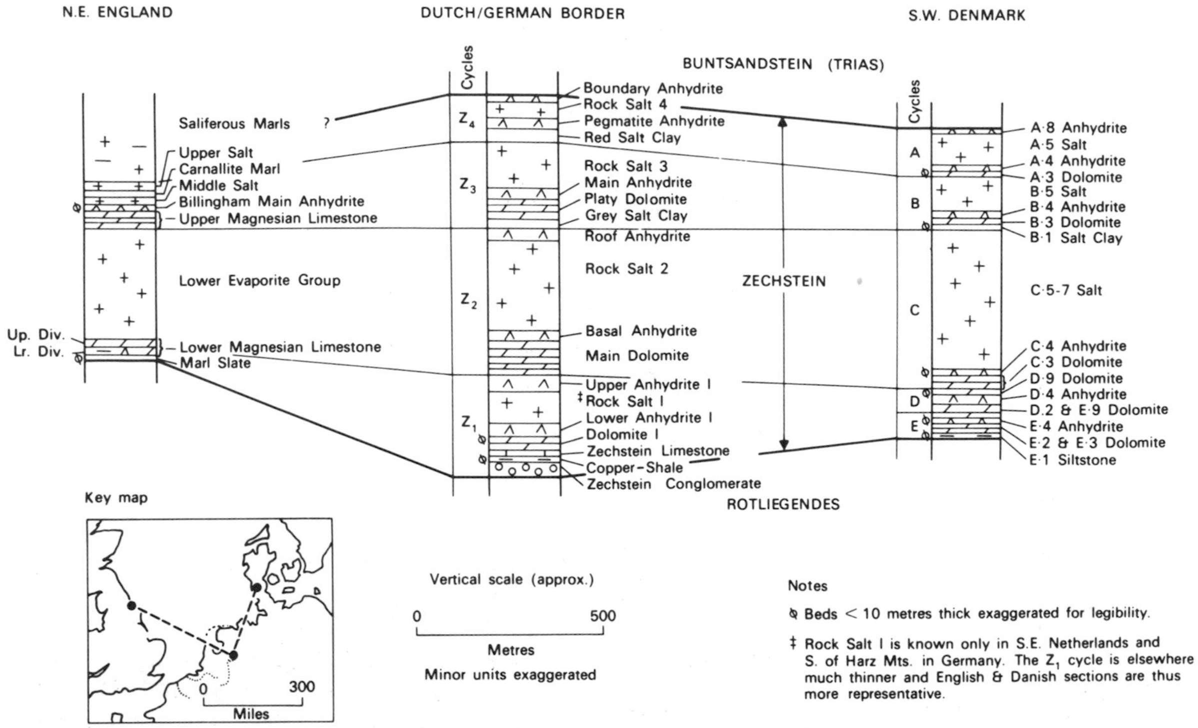

N.E. ENGLAND

DUTCH/GERMAN BORDER

S.W. DENMARK

BUNTSANDSTEIN (TRIAS)

Cycles

Saliferous Marls ?
Upper Salt
Carnallite Marl
Middle Salt
Billingham Main Anhydrite
Upper Magnesian Limestone
Lower Evaporite Group
Up. Div.
Lr. Div.
Lower Magnesian Limestone
Marl Slate

Z_4
Z_3
Z_2
Z_1
Boundary Anhydrite
Rock Salt 4
Pegmatite Anhydrite
Red Salt Clay
Rock Salt 3
Main Anhydrite
Platy Dolomite
Grey Salt Clay
Roof Anhydrite
Rock Salt 2
Basal Anhydrite
Main Dolomite
Upper Anhydrite I
‡Rock Salt I
Lower Anhydrite I
Dolomite I
Zechstein Limestone
Copper-Shale
Zechstein Conglomerate

ZECHSTEIN

ROTLIEGENDES

A
B
C
D
E
A·8 Anhydrite
A·5 Salt
A·4 Anhydrite
A·3 Dolomite
B·5 Salt
B·4 Anhydrite
B·3 Dolomite
B·1 Salt Clay
C·5-7 Salt
C·4 Anhydrite
C·3 Dolomite
D·9 Dolomite
D·4 Anhydrite
D.2 & E·9 Dolomite
E·4 Anhydrite
E·2 & E·3 Dolomite
E·1 Siltstone

Key map

0 300
Miles

Vertical scale (approx.)

0 500

Metres

Minor units exaggerated

Notes

Beds < 10 metres thick exaggerated for legibility.

‡ Rock Salt I is known only in S.E. Netherlands and S. of Harz Mts. in Germany. The Z_1 cycle is elsewhere much thinner and English & Danish sections are thus more representative.

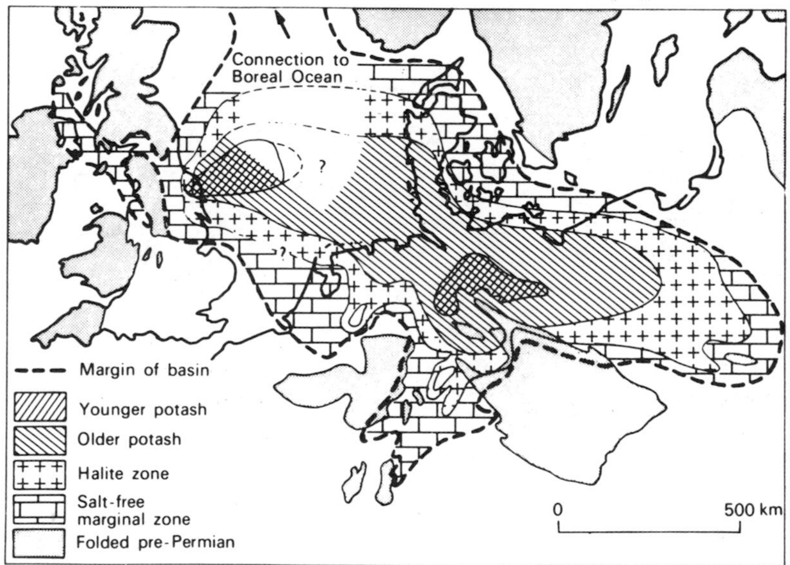

Fig. 6.32. Lithofacies distribution in the Zechstein Basin of western Europe (Upper Permian), and a cross section through the deposits (Borchert and Muir, 1964; Brunstrom and Walmsley, 1969; Reading, 1978).

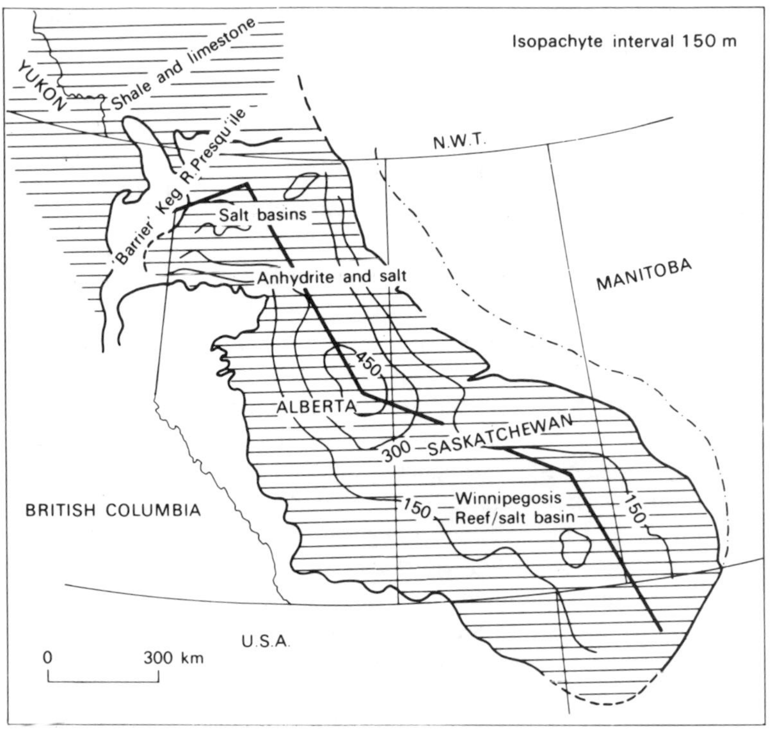

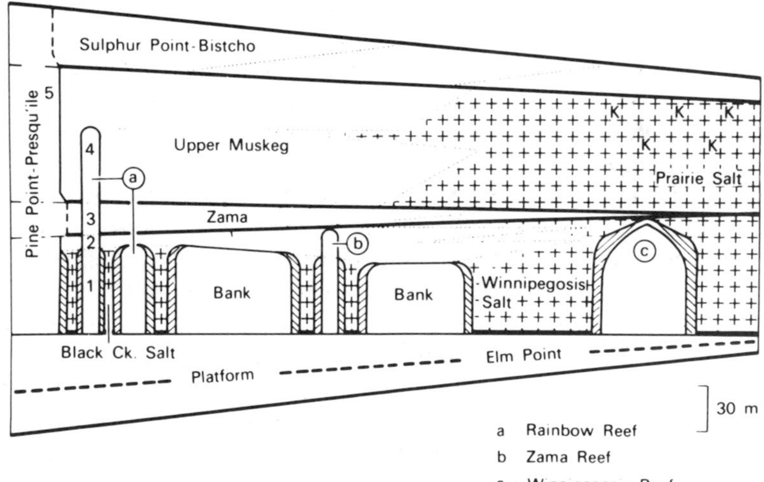

Fig. 6.33. The Upper Elk Point Basin of western Canada (Middle Devonian) and a diagrammatic cross section. Anhydrite flanks reefs and bank edges (oblique hatching), halite is shown by crosses (Fuller and Porter, 1969; Reading 1978).

be repeated several or many times as sea levels fluctuate (see also section 4.6.8 and discussion of Figs. 8.22 and 8.23 in Chap. 8).

6.7 Continental slope, rise and basin clastic depositional systems

During about the last twenty years the techniques of geophysics and oceanography have provided us with an immense wealth of bathymetric, seismic and core data on the world's continental margins. Emery (1980), in a major classification of margin types, listed 178 regional studies as his source material, the oldest published in 1964. Needless to say this work reveals a great variability in morphology and process, reflecting variations in margin configuration, shelf width, amount of sediment discharge from adjacent rivers, strength of oceanic currents, rate of subsidence and sea level change and, in some ways most important of all, plate tectonic environment.

Sedimentological studies of modern and ancient slope and basin deposits (summarized in section 4.6.9) have tended to concentrate on the nature of sedimentary processes (sediment gravity flows, contour currents) and the individual elements of the depositional system, particularly submarine fans. Very few studies have adopted a large-scale depositional systems approach for the study of an ancient slope-basin deposit, in which all the controls listed above have been evaluated. The debate over submarine fan facies models has perhaps obscured many fascinating broader questions about the evolution of ancient continental margin basins. A possible reason for this is that margins still undergoing sedimentation are largely innaccessible to normal sedimentological basin analysis study, whereas those margins that have been brought to the surface owe their exposure to compression and uplift in a suture zone, in which stratigraphic relationships are obscured by intense structural deformation (e.g. Lower Paleozoic of Appalachians, Innuitian Orogen and Scottish Southern Uplands; Alpine Cenozoic). What follows is therefore an attempt to evaluate slope and basin depositional systems using a mixture of modern and ancient examples. As with most marine environments seismic stratigraphy has become an invaluable tool in this process of synthesis.

The principal depositional components of continental margins are as follows:

1. submarine canyons
2. ephemeral gullies
3. continental slope
4. base-of-slope sediment apron
5. submarine fan
6. basin plain.

Environments 4 and 5 constitute the continental rise. Not all these elements are everywhere present. For example there are (as noted in section 4.6.9) margins characterized by broad shelves cut by submarine canyons (e.g. Atlantic margin of the United States) and those with narrow margins and submarine fans fed directly by major rivers (e.g. Mississippi, Amazon, Nile). Contour currents may be present and modify sediment distribution and dispersal. The margin may flank a major ocean, with one-sided sediment input (Atlantic), or a small ocean with centripetal inflow (Gulf of Mexico) or longitudinal sediment dispersal (Labrador Sea, Bay of Bengal). Rise and fall of sea level has a profound effect on margin evolution and its depositional systems.

Most of the coarse sediment (gravel, sand) which reaches the deep ocean is transported there by submarine canyons. These are major valleys up to 2000 m deep. They are eroded partly subaerially by rivers when the continental shelves are exposed during low sea level stands, and partly subaqueously by sediment gravity flows (Shepard, 1981). Most reveal a complex cut-and-fill stratigraphy on seismic records (Figs. 6.19, 6.34). The upper reaches of canyons are widened and deepened during subaerial exposure at times of low sea level, but also are eroded by submarine processes (particularly slumping) during high sea level stands. Erosion and entrenchment may occur down to the mouth of the canyon if the sediment supply is largely cut off. This may occur immediately following a rise in sea level, when the sediment supply may be trapped in deltaic sinks. Sediment onlap tends to follow this period of downcutting, as the sediment supply is reestablished. Eventually this may change to an offlap configuration, when abundant deltaic or shelf detritus is fed into the canyon by shelf progradation or by fluvial rejuvenation during the next phase of sea level lowering. McGregor (1981) showed that in Wilmington Canyon, off Dela-

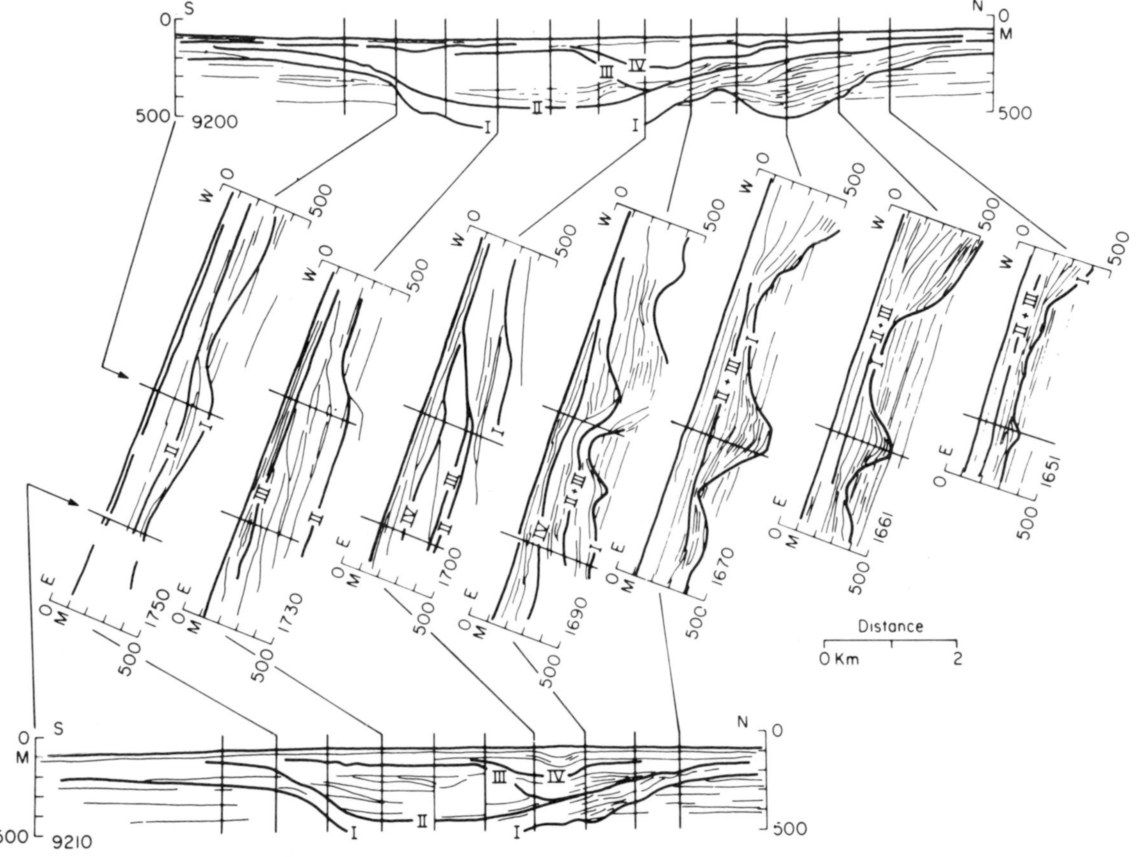

Fig. 6.34. Network of seismic cross sections through the head of Wilmington Canyon, U.S. Atlantic shelf. Lines 9200 and 9210 are oriented across the canyon, and the other sections are oriented down canyon. The canyon fill consists of a series of erosion surfaces and onlapping sediments of Miocene to Recent age (McGregor, 1981).

ware, this alternation of cut and fill has occurred at least four times since the end of the Miocene (Fig. 6.34). Ryan et al. (1978) found evidence of other phases of cut and fill extending back to the Cretaceous in New England canyons. Submarine canyons can thus be very long-lived elements of the continental margin environment, particularly on divergent margins, where they may evolve for tens of millions of years undisturbed by tectonic deformation.

The cut-and-fill stratigraphy of a canyon fill would be difficult to decipher from outcrop or well records without extremely good biostratigraphic control. This probably is why so few good descriptions of them are available in the geological literature (Picha, 1979; Von der Borch et al., 1982).

The sedimentology of submarine fans is discussed briefly in section 4.6.9 and it is concluded that fan facies models are in an active stage of evolution. Summary diagrams such as that illustrated in Figure 4.65 may be useful guides to interpreting modern and ancient fans, but work on modern fans reveals an enormous variability in size, shape, thickness, grain size, structural setting and associated processes such as slumping and bottom currents. The practical utility of any single facies model is therefore limited, and there is a danger that their uncritical use can lead to wrong interpretations. Three Cenozoic fans are illustrated to demonstrate this point.

The Nile Cone is an arcuate bulge of sediment 600 km long and 220 km wide, fed by the Nile River. The fan is fed by two major delta distributaries, one of which extends seaward into a shallow submarine canyon (Alexandria Canyon). The Cone consists of two sub-fans, Rosetta Fan and Levant Platform. These merge laterally.

Maldonado and Stanley (1979) and Stanley and Maldonado (1979), who have studied the Nile Cone extensively using seismic profiles and 65 shallow piston cores, have mapped lithofacies patterns and erected a near-surface stratigraphy. Pliocene to Recent sediments reach 3000 m in the cone. Seismic profiles (Fig. 6.35) show the depositional topography, presence of active faults, and the effects of salt tectonics (uplap reflection terminations) in the Levant Platform. Eight lithofacies types have been recognized; their distribution is not particularly reminiscent of any published fan model (Fig. 6.36). Sand constitutes only 5% of the section, the remainder being silt and mud. Well-sorted sands of terrigenous origin occur in beds up to 9 m thick, whereas more immature sands form thin-bedded turbidite sequences up to 50 cm thick. Both lithofacies are concentrated in a channelized region in the middle and outer part of the fan. No suprafan lobes are present. Sedimentation rates (calculated from radiocarbon dates) locally exceed 30 cm/1000 a.

The Frigg Fan is part of an early Tertiary succession that prograded into the Viking Graben of the North Sea during a period of intermittent fault movement. The fan is a stratigraphic trap for a giant gas field and has therefore been extensively studied (Heritier et al., 1980). Figure 6.37 is a structure contour map of the main fan, based on seismic data. The detail evident in this map attests to the remarkably refined interpretations that can now be made from high-quality seismic records. Note that the fan is not fan-shaped. The lobate nature reflects the effect of basement structure and depositional bottom topography on dispersal patterns. The latter effect is illustrated in Figure 6.38. Heritier et al. (1980) found that because channel and levee deposits, being sand rich, compact less than interchannel muds, a build-up of channel deposits created a marked topographic feature within the basin. Successive sand bodies therefore tend to be offset laterally from each other. This process occurred on all scales from individual channels to entire fan lobes, and is analogous to the distributary and lobe switching that occurs in deltas and alluvial fans.

Lastly, Figure 6.39 illustrates the Bengal Cone, the largest submarine fan in the world and a depositional system 3000 km in length, within which up to 10 km of sediment have been deposited since the mid-Eocene (Curray and Moore, 1974). This is only termed a fan because it has a single major source (Ganges–Brahmaputra River), and is constrained between India and the Andaman–Nicobar Ridge, which imposes a fan-like outline. In other respects the Bengal Fan resembles a longitudinally filled basin such as those discussed below.

Numerous other examples of submarine fan variation and complexity could be cited; for example Graham and Bachman (1983) used seismic data to demonstrate the complex architecture of the La Jolla fan, and its dependence on syndepositional tectonism.

Submarine fans may prograde rapidly, giving rise to clinoform seismic patterns (Fig. 6.40). Onlap and offlap may occur in response to changes in sea level or sediment supply, just as occurs in submarine canyons (discussed above). Similar effects may result from lobe switching, which gives rise to hummocky, mounded or lensoid reflections in strike section (Fig. 6.38). Reflection continuity may be poor near channels, but excellent elsewhere, because of the lateral persistence of sediment gravity flows.

Submarine fans are not the only way by which continental slopes prograde. Thermal stratification of the oceans can be disturbed by seasonal temperature changes, causing upwelling at the continental margins and, conversely, spillover whereby dense, sediment-laden shelf waters flow over the shelf-slope break (Southard and Stanley, 1976; Bouma, 1979). Small gullies may channel some of this movement. As discussed in section 4.6.8 continental slopes beneath carbonate banks are particularly active because of their rapid seaward progradation. Present-day continental slopes contain evidence of numerous periods of erosion, reflecting changes in sea level. Some of the slope sediments may be deltaic sediments dumped at the shelf edge during periods of low sea level. However, Doyle et al. (1979) were able to demonstrate that on the U.S. Atlantic margin the slope receives large quantities of silt and mud, and some sand, by hemipelagic settling related to the shelf-slope break spillover process. Intercanyon sedimentation rates may exceed 20 cm/1000 a. Figure 6.41 is a typical seismic section across the outer continental shelf and slope off Georgia. It shows the complex Cretaceous–

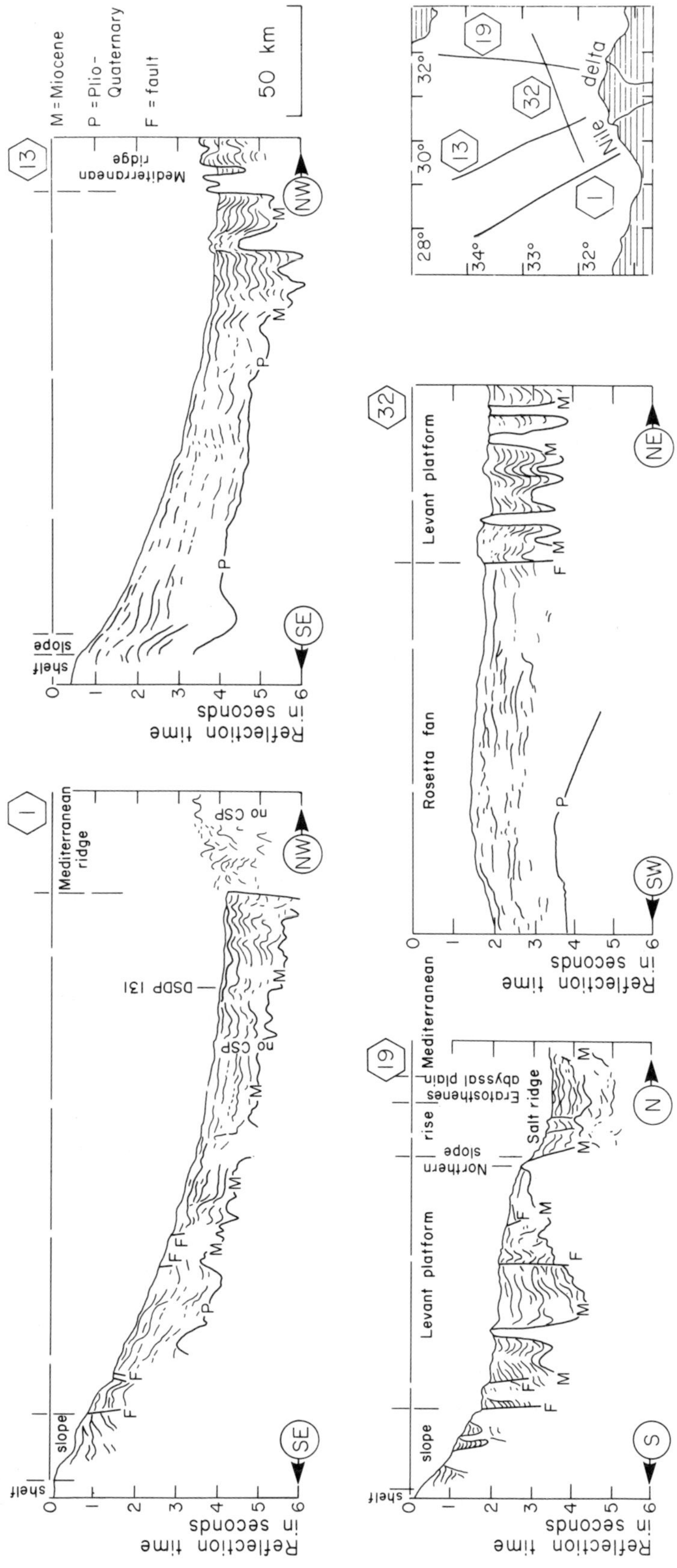

Fig. 6.35. Seismic profiles across the Nile Cone. Lines 1 and 13 traverse the Rosetta Fan, and line 19 the Levant Platform. This line and the east end of line 32 show the effects of salt tectonics in the Levant Platform area (Maldonado and Stanley, 1979).

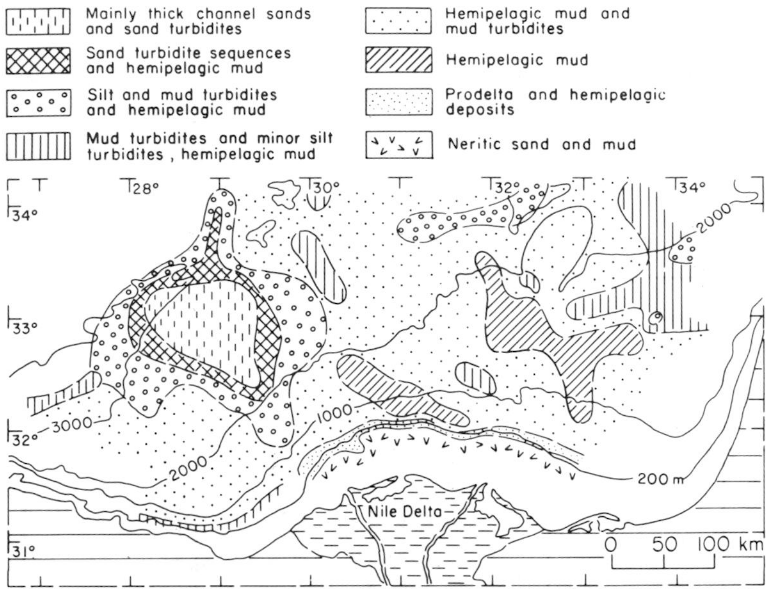

Fig. 6.36. Lithofacies distribution on Nile Cone (Maldonado and Stanley, 1979).

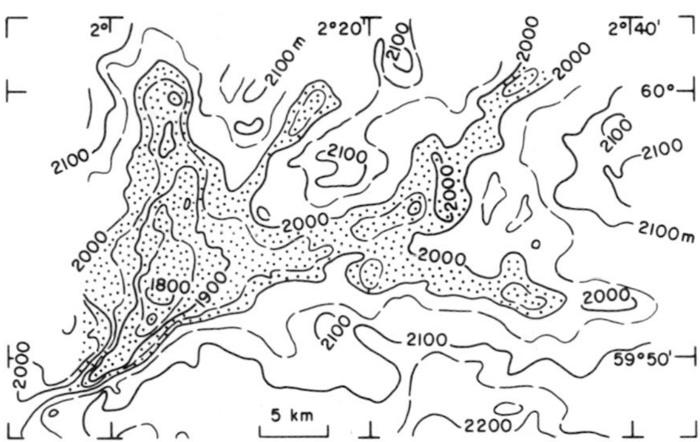

Fig. 6.37. Seismic structure map drawn on top of Frigg sand, Frigg Field, North Sea. The distribution of the main fan sand is shown by stippling (Heritier et al., 1980).

Cenozoic stratigraphy, including the prograding clinoforms of the fine-grained slope deposits (Buffler et al., 1979).

Slumps, slides, olisthostromes and debris flows are an integral part of continental slopes. Slumped sediment masses may have volumes of several cubic kilometers (Rupke, 1976) and their movement may set off major sediment gravity flows, such as the famous turbidity current generated by the 1929 Grand Banks earthquake (Heezen and Hollister, 1971). Embley (1976) reported a slide off the Spanish Sahara that displaced more than 600 km^3 of sediment on a slope of less than 1.5°. The debris flows, consisting of angular pebbly fragments in a mud matrix, traveled for 500 km and cover an area of 30,000 km^2 (Fig. 6.42). Such deposits are common on the shelf and slope off Antarctica, where they represent reworked till (Kurtz and Anderson, 1979).

Contourites may form thick sequences on some continental slopes. Bein and Weiler (1976) described a Cretaceous unit in Israel that is up to 3000 m thick. Petrographic analysis revealed that about 70% of the detritus is carbonate material derived from an adjacent shelf. It was transported down the continental slope by shelf spillover and turbidity currents, but evidence such as that discussed in section 4.6.9 suggests that the final transportation and sedimentation of the detritus was by contour currents.

On many divergent continental margins sedimentation is markedly affected by salt tectonics, and this must be taken into account in any consideration of depositional systems, because diapir movement occurs at the same time as sedimentation. The association of evaporites with divergent margins is explained in Chapter 9. Diapirs are abundant and have been particularly well studied in the Gulf of Mexico (Worzel and Burk, 1979; Stuart and Caughey, 1977). There the salt layer is Upper Jurassic. Movement of the diapirs took place at least from latest Cretaceous time as indicated by thinning of layers adjacent to the diapirs and change of facies from turbidite silts and muds to pelagic deposits in flanking and

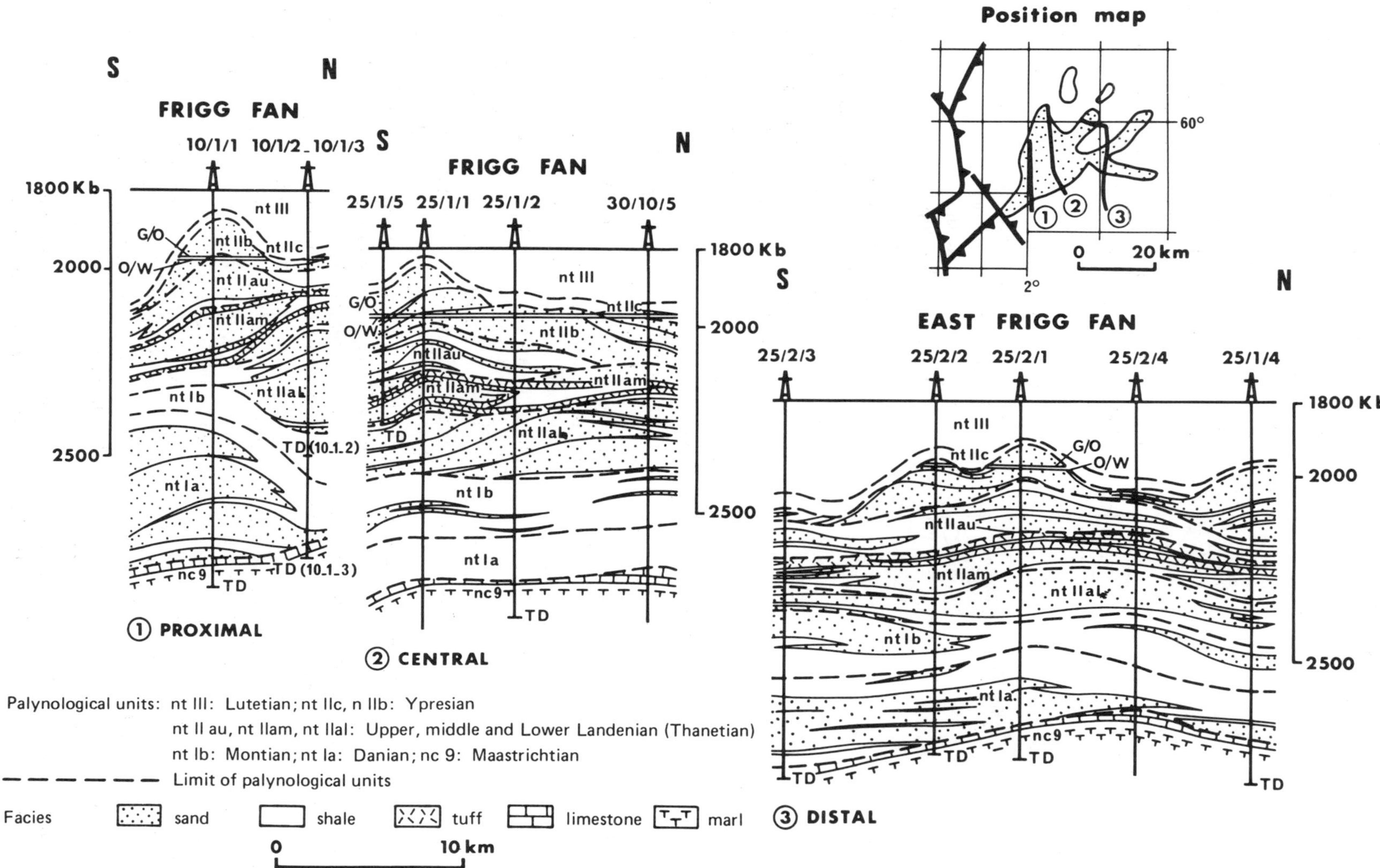

Fig. 6.38. Cross sections through Frigg fan showing deposition of thick sand lobes and channel fill sequences on flanks of earlier depositional lobes (Heritier et al., 1980).

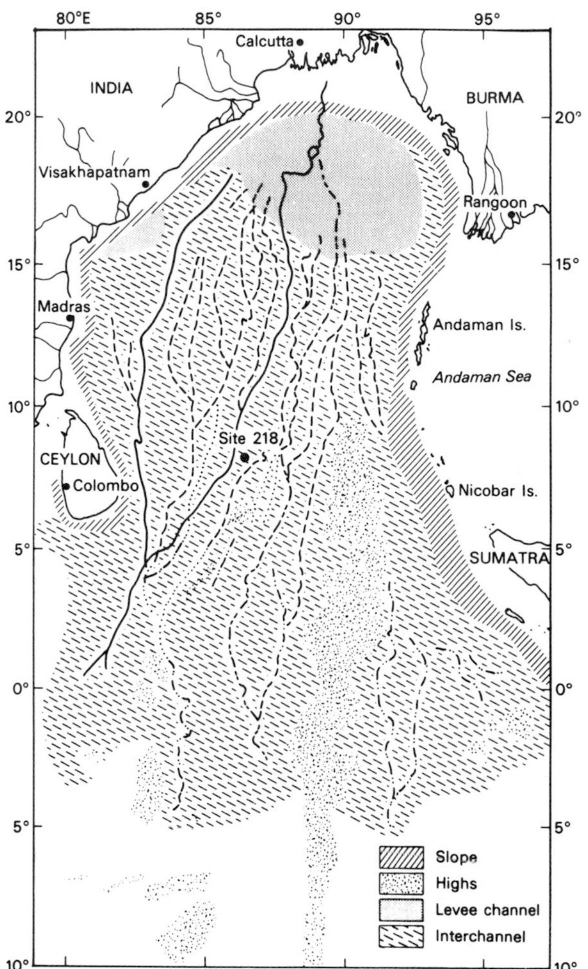

Fig. 6.39. The Bengal Cone, a giant submarine fan (Curray and Moore, 1974; Reading, 1978).

overlying beds. This demonstrates that the diapirs formed elevated areas on the sea floor, as they still do at the present day (e.g. Challenger Knoll).

Although many ancient continental margin deposits have been described, most clastic sequences are interpreted in terms of the submarine fan model. Undoubtedly there are many intercanyon slope sequences wating to be described—all that is needed is a suitably attractive facies model. This constrasts with the more extensive literature now available on carbonate slopes (section 6.6). Galloway and Brown (1973) described a Pennsylvanian shelf-slope system in which laterally coalesced fans were deposited from sediment "funneled through the shelf-edge bank complex in one or more restricted channels" and dumped in slope troughs. A cross section through this deposit is given in Figure 6.18.

It remains to describe the basin plain environment. Excellent summaries of depositional processes are given by Jenkyns (1978), Rupke (1978) and Pilkey et al. (1980), and will not be repeated here. Basin plains are floored by oceanic crust and are thus not likely to be exposed on continents unless raised tectonically during plate suturing. Some well-exposed Cenozoic deposits in the Alpine region are of this type (e.g. Hesse, 1974, 1975; Ricci-Lucchi and Valmori, 1980) but, in most cases, considerable tectonic deformation is to be expected. Basin plain depositional systems are characterized by sheet-like turbidite deposits and pelagic and hemipelagic muds. The deposits blanket preexisting topography, such as fault blocks, and develop extremely flat basin floors. Individual turbidite units may be traceable for up to 500 km, as determined by Pilkey et al. (1980) in the modern Hatteras Basin. Hesse (1974) traced units for 200 km in the Cretaceous of the Alps. Ricci-Lucchi and Valmori (1980) traced gradual facies changes in a Miocene basin over a distance of 120 km (Fig. 3.6). A suite of typical basin plain turbidites is illustrated in Figure 6.43.

The thickness, extent and paleocurrent patterns of basin plain deposits depend largely on tectonic setting (Pilkey et al., 1980). Most such plains are fed by submarine canyons or by turbidity currents extending beyond the outer fringes of submarine fans. Thicknesses of up to 2 km have been recorded in modern basins such as the western Atlantic (Collette et al., 1969), but where the sediment is ponded by trenches, volcanic arcs or faults an oceanic basin may be starved and receive little but a thin blanket of pelagic sediment. The thickness and frequency of turbidite units also depends on tectonic setting. They are more numerous but relatively thin in active tectonic areas such as convergent margins or wrench-fault basins, where triggering earthquakes are more frequent (Pilkey et al., 1980). Typical basin plain seismic facies are parallel bedded, with excellent reflection continuity. Reflection amplitude varies from strong to weak, depending on the abundance of good turbidite sand–silt reflectors (Stuart and Caughey, 1977). Paleocurrents may be variable. They can be reversed by tilting of the basin and may reflect turbidity currents from more than one source that can flow freely in virtually any direction over the

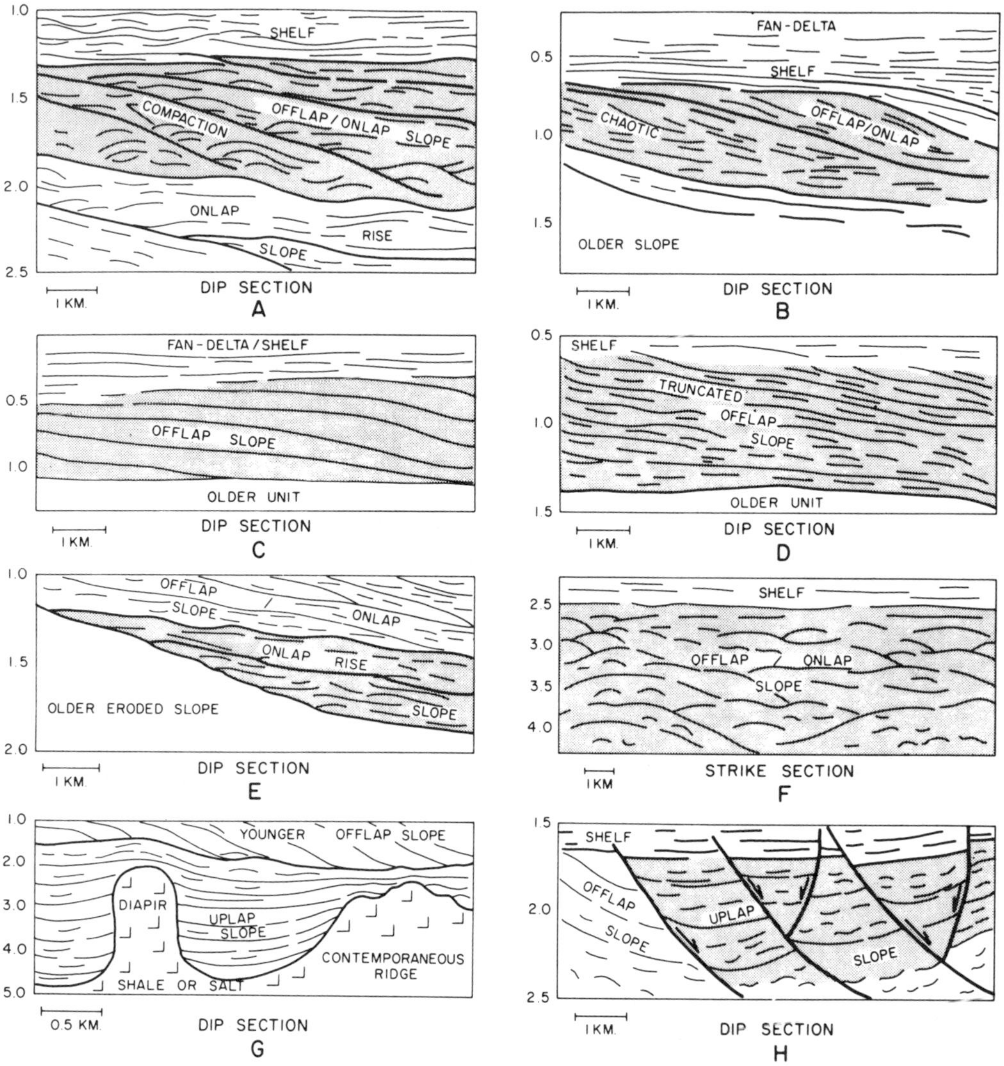

Fig. 6.40. Seismic facies patterns showing examples of continental slope and rise depositional systems (Brown and Fisher, 1977).

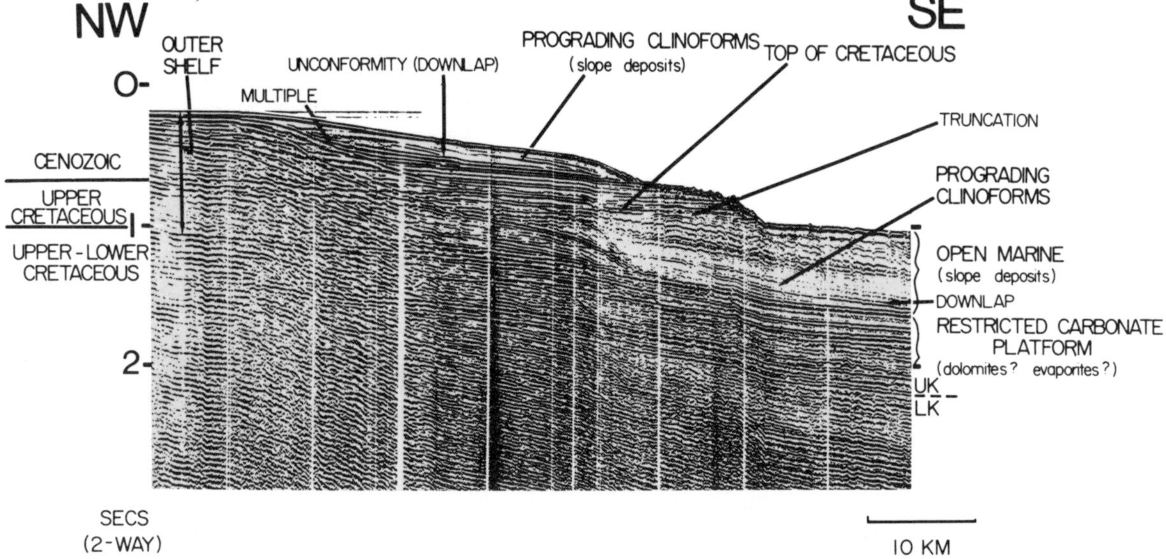

Fig. 6.41. Typical seismic section across a modern continental slope, off Georgia (Buffler et al., 1979).

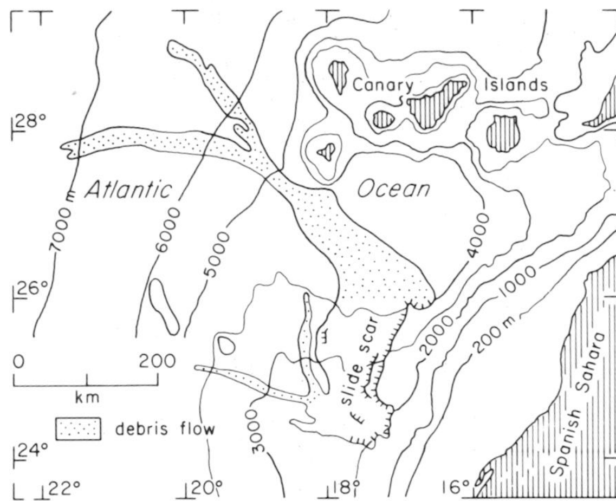

Fig. 6.42. A recent debris flow, Spanish Sahara (Embley, 1976).

nearly flat floor. Longitudinal current patterns, parallel to the continental margin, are common. Tectonic control of these variations is discussed in Chapter 9.

6.8 Depositional systems and stratigraphic nomenclature

Handford and Dutton (1980) discussed the evolution of a mixed carbonate–clastic shelf and slope succession of Pennsylvanian–Permian age in northern Texas. At the conclusion of their paper they wrote:

> The task of documenting regional Pennsylvanian and Permian depositional history of the Palo Duro basin was made easier by the lack of formal stratigraphic units because few lithologic units are laterally consistent enough for regional correlation through the basin. Hence our task was unencumbered by what otherwise could have been a column consisting of locally established stratigraphic units that offered little utility toward unraveling the depositional history of the basin.

In Chapters 1 and 3 it is suggested that the definition of formal statigraphic units should be left to the later stages of a basin analysis. In frontier basins this is simple to do, but in older, more mature or well-explored basins there will inevitably exist a suite of stratigraphic names, many of which will have been erected before the origins of the rocks were fully understood. In such cases there is a great need for a radical reappraisal of basin stratigraphy using the facies analysis and depositional system methodologies. The analyst must be clear headed and determined enough to cut through the clutter of names, retain only those lithostratigraphic and biostratigraphic correlations that are firmly based on independent evidence, and rebuild the stratigraphy from scratch. Old formation names can then be revised, if necessary, or a new nomenclature erected to reflect the improved understanding of the rocks. Two examples of this procedure are illustrated.

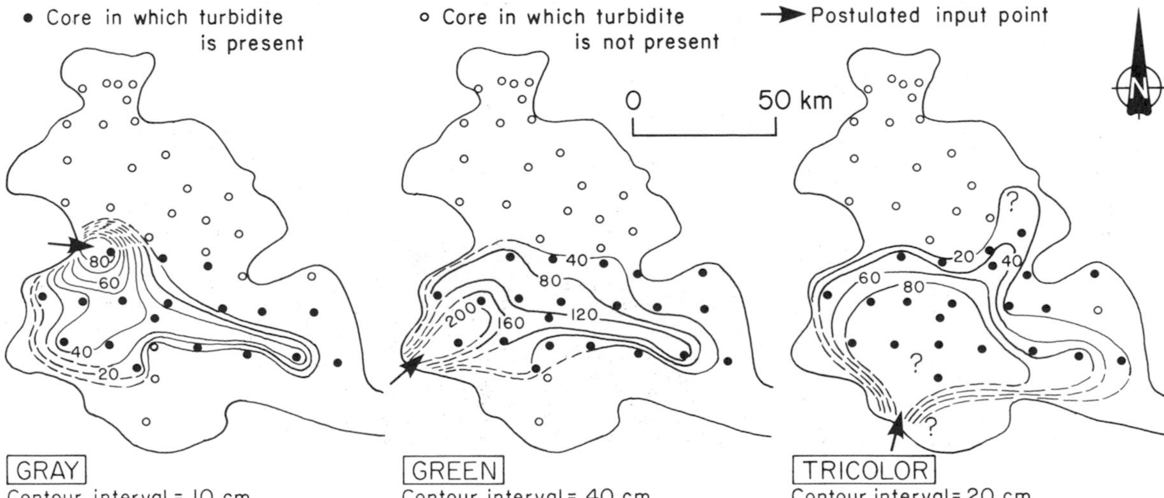

Fig. 6.43. Isopach maps of three superimposed turbidity current deposits on the floor of the Hispaniola–Caicos Basin. The middle (Green) unit has a volume of $30.8 \times 10^8 m^3$ (Bennetts and Pilkey, 1976).

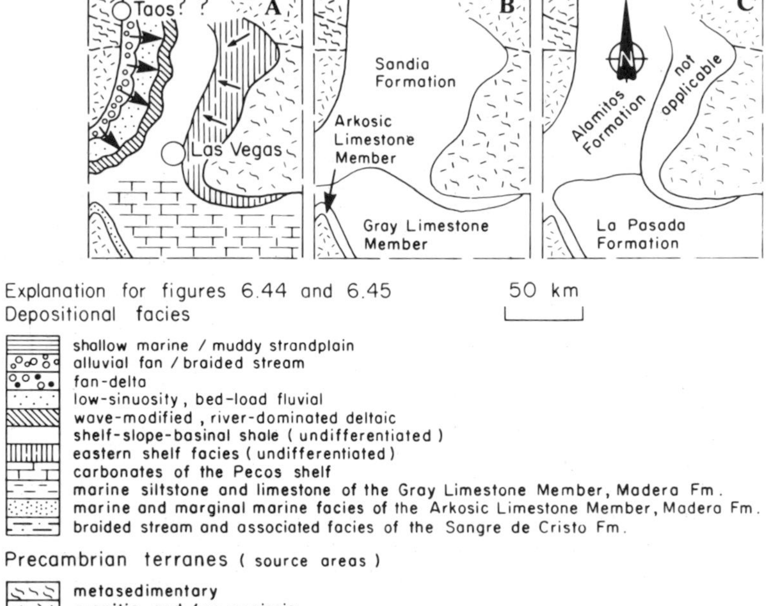

Fig. 6.44. A Depositional systems in Taos Trough, New Mexico, in Middle Desmoinesian time; **B** and **C** two systems of stratigraphic nomenclature that have been applied to the same rocks (Casey, 1980).

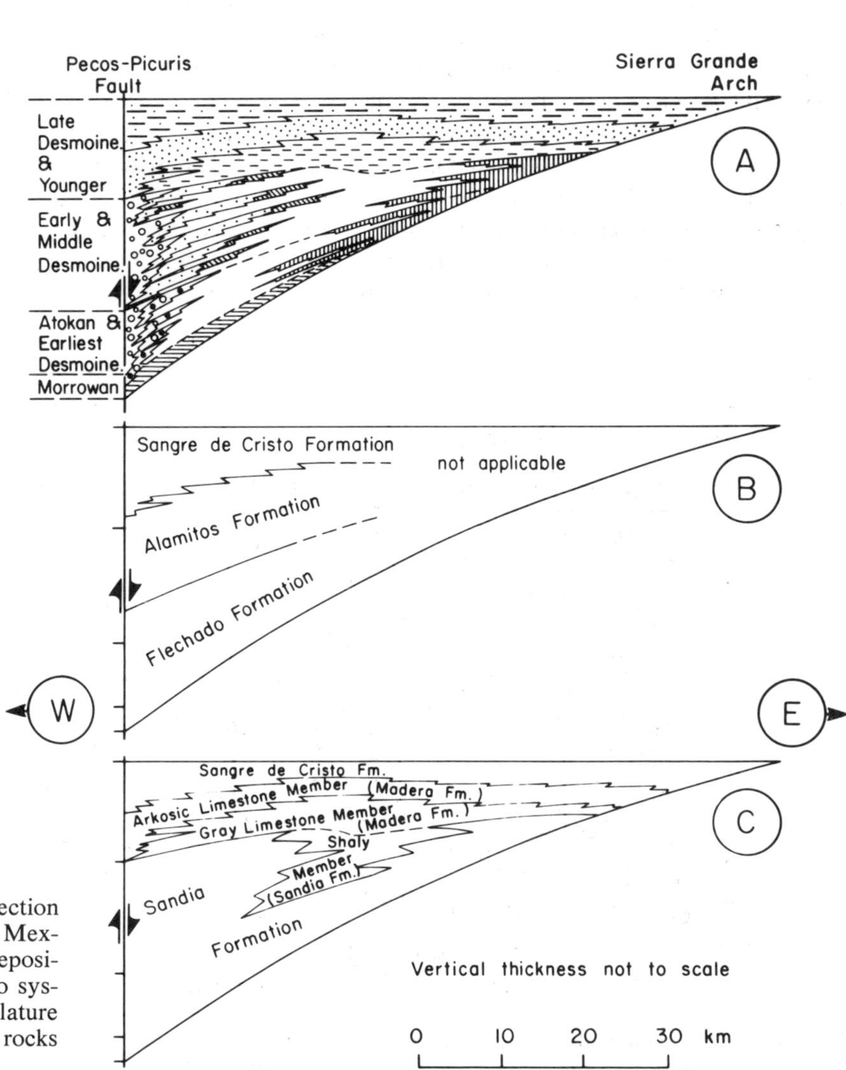

Fig. 6.45. A Cross section through Taos Trough, New Mexico, showing evolution of depositional systems; **B** and **C** two systems of stratigraphic nomenclature that have been applied to the rocks (Casey, 1980).

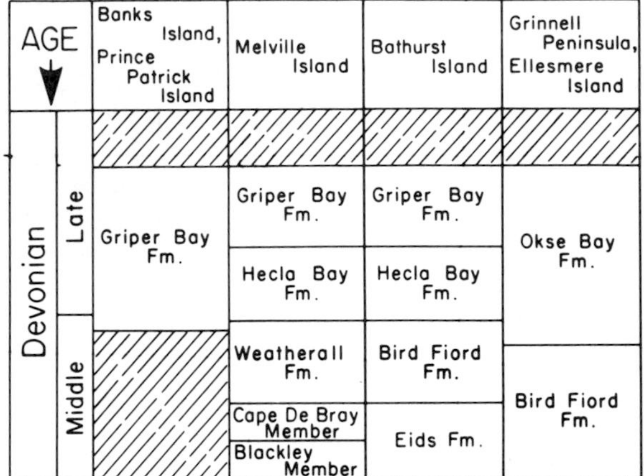

AGE		Banks Island, Prince Patrick Island	Melville Island	Bathurst Island	Grinnell Peninsula, Ellesmere Island
Devonian	Late				
		Griper Bay Fm.	Griper Bay Fm.	Griper Bay Fm.	Okse Bay Fm.
			Hecla Bay Fm.	Hecla Bay Fm.	
	Middle		Weatherall Fm.	Bird Fiord Fm.	Bird Fiord Fm.
			Cape De Bray Member	Eids Fm.	
			Blackley Member		

Fig. 6.46. Table of stratigraphic nomenclature for the Middle–Upper Devonian clastic wedge of the Canadian Arctic Islands (Thorsteinsson and Tozer, 1970).

Casey (1980) described the Late Paleozoic evolution of Taos Trough in northern New Mexico, based on a depositional systems analysis. He produced a series of six paleogeographic maps for the area, showing the relationship of the various depositional environments to the older stratigraphic nomenclature at different time periods. One of these is shown in Figure 6.44. A cross section through the basin illustrates the complex facies relationships (Fig. 6.45) and Casey demonstrated how the same rocks have been subdivided in two earlier systems of nomenclature.

The Middle and Upper Devonian clastic wedge of the Canadian Arctic Islands consists of up to 5 km of strata extending in the surface and subsurface over an area exceeding 200,000 km^2. During the nineteen fifties and sixties the stratigraphy of the sequence was worked out by several officers of the Geological Survey of Canada working more or less independently on the surface geology of different islands or island groups, resulting in the development of two sets of stratigraphic nonenclature, one for the western Arctic and one for the eastern Arctic (Fig. 6.46). No detailed sedimentological analyses were attempted. Embry and Klovan (1976) carried out a depositional systems analysis of these rocks, incorporating new subsurface data that had become available. With a few judicious revisions of the Survey nomenclature and a unifying concept of the origin of the rocks the stratigraphy of this major succession sprang to life. Figure 6.47 is their generalized cross section through the clastic wedge, showing it to consist of a prograding fluvial–deltaic–slope–submarine fan complex. Three periods of progradation were recognized, based on sedimen-

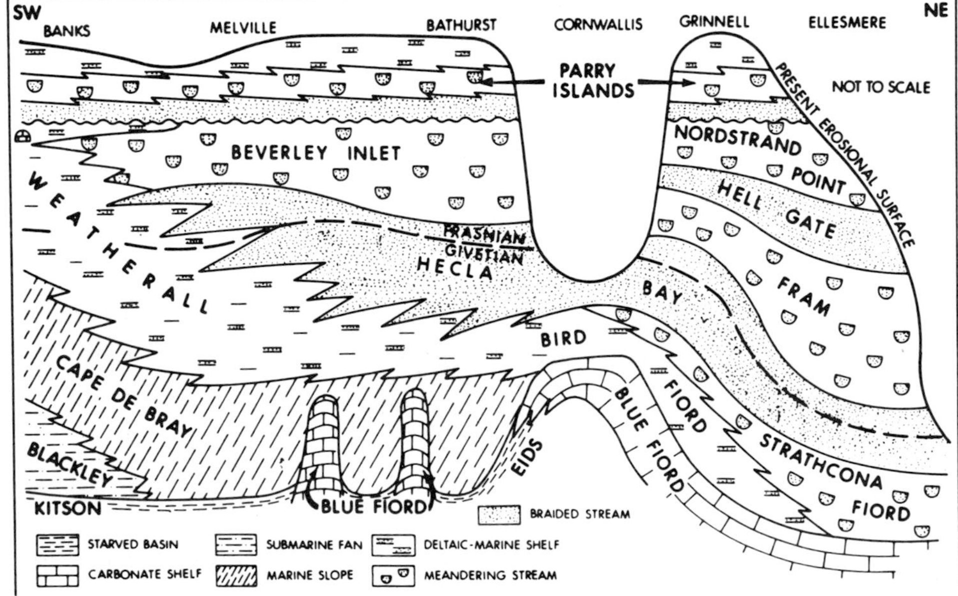

Fig. 6.47. Cross section through the Middle–Upper Devonian clastic wedge of the Canadian Arctic Islands, based on a depositional systems analysis (Embry and Klovan, 1976).

tological and palynological data. Many of the older formation and member names were retained, but their meaning and usefulness are now more readily understood.

References

BEIN, A. and WEILER, Y., 1976: The Cretaceous Talme Yafe Formation: a contour current shaped sedimentary prism of calcareous detritus at the continental margin of the Arabian Craton; Sedimentology, v. 23, p. 511–532.

BENNETTS, K.R.W. and PILKEY, O.H. Jr., 1976: Characteristics of three turbidites, Hispaniola–Caicos Basin; Geol. Soc. Am. Bull., v. 87, p. 1291–1300.

BLATT, H., MIDDLETON, G.V. and MURRAY, R., 1980: Origin of sedimentary rocks; 2nd. ed., Prentice-Hall, Englewood Cliffs, New Jersey, 782 p.

BLUCK, B.J., 1978: Sedimentation in a late orogenic basin: the Old Red Sandstone of the Midland Valley of Scotland, *in* D.R. Bowes and B.E. Leake, eds., Crustal Evolution in northwestern Britain and adjacent regions; Geol. Jour. Spec. Issue 10, p. 249–278.

BORCHERT, H. and MUIR, R.O., 1964: Salt deposits. The origin, metamorphism and deformation of evaporites; Van Nostrand, London, 338 p.

BOUMA, A., 1979: Continental slopes, *in* L.J. Doyle and O.H. Pillkey Jr., eds., Geology of continental slopes; Soc. Econ. Paleont. Mineral. Spec. Pub. 27, p. 1–16.

BRENNER, R.L., 1978: Sussex Sandstone of Wyoming—example of Cretaceous offshore sedimentation; Am. Assoc. Petrol. Geol. Bull., v. 62, p. 181–200.

BRENNER, R.L., 1980: Construction of Process–Response models for ancient epicontinental seaway depositional systems using partial analogs; Am. Assoc. Petrol. Geol. Bull., v. 64, p. 1223–1244.

BRENNER, R.L. and DAVIES, D.K., 1973: Storm-generated coquinoid sandstone: genesis of high energy marine sediments from the Upper Jurassic of Wyoming and Montana; Geol. Soc. Am. Bull., v. 84, p. 1685–1698.

BRENNER, R.L. and DAVIES, D.K., 1974: Oxfordian sedimentation in Western Interior United States; Am. Assoc. Petrol. Geol. Bull., v. 58, p. 407–428.

BROWN, L.F. Jr. and FISHER, W.L., 1977: Seismic–stratigraphic interpretation of depositional systems: examples from Brazilian rift and pull-apart basins, *in* C.E. Payton, ed., Seismic stratigraphy—applications to hydrocarbon exploration; Am. Assoc. Petrol. Geol. Mem. 26, p. 213–248.

BRUNSTROM, R.G.W. and WALMSLEY, P.J., 1969: Permian evaporites in the North Sea Basin; Am. Assoc. Petrol. Geol. Bull., v. 53, p. 870–883.

BUBB, J.N. and HATLELID, W.G., 1977: Seismic stratigraphy and global changes of sea level, Part ten: Seismic recognition of carbonate buildups, *in* C.E. Payton, ed., Seismic stratigraphy—applications to hydrocarbon exploration, Am. Assoc. Petrol. Geol. Mem. 26, p. 185–204.

BUFFLER, R.T., WATKINS, J.S. and DILLON, W.P., 1979: Geology of the offshore southeast Georgia Embayment, U.S. Atlantic Continental Margin, based on multichannel seismic reflection profiles, *in* J.S. Watkins, L. Montadert, and P.W. Dickerson, eds., Geological and geophysical investigations of continental margins; Am. Assoc. Petrol. Geol. Mem. 29, p. 11–26.

CASEY, J.M., 1980: Depositional systems and basin evaluation of the Late Paleozoic Taos Trough, northern New Mexico; Texas Petrol. Res. Committee, Austin, Texas, Report UT 80–1.

CLARK, T.H. and STEARN, C.W., 1968: The geological evolution of North America, 2nd ed., New York, Ronald Press, 570 p.

COLEMAN, J.M. and WRIGHT, L.D., 1975: Modern river deltas: variability of processes and sand bodies, *in* M.L. Broussard, ed., Deltas, models for exploration; Houston Geol. Soc., p. 99–149.

COLLETTE, B.J., EWING, J.I., LAGAAY, R.A. and TRUCHAN, M., 1969: Sediment distribution in the oceans: the Atlantic between 10° and 19° N; Mar. Geol., v. 7, p. 279–345.

COOK, H.E. and ENOS, P., eds., 1977: Deep-water carbonate environments; Soc. Econ. Paleont. Mineral. Spec. Pub. 25.

CURRAY, J.R., 1965: Late Quaternary history, continental shelves of the United States, *in* H.E. Wright and D.G. Frey, eds., The Quaternary of the United States, Princeton Univ. Press, New Jersey, p. 723–735.

CURRAY, J.R. and MOORE, D.G., 1974: Sedimentary and tectonic processes in the Bengal deep-sea fan and geosyncline, *in* C.A. Burk and C.L. Drake, eds., The Geology of continental margins; Springer-Verlag, New York, p. 617–627.

CURTIS, D.M., 1970: Miocene deltaic sedimentation, Louisiana Gulf Coast, *in* J.P. Morgan, ed., Deltaic sedimentation modern and ancient; Soc. Econ. Paleont. Mineral. Spec. Pub. 15, p. 293–308.

DONOVAN, R.N., 1975: Devonian lacustrine limestones at the margin of the Orcadian Basin, Scotland; J. Geol. Soc. London, v. 131, p. 489–510.

DOYLE, L.J., PILKEY, O.H. Jr. and WOO, C.C., 1979: Sedimentation on the eastern United States continental slope, *in* L.J. Doyle and O.H. Pilkey, Jr., eds., Geology of continental slopes; Soc. Econ. Paleont. Mineral. Spec. Pub. 27, p. 119–130.

EDWARDS, M.B., 1976: Growth faults in Upper Triassic deltaic sediments, Svalbard; Am. Assoc. Petrol. Geol. Bull., v. 60, p. 341–355.

EMBLEY, R.W., 1976: New evidence for occurrence of debris flow deposits in the deep sea; Geology, v. 4, p. 371–374.

EMBRY, A. and KLOVAN, J.E., 1976: The Middle–Upper Devonian clastic wedge of the Franklinian Geosyncline; Bull. Can. Petrol. Geol., v. 24, p. 485–639.

EMERY, K.O., 1980: Continental margins—classification and petroleum prospects; Am. Assoc. Petrol. Geol. Bull., v. 64, p. 297–315.

EVANS, W.E., 1970: Imbricate linear sandstone bodies of Viking Formation in Dodsland-Hoosier area

of southwestern Saskatchewan, Canada; Am. Assoc. Petrol. Geol. Bull, v. 54, p. 469–486.
FISHER, W.L. and McGOWEN, J.H., 1967: Depositional systems in the Wilcox Group of Texas and their relationship to occurrence of oil and gas; Gulf Coast Assoc. Geol. Soc. Trans., v. 17, p. 105–125.
FRAZIER, D.E., 1967: Recent deltaic deposits of the Mississippi Delta: their development and chronology; Gulf Coast Assoc. Geol. Soc. Trans., v. 17, p. 287–315.
FRAZIER, D.E., 1974: Depositional episodes: their relationship to the Quaternary stratigraphic framework in the northwestern portion of the Gulf Basin; Bureau Econ. Geol., Texas, Geol. Circ. 74–1.
FULLER, J.G.C.M. and PORTER, J.W., 1969: Evaporite formations with petroleum reservoirs in Devonian and Mississippian of Alberta, Saskatchewan and North Dakota; Am. Assoc. Petrol. Geol. Bull., v. 53, p. 909–926.
GALLOWAY, W.E., 1975: Process framework for describing the morphologic and stratigraphic evolution of the deltaic depositional systems; in M.L. Broussard, ed., Deltas, models for exploration; Houston Geol. Soc., p. 87–98.
GALLOWAY, W.E., 1981: Depositional architecture of Cenozoic Gulf Coastal Plain fluvial systems, *in* F.G. Ethridge and R.M. Flores, ed., Recent and ancient nonmarine depositional environments: models for exploration; Soc. Econ. Paleont. Mineral. Spec. Publ. 31, p. 127–156.
GALLOWAY, W.E. and BROWN, L.F. Jr., 1973: Depositional systems and shelf–slope relations on cratonic basin margin, uppermost Pennsylvanian of north-central Texas; Am. Assoc. Petrol. Geol. Bull., v. 57, p. 1185–1218.
GARY, M., McAFEE, R. Jr. and WOLF, C.L., 1972: Glossary of geology; Am. Geol. Inst., 805 p.
GLENNIE, K.W., 1970: Desert sedimentary environments; Developments in Sed., Elsevier, no. 14.
GLENNIE, K.W., 1972: Permian Rotliegendes of northwest Europe interpreted in the light of modern desert sedimentation studies; Am. Assoc. Petrol. Geol. Bull., v. 56, p. 1048–1071.
GRAHAM, S.A., and BACHMAN, S.B., 1983: Structural controls on submarine-fan geometry and internal architecture: Upper La Jolla fan system, offshore southern California; Am. Assoc. Petrol. Geol. Bull., v. 67, p. 83–96.
HANDFORD, C.R. and DUTTON, S.P., 1980: Pennsylvanian–Early Permian depositional systems and shelf–margin evolution, Palo Duro Basin, Texas; Am. Assoc. Petrol. Geol. Bull., v. 64, p. 88–106.
HEEZEN, B.C. and HOLLISTER, C.D., 1971: The face of the deep; Oxford Univ. Press, New York, 659 p.
HERITIER, F.E., LOSSEL, P. and WATHNE, E., 1980: Frigg Field—large submarine fan trap in Lower Eocene rocks of the Viking Graben, North Sea, *in* M.T. Halbouty, ed., Giant Oil and gas fields of the decade 1968–1978; Am. Assoc. Petrol. Geol. Mem. 30, p. 59–80.
HESSE, R., 1974: Long-distance continuity of turbidites: possible evidence for an Early Cretaceous trench-abyssal plain in the east Alps; Geol. Soc. Am. Bull., v. 85, p. 859–870.
HESSE, R., 1975: Turbiditic and non turbiditic mudstone of Cretaceous flysch sections of the East Alps and other basins; Sedimentology, v. 22, p. 387–416.
HOUBOLT, J.J.H.C., 1968: Recent sediments in the southern Bight of the North Sea; Geol. Mijnb., v. 47, p. 245–273.
JENKYNS, H.C., 1978: Pelagic environments, *in* H.G. Reading, ed., Sedimentary environments and facies; Blackwell, Oxford, p. 314–371.
KENDALL, A.C., 1979a: Facies models 13: Continental and supratidal (sabkha) evaporites, *in* R.G. Walker, ed., Facies models, Geosci. Can. Reprint Series 1, p. 145–157.
KENDALL, A.C., 1979b: Facies models 14: Subaqueous evaporites, *in* R.G. Walker, ed. Facies models.; Geosci. Can. Reprint Series 1, p. 159–174.
KING, P.B., 1948: Geology of the southern Guadalupe Mountains, Texas; U.S. Geol. Surv. Prof. Paper 215.
KLEIN, G. deV., 1977: Tidal circulation model for deposition of clastic sediment in epeiric and mioclinal shelf seas; Sediment. Geol., v. 18, p. 1–12.
KLEIN, G. deV. and RYER, T.A., 1978: Tidal circulation patterns in Precambrian, Paleozoic, and Cretaceous epeiric and mioclinal shelf seas; Geol. Soc. Am. Bull., v. 89, p. 1050–1058.
KOCUREK, G., 1981: Erg reconstruction: the Entrada Sandstone (Jurassic) of northern Utah and Colorado; Palaeogeog., Palaeoclim., Palaeoec., v. 36, p. 125–153.
KREBS, W. and MOUNTJOY, E.W., 1972: Comparison of central European and Western Canadian Devonian reef complexes; 24th Internat. Geol. Congr., Montreal, Sect. 6, p. 294–309.
KURTZ, D.D. and ANDERSON, J.B., 1979: Recognition and sedimentologic description of recent debris flow deposits from the Ross and Weddel Seas, Antarctica; J. Sediment. Petrol., v. 49, p. 1159–1170.
MALDONADO, A. and STANLEY, D.J., 1979: Depositional patterns and late Quaternary evolution of two Mediterranean submarine fans: a comparison; Mar. Geol. v. 31, p. 215–250.
McGREGOR, B.A., 1981: Ancestral head of Wilmington Canyon; Geology, v. 9, p. 254–257.
McKEE, E.D., ed., 1979: A study of global sand seas; U.S. Geol. Surv. Prof. Paper 1052.
MEISSNER, F.F., 1972: Cyclic sedimentation in Middle Permian strata of the Permian basin, West Texas and New Mexico, *in* J.C. Elam and S. Chuber, eds., Cyclic sedimentation in the Permian Basin; 2nd ed., West Texas Geol. Soc., p. 203–232.
MIALL, A.D., 1978: Tectonic setting and syndepositional deformation of molasse and other nonmarine-paralic sedimentary basins; Can. J. Earth Sci., v. 15, p. 1613–1632.
MIALL, A.D., 1981: Alluvial sedimentary basins: tectonic setting and basin architecture, *in* A.D. Miall, ed., Sedimentation and tectonics in alluvial basins; Geol. Assoc. Can. Spec. Paper 23, p. 1–33.
MITCHUM, R.M. Jr., 1977: Seismic stratigraphy and

global changes of sea level, Part eleven: Glossary of terms used in seismic stratigraphy, *in* C.E. Payton, ed., Seismic stratigraphy—applications to hydrocarbon exploration; Am. Assoc. Petrol. Geol. Mem. 26, p. 205–212.

MITCHUM, R.M. Jr., VAIL, P.R. and SANGREE, J.B., 1977: Seismic stratigraphy and global changes of sea level, Part six: Stratigraphic interpretation of seismic reflection patterns in depositional sequences, *in* C.E. Payton, ed., Seismic stratigraphy—applications to hydrocarbon exploration; Am. Assoc. Petrol. Geol. Mem. 26, p. 117–133.

MULLINS, H.T. and NEUMANN, A.C., 1979: Deep carbonate bank margin structure and sedimentation in the northern Bahamas, *in* L.J. Doyle and O.H. Pilkey, Jr., eds., Geology of continental slopes; Soc. Econ. Paleont. Mineral. Spec. Publ. 27, p. 165–192.

NEWELL, N.D., RIBGY, J.K., FISCHER, A.G., WHITEMAN, A.H., HICKOX, J.E. and BRADLEY, J.S., 1953: The Permian reef complex of the Guadalupe Mountains region, Texas and New Mexico; W.H. Freeman and Co., San Francisco, 236 p.

OFF, T., 1963: Rhythmic linear sand bodies caused by tidal currents; Am. Assoc. Petrol. Geol. Bull., v. 47, p. 324–341.

PICHA, F., 1979: Ancient submarine canyons of Tethyan continental margins, Czechoslovakia; Am. Assoc. Petrol. Geol. Bull., v. 63, p. 67–86.

PILKEY, O.H. Jr., LOCKER, S.D., and CLEARY, W.J., 1980: Comparison of sand-layer geometry on flat floors of 10 modern depositional basins; Am. Assoc. Petrol. Geol. Bull., v. 64, p. 841–856.

READING, H.G., ed., 1978: Sedimentary environments and facies; Blackwell Scientific Publications, Oxford, 557 p.

RICCI-LUCCHI, F. and VALMORI, E., 1980: Basin-wide turbidites in a Miocene over-supplied deep-sea plain: a geometrical analysis; Sedimentology, v. 27, p. 241–270.

RUPKE, N.A., 1976: Large-scale slumping in a flysch basin, southwestern Pyrenees; J. Geol. Soc. London, v. 132, p. 121–130.

RUPKE, N.A., 1978: Deep clastic seas, *in* H.G. Reading, ed., Sedimentary environments and facies; Blackwell, Oxford, p. 372–415.

RUST, B.R., 1979: Facies models 2: Coarse alluvial deposits, *in* R.G. Walker, ed., Facies models; Geosci. Can. Reprint Series 1, p. 9–21.

RYAN, W.B.F., et al., 1978: Bedrock geology in New England submarine canyons; Oceanologica Acta, v. 1, p. 233–254.

SANFORD, B.V. and NORRIS, A.W., 1973: Hudson Platform, *in* R.G. McCrossan, ed., The future petroleum provinces of Canada—their geology and potential; Can. Soc. Petrol. Geol. Mem. 1, p. 387–410.

SHEPARD, F.P., 1981: Submarine canyons: multiple causes and long-time persistence; Am. Assoc. Petrol. Geol. Bull., v. 65, p. 1062–1077.

SOUTHARD, J.B. and STANLEY, D.J., 1976: Shelf-break processes and sedimentation, *in* D.J. Stanley and D.J.P. Swift, eds., Marine sediment transport and environmental management; Wiley, New York, p. 351–377.

STANLEY, D.J. and MALDONADO, A., 1979: Levantine Sea—Nile Cone lithostratigraphic evolution: quantitative analysis and correlation with paleoclimatic and eustatic oscillations in the late Quaternary; Sediment. Geol., v. 23, p. 37–65.

STUART, C.T. and CAUGHEY, C.A., 1977: Seismic facies and sedimentology of terrigenous Pleistocene deposits in northwest and central Gulf of Mexico, *in* C.E. Payton ed., Seismic stratigraphy—applications to hydrocarbon exploration; Am. Assoc. Petrol. Geol. Mem. 26, p. 249–276.

SWIFT, D.J.P., DUANE, D.B. and McKINNEY, T.F., 1973: Ridge and swale topography of the Middle Atlantic Bight, North America: secular response to the Holocene hydraulic regime; Mar. Geol., v. 15, p. 227–247.

THORSTEINSSON, R., and TOZER, E.T., 1970: Geology of the Arctic Archipelago, *in* R.J.W. Douglas, ed., Geology and Economic minerals of Canada; Geol. Surv. Can. Econ. Geol. Rept 1, p. 548–590.

TILL, R., 1978: Arid shorelines and evaporites, *in* H.G. Reading, ed., Sedimentary environments and facies, Blackwell, Oxford, p. 178–206.

TIZZARD, P.G. and LERBEKMO, J.F., 1975: Depositional history of the Viking Formation, Suffield area, Alberta, Canada; Bull. Can. Petrol. Geol., v. 23, p. 715–752.

TOOMEY, D.F., MOUNTJOY, E.W. and MACKENZIE, W.S., 1970: Upper Devonian (Frasnian) algae and foraminifera from the Ancient Wall carbonate complex, Jasper National Park, Alberta, Canada; Can. J. Earth Sci., v. 7, p. 946–981.

TOWNSON, W.G., 1975: Lithostratigraphy and deposition of the type Portlandian; J. Geol. Soc. London, v. 131, p. 619–638.

VON DER BORCH, C.C., SMIT, R., and GRADY, A.E., 1982: Late Proterozoic submarine canyons of Adelaide Geosyncline, South Australia; Am. Assoc. Petrol. Geol., v. 66, p. 332–347.

WALKER, R.G., 1979: Facies models 7: Shallow marine sands, *in* R.G. Walker, ed., Facies models; Geosci. Can. Reprint Series 1, p. 75–90.

WALKER, R.G. and MIDDLETON, G.V., 1979: Facies models 4. Eolian sands, *in* R.G. Walker, ed., Facies models; Geosci. Can. Reprint Series 1, p. 33–41.

WEBER, K.J., and DAUKORU, E., 1975: Petroleum geology of the Niger Delta; Proc. 9th World Petrol. Conf., Tokyo, v. 2, p. 209–222.

WILSON, I.G., 1973: Ergs; Sediment. Geol., v. 10, p. 77–106.

WILSON, J.L., 1975: Carbonate facies in geologic history; Springer-Verlag, New York, 471 p.

WILSON, J.L., and D'ARGENIO, B., 1982: Penrose conference report: controls on carbonate platforms and basin systems development; Geology, v. 10., p. 659–661.

WORZEL, J.L. and BURK, C.A., 1979: The margins of the Gulf of Mexico, *in* J.S. Watkins, L. Montadert and P.W. Dickerson, eds., Geological and geophysical investigations of continental margins; Am. Assoc. Petrol. Geol. Mem. 29, p. 403–419.

CHAPTER 7

Burial history

7.1 Introduction

The quantification of the amount and rate of basin subsidence is a vital part of basin analysis. Geophysicists are finding that certain subsidence styles are characteristic of particular tectonic settings, and are using these data to develop quantitative models of subsidence mechanisms. The details of basin subsidence may provide important information on the tectonic history of an area, yielding useful comparisons with the orogenic or igneous history of adjacent deformed belts. Most important of all from the economic perspective, the burial history of a basin is the primary control on the thermal history of its contained sediments. The generation and migration of oil and gas, and the rank of coal, are directly dependent on the temperatures achieved in the host sediments and the time over which these temperatures were maintained.

There are two main approaches to studying basin burial history. The first simply documents stratigraphic thicknesses, making allowance for various corrections (section 7.2). The second approach makes use of a variety of petrographic or geochemical techniques to reconstruct thermal history and to deduce burial history from models of diagenetic change (section 7.3). Both have their advantages and drawbacks. Geochemical methods have evolved rapidly in the last dozen years and are the subject of active research at the present, with the appearance of many fascinating publications on regional applications (section 7.4).

7.2 Stratigraphic analysis

The use of thickness and facies data to produce curves of burial history was termed "geohistory analysis" by Van Hinte (1978). This singularly ambiguous phrase was intended by Van Hinte to apply to the method of constructing time–depth plots for continuous stratigraphic sections. An example is illustrated in Figure 7.1, and its method of construction is as follows. The chronostratigraphy of the section is determined as accurately as possible. Well-dated marker beds (*a, b, c,* etc. in Fig. 7.1) are indicated on the time scale along the top of the graph. In the example illustrated, an unconformity occurs at 700 m depth, and this must be accounted for in the final graph. The paleobathymetry is then determined for each marker bed and a data point inserted on the graph in the correct time–depth position for each marker bed (closed triangles). Extrapolation may be necessary to find the correct position of the points for total depth and for the age of the surfaces above and below the unconformity (open triangles). This can be done by estimating sedimentation rates from adjacent well-dated parts of the section, making all due allowance for the variations in sedimentation rate shown by different lithologies deposited in different depositional environments. A curve joining the data point triangles defines the changes in water depth with time.

Cumulative sediment thicknesses are then determined for each marker bed, and these values are plotted below the corresponding point on the water depth curve (closed circles). A curve joining these points defines the subsidence history of the total depth point in the section. Two curves are shown in Figure 7.1. The *uRs* curve is for "uncorrected" subsidence, whereas the *Rs* curve is for subsidence corrected according to compaction. Van Hinte (1978) provided a method for determining compaction corrections but Keen (1979) argued that it is not important to make this correction unless the section contains thick intervals of mudstone.

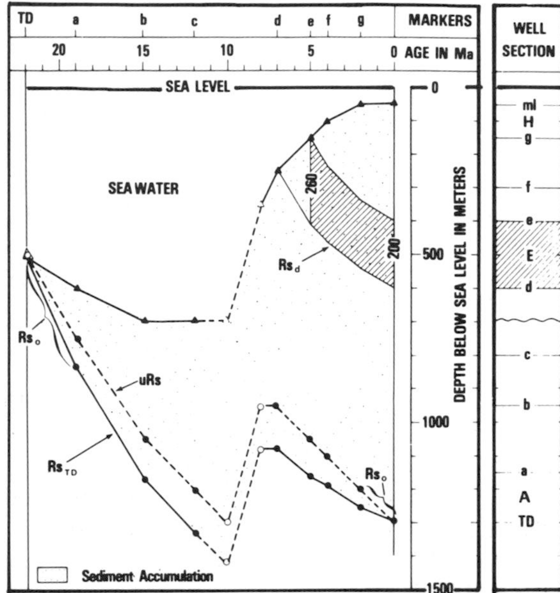

Fig. 7.1. Sea level and sediment accumulation curves for a hypothetical well. a to g are accurately dated marker horizons. Rs = corrected subsidence curve, uRs = uncorrected curve (Van Hinte, 1978).

The unconformity in the section has been accounted for simply by drawing dashed lines between the open circles and triangles. This suggests that sedimentation simply ceased while water depths shallowed by about 350 m due to tectonic uplift or eustatic sea level change. In fact, of course, an unconformity is normally a much more complicated event involving uplift and erosion, but it is usually not possible to depict this accurately because the very nature of an unconformity involves a loss of part of the record.

If the TD point represents top of basement the subsidence curve depicts the overall subsidence of the basin relative to sea level. Much depends on the accuracy of the biostratigraphic and paleoecological control. For the Cenozoic, bathymetric determinations from micropaleontological evidence should be accurate to within a few hundred meters, but become much less accurate at deep bathyal and abyssal (>2000m) depths. Precision drops off rapidly for older rocks (Van Hinte, 1978). An example of the type of evidence used in reconstructing water depths is discussed in Section 4.5.7 (Fig. 4.43).

As described at some length in Chapter 8, global sea level has varied considerably at least during the Phanerozoic, on time scales ranging from 10^5 to 10^8 a. Time–depth plots such as those illustrated in Figures 4.43 and 7.1 do not show this very effectively. For bathyal or abyssal depths eustatic sea level changes of a few hundred meters or less would be hard to detect using available paleoecological criteria. If a sea level change could be recognized, it would produce anomalous changes in slope of the subsidence curve. Even these would be hard to analyze, because changes in sea level produce changes in sedimentation rate through their effects on the sediment supply (Chapter 8). At shallower depths sea level changes produce rather more obvious changes in microfauna, and the sediment record may contain clear evidence of transgressions and regressions, and possibly several unconformities.

Van Hinte (1978) provided a hypothetical example of the reasoning that can be used to recognize an eustatic change in sea level from stratigraphic data. Two wells are shown in Figure 7.2. In both a paleobathymetric break has been recognized at about 2.7 Ma. Calculation of sedimentation rates immediately above and below the break in each well suggests a time gap of 30,000 a in well 1 and 600,000 a in well 2. The break in well 2 clearly represents an unconformity, whereas that in well 1 may only reflect inaccuracies introduced by changes in sedimentation rate. A simple explanation is that the paleobathymetric break represents a drop in sea level. Estimates from well 1 suggest that it amounted to about 700 ft (213 m).

7.3 Petrographic and geochemical analysis

Many physical and chemical changes are induced in sediments by the rise in temperature and pressure during burial. For some of these the changes can be quantified with varying degrees of accuracy and related to burial history. Measurement of such effects in research laboratories is now routine in the coal and petroleum industries, because the composition and quality of hydrocarbons in sediments is directly dependent on this process of diagenesis. A survey of the area by Heroux et al. (1979) included most of the following measurable parameters:

Coal petrology
- coal rank
- percent volatile matter (in vitrinite)
- total organic carbon (in vitrinite)

calorific content
light reflectance (vitrinite)
Petroleum (these parameters dependent on kerogen type)
atomic ratio
carbon ratio
Mineralogy
clay mineral content
illite crystallinity
pre-metamorphic mineral assemblages (zeolites, etc.)
fluid inclusion thermometry
Micropaleontology—palynology
spore-pollen color
conodont color

The literature dealing with the theory, laboratory procedures and interpretation of these various parameters is extensive and could fill a book on its own. The following is a brief discussion of some of the more widely used techniques.

Coal rank. This is one of the oldest measures of thermal alteration or "maturity". The standard scale of coal ranks shown in Figure 7.3 is based on that of Suggate (1959). Earlier workers classified coals into the appropriate rank on the basis of such properties as ash content, moisture content, coking power, volatile content, heat value, hardness or grindability. However, nowadays the most common method is that of **vitrinite reflectance**.The advantage of the method is that it uses a relatively simple laboratory procedure and can be carried out on very small samples, enabling rank determinations to be made on scattered plant fragments rather than requiring large samples from an actual coal seam. Polished samples are examined under a reflected-light microscope using standard lighting conditions, and a microphotometer measures the amount of light reflected from the surface. This increases in a predictable way with rank, from reflectance values of $R_o < 0.4\%$ for lignite to $R_o > 2.5\%$ for anthracite. Large numbers of individual measurements are made and the data are treated statistically. The technique was developed in Germany (e.g. see McCartney and Teichmuller, 1972) and is widely used for measuring maturity in petroleum-producing areas (e.g. Hacquebard, 1977). Problems with the method include the fact that it only works on the vitrinite component, which is absent in pre-Carboniferous rocks, is unreliable for values less than 0.3% R_o, and is affected by the lithology of the sample host rock (Heroux et al., 1979).

One of the main changes that takes place in coal with increasing burial is the emission of volatile components (water, carbon dioxide, methane etc.) and the increase in carbon content. This increase in "carbonization" is responsible for the changing reflectance, and is also the controlling factor behind another major maturity indicator, that of **spore color** and preservation. Gutjahr (1966) proposed a color scale of yellow through brown to black. Staplin (1969) erected a "thermal alteration index" (TAI) based on microscopic observations of color and structural alteration of organic debris, and Correia (1967) used "state of preservation" and light absorption. The color and TAI scales have been correlated to coal rank, as shown in Figure 7.3.

The study of clay mineralogy has provided several useful diagenetic tools. One of the most widely used is the **illite crystallinity index**. This is defined as the width at half-height of the peak located around 10 nm on an X-ray diffractogram (Kübler, 1968). Crystallinity increases with depth, but reproducible results can only be obtained by careful standardization of instruments and laboratory procedures and making allowance for inherent (pre-burial) characteristics of the clay mineral assemblage. Numerous other diagenetic changes in clay have been reported, for example, the shift in diffractogram reflection of smectite with burial due to a loss of interlayer water and changes in the proportions of illite and other clays in the assemblage. Heroux et al. (1979) listed the key references to this work.

Such maturity indicators as coal rank and illite crystallinity are not suitable for use in pure carbonate sediments (although they may be applied to interbedded clastic units). However, Epstein et al. (1977) demonstrated that conodonts, which occur mainly in Paleozoic carbonate sequences, also can yield valuable maturity information. Field and laboratory studies demonstrated that with increasing temperature carbon is fixed in the conodonts and color changes from pale yellow through brown to black, then to colorless crystalline. The major color changes occur over the range 50 to 550°C. This contrasts with the most useful range of color alteration in palynomorphs, which is 60 to 150°C. Where both fossil types are present in a stratigraphic section the palynomorphs are therefore more useful for younger,

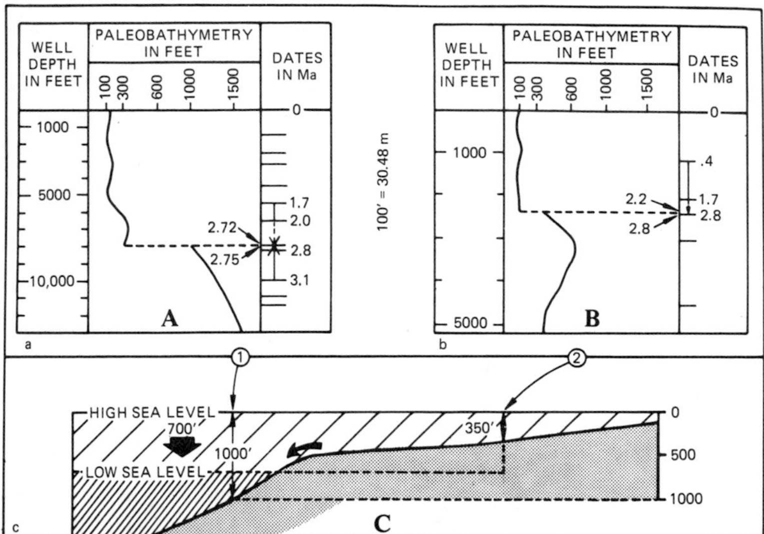

Fig. 7.2. Paleobathymetry of two hypothetical wells (**A**,**B**) showing a paleobathymetric break in each, and estimation of age of sediments above and below the break by extrapolation of sedimentation rates. In **C** the data are interpreted in terms of an eustatic sea level drop. See text for further explanation (Van Hinte, 1978).

less altered sediments and conodonts for the more deeply buried part of the section. As with most other temperature indicators, the temperature–alteration relationship for conodonts is not without some complications, but varies somewhat with the structure and size of the conodont element and with host-rock lithology. Color changes have been converted into a semiquantitative scale termed the Color Alteration Index (CAI); it ranges in value from 1 to 8 (Epstein et al., 1977).

Hood et al. (1975) discussed the need for a unified metamorphic scale for studying burial history. They studied the standard coal rank scale and compared it with Staplin's (1969) thermal alteration index and the vitrinite reflectance scale. Their proposal was for a scale called Level of Organic Metamorphism (LOM), with arbitrary values 1 to 20. Samples could be assigned to the correct level by using any one or more of the organic scales, and correlating the results using the chart reproduced as Figure 7.3.

Unfortunately, neither LOM values nor any other metamorphic scale provide a simple measure of maximum burial depth, because of variations in geothermal gradient and because heating time is as important as heating temperature in bringing about the organic changes. For example Hood et al. (1975) quoted two examples of samples exhibiting LOM values of 11. One was Paleozoic rock from west Texas, with a maximum burial temperature of 105°C, the other was a Tertiary sample from California, heated as high as 205°C. Hood et al. (1975) determined that the time during which a rock had been within 15°C of its maximum temperature (T_{max}) "represents a reasonably suitable, though somewhat arbitrary, definition of effective heating time" (t_{eff}). For example, consider a sediment formed 150 Ma ago, heated to 120°C by 50 Ma and then to 135° somewhat later. Assuming the temperature up to the present has not dropped back below 120°, then the last 50 Ma is counted as t_{eff}. Figure 7.4 shows the relationships determined between t_{eff}, T_{max} and LOM.

The principal importance of the various metamorphic scales is in predicting petroleum generation. It has been found that oil is released from source rocks at LOM values between 8 and 12, followed by condensate and wet gas up to LOM = 13.5, after which only high-temperature dry gas is released. A complete discussion of this topic is beyond the scope of the present book but is considered at greater length by Tissot and Welte (1978).

7.4 Selected case studies

Apart from the by now widespread use of metamorphic indicators to study petroleum generation and migration, burial history data can be of assistance in three main areas of basin analysis:

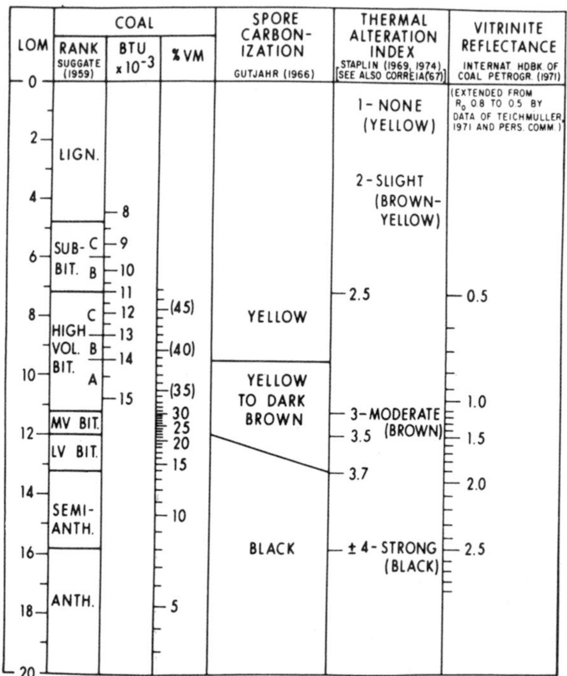

Fig. 7.3. Correlation of some scales of organic metamorphism. LOM = level of organic metamorphism, BTU = heat value in British thermal units; %VM = percent volatile matter (Hood et al., 1975).

1. Subsidence curves or time–depth plots drawn for continuous stratigraphic sections can be used to test geophysical models of basin subsidence.
2. Variations in metamorphism with depth can indicate amounts of section missing as a result of erosion. This can be of considerable assistance in reconstructing the former paleogeographic extent of a basin or of particular depocenters.
3. Patterns of burial history when reconstructed in two or three dimensions across tectonically deformed areas can yield information on the relationship of burial metamorphism to tectonism.

A few example of these applications follow.

Keen (1979) studied the subsidence history of offshore eastern Canada using well records. She corrected subsidence for eustatic sea level changes and paleobathymetry assuming simple isostatic behaviour. Tectonic ("true") subsidence, y, is given by the equation

$$y = \frac{(\rho_m - \rho_s)}{(\rho_m - \rho_w)} S + h_w - \frac{\rho_m}{\rho_m - \rho_w} E$$

where ρ_m, ρ_s and ρ_w are densities of mantle, sediment and water, S is sediment thickness, h_w is paleo-water depth and E is eustatic height of sea level above present sea level. Calculations were

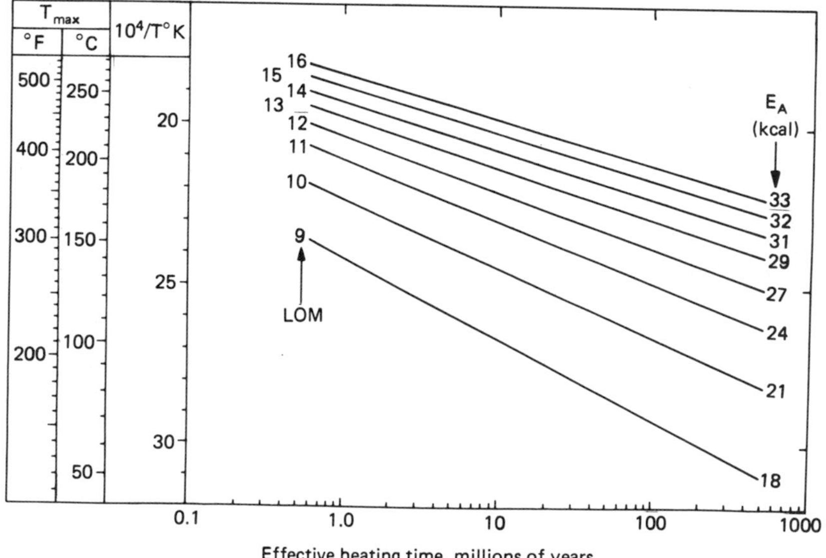

Fig. 7.4. Relation of LOM to maximum temperature and effective heating time (Hood et al., 1975).

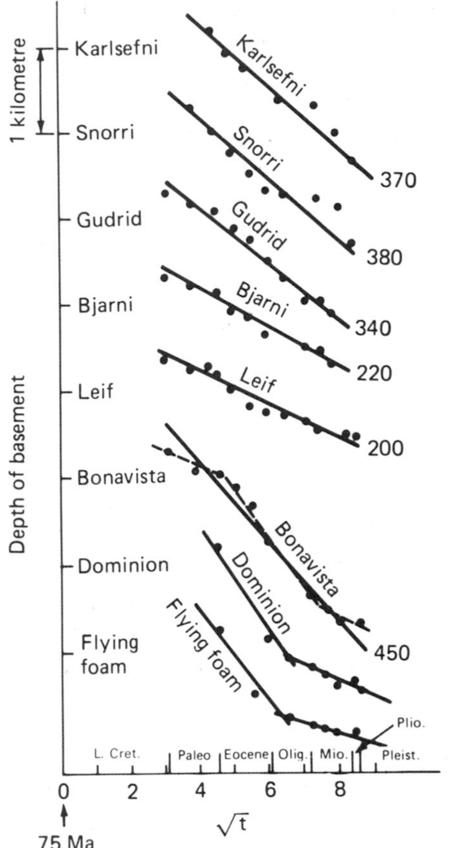

Fig. 7.5. Tectonic subsidence of wells off Labrador plotted against $t^{1/2}$ since subsidence began. Each curve is plotted relative to zero subsidence at points marked on the vertical axis. Numbers after each line are slopes in m $Ma^{-1/2}$ (Keen, 1979).

made for successive time increments, adjusting for sediment densities. Paleobathymetric data were provided by micropaleontological and palynological studies, and eustatic sea level values were taken from those deduced by Vail et al. (1977) and Pitman (1978) (this latter topic is discussed at length in Chapter 8).

Keen (1979) was interested in testing subsidence history against a thermal cooling model for divergent continental margins which predicted that subsidence is a linear function of $t^{1/2}$, t being the age of the basement (see Chapters 8 and 9 for some additional discussion of subsidence mechanisms and basin models). Accordingly, she plotted her subsidence data against the square root of time since subsidence began. Results for wells off Labrador are shown in Figure 7.5. For most of the wells the data can be fitted convincingly to a straight line when allowance is made for all the possible sources of error in the input data.

The removal by erosion of significant thicknesses of sediment along the margin of a basin may be clearly indicated by truncation of depositional systems or by anomalous relationships of facies mapping patterns (e.g. isopachs, sand/mud ratios, paleocurrent trends) to the present basin margin. Similar conclusions may be drawn from a few very simple observations of organic metamorphic changes. For example, a surface exposure of anthracite means thousands of meters of overburden must have been removed. Using the techniques described in the preceding section, it is possible to obtain approximate quantitative estimates of this erosion.

An example of such an estimate was given by Friedman and Sanders (1982) from Middle Devonian strata of eastern New York State. Anthracite samples have a LOM of 16 while kerogen particles show a TAI of 4. Conodont CAI values of 4 were obtained, indicating maximum temperatures of 190 to 240°C. The presence of pore-filling authigenic chlorite and sericite is interpreted as greenschist-facies metamorphism and a temperature in the range of 200–300°C. The rocks are about 350 Ma old but, to allow time for subsidence to the realm of highest temperatures and subsequent uplift to the present land surface, the effective heating time must have been some fraction of this. According to Figure 7.4 a LOM of 16 will develop at a t_{max} of 200°C over a t_{eff} of about 100 Ma, which seems reasonable. Assuming a geothermal gradient similar to that at the present, 26°C km^{-1}, and a surface temperature of 15°C, this indicates a burial depth of 7.1 km. Friedman and Sanders (1982) arrived at a figure of 6.5 km using $t_{max} = 200$ Ma and $t_{eff} = 190$°C. These burial figures are surprising, but can only be reduced by suggesting a much higher geothermal gradient than at present, which would imply much higher levels of metamorphism in underlying rocks than has in fact been observed. The conclusion is that a pile of sediment at least 6.5 km thick has been removed by erosion from eastern New York State. This is comparable to the thickness of Carboniferous strata preserved along strike in Pennsylvania, and indicates that the coal basin there formerly extended much further to the north.

Legall et al. (1981) carried out a similar study of the lower Paleozoic strata of southern Ontario

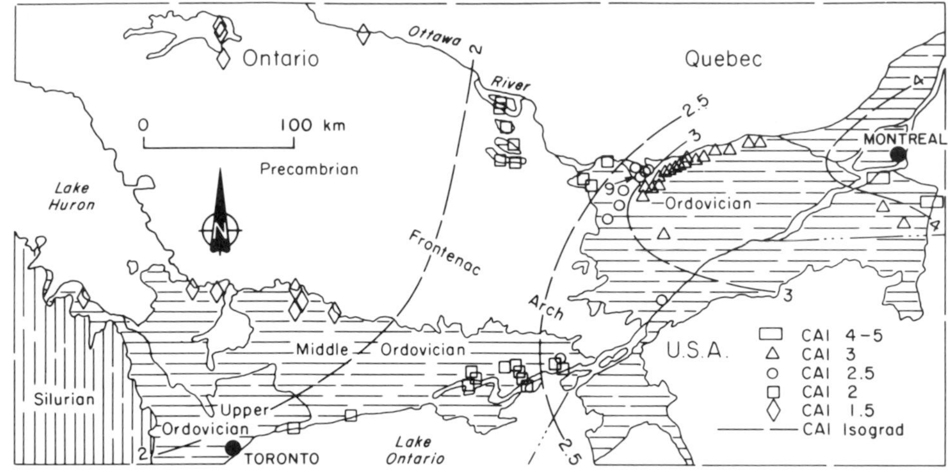

Fig. 7.6. Conodont CAI variations in Ordovician sediments of southern Ontario (Legall et al., 1981).

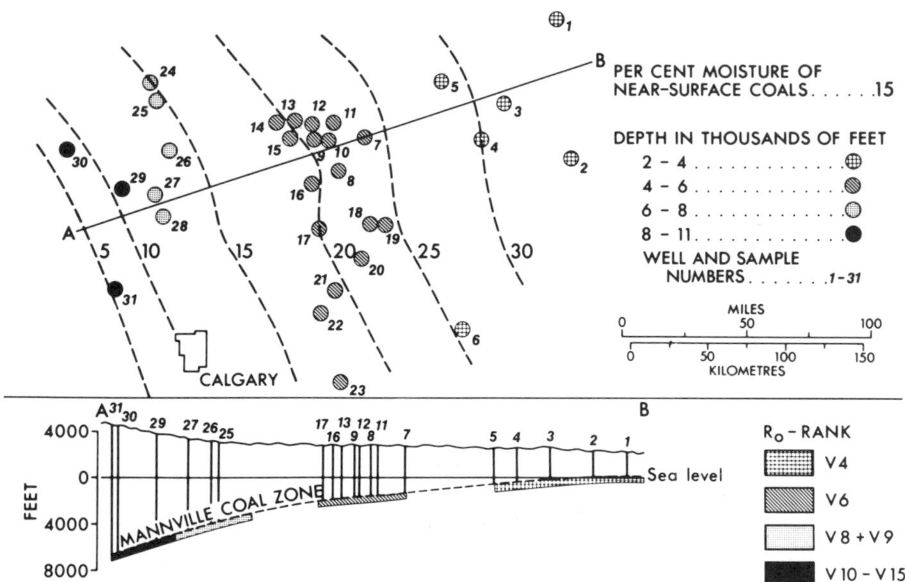

Fig. 7.7. Location, depth and rank of Mannville coal samples, and isomoisture contours of near surface coals of Edmonton and Paskapoo Formations, Alberta (Hacquebard, 1977).

and western Quebec. They were able to suggest from burial data the former existence of several hundred meters of younger Paleozoic or Mesozoic strata. In addition, the broad regional distribution of CAI values below 2 in southwestern Ontario indicates a regular burial pattern. The Algonquin Arch is a distinct cratonic structure trending southwest through the area toward Detroit, but the maturation data indicate that it had no major control on the thermal history of the sequence. Likewise, the Frontenac Arch exposes the Canadian Shield along an irregular inlier extending southeast across the project area and leaves the Paleozoic rocks as remnant outliers, but the maturity data can be contoured as if the arch did not exist (Fig. 7.6). Clearly, these cratonic structures are entirely post-depositional (post-lower Paleozoic) in origin.

Hacquebard (1977) carried out a detailed study of the burial history of Cretaceous and Tertiary coals in Alberta. Figure 7.7 shows a contour map of moisture content of near-surface coals, and rank variations in buried (Mannville Group) coals, data for which were obtained from cores in

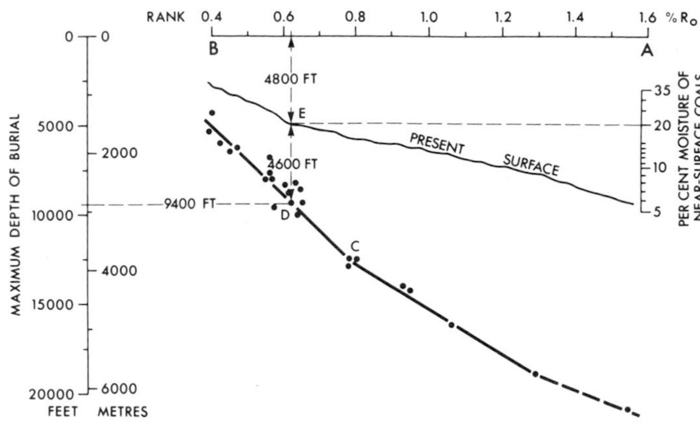

Fig. 7.8. Determination of rank and original burial depths of Mannville coals. See text for explanation (Hacquebard, 1977).

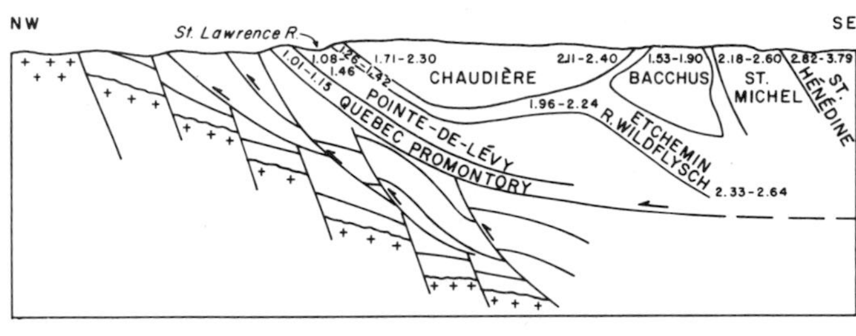

Fig. 7.9. Schematic structural cross section through the Taconic orogen, Quebec (not to scale) showing reflectance values (Ogunyami et al., 1980).

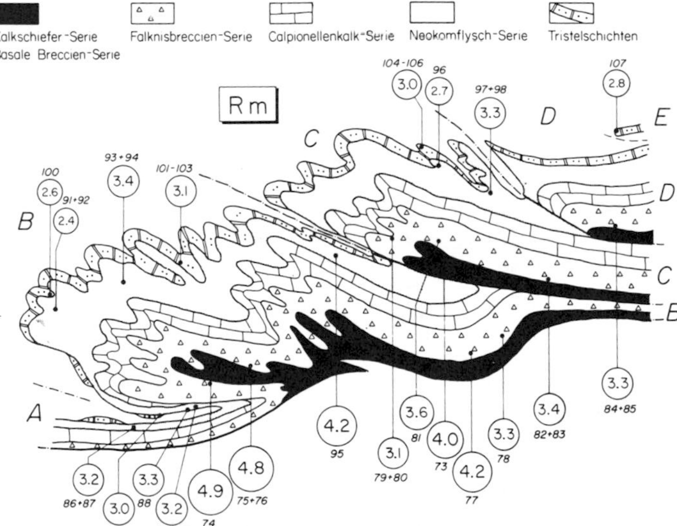

Fig. 7.10. Variation in reflectance values across the Falknis nappe, Switzerland (Frey et al., 1980).

petroleum exploration wells. Rank is shown by codes V4 (= R_o, 0.40 to 0.49%), V5 (= R_o, 0.50 to 0.59%), etc. The moisture content of coal shows a linear relationship to maximum burial depth, and so the former depths of the near-surface samples can readily be plotted, permitting a precise estimate of the maximum burial depth of Mannville coals. The procedure is shown in Figure 7.8. A sample of near-surface coal contains 20% moisture, indicating a maximum burial depth of 1463 m (4800 ft). Mannville coal D now lies at 1402 m (5600 ft) below the surface, but must formerly have lain at a maximum depth of 1402 + 1463 = 2865 m (9400 ft). Rank of the Mannville coals is along the top, and the data show a curved relationship between coal rank and depth.

Several studies have been carried out of the low

grade metamorphism of fold-thrust belts in order to investigate the relationship of metamorphism to tectonism. Ogunyami et al. (1980) studied the Lower Paleozoic (Taconic) orogen near Quebec City, and Frey et al. (1980) investigated several cross sections through the nappes of the Swiss Alps. Both studies tackled the question of whether metamorphic trends paralleled the stratigraphy or whether they were related to present structural depth and therefore cut across stratigraphy. In the first case metamorphism must be related entirely to pre-tectonic burial, whereas in the second case metamorphism can be related to tectonic loading. Several other complications occur.

Ogunyami et al. (1980) found that metamorphism increased toward the center of the orogen and that the contours of illite crystallinity, vitrinite reflectance and other parameters did not coincide with structural (nappe) boundaries. This is illustrated by the reflectance measurements shown on a regional cross section in Figure 7.9, and is interpreted as the result of regional heating during the Tacónic Orogeny. In addition, they noted that within the outermost nappes the degree of thermal alteration increases downward. Using similar arguments to those of Friedman and Sanders (1982), quoted above, they demonstrated that several kilometers of sediment must have been removed, and attempted to determine whether this was a sedimentary or tectonic load. In the case of Cambrian sediments at the base of the Quebec Promontory Nappe stratigraphic reconstructions indicate a sediment pile of 6 to 7 km accumulated prior to tectonism in the Ordovician. Thermal alteration of these sediments was therefore probably due entirely to stratigraphic burial. Younger sediments in the Quebec Promontory Nappe and the Pointe-de-Lévy Nappe are not known to have been deeply buried prior to tectonic deformation, and their thermal alteration is probably the result of tectonic loading by nappe emplacement.

Frey et al. (1980) measured illite crystallinity, coal rank and fluid inclusion temperatures and pressures along several transects through the Swiss Alps. Their conclusions were similar to those reported from Quebec. However, a particularly detailed study of one nappe revealed the complexities involved in this type of work. Figure 7.10 illustrates reflectance values measured in five thrust slices or schuppen (A to E) of the Falknis Nappe in eastern Switzerland. Rather consistent reflectance values in the two youngest units, the Tristelschichten and Neokomflysch-Serie, seem to suggest that coalification is pre-tectonic. However, measurements made in the Falknisbreccien-Serie indicate an increase in thermal alteration going from schuppen C and D to the deeper structural level of schuppe A. This would suggest that coalification is related to tectonic loading. A further complication is the inversion of reflectance values between schuppen B and A. Two possible interpretations present themselves. Either the thrusting of schuppen B to E onto schuppe A is a very late event, or the increased coalification level of the Falknisbreccien-Serie in schuppe B is due to internal shearing. Illite crystallinity and fluid inclusion data do not resolve these problems and, indeed, introduce their own complications. These results should serve as a warning against drawing sweeping conclusions from a few data points, and they indicate that we still have a lot to learn about the physics and chemistry of the thermal alteration processes.

References

CORREIA, M., 1967: Relations possibles entre l'état de conservation des élements figures de la matière organique (microfossiles palynoplanctologiques) et l'existence de gisements d'hydrocarbures; Inst. Français Pétrole Rev., v. 22, p. 1285–1306.

EPSTEIN, A.G., EPSTEIN, J.B., and HARRIS, L.D., 1977: Conodont color alteration—an index of organic metamorphism; United States Geol. Surv., Prof. Paper 995.

FREY, M., TEICHMULLER, M., TEICHMULLER, R., MULLIS, J., KUNZI, B., BREITSCHMID, A., GRUNER, U. and SCHWIZER, B., 1980: Very low-grade metamorphism in external parts of the central Alps: illite crystallinity, coal rank and fluid inclusion data; Eclogae Geol. Helv., v. 73, p. 173–203.

FRIEDMAN, G.M. and SANDERS, J.E., 1982: Time-temperature-burial significance of Devonian anthracite implies former great (~6.5 km) depth of burial of Catskill Mountains, New York; Geology, v. 10, p. 93–96.

GUTJAHR, C.C.M., 1966: Carbonization of pollen grains and spores and their application; Leidse Geol. Meded., v. 38, p. 1–30.

HACQUEBARD, P.A., 1977: Rank of coal as an index of organic metamorphism for oil and gas in Alberta, *in* The origin and migration of petroleum in the Western Canada Sedimentary Basin, Alberta; Geol. Surv. Can Bull. 262, p. 11–22.

HEROUX, Y., CHAGNON, A., and BERTRAND,

R., 1979: Compilation and correlation of major thermal maturation indicators; Am. Assoc. Petrol. Geol. Bull., v. 63, p. 2128–2144.

HOOD, A., GUTJAHR, C.C.M., and HEACOCK, R.L., 1975: Organic metamorphism and the generation of petroleum; Am. Assoc. Petrol. Geol. Bull., v. 59, p. 986–996.

KEEN, C.E., 1979: Thermal history and subsidence of rifted continental margins—evidence from wells on the Nova Scotian and Labrador Shelves; Can. J. Earth Sci., v. 16, p. 505–522.

KÜBLER, B., 1968: Evaluation quantitative du metamorphisme par la cristallinité de l'illite. Etat des progrès réalisés ces dernières années; Centre Recherche Pau Bull., v. 2, p. 385–397.

LEGALL, F.D., BARNES, C.R., and MACQUEEN, R.W., 1981: Thermal maturation, burial history and hotspot development, Paleozoic strata of Southern Ontario–Quebec, from conodont and acritarch colour alteration studies; Bull. Can. Petrol. Geol., v. 29, p. 492–539.

McCARTNEY, J.T., and TEICHMULLER, M., 1972: Classification of coals according to degree of coalification by reflectance of the vitrinite component; Fuel, v. 51, p. 64–68.

OGUNYAMI, O., HESSE, R., and HEROUX, Y., 1980: Pre-orogenic and synorogenic diagenesis and anchimetamorphism in Lower Paleozoic continental margin sequences of the northern Appalachians in and around Quebec City, Canada; Bull. Can. Petrol. Geol., v. 28, p. 559–577.

PITMAN, W.C. III, 1978: Relationship between eustacy and stratigraphic sequences of passive margins; Geol. Soc. Am. Bull., v. 89, p. 1389–1403.

SUGGATE, R.P., 1959: New Zealand coals. Their geological setting and its influence on their properties; New Zealand Dept. Sci. Indust. Res. Bull. 134.

STAPLIN, F.L., 1969: Sedimentary organic matter, organic metamorphism, and oil and gas occurrence, Bull. Can. Petrol. Geol., v. 17, p. 47–66.

TISSOT, B.P., and WELTE, D.H., 1978: Petroleum formation and occurrence; Springer-Verlag, Berlin, 538 p.

VAIL, P.R., MITCHUM, R.M., and THOMPSON, S., 1977: Seismic stratigraphy and global changes of sea level, Part 4, global cycles of relative changes of sea level, *in* C.E. Payton, ed., Seismic stratigraphy—applications to hydrocarbon exploration; Am. Assoc. Petrol. Geol. Mem. 26, p. 83–97.

VAN HINTE, J.E., 1978: Geohistory analysis—applications of micropaleontology in exploration geology; Am. Assoc. Petrol. Geol. Bull., v. 62, p. 201–222.

CHAPTER 8

Regional and global stratigraphic cycles

8.1 Back to the layer cake

One of the achievements of modern sedimentology has been to demonstrate the almost ubiquitous presence of major lateral facies changes in sedimentary rocks, with the result that nobody uses lithostratigraphy nowadays as a basis for chronostratigraphic correlation. However, it has long been recognized that on a continental and even an intercontinental and global scale, there seem to have been a number of synchronous, correlatable stratigraphic events. These include the virtually worldwide basal Cambrian unconformity (Matthews and Cowie, 1979), major transgressive episodes in the Ordovician (McKerrow, 1979; Vail et al., 1977b) and the Cretaceous (Hallam, 1963; Hancock and Kauffman, 1979; Vail et al., 1977b), and the post-Cretaceous regression (Hallam, 1963). Glacial episodes of hemispherical or global scope have been recognized in the Early Proterozoic, latest Precambrian (Eocambrian), Late Ordovician and Early Carboniferous to mid-Permian (Crowell, 1978; Edwards, 1978; Harland, 1981). Other widespread cyclic events have been reported by Hallam (1978), Ager (1981) and many others. The Late Cretaceous transgression was first recognized as being exceptionally prominent by Suess. He suspected that it was only one of several such events in the Phanerozoic related to worldwide changes in sea level, and he named such movements "eustatic" (Suess, 1906).

To these can be added many events that are only slightly more parochial in scope, such as the distinctive "Coal Measure" cyclothems of Carboniferous age in North America and Europe (Wanless and Shepard, 1936; Moore, 1964; Crowell, 1978; Ramsbottom, 1979) and the late Paleozoic arid facies of the same two continents. Stratigraphic and tectonic comparisons between the Gondwana continents of South America, southern Africa, India, Australia and Antarctica have long been evident (du Toit, 1937). Many attempts at global correlation have not withstood the application of rigorous biostratigraphic or radiometric testing, such as the use of such essentially European terms as Caledonian, Hercynian and Alpine for orogenic episodes in North America, the Soviet Union and elsewhere; but the list of global or hemispherical stratigraphic events is nevertheless impressive.

The wealth of subsurface data in North America enabled Sloss et al. (1949) to recognize four continent-wide stratigraphic sequences in the Cambrian to Jurassic. Later this list was expanded to six to include the Jurassic to Recent (Sloss, 1963). Subsequently detailed comparisons have been made between the Phanerozoic record of Western Canada and the Russian Platform (Sloss, 1972), and between selected intracratonic and pericratonic basins on these continents (Sloss, 1978). Soares et al. (1978) have recognized a similar chronology of stratigraphic sequences in the intracratonic basins of Brazil. The fact that these sequences have now been documented in the cratonic and pericratonic regions of three continents would suggest that they are in fact global in extent, and this conclusion has recently received strong support from the Exxon work on subsurface stratigraphy using seismic methods (Vail et al., 1977a, 1977b). As discussed later the development of these sequences depended in part on global changes in sea level, for which a mechanism has been discovered, based on varying rates of sea-floor spreading.

How are these sequences defined in the rock record? Sloss (1963) defined them as "rock stratigraphic units of higher rank than group, megagroup or supergroup, traceable over major areas of a continent and bounded by unconformities of

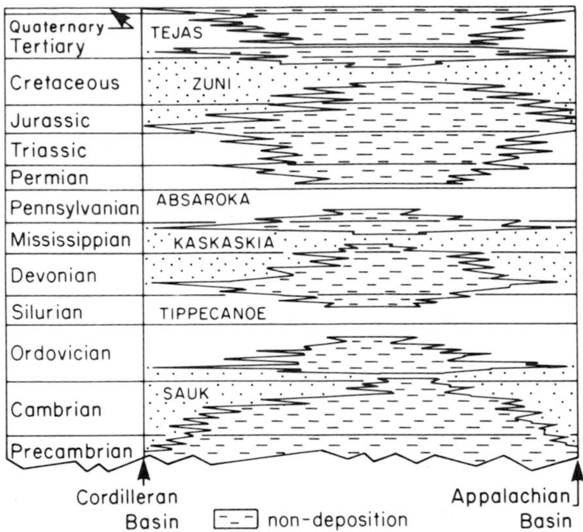

Fig. 8.1. The six sequences of Sloss (1963). Black areas represent non-deposition. (Reproduced, with permission, from L.L. Sloss, *Geological Society of America Bulletin*, v. 74, pp. 93–113).

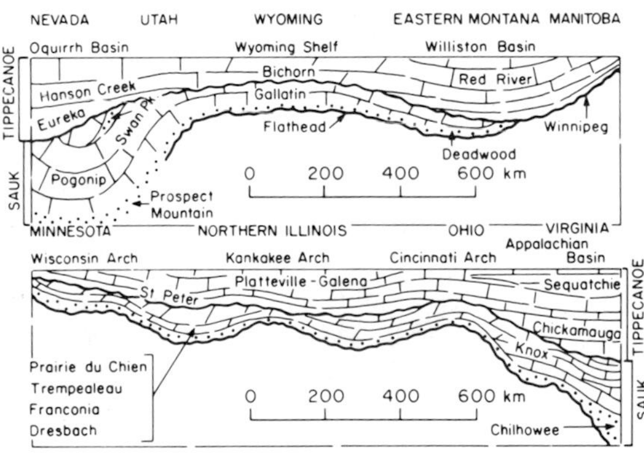

Fig. 8.2. Regional cross sections through the interior of the United States to illustrate the extent and relationships of the Sauk (white) and lower Tippecanoe (black) sequences (Sloss, 1963).

interregional scope". Their recognition can only be achieved by studying subsurface stratigraphic successions in the centers of sedimentary basins, where the record can be assumed to be relatively complete. At the basin margins and on the flanks of positive elements, there commonly are many local unconformities, which may be of only local significance but which, nevertheless, obscure broader stratigraphic relationships (Fig. 2.1). Research carried out only on surface outcrops will, of necessity, be biased by this effect, particularly in the case of the older units which may only crop out at the basin margins. Sloss (1963) pointed out that there is no apparent relationship between the prominence of an unconformity and its geographic extent or regional importance. Spectacular angular unconformities may be the product of very localized syndepositional tectonism, the evidence of which disappears completely within a few kilometers (Riba, 1976; Miall, 1978). By contrast regional unconformities may be the product of sea level change unaccompanied by tectonic tilting. They are characterized by very low structural discordance and, in many cases, little erosional relief. Very careful facies studies or biostratigraphic zonation may be required to recognize this kind of unconformity.

Sloss (1963) recognized six major unconformities within the North American continent. These are most readily recognized within the craton and become obscured within the mobile belts, on the margins of the continent. Each unconformity was followed by a major transgression and onlap, beginning on the craton margins, extending into the major intracratonic basins and, eventually, onto the margins of the Canadian Shield. Sloss noted that the transgressive phase of each sequence tends to be better preserved than the regressive phase at the end of the cycle, because the transgressive units were protected by successively overlapping sediments. By contrast the top of each cycle was immediately exposed to erosion as the sequence was terminated by regression.

The six sequences are shown diagrammatically in Figure 8.1 and a regional cross section through the lower two is illustrated in Figure 8.2. They are as follows:

Tejas Sequence
- sub-Tejas unconformity: Paleocene, c. 64 Ma

Zuni Sequence
- sub-Zuni unconformity: Early Jurassic, c. 172 Ma

Absaroka Sequence
- sub-Absaroka unconformity: late Early Carboniferous, c. 325 Ma

Kaskasia Sequence
- sub-Kaskasia unconformity: Early Devonian, c. 385 Ma

Tippecanoe Sequence
- sub-Tippecanoe unconformity: Early Ordovician, c. 480 Ma

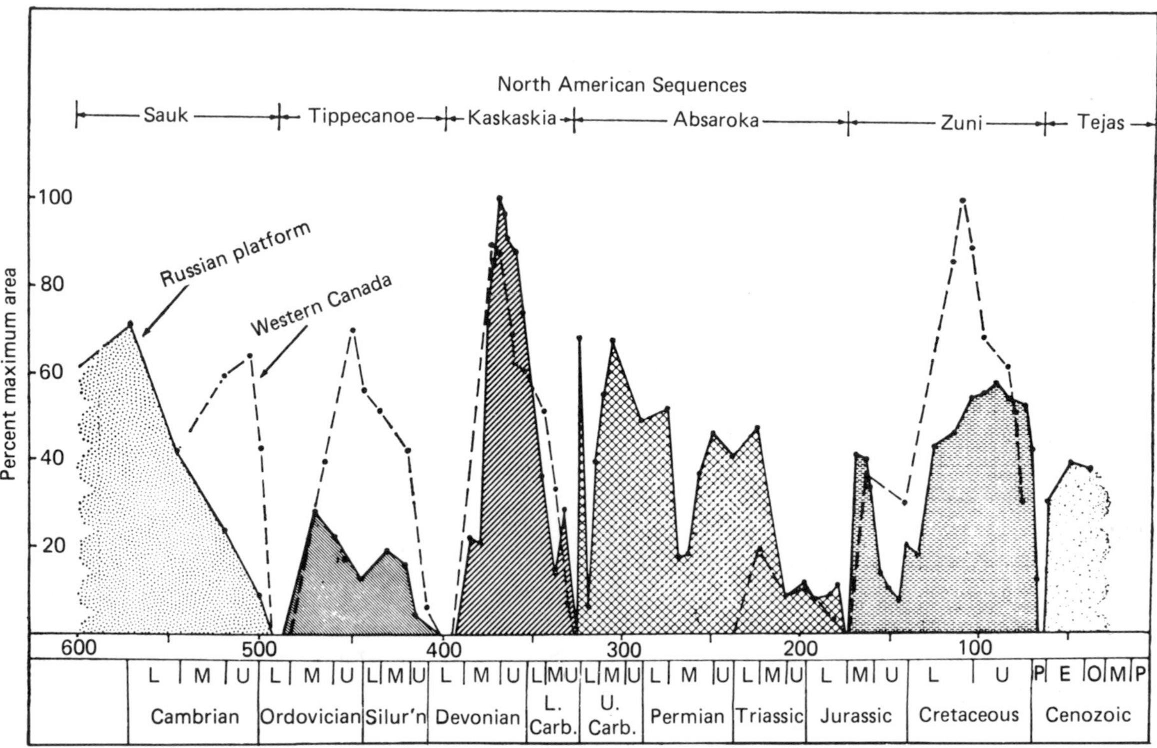

Fig. 8.3. Areas of preservation of units in western Canada and the Russian platform, showing relationship to the six sequences of Sloss (1963) (Sloss, 1972).

Sauk Sequence
sub-Sauk unconformity: Late Precambrian c 600 Ma

The names of the sequences were derived by Sloss et al. (1949) and Sloss (1963) from North American Indian tribal names. Figure 8.1 shows the approximate time ranges of the sequences as they vary from west to east across the continent. The diagram does not extend into either the Cordilleran or Appalachian regions where stratigraphic relationships have been complicated by tectonics. Note that the Sauk and Zuni sequences are characterized by slow transgression and rapid regression, whereas the Absaroka and Tejas sequences, which were formed during times of marked cratonic mobility, show a rapid basal transgression and a long regressive phase.

The inital recognition of the six cratonic sequence was based on careful lithostratigraphic and biostratigraphic correlation. In later papers Sloss (1972, 1978) took a different approach, and showed that a similar sequence chronology could be recognized in Europe and Russia. In his 1972 paper Sloss reported on an analysis of detailed isopach and lithofacies maps of the Western Canada Sedimentary Basin and the Russian Platform. The data source included 29 Canadian maps (McCrossan and Glaister, 1964) and 62 of Russia (Vinogradov and Nalivkin, 1960; Vinogradov et al., 1961). Each map was divided into a grid with intersection points spaced about 60 km apart, and the thickness and lithofacies were recorded for each point. From these data the areal extent and volume of each of the mapped units could be calculated and compared. This approach is subject to possibly serious error because of the high probability of inter-sequence and even intra-sequence erosion. The presence of a single isolated outlier or fault block beyond the edge of the main basin can change the interpreted former area of extent of a map unit by hundreds of square kilometers. Nevertheless the data from the two areas show remarkable similarities (Fig. 8.3), and the detailed statistical documentation confirms that Sloss's six sequences can be recognized in two widely separated continents that formerly would have been assumed to have undergone a quite different geological history.

In order to reduce the error inherent in the analysis of thin feather-edge cratonic remnants Sloss (1978) turned to intracratonic and pericratonic basins, as representing loci of more continuous subsidence. Figure 8.4 shows a plot of the volume of sediment per unit time for six Triassic to Recent basins on four continents, and Figure 8.5 shows his thickness per unit time plot for three basins of Ordovician to Devonian age in Russia and North America (see Sloss, 1978, for data sources). The basins are in very different plate tectonic settings, and yet some strong similarities are evident within the two plots. The sub-Zuni unconformity is represented by breaks in sedimentation or by volume minima in four of the six basins in Figure 8.4, and the sub-Tejas unconformity is similarly recognizable in all six basins. Volume maxima may be correlated between several of the basins. Particularly noticeable is a mid-Cretaceous peak between about 120 and 80 Ma. The similarities between the three plots in Figure 8.5 are even more marked. The two main thickness peaks correspond to the Tippecanoe and Kaskasia sequences, although it is of interest to note that the sedimentary break between them is at about 420 Ma, considerably older than the 385 Ma figure suggested by Sloss (1972).

Soares et al. (1978) reported a stratigraphic analysis of the three major intracratonic basins in Brazil, namely the Amazon, Parnaiba and Parana Basins, all of which contain successions spanning most of the Phanerozoic. Their interpretation of the geomorphic behaviour of these basins is given in Figure 8.6. They recognized seven sequences, which correlate reasonably closely with those of Sloss, as shown in Figure 8.7.

These Phanerozoic cycles clearly are global in scope. What is their origin? The simplest explanation is that given by Hallam (1963) to explain Late Cretaceous and Cenozoic events, namely global changes in sea level in response to volume changes of oceanic spreading centers. As discussed later, this mechanism has now been examined in detail and found to be probably the correct one, but eustatic sea level changes cannot explain all the features of the major sequences. They commonly are separated by angular unconformities (Figs. 8.2, 8.6) and in many cases the sequences contain thick continental deposits. A passive rise in sea level would terminate widespread nonmarine deposition, and therefore more must be involved.

Soares et al. (1978) described an epeirogenic cycle consisting of five phases, which explained stratigraphic events in the Brazilian basins:

1. Initial rapid basin subsidence with development of nonmarine facies and numerous local unconformities.
2. Basin subsidence slower, with deepening basin centers, marine transgression and differentiation of central marine and marginal nonmarine facies belts.
3. Development of intrabasin uplifts and local downwarps, much local facies variability.
4. Renewed basin-wide subsidence, time of maximum transgression, generally fine-grained deposits.
5. Broad cratonic uplift, return to nonmarine sedimentation.

Because the Brazilian cycles can be correlated with those in Russia and North America it suggests that the tectonic events outlined above might be intercontinental in scope. This point will be returned to later.

In conclusion, it has been demonstrated that between Russia, Europe, North America and a significant part of South America there are broadly correlatable intracratonic and pericratonic Phanerozoic stratigraphic sequences. Similar detailed analyses are not yet available for most of Africa, southern or eastern Asia or Australia, although in the next section it is shown how these gaps have been partially filled by seismic interpretation. A major control of stratigraphic events by eustatic sea level changes seems indicated but is clearly inadequate to explain all the observed events.

8.2 The contribution of seismic stratigraphy

8.2.1 New developments and remaining problems

Seismic reflections result from acoustic velocity impedance contrasts along bedding planes or unconformities, and can therefore be used as stratigraphic markers. Modern seismic processing techniques may permit resolution of stratigraphic relationships to depths of 8 or 10 km, and gross structural features even deeper. Long, deep seis-

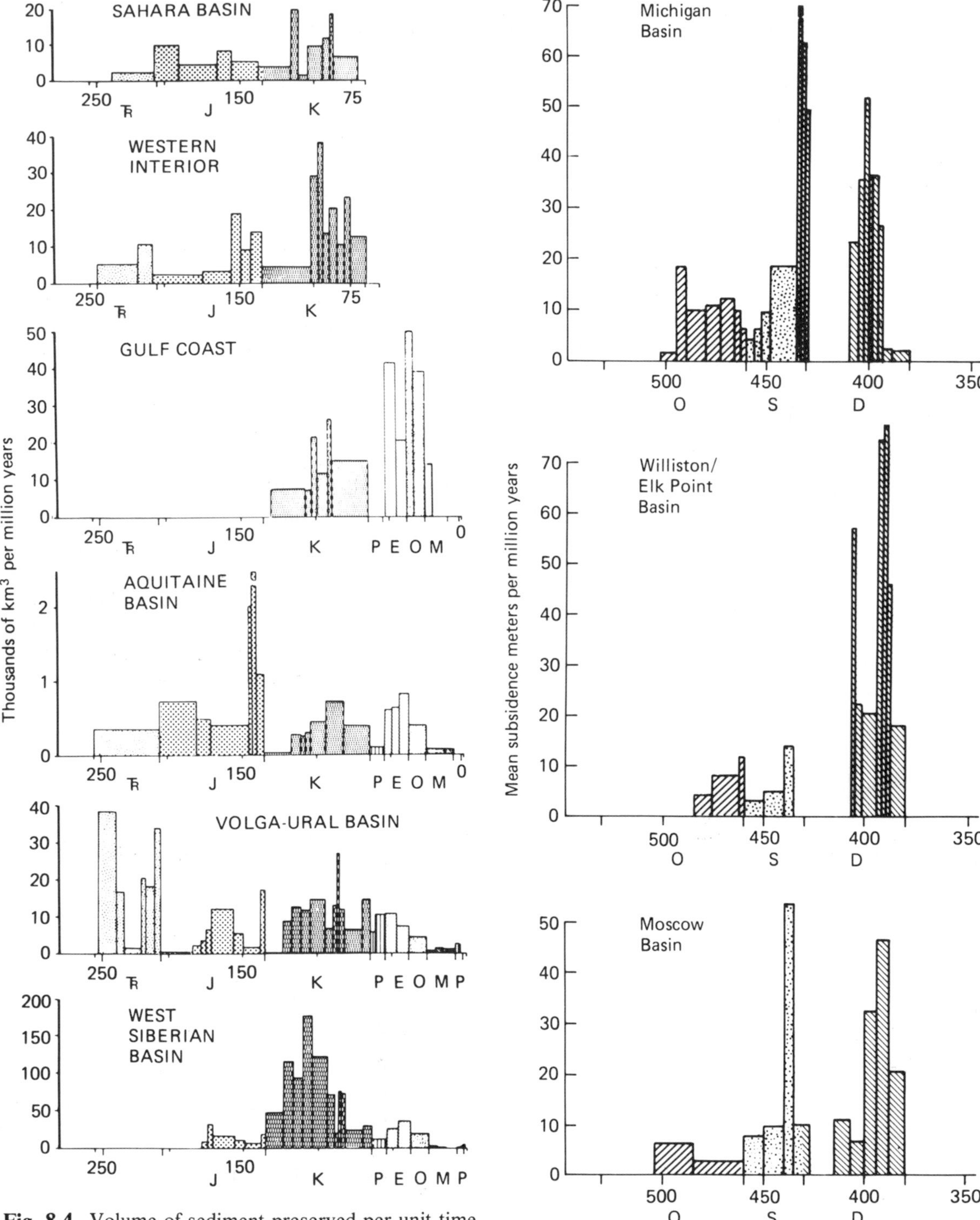

Fig. 8.4. Volume of sediment preserved per unit time in six Mesozoic–Cenozoic basins (Sloss, 1978).

Fig. 8.5. Mean thickness per unit time of stratigraphic units preserved in three mid-Paleozoic basins (Sloss, 1978).

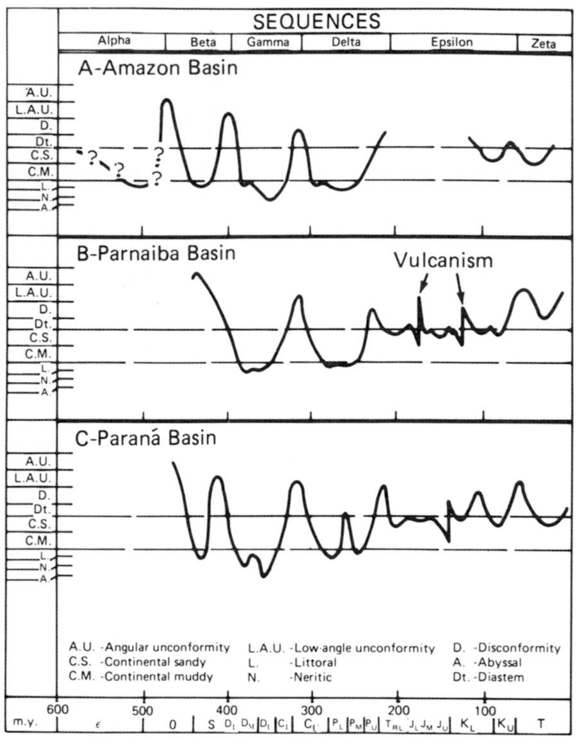

Fig. 8.6. Geomorphic expression of oscillatory movements in three Brazilian basins (Soares et al., 1978).

mic sections across a depositional basin therefore provide a picture of basin architecture that is available in no other way.

Seismic interpretation is more reliable in many cases than detailed log correlation, because the latter depends on recognition and tracing of facies variations within a succession, and these may be markedly diachronous. Figures 8.8 and 8.9, reproduced from Vail, Todd and Sangree (1977), illustrate this point using a Tertiary example from South America. The electric logs show a section consisting of interbedded sandstone, siltstone and mudstone with the sandstone beds thinning toward the south. Correlation lines suggest that the entire section between the basal unconformity and the upper datum thickens toward the south. Figure 8.9 is an interpreted seismic line along the same line of section. Seismic reflections are interpreted as high: 3300 to 4200 m/sec, mainly sandstone; medium: 2700 to 3300 m/sec, siltstone; and low: 2200 to 2700 m/sec, mudstone. The distribution of lithologies in this section is essentially the same as in Figure 8.8, but the seismic reflections suggest a slightly different pattern of correlations for the lowermost unit. The onlap of the basal sandstone onto the unconformity and

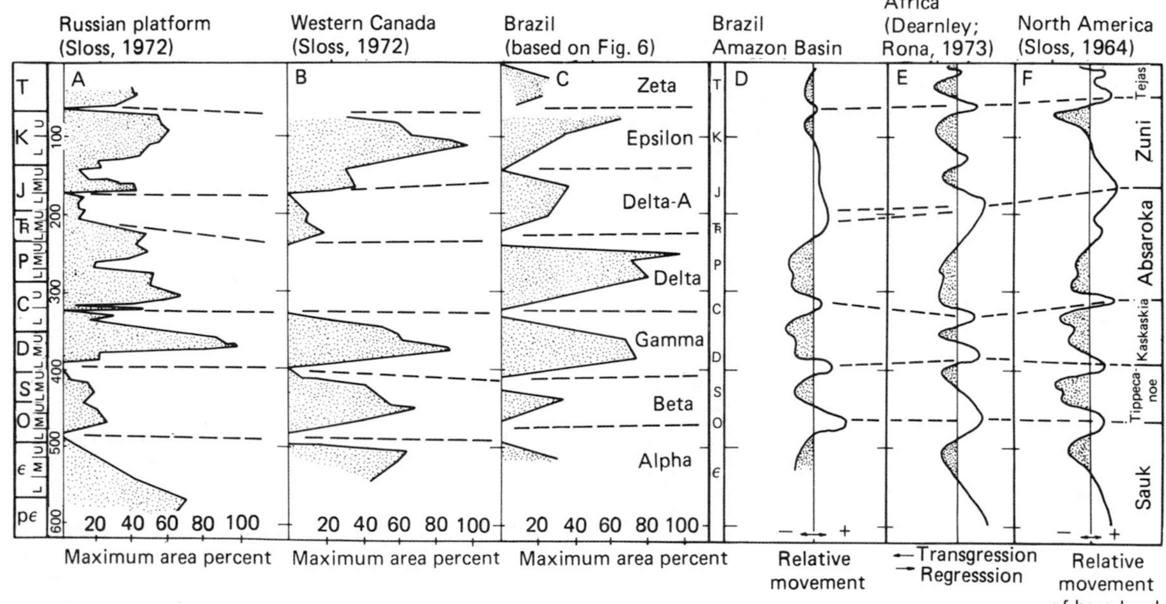

Fig. 8.7. Correlation of sequences of oscillatory movement in North American, Brazilian, European and African cratons. **A**, **B**, **C** are based on preserved sediments; **D**, **E**, **F**, on relative base-level movements (Soares et al., 1978).

Fig. 8.8. An example of electric log correlation stressing broad lithologic similarities (Tertiary, South America; Vail, Todd and Sangree, 1977).

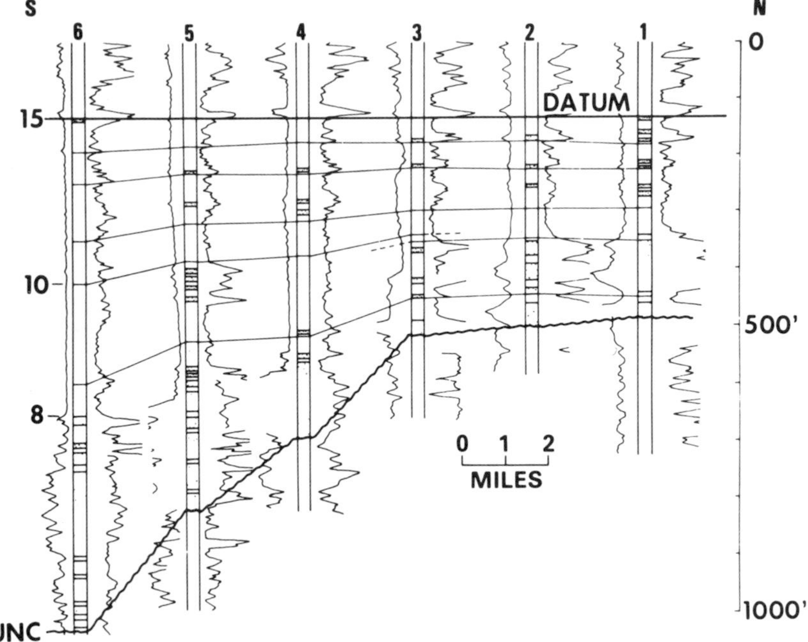

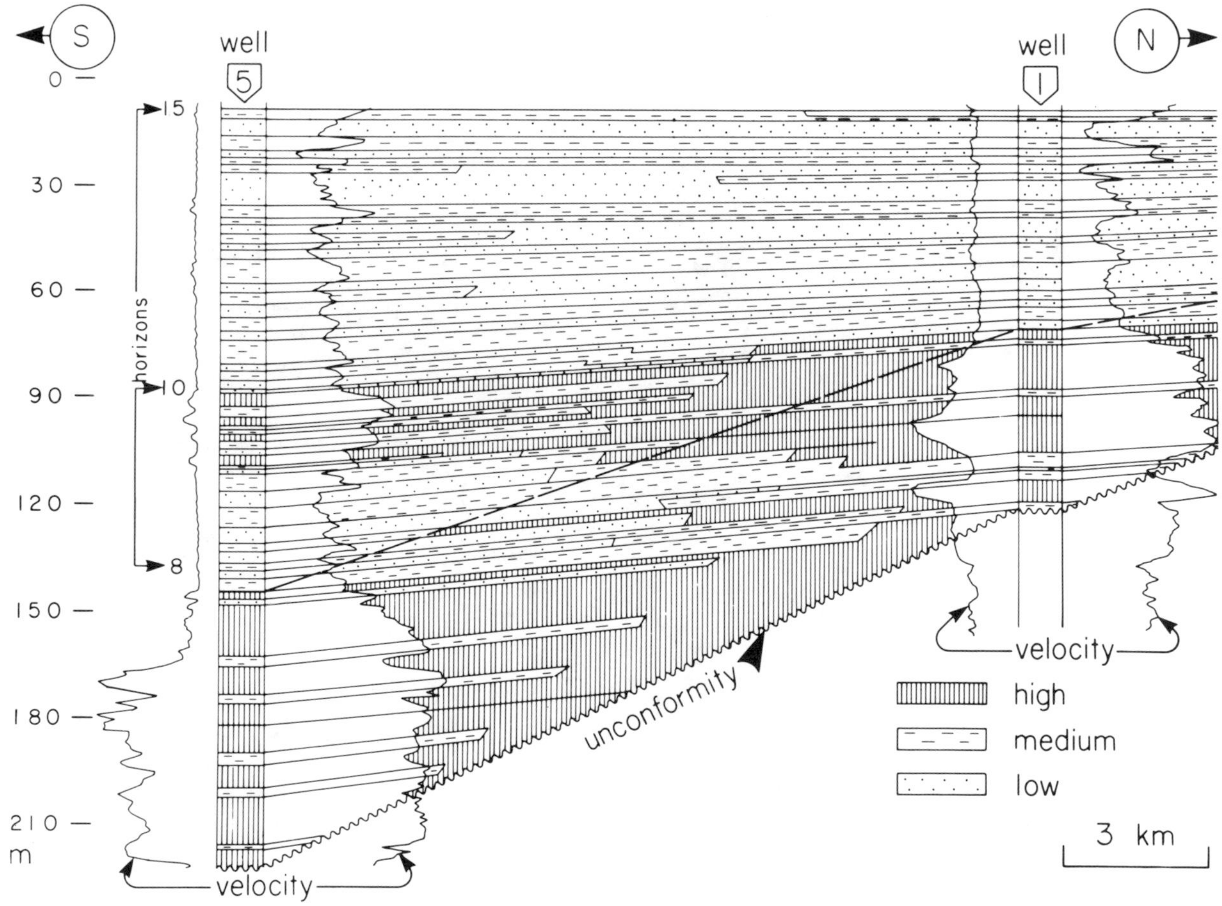

Fig. 8.9. Seismic velocity correlation between wells 1 and 5 of Fig. 8.8. Dashed line extending upward from horizon 8 at left marks top of sandy beds (Vail, Todd and Sangree, 1977).

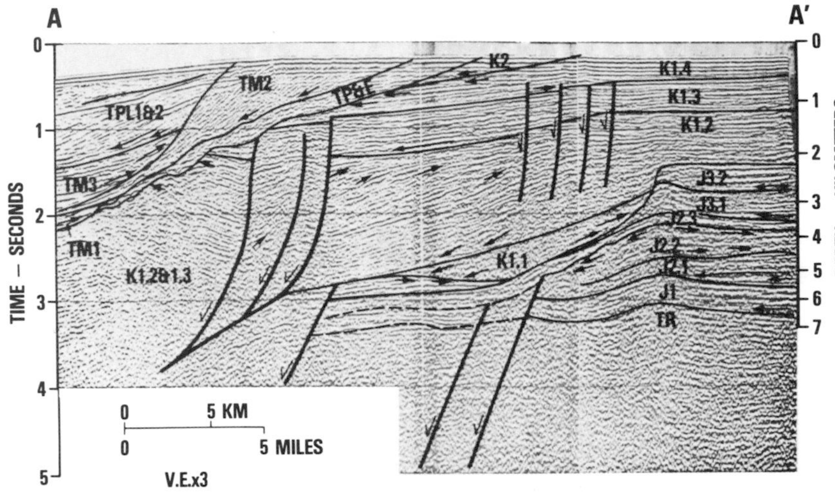

Fig. 8.10. A seismic section across the continental shelf off northwest Africa showing sequences defined by seismic reflections. Arrows indicate downlap, onlap, truncation and toplap, as defined in Chapter 6 (Mitchum et al., 1977).

Fig. 8.11. Scale of a typical seismic wave form as compared to an outcrop (based on an idea from A.E. Pallister and A.E. Wren).

its disappearance near well 3 are clearly indicated. The diagonal line drawn across the section from horizon 8 corresponds to the top of the basal sandstones in the sequence and, with only a few wells available, this line would almost certainly have been drawn in as a stratigraphic correlation by the subsurface geologist. It can be seen from Figure 8.9 that this line does not correspond to any physical changes in the succession. At best it can be viewed as a smoothed approximation of the sandstone–mudstone facies change.

Figure 8.10 illustrates a seismic section across a portion of offshore northwest Africa from Mitchum et al. (1977). It has been interpreted in terms of three major unconformity-bounded sequences and contains several major normal faults. Clearly, seismic sections of this type offer a powerful tool for testing and extending the concept of regional continental or global sequences, especially where the sections coincide or can be tied in with wells that can provide accurate biostratigraphic control.

Some words of caution are in order here, because the clarity and detail of seismic lines such as that in Figure 8.10 often blind the user to some common problems of seismic interpretation, particularly where the seismic records have not been collected and processed using the most sophisticated of available modern methods. (The illustrated section is by now a classic, having been reproduced numerous times because of its high quality. It is not necessarily a typical example.) Resolution of bedding units is still relatively crude, as illustrated in the comparison between typical seismic waveforms and the scale of actual bedding units shown in Figure 8.11. The reflections may contain multiples (analogous to echoes) which overlay and obscure deep structure, and may also be distorted by refractions from steeply dipping events such as faults or buried cliffs (section 5.4.1). Tracing of reflections must be caried out with great care, because miscorrelations frequently occur across faults or where the reflection character alters due to facies changes. Nevertheless, seismic methods have now reached a level of development at which they are having a profound impact on the science of stratigraphy.

8.2.2 Reconstructing stratigraphic sequences

A detailed analysis such as that given in Figure 8.10 must be based on a grid of seismic sections and on biostratigraphic control from one or more deep wells. Such data are now available for most major sedimentary basins in at least the western world, and can therefore provide a thorough test of Sloss's sequence concept.

The procedure described below (based on Vail, Mitchum and Thompson, 1977a, 1977b) explains how to analyze these data in order to deduce a sequence chronostratigraphy. In the next section it will be shown how the method has been used to build up a picture of globally correlatable sequences. These developed in response to eustatic sea level changes and provide ample support for Sloss's ideas.

The first step is to define the sequence-bounding unconformities. On good seismic records this should be a relatively simple step, except within tectonically very stable areas such as cratons, where the low angularity of the unconformities may inhibit recognition, or in basin centers where there may be insufficient acoustic impedance between beds of similar facies. Prominent seismic markers should then be drawn in, in order to emphasize the internal architecture of each sequence. These steps are illustrated in Figure 8.12a. The details of onlap, offlap, truncation, etc., are related to depositional environment and subsidence history and are considered in detail in Chapter 6.

The sequences must then be dated using the best available biostratigraphic and (if available) radiometric data. Biostratigraphic ages should be converted to values in years using a standard scheme such as that compiled by Van Eysinga (1975). For Cenozoic marine strata a sensitive time scale is available using a correlation of zones based on marine planktonic organisms with the sequence of magnetic reversals preserved in oceanic crust (e.g. Berggren et al., 1978; Vail, Mitchum and Thompson, 1977b). Van Hinte (1976a, 1976b) provided Jurassic and Cretaceous time scales, and other periods are discussed by Harland et al. (1964). However, the older the rocks, the greater the range of error of age determinations (this topic is discussed in some detail in Chapter 3). Several age values in millions of years are given for the sequences shown in Figure 8.12a. These then permit the construction of a chronostratigraphic chart as shown in Figure 8.12b. The seismic reflectors within each sequence can be assumed to be time markers, and can be straightened out to permit plotting of the section on a time ordinate. This shows that the

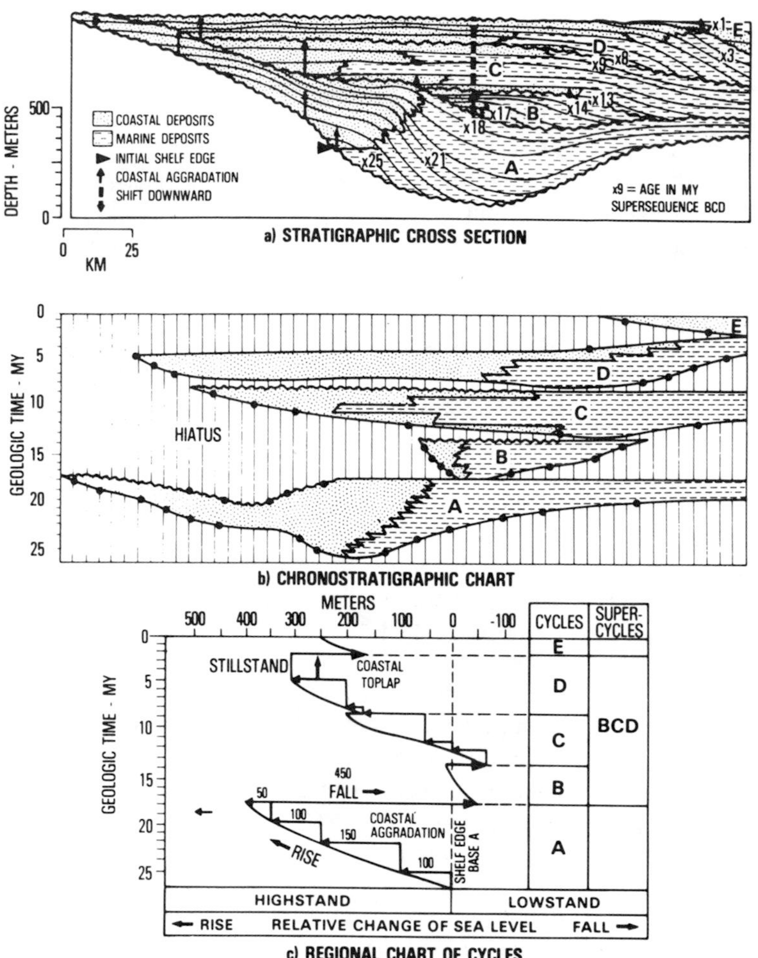

Fig. 8.12. Procedure for constructing charts of relative changes in sea level. See text for explanation (Vail et al., 1977a).

boundaries of some of the sequences are strongly diachronous, indicating slow transgression and filling of the basin up to contemporaneous base level. It also shows that the bounding unconformities are not time planes but vary in age across the section as a result of differential tilting or slight tectonic disturbance. Figure 8.12b should be compared with Figure 8.1 which shows a similar pattern developed on a much broader scale (although the two diagrams were derived by entirely different methods).

8.2.3 Global seismic correlation and eustatic sea level changes

The seismic method discussed above could be used to compile and compare chronostratigraphic sequence charts around the world and, indeed, this has been done. Figure 8.13 shows the location of regional seismic sections used by Vail, Mitchum and Thompson (1977b) to derive a global pattern of sequences. Note that most of their locations are on craton–margin or divergent, trailing-edge plate margins. This is an important point and will be returned to later.

These authors have used seismic sections and chronostratigraphic charts such as those shown in Figure 8.12a,b, to construct charts showing relative changes in sea level, as given in Figure 8.12c. Such charts have been drawn by geologists for many years based on conventional stratigraphic and sedimentological analysis (e.g. Figure 8.6), but Vail et al. (1977a) have devised a new technique for constructing these charts which purports to generate semiquantitative information about sea level changes. The method promises to become a standard one but, as noted below, it should be used with care.

Fig. 8.13. Location of regional studies of seismic stratigraphy used in construction of the global charts shown in Figs. 8.16, 8.17 and 8.18 (Vail et al., 1977b).

Figure 8.12c is a graph of relative sea level change constructed for the section shown in Figure 8.12a. Vertical aggradation and onlap of marginal marine sediments is taken to indicate a relative rise in sea level. This could be either a transgression over a stable continent (an actual sea level rise) or basin subsidence without a change of sea level. From local stratigraphic data alone it is impossible to tell which is the correct interpretation. A relative fall in sea level is indicated by offlapping sequences or by a downward shift in coastal onlap.

Increments of sea level rise and fall can be measured on the seismic section as shown in Figure 8.12. The starting point of the analysis is the beginning of a major cycle of rising sea level at the base of sequence A. Measurements are made from the assumed position of the "shelf edge" at the facies contact between "coastal" and "marine" deposits. In one million years the initial deposits of sequence A aggrade vertically by 100 m (see vertical arrow at "initial shelf edge" in Fig. 8.12a). This can be plotted as a corresponding relative sea level rise, as shown in Figure 8.12c. Successive increments plotted the same way indicate a total relative sea level rise of 400 m during the deposition of sequence A. There is then a dramatic fall in sea level before the beginning of sequence B, the total amount is indicated by the heavy dashed arrow in Figure 8.12a, and is 450 m, indicating that sea level dropped to below its starting position at the beginning of sequence A. This is clear from Figure 8.12c. Remaining sea level changes are plotted in the same way. The resulting curve is a graphical portrayal of changing relative sea level. It includes a period of stillstand at the end of sequence D time when, because of the unchanging position of sea level, sediments aggraded laterally rather than vertically.

Figure 8.12 is an hypothetical example used by Vail, Mitchum and Thompson (1977a) to illustrate the principles of seismic stratigraphic analysis and sea level change curve construction. However, it also illustrates a number of problems with the methodology which are not discussed by these authors. They state that measurements of coastal aggradation should be made as closely as possible to the underlying unconformity to minimize the effect of differential basin subsidence. However, very little differential subsidence is indicated in their section. Seismically-defined time lines are parallel for much of sequences C and D and nearly so for much of sequences A and B. In practice this is rarely the case. Stratigraphic units almost invariably thicken toward a basin center because of differential subsidence, and so the measurements of relative sea level change would show quantitative differences depending on where the analysis was carried out within the basin. Secondly, and related to this problem, units may not show the dramatic onlap relationships illustrated in Figure 8.12a but may all taper to a feather edge near a continental hinge line, so that the method described above could not be used. Thirdly, it should be pointed out that "coastal deposits" may include significant thicknesses of nonmarine sediments deposited above sea level. These must clearly be separated out prior to the analysis, but it may be very difficult to do this from seismic data alone. Seismic facies analysis (Chaps. 5, 6) may provide some clues.

Many onlap relationships are developed by deep marine (e.g. submarine fan) deposits building up the continental slope (e.g. Figs. 6.19, 6.34, 6.40) in a very similar pattern to that shown in Figure 8.12a. Care must be taken to distinguish these relationships because, of course, deep marine onlap provides no useful information about sea level.

Figure 8.12c and, indeed, all the curves of sea level change published by Vail et al. are asymmetric. Rises in sea level are shown to be relatively slow, whereas falls are shown to be very rapid. In fact in all the curves falls are indicated by horizontal lines, suggesting that they took place instantaneously. Obviously a change in sea level takes longer than this but, nevertheless, Vail et al. claim that in all the seismic records examined the same pattern is apparent. This seems unlikely to many geologists (Kerr, 1980), and independent work by Pitman (1978) indicated that falls in sea level may be just as slow as the rises. A slow fall in sea level should be recognized in the stratigraphic record by offlap, but Vail et al. (1977a, p. 72) stated that they have not observed this on seismic data. A partial explanation may be that a fall in sea level results in subaerial exposure and erosion, so that the evidence of offlap may be lost. This is illustrated in Figure 8.14. Units 1 to 4 are deposited during a rise in sea level from A to B. Sedimentation is accompanied by differential subsidence. During a subsequent slow fall in sea level from B to C units 5, 6 and 7 are deposited with an offlap relationship to each other and to unit 4. Subaerial erosion would remove the landward edge of these offlap wedges, possibly down to base level C. Without independent facies or bathymetric evidence the end result might be interpreted as seaward progradation taking place

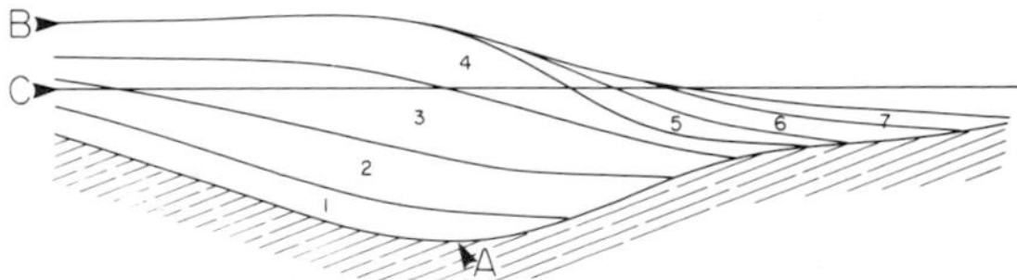

Fig. 8.14. Development of an offlap relationship obscured by subsequent erosion. See text for explanation.

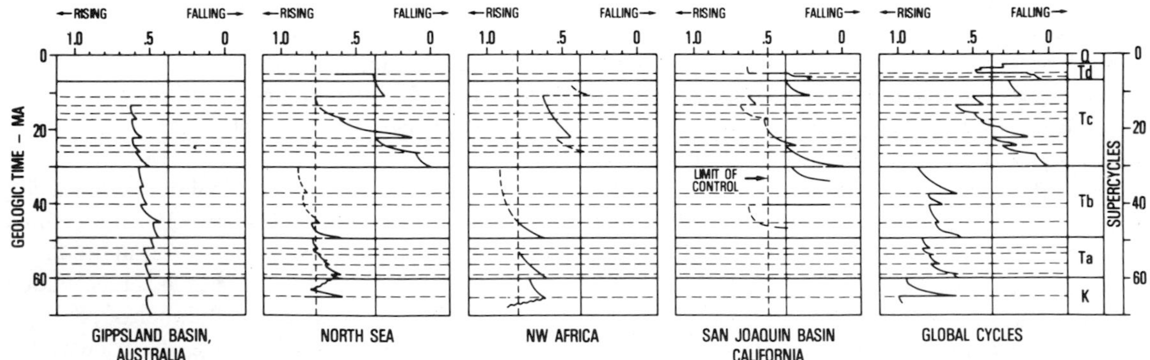

Fig. 8.15. Correlation of regional cycles of relative change of sea level from four continents, showing how these have been averaged to produce the global cycle chart (Vail et al., 1977b).

during a sea level stillstand, as was unit D in the hypothetical example in Figure 8.12. Kerr (1980) summarized other views on this controversy.

The major thesis of the Vail et al. (1977b) work is that cycles of sea level change such as those illustrated in Figure 8.12 can be correlated around the world, indicating that they are not a response to local tectonic events but the result of global or eustatic sea level changes. It is stated that much evidence has been amassed to demonstrate this idea, but the only evidence actually offered is a correlation between five Cretaceous to Recent basins in a chart reproduced here as Figure 8.15. A series of major and minor sea level changes have been correlated between these basins, producing what is at first sight a convincing proof of the eustatic sea level model. The model probably is valid—it is supported by the independent research of Sloss, that was discussed earlier (perhaps not entirely independent. Vail was a student of Sloss!), plus that of many others working in individual basins (see below), but the evidence as presented here seems suspect. In Figure 8.15 correlation lines are all based on sudden sea level falls. The questionable validity of this interpretation has been discussed above. Secondly, it is rare in geology for the evidence to permit such precise correlation of events that they can be indicated by the straight lines encompassing the globe shown in Figure 8.15. There is still much disagreement over the exact age of magnetic anomalies (e.g. see Fig. 8.16 and Chapter 3), and the chronostratigraphic age of biostratigraphic zones, even in the Cenozoic, is characterized by a certain experimental error which increases with the age of the sediments (Chapter 3). Yet there is no indication of these possible errors in Figure 8.15. If such data are not provided it is difficult to allow for revisions and refinements.

Global correlations such as that given in Figure 8.15 permitted Vail et al. (1977b) to construct a chart showing sea level changes relative to magnetic reversal and biostratigraphic events for the Cenozoic (Fig. 8.16), a less detailed chart for

Table 8.1. Global highstands and lowstands of sea level and associated major interregional unconformities during Phanerozoic time

SEA LEVEL HIGHSTANDS	MAJOR GLOBAL SEA LEVEL FALLS	SEA LEVEL LOWSTANDS
	PRE-LATE PLIOCENE & PRE-PLEISTOCENE (3.8 & 2.8 MA)	LATE PLIOCENE - EARLY PLEISTOCENE
EARLY & MIDDLE PLIOCENE		
	PRE-LATE MIOCENE & PRE-MESSINIAN (10.8 & 6.6 MA)	LATE MIOCENE
MIDDLE MIOCENE		
	PRE-MIDDLE LATE OLIGOCENE (30 MA)	MIDDLE LATE OLIGOCENE
LATE MIDDLE EOCENE & EARLY OLIGOCENE		
	PRE-MIDDLE EOCENE (49 MA)	EARLY MIDDLE EOCENE
LATE PALEOCENE - EARLY EOCENE		
	PRE-LATE PALEOCENE (60 MA)	MID-PALEOCENE
CAMPANIAN & TURONIAN		
	PRE-MIDDLE CENOMANIAN (98 MA)	MID-CENOMANIAN
ALBIAN - EARLIEST CENOMANIAN		
	PRE-VALANGINIAN (132 MA)	VALANGINIAN
EARLY KIMMERIDGIAN		
	PRE-SINEMURIAN (190 MA)	SINEMURIAN
NORIAN & MIDDLE GUADALUPIAN		
	PRE-MIDDLE LEONARDIAN (270 MA)	MID-LEONARDIAN
WOLFCAMPIAN & EARLIEST LEONARDIAN		
	PRE-PENNSYLVANIAN (324 MA)	EARLY PENNSYLVANIAN
OSAGIAN & EARLIEST MERAMECIAN		
	PRE-DEVONIAN (406 MA)	EARLY DEVONIAN
MIDDLE SILURIAN		
	PRE-MIDDLE ORDOVICIAN (490 MA)	EARLY MIDDLE ORDOVICIAN
LATE CAMBRIAN & EARLY ORDOVICIAN		
		EARLY CAMBRIAN & LATEST PRECAMBRIAN

the Late Triassic to Recent (Fig. 8.17) and a generalized chart for the entire Phanerozoic (Fig. 8.18). These diagrams promise to become amongst the most widely reproduced illustrations in the history of geology, because they provide a key as important as that of plate tectonics for understanding worldwide stratigraphic events. In these charts the rises and falls for individual regions have been averaged to produce a global curve, and this is given in terms of a relative change, because it differs in amount from region to region depending on local tectonic events (e.g. Fig. 8.15).

The pre-Jurassic cycles are constructed mainly from North American data (Vail et al., 1977b, p. 88). For rocks of this age range correlations with oceanic events are not available (the oldest oceanic crust is probably Middle Jurassic), and so Vail and his co-workers presumably used much the same data as did Sloss, namely subsurface information from the continental interior. It is therefore hardly surprising that the first four of Sloss's (1963) sequences (Fig. 8.1) are almost identical to the corresponding "supercycles" in Figure 8.18. Younger supercycles differ considerably from those of Sloss because of the present day availability of a wholly new and more accurate data base.

Table 8.1 is a listing of the major high and low sea level stands and times of rapidly falling sea level in the Phanerozoic, as determined by Vail et al. (1977b). It is instructive to compare this with Sloss's (1963) table for North America (section 8.1), which was based on much less precise data, particularly for the Jurassic and younger rocks. In light of the reservations expressed above about the chronostratigraphic accuracy of the seismic work it is also instructive to compare Table 8.1 and the sea level charts with the independent stratigraphic documentation of Hallam (1978), Ager (1981) and others. Some nagging discrepancies are apparent, which should warn against too facile an acceptance of the validity of any such tabulation.

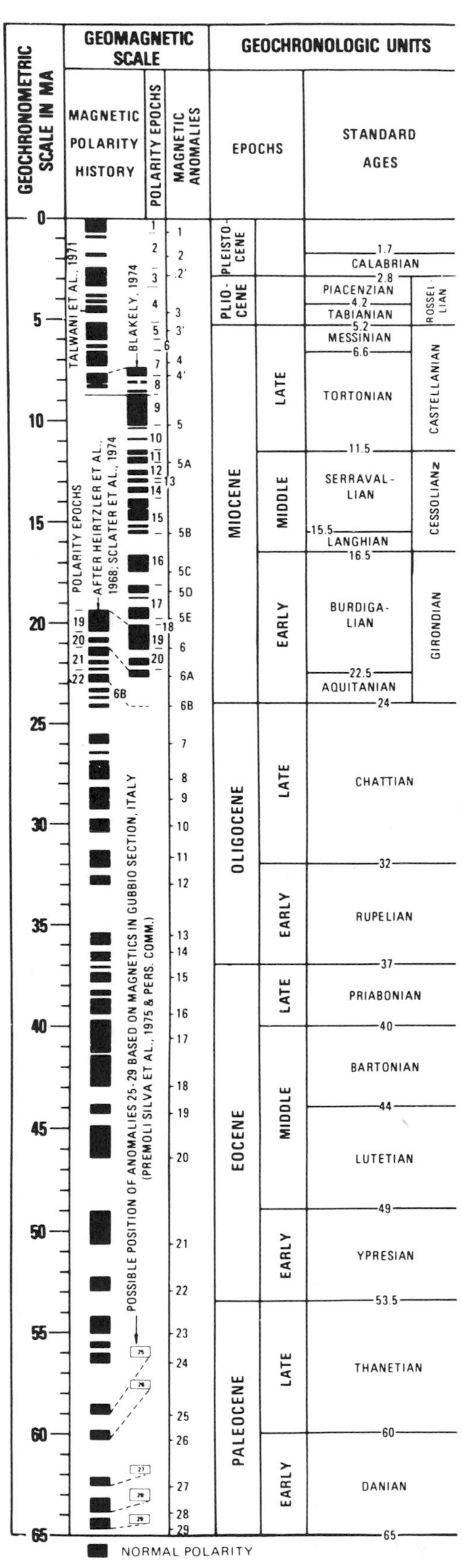

▷

Fig. 8.16. Global cycles of relative change of sea level during the Cenozoic (Vail et al., 1977b).

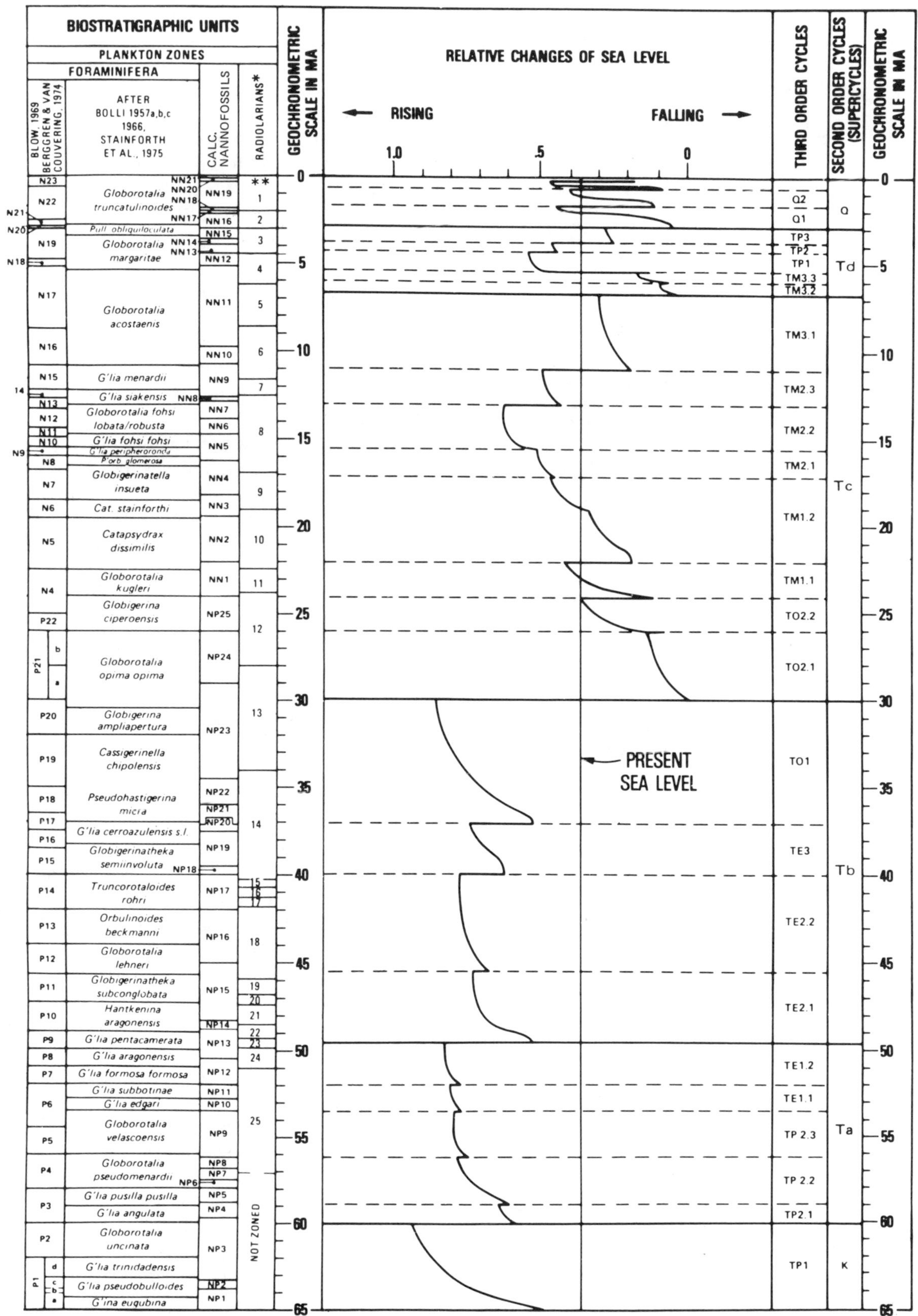

BIOSTRATIGRAPHIC UNITS
PLANKTON ZONES
FORAMINIFERA
BLOW 1969 BERGGREN & VAN COUVERING, 1974
AFTER BOLLI 1957a,b,c 1966, STAINFORTH ET AL., 1975
CALC. NANNOFOSSILS
RADIOLARIANS*
GEOCHRONOMETRIC SCALE IN MA
RELATIVE CHANGES OF SEA LEVEL
RISING
FALLING
1.0
.5
0
THIRD ORDER CYCLES
SECOND ORDER CYCLES (SUPERCYCLES)
GEOCHRONOMETRIC SCALE IN MA
PRESENT SEA LEVEL
Globorotalia truncatulinoides
Pull. obliquiloculata
Globorotalia margaritae
Globorotalia acostaensis
G'lia menardii
G'lia siakensis
Globorotalia fohsi lobata/robusta
G'lia fohsi fohsi
G'lia peripheroronda
P'orb. glomerosa
Globigerinatella insueta
Cat. stainforthi
Catapsydrax dissimilis
Globorotalia kugleri
Globigerina ciperoensis
Globorotalia opima opima
Globigerina ampliapertura
Cassigerinella chipolensis
Pseudohastigerina micra
G'lia cerroazulensis s.l.
Globigerinatheka semiinvoluta
Truncorotaloides rohri
Orbulinoides beckmanni
Globorotalia lehneri
Globigerinatheka subconglobata
Hantkenina aragonensis
G'lia pentacamerata
G'lia aragonensis
G'lia formosa formosa
G'lia subbotinae
G'lia edgari
Globorotalia velascoensis
Globorotalia pseudomenardii
G'lia pusilla pusilla
G'lia angulata
Globorotalia uncinata
G'lia trinidadensis
G'lia pseudobulloides
G'ina eugubina
NOT ZONED
Q2
Q1
Q
TP3
TP2
TP1
TM3.3
TM3.2
Td
TM3.1
TM2.3
TM2.2
TM2.1
TM1.2
TM1.1
TO2.2
TO2.1
Tc
TO1
TE3
TE2.2
TE2.1
Tb
TE1.2
TE1.1
TP 2.3
TP 2.2
TP2.1
Ta
TP1
K

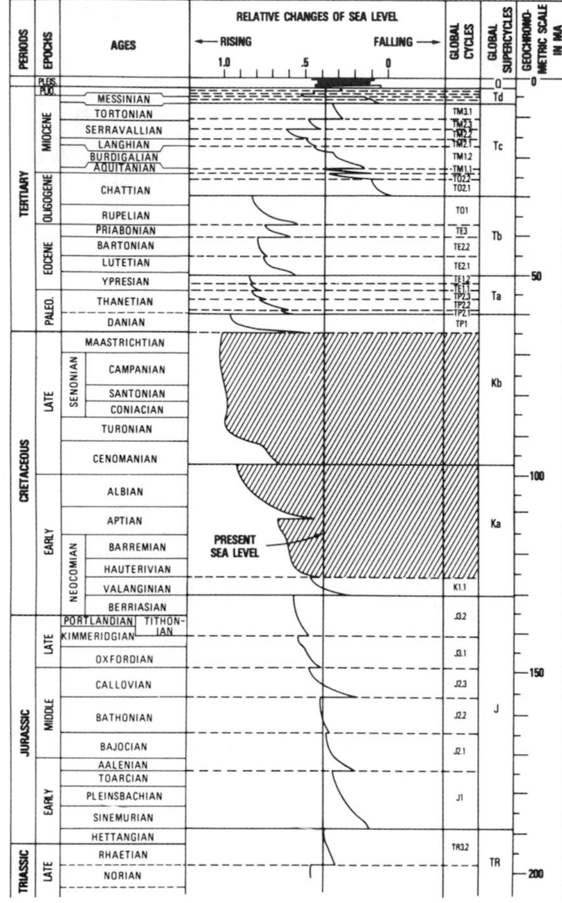

Fig. 8.17. Global cycles of relative change of sea level during the Jurassic–Tertiary (Vail et al., 1977b). Details of Cretaceous cycles have not been released for publication.

8.3 Cycles within cycles

8.3.1 Time scales of stratigraphic sequences

There are at least three, probably four, scales of sequence development in the stratigraphic record, related to a range of time scales shown by sea level changes.

First order cycles, as defined by Vail et al. (1977b), are illustrated in Figure 8.18. They include the two extended periods of maximum marine transgression in the Phanerozoic, the Late Cambrian to Mississipian, and the Cretaceous, and a period of maximum regression in the Pennsylvanian to Jurassic. A glance at the geological map of North America confirms the importance of these broad changes. The Canadian Shield is flanked by Cretaceous rocks resting unconformably on a Cambrian or Ordovician to Devonian sequence over wide areas of the craton, from the Great Lakes region, across the Prairies into the Beaufort–Mackenzie region and the Arctic Platform. Rocks of Pennsylvanian to Jurassic age are largely confined to intracratonic basins and the mobile belts flanking the cratonic interior.

Second order cycles are termed supercycles by Vail et al. (1977b). They correspond in part to the sequences defined by Sloss (compare Figs. 8.1 and 8.18). Supercycles are equivalent in importance to periods, and range in length from about 10 to 100 Ma.

Third order cycles vary in length from less than 1 Ma to about 10 Ma. Global correlation by seismic stratigraphy has resulted in a symbolic terminology (as illustrated in Figs. 8.16 and 8.17), for example, subdivision of the mid-Cenozoic supercycle Tb into third order cycles TE2.1, TE2.2, TE3 and TO1 (Fig. 8.16). The use of this terminology is illustrated in the annotation of seismic sections such as that shown in Figure 8.10. Abbreviated notation of this type also has obvious advantages for input into computer processible data files of stratigraphic depth and thickness data.

A fourth order of cyclicity can be erected for stratigraphic events corresponding to the relatively very rapid changes of sea level during glacial advance and retreat. These would have durations in the order of 10^4 to 10^6 years. This is suggested by the sea level change graph at the top of Figure 8.16, which shows rapid rises and falls during the last million years. Changes of this rapidity may leave little permanent stratigraphic mark, because they are comparable in rate to that of isostatic visco-elastic crustal adjustment and to rates of sedimentation. Pleistocene sequences that can be related to such cycles are widely developed, but these are surficial sediments and their long term preservation potential is debatable. Some possible ancient examples are discussed in the next section.

As discussed in section 8.4 there are several causes for this wide range in time scales. Second, third and fourth order cycles are those of most concern to practicing basin analysts. Second order cycles have been discussed in preceding

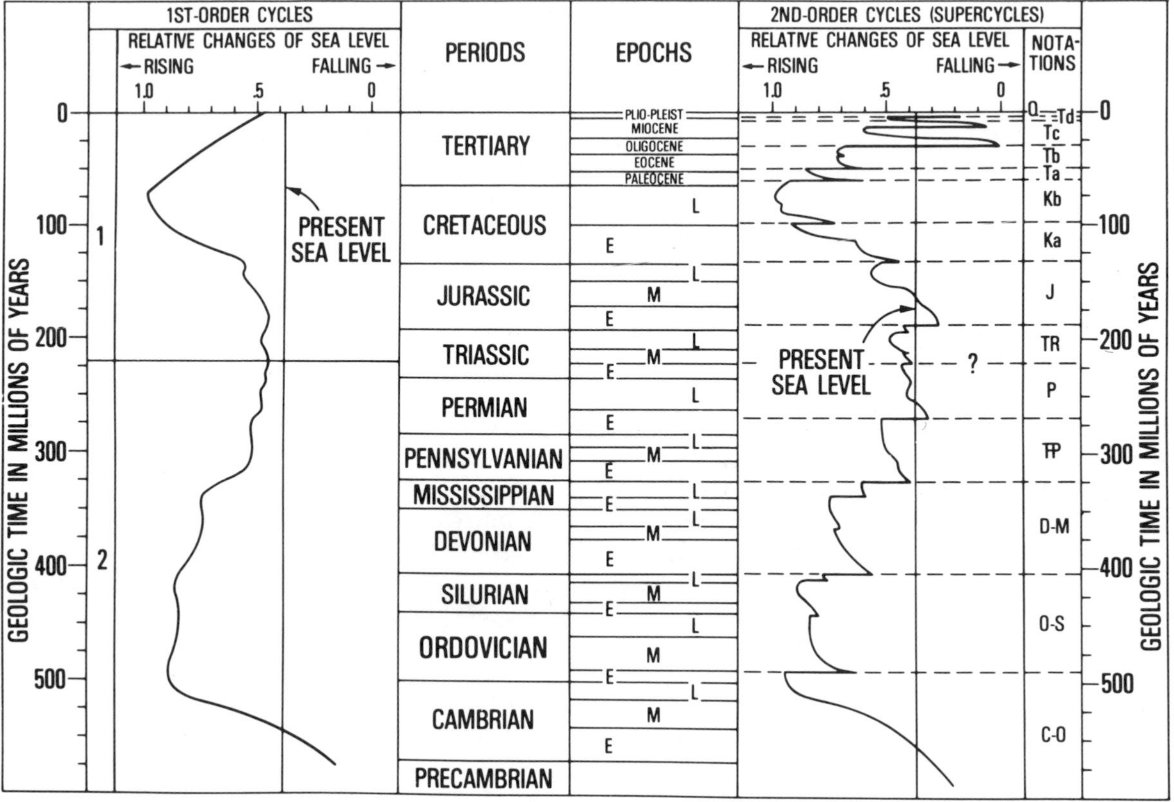

Fig. 8.18. First and second order cycles during the Phanerozoic (Vail et al., 1977b).

sections of this chapter, and the next section provides some examples of third and fourth order cycles.

8.3.2 Examples of third and fourth order cycles

It is the function of biostratigraphers to examine correlations on a regional, continental or global scale, and it is therefore not surprising that recognition of many of the widespread stratigraphic events in the Phanerozoic has stemmed from their work. In this section we shall survey briefly three examples of attempted continental or global cyclic correlation which rely strongly on detailed lithostratigraphic analysis and biostratigraphic zonation. They show how seismic stratigraphy may be amplified and clarified by careful examination of outcrops and cores.

Sea level changes in the Ordovician and Silurian. The most sensitive indicators of sea level changes are the rocks formed on stable cratons, beyond the influence of tectonic uplift or downwarp and free of the effects of major detrital influxes. Sea level changes in shallow epeiric seas may result in the shifting of lithofacies belts and ecological zones (biofacies) over large distances, generating marked local facies and faunal changes. Typical shelf facies are carbonate, evaporite and mature or very fine grained clastic sediments (quartzose, sandstone, siltstone, mudstone). Klein (1982) has provided a general model for cratonic cyclicity, as discussed in section 9.3.6.

In the Ordovician and Silurian marine shelf, faunas consisted mainly of brachiopods, trilobites, corals, stromatoporoids and bryozoans. These comprised what is informally termed the "shelly" fauna. It contrasts with the graptolitic fauna of deeper water, continental slope and abyssal oceanic environments.

Several authors have attempted to subdivide lower Paleozoic shelf faunas into depth-controlled communities. Ziegler (1965) recognized five brachiopod-dominated assemblages in the early Silurian, which he named after typical genera. They are *Lingula, Eocoelia, Pen-*

tamerus, Stricklandia and *Clorinda*, in order of increasing water depth. These communities map out in bands parallel to the shore in shelf sequences in Wales, the Appalachian Basin, New Brunswick and Iowa (McKerrow, 1979). The communities are not related to distance from shore as the shelf width varies from 5 to more than 100 km, and they are not related to sediment character, as each community occurs in a variety of rock types. Similar faunal differentiation has been established for the Upper Ordovician and the remainder of the Silurian in a few areas. Sea level changes over the shelf should be accompanied by lateral shifts in these communities, which should be recognizable in vertical sections through the resulting sediments. McKerrow (1979) used this approach, plus supplementary facies data, to construct depth change curves for the Middle Ordovician to Early Devonian in thirteen locations in Europe, Africa, North and South America. The result is shown in Figure 8.19. Some of the depth changes can be correlated between many or all of the regions examined, and undoubtedly reflect eustatic sea level changes. Others are more local in scope and probably were caused by regional tectonic events.

McKerrow (1979) distinguished two types of eustatic depth change, slow and fast. Slow changes include the rise in sea level during the Llandovery and the fall in the Ludlow and Pridoli. Fast changes include the rapid rise in latest Llandeilo and earliest Caradoc time, a short-lived fall at the end of the Ashgill, and a rise and fall at the beginning of the Upper Llandovery. The slow rise and fall during the Silurian took about 40 Ma to complete and is of second order or supercycle rank. It is recognizable as a secondary peak on the Vail et al. curve (Figure 8.18) and encompases the upper part of the Tippecanoe sequence of Sloss (1963). The various rapid changes that took place within this cycle are third or fourth order cycles of 1 to 2 Ma duration.

At least one of these rapid fluctuations in sea level is attributed to withdrawal of water from the sea during a major ice age. Beuf et al. (1966) and Bennacef et al. (1971) documented a major continental glaciation during the Late Ordovician in the Algerian Sahara, and it has subsequently been reported in Europe, South Africa and South America (Lenz, 1976; Hambrey and Harland, 1981). It ranged in age from late Caradoc to Llandovery, possibly early Late Llandovery, a time span approaching 20 Ma. A major continental ice phase seems certainly to have occurred at the end of Ashgill time, as this period is represented by a widespread, if short-lived, shallowing in most of the sections studied by McKerrow (Fig. 8.19), and by a widespread disconformity in platform sequences throughout much of Western and Arctic Canada, as shown in Figure 8.20 (from Lenz, 1976). Berry and Boucot (1973) also reported widespread changes in marine benthic faunas at this time.

Johnson and Campbell (1980) studied detailed biofacies variations in Llandoverian carbonate strata of northern Michigan, which onlap the basal Silurian unconformity, and demonstrated the existence of six cycles averaging about 3 Ma in length which correlate well with those documented in Iowa and New York (Fig. 8.19). Johnson and Campbell (1980) defined three depth-related facies:

1. A shallow water, possibly intertidal facies characterized by laminated, stromatolitic dolomite with fucoids and ostracodes. Desiccation cracks and synaeresis structures attest to intermittent exposure and salinity changes.
2. A subtidal shelf facies containing abundant tabulate corals, stromatoporoids and stromatolites, probably located above wave base.
3. A deeper shelf environment characterized by thin-shelled Pentamerid brachiopods, which normally preferred relatively quiet water conditions below wave base.

Recurrence of these biofacies through the nearly 200 m of Llandovery section defines cyclic sea level changes, as shown in Figure 8.21. These short-term fluctuations may reflect glacial and interglacial episodes, though they indicate sea level changes in the order of only a few tens of meters, considerably less than the 100 to 200 m associated with major ice ages such as that affecting the northern hemisphere during the Plio-Pleistocene.

A quite different type of sedimentary cyclicity is demonstrated by the Upper Silurian (Ludlow–Pridoli) carbonate–evaporite rocks of the central Michigan Basin. These consist of alternating micritic carbonates and halite–anhydrite beds. Pinnacle reefs developed within the carbonate layers and these are surrounded by the evaporites (Fig. 8.22). There has been much argument in the literature about whether or not the pinnacle reefs

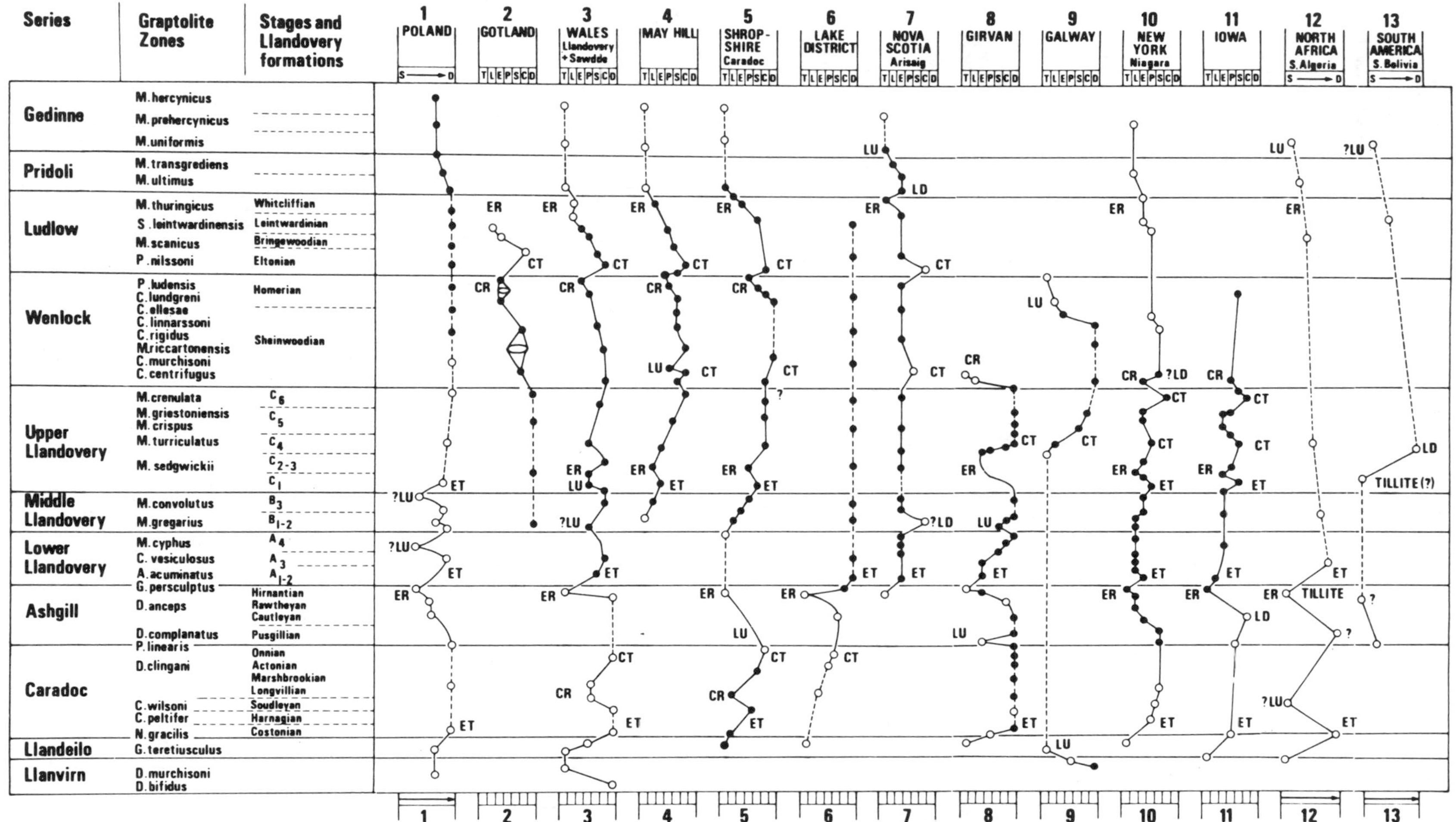

Fig. 8.19. Depth changes on three continents during the Ordovician and Silurian. Symbols at head of column: S, D = shallow, deep; TLEPSCD = terrestrial, **Lingula**, **Eocoelia**, **Pentamerus**, **Stricklandia**, **Clorinda**, deepening (see text for explanation). Symbols on graphs: E = eustatic, C = continental, L = local, T = transgression, R = regression, U = uplift, D = deepening. Open circles indicate uncertainty of depth and/or age. Dashed lines indicate terrestrial, unfossiliferous or very deep environments which do not yield good depth control (McKerrow, 1979).

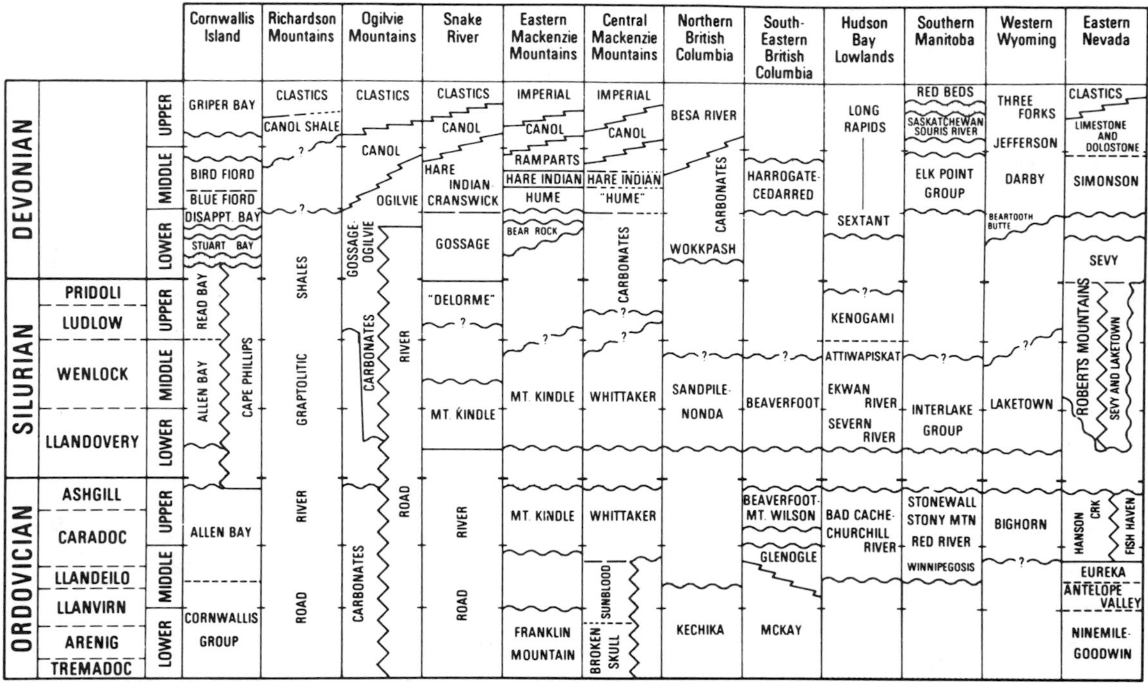

Fig. 8.20. Correlation of Ordovician to Devonian sections in selected areas of western and Arctic North America, illustrating the widespread Late Ordovician—Early Silurian stratigraphic break. The Road River and Cape Phillips Formations, which straddle this break, are basinal graptolitic mudrocks. The remainder are mainly platform deposits, predominantly carbonates (Lenz, 1976).

were formed contemporaneously with the evaporites. Contemporaneous deposition would require that a coral–crinoidal–algal community was capable of withstanding saline waters and it would indicate that there must be a complete lateral facies change between the reefs and the evaporites, which cannot, in fact, be demonstrated.

Mesolella et al. (1974) developed an alternative model based on cyclical changes in sea level (Fig. 8.23). Pinnacle reef and inter-reef carbonates formed during high sea level stands, while evaporites were deposited during periods of low sea level and basin restriction. The pinnacle reefs may have been exposed to the atmosphere during this second stage. They provided a base for renewed colonization and reef growth when sea level rose again.

Fossils are sparse in these rocks, and so the cycles cannot easily be correlated with McKerrow's (1979) curves. Pinnacle reef growth stopped early in the Pridoli, and evaporites occupied a progressively larger proportion of the basin, suggesting a gradual overall shallowing. This, at least, is compatible with the general lowering of sea level indicated in Figure 8.19.

Carboniferous cyclothems and mesothems. The word **cyclothem** was coined by Wanless and Weller (1932) for spectacular coal-bearing cyclic sequences of Pennsylvanian and Permian age in the United States. The cycles generally are each a few tens of meters in thickness and typically are repeated dozens of times in any one basin. For example more than 100 repetitions have been mapped in Kansas (Moore, 1964). These cycles were first recognized in the mid-nineteenth century and are known from many parts of the central and eastern United States, Europe, Russia and North Africa. There is a vast literature discussing the nature and origin of the cycles, a summary of which is beyond the scope of this book.

A diagrammatic section of two typical cyclothems in Kansas is shown in Figure 8.24, with interpretations indicated (from Moore, 1964; Crowell, 1978). Some individual beds within the cyclothems can be traced for more than 300 km. Figure 8.25 illustrates how a typical cyclothem in Illinois varies as individual components thicken and thin. Thick sandstones and shales represent deltaic units deposited in regions of local subsidence.Nonmarine sandstones may rest on an

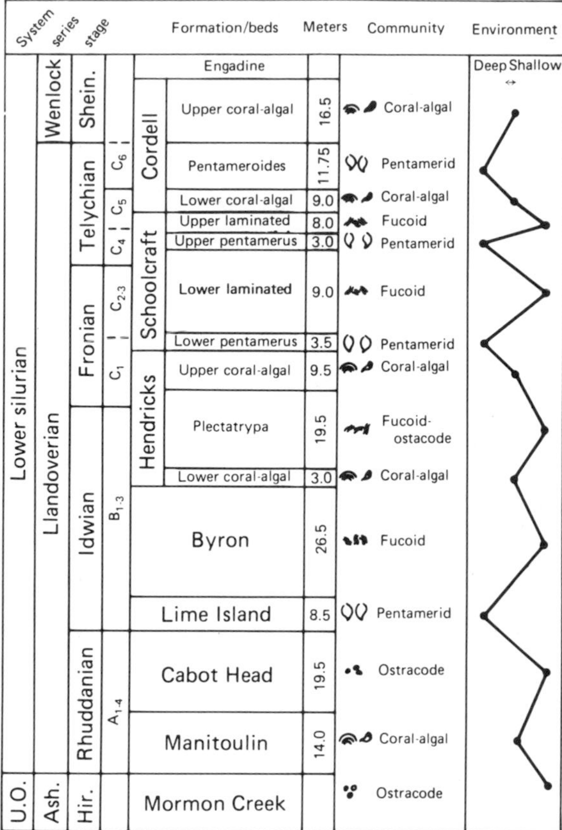

Fig. 8.21. Composite stratigraphic section of Lower Silurian rocks of northern Michigan, illustrating variations in biofacies and interpreted fluctuations in water depth (Johnson and Campbell, 1980).

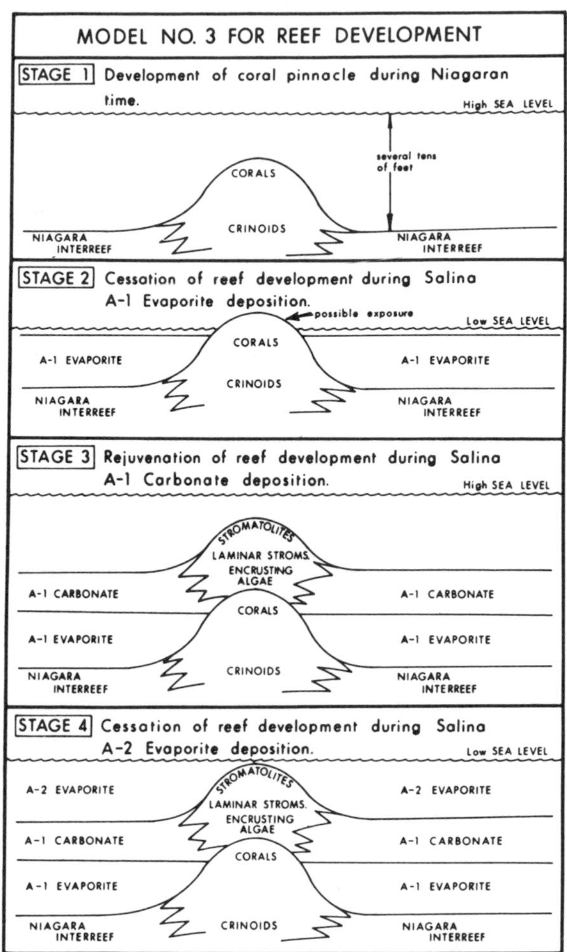

Fig. 8.23. A model for reef growth and evaporite deposition based on fluctuating sea level (Mesolella et al., 1974).

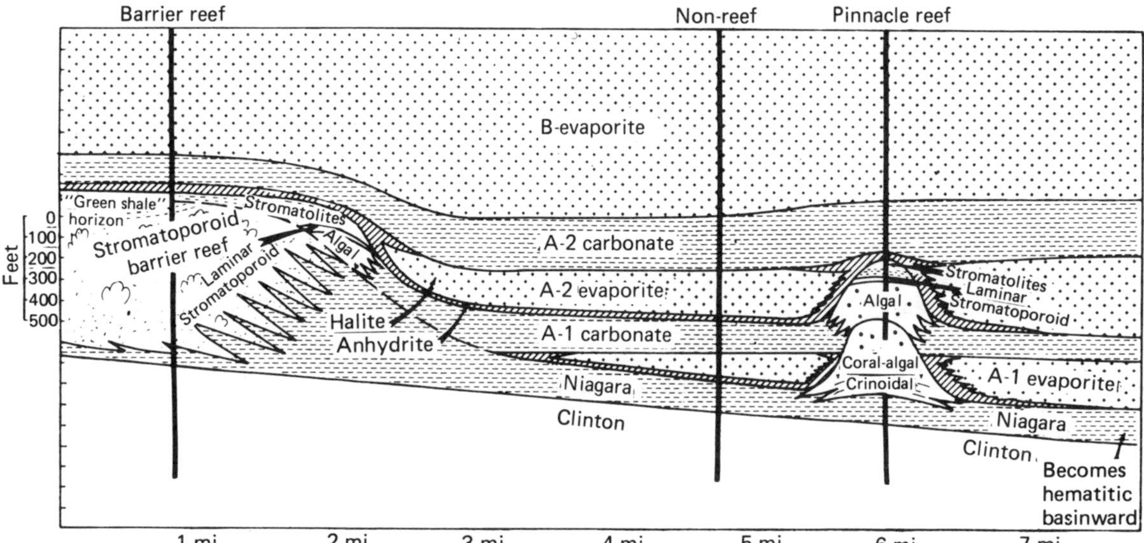

Fig. 8.22. Generalized cross section through Silurian reefs of northern Michigan Basin, showing relationship to evaporites (Mesolella et al., 1974).

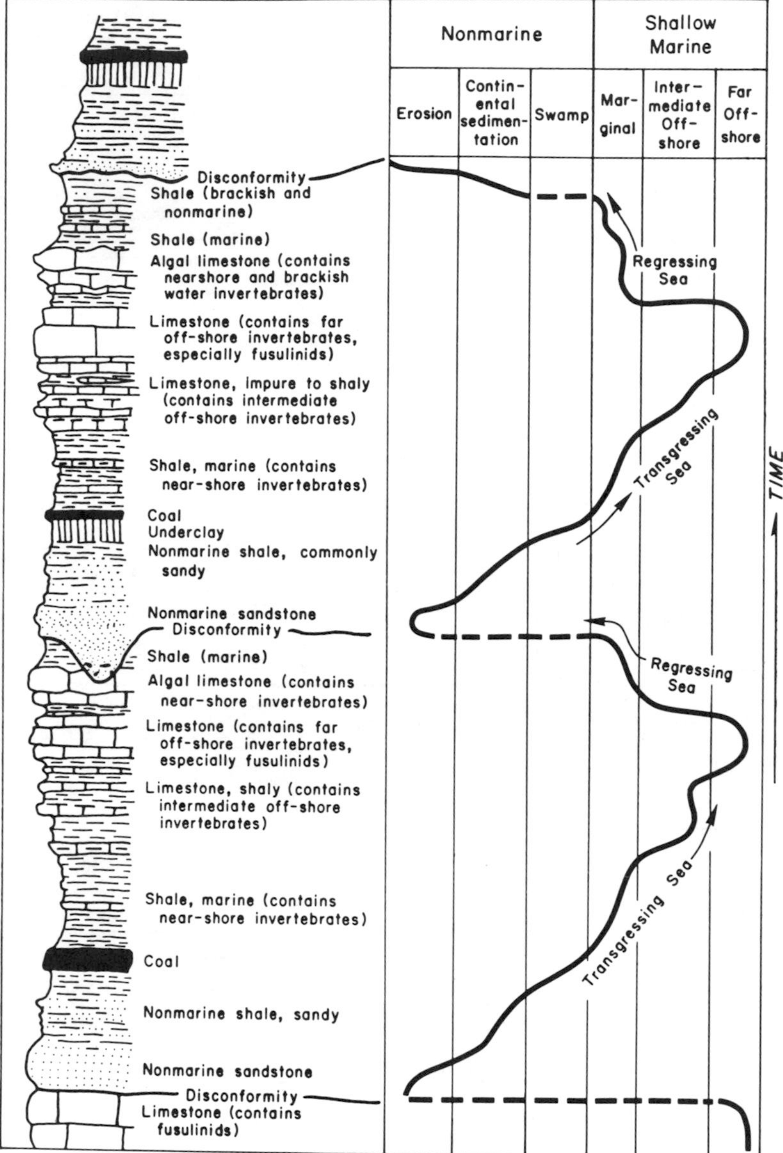

Fig. 8.24. Two typical Carboniferous cyclothems, showing interpreted development in terms of changing water level (Moore, 1964, with interpretation by Crowell, 1978).

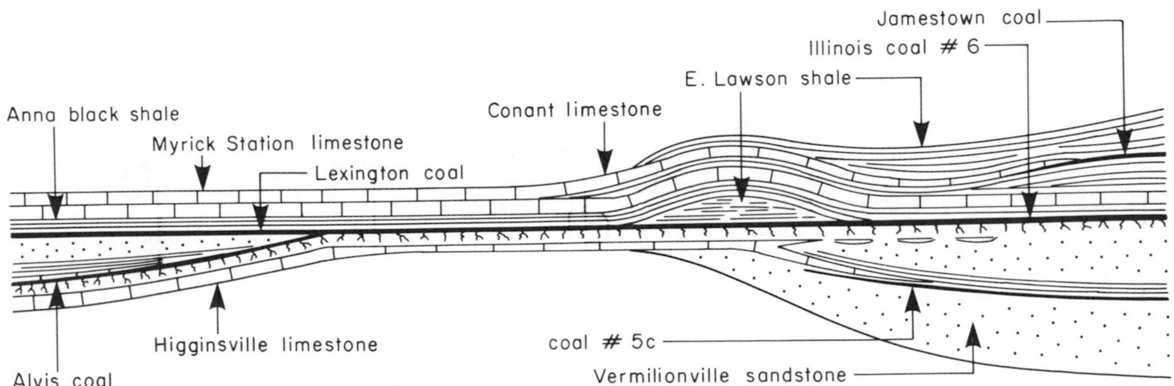

Fig. 8.25. Generalized cross section through a cyclothem, illustrating lateral variations (Wanless, 1964).

erosional, channelized unconformity at the top of the underlying marine units. Coals, underclays, black shales and marine limestones typically retain considerable uniformity over hundreds of kilometers (Wanless, 1964). In cratonic areas away from detrital sources the nonmarine clastic units may thin to zero, so that the marine limestones rest on each other.

In Europe cyclothems may be correlated using widespread marker beds and "marine bands", which contain goniatite faunas. Ramsbottom (1979) has recognized that cyclicity occurs on three scales, of which the cyclothem is the smallest. The cyclothems themselves show rhythmic variations, enabling them to be grouped into what Ramsbottom (1979) termed **mesothems**, and he adopted the term **synthem** for the largest scale of cyclicity, which would correspond to second order cycles in the Vail et al. (1977b) terminology (Fig. 8.26). There is a possibility for

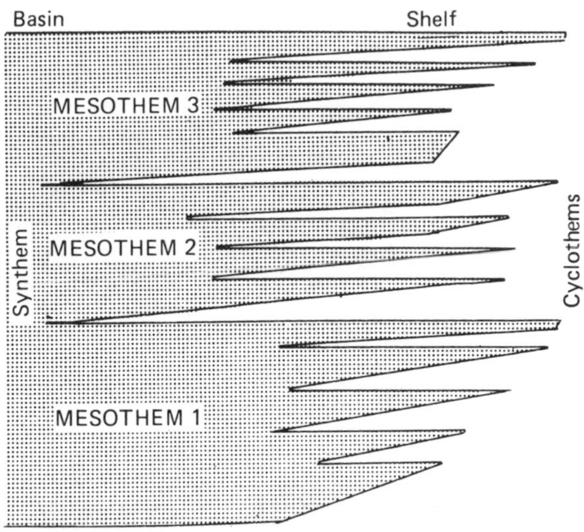

Fig. 8.26. Three scales of cyclicity in the Carboniferous of northwest Europe (Ramsbottom, 1979).

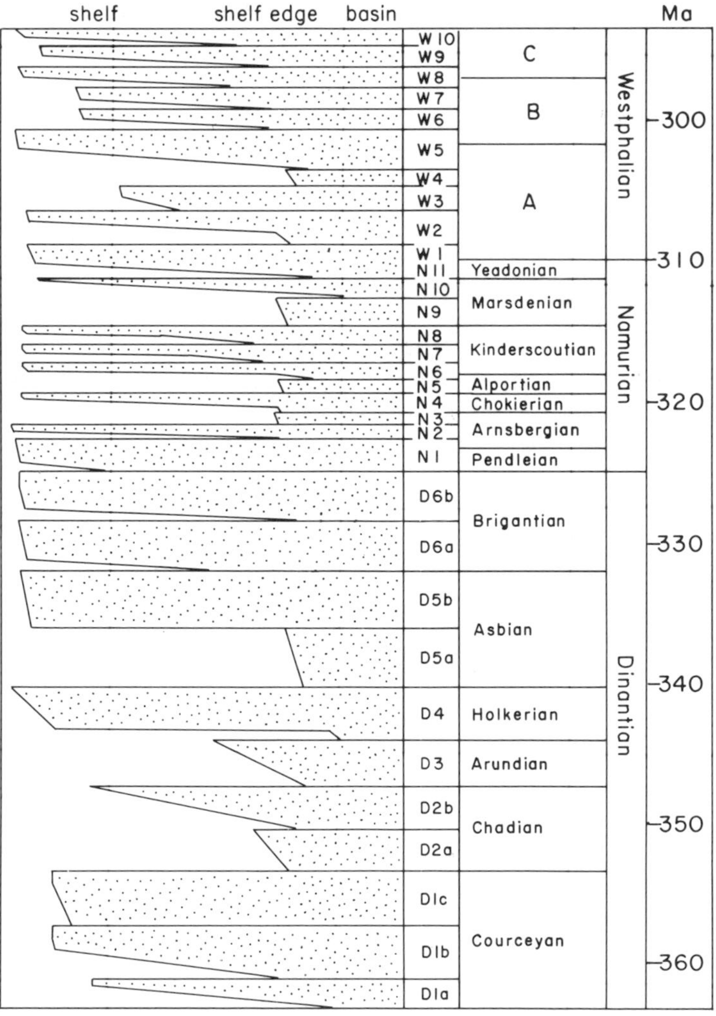

Fig. 8.27. Extent and duration of Carboniferous mesothems, northwest Europe (Ramsbottom, 1979).

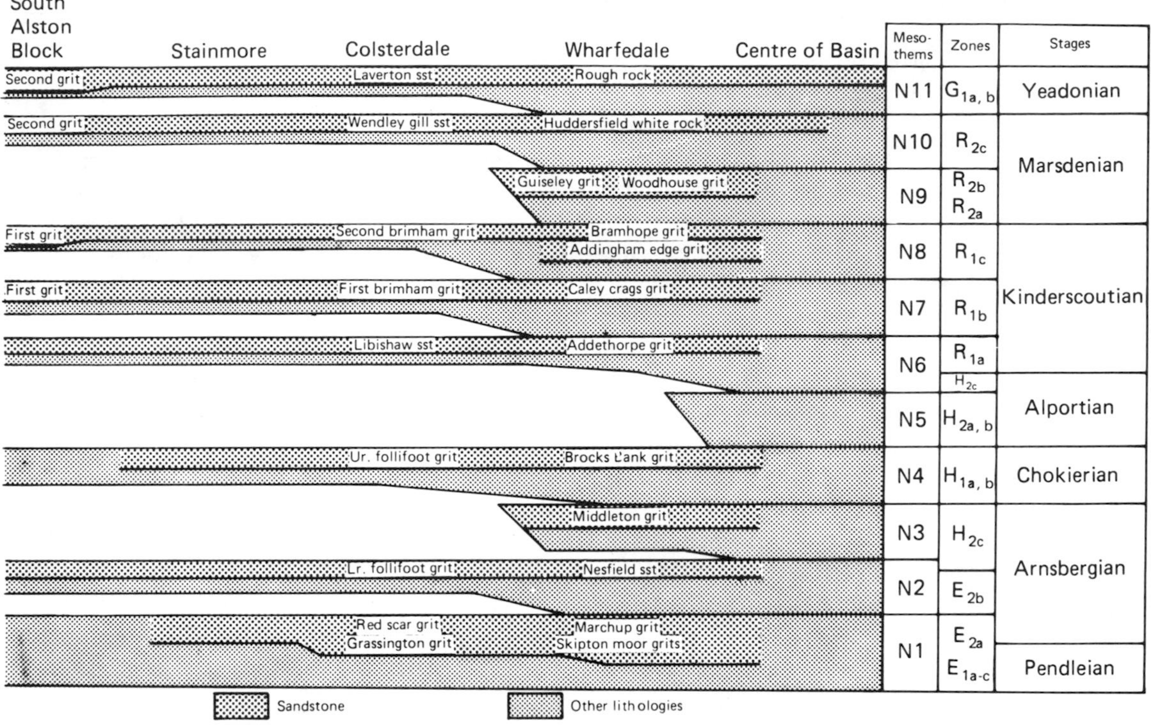

Fig. 8.28. Chronostratigraphic section through Namurian mesothems of northern England. Grit is a local term for sandstone. Each sandstone is followed by a disconformity on the shelf (left side of diagram) but sedimentation is continuous in basin centre (Ramsbottom, 1979).

confusion here, as Hedberg (1976) employed the word synthem as a general term for unconformity-bounded sequences. Ramsbottom (1979) showed that in northwest Europe the Dinantian to Westphalian-C can be subdivided into 34 mesothems. These ranged in duration from 1.1 Ma in the Namurian to an average length of 3.6 Ma in the Dinantian (Fig. 8.27). They are therefore typical third order cycles. Cyclothems ranged in duration from 225,000 to 500,000 years.

In the Namurian each mesothem consists of a muddy sequence at the base containing several cyclothems, followed by one or more sandy cyclothems. The lower, muddy parts of the mesothems are broadly transgressive, although the transgressions were slow and pulsed, each cyclothemic transgression reaching further than its predecessor out from the basin onto the shelf (Fig. 8.28). The regression at the end of each mesothem appears to have occurred rapidly. The sandy phase commonly commenced with turbidites, and is followed by thick deltaic sandstones (commonly called Grits in the British Namurian; see Fig. 8.28). The cycles may be capped by coal. On the shelf areas each mesothem is bounded by a disconformity (Fig. 8.28), but sedimentation probably was continuous in the basins. Deltaic progradation was rapid, approaching the growth rate of the modern Mississippi delta.

Ramsbottom (1979) noted that the Namurian mesothems were of the shortest duration, and many do not extend up onto the shelf. This point in time, near the Mississippian–Pennsylvanian boundary, is interpreted as the boundary between the Kaskasia and Absaroka Sequences in North America (Sloss) and between two of the Vail et al. (1977b) global supercycles (Figs. 8.1, 8.18). Saunders et al. (1979) claimed that an incomplete record of the mid-Carboniferous mesothems can be recognized in the Ozarks, Arkansas.

There is by no means universal agreement that cyclothems and mesothems may be correlated over such wide areas as was claimed by Ramsbottom (1979) and Saunders et al. (1979). For example George (1978) provided a detailed critique of Ramsbottom's work, based in part on a meticulous examination of the biostratigraphic record. Some geologists reject the concept of the cyclothem entirely. They note the numerous variations in the cyclothem sequence (e.g. Fig. 8.25) and maintain that the concept of an "ideal" or "model" cyclothem is a dubious one (e.g. Duff et al., 1967). The development of the facies model methodology during the last twenty or so years has given geologists an entirely new way of analyzing cyclic sequences. Application of the "process" approach, using vertical profiles and Walther's Law, has shown that many cyclic relationships are the result of such processes as point-bar lateral accretion, or deltaic progradation, or the shoaling of tidal flats (section 4.5.8). Ferm (1975) showed that most variations in the cyclothem sequence could be explained by the progradation and abandonment of deltas. He interpreted each cyclothem as a complex clastic wedge, which he termed the "Allegheny Duck" model, because of the fancied resemblance of the model cross section to a flying duck. (Ferm also expressed doubts about his own model in the 1975 paper, but it seems a good one to the writer.) That cyclothems can now be explained in modern sedimentological terms does not mean that the concept and term "cyclothem" are no longer valid. The widespread nature of these distinctive sequences, their restriction to rocks of Pennsylvanian and Permian age and, in spite of the doubts of George and others, the fact that individual cyclothems and mesothems can be correlated for considerable distances, all call for a special interpretation.

It seems likely that cyclothems are the product of fourth order eustatic sea level fluctuations induced by repeated glaciation. This is an old idea (Wanless and Shepard, 1936; Wanless, 1950, 1972), and has recently been revived by Crowell (1978). He showed that the Carboniferous–Permian glaciation of Gondwana had the same time span as the cyclothem-bearing sequences of the northern hemisphere. The lower part of each cyclothem was formed during a transgression, which drowned clastic coastal plain complexes and developed a sediment-starved shelf on which carbonates were deposited (Fig. 8.24). This phase represents the melting of continental ice caps. With renewed cooling and ice formation regression began, with the initiation of rapid deltaic progradation. The erosion surface at the base of the nonmarine sandstone (Fig. 8.24) may represent local deltaic channeling or widespread subaerial erosion. Variations in thickness and composition of the cyclic sequences were caused by local tectonic adjustments (epeirogenic warping, movement on fault blocks, etc.).

Much work needs to be done on the chronology of the Gondwana glaciation to determine if its stratigraphy can, indeed, be related to the cyclothem–mesothem record. Until this is documented the proposed genetic link will remain only as an intriguing possibility, and many sedimentologists will remain unconvinced that there is anything unique about the cyclothems themselves.

Cretaceous cycles. The great eustatic sea level rise in the Cretaceous created vast epicontinental seas throughout the world. Superimposed on this first order cycle are two second order cycles and several third order cycles. These can readily be identified in the sedimentary record of cratonic areas, where changes in sea level caused large shifts in the coastline.

A typical cyclic sequence is shown in Figure 8.29, and Figure 8.30 illustrates the occurrence of two such cycles in the Colorado–Kansas area based on stratigraphic work by Weimer (1960), Hattin (1964) and Kauffman (1969). Each of the two transgressive–regressive cycles took place over about 7 Ma, and therefore they rank as third

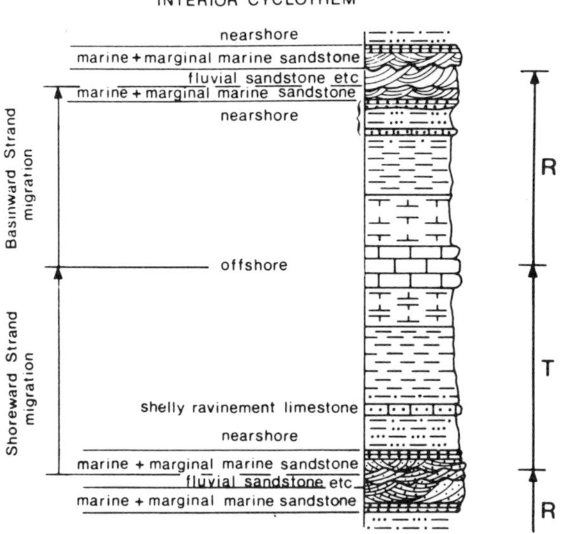

Fig. 8.29. Typical Upper Cretaceous mesothem and its interpretation as the product of transgression (T) and regression (R) (Kauffman, 1969; Hancock and Kauffman, 1979).

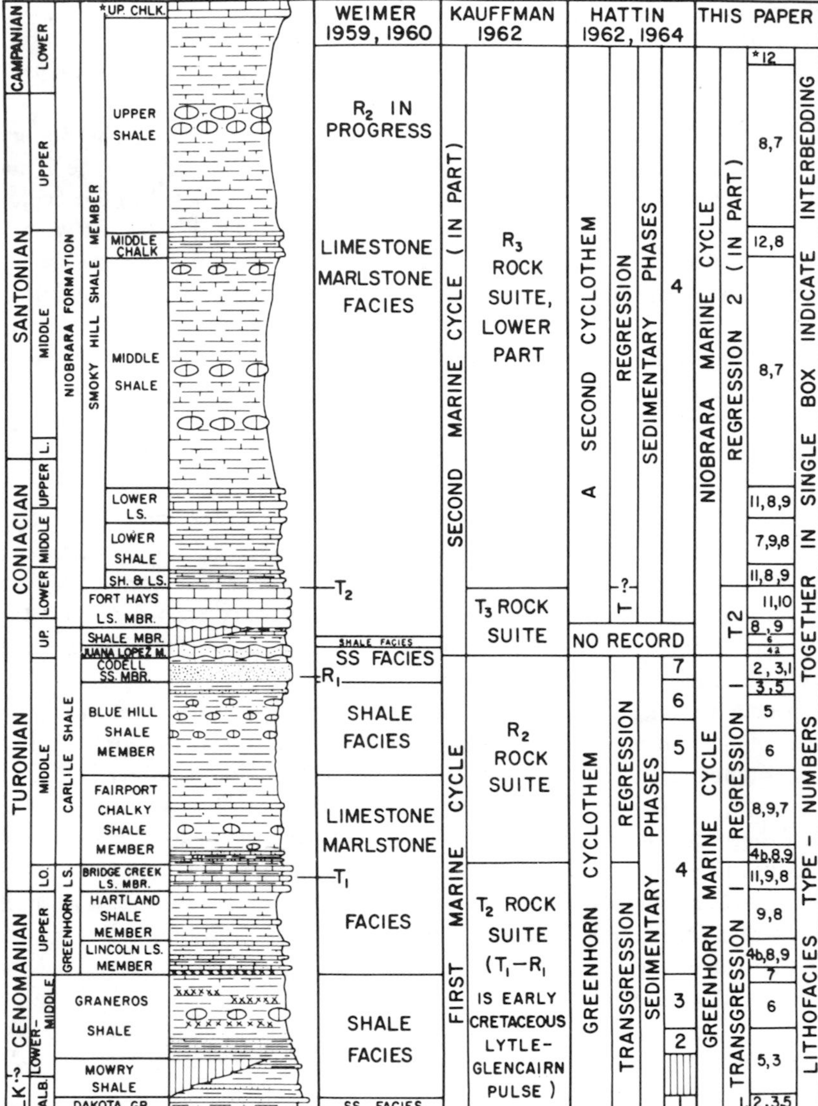

Fig. 8.30. Two mesothems in the Upper Cretaceous of the Colorado-Kansas area. Lithofacies numbers (last column) are shown in Fig. 8.31 and are discussed in text. Lithofacies type-numbers shown together in a single box indicate interbedding (Kauffman, 1969).

order cycles or mesothems, and not cyclothems, as stated by Hancock and Kauffman (1979). Approximately 300 m of section are shown in Figure 8.30, but the total Cretaceous section in the Western Interior may exceed 5000 m (Weimer, 1960), and contains five third order cycles between the Albian and the Maastrichtian (Hancock and Kauffman, 1979).

Kauffman (1969) erected a generalized model for the Cretaceous cycles, which is shown in Figure 8.31. It consists of 12 lithofacies with accompanying faunas, which can be grouped into four main cyclic phases (or lithofacies assemblages):

A. Lithofacies 1: coarse clastics and muds, local facies changes, deposited in fluvial, lagoonal and estuarine settings.

B. Lithofacies 2, 3, 4a: coarse clastics with sublittoral marine faunas, deposited in deltaic, beach-barrier and shallow subtidal environments.

C. Lithofacies 5, 6, 7: mudstones with laminations, small-scale sedimentary structures, deeper-water marine faunas, deposited in a mid-basin setting.

D. Lithofacies 4b, 8, 9, 10, 11, 12: calcareous mudstones and various carbonates, indicating

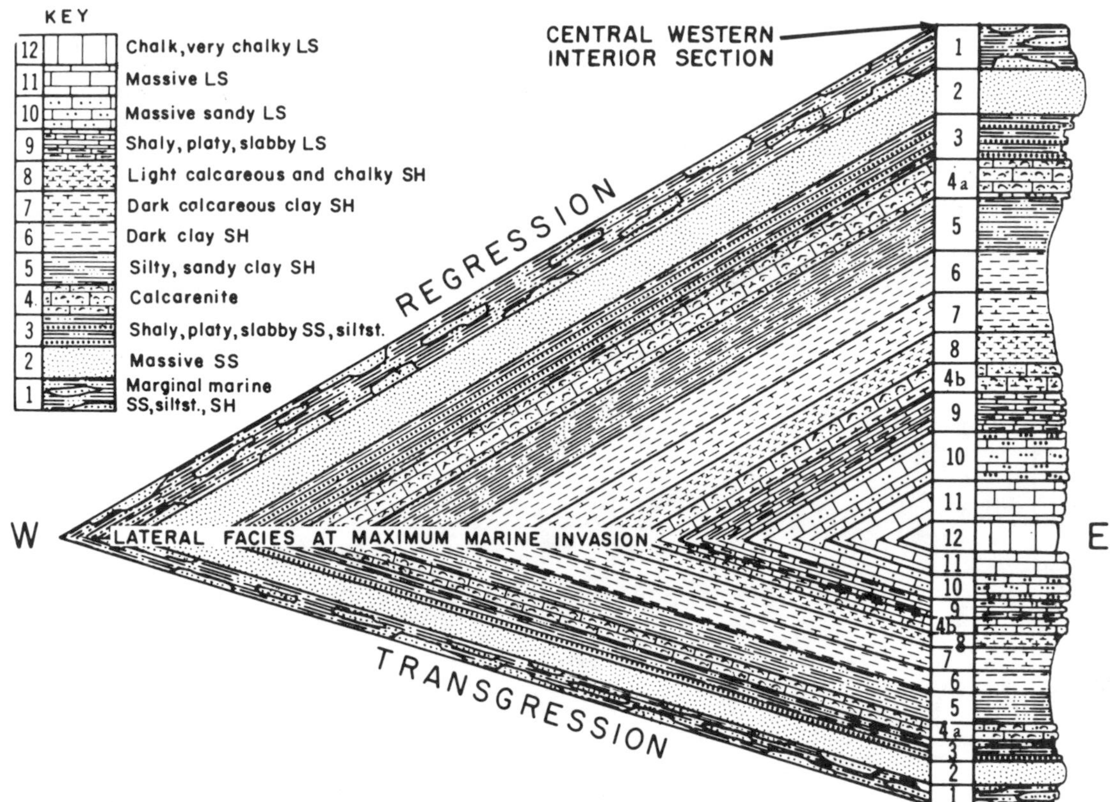

Fig. 8.31. Model of cyclic Cretaceous sedimentation in the Western Interior (Kauffman, 1969).

shallow to deep shelf environments distant from sediment sources.

The model is highly generalized and, like those erected for other cyclic sequences discussed earlier, can contain complications introduced by local coastal progradation and local tectonic effects.

Hancock and Kauffman (1979) compared the third order cycles of the Western Interior with those of northwest Europe (Fig. 8.32) and showed that there is a considerable amount of similarity between the two areas. Many lithofacies are common to the two continents, in spite of the differences in tectonic setting (Hancock, 1975).

Figure 8.33 illustrates that such correlations cannot be made everywhere. Jeletzky (1978) provided detailed biostratigraphic documentation for sea level changes in five Cretaceous basins in Canada (first five graphs in Fig. 8.33) which he compared to the curves of Hancock and Kauffman. The latter were redrawn by Jeletzky using equal time subdivisions for the Cretaceous stages and ignoring the available data on their radiometric age, which he regards as unreliable. Jeletzky offers this illustration as evidence that sea level changes cannot be correlated from basin to basin once the detailed biostratigraphic evidence is considered and, indeed, the lack of correlation is convincing, although certain events do

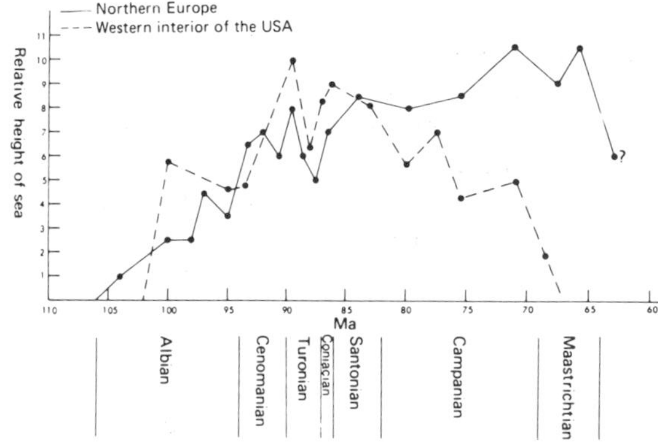

Fig. 8.32. Oscillations of sea level in northern Europe and Western Interior, based on facies variations (Hancock and Kauffman, 1979).

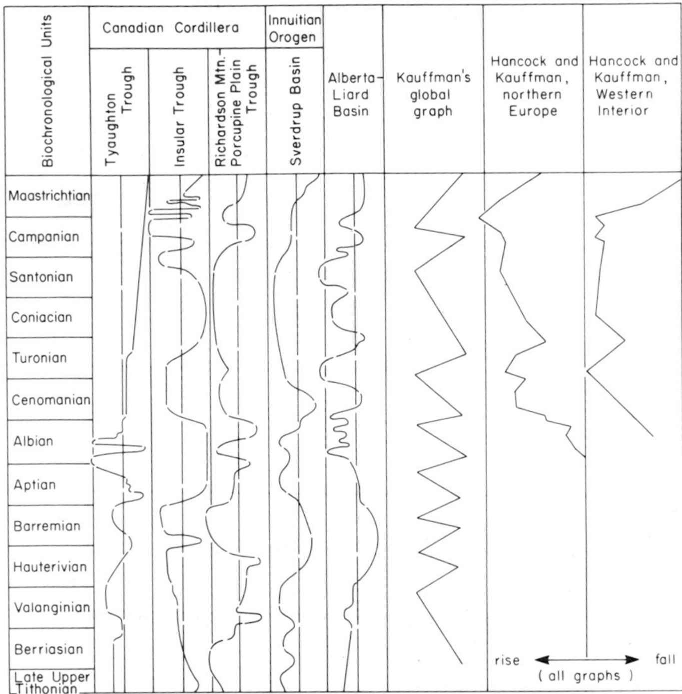

Fig. 8.33. Sea level changes in five Cretaceous basins in Canada, compared to the curves of Hancock and Kauffman. The latter have been replotted to correspond to equal-stage time subdivisions (Jeletzky, 1978).

seem to occur nearly simultaneously in several of the basins (early Turonian transgression, late Santonian transgression, mid-Campanian regression, late Maastrichtian regression). Jeletzky argued that the evidence does not support the eustatic control hypothesis but, unfortunately, his selected basins are all in or marginal to mobile belts where tectonic overprinting would be expected to mask the passive type of sea level change discussed here.

8.3.3 Other evidence of cyclicity

As noted in section 8.2.3, the charts of sea level change published by the Exxon group have been widely reproduced in recent years, as geologists have attempted to fit their data to this new model. Apart from the time-specific studies such as those reported in section 8.3.2 some important generalizations are starting to appear. For example, as noted in section 6.7, erosional and depositional events in submarine canyons and fans are markedly affected by sea level. During high stands of the sea, continental detritus tends to be trapped along the shorelines in deltas and coastal plain complexes. Shelf progradation rates are high, but the continental rise tends to be starved. Clinoform shelf–slope architecture and offlapping sequences result. Conversely, during periods of low sea level, canyon erosion is active and much detritus is fed directly to the submarine canyons at their mouths. The result is deep marine onlap of thick fan sequences (Fig. 6.40). Because of the increased sediment supply to the deep oceans during periods of low sea level contourites are also likely to be more common at these times. Shanmugam and Moiola (1982) have attempted to document the major occurrences of thick submarine fan and contourite deposits in the stratigraphic record, and they claim that most correspond to periods of minimum sea level on the Vail et al. (1977b) curves. Conversely G. deV. Klein (oral communication, 1982) suggested that the major tidalite sequences of the Phanerozoic correspond to periods of high sea level, when shelf widths and tidal effects were at a maximum.

Several recent studies have suggested that many physical, chemical and biological events in the continents and oceans are correlated with each other and that they change cyclicly over periods of 10^6 to 10^7 a. Funnell (1981) termed this "autocorrelation". The subject was explored in depth by Fischer and Arthur (1977), who studied the Mesozoic and Cenozoic record and, later, by Leggett et al. (1981), who carried out a similar analysis for the early Paleozoic. The main control on autocorrelation seems to be sea level. During periods of high sea level the world's oceans tend to be warm, with much reduced latitudinal and vertical temperature gradients. Oceanic circulation is thereby reduced, leading to increased stratification and severe oxygen depletion at depth. Deposition of organic-rich sediments in the deep ocean becomes widespread, the carbonate compensation depth rises and faunal diversity increases. These are the times of global "anoxic events" when widespread black shales are developed around the world.

During periods of low sea level global climates are more variable, the oceans are, in general, cooler and better oxygenated because of better circulation. Submarine erosion may be intensified. Initiation of these episodes of lower sea level may be a cause of biotic crises, in which faunal diversity is sharply reduced by major extinctions.

Fischer and Arthur (1977) encapsulated these faunal variations by applying the terms **polytaxic** and **oligotaxic** to periods of, respectively, high and low faunal diversity. They demonstrated that since the Triassic the oceans have fluctuated between these two modes with a periodicity of approximately 32-Ma. These cycles correlate reasonably well with the sea level changes of Figure 8.17, but there are discrepancies, which raises the question of whether the 32-Ma periodicity or the correlation with the seismic stratigraphic record has the greater significance. There is a natural human tendency to be intrigued by regularities in the earth's record, but there is nothing known about earth history that would explain it. On the contrary, as discussed in the next section, the causes of sea level change are varied, and no regularity is to be expected. Perhaps this is a case where a cycle is being forced to fit too rigid and artificial a time constraint.

8.4 Causes of eustatic sea level change

Many causes have been proposed for changes in global sea levels (Fairbridge, 1961; Donovan and Jones, 1979; Pitman, 1978):

1. differentiation of lithosphere
2. sediment infill of ocean basins
3. crustal shortening during orogeny
4. volume changes of oceanic spreading ridges
5. desiccation of small ocean basins
6. growth and decay of continental ice sheets
7. geoid changes
8. changes in the volume of the hydrosphere
9. changes in ocean temperature
10. changes in atmospheric moisture content
11. vertical tectonic movements.

These have the effect of either changing the total volume of sea water (6, 8, 9, 10) or changing the total volume of the ocean basins (1, 2, 3, 4, 5, 7, 11). The critical questions are, which of these mechanisms can cause sea level changes of up to several hundred meters, and which can cause repeated rises and falls at geologically fast rates?

Calculations show that only two or, possibly, three of the mechanisms listed are likely to have been of importance, and that these two can, in fact, account for most of the cyclic changes discussed earlier in this chapter (Table 8.2). Volume changes of oceanic spreading ridges are caused by variations in spreading rates; they probably account for the second order cycles and, possibly, the third order cycles. Changes in the volume of land ice can account for fourth order cycles and, possibly, third order cycles, at least during those periods of earth history when there were major continental glaciations. Most of the remaining mechanisms have been shown to have had insignificant effects on sea level or to be incapable of causing changes fast enough to account for the oscillations deduced from the sedimentary record.

8.4.1 Volume changes of oceanic spreading ridges

The oceanic lithosphere formed at a spreading center is intially hot, and cools as it moves away from the axis. Cooling is accompanied by thermal contraction and subsidence (Sclater et al., 1971). The age versus depth relationship is constant, regardless of spreading history, and follows a time-dependent exponential cooling curve (McKenzie and Sclater, 1969), as shown in Figure 8.34.

Hallam (1963), Russell (1968), Valentine and Moores (1970, 1972) and Rona (1973) argued that eustatic sea level oscillations would be caused by variations in spreading rates. Low stands of sea level would occur during episodes of slow spreading, during which relatively small

Table 8.2. Stratigraphic cycles and their causes

Type (Vail et al., 1977b)	Other terms	Duration Ma	Probable cause
First order	—	200–400	major eustatic cycles caused by formation and breakup of super-continents
Second order	supercycles (Vail et al., 1977b) sequence (Sloss, 1963) synthem (Ramsbottom, 1979)	10–100	eustatic cycles induced by volume changes in global mid-oceanic spreading ridge system
Third order	mesothem (Ramsbottom, 1979)	1–10	possibly produced by ridge changes and/or continental ice growth and decay
Fourth order	cyclothem (Wanless and Weller, 1932)	0.2–0.5	rapid eustatic fluctuations induced by growth and decay of continental ice sheets, growth and abandonment of deltas

volumes of hot oceanic lithosphere are being generated. Conversely, episodes of fast spreading would raise sea levels by increasing the ridge volume. Using the data of Sclater et al. (1971), Pitman (1978) modeled volume changes in a hypothetical ridge, as shown in Figure 8.35. The elevation of any part of a ridge can be calculated by converting age to depth, using an appropriate spreading rate. Pitman (1978) showed, for example (Fig. 8.35), that a ridge spreading at 6 cm/a will have three times the volume of one spreading at 2 cm/a provided these rates last for 69 Ma. This is the time taken for the oldest (outermost) part of the ridge to subside to average oceanic abyssal depths of 5.5 km, by which time the ridge has achieved an equilibrium profile.

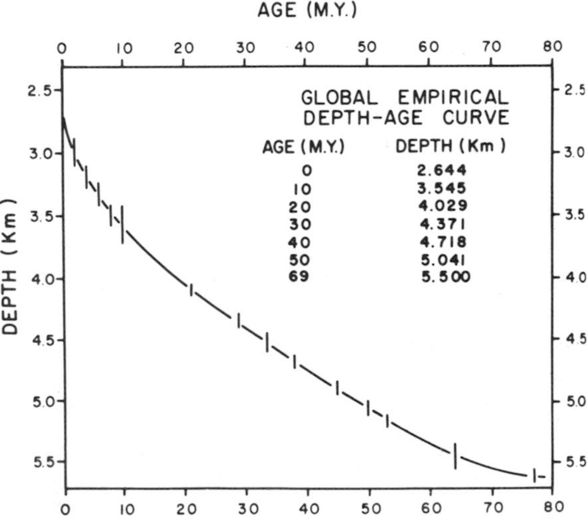

Fig. 8.34. The standard age versus depth curve for oceanic crust. Vertical bars show standard deviation for each data set (Sclater et al., 1971; Pitman, 1978).

The total length of the world mid-oceanic ridge system is about 45,000 km (Hays and Pitman, 1973; Pitman, 1978) and Pitman (1978) showed that, allowing for the shape of the continental margins, measured spreading rates can account for eustatic sea level changes of the order of 1 cm/1000 a. This is fast enough to accomodate second order cycles.

However, a rise or fall in sea level is not the same thing as a transgression or regression. As the rifted margins of a continent move away from a spreading center they subside as a result of thermal contraction, crustal attenuation and, possibly, phase changes (Sleep, 1971; Watts and Ryan, 1976). Sediments deposited on the subsiding margin cause further isostatic subsidence. Immediately after rifting and the appearance of oceanic crust the margins may subside at rates of the order of 20 cm/1000 a, decreasing after 20 to 30 Ma to about 2–4 cm/1000 a (Rona, 1973; Watts and Ryan, 1976; Pitman, 1978). Shelf-edge subsidence appears to be always faster than the rate of eustatic sea level change, and therefore

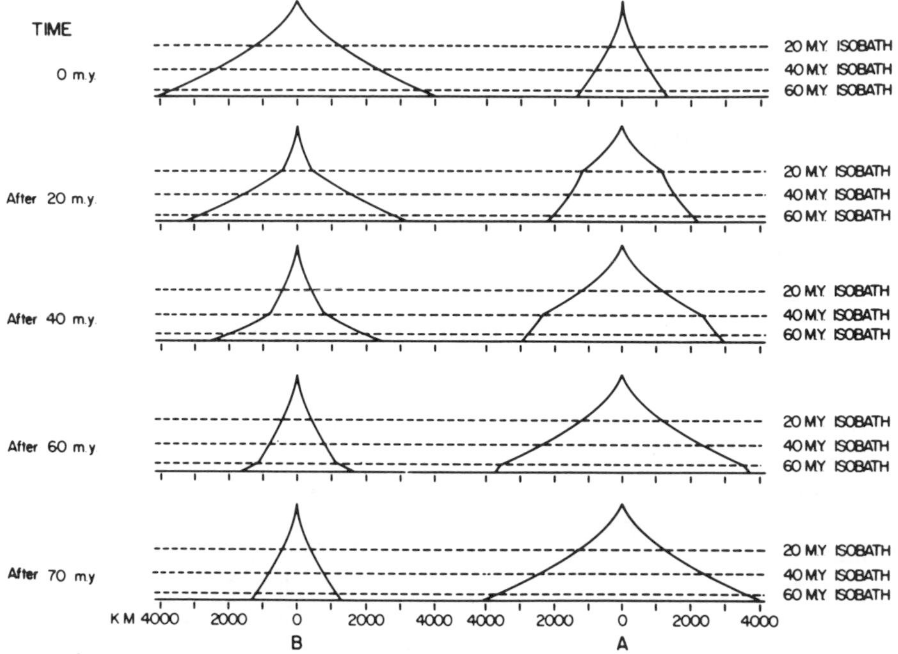

Fig. 8.35. Profiles of spreading ridges showing the effect of different spreading rates on volume at 20, 40, 60 and 70 Ma after initial condition. **A** profile of a ridge that has been spreading at 2 cm/a for 70 Ma and changes to 6 cm/a at time 0; **B** a ridge that has been spreading at 6 cm/a for 70 Ma and changes to 2 cm/a at time 0. After 70 Ma ridge **A** has three times the volume of ridge **B** (Pitman, 1978).

transgressions can actually occur during periods of falling sea level, if the rate of lowering is sufficiently slow. Conversely, regressions can occur locally during periods of rising sea level if there is an adequate sediment supply. Growth of large deltas such as the Mississippi following the Holocene post-glacial sea level rise is adequate demonstration of this. However, such local regressions are not relevant to the consideration of global stratigraphic cycles.

The spreading histories of the world oceans are now reasonably well understood (with the exception of the Arctic Ocean), based on deep-sea drilling data and magnetic reversal stratigraphy. Knowledge of worldwide spreading rates enabled Hays and Pitman (1973) to calculate ridge volume changes and a sea level curve for the last 110 Ma. A revised version of this curve was recalculated for the period 85 to 15 Ma by Pitman (1978), based on refinements in oceanic data (Fig. 8.36). It shows a gradual drop in sea level of 300 m at an average rate of 0.4 cm/1000 a. Sea levels rose to a maximum during a period of fast spreading between 110 and 85 Ma (Larson and Pitman, 1972), and the subsequent drop reflects slower spreading rates. By calculating the relationship between spreading rates, subsidence rates and falling sea level Pitman (1978) was able to model a major global transgression during the Eocene, as actually documented from stratigraphic evidence by Hallam (1963), and a second transgression during the Miocene (Fig. 8.36). The

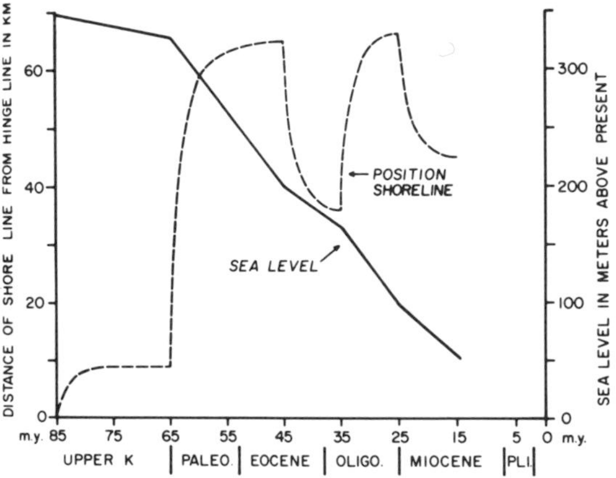

Fig. 8.36. Calculated fall in global sea level during the Late Cretaceous to Miocene. Dashed line shows position of shoreline relative to continental edge hinge line (left-hand axis), demonstrating that a transgression took place during the Eocene during a period of slow sea level fall (Pitman, 1978).

Eocene transgression does not appear on the Vail et al. (1977b) curve (Fig. 8.16), which remains an anomaly to be resolved.

Much work remains to be done on this problem. There are continuing disagreements over the magnetic reversal chronology (e.g. see Fig. 8.16 and Chapter 3) based on disparities between biostratigraphic and radiometric data (a problem discussed at length by Jeletzky, 1978), and this hampers attempts at global cycle correlation and relating continental and oceanic events.

8.4.2 Plate movements and major cycles

During the Phanerozoic there were high sea level stands between the Late Cambrian and Mississippian, and again during the Cretaceous, and generally low stands during the Pennsylvanian to Jurassic (Fig. 8.18). These broad changes can be related to the formation and breakup of supercontinents. High global rates of sea-floor spreading are accompanied by continental breakup, active subduction, plutonism and arc volcanism on convergent plate margins, and by major transgressions on cratons and divergent margins. Low average rates of spreading may occur during ridge reordering following major continental collision and suturing events. Suturing might therefore be expected to correlate with, or precede, times of low sea level (Valentine and Moores, 1970, 1972; Larson and Pitman, 1972; Vail et al., 1977b; Schwan, 1980).

With these generalizations in mind it is instructive to examine briefly the Phanerozoic record. The Cambrian transgression may reflect the breakup of the Eocambrian supercontinent (Matthews and Cowie, 1979; Donovan and Jones, 1979). It has been suggested that the late Precambrian Sparagmite sequence of Norway was formed in rifts representing the incipient Iapetus (proto-Atlantic) Ocean (Bjørlykke et al., 1976). The exceptionally high sea level stand during the Ordovician might then relate to the rapid widening of Iapetus Ocean (and possibly other world oceans).

Johnson (1971) showed by a detailed stratigraphic analysis that the Antler, Ellesmerian and Acadian orogenies of North America coincided with the major transgression of the Kaskasia sequence (Fig. 8.37). These orogenies were followed by the sub-Absaroka unconformity, suggesting a post-orogenic reordering of spreading axes. Johnson (1971) extended this analysis to other North American orogenies, as shown in Figure 8.37.

Sea level lowering occurred during the Caledonian–Acadian and Hercynian–Appalachian suturing of Pangea between the Devonian and Permian (Schopf, 1974). The production of this supercontinent may well have been the cause of increasingly "continental" type climates, leading to the Carboniferous–Permian glaciation of Gondwana (Crowell, 1978) and the widespread Pennsylvanian to Jurassic eolian facies of the United States and Europe.

Worldwide transgressions during the Jurassic and Cretaceous probably reflect the progressive splitting of Pangea. North America and Africa rifted apart in the mid-Jurassic, South America and Africa in the Early Cretaceous; North America and Britain split in the mid-Cretaceous, as did Africa and Antarctica; India and Madagascar rifted apart in the Late Cretaceous, and the North Atlantic split extended northward between Greenland and Scandinavia in the early Paleo-

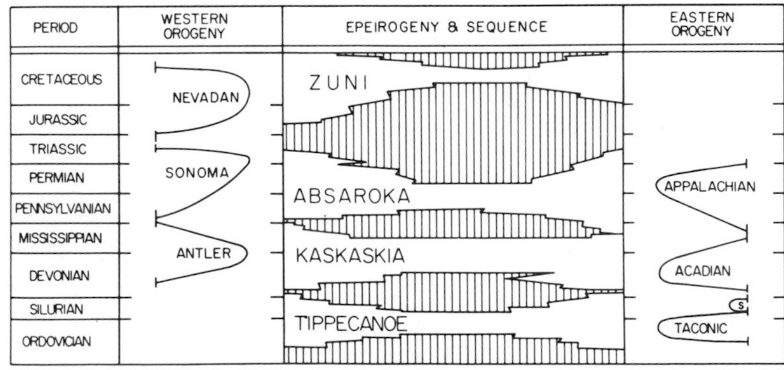

Fig. 8.37. Timing of North American orogenic episodes relative to Sloss's cratonic sequences (Johnson, 1971).

cene (summary and data sources in Bally and Snelson, 1980, p. 31).

Schwan (1980) compiled data showing that worldwide orogenic activity in the Late Jurassic to Recent, including the development of unconformities and compressional structures (folds, thrust faults), and metamorphic and plutonic episodes, could be correlated with major readjustments in the world plate pattern. The latter can now be documented in detail using the oceanic magnetic anomaly pattern. The method of analysis is similar to that of Johnson (1971) but the availability of data from the oceanic crust permits a much more refined approach. The periods of peak global tectonic activity are, according to Schwan (1980), at 148, 115–110, 80–75, 63, 53, 42–38, 17 and 10–9 Ma. Most of these correlate with episodes of rising sea level on the Vail et al. (1977b) curve (Fig. 8.17).

8.4.3 Growth and decay of continental ice sheets

There is no doubting the efficacy of glacial advance and retreat as a mechanism for changing sea level; we have the evidence of the Pleistocene glaciation at hand throughout the northern hemisphere. It has been calculated that, during the period of maximum Pleistocene ice advance, the sea was lowered by about 100 m (Donovan and Jones, 1979). Melting of the remaining ice caps would raise present sea level by about 40 to 50 m (Donovan and Jones, 1979). Recent history thus demonstrates a mechanism for changing sea level by at least 150 m at a rate of about 1 cm/a. This is fast enough to account for fourth order cycles.

Evidence for glacial episodes during the Phanerozoic now reveals two major ice ages, each far longer than that of the Late Cenozoic (suggesting that the latter may only just be beginning!). Bennacef et al. (1971) have documented a late Ashgill glaciation in Algeria, which may have occupied much of the southern hemisphere and may have extended until the Late Llandovery, as tillites of this age have been recorded in Argentina (Berry and Boucot, 1972). Crowell (1978) summarized evidence for the Gondwana glaciation, which occupied various parts of that continent between the early Carboniferous and mid-Permian. The Cenozoic glaciation probably started in the Antarctic in the early Oligocene (Shackleton and Kennett, 1975).

Glacials and interglacials, as we understand them from the Pleistocene record, are geologically short-term events whose duration and magnitude suggest a correlation with the transgressions and regressions which generated cyclothems. At present we have no hypotheses to explain the third order cycles or mesothems of Ramsbottom (1979). Research into the details of the well-preserved Gondwana glaciation may well reveal longer-term advance and retreat phases capable of generating third order cycles. However, this still leaves unexplained the third order cycles of times when there appear to have been no major continental ice sheets, such as during the Jurassic and Cretaceous.

8.4.4 Desiccation of small ocean basins

It has been suggested that the Mediterranean Sea became isolated from the world ocean and dried out during the Late Miocene (Hsü et al., 1973; Adams et al., 1977). The Mediterranean contains 0.28% of the volume of the world ocean, and Berger and Winterer (1974) calculated that when this water was added to the main body of the ocean it would have raised sea level by about 10–15 m.

Did this effect occur at any other time in geological history? Berger and Winterer (1974) suggested that the incipient South Atlantic Ocean might have been similarly isolated, before separation between Africa and South America was complete in the Early Cretaceous. This is a much larger, deeper basin and its desiccation could have caused a eustatic rise of about 60 m. There is no evidence that this did, indeed, occur, but it is an idea worth exploring. Many oceans develop evaporite basins during the incipient sea-floor spreading stage, suggesting a possible desiccation phase, and therefore eustatic rises might be sought during times of continental breakup. However, this immediately precedes the time of most active sea-floor spreading, when eustatic rises are to be expected in response to expansions in the world mid-oceanic ridge system. Separation of various causes of eustasy might be difficult.

It seems likely that this mechanism is not one that can be called upon to explain regular, frequent eustatic sea level changes, but might have added its effects spasmodically during times of continental breakup.

8.5 The sequence concept in basin exploration

It might at first sight seem difficult to imagine the practical utility of the material presented in this chapter. How do global correlations of plate movements, transgressions and orogenies help us find more gas in the Deep Basin of the Western Interior? What use are cyclothems and mesothems in exploring for oil in the offshore frontier regions of the world?

The answer is that the pattern of sequences and cycles now emerging through application of seismic stratigraphic, biostratigraphic, magnetic and radiometric methods is providing a detailed stratigraphic pattern to which new observations can be fitted. We now know most of the second order cycles and many of the third order cycles which have occurred during the Phanerozoic. These are potentially invaluable keys for exploration in frontier basins or for re-evaluating the data from mature areas, because the pattern they demonstrate is unique and, with certain restrictions, should be recognizable worldwide. In fact, as Vail et al. (1977b) pointed out, the pattern of cycles has great potential as a standard for Phanerozoic time. Charts such as those given in Figures 8.16, 8.17 and 8.18 can be used to calibrate the stratigraphic record in poorly understood areas. Local charts of sea level change derived from seismic stratigraphic interpretation can be compared to the global standard and predictions about geological age of stratigraphic units and unconformities can be made before the area has been explored by drilling. Where good biostratigraphic control is available from well data departures from the global standard might be apparent, and these reveal local tectonic disturbances which may be of considerable importance in local correlation and interpretations of basin history.

The record is, of course, particularly good for the Late Jurassic to Recent. For this period seismic records are available for areas flanking ocean basins that have evolved essentially undisturbed by convergent plate tectonics; their records can be correlated with the sea-floor spreading history of the oceanic crust. For the Cambrian to Middle Jurassic the cyclic record is based entirely on continental data, notably the sequences of Sloss (1963, 1972, 1978) and the various third order cycles, examples of which were discussed in section 8.3.2.

There are, of course, dangers inherent in this approach. The search for specific cycles or global unconformities in new areas is liable to produce many self-fulfilling prophesies and circular arguments. The geologist should tackle the problem with the most rigorous of methods, taking care always to state clearly the limits of accuracy of available biostratigraphic and radiometric data, and pointing out alternative interpretations. The method of developing multiple working hypotheses is as important in this area of geology as in any other.

Successful application of the global stratigraphic standard is most likely to be accomplished in basins located on cratons and divergent continental margins. On convergent margins subduction and oblique-slip generate numerous local tectonic events, with consequent uplifts and downwarps of sedimentary basins that mask the eustatic cycles discussed in this chapter (sections 9.3.2, 9.3.3). This is particularly true of suture zones, where progressive closure of irregular continental margins or collisions between microplates may result in an exceedingly complex tectonic and sedimentary history (sections 9.3.4., 9.3.5). That is not to say that tectonism does not occur elsewhere. Epeirogenesis is an important process governing cratonic and divergent margin basins, as is discussed in section 9.3.6.

Can these global cycle concepts be applied to the Precambrian? The difficulty is that chronostratigraphic correlation of Precambrian sediments is very crude compared to that of the Phanerozoic (Chap. 3). Some attempts have been made at regional correlation. For example Young (1979) compared stratigraphic sections of Proterozoic age across northern North America, and suggested the existence of three unconformity-bounded sequences spanning the period 1.4 to 0.7 Ga. (see Fig. 3.10). These would compare to the first order cycles discussed here. A more detailed subdivision of the Huronian sediments of the southern Canadian Shield is possible. Robertson (1973) divided the Huronian Supergroup into four unconformity-bounded groups. The Huronian was deposited between about 2.4 and 2.2 Ga, and therefore each cycle represents approximately 50 Ma. They are therefore comparable to second order cycles. However, the regional extent of the unconformities is not known, and these

Huronian cycles have not yet been correlated with any others elsewhere on the North American continent, and so they may be of only local significance.

8.6 Conclusions

The debate about stratigraphic cycles and their causes exemplifies a classic dichotomy amongst geologists. On the one hand there are individuals such as Sloss, Vail and his co-workers, Ramsbottom, Fischer, Kennett, Hancock and Kauffman, who believe that there is a series of global stratigraphic events which can be recognized worldwide. Their philosophy seems to be that the evidence in support of this stratigraphic cyclicity is so solid and the causes of the cyclicity so self-evident that any negative evidence must be due to imperfections in the record or to local biostratigraphic or tectonic complications. On the other hand there are the doubters: George, Jeletzky and some others, many of them biostratigraphers, who are very much aware of the shortcomings of the fossil record, who point out the weaknesses in our paleomagnetic and radiometric scales, and conclude that the case for global eustatic control has not been made.

The dichotomy is analogous to that between the "lumpers" and the "splitters" in taxonomic studies or, in the nineteen sixties, that between pro- and anti-continental drifters. It represents a basic difference in the way scientists interpret data. There are those who try to generalize, synthesize and build models, and those who believe nothing and constantly call for more data. The tension between these two groups is a healthy one; science develops mainly because of the model builders, yet they need the second group to keep them honest.

References

ADAMS, C.G., BENSON, R.H., RIDD, R.B., RYAN, W.B.F., and WRIGHT, R.C., 1977: The Messinian salinity crisis and evidence of late Miocene eustatic changes in the world ocean; Nature, v. 269, p. 383–386.

AGER, D.V., 1981: Major marine cycles in the Mesozoic; J. Geol. Soc. London, v. 138, p. 159–166.

BALLY, A.W., and SNELSON, S., 1980: Realms of subsidence, *in* A.D. Miall, ed., Facts and principles of world petroleum occurrence; Can. Soc. Petrol. Geol. Mem. 6, p. 9–94.

BENNACEF, A., BEUF, S., BIJU-DUVAL, B., DE CHARPAL, O., GARIEL, O., and ROGNON, P., 1971: Examples of cratonic sedimentation: Lower Paleozoic of Algerian Sahara; Am. Assoc. Petrol. Geol. Bull., v. 55, p. 2225–2245.

BERGER, W.H., and WINTERER, E.L., 1974: Plate stratigraphy and the fluctuating carbonate line, *in* K.J. Hsü and H.C. Jenkyns, eds., Pelagic sediments: on land and under the sea; Internat. Assoc. Sedimentologists Spec. Publ. 1, p. 11–48.

BERGGREN, W.A., McKENNA, M.C., HARDENBOL, J., and OBRADOVICH, J.D., 1978: Revised Paleogene polarity time scale; J. Geol., v. 86, p. 67–81.

BERRY, W.B.N., and BOUCOT, A.J., 1973: Glacio–eustatic control of Late Ordovician–Early Silurian platform sedimentation and faunal changes; Geol. Soc. Am. Bull., v. 84, p. 275–284.

BEUF, S., BIJU-DUVAL, B., STEVAUX, J., and KULBICKI, G., 1966: Ampleur des glaciations "siluriennes" au Sahara; leurs influences et leurs conséquences sur la sédimentation; Inst. Français Pétrole Rev., v. 21, no. 3, p. 363–381.

BJØRLYKKE, D., ELVSBORG, A., and HØY, R., 1976: Late Precambrian sedimentation in the central Sparagmite basin of south Norway; Norsk Geol. Tidsk., v. 56, p. 233–290.

CROWELL, J.C., 1978: Gondwanan glaciation, cyclothems, continental positioning, and climate change; Am. J. Sci., v. 278, p. 1345–1372.

DONOVAN, D.T., and JONES, E.J.W., 1979: Causes of world-wide changes in sea level; J. Geol. Soc. London, v. 136, p. 187–192.

DUFF, P.M.D., HALLAM, A., and WALTON, E.K., 1967: Cyclic sedimentation; Develop. in Sedimentology, v. 10, Elsevier, Amsterdam, 280 p.

du TOIT, A.L., 1937: Our wandering continents; Oliver and Boyd, Edinburgh, 366 p.

EDWARDS, M.B., 1978: Glacial environments, *in* H.G. Reading, ed., Sedimentary environments and facies; Blackwell, Oxford, p. 416–438.

FAIRBRIDGE, R.W., 1961: Eustatic changes in sea level, *in* L.H. Ahrens, F. Press, K. Rankama and S.K. Runcorn, eds., Physics and chemistry of the Earth; Pergamon Press, London, p. 99–185.

FERM, J.C., 1975: Pennsylvanian cyclothems of the Appalachian Plateau, a retrospective view; U.S. Geol. Survey Prof. Paper 853, p. 57–64.

FISCHER, A.G., and ARTHUR, M.A., 1977: Secular variations in the pelagic realm, *in* H.E. Cook and P. Enos, eds., Deep water carbonate environments; Soc. Econ. Paleont. Mineral. Spec. Publ. 25, p. 19–50.

FUNNELL, B.M., 1981: Mechanisms of autocorrelation; J. Geol. Soc.London, v. 138, p. 177–182.

GEORGE, T.N., 1978: Eustasy and tectonics: sedimentary rhythms and stratigraphical units in British Dinantian correlation; Proc. Yorks. Geol. Soc., v. 42, p. 229–253.

HALLAM, A., 1963: Major epeirogenic and eustatic changes since the Cretaceous and their possible relationship to crustal structure; Am. J. Sci., v. 261, p. 397–423.

HALLAM, A., 1978: Eustatic cycles in the Jurassic; Palaeogeog., Palaeoclim., Palaeoec., v. 23, p. 1–32.

HAMBREY, M.J. and HARLAND, W.B., eds., 1981: Earth's pre-Pleistocene glacial record; Cambridge Univ. Press, 1004 p.

HANCOCK, J.M., 1975: The sequence of facies in the Upper Cretaceous of northern Europe compared with that in the Western Interior, in W.G.E. Caldwell, ed., The Cretaceous System in the Western Interior of North America; Geol. Assoc. Can. Spec. Paper 13, p. 83–118.

HANCOCK, J.M., and KAUFFMAN, E.G., 1979: The great transgressions of the Late Cretaceous; J. Geol. Soc. London, v. 136, p. 175–186.

HARLAND, W.B., 1981: Chronology of earth's glacial and tectonic record; J. Geol. Soc. London, v. 138, p. 197–203.

HARLAND, W.B., SMITH, A.G., and WILCOCK, B., eds. 1964: The Phanerozoic time scale; Geol. Soc. London, Supplement to Quart. J. Suppl., v. 120, 458 p.

HATTIN, D.E., 1964: Cyclic sedimentation in the Colorado Group of west–central Kansas; Kansas Geol. Surv. Bull. 169, p. 205–217.

HAYS, J.D., and PITMAN, W.C., III, 1973: Lithospheric plate motion, sea level changes and climatic and ecological consequences; Nature, v. 246, p. 18–22.

HEDBERG, H.D., ed., 1976: International stratigraphic guide; John Wiley and Sons, New York, 200 p.

HSÜ, K.J., RYAN, W.B.F., and CITA, M.B., 1973: Late Miocene desiccation of the Mediterranean; Nature, v. 242, p. 240–244.

JELETZKY, J.A., 1978: Causes of Cretaceous oscillations of sea level in Western and Arctic Canda and some general geotectonic implications; Geol. Surv. Can. Paper 77–18.

JOHNSON, J.G., 1971: Timing and coordination of orogenic, epeirogenic, and eustatic events; Geol. Soc. Am. Bull., v. 82, p. 3263–3298.

JOHNSON, M.E., and CAMPBELL, G.T., 1980: Recurrent carbonate environments in the lower Silurian of northern Michigan and their inter-regional correlation; J. Paleont., v. 54, p. 1041–1057.

KAUFFMAN, E.G., 1969: Cretaceous marine cycles of the Western Interior; Mountain Geologist, v. 6., p. 227–245.

KERR, R.A., 1980: Changing global sea levels as a geologic index; Science, v. 209, p. 483–486.

KLEIN, G. deV., 1982: Probable sequential arrangement of depositional systems on cratons; Geology, v. 10, p. 17–22.

LARSON, R.L., and PITMAN, W.C., III, 1972: World-wide correlation of Mesozoic magnetic anomalies and its implications; Geol. Soc. Am. Bull., v. 83, p. 3645–3662.

LEGGETT, J.K., McKERROW, W.S., COCKS, L.R.M., and RICKARDS, R.B., 1981: Periodicity in the Lower Paleozoic marine realm; J. Geol. Soc. London, v. 138, p. 167–176.

LENZ, A.C., 1976: Late Ordovician–Early Silurian boundary in the northern Canadian Cordillera; Geology, v. 4, p. 313–317.

MATTHEWS, S.C., and COWIE, J.W., 1979: Early Cambrian transgression, J. Geol. Soc. London, v. 136, p. 133–135.

McCROSSAN, R.G., and GLAISTER, R.P., eds., 1964: Geological history of Western Canada; Alberta Soc. Petrol. Geol., 232 p.

McKENZIE, D., and SCLATER, J.G., 1969: Heat flow in the eastern Pacific and sea-floor spreading; Bull. Volcanology, v. 33, p. 101–118.

McKERROW, W.S., 1979: Ordovician and Silurian changes in sea level; J. Geol. Soc. London, v. 136, p. 137–146.

MESOLELLA, K.J., ROBINSON, J.D., McCORMICK, L.M., and ORMISTON, A.R., 1974: Cyclic deposition of Silurian carbonates and evaporites in the Michigan Basin; Am. Assoc. Petrol. Geol. Bull., v. 58, p. 34–62.

MIALL, A.D., 1978: Tectonic setting and syndepositional deformation of molasse and other nonmarine–paralic sedimentary basins; Can J. Earth Sci., v. 15, p. 1613–1632.

MITCHUM, R.M. Jr., VAIL, P.R., and THOMPSON, S., III, 1977: Seismic stratigraphy and global changes of sea level, part two: the depositional sequence as a basic unit for stratigraphic analysis; Am. Assoc. Petrol. Geol. Mem. 26, p. 53–62.

MOORE, R.C., 1964: Paleoecological aspects of Kansas Pennsylvanian and Permian cyclothems, *in* D.F. Merriam, ed., Symposium on cyclic sedimentation; Kansas Geol. Surv. Bull. 169, p. 287–380.

PITMAN, W.C., III, 1978: Relationship between eustacy and stratigraphic sequences of passive margins; Geol. Soc. Am. Bull., v. 89, p. 1389–1403.

RAMSBOTTOM, W.H.C., 1979: Rates of transgression and regression in the Carboniferous of NW Europe; J. Geol. Soc. London, v. 136, p. 147–153.

RIBA, O., 1976: Syntectonic unconformities of the Alto Cardener, Spanish Pyrenees: a genetic interpretation; Sediment. Geol., v. 15, p. 213–233.

ROBERTSON, J.A., 1973: A review of recently acquired geological data, Blind River—Elliot Lake area, *in* G.M. Young, ed., Huronian stratigraphy and sedimentation; Geol. Assoc. Can. Spec. Paper 12, p. 169–198.

RONA, P.A., 1973: Relations between rates of sediment accumulation on continental shelves, sea-floor spreading and eustacy inferred from the central North Atlantic; Geol. Soc. Am. Bull., v. 84, p. 2851–2872.

RUSSELL, L.K., 1968: Oceanic ridges and eustatic changes in sea level; Nature, v. 218, p. 861–862.

SAUNDERS, W.B., RAMSBOTTOM, W.H.C., and MANGER, W.L., 1979: Mesothemic cyclicity in the mid-Carboniferous of the Ozark shelf region?; Geology, v. 7, p. 293–296.

SCHOPF, T.J.M., 1974: Permo-Triassic extinctions: relation to sea-floor spreading; J. Geol., v. 82, p. 129–143.

SCHWAN, W., 1980: Geodynamic peaks in Alpinotype orogenies and changes in ocean-floor spreading during Late Jurassic—Late Tertiary time; Am. Assoc.

Petrol. Geol., v. 64, p. 359–373.

SCLATER, J.G., ANDERSON, R.N., and BELL, M.L., 1971: The elevation of ridges and the evolution of the central eastern Pacific; J. Geophys. Res., v. 76, p. 7888–7915.

SHACKLETON, N.J., and KENNETT, J.P., 1975: Paleotemperature history of the Cenozoic and the initiation of Antarctic glaciation: oxygen and carbon isotope analysis in DSDP sites 277, 279 and 281; Initial Rep. D.S.D.P., v. 29, p. 743–755.

SHANMUGAM, G., and MOIOLA, R.J., 1982: Eustatic control of turbidites and winnowed turbidites; Geology, v. 10, p. 231–235.

SLEEP, N.H., 1971: Thermal effects of the formation of Atlantic continental margins by continental breakup; Geophys. J. Roy. Astron. Soc., v. 24, p. 325–350.

SLEEP, N.H., 1976: Platform subsidence mechanisms and "eustatic" sea-level changes; Tectonophysics, v. 36, p. 45–56.

SLOSS, L.L., 1963: Sequences in the cratonic interior of North America; Geol. Soc. Am. Bull., v. 74, p. 93–113.

SLOSS, L.L., 1972: Synchrony of Phanerozoic sedimentary–tectonic events of the North American craton and the Russian platform; 24th Internat. Geol. Congr., Montreal, Sect. 6, p. 24–32.

SLOSS, L.L., 1978: Global sea level changes: a view from the craton, *in* J.S. Watkins, L. Montadert and P.W. Dickerson, eds., Geological and geophysical investigations of continental margins; Am. Assoc. Petrol. Geol. Mem. 29, p. 461–468.

SLOSS, L.L., KRUMBEIN, W.C., and DAPPLES, E.C., 1949: Integrated facies analysis, *in* C.R. Longwell, ed., Sedimentary facies in geologic history; Geol. Soc. Am. Mem. 39, p. 91–124.

SOARES, P.C., LANDIM, P.M.B., and FULFARO, V.J., 1978: Tectonic cycles and sedimentary sequences in the Brazilian intracratonic basins; Geol. Soc. Am. Bull., v. 89, p. 181–191.

SUESS, E., 1906: The face of the earth; Clarendon Press, Oxford.

VAIL, P.R., MITCHUM, R.M. Jr., and THOMPSON, S., III, 1977a: Seismic stratigraphy and global changes of sea level, part three: relative changes of sea level from coastal onlap; Am. Assoc. Petrol. Geol. Mem. 26, p. 63–82.

VAIL, P.R., MITCHUM, R.M. Jr., and THOMPSON, S., III, 1977b: Seismic stratigraphy and global changes of sea level, part four: global cycles of relative changes of sea level; Am. Assoc. Petrol. Geol. Mem. 26, p. 83–98.

VAIL, P.R., TODD, R.G., and SANGREE, J.B., 1977: Seismic stratigraphy and global changes of sea level, part five: chronostratigraphic significance of seismic reflections; Am. Assoc. Petrol. Geol. Mem. 26, p. 99–116.

VALENTINE, J.W., and MOORES, E., 1970: Plate tectonic regulation of faunal diversity and sea level; Nature, v. 228, p. 657–669.

VALENTINE, J.W., and MOORES, E., 1972: Global tectonics and the fossil record; J. Geol., v. 80, p. 167–184.

VAN EYSINGA, F.W.B., 1975: Geological time table; 3rd ed., Elsevier, New York.

VAN HINTE, J.E., 1976a: A Jurassic time scale; Am. Assoc. Petrol. Geol. Bull., v. 60, p. 489–497.

VAN HINTE, J.E., 1976b: A Cretaceous time scale; Am. Assoc. Petrol. Geol. Bull., v. 60, p. 498–516.

VINOGRADOV, A.P., and NALIVKIN, V.D., eds., 1960: Atlas of lithopaleogeographical maps of the Russian Platform and its geosynclinal framing, Part I—Late Precambrian and Paleozoic; Acad. Sci. U.S.S.R., Moscow.

VINOGRADOV, A.P., RONOV, A.B., and KHAIN, V.E., eds., 1961: Atlas of lithopaleogeographical maps of the Russian Platform and its geosynclinal framing, Part II—Mesozoic and Cenozoic; Acad. Sci. U.S.S.R, Moscow.

WANLESS, H.R., 1950: Late Paleozoic cycles of sedimentation in the United States; Internat. Geol. Congr. 18th, Algiers, Pt. 4, p. 17–28.

WANLESS, H.R., 1964: Local and regional factors in Pennsylvanian cyclic sedimentation; Kansas Geol. Surv. Bull. 169, p. 593–606.

WANLESS, H.R., 1972: Eustatic shifts in sea level during the deposition of Late Paleozoic sediments in the central United States, *in* J.G. Elam and S. Chuber, eds., Cyclic sedimentation in the Permian Basin; West Texas Geol. Soc. Symp., p. 41–54.

WANLESS, H.R., and SHEPARD, E.P., 1936: Sea level and climatic changes related to Late Paleozoic cycles; Geol. Soc. Am. Bull., v. 47, p. 1177–1206.

WANLESS, H.R., and WELLER, J.M., 1932: Correlation and extent of Pennsylvanian cyclothems; Geol. Soc. Am. Bull., v. 43, p. 1003–1016.

WATTS, A.B., and RYAN, W.B.F., 1976: Flexure of lithosphere and continental margin basins; Tectonophysics, v. 36, p. 25–44.

WEIMER, R.J., 1960: Upper Cretaceous stratigraphy, Rocky Mountain area; Am. Assoc. Petrol. Geol. Bull., v. 44, p. 1–20.

YOUNG, G.M., 1979: Correlation of middle and upper Proterozoic strata of the northern rim of the North Atlantic craton; Trans. Roy. Soc. Edin., v. 70, p. 323–336.

ZIEGLER, A.M., 1965: Silurian marine communities and their environmental significance; Nature, v. 207, p. 270–272.

CHAPTER 9

Sedimentation and plate tectonics

9.1 The basin model concept

Until the nineteen sixties sedimentary basins were explained and categorized in terms of geosynclinal theory (Dott, 1974, 1978; Mitchell and Reading, 1978). Such classic books as those by Kay (1951), Krumbein and Sloss (1963) and Aubouin (1965) had a profound impact on geologists and formed the basis for all large-scale interpretations. However, we can now see that these and other studies, although meticulously descriptive, could not ultimately explain why or how most basins formed or why there were recurrent structural styles or lithofacies assemblages. With the development of plate tectonics much has become clear. The kinematics of modern plate movements have been documented in some detail, and have provided geologists and geophysicists with a reliable data bank from which to build and constrain models of deep crustal behavior. Most sedimentary basins can now be explained in terms of plate margin or plate interior processes, and their structure and stratigraphy have become more comprehensible. This has led to the growth, in the last ten years, of a brand new terminology for types of basin. For miogeosyncline we now have divergent margin basin, for exogeosyncline we have retroarc basin, and so on. In fact, it is recommended that the old terminology be entirely abandoned. Such terms as eugeosyncline and geanticline now serve only to confuse, because we can study many features so described and recognize that these terms fail to distinguish between several subtly, but importantly, different types of plate setting. Other general terms, including intermontane basin and successor basin, are also so imprecise as to be of little value. The papers by Dickinson (1974) and Bally and Snelson (1980) provide the best general basin descriptions and classifications using this new terminology. They are referred to extensively in this chapter.

The importance of plate tectonics is by now, of course, universally recognized, and we do not need to congratulate ourselves on our revolution any further. The point of this chapter is to apply plate tectonic principles to the details of basin stratigraphy and structure. The work of Dickinson, Bally and Snelson was concerned mainly with causes of basin subsidence, and paid far less attention to the details of basin fill. It is suggested here that we can systematize the descriptions of depositional systems, structural geology, petrology and tectonic setting into a series of **basin models**, for the purpose of interpreting modern and ancient sedimentary basins. In Chapter 4 we discuss the methods of facies analysis, a technique used to systematize the welter of sedimentological detail visible in outcrops and cores. The evolution of facies models, using the facies analysis methodology, has been one of the great achievements of sedimentology. We can now attempt to apply the same skills of synthesis and judicious simplification to the erection of all-encompassing basin models, based on a plate tectonics foundation. Dickinson (1980, 1981) used the term "petrotectonic assemblages" with the same meaning. These basin models are then a powerful tool for interpreting regional plate tectonic history.

This process has been underway in a piecemeal manner for several years. For example Mitchell and McKerrow (1975) and Graham et al. (1975) attempted to explain the complex Paleozoic stratigraphy of southern Scotland and the Appalachians–Ouachitas, respectively, with reference to modern plate interactions in Burma and India. Their references to modern analogues and the erection of local basin models is exactly the same method as that used in facies analysis. There are

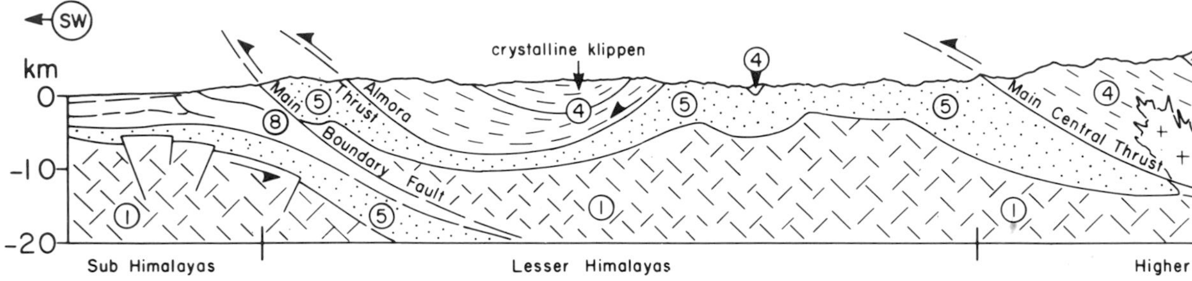

Fig. 9.1. Structural cross section across the Himalayas to illustrate the juxtaposition in a suture zone of sediments formed in more than one plate setting. 1. Precambrian Indian Shield, 2. post suture granites, 3. ophiolites, 4. Central Crystallines (Tethyan basement), 5. Tethyan Precambrian–Paleozoic sediments, 6. Tethyan upper

enough such local studies available that we can now develop a set of useful generalizations.

Many workers have attempted to adapt data and ideas for this purpose from the work that predates plate tectonics. The definition and interpretation of flysch (Lajoie, 1970; Reading, 1972) and molasse (Van Houten, 1973; Miall, 1981a) are good examples of such adaptation. They are well-known terms that describe readily recognizable sedimentary–tectonic associations. However, this is as far as the adaptation has gone, because modern plate tectonic theory has shown that they represent a much wider variety of tectonic settings than formerly thought, and there is no agreement as to how the two terms should now be used.

Basin analysts have much work to do to flesh out the new basin models with the details of facies, paleoslope, vertical succession, lateral variation, sedimentation rate, and so on, that categorize particular plate tectonic settings. They cannot work alone, because they require the services of seismic experts to reveal basin architecture and geophysical theoreticians to model mechanisms of basin subsidence. Nor can they ignore the rest of geology. Plutonism, volcanism, metamorphism and metallogeny all reflect plate tectonic setting, and integration of all these types of data may be essential for arriving at correct local interpretations and for ultimately developing satisfactory basin models. This is particularly the case in areas of accretionary terranes, as discussed in section 9.3.5. New types of data are constantly changing our ideas about crustal structure and the ways in which plates interact. The discoveries of the Consortium for Continental Reflection Profiling (COCORP) in the United States are the most spectacular illustration of this. Deep seismic reflection profiling has revealed major thrust faults in the Wind River Uplift, Wyoming and the southern Appalachians, the latter demonstrating that at least 400 km of crustal shortening occurred during Paleozoic suturing (see Oliver, 1982, for a review of the COCORP projects). At the same time there is a continuing controversy over the driving force or forces of plate tectonics. Earlier ideas emphasized mantle upwelling as a primary mechanism, forcing plates apart, and the gravitational effects of subduction, where oceanic crust was supposedly sucked down. However, nowadays there is a tendency to view sea-floor spreading and other processes as a response to the drift of plates, not a cause of it. Plate movement may be driven in part by torques arising from gyroscopic effects of the earth's spin (Hamilton, 1979). Such developments require a revaluation of many of our ideas about the crustal development of structurally complex areas, and demonstrate the need for a very broadly based, multivariate approach to sedimentary basin analysis. Having said this, we concentrate here on the sediments, about which there is quite enough to say to fill the remainder of the book. A broad familiarity with plate tectonic principles at an introductory undergraduate level is assumed for the readers of this chapter.

9.2 Basin classification

Most sedimentary basins can now be readily classified in terms of three criteria:

1. the type of crust on which the basin rests
2. the position of the basin relative to plate margins

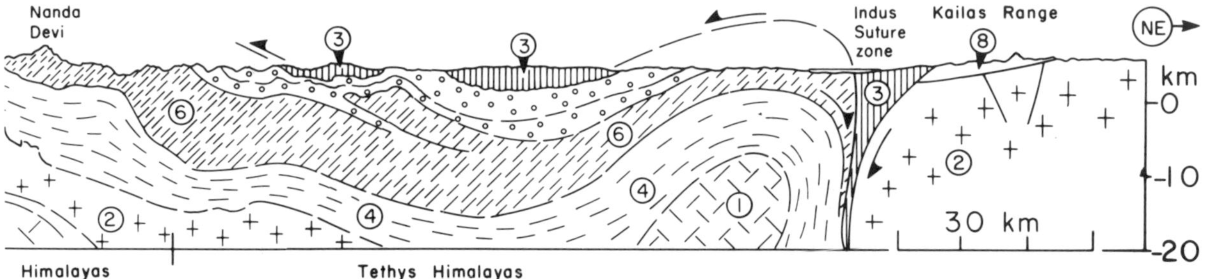

Precambrian–Paleozoic sediments, 7. Tethyan Upper Paleozoic–Mesozoic, 8. post-suture Tertiary (Geology of the Himalayas, A. Gansser, © 1964 Interscience, London. Reprinted by permission of John Wiley & Sons, Ltd.)

3. where the basin lies close to a plate margin, the type of plate interaction occurring during sedimentation.

Plate tectonics theory has shown us that all three of these parameters can change with time. This is a fundamental discovery, because it means that basins having several or many different origins normally are juxtaposed in the same orogenic belt. Many of the confusions of geosyncline theory can be resolved using this "mobilist" approach. For example the collision of India and Asia has juxtaposed shield and cratonic cover of two continental plates and has involved, in the intense deformation of the suture zone, thick Phanerozoic sediments originally developed as divergent (or "passive") margin wedges on the flanks of the Tethyan Ocean, remnants of volcanic arcs plus oceanic crust and overlying pelagic sediments (ophiolites), plus immense thicknesses of detrital sediments derived from uplifts generated during and following plate collision (Gansser, 1964). The suture zone thus contains at least four entirely different types of sedimentary basin (Fig. 9.1), and neither basin classification nor interpretation will be meaningful unless the entire tectonic history can be unraveled and each basin placed in its correct plate context.

Mitchell and Reading (1978) have pointed out that many basins as old as Jurassic can be regarded as "modern" in the sense that they still occupy essentially the same plate tectonic situation as they did when formed. For example the basins flanking the Atlantic Ocean, developed at various times between the Jurassic and Tertiary, can be so classified because the Atlantic is still, as it was then, an actively spreading ocean. Conversely, all but the most recent basins in the North American Cordillera are "ancient", in the sense that plate tectonic patterns there have undergone several profound changes since the Triassic and more are to be expected as the East Pacific Rise continues to interact with the westward-drifting North American plate. Numerous ancient basin types are therefore present in the Cordilleran region.

As discussed elsewhere in this book, geological interpretations of Jurassic and younger sediments are considerably facilitated by the evidence of sea-floor spreading preserved in magnetic anomalies and fracture zones on the ocean floors. Analysis and interpretation of this evidence have provided historical data on position of rotation poles, and direction and rate of plate movement, all of which considerably simplifies basin analysis. The superb study of Indonesian tectonics by Hamilton (1979) is an excellent example of this approach. For the most recent events this methodology can be supplemented by the development of fault plane solutions from earthquake records. For ancient basins the same procedure is sometimes attempted in reverse. For example Phillips et al. (1976) deduced plate vectors for the British area during the Caledonian closure of Iapetus Ocean, based on orientation of structural surfaces, volcanic centers, age of magmatic activity, and so on. It is the interpretation of complex ancient sutures such as this that offers the greatest challenge to basin analysts, and it is here that the availability of basin models should be the most useful.

The first of the three criteria for basin classification deals with the type of basement below the sedimentary basin. Oceanic crust normally sits at between 2.5 and 6 km below sea level, depending on its age (the thermal contraction of oceanic

Table 9.1. Basin classification*

1. Basins Located on the Rigid Lithosphere, Not Associated with Formation of Megasutures
- 11. Related to formation of oceanic crust
  - 111. *Rifts*
  - 112. *Oceanic transform fault associated basins*
  - 113. *Oceanic abyssal plains*
  - 114. *Atlantic-type passive margins (shelf, slope & rise) which straddle continental and oceanic crust*
    - 1141 Overlying earlier rift systems
    - 1142 Overlying earlier transform systems
    - 1143 Overlying earlier backarc basins of (321) and (322) type
- 12. Located on pre-Mesozoic continental lithosphere
  - 121. *Cratonic basins*
    - 1211. Located on earlier rifted grabens
    - 1212. Located on former backarc basins of (321) type

2. Perisutural Basins on Rigid Lithosphere Associated with Formation of Compressional Megasuture
- 21. Deep sea trench or moat on oceanic crust adjacent to B-subduction margin
- 22. *Foredeep and underlying platform sediments*, or moat on continental crust adjacent to A-subduction margin
  - 221. Ramp with buried grabens, but with little or no blockfaulting
  - 222. Dominated by block faulting
- 23. *Chinese-type* basins associated with distal blockfaulting related to compressional or megasuture and without associated A-subduction margin

3. Episutural Basins Located and Mostly Contained in Compressional Megasuture
- 31. Associated with B-subduction zone
  - 311. *Forearc basins*
  - 312. *Circum Pacific backarc basins*
    - 3121. Backarc basins floored by oceanic crust and associated with B-subduction (marginal sea sensu stricto).
    - 3122. Backarc basins floored by continental or intermediate crust, associated with B-subduction
- 32. Backarc basins, associated with continental collision and on concave side of A-subduction arc
  - 321. On continental crust or *Pannonian-type* basins
  - 322. On transitional and oceanic crust or *W. Mediterranean-type* basins
- 33. Basins related to episutural megashear systems
  - 331. *Great basin-type* basin
  - 332. *California-type* basins

*Bally and Snelson, 1980.

crust as it retreats from a spreading centre is discussed in Chapter 8; see Fig. 8.34), dropping to as much as 11 km in trenches at subduction zones. Sediments deposited on oceanic crust are therefore all of deep marine facies. Continental crust ranges from a few hundred meters below sea level up to the highest mountains, although the depth of sedimentary basement varies considerably depending on sediment thickness and structural foreshortening. Basins floored by continental crust are therefore typically shallow marine to nonmarine. However, this simple subdivision breaks down at plate margins, particularly at divergent margins, where continental crust may subside due to attenuation, listric normal faulting or ductile creep (see section 9.3.1) and, conversly, sedimentation can add as much as 15 km to crustal thickness (usually compensated in part by subcrustal erosion or creep).

Basin position relative to plate margin (the second of our classification criteria) is one of the primary criteria used by Bally and Snelson (1980) in their classification of sedimentary basins (Table 9.1). They distinguish basins on rigid lithosphere from those occurring in mobile belts. The latter group are subdivided into those at the margins of the belt (perisutural basins) and those within the belt (episutural basins).

Although this is a particularly comprehensive system of basin classification it groups together basins formed by different plate margin processes. Since all sedimentary basins are shaped by such processes, except those located on cratonic basement, it is perhaps not the most satisfactory approach for use by basin analysts.

In this chapter basin classification is based on the Wilson cycle of opening and closing oceans, and this emphasizes plate margin behaviour as the

Table 9.2. Classification of the types of sedimentary basin discussed in this chapter

1. Divergent margin basins
   - Rift basins
     - Rifted arch basins
     - Rim basins
   - Ocean margin basins
     - Red Sea type ("youthful")
     - Atlantic type ("mature")
   - Aulacogens and failed rifts
2. Convergent margin basins
   - Trenches and subduction complexes
   - Forearc basins
   - Interarc and backarc basins
   - Retroarc (foreland) basins
3. Transform and transcurrent fault basins
   - Basin setting:
     - Plate boundary transform fault
     - Divergent margin transform fault
     - Convergent margin transcurrent fault
     - Suture zone transcurrent fault
   - Basin type:
     - Basins in braided fault systems
     - Fault termination basins
     - Pull-apart basins in en echelon fault systems
4. Basins developed during continental collision and suturing
   - Foreland basins
   - Peripheral (foredeep) basins
   - Intra-suture embayment basins (remnant ocean basins)
   - Associated transcurrent fault basins†
5. Cratonic basins

†this type of basin may be classified under two headings

primary classification criterion. Basins can be conveniently divided into five categories (cf. Dickinson, 1974):

1. Divergent margin basins (section 9.3.1): intracratonic rifts that precede sea-floor spreading, divergent margin wedges, aulacogens and failed rifts.
2. Convergent margin basins (section 9.3.2): basins related to volcanic arcs, including trenches, forearc, interarc, backarc and retroarc basins.
3. Basins associated with transform plate margins and megashears (section 9.3.3).
4. Basins generated during the process of continental collision and suturing (section 9.3.4): foreland, foredeep or peripheral basins and some graben and wrench fault basins. Several mobile belts consist of accreted terranes or microplates, and these raise particular problems of basin analysis (section 9.3.5).
5. Cratonic basins (section 9.3.6): this is the only category of basins whose development appears unrelated to contemporaneous plate margin tectonics and may reflect long-term, deep-seated intraplate processes or reactivation of ancient plate margin lineaments.

A summary of the basin types discussed here is listed in Table 9.2.

9.3 Basin models

This section, comprising the main body of the chapter, presents a systematic description of the major basin models. For each basin type the following essential information is discussed:

1. plate tectonic process generating the basin
2. mechanism of crustal subsidence
3. structural geology of the basin
4. typical evolutionary development of depositional systems.

The discussion is based primarily on modern examples (in the sense defined in section 9.2), and selected examples of interpreted ancient equivalents are briefly described or referred to. Because of the special problems of plate tectonic interpretations in the Precambrian and the controversy surrounding them this topic is dealt with separately (section 9.5). Theoretical, thermal, mechanical and other models of basin subsidence are described briefly, but space does not permit an extended treatment of this topic.

9.3.1 Divergent margin basins

A distinctive suite of basins develops where plates rift and separate as a result of sea-floor spreading. Continental margins under these conditions are referred to as "divergent" or "passive" margins. The latter term is intended to suggest a tectonic contrast with the highly active character of convergent margins, but it is a misleading comparison. Several long-term, deep-seated tectonic processes occur on so-called passive margins. These have a profound effect on basin formation and evolution, and so the term is a poor one.

In this section we deal with basins formed during orthogonal spreading. Several important classes of basin also form by oblique spreading, controlled by transform faults. These are discussed in section 9.3.3.

Rift basins. Bott (1978, 1982), Sengör and Burke (1978), Sheridan (1981), Le Pichon et al. (1982) and many others have discussed the process of continental splitting. Veevers (1981) provided a detailed analysis of African, Arabian and Australian rifted margins and developed a basin model that generalized information from these areas. Pertinent points regarding rift generation to be discussed here include:

1. location of rift
2. mechanism of rifting
3. mechanism of subsidence
4. rift kinematics (extent of rifting, rift propagation, relief generation).

Sengör and Burke (1978) and Baker and Morgan (1981) have divided rifts into two categories, those resulting from differential stresses during two-dimensional plate evolution ("passive mantle hypothesis") and those produced by convective upwelling in the mantle ("active mantle hypothesis"). The first category is probably more common.

Why do rifts occur where they do? Intraplate stresses may generate rifts along old lines of weakness, possibly because the crust is thinner there and clearly because the crustal weakness (previously existing rift or suture) still exists. The reopening of the modern Atlantic Ocean close to the Iapetus suture (Wilson, 1966) is a good example of this. Similarly, southeast Africa separated from Antarctica in the Cretaceous (Hallam, 1980) along the site of a Paleozoic failed rift (Natal Embayment) (Hobday and Von Brunn, 1979; Tankard et al., 1982). McConnell (1977, 1980) demonstrated that the Cenozoic East African Rift System follows a Precambrian taphrogenic lineament that may be Archean in origin. Baffin Bay, an incipient ocean, parallels Precambrian (Hadrynian) dike swarms in Baffin Island (Fahrig et al., 1973), suggesting a great age for this lineation also. Veevers (1981) cited other examples.

Rifting may also be initiated by the upwelling of hot mantle. Wilson (1963) suggested that this was a localized process, occurring in what he called "hot spots". Morgan (1971, 1972) developed this idea into the mantle plume hypothesis. Upwelling causes uplift along the incipient rift, particularly over the mantle plumes, where the uplift has a domal shape with a relief of up to 4 km. Basic volcanism precedes and follows the uplift. Uplift generates horizontal tensional stresses, and it is these that lead to the failure of the crust by extensional faulting. Burke and Dewey (1973) suggested that most triple-point plate junctions are initiated over these uplifts, and that sea-floor spreading proceeds by the formation of spreading centers and transform faults linking these triple points.

Various mechanisms have been proposed to account for the generation of graben along the rifts. They may subside to depths of up to 5 km, permitting the accumulation of significant quantities of sediment. Bott (1978, 1982) suggested that the upper 10–20 km of crust deforms by brittle failure, whereas the lower part of the crust responds to tensional stress and sediment loading by ductile creep, leading to crustal attenuation. Significant extension can be accounted for by these processes. The only surface expression may be steeply dipping normal faults, but these typically are listric, that is, they flatten out at depth.

In the rift phase of continental splitting the graben floors remain above sea level, and are floored by nonmarine sediments and volcanics. The duration of this phase seems to vary considerably, depending on regional spreading patterns. Veevers (1981) suggested that true oceanic crust may be recognizable from ophiolite generation after 20 Ma or less. However, where the plates are constrained by surrounding active spreading centers, as in the case of Africa at the present day, the rift phase may persist for a considerable period. Rift basins bordering Baffin Bay span the Barremian to Eocene (~60 Ma) (Miall et al., 1980; McWhae, 1981), while those on East Greenland were intermittently active for about 170 Ma, from Early Triassic until the first generation of oceanic crust in the Paleocene (Surlyk et al., 1981). Fourteen phases of rifting have been identified there, some of which can be correlated with rifting episodes in the North Sea. These record long continued intraplate movements, and it seems clear that here sea-floor spreading is a passive response to plate movement rather than the driving force, and the rifting is of the "passive mantle" variety.

The East Greenland basins record a history of slightly oblique spreading, with marine conditions

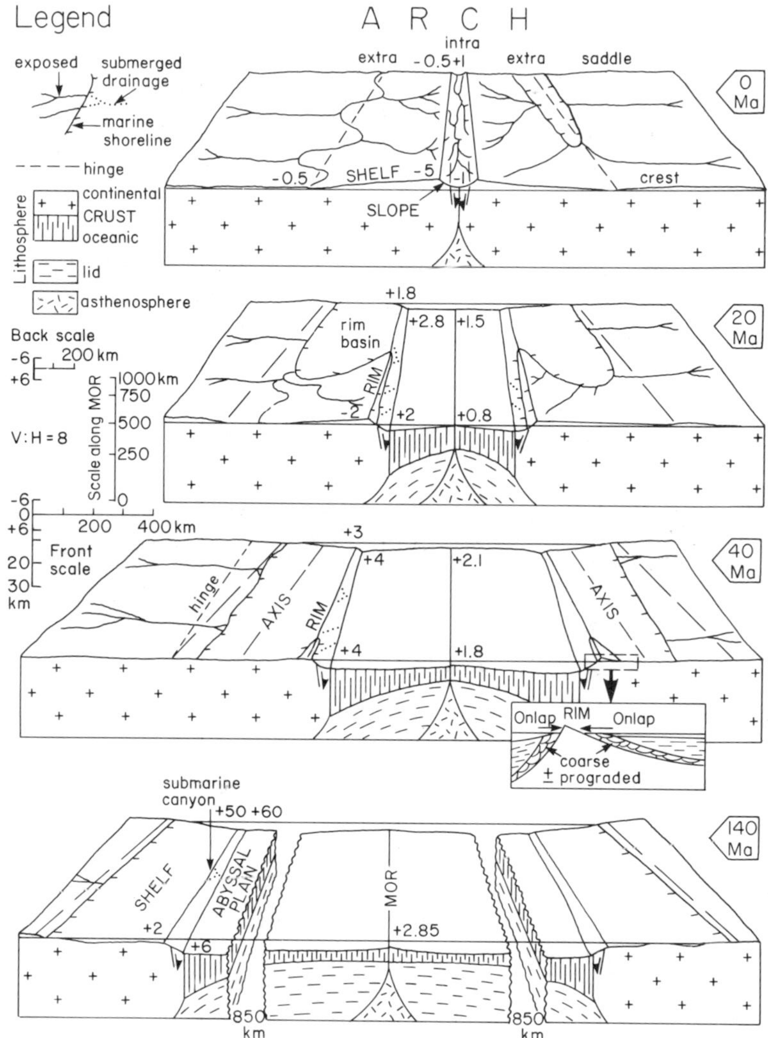

Fig. 9.2. Block diagram model of the development of a rifted arch system during continental break-up. Longitudinal topographic variation is shown as crest (elevated) and saddle (depressed) areas (Veevers, 1981).

appearing first in the south and gradually extending northward. It is of interest to note that the global stratigraphic cycles discussed in Chapter 8 could not be recognized there by Surlyk et al. (1981).

Veevers (1981) presented a series of block diagrams illustrating the evolution of rift basins, based on East Africa, the Red Sea and western Australia (Fig. 9.2). The initial rift is formed at or near the crest of the uplift and is called a rifted arch basin. It commonly is about 50 km wide. Continued faulting develops a series of graben, half-graben and tilted fault blocks as shown in the model experiments by Le Pichon et al. (1982), reproduced here as Figure 9.3. Note the considerable crustal extension achieved by this pattern of deformation. Such faulting may extend over a belt up to 1000 km wide, based on subsurface data from Atlantic rifted margins. The rifted margin of the original arch may remain as an uplifted rim, which Veevers (1981) suggested could survive as an internal sediment source for about 40 Ma after rift initiation (see inset below 40 Ma block model in Fig. 9.2). Behind this rim there may be a shallow rim basin. The rift system also has a longitudinal relief, as it is elevated several kilometers higher over the domal uplifts than in the saddles between them.

The architecture of rift-basin fills is well known, based on study of present day sedimentation patterns in East Africa and such ancient basins as the Triassic graben of northeast North America (Hubert et al., 1976) and the Triassic to Paleocene graben of East Greenland (Surlyk et

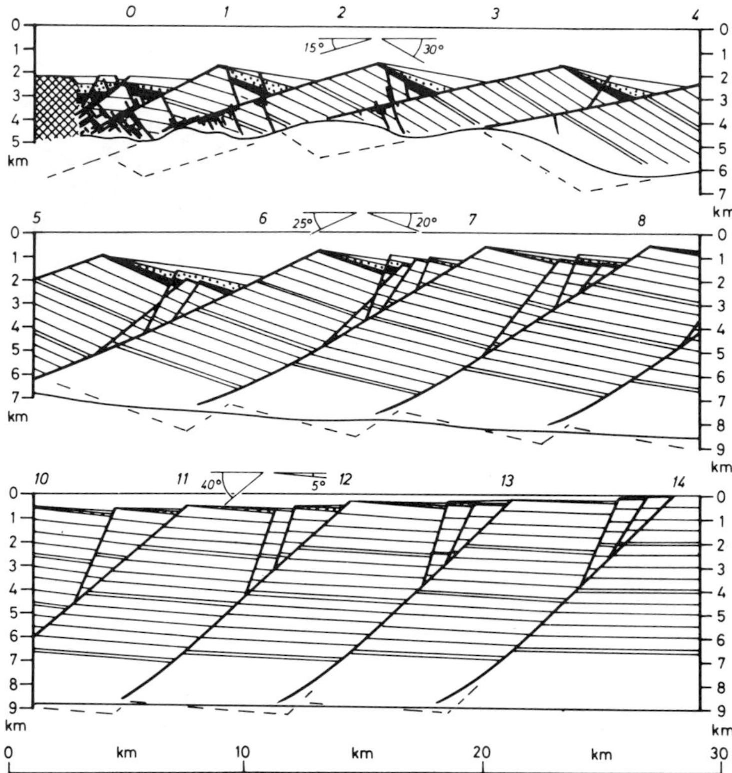

Fig. 9.3. Model of structural development of a continental margin. Brittle deformation occurs in upper levels; faults remain planar. Plastic deformation may occur in lower levels; faults may resemble curved listric faults (Le Pichon et al., 1982).

al., 1981). River systems drain transversely across fault scarps, possibly forming alluvial fans, and longitudinal trunk rivers occupy basin axes. In different parts of its course the Nile River demonstrates both patterns in Uganda (Miall, 1981b, Fig. 4), whereas in Egypt and Sudan it is a longitudinal river in an "extra-arch" or rim basin. Saddle areas along the rift may be occupied by lakes, containing clastic or chemical sediments, depending on sediment supply and climate. Basin fill patterns 1 and 6 of Miall (1981b) are typical (Fig. 6.5), and patterns 7 and 8 may occur where the rift has generated (or captured) a major trunk river that deposits deltas at the head of an embryonic ocean beginning to occupy the rift. With continued subsidence, transverse river valleys may become submarine canyons and alluvial fans may continue to form as submarine fans. This general sequence of events is well documented in East Greenland. Figure 9.4 illustrates the pattern of block faults there, and Figure 9.5 shows a typical paleogeographic reconstruction.

Subsidence and sedimentation rates are rapid in rift basins. Tiercelin and Faure (1978) reported fault displacements and accumulation of fluvial–lacustrine sediments in the Ethiopian Rift and Danakil Depression at comparable rates of 0.2–1.0 m/1000 a. Clastic sedimentation rates locally reach 3.5 m/ 1000 a, and accumulation rates are very high for lacustrine halite precipitation, possibly as much as 10 to 35 m/1000 a.

Rift basin sediments typically are intimately associated with volcanic flows, which may reach cumulative thicknesses of hundreds of meters. Space does not permit a discussion of these here. Crossley (1979) discussed the sedimentary and volcanic stratigraphy of the South Kenya rift valley. Mantle plumes generate immense volumes of flood basalt in rifts, for example in the domal uplifts of the East African system, east Greenland and west Greenland. Burke and Kidd (1980) provided a useful review of these occurrences.

Mature ocean margin basins. If true oceanic crust is generated in rift basins and sea-floor spreading continues for at least a few tens of millions of years basin subsidence mechanisms and sedimentary style undergo a gradual change. The generation of a spreading center creates a

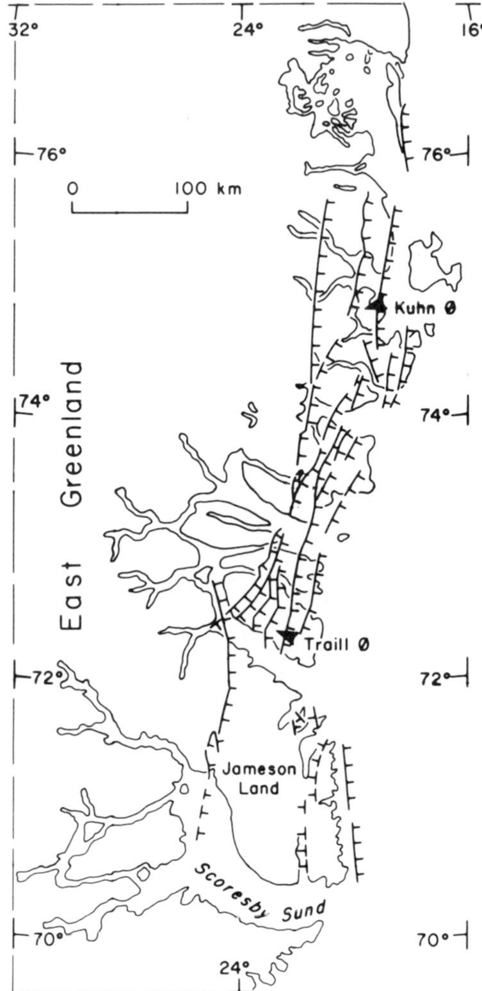

Fig. 9.4. Extensional faults on the East Greenland divergent continental margin (Surlyk et al., 1981).

thermal uplift at the center of the basin, and this effectively separates the two sides from each other in terms of basin floor transport processes.

The various processes of crustal extension plus the effects of thermal uplift discussed in the preceding section may by this time have generated a relief of 4 to 5 km between the rifted continental margin and the basin floor, with the latter situated 2 to 3 km below sea level. For the next 80 Ma or more, as the basin evolves from a narrow ocean (otherwise termed an incipient or youthful ocean) to a fully developed (mature) ocean, subsidence mechanisms are dominated by thermal contraction. As discussed in section 8.4.1, Sclater et al. (1971) demonstrated that the oceanic crust subsides as it moves away from a spreading center and cools. The rate of subsidence shows an exponential decrease, from about 20 cm/1000 a at first to about 2–4 cm/1000 a after 20–30 Ma (Fig. 8.34). Sleep (1971) demonstrated that the same process affects the adjacent continental crust. Keen (1979) analyzed subsurface stratigraphic data on the continental shelf of eastern Canada and showed that the thermal model can account with some accuracy for the actual subsidence history if the stratigraphic effects of eustatic sea level change plus the isostatic effect of sediment loading are removed. The rate of subsidence is proportional to $t^{1/2}$, where t is the time since subsidence began.

Three distinctive tectonic styles are typical of fully developed divergent margin basins:

1. extensional faulting, including growth faults
2. giant slumps
3. salt tectonics.

Faulting is less widespread than during the rift stage, the basin deforming mainly by downwarping. However some of the major rift faults may remain active. Growth faults are particularly common in areas of high sedimentation rate, such as major deltas (section 6.4). A typical structural style is that of tilted fault blocks and roll-over anticlines. In plan view most faults have a curved trace and intersect at angles of less than 45°. Figure 6.12 illustrates this style in the Niger Delta; Figure 9.6 is a detailed structure map showing part of the northeast side of the Gulf of Suez (Brown, 1980), a divergent margin that has been developing for about the last 20 Ma. The Red Sea is toward the lower left in Figure 9.6. Note that most, but not all of the faults are downthrown in that direction.

Giant slumps have been described from the continental slope and rise off southwestern and southern Africa by Dingle (1980). These are enormous features up to 250 m in thickness and 700 km long. Dingle identified two phases of slumping that occurred during the Cretaceous, probably as a result of slope instability during rapid sedimentation. Two additional cycles of slumping took place in the Cenozoic. These are not associated with rapid sediment loading but may have been initiated by seismic shock, with failure aided by high fluid pore pressures. Figure 9.7 shows a map and seismic section through a major slump near

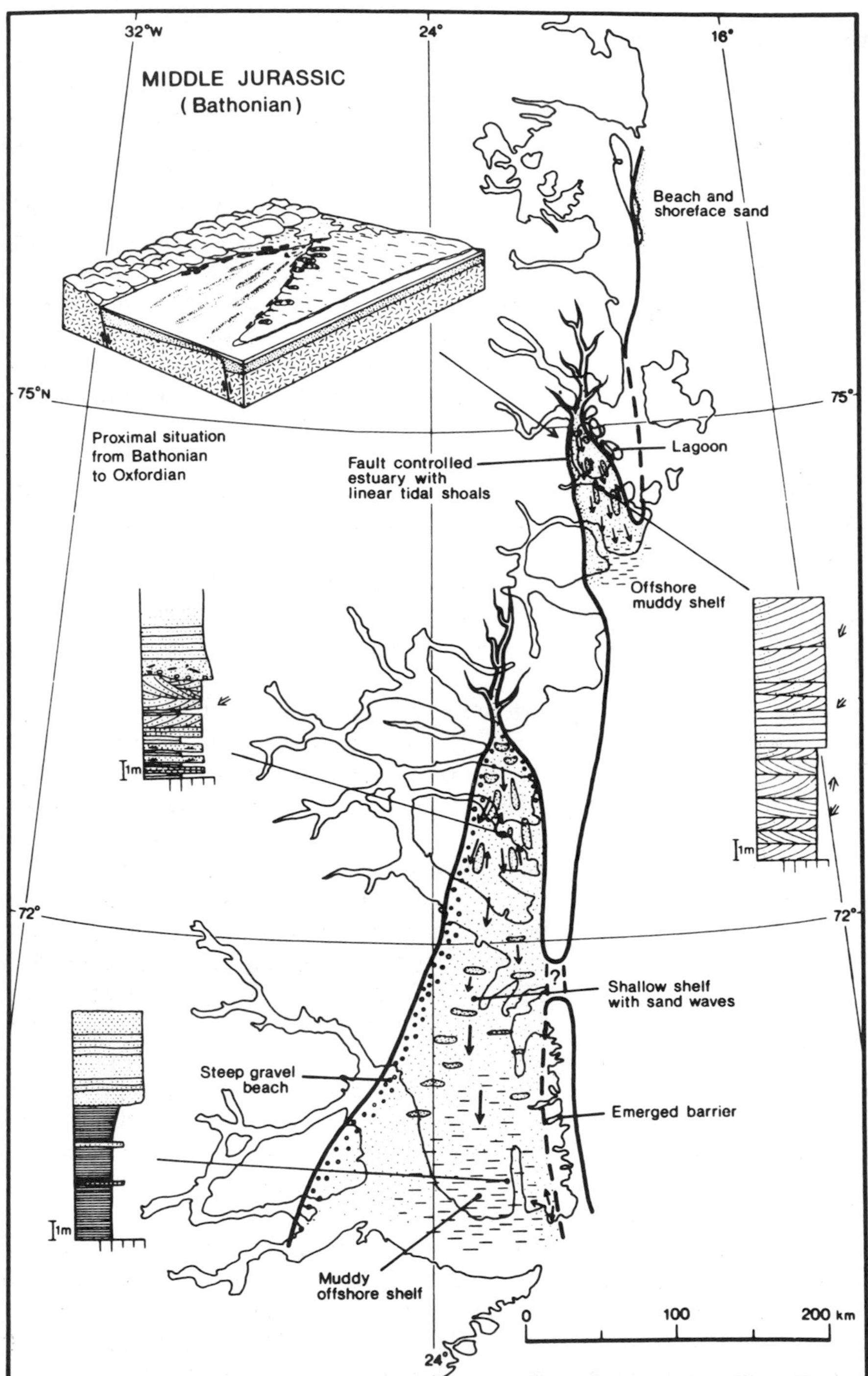

Fig. 9.5. Paleogeography of the East Greenland continental margin during the Bathonian. Embayments are probably extra-arch basins (Surlyk et al., 1981).

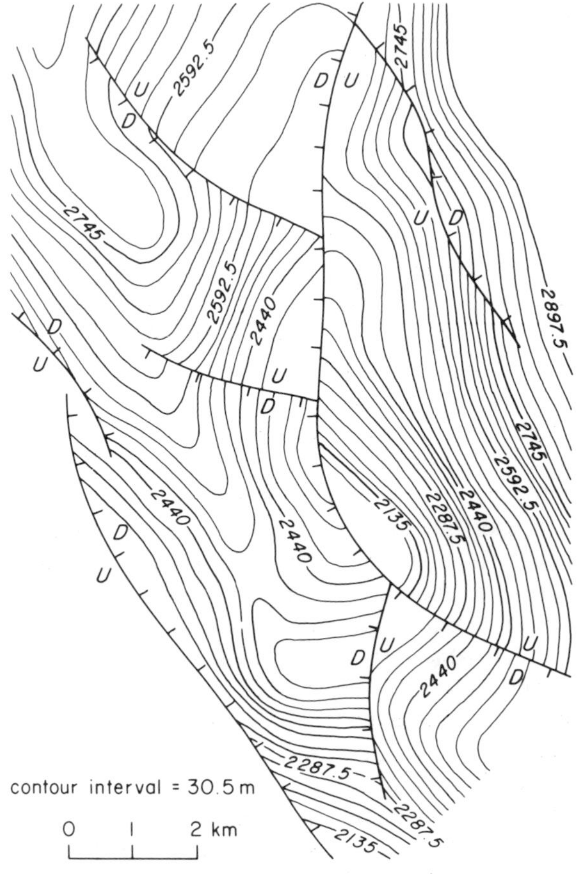

Fig. 9.6. Structure contour map of typical divergent continental margin, showing curved normal faults and gentle warps. Based on seismic data. July Field, Gulf of Suez. Red Sea to southwest (Brown, 1980).

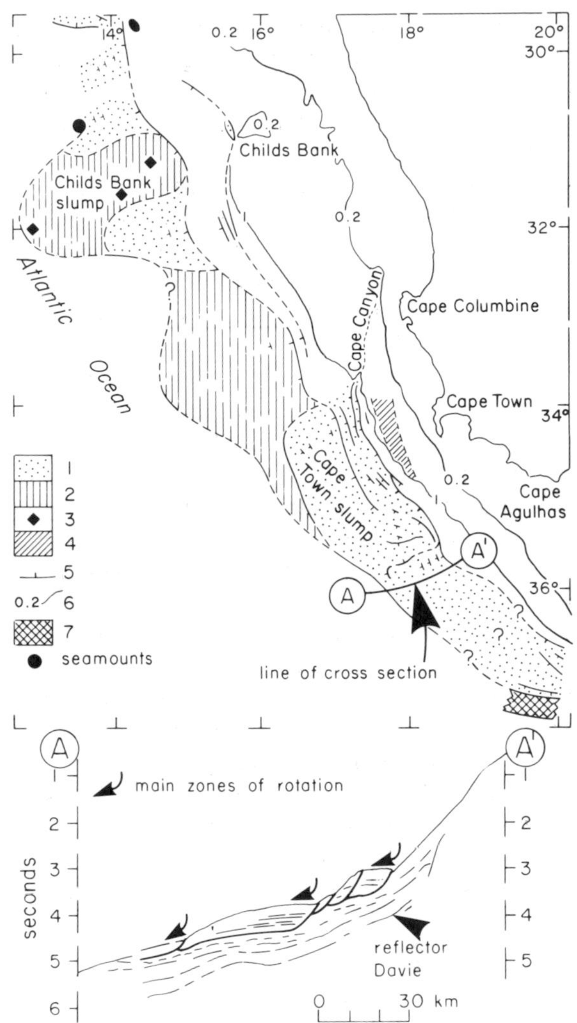

Fig. 9.7. Neogene slumps, Atlantic margin near Cape Town. 1. large rotated slump blocks with coherent bedding; 2. incoherent or horizontal bedding; 3. areas with detailed surface microrelief; 4. fissured zone; 5. main glide-plane traces; 6. isobaths, in km; 7. basement outcrops (Dingle, 1980).

Cape Town. Note the curved glide plane and rotated slump blocks. The presence of such features would play havoc with subsurface stratigraphic correlation in an ancient divergent margin basin, unless they could be identified on seismic records. They are by no means a ubiquitous feature of divergent continental margins, but seem to be particularly common off southern Africa.

As noted below, evaporites are a particularly common depositional facies in the incipient ocean phase of divergent margin development. They are invariably buried by several kilometers of sediment as the basin subsides, and such loading initiates diapirism. Diapir growth may continue for millions of years as burial continues, and they may project 10 km or more up to the sea bottom, where they form local topographic highs affecting sedimentation patterns. Such is the case in parts of the modern Gulf Coast which is underlain by Tertiary evaporites (Worzel and Burk, 1979). Onlap of sediments against these structures is recognized as "uplap" on seismic records (Fig. 6.40). The mechanics and structural geology of diapirs are beyond the scope of this book but are discussed in a useful review by Woodbury et al. (1980).

The stratigraphic evolution of divergent margins follows a rather predictable pattern. Veevers (1981) illustrated this using the south coast of Australia as an example (Fig. 9.8). During the

Lower Cretaceous fluvial and lacustrine deposits accumulated in graben and half-graben. The initial arch rifted apart, forming a rifted arch basin and a rim basin in which deltaic and other marginal marine facies developed. These gradually onlapped the rim as suggested in Figure 9.2. Continued differential subsidence caused the elevated rim to be planed off by subaerial erosion, and succeeding marine Tertiary deposits formed clinoform sets prograding unconformably across successively older units to rest on Lower Cretaceous strata over the site of the original arch (stratigraphic data from Boeuf and Doust, 1975; Talwani et al., 1978). Veevers (1981) pointed out that onlap in rim basin sediments will, with continued differential subsidence, become obscured as the distinctiveness of the rim disappears and may be interpreted as downlap on seismic records.

At the stage of the incipient or youthful ocean the basin may continue to accumulate clastic sediments under progressively deeper water conditions, or it may contain one of three other facies assemblages that are particularly characteristic of this stage of basin development. In all cases, sedimentation is now extending on to oceanic crust and further obscuring the boundary between oceanic and continental crust. The three facies are:

1. evaporites
2. black, organic shales
3. starved basin carbonates and pelagics.

In the early stages of marine inundation the connection with the sea may be spasmodic and subject to being cut off by eustatic sea level changes, movement of fault blocks or volcanic barriers. Under such circumstances conditions similar to those of the Mediterranean in the Miocene may develop, namely periodic complete desiccation and the deposition of thick evaporite sequences. Alternatively evaporites may be deposited from concentrated brines in deep, isolated basins (see discussion in sections 4.6.6 and 6.6). There are numerous examples of this in Atlantic margin basins (Rona, 1982), and dating of the evaporites has helped determine time of final continental rifting. For example, Jurassic evaporites and diapirs are present off Nova Scotia and Newfoundland (McWhae, 1981; Wade, 1982; see Fig. 9.10) and Lower Cretaceous evaporites occur off Brazil and west Africa from Nigeria to Angola (Ponte et al., 1980; Pautot et al., 1973; Simpson, 1977). The Red Sea, which dates back to the Miocene, is underlain by as much as 7.5 km of evaporites (Lowell and Genik, 1972). Rona (1982) pointed out that evaporites in the Atlantic region occur only in those parts of divergent margins that are oriented more or less perpendicular to oceanic fracture zones. This is presumably because the sheared margins (e.g. north Brazil) did not open into sizable basins until oceanic circulation was fully established.

At several times during the Phanerozoic extensive deposits of black organic shale (2–15% weight percent of organic carbon) have formed in the world's oceans. These include beds of Lower Paleozoic, Jurassic and Cretaceous age. The causes of this unusual depositional environment include high organic productivity and restricted marine circulation. Youthful oceans, where a volcanic ridge or upfaulted basement sill is present to inhibit current movement, are an ideal environment for this type of sedimentary accumulation, and Arthur and Schlanger (1979) showed that many of the Cretaceous deposits were located in such a tectonic setting. However, this facies is by no means restricted to young oceans; for example the best modern analogue for the euxinic environment is the Black Sea, an oceanic embayment within a suture zone. Arthur and Schlanger (1979) argued that, at least during the Cretaceous, "equable climates and low latitudinal thermal gradients would have led to sluggish circulation and therefore slow renewal of oxygen to deep water masses." This could help explain black shale development in other tectonic settings.

Where continental margin relief is low, sediment supply is minimal, and sedimentation may be dominated by platform and pelagic carbonates and other fine-grained, pelagic facies. Facies distribution is controlled by basement topography. Uplifted fault blocks and volcanic seamounts are the site of shallow water carbonate sedimentation, which may show rapid transitions into deep water facies in intervening faulted basins. Early sediments rest directly on oceanic crust. The modern Bahamas are a good example of this style of divergent margin. The carbonates there have built up from continental margin fault blocks that probably were first formed during the Triassic–Jurassic Atlantic rifting episode. Their sedimentology is discussed elsewhere in this book (see

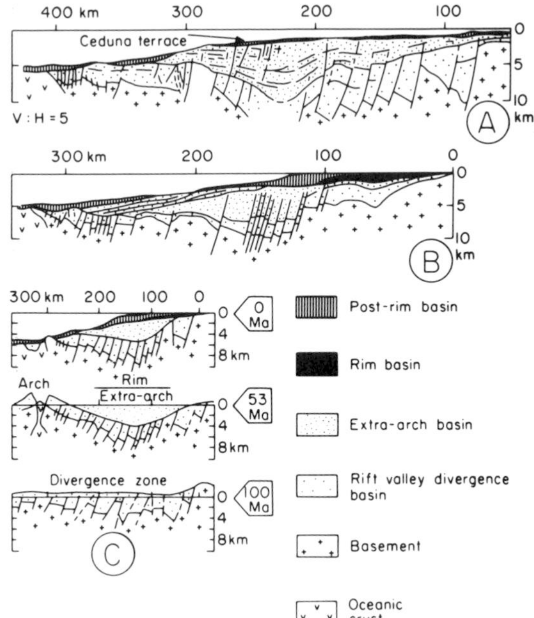

Fig. 9.8. Typical divergent margins, Australia. **A** Great Australian Bight; **B** Otway Basin; **C** schematic development of south coast, showing development of basin types based on model shown in Fig. 9.2 (Veevers, 1981).

Figs. 4.9, 6.31). The southern margin of the Tethyan Ocean in Greece, Italy (Bernoulli and Jenkyns, 1974) and Oman (Searle and Graham, 1982) provides many well-exposed ancient examples. The rocks range in age from Permian to Cretaceous. Those in Oman are mainly reef and lagoonal facies and are thought to have been deposited on volcanic seamounts similar to those of the modern Mauritius and Reunion Islands. The European examples occur mainly on continental basement. Sedimentation during the Jurassic was slow, resulting in various starved pelagic facies (Figure 9.9).

The fully mature divergent margin that evolves about 10^7 a after the start of sea-floor spreading is characterized by a decrease in active faulting (except for growth faults in deltas) and an increase in the scale of continental margin depositional systems. The early fault basins are buried by immense wedges of carbonate and/or clastic sediment, which usually show as much evidence of lateral progradation as vertical upbuilding. Much of the information we now have on large-scale depositional systems is derived from basins of this type (e.g. Gulf of Mexico, Brazil) and the reader is referred to Chapter 6 for a discussion. Similarly, the best evidence for the existence of regional and global stratigraphic cycles is derived from divergent margins, as described in Chapter 8.

A generalized basin model for divergent margins is summarized in Table 9.3 and illustrated in Figures 9.8 and 9.10. The three evolutionary stages defined in Table 9.3 can be readily identified in most of these examples. Variations in stratigraphy are summarized in Figure 9.11.

In the discussion up to this point we have assumed simple, continuous orthogonal spreading, and the examples illustrated essentially reflect this pattern. However, the structure and stratigraphy of some margins reflect important departures from the model:

1. oblique spreading, where the spreading ridge is offset by transform faults
2. intersection of oceanic transform fracture zones with continental margin
3. changes in position of the spreading center
4. fragmentation of the continental margin into major fault blocks
5. incomplete or aborted spreading
6. formation of triple junctions and failed rifts.

With regard to the first two points Franchetau and Le Pichon (1972) noted that coastal basins on the Atlantic margins of Argentina are oriented perpendicular to the continental margin and are bordered by structural highs on the prolongation of oceanic transform facture zones. They suggested that the development of these basins was influenced by thermal contrasts on either side of the transform fault. Wilson and Williams (1979) reviewed the relationship of Atlantic fracture zones to marginal sedimentary basins and pointed out that spreading directions ranged between parallel and perpendicular to continental margins and to basin axes in the Atlantic area. Most basins probably were influenced to some degree by wrench tectonics (as discussed further in section 9.3.3), although where continental margin and basin axis are perpendicular to transform direction as on the U.S. Atlantic border and southwestern Africa, the effects were small, and the basin model discussed here is the most appropriate one.

The third case is exemplified by the Blake Plateau area of the U.S. Atlantic Coast. Sheridan

Table 9.3. Basin model for divergent continental margins*

Evolutionary stage	Structure-Stratigraphy	Sedimentology
3	Few active faults (except growth faults in deltas), seaward-dipping prograding (clinoform) wedges, regional unconformities and onlap-offlap relationships reflecting major stratigraphic (eustatic) cycles. Progradation may extend far on to oceanic crust.	A. Continental coast-shelf-slope-rise clastic depositional systems, or: B. Carbonate platform or ramp depositional systems, or: C. Mixed carbonate-clastic systems showing crude cyclicity
2	Some active faults, blanket deposits in basins, wedge-out and drape over horsts. Sedimentation on oceanic or continental basement.	A. Marginal fluvial, lacustrine, fan delta deposits with basin centre containing either A1: thick evaporites or A2: black, organic shales; or: B. Carbonate platform and basinal pelagic deposits, some starved sequences.
1	Rifted arch and rim basins, many active faults	Fluvial and lacustrine sediments, interbedded with basic volcanics

*for orthogonal spreading only. See section 9.3.3 for discussion of oblique spreading.

et al. (1981) pointed out that the gravity and magnetic anomaly pattern for the outer part of the Plateau indicate thickened continental crust. The continental shelf is also unusually wide in this area. They explain these data by a shift in spreading center in the mid to Late Jurassic, as suggested in Figure 9.12.

The continental shelf around the British Islands is a good example of the fourth type of pattern. The shelf consists of a series of fault-bounded troughs (Hatton, Rockall, Porcupine) with intervening highs, some of which emerge as small islands (Rockall, Faroes). Short-lived spreading centers probably existed between some of these highs (the troughs therefore qualify as failed rifts), and the northern part of the area was markedly affected by Tertiary volcanism of the Icelandic or Thulean plume (Hall, 1981; Wilson, 1981). Shallow to deep water clastic sediments drape Precambrian to Paleozoic basement. This particular style of margin development is essentially a more complex example of "case 2" in the list above.

The Sverdrup Basin in Arctic Canada is a possible example of incomplete or aborted rifting. The basin shows virtually all the stratigraphic and structural characteristics listed in Table 9.3 (Balkwill, 1978), yet it is floored by Paleozoic continental crust.

The value of these models is that they permit the interpretation of plate setting for ancient basins that have been subsequently modified by quite different plate tectonic processes. For example we can identify a divergent margin in western North America by the westward-thickening Proterozoic and Paleozoic sequences extending from Nevada to Alaska (Douglas et al., 1970; Stewart, 1972). On this margin have now been welded all the accretionary terranes comprising the Western Cordillera (section 9.3.5). Similarly the Cambro–Ordovician divergent margin of Iapetus Ocean is now preserved in parts of the eastern United States, Quebec, Newfoundland and Scotland, along the fringe of the Appalachian–Caledonian suture (Williams, 1978), and Schenk (1981) argued that the Meguma Group of Nova Scotia represents a fragment of the divergent margin of northwest Africa.

Aulacogens and failed rifts. Burke and Dewey (1973) showed how divergent margins develop from triple-point junctions. The basic kinematics are now well understood and need not be repeated here. One arm of a ridge–ridge–ridge (RRR) junction commonly "fails" after a few million years, in the sense that spreading is aborted shortly before or after the development of true oceanic

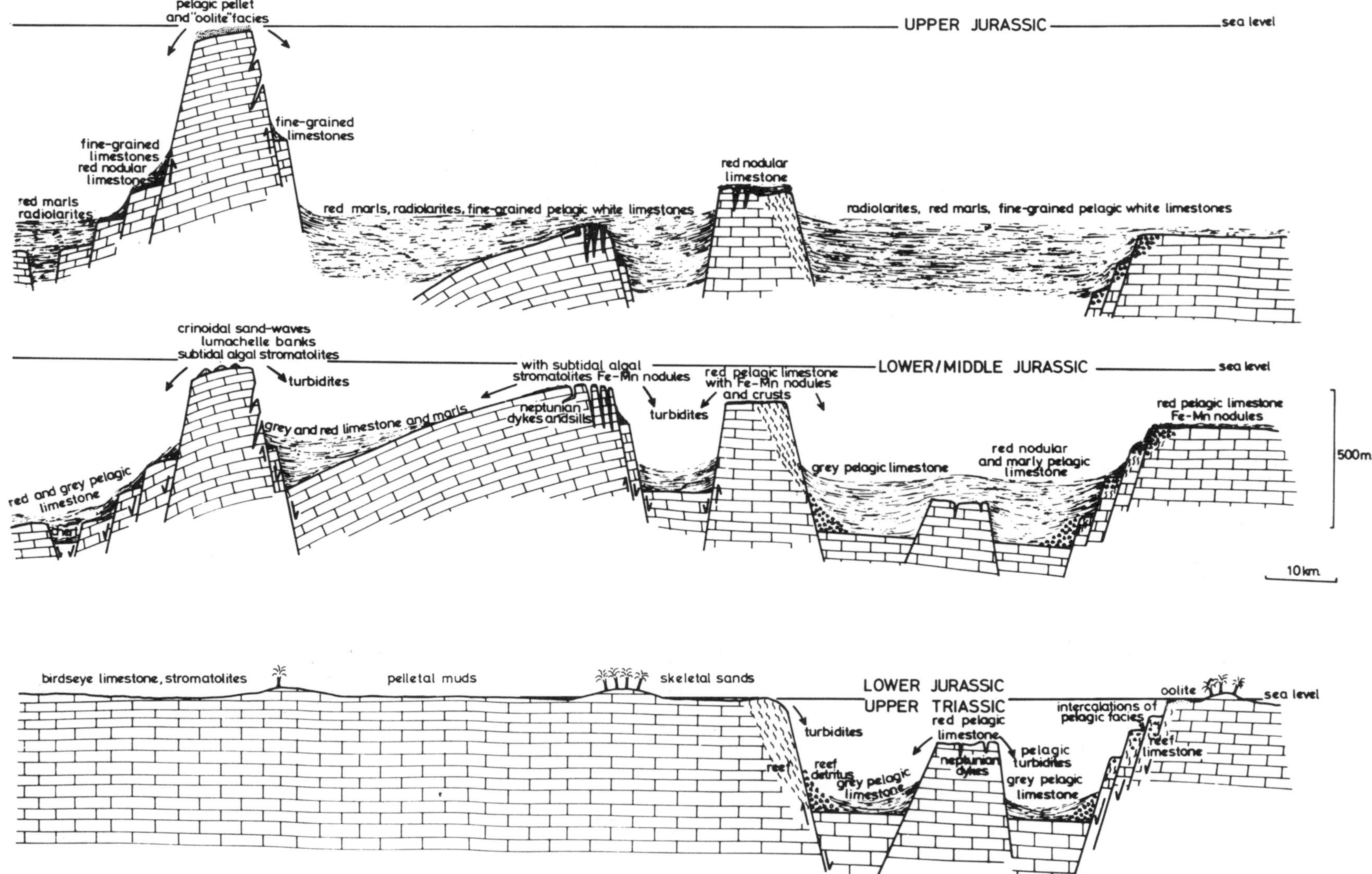

Fig. 9.9. Paleogeographic evolution of part of the southern continental margin of Tethys, showing the gradual foundering of a carbonate platform and the absence of major detrital depositional systems (Bernoulli and Jenkyns, 1974).

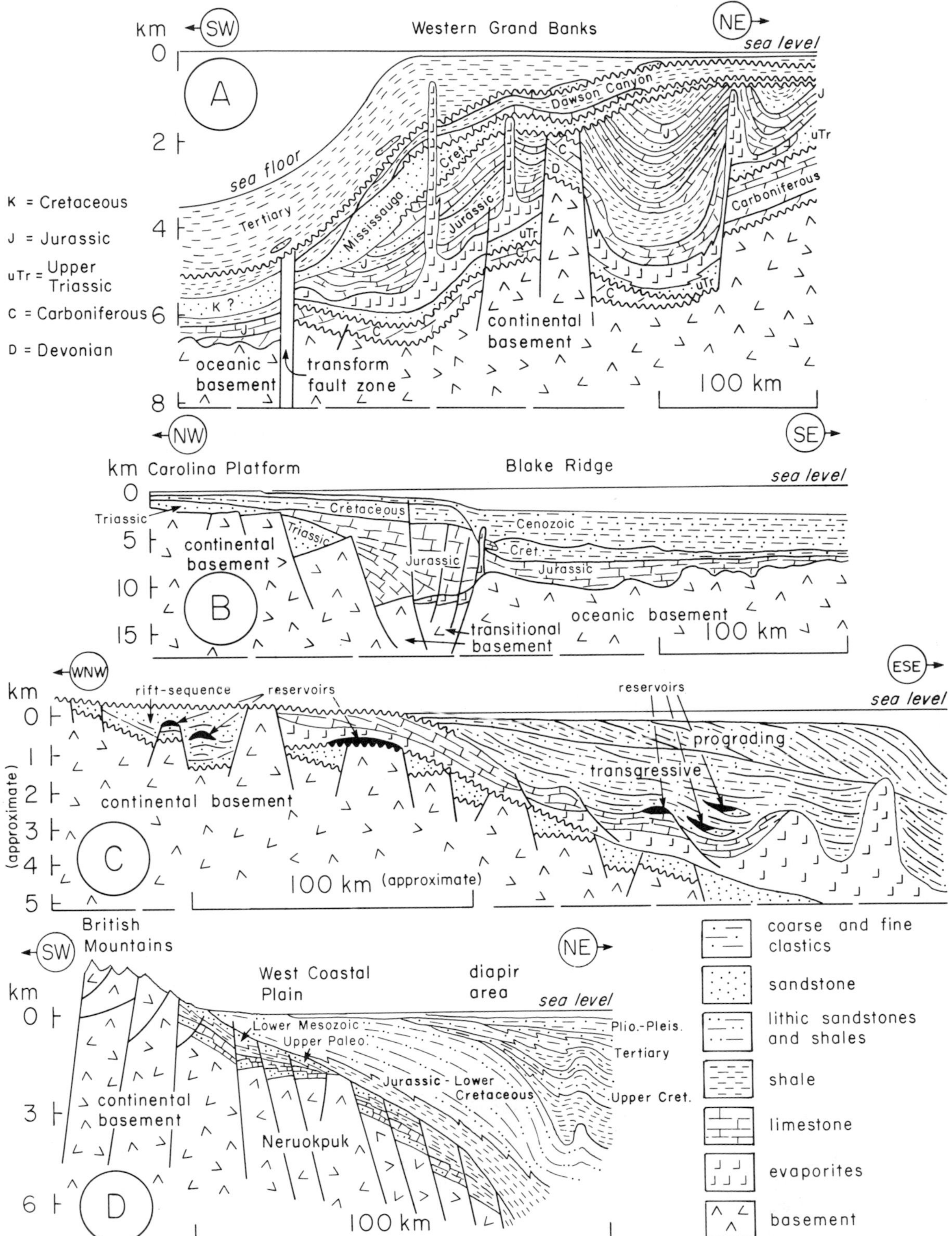

Fig. 9.10. Examples of typical fully developed divergent continental margins. **A** Grand Banks, Newfoundland (McWhae, 1981); **B** Caroline Trough, off Cape Fear, North Carolina (Sheridan et al., 1981); **C** idealized section across Brazilian continental margin (Ponte et al., 1980); **D** Beaufort Mackenzie Basin, northern Canada (Hawkings and Hatlelid, 1975).

Fig. 9.11. Examples of large-scale stratigraphic variations within divergent margin basins reflecting variations in sediment supply, boundary currents and presence of evaporites (Sheridan, 1981).

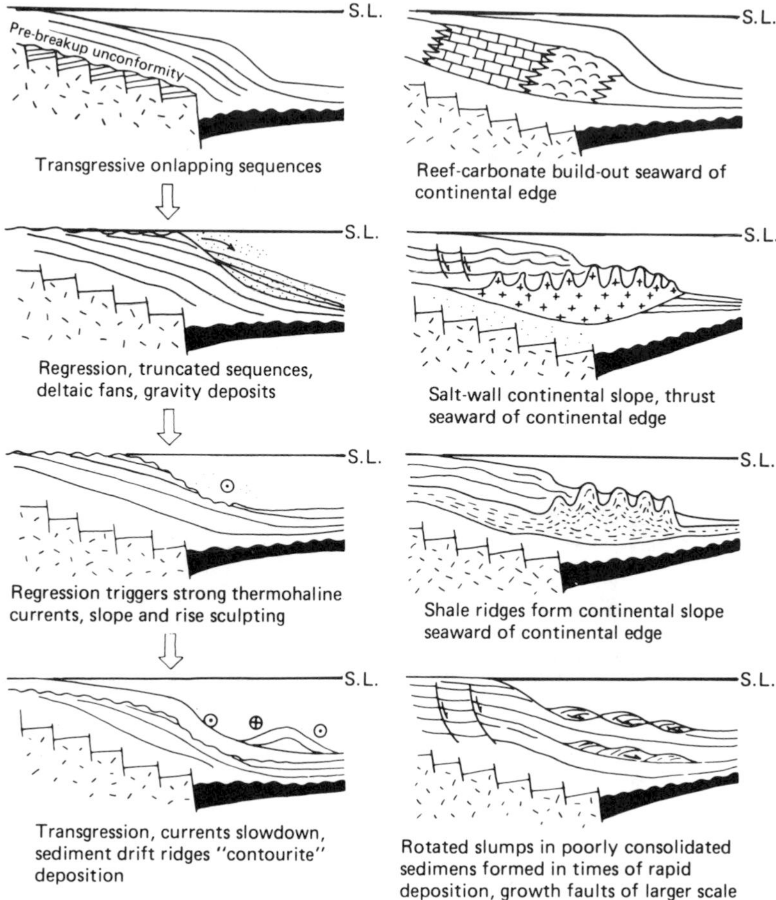

crust, while complete continental separation occurs on the other two arms. The "failed" arms form deep troughs extending at a high angle away from the continental margin. The continental margin may later collide with another continental plate forming a suture. This event may reactivate structures in the failed rift, producing a mildly deformed trough trending obliquely from the suture. These types of rifts are now called **aulacogens**, a term coined by the Russian geologist N. Shatski long before the development of plate tectonics. The term **failed rift** is retained for rifts striking from oceans into continents (see review by Burke, 1977). Hoffman (1973) and Hoffman et al. (1974) described the structural and stratigraphic evolution of aulacogens, and Burke (1977) summarized the structural complexities of various basin types. Brewer et al. (1983) pointed out that failed rifts and aulacogens may, like other types of rift, be located along earlier lines of weakness.

The structural and stratigraphic evolution of failed rifts and aulacogens parallels that of divergent margins, except that the basin continues to be floored by continental crust and therefore does not subside to abyssal depths. A genetic model of crustal extension and upwelling of hot asthenosphere was suggested for the Athapuscow Aulacogen (Proterozoic, northern Canada) and the Oklahoma Aulacogen (Proterozoic–Permian) by Hoffman et al. (1974) and for the Viking–Central Graben of the North Sea (Triassic–Cretaceous) by Sclater and Christie (1980). This model provides for a simple two-phase development, an initial graben subsidence during crustal extension and a downwarping phase during thermal contraction and loading. In Figure 9.13 these two phases are shown, together with a transitional phase and a final compressive phase caused by a change in spreading patterns. The intrusion and loading effect of dike swarms may be important in some aulacogens (Burke, 1976a, 1977) but does

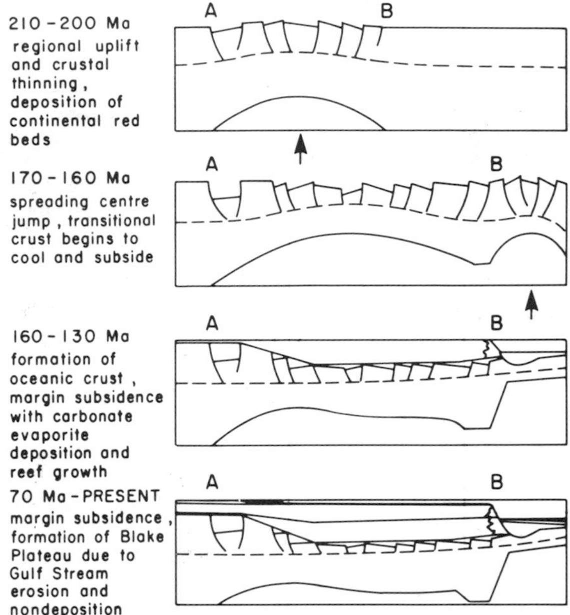

Fig. 9.12. Model of the development of Blake Plateau, underlain by a sliver of "African" crust as a result of a jump in the spreading center. Distance between **A** and **B** increases with time due to intrusion of basaltic dykes (Sheridan et al., 1981).

not seem to have occurred in the Athapuscow or North Sea examples. The duration of these phases is partly controlled by the constraints of the mantle upwelling and cooling model, and partly by regional stress patterns imposed by plate spreading vectors. For example the Viking–Central Graben (Fig. 9.14) shows several periods of rifting (Ziegler, 1980), which can be correlated with rifting episodes in the east Greenland divergent margin (Surlyk et al., 1981; see earlier disscusion of this area). These probably reflect changes in plate movement as the Atlantic Ocean opened.

Aulacogens and failed rifts do not appear to be characterized by any distinctive lithofacies assemblage or sequence. Sediments may be shallow or deep marine, carbonate or clastic. A transition from nonmarine in the graben stage to marine in the downwarp stage seems typical. There may be transverse alluvial fans, submarine fans (e.g. Frigg area of North Sea, see Fig. 6.37, 6.38) or carbonate platforms (Fig. 9.13) along the flanks, with longitudinal rivers and deltas or a basin plain turbidite trough along the axis. Up to 10 km of sediment may accumulate. These typically show ideal levels of organic maturity for petroleum generation and are thus of considerable economic importance (e.g. North Sea).

A note of caution now seems necessary in the interpretation of basins as failed rifts or aulacogens. Divergent margins typically contain a variety of basins showing a wide range of orientations. It is a simple matter to draw in hypothetical spreading axes and triple-point junctions, and label any basin extending inland at a high angle to the continental margin as a failed rift. For many ancient basins, particularly those flanking sutures where all evidence of plate kinematics has been lost, this may be the only type of analysis possible. However, for basins flanking modern oceans detailed analysis of sea-floor spreading patterns permits a rather more sophisticated approach. Burke (1976a) interpreted the Atlantic margins largely in terms of the triple-point model, although recognizing the importance of transform faulting between northern Brazil and adjacent parts of Africa. As a result of this analysis the Benue Trough in Nigeria has become widely regarded as a classic example of a failed rift (see also Hoffman et al., 1974). An important modification of this interpretation was offered by Wilson and Williams (1979) and Benkhelil (1982) who noted the presence of three Atlantic fracture zones striking into the trough. Benkhelil (1982) suggested that the trough is a complex pull-apart basin, and in support of this analysis he presented evidence of structural patterns inconsistent with simple orthogonal extension. This point is taken up again in Section 9.3.3.

Burke and Dewey (1973) and Burke (1977) discussed a variety of possible ancient failed rifts and aulacogens of Phanerozoic and Precambrian age. Hobday and Von Brunn (1979) described the alluvial fill of Natal Embayment, a failed rift that occupied the present southeast coastal area of Africa in the Paleozoic and was subsequently the site of one of the major rifts through Gondwana in the Cretaceous. All these interpretations must be reevaluated in the light of the caution expressed above.

9.3.2 Convergent margin basins

These are basins related to active magmatic arcs and subduction zones. Many modern arcs in the Pacific have now been studied extensively by

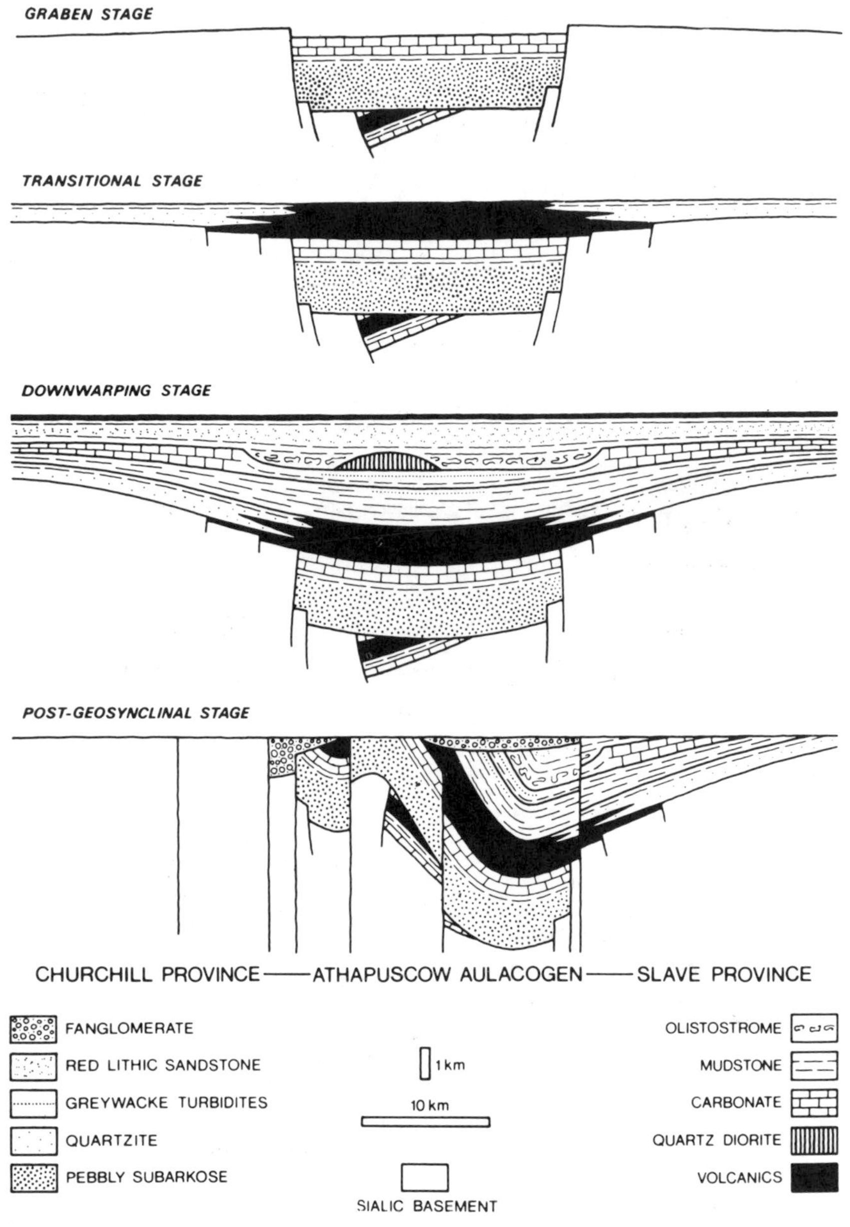

Fig. 9.13. Model of the development of an aulacogen, based on Athapuscow Aulacogen, northern Canada (Hoffman et al., 1974).

geophysical methods, and some scattered stratigraphic–sedimentological information is available from relatively shallow DSDP cores (Von Huene, 1981). However, these basins have not been as extensively explored by deep petroleum exploration drilling as have divergent margins. Good sedimentological descriptions of modern arcs are few, and many of our best ideas are derived from studies of interpreted ancient convergent margin basins, such as the Ordovician–Silurian subduction complex of the Southern Uplands, Scotland, the Jurassic to Paleogene forearc basin of the Great Valley, California, and the associated Franciscan subduction complex. These are discussed below.

Tectonic review. The classification and terminology of convergent margin morphologic and petrotectonic features have been described and illustrated by Dickinson (1974), Dickinson and Seely (1979), Bally and Snelson (1980) and others. Some of the terms used by those authors are illustrated in Figures 9.15 and 9.16. Slightly different terminologies are employed by Karig and Shar-

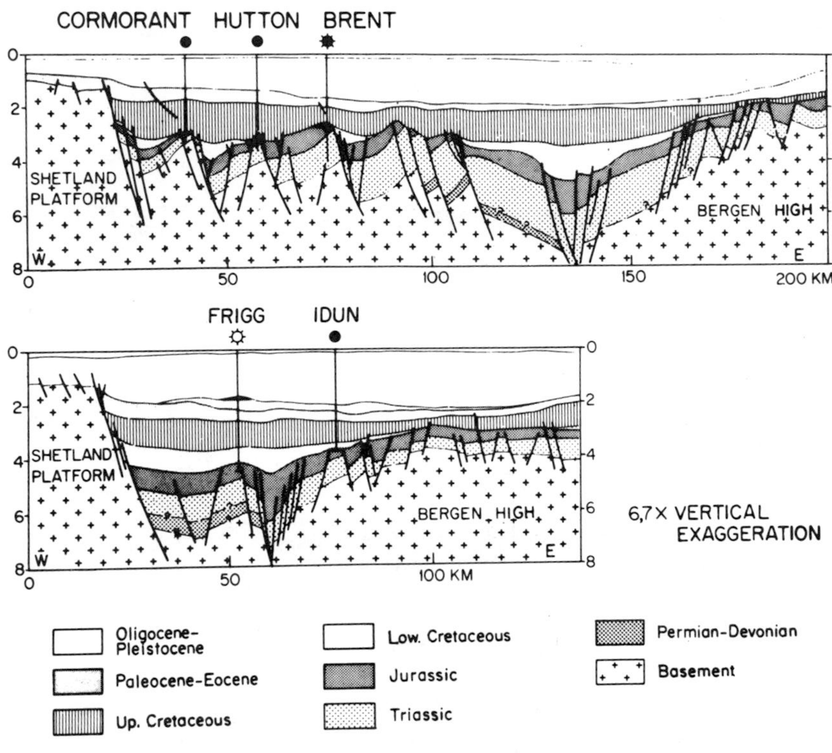

Fig. 9.14. Structural cross sections through a typical failed rift, Viking Graben, North Sea (Ziegler, 1980).

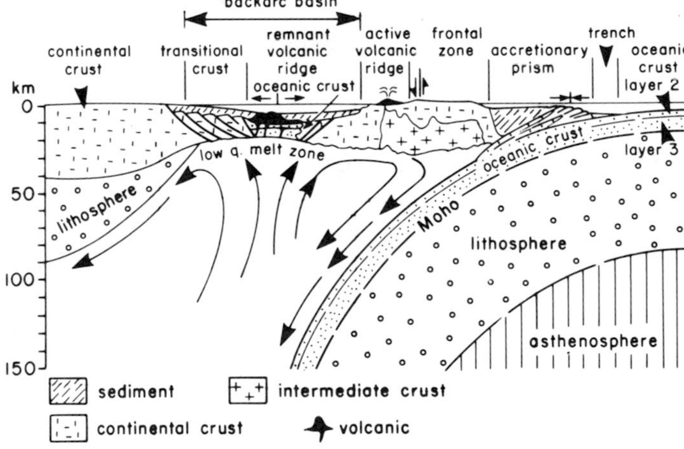

Fig. 9.15. Terminology of a convergent margin involving a continental overriding plate. A model for the development of a backarc basin is shown, based on Toksoz and Bird (1977; see also Bally and Snelson, 1980).

man (1975) and Hamilton (1977, 1979). The main components of convergent systems are as follows (in order from subducting to overriding plate):

1. Arch or outer rise: an upwarp in the oceanic crust elevated as much as a kilometer or more above the abyssal plain and up to several hundred kilometers wide. The arch is probably generated by bending of the crust above the subduction zone.

2. Trench: a trough up to 11 km deep in front of the arc, and the final repository of most arc-derived sediment. Turbidites and pelagics typically are present here, but are deformed and overriden by the subduction complex. Many trenches are essentially empty of sediments because of the absence of any adjacent sediment supply.

3. Subduction complex: a highly complex zone formed by the shearing between subducting and

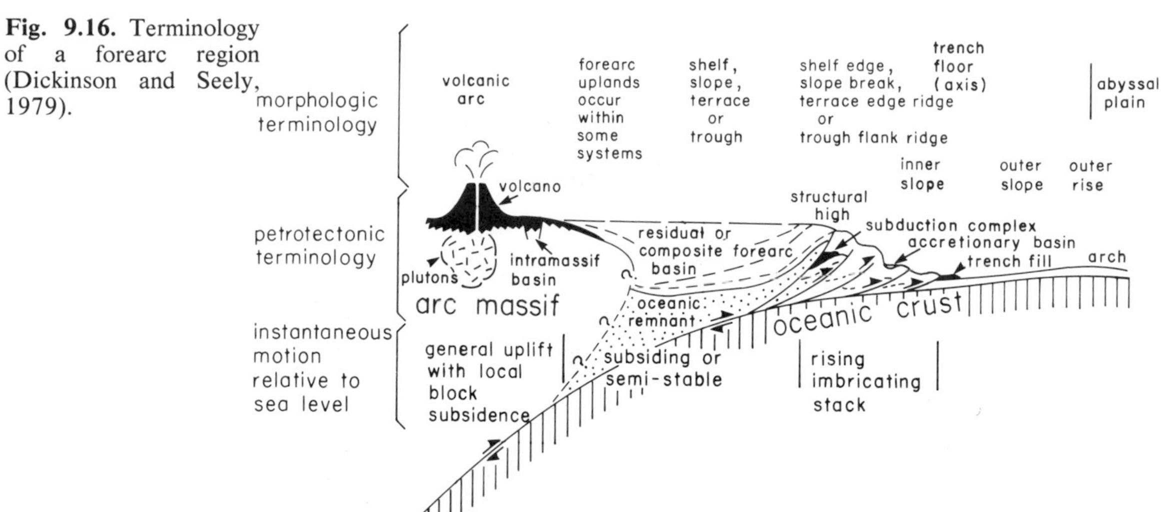

Fig. 9.16. Terminology of a forearc region (Dickinson and Seely, 1979).

overriding plates. The composition and structure of the zone varies widely. Some subduction complexes form a prominent topographic high and emerge as an "outer accretionary arc" or "non-volcanic outer arc" parallel to the volcanic arc. The subduction zone dips at an angle of between 10 and 85°, and can be traced to depths of up to 600 km by its active seismicity. It is commonly called a Benioff Zone (hence "B-subduction").

4. Forearc basin: Where the subduction complex forms a terrace or ridge in front of the volcanic arc, a topographic depression will occur, and this acts as a sediment trap in which great thicknesses of sediment may occur.

5. Magmatic arc: Where the subduction zone dips to a depth of between about 100 and 150 km partial melting of the overriding plate and possibly of the subducting plate generates magma that feeds an active magmatic arc. The arc lies at least 100 km landward of the trench, depending on the dip of the subduction zone. Volcanism is typically andesitic, but varies widely depending mainly on the composition of the overriding plate. Uplift of this arc provides the major sediment source on convergent margins.

6. Backarc region: The overriding plate may consist either of continental or oceanic crust. In the case of continental crust the volcanic arc may be flanked by small extensional basins, and typically there is a major fold–thrust belt adjacent to a foreland or retroarc basin overlying the cratonic hinterland. Bally and Snelson (1980) defined the fold-thrust belt as A-subduction (after Ampferer). This is explained in more detail in a later section.

Backarc areas may also be floored by oceanic or transitional crust and are covered by the sea (Fig. 9.15). These are termed marginal seas and are typically extensional in character; their continental side may to some extent mimic a divergent plate margin. There is some controversy over the origin of backarc basins, as discussed later.

Arc geology encompasses a daunting array of complexities that reflect variations in the age and composition of the converging plates and variations in plate trajectories. Indonesia provides examples of most types of convergent margin behavior. The plate tectonics of this region have been summarized by Hamilton (1977) and described in greater detail in a masterpiece of regional geology by Hamilton (1979).

Throughout the Cenozoic, Indonesia and the Philippines have undergone a highly complex history as a result of the convergence of three major plates. The Indian–Australia plate is moving northward and approaching Asia at a velocity of about 10 cm/a, while the Pacific plate is moving relatively westward. The Indonesia–Philippine region

has responded with rapidly changing patterns of subduction, spreading, migration, strike–slip faulting, oroclinal folding, and rotation. After arcs have collided with each other or with continents, convergence continues at a new subduction zone on the outside of the

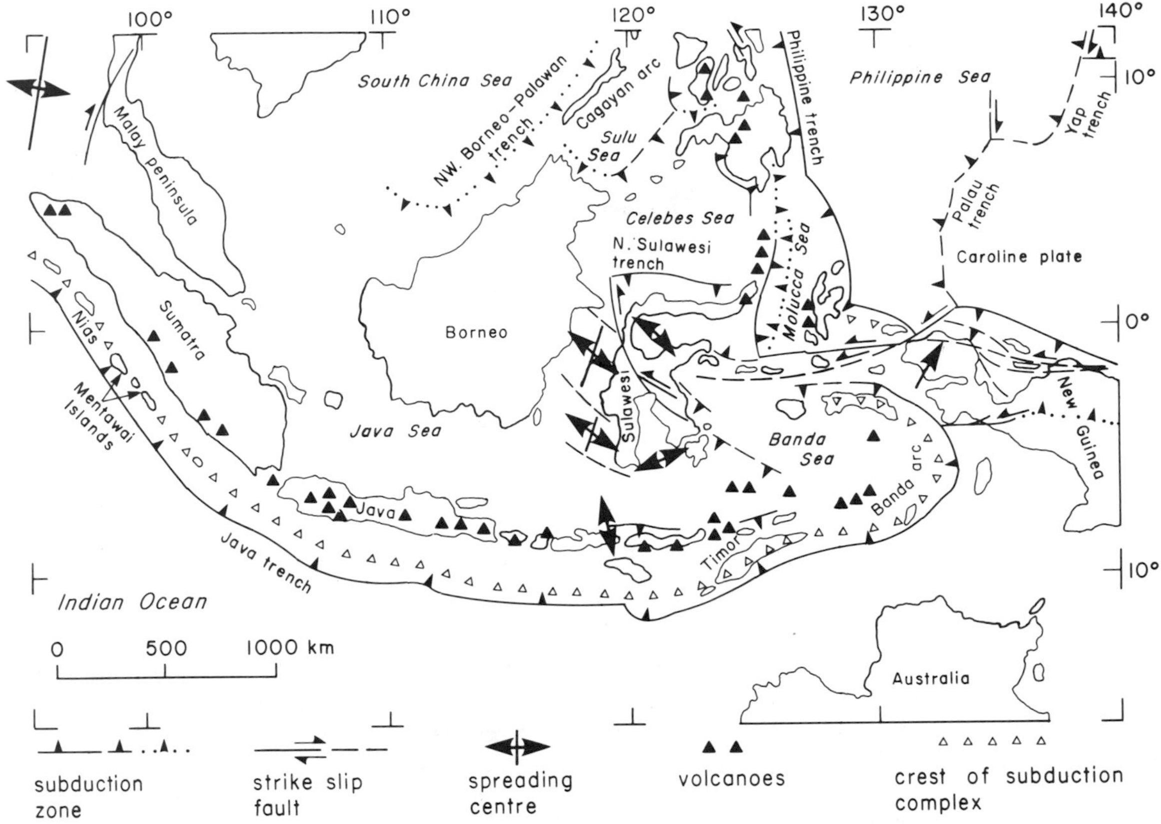

Fig. 9.17. Major Cenozoic tectonic elements in the Indonesian region (Hamilton, 1977, 1979).

collision aggregate. Some arcs die without collision, some arcs reverse polarity, and others migrate and change shape as spherical plates inflect through them (Hamilton, 1977).

Some of the major Cenozoic tectonic elements in the area are shown in Figure 9.17. The Java–Sumatra arc is a classic example of a convergent margin, showing an active volcanic arc, a foreland floored by continental crust to the north, a deep marine forearc basin, a well-developed outer accretionary arc (uplifted off Sumatra as the Mentawai Islands), and a deep trench partly filled with turbidites of Bengal Fan. Toward the east the forearc region crosses the island of Timor, which consists mainly of a subduction complex containing Permian to Paleogene shallow-water strata representing the northernmost edge of the Australian continental plate. The arc has collided with a major continental mass and it is likely that the subduction zone will shortly "flip", or reverse polarity, with a new southward-dipping subduction zone appearing to the north of the present volcanic arc.

This arc continues eastward as the Banda arc, which curves northward against New Guinea. Hamilton (1977; 1979, Fig. 77) suggested that the curvature resulted from a collision with New Guinea coupled with a continued tendency for the subduction hinge (the bend in the oceanic crust at the outer rise) to migrate over the oceanic plate.

The Sangihe and Halmahera island arcs face each other in the Molucca Sea (Fig. 9.17) and are in the process of colliding. Collision is complete in the north, in Mindanao, but incomplete in the south. North Sulawesi is undergoing clockwise rotation bounded by an active sinistral strike-slip fault to the west and a subduction zone to the north. Convergence on this zone decreases eastward to zero at an orocline or pole of rotation. Borneo and Sulawesi rifted apart along a mid-Tertiary spreading center underlying Makassar Strait. East Sulawesi is a continental fragment of New Guinea sutured with Sulawesi in the Miocene following westward subduction and strike-slip faulting of the Australia–New Guinea plate. Numerous other complications could be described. They are all superimposed on earlier, quite different patterns.

Hamilton (1977, 1979) suggested that many megasutures, such as the Appalachians and the Alpine–Himalayan system would have shown Indonesian-style tectonics in early stages of collision. The difficulties inherent in resolving Indonesian geology, where the tectonic elements are still relatively spread out and undeformed, pale by comparison with the task of sorting out a major suture zone where thousands of square kilometers of microplates, arc complexes and oceanic crust have been compressed into narrow, linear orogenic belts.

We discuss suture zones and microplates in greater detail in sections 9.3.4 and 9.3.5. The purpose of introducing the topic at this stage is to make the point that convergent margins cannot be properly understood on the basis of a simple two-dimensional analysis such as the cross sections of Figures 9.15 and 9.16. These imply orthogonal convergence, which is practically never the case, and tell us nothing about variations in subduction behavior in response to sediment supply and sediment loading, highly oblique convergence, variations in age of oceanic crust, presence of small (but still subductable) masses such as volcanic seamounts on the subducting plate, the effects of subducting spreading ridges or oceanic fracture zones (old transform faults), or the nature of backarc spreading. Many writers have tackled these problems, including Wilson and Burke (1972), Toksöz and Bird (1977), Uyeda (1981, 1982), Molnar and Atwater (1978), Dickinson and Seely (1979), Dewey (1980) and Cross and Pilger (1982).

The kinematics of convergence have been examined by Dewey (1980b) in an elegant exercise of vector analysis. Some of his major conclusions are summarized here. He recognized, following Molnar and Atwater (1978), that convergent margins in the western Pacific are subducting mainly Mesozoic crust older than about 100 Ma, and that most show a history of present or relatively recent backarc spreading. Conversely in the eastern Pacific subducted crust is mainly younger than about 50 Ma and the arcs there are located on continental crust with active backarc compression (metamorphic core complexes, fold-thrust belts). Documentation of Pacific arcs yields an additional point that in general the older the subducted crust, the steeper the dip of the Benioff Zone. This is attributed to the thermal properties of the downgoing slab: the older it is the colder and more dense it will be, and hence the faster its rate of sinking. It is suggested that where the subducting crust is old it sinks faster than the adjacent plates can converge. The subducting hinge rolls back (oceanward), carrying with it the forearc region of the overriding plate. This leads to extension and backarc spreading in the overriding plate. Bally and Snelson (1980) summarized alternative explanations of backarc spreading, such as that of Toksöz and Bird (1977) and McKenzie (1978) in which the spreading, when it occurs, is caused by secondary mantle convection induced by the downward motion of the slab. However, the model of Molnar and Atwater (1978), as elaborated by Dewey (1980), seems to be capable of explaining most varieties of arc behavior, and is the one preferred here.

Dewey (1980) established five models for convergent margins.These are illustrated in Figure 9.18 and summarized below. Vectors in Figure 9.18 are shown with reference to "fixed" asthenosphere and are therefore "absolute" motions. Models 1 and 2 are cases where the rate of retreat of the subduction hinge (V_r) is faster than the component of oceanward movement of the overriding plate. This would be particularly likely to occur where the overriding plate is moving obliquely to the trench line, and such a situation would arise if the subduction zone developed along an old line of weakness (e.g. a fracture zone) oriented at an oblique angle to V_o These two models, particularly model 2, are characterized by backarc extension. Subduction complexes may accrete in the forearc region but because of the tendency of the arc to migrate oceanward it will become isolated from continental sediment sources. The Mariana and Tonga arcs are examples of these **extensional arcs**.

Model 3 is that of a **neutral arc**, where oceanward movement of the overriding plate is balanced by the oceanward retreat of the subduction hinge. A strike-slip fault may be present along the arc (where the crust is weakest) to accommodate oblique convergence (Fitch, 1972), but backarc spreading does not occur except where this fault is offset. The Alaska–Aleutian and Sumatra arcs are of this type, and are characterized by well-developed subduction complexes. The Andaman Sea is an example of backarc spreading developed as a result of transform offset in a neutral arc.

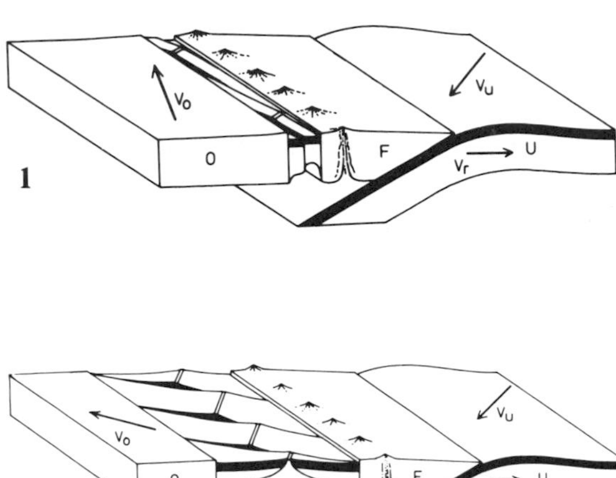

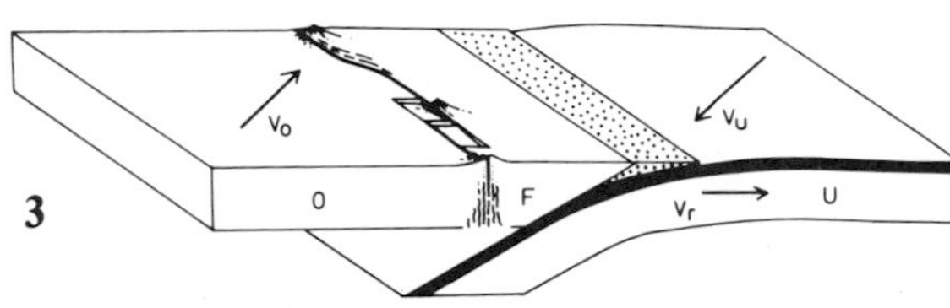

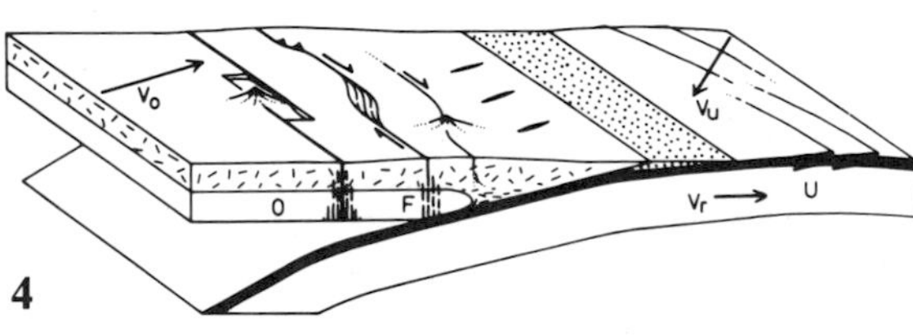

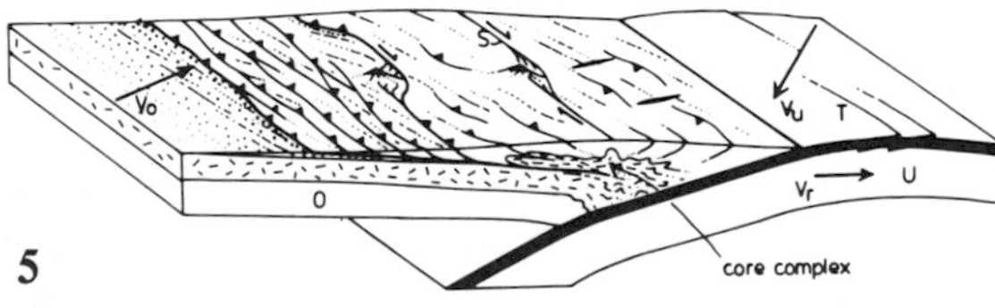

Fig. 9.18. Five models for the kinematics of a convergent margin, based on a vector analysis. V_o = velocity of overriding plate, O; V_u = velocity of underriding plate, U; V_r = velocity of migration of subduction hinge; F = frontal (forearc) part of arc (Dewey, 1980).

Models 4 and 5 are **compressional arcs** and are thought to occur where the subducting crust is young. The hinge line roll-back (V_r) is therefore slow, and the forearc region is under active compression. Tectonic erosion of the overriding plate may occur in the subduction zone, and some of the shear stress is accommodated by thrusting in the oceanic plate. In model 4 a discrete forearc region is maintained, and oblique convergence gives rise to strike-slip movement within the arc. In model 5 the whole frontal region of the overriding plate is involved in compressional tectonics. Cordilleran-type metamorphic core complexes, nappes and foreland A-subduction occur. Peru and the Canadian-western U.S. Cordillera fit this model at certain stages of their development.

A convergent margin may show variations in style along strike if the age of the crust varies. For example the subduction zone will be segmented where it intersects a fracture zone (Fig. 9.19). Subduction of older crust on one side of the zone may develop an extensional margin with backarc spreading, whereas the arc facing younger crust may be neutral or compressional. This model illustrates that backarc spreading can be terminated along strike, particularly at a cross-cutting transform fault. This in fact occurs at several places in the western Pacific (Dewey, 1980; Hamilton, 1979). Other variations in style are caused by subduction of aseismic ridges or seamounts, and by differences in the volume of sediment fed to the trench (Cross and Pilger, 1982).

Dewey (1980) showed that arc behaviour may also vary with time along a given convergent margin, as the age of the subducted crust becomes progressively older or younger. The latter case is illustrated in Figure 9.20. This adds much useful new detail to the concept of the Wilson cycle of opening and closing oceans. It demonstrates how backarc basins may open and close and accounts for the episodic nature of subduction–accretion or orogenic activity in the foreland. There is not space to dwell here on the parallel implications for volcanism and plutonism.

With this introduction to the complexity of arc tectonics we shall now examine the main types of convergent margin basin listed in Table 9.2. Unfortunately, as noted earlier, the acquisition of stratigraphic–sedimentologic data has not kept pace with the production of seismic cross section or the development of theoretical tectonic models.

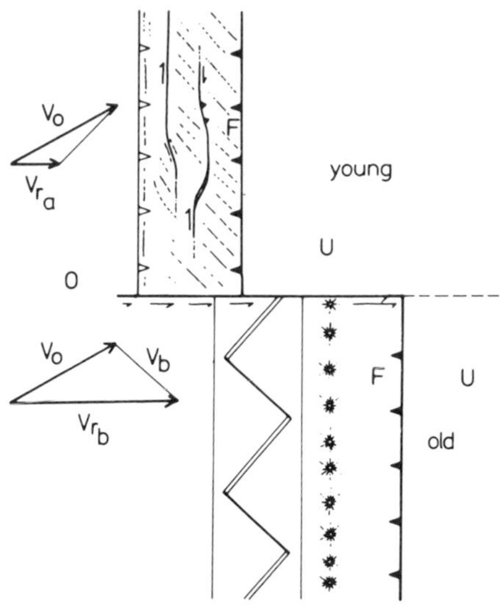

Fig. 9.19. Segmentation of a subduction zone where young and old oceanic crust are juxtaposed along a fracture zone (Dewey, 1980).

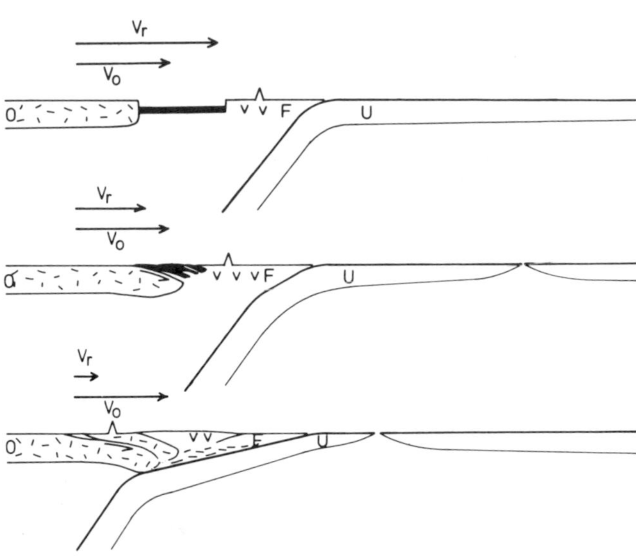

Fig. 9.20. Possible sequence of events during a Wilson cycle as progressively younger oceanic crust is subducted. A backarc basin is first opened and then closed (Dewey, 1980).

For example a recent review of DSDP data from convergent margins by Von Huene (1981) does not mention either the various hypotheses of backarc spreading or the different styles of convergent margin. Therefore it is not yet possible to document the sedimentological implications of all the complexities discussed in this section.

Trenches and subduction complexes. In complex accreted terranes or megasutures the identification of subduction complexes is one of the surest indicators of the previous existence of a convergent margin, as their structural and stratigraphic character are unique. If the arc magmatic belt can also be recognized and the two proved to be contemporaneous the polarity of the subduction zone is then known. The existence of paired metamorphic belts (high P, low T and low P, high T) in arcs is also a well-known criterion (Miyashiro, 1980).

Dickinson and Seely (1979) identified "three distinct structural styles (that) may be formed as discrete belts or slabs within the subduction complex":

1. Bedded sequences, which may be isoclinally folded, in which stratification is largely preserved and metamorphic fabrics are not widely developed, although metamorphic minerals may largely replace original constituents;
2. Metamorphic tectonites, schistose or semischistose, in which a pervasive metamorphic fabric that largely supplants original textures exerts the dominant control on structural relations; and
3. Chaotic melanges, in which pervasive mesoscopic shear fractures have disrupted the original rock into isolated tectonic inclusions immersed in a sheared matrix (Dickinson and Seely, 1979, p. 18).

Important constituents in any of these types may include slivers of oceanic crust (ophiolites) or giant slide masses (olistostromes). Some variations in the relative scale and style of subduction complexes and other forearc elements are shown in Figure 9.21.

Some subduction complexes are dominated by folded, bedded sequences. Modern examples include the Shikoku subduction zone off southwestern Japan (J.C. Moore and Karig, 1976) and the Aleutian Kodiak–Kenai subduction complex of Alaska (J.C. Moore and Wheeler, 1978; Von Huene, 1978; Fisher and Von Huene, 1980). An ancient example is the Ordovician–Silurian subduction complex of the Southern Uplands, Scotland (Leggett, 1980). Other complexes consist largely of chaotic melanges such as the Mentawai Islands complex of the Sunda arc, off Sumatra (Karig et al., 1978; G.F. Moore and Karig, 1980) and much of the Jurassic–Paleogene Franciscan complex of California.

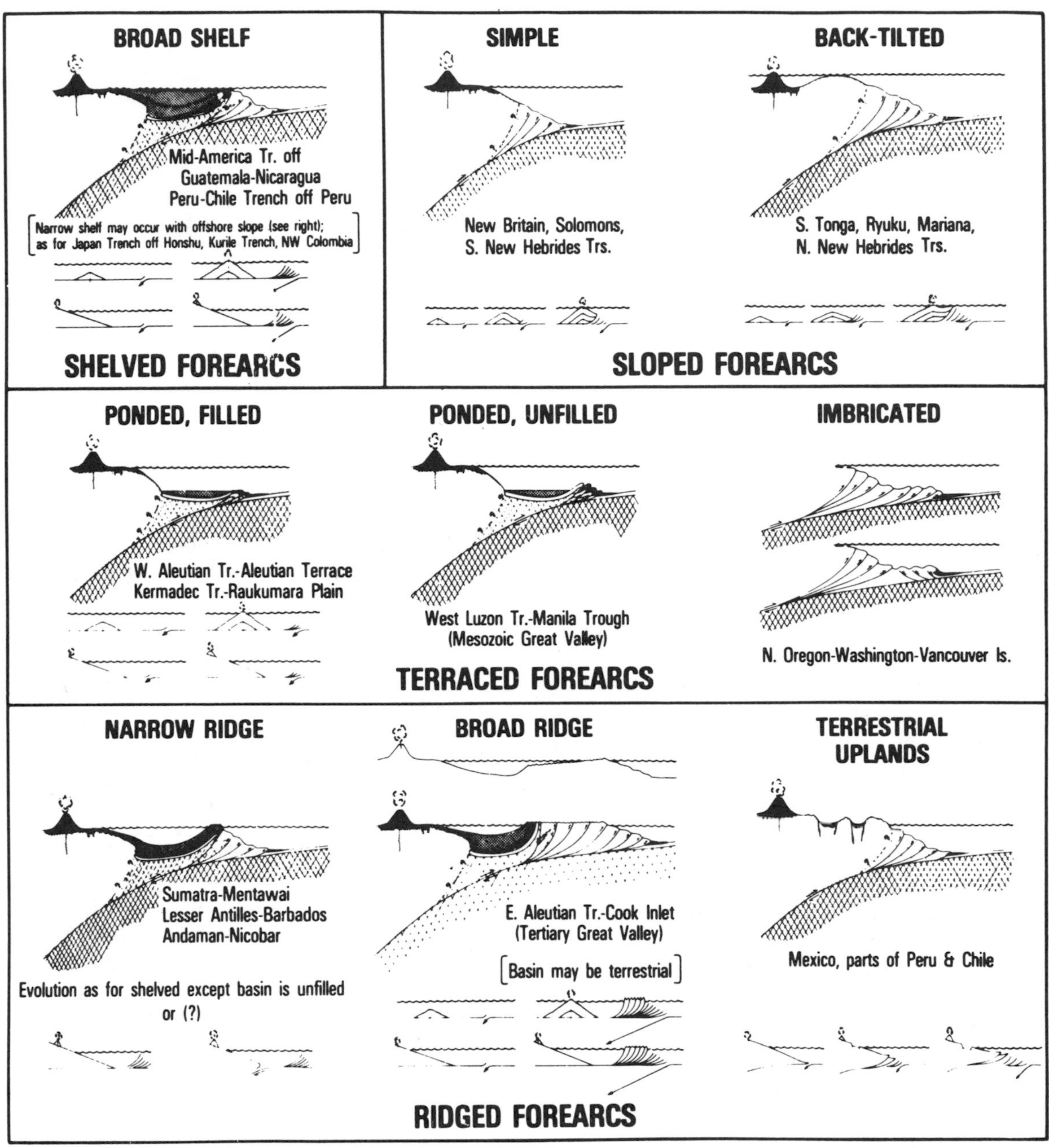

Fig. 9.21. Configuration of modern forearc regions (Dickinson and Seely, 1979).

It is not yet clear whether Dewey's (1980) arc models can be extended to embrace differences in style of subduction complex. G.F. Moore and Karig (1980) suggested that pervasive shearing would result from rapid dewatering under high stress. In the Shikoku subduction zone folded sediments in the lower part of the subduction complex are over-consolidated as a result of tectonic dewatering (J.C. Moore and Karig, 1976). Hamilton (1979) reported the existence of active mud volcanoes on top of several Indonesian subduction complexes, attesting to rapid dewatering. Possibly differences in the intensity of internal deformation, from moderate (recumbent or isoclinal folding, thrust faulting) to intense (shearing and production of chaotic melange) can be related to the rate of dewatering, and this to arc style, be it extensional, neutral or compressional.

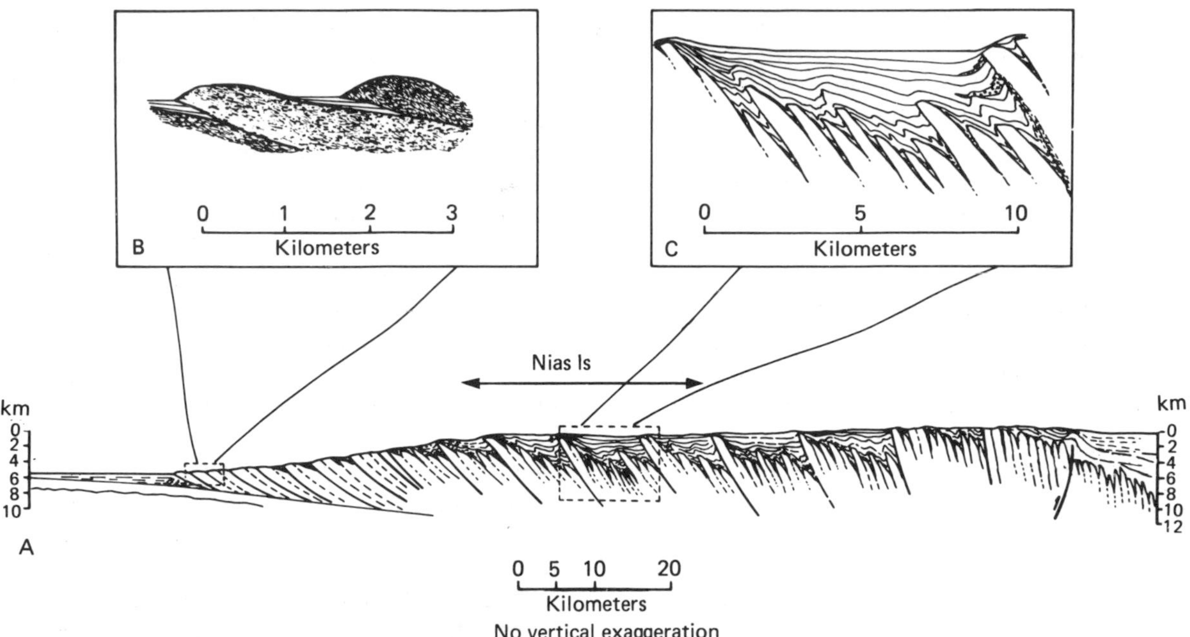

Fig. 9.22. Seismic profiles across the subduction complex of Sunda arc. Section is entirely below sea level, but approximate along-strike position of Nias Island is shown in profile **A** (G.F. Moore and Karig, 1976; Karig et al., 1978).

In order to illustrate the generation of subduction complexes, that of the Sunda arc off Sumatra is now described. This area has long served as a classic illustration of an island arc because of the work there in the nineteen thirties and forties, mainly by Dutch geologists (van Bemmelen, Umbgrove, Vening Meinesz). Recently detailed geological and geophysical studies by G.F. Moore and Karig (1976, 1980) and Karig et al. (1978) have provided a detailed picture of the unique geological style of an outer accretionary arc. Figure 9.22 is an interpreted seismic reflection profile, showing the position of Nias Island, one of the Mentawai Islands, which owe their existence to the uplift of the subduction complex. The structure is dominated by imbricate thrust faults developed in a melange of pervasively sheared sedimentary rock debris. The matrix consists of comminuted sandstone and siltstone plus sheared mudstone, and contains variously shaped blocks from millimeters to tens of meters in length. Some inclusions are fractured or have slickensides on the surface. The included blocks consist mainly of turbidite sandstones and siltstones, together with mudstone, rare conglomerate, clasts of chert, red shale and limestone. Up to 20% of the inclusions are basalt fragments, including pillow lavas and, very rarely, other ophiolite fragments such as peridotite, dunite and serpentinite.

The cross section shows that the thrust slices steepen progressively away from the trench. It is now generally accepted that this is the result of continuous underthrusting by the oceanic plate which tilts the thrust slices landward. As tectonic compaction increases the resistance of the rocks to shearing a new thrust slice is initiated in the trench, and in this way the subduction zone gradually expands seaward. On Nias Island the melange is dated by stratigraphic relationships as no younger than early Miocene, and possibly as old as Eocene, and there is probably a continuous age range to that of newly formed melange units in the present-day trench.

Clastic sedimentary rock fragments which comprise the bulk of the melange are mainly derived from the Bengal Fan. These have been transported longitudinally along the trench since the Eocene. Because they increase in volume toward the source the subduction complex becomes thicker and the trench shallower toward the northwest. Off Java the outer accretionary arc is 1–3 km below sea level and the trench is up to 7 km deep. Off northern Sumatra the arc is partly above sea level and the trench is less than 4 km deep (Hamilton, 1979). Arc sediments may have contributed to the trench early in the formation of the complex, before this source was cut off by the growth of the complex into a structural high. Nonclastic sedimentary fragments in the melange (chert, red shale, limestone) are pelagic sediments that covered the oceanic crust before it was carried like a conveyor belt into the subduction zone. The oceanic crust itself is represented by the mafic and ultramafic rock fragments.

As shown in Figure 9.22, resting on and folded into the melange are basin shaped masses of sediment a few kilometers across. On Nias Island these are Miocene to Pliocene in age and consist of coarsening upward sequences 3 to 5 km thick. At the base are thin-bedded, carbonate-poor siltstones and shales. These are overlain by thin bedded-calcareous and terrigenous turbidites, and then by calcareous and terrigenous sandstones and conglomerates. The beds are folded into tight, asymmetric, chevron folds. They have sheared but structurally nearly conformable contacts with the melange on the southwest side of each basin, and thrust contacts on the northeast. They are interpreted as the fill of sedimentary basins which developed as structural depressions between thrust masses, and were then themselves mildly deformed before each thrust mass became inactive. The stratigraphic sequence reflects gradual uplift to above the carbonate compensation depth and increasing relief. Dickinson and Seely (1979) named them "accretionary" basins. "Trench-slope basins" is another common term.

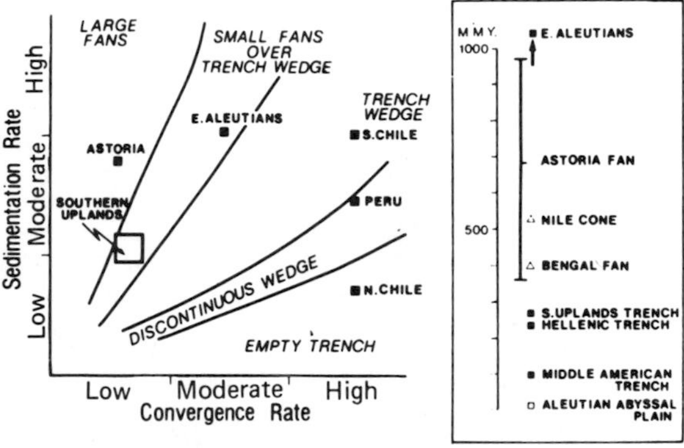

Fig. 9.23. Sedimentary style in the trench and subduction complex and its relationship to sedimentation rate and convergence rate. Small squares show estimated positions of some modern examples, and larger open square shows approximate position of Paleozoic subduction complex of the Southern Uplands, Scotland (Leggett, 1980).

The detailed geology of other subduction complexes show many differences, but the basic theme is the same. For example Leggett (1980) showed that the thickness of the sedimentary assemblage reflects a balance between convergence rate and sedimentation rate (Fig. 9.23). High sedimentation rates may lead to the development of major submarine fans prograding across the inner trench slope, such as the Astoria Fan on the U.S. west coast (Kulm and Fowler, 1974). Broad subduction complexes such as that south of Kenai Peninsula, Alaska, may contain large, shallow accretionary basins. Those off Alaska are at least 100 km wide and are accumulating sediment in shelf and coastal environments (Von Huene, 1978; Dickinson and Seely, 1979). In Hawke Bay, New Zealand an ac-

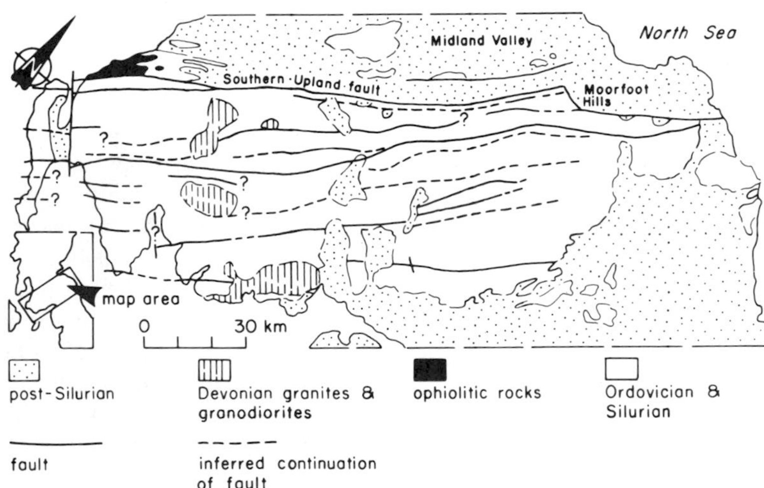

Fig. 9.24. Ordovician–Silurian outcrops of the Southern Uplands, showing location of the major thrust faults within the subduction complex. Thrust slices mainly dip northwest. Iapetus suture is located approximately along southeast edge of map (Leggett, 1980).

Table 9.4. The differences between melange and olistostrome*

Criterion	Melange	Olistostrome
clast character	angular, fractured, sheared, may be deformed into boudins or smeared phacoids	angular to rounded, may be fractured but not sheared
clast source	overlying and/or underlying unit	overlying unit only
matrix	sheared, plastic intrusion into cracks	not necessarily sheared
contacts with other units	sheared	may have sedimentary contacts with "normal" slope or trench sediments

*data from Hsü, 1974.

cretionary basin is accumulating coastal and fluvial sediments (Ballance, 1980).

Leggett's (1980) analysis of the Southern Uplands of Scotland is an excellent example of the application of a well-defined basin model to a hitherto incomprehensible piece of geology (Fig. 9.24). Following Mitchell and McKerrow (1975) this is now recognized as a subduction complex developed above a northwest-dipping Benioff zone immediately prior to the closure of Iapetus Ocean (the suture lies near the southeast edge of the area shown in Fig. 9.24). Each of the major thrust slices youngs northward, whereas the slices as a whole becomes younger southward. Melange is absent in this complex, the rocks consisting of turbidites and associated sediments, most of which were transported longitudinally southwestward into the trench. Some sediment derivation from the rising accretionary arc itself can be demonstrated, just as younger (Pliocene) beds in Nias Island contain material derived from the uplifted melange. This uplift also contributed sediment northward to the forearc basin in both cases. The Southern Uplands complex contains a major slice of oceanic rocks, the Ballantrae Ophiolite, which was incorporated into the complex during an early stage of subduction (Early Ordovician).

In other subduction complexes, such as many of those in Indonesia, ophiolitic material is more abundant and the rocks show evidence of high-pressure, relatively low-temperature metamorphism (blueschist facies). Glaucophane, jadeite, and prehnite–pumpellyite are typical metamorphic minerals (Hamilton, 1979; Blake and Jones, 1974, 1981).

The term melange has two meanings, and it is important to distinguish them. As used here it implies a pervasively sheared rock mass, but some authors extend the term to include ill-sorted deposits that have been derived by submarine slumping. The correct term for these is olistostromes. Hsü (1974) provided a useful discussion of the differences between tectonic melange and sedimentary olistostromes. These are summarized in Table 9.4. Olistostromes result from the failure of oversteepened slopes. Rapid sedimentation or advancing thrust sheets could trigger major slope failure, and both are typical of active subduction zones. D.G. Moore et al. (1976) described a giant olistostrome on the north end of the Sunda arc in the Bay of Bengal which they interpret as having been caused by rapid sedimentation on the Irrawaddy Delta. Individual blocks (olistoliths) are up to 2.8 km long and the slide is locally at least 30 km across. D.G. Moore et al. (1976) stated that based on numerous geophysical transects this is the only such slide mass on the Sunda arc, and therefore major olistostromes may not be common in subduction complexes. However, Field and Clarke (1979) showed that active continental slopes may contain numerous small slides a few hundreds of meters or less in width, and that these are beyond the resolving power of conventional seismic equipment.

Distinguishing a melange from an olistostrome in any given outcrop may not always be easy, because some rock masses could be the product of both processes. An olistostrome could be produced by slumping of a melange from an active thrust sheet, and then itself reincorporated in the melange by continued thrusting (D.G. Moore et al., 1976). Resolving the age of a melange may also be very difficult because it may contain only derived fauna. An olistostrome could contain blocks that are older than the matrix, as determined by contained fossils, whereas the reverse is only possible in a melange. For example some Swiss melange blocks containing an early Tertiary nummulitic fauna are set in a matrix of shale yielding a Late Cretaceous fauna (Hsü, 1974). Normally any contained fauna in a melange will only indicate a maximum age for the tectonism.

Forearc basins. Dickinson and Seely (1979) recognized three main types of forearc basin. The first, which they called "accretionary" basins, are the relatively small basins of ponded sediment on the inner trench slope such as are exposed on Nias island (Fig. 9.22). These are discussed in the preceding section and will not be dealt with further.

The second type are extensional intra-massif basins that occur within the magmatic arc. These basins may be of considerable size, particularly where a broad forearc region has developed over continental crust, as in Chile. They are typically bounded by normal faults, the orientation of which is not necessarily parallel to arc trends but may be determined by preexisting basement grain. Such is the case in Chile, where bounding faults and the alignment of andesitic arc eruptions are oriented either northwest or northeast, reflecting preexisting Gondwana architecture (Katz, 1971). The Osorno Basin is one of the larger of these basins. It is over 200 km long, and locally contains 1100 m of Tertiary strata or 1300 m of Quaternary sediments, although not in the same location because of differential movement. The strata are mainly nonmarine, deposited by rivers whose alignment is determined by the same structural grain. However, repeated block movements related to subduction have caused local facies changes, and some of the strata are shallow marine (Katz, 1971).

The third type of basin includes the largest and most important of the forearc basins. These are located between the magmatic arc and the outer accretionary arc and sediment is ponded in them to the extent that the outer arc is uplifted to form a linear (subaqueous to subaerial) topographic feature. Dickinson and Seely (1979) subdivided them into two subtypes, "residual" basins floored by attenuated continental crust or a fragment of obducted oceanic crust, and "constructed" basins floored by the subduction complex. Distinction of the two subtypes is difficult from geophysical data and may not be possible from outcrop information in ancient, uplifted basins. Most basins probably are composites of the two subtypes. They show essentially the same style of structural and stratigraphic evolution and can be dealt with together.

Forearc basins tend to become wider and shallower with time. Initially the forearc region may consist of a simple seaward slope, depending on the location of the arc (old oceanic fracture zone, continental margin, etc). Sediments are carried directly (transversely) into the trench and feed a growing subduction complex. As ponding commences water depths decrease from abyssal to shallow, and environments may becomes nonmarine before the basin ceases to be active. At the same time the basin becomes wider as the subduction complex advances oceanward. The magmatic arc may tend to retreat in the oposite direction, particularly if the angle of subduction decreases over progressively younger crust. Both these processes have been well documented in the Great Valley of California, a Late Jurassic to early Tertiary forearc basin (Ingersoll, 1979).

The forearc fill onlaps the magmatic arc, although the contact may be faulted, and has a sedimentary or partly sheared contact with the subduction complex. Details of these relationships in the Paleogene to Recent forearc basin off Sumatra are shown in Figure 9.25.

Basal sediments in a "residual" forearc basin are invariably deep marine, consisting of pelagic trench deposits resting on ophiolite. "Constructed" basins may be somewhat shallower. In both cases the first major depositional systems are submarine fans that build transversly across the basin from the magmatic arc. In the case of intraoceanic arcs which are largely submerged (e.g. Mariana arc) the derived deposits will be entirely deep marine, whereas arcs built on continental crust may contain a significant proximal shallow marine to nonmarine component (e.g. west coast of Oregon, Washington). In general the submarine fans show a gross upward coarsening indicating overall progradation, but in detail both coarsening- and fining-upward trends are present as the fans switch position due to the interplay of active tectonics, sea level change and sedimentation (Ingersoll, 1978a, 1979). Sediment thicknesses typically are considerable. The modern Sumatra trough contains at least 6 km of basin fill (Karig et al., 1978), whereas the Great Valley

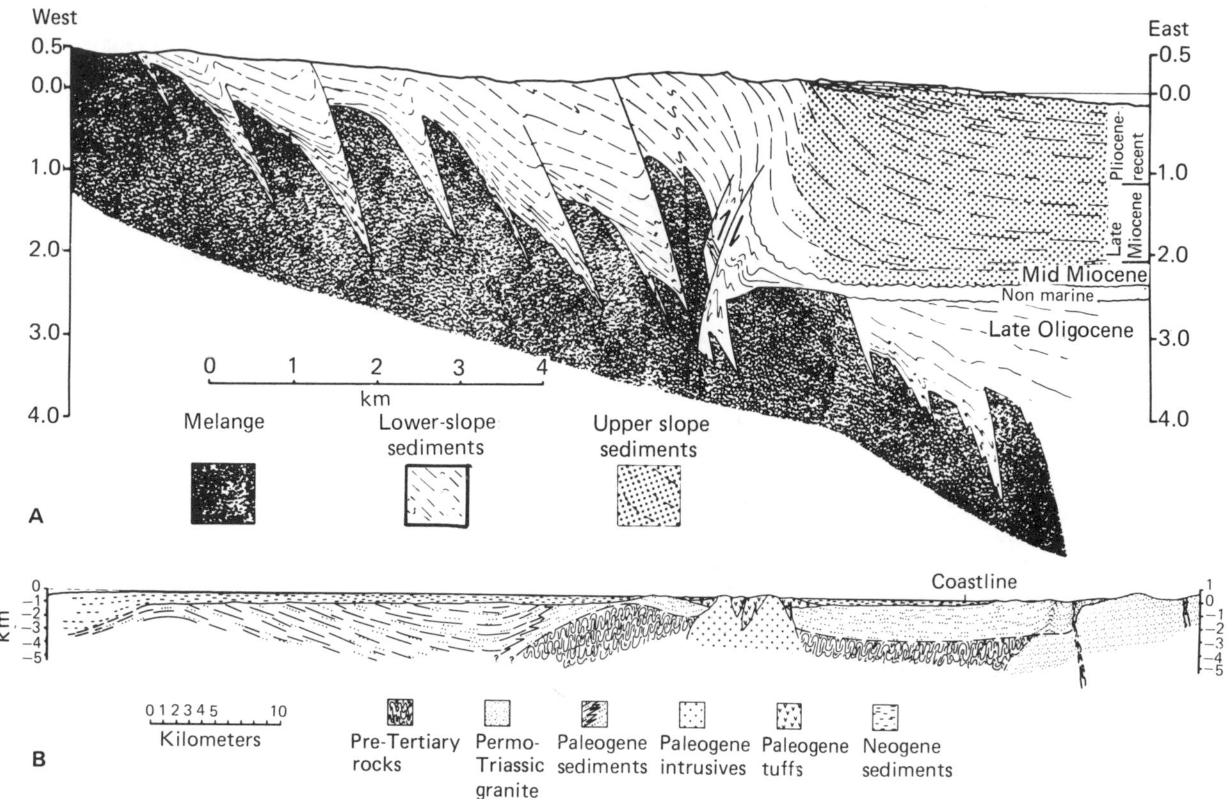

Fig. 9.25. Seismic reflection profiles across the Sunda forearc basin near Nias Island. **A** shows relationship between forearc basin and subduction complex; **B** onlap of forearc basin on to pre-Tertiary basement and magmatic arc (Karig et al., 1978).

Group is about 15 km thick (Ingersoll and Dickinson, 1981). Sedimentation rates in the latter ranged from about 0.2 to 0.4 m/1000a.

The growth of the outer accretionary arc gradually changes the shape of the basin to a two-sided trough and turbidity currents then turn 90° at the fringe of the submarine fan wedge to flow longitudinally down the basin plain. This is well documented in the Great Valley Group (Ingersoll, 1979). The outer arc may itself become a sediment source as in the Silurian forearc basin north of the Southern Uplands Fault, Scotland (Leggett, 1980).

The youngest sediments in a forearc basin may be shallow marine or nonmarine. For example in the Great Valley the Cretaceous–Tertiary transition was marked by the progradation of deltas in the Sacramento Valley, whereas in the adjacent San Joaquin Valley turbidite sedimentation continued (Ingersoll and Dickinson, 1981).

Kuenzi et al. (1979) provided an interesting study of fluvial–deltaic sedimentation on the Guatemala coastal plain. Fluvial styles here seem unusual, but may in fact be typical of nonmarine forearc (or backarc) regions characterized by active intermediate volcanism. Explosive volcanic eruptions coupled with very high rainfall lead to the catastrophic transportation and sedimentation of large quantities of coarse volcaniclastic debris. This may be confined to the main trunk rivers, and the result is that tributaries are dammed at their junction with the main river, and back up to form lakes. These then fill with muds and with small Gilbertian deltas (Fig. 9.26). Lobate, river-dominated deltas form at the coast when sediment supply is high, but between major eruptions these may be reworked by waves into narrow arcuate deltas, as in Figure 9.26.

Forearc styles undergo dramatic change along strike if the convergent plate boundary runs from

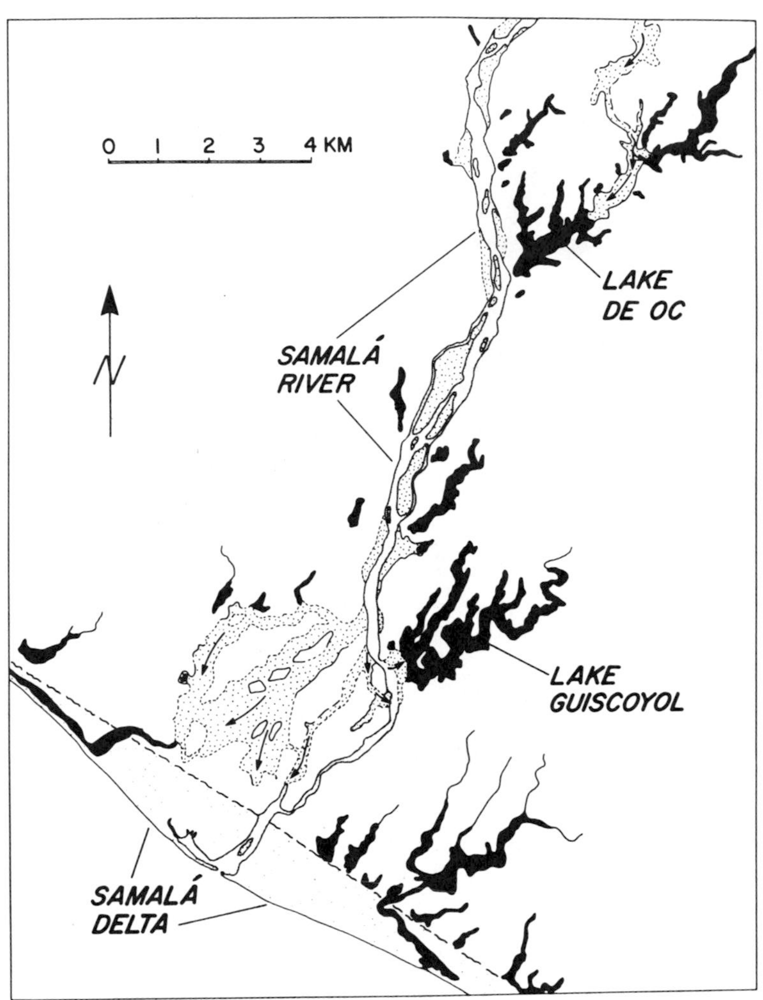

Fig. 9.26. Fluvial style in a volcaniclastic setting, Guatemala forearc basin. See text for discussion (Kuenzi et al., 1979).

oceanic to continental crust.There are two well-documented Tertiary to modern examples of this—Cook Inlet, Alaska (Hayes et al., 1976; Fisher and Magoon, 1978; Hayes and Michel, 1982), and Western Trough, Burma (Mitchell and McKerrow, 1975; Curray et al., 1978; Bannert and Helmcke, 1981). In both cases the subduction complex increases considerably in thickness along strike toward land, and has become uplifted to form a major landmass. The availability of a greater volume of sediment to fill the trench is the main reason for this. Secondly, the continental part of the forearc basin has become a major alluvial valley receiving sediment both from the magmatic arc and the subduction complex. Rivers drain transversly into the basin from both sides, feeding a longitudinal river that empties into the marine part of the forearc basin along strike.

In the case of Cook Inlet, although fluvial sedimentation has probably been continuously active since the Tertiary (Hayes et al., 1976) the basin is still not filled up. At present it is an estuary receiving sandy sediment from both the arc and the subduction complex, and this is deposited as spits and deltas (Fig. 9.27) in the pattern of Miall's (1981b) basin model 6 (Fig. 6.5). By contrast the forearc basin in Burma (the Western Trough in Fig. 9.28, 9.29) contains up to 17 km of Tertiary marine and fluvial sediments derived from Burman and Tibetan highlands to the north and east and the Indo–Burman ranges (subduction complex) to the west. The basin is followed by the longitudinal Irrawaddy River, which has built a major tide-dominated delta out over the continental margin (basin model 8 of Miall, 1981b; see Fig. 6.5).

An attempt at a summary basin model for

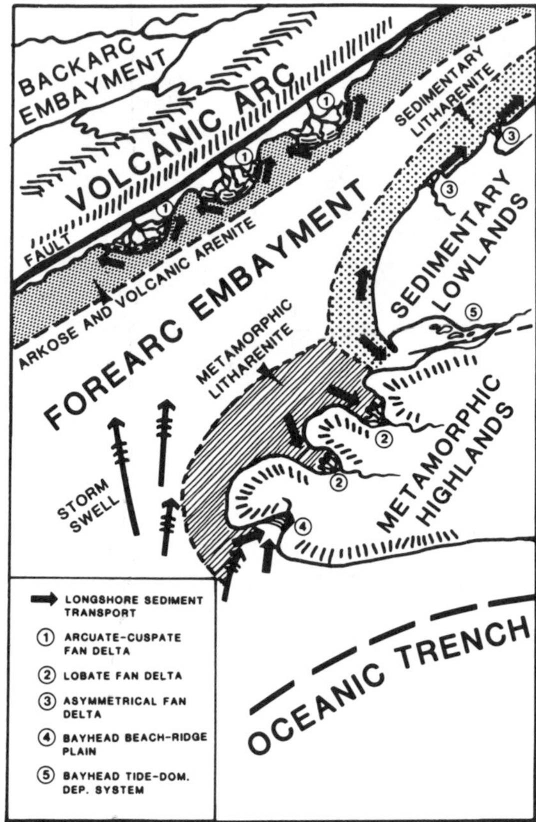

Fig. 9.27. Depositional environments in a forearc embayment, Cook Inlet, Alaska (Hayes and Michel, 1982).

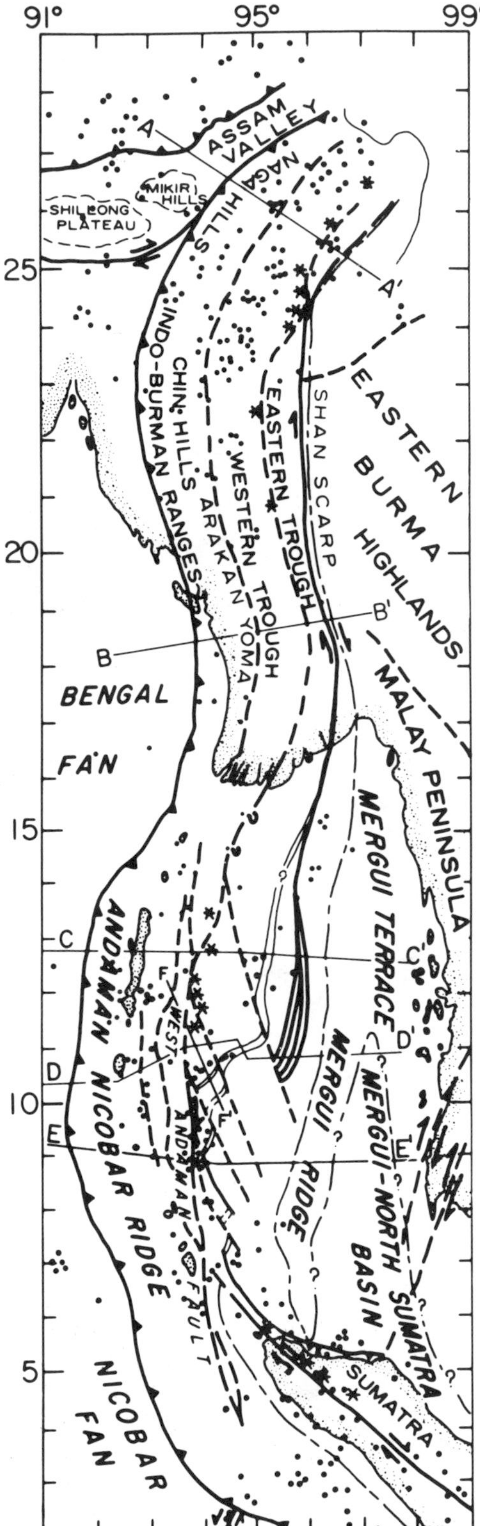

Fig. 9.28. Tectonic map of Burma and Andaman Sea (Curray et al., 1978). ▲ thrust fault; ⥂ active fault, ---inactive fault; ● epicenters; * volcano; = spreading ridge; == inactive spreading ridge; —·— edge of continental crust; A—A′ line of section.

forearc regions is presented in Table 9.5. Generalizations of the type given in the table have been used by several authors in recent years to interpret various ancient deformed sequences and their relationship to adjacent igneous and metamorphic terranes. For example Smellie (1981) and Hyden and Tanner (1981) interpreted the Antarctic Peninsula as a Jurassic–Tertiary forearc basin and subduction complex. Smellie (1981) suggested that the Karoo Basin of South Africa might be the retroarc counterpart of this system, now separated from Antarctica by the Atlantic seafloor spreading.

Interarc and backarc basins on oceanic or transitional crust. Backarc spreading accompanies the rifting away of a magmatic arc from a continental margin in an "extensional" arc setting (Dewey, 1980). In complex areas such as the western Pacific many basins are located between two island arcs and have been described as in-

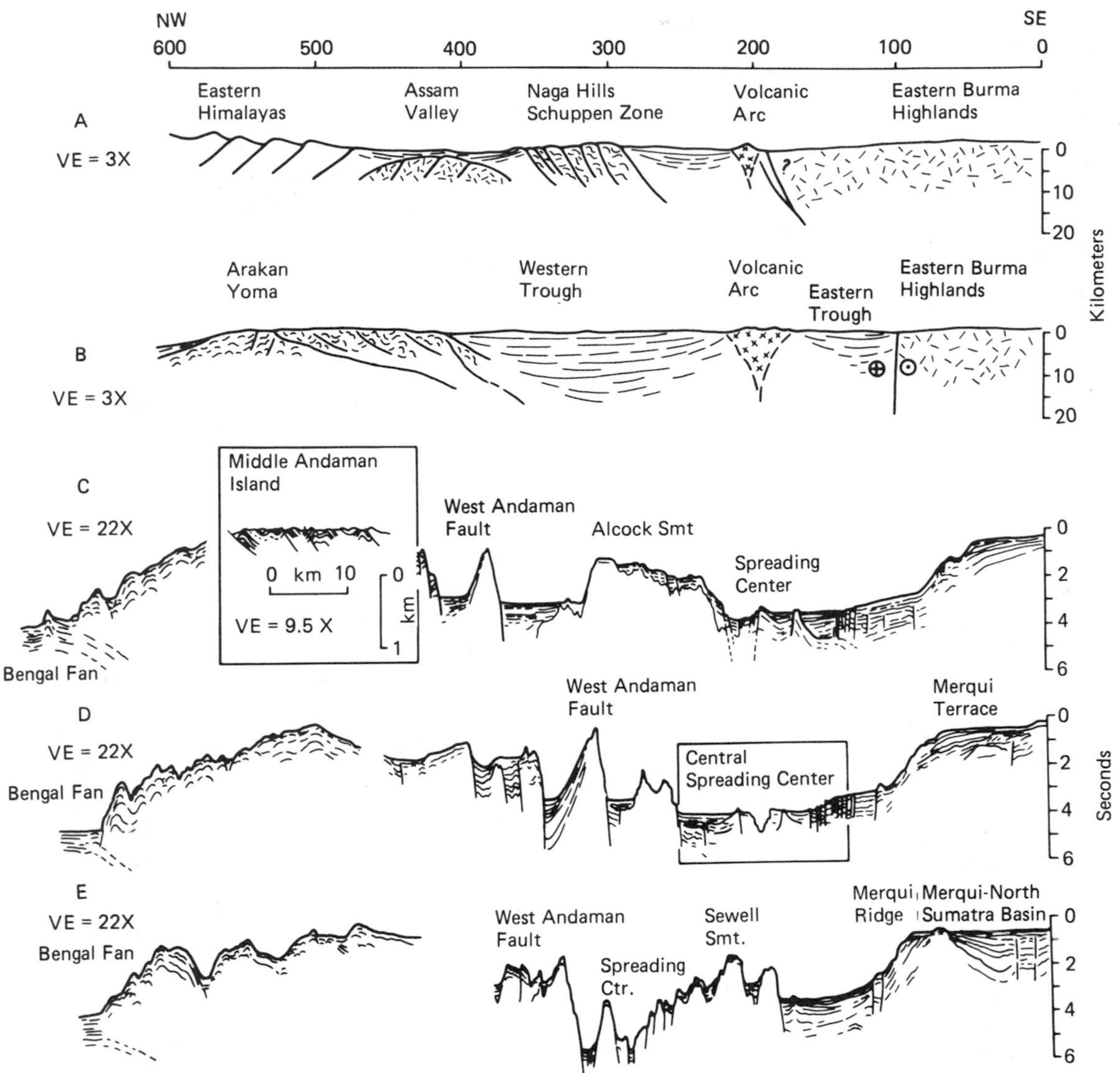

Fig. 9.29. Structural cross sections across Burma and Andaman Sea. Locations are shown in Fig. 9.28. **A**, **B** based on outcrop data, **C**, **D**, **E** based on seismic data (compiled by Curray et al., 1978).

terarc basins. However, it is rare for there to be two active arcs with the same polarity in close proximity, except in areas of complex microplate interaction. In most cases there is only one active arc, and the basin behind this is better described as a backarc basin. What happens is that arcs migrate suddenly to accommodate very rapid oceanward rollback of the subduction hinge, creating a basin bounded by the active arc and an older, remnant arc.

The clearest example of arc migration is the Philippine Sea (Fig. 9.30), where the subduction zone has moved eastward twice since the Cretaceous, according to Karig (1971a, 1971b), Karig and Moore (1975) and Hamilton (1979). At the surface the arc rifted along the line of volcanic activity, the backarc region was abandoned and became inactive, while the forearc region migrated rapidly oceanward into a new, temporarily more stable position. The Kyushu–Palau Ridge was thus split in the mid-Tertiary and the West Mariana Ridge in the Pliocene. The currently active Mariana Ridge now contains volcaniclastics of Eocene to Miocene age, indicating that the

Table 9.5. Basin Model for forearc regions

Parameter	Subduction complex	Forearc basin
lower contact with	Ophiolites, may be sheared	To seaward: subduction complex, may be faulted. To landward: magmatic arc, may be faulted.
oldest sediments	Pelagic chert, abyssal limestone, on oceanic crust	similar to younger sediments
sediments from transverse sources	Arc derived volcaniclastics, may give way to reworked melange as complex grows. Olistostromes. Possible fringing reefs	Arc derived volcaniclastics. Some material from subduction complex as latter is uplifted
sediments from longitudinal sources	Depends on proximity to continental source. Fluvial and/or submarine fan detritus may give rise to thick trench fill and large complex	As above, redirected longitudinally by topography. Major terrigenous input if close to continent.
depositional environments	Abyssal oceanic to nonmarine; typically shallows upward	Deep marine to nonmarine, typically shallows upward
structure*	Imbricate thrust sheets, dipping toward arc but sheets as a whole young in opposite direction. Sediments may be moderately to intensly folded or sheared to melange	Simple, open syncline
metamorphism*	High pressure: blueschist facies (glaucophane etc.)	Negligible
associated rocks and structures	May incorporate sheared slabs or thrust masses of ophiolite	Sediments may onlap arc volcanics or eroded plutons.

*This assumes that the rocks have not been further deformed in a suture zone.

older part of the arc has indeed migrated oceanward from the Kyushu–Palau Ridge.

Many areas of backarc spreading have now been identified in the region of complex tectonics between the Asian, Indian–Australian and Pacific plates. The Andaman Sea (Figs. 9.28, 9.29) lies over an active backarc ridge–transform spreading system that accommodates the highly oblique convergence of the Indian Ocean against Burma and Malaya (Curray et al., 1978; Hamilton, 1979). The Okinawa Trough, northeast of Taiwan, is a backarc basin behind the active, west-dipping Ryukyu subduction zone (Herman et al., 1978). Sulu Sea contains a now inactive backarc basin and west-facing volcanic arc that was active during the Tertiary (Mascle and Biscarrat, 1978; Hamilton, 1979). The Japan Sea formed as a backarc basin as the Japanese arc rifted away from Asia and migrated relatively eastward during the Cretaceous and early Tertiary. Spreading there has now ceased (Uyeda and Miyashiro, 1974). Many other examples can be identified from a study of Hamilton's (1979) tectonic map of the Indonesian region. The Aegean Sea is a backarc basin formed behind the East Mediterranean Rise (subduction zone), and has been active since Neogene time (McKenzie, 1978).

The depositional environments in backarc basins are deep marine, except along their margins. Stratigraphic–sedimentologic information from modern basins is therefore limited to seismic data and scattered DSDP holes. Karig and Moore (1975) presented a model based on these sources for western Pacific basins (Fig. 9.31). A volcaniclastic submarine fan apron progrades into the basin from the arc, and interfingers with pelagic clays. Calcareous organic debris may be important where the volcanic detritus decreases in abundance toward the mainland side of the basin.

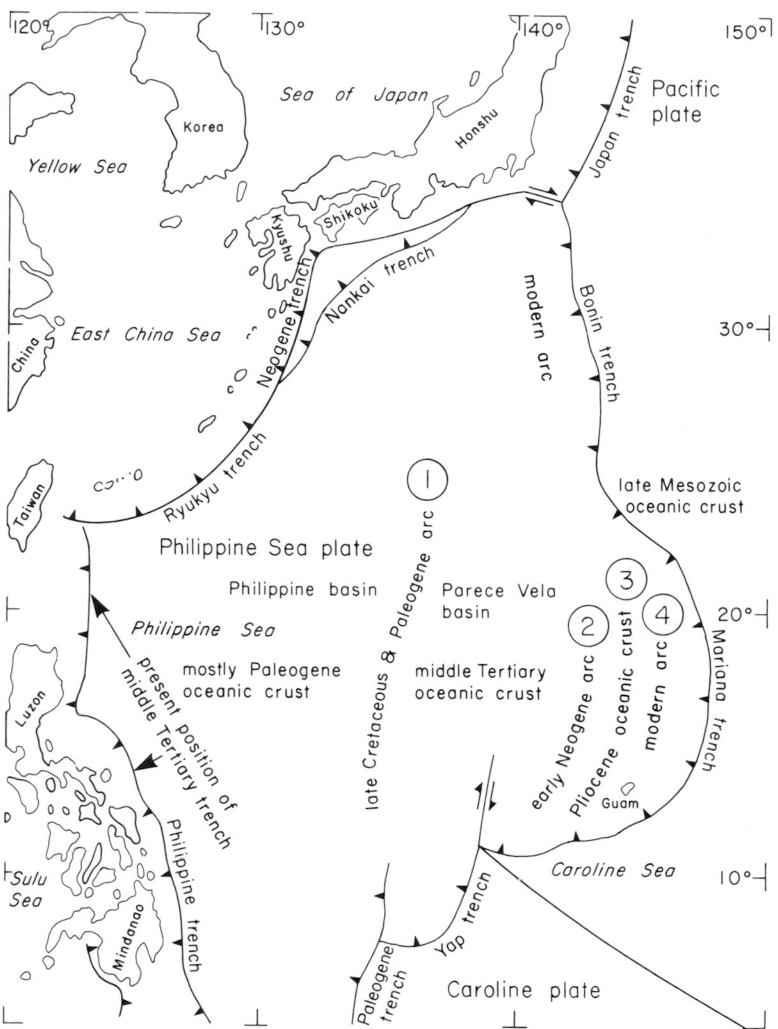

Fig. 9.30. Geology of Philippine Sea, showing evolution of backarc basins by migration of subduction hinge (Hamilton, 1979). 1. Kyushu–Palau Ridge; 2. West Mariana Ridge; 3. Mariana Trough; 4. Mariana Ridge.

As the basin subsides the floor will eventually drop below the carbonate compensation depth, and younger basin fill will be carbonate free. Where the basin is developed against a continental margin there will be a wedge of continent-derived sediment showing deep to shallow marine transitions in depositional environments. This should display an extensional structure and a stratigraphy very similar to that of divergent continental margins (not shown in Fig. 9.31) because the origins are similar.

Klein and Lee (1983) synthesized all available DSDP data from backarc basins in the western Pacific and showed that when each basin is examined in detail the sedimentation patterns, and their relationship to episodes of arc magmatism and rifting, vary considerably. They identified nine types of depositional system: submarine fan, debris flow, silty basinal turbidite, biogenic pelagic carbonate, resedimented carbonate, biogenic pelagic silica, pyroclastics, hemipelagic and turbid layer clay, and pelagic clay. They stated:

> Debris flows and submarine fan systems tend to accumulate after uplift of andesitic volcanic arc sources following development of a regional drainage network. Biogenic pelagic systems tend to be controlled by biological productivity and are indicators of back-arc basin movement through such productivity zones by regional crustal shift. Hemipelagic and turbid-layer clay systems are derived from terrigenous sources and their volume is dependent on climatically controlled sediment yield from such sources. Resedimented carbonate systems tend to occur during times of active basinal faulting.

In modern mature backarc basins sediment thicknesses typically are in the order of 2–3 km in the basin center, except where there is a par-

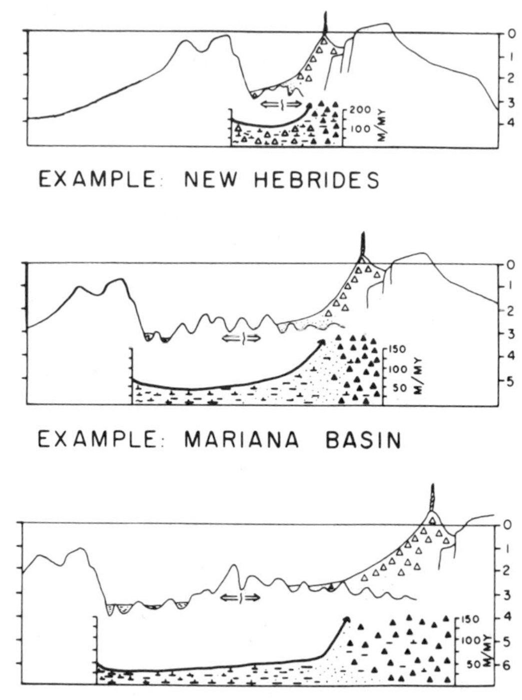

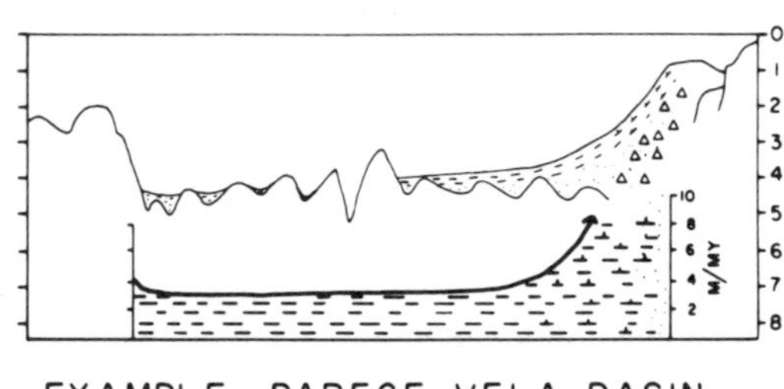

Fig. 9.31. Depositional model for the evolution of backarc basins, based on four basins in various stages of development from oldest (bottom) to youngest (top). At base of each section is a diagram showing sedimentation rate and lithofacies, assuming no continental sediment sources (Karig and Moore, 1975).

ticularly active local sediment source. For example the north end of the Andaman Sea is underlain by the submarine fan equivalent of the Irrawaddy Delta, and sediment thicknesses reach at least 6 km (Curray et al., 1978). Similar or greater thicknesses may be expected in the volcaniclastic wedge, depending on the duration of active backarc spreading. The contribution of volcanic and sedimentary rocks from the backarc spreading center itself is not clear. Curray et al. (1978) provided a seismic section across the Andaman Sea spreading center which shows it to consist of a deep rift valley with gently dipping sediments ponded between fault blocks (Fig. 9. 29).

The identification of ancient backarc basins in a suture zone may be difficult, because the elements that define them tend to be deformed out of existence. For example an arc–continent collision would probably result in reversal of arc polarity and the subduction of backarc oceanic crust. This topic is discussed further in section 9.3.4.

Backarc basins on continental crust. Where the arc is located on the leading edge of a continental plate there may or may not be backarc spreading, depending on whether the arc is extensional or compressional. According to Dewey's (1980) model this depends mainly on the age of the subducted crust. Thus the Andes do not show backarc spreading, although small, superficial, extensional intermontane basins are present, probably as a result of warping during subduction.

However, some other areas of the world show what might be termed incipient or aborted backarc spreading, which may or may not lead to full-scale extension and the development of oceanic crust. One of these is the Basin and Range province of the western United States and northern Mexico (Fig. 9.32). This area consists of a series of tilted fault blocks and half-graben basins filled with up to 3000 m of nonmarine Late Cenozoic sediment (Fig. 9.33). Fault-generated relief between uplifted blocks and the basement floor of the basins is 2–5 km and the entire area has been elevated by 2 to 3 km (Stewart, 1978). The province contains abundant early Cenozoic calc-alkaline and silicic volcanics, whereas later volcanics are basaltic, including the voluminous Columbia Plateau basalts, of Miocene age. The tectonic style indicates considerable crustal extension, but there is no agreement as to the exact amount. Stewart (1978) summarized earlier views, which ranged from 8 to 100%. Estimates partly depend on the interpretation of fault style, and recent workers (Stewart, 1978; Bally and Snelson, 1980) seem to favor the idea that the faults are listric in character (e.g. Fig. 9.33) which would indicate greater rather than lesser amounts of extension.

Hypotheses of the origin of the Great Basin involve mantle upwelling over a subducted East Pacific Rise, or a wide zone of transcurrent faulting in a dextral megashear, or backarc spreading.

Scholz et al. (1971) and Stewart (1978) favored the last of these. Rifting and volcanism began in the early Cenozoic, but active extension and basin subsidence did not commence until the mid to late Miocene. It is possible that the change in volcanic composition and the initiation of large-scale extension are both related to the reduction in regional compression following the collision of the East Pacific Rise with the western continental margin in the late Oligocene. After this time interplate stresses were mainly taken up by transcurrent movement on the San Andreas transform system (Stewart, 1978; Dickinson, 1981).

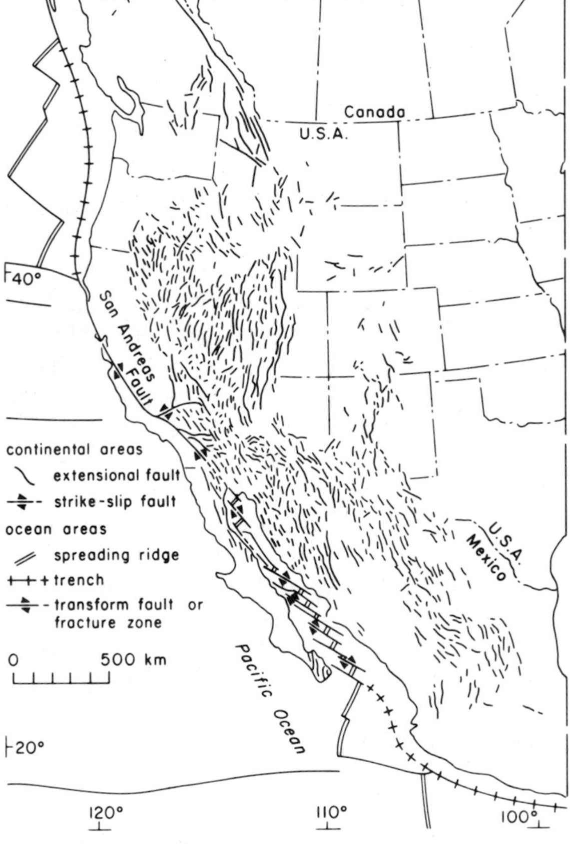

Fig. 9.32. Distribution of extensional faults in the western Cordillera, defining the area of the Great Basin. The San Andreas transform fault and other elements of the North American plate boundary are also shown (Stewart, 1978).

The Basin and Range Province seems to be a special case, and may not be a good analogue on which to build a general model. Thus Ingersoll (1982) has developed an alternative model based on rotation and shear along the San Andreas transform system as the triple junction at its northern end migrated northward. Another complex area is the Pannonian Basin of the Carpathian region in southeast Europe, which Bally and Snelson (1980) termed a backarc basin. Burchfiel and Royden (1982) confirmed this interpretation but showed that microplate collision and strike-slip faulting complicated the history of basin development. The Aegean Sea may be an example of a backarc basin in an intermediate state of development between a continental (Great Basin) stage and a fully oceanic (western Pacific) stage. It is partly underlain by oceanic crust, but also contains a series of elongate graben filled with 600–1000 m of nonmarine to shallow marine Pliocene to Quaternary clastic sediment (McKenzie, 1978; Biju-Duval et al., 1978).

An alternative approach for developing a general model is to examine various ancient mobile belts. Ensialic backarc basins may be expected to have survived somewhat more intact than those developed over more readily subductable oceanic crust. Three possible examples are the Cambrian to Silurian (Caledonian) Welsh Basin, Devonian to Carboniferous (Hercynian) basins in western Europe and the Lower Paleozoic Hazen Trough of the Arctic Islands.

The Welsh basin developed behind a volcanic arc trending northeastward through North Wales and contains at least 12 km of volcaniclastics and continent-derived turbidites and mudstones. It shows a transition to shallow marine continental shelf facies toward the southeast (Phillips et al., 1976; Siever and Hager, 1981). Middle Devonian to Carboniferous rocks along a trend from South Devon, England to the Ardennes are characterized by pelagic mudstones and car-

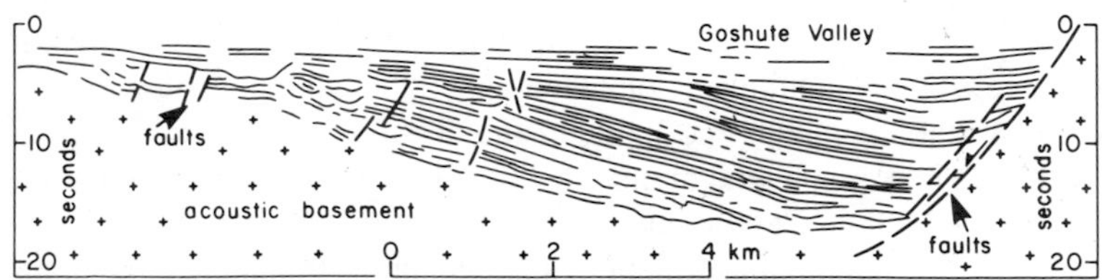

Fig. 9.33. Seismic cross section through a typical fault-bounded basin in the Great Basin, Goshute Valley, Nevada (Bally and Snelson, 1980).

bonates which Leeder (1976, 1982) interpreted as the fill of a backarc basin above a north-dipping subduction zone. Northward in England these pass into block faulted basins, the outlines of which are strongly controlled by earlier lines of weakness. Lastly, Hazen Trough in the Arctic Islands contains a fill of radiolarian chert and thin-bedded turbidites derived (transversly) from a magmatic arc to the northwest. The southeast margin of the basin is well exposed, and shows a complete transition through slope facies to shelf carbonates (Trettin et al., 1972). Similarities between these three examples include the following, which may be regarded as the essential elements of a basin model:

1. a deep marine basin fill derived from either or both sides of the basin
2. a transition into slope and shelf facies on the continental margin of the basin, analagous to that of divergent plate margins
3. a magmatic arc on the other side of the basin
4. little evidence of contemporaneous tectonics, except normal faulting
5. a basement of normal or attenuated continental crust.

Retroarc (foreland) basins. In compressional arcs metamorphic core complexes are developed deep within the magmatic arc, and the edge of the continental plate is deformed into a fold-thrust belt (Fig. 9.18, model 5, after Dewey, 1980). The crust may be shortened and thickened by a factor of two, giving rise to a major mountain belt exposing the metamorphic and plutonic core rocks of the arc, as well as the upthrusted masses of older rocks comprising the local continental basement. The loading of the craton with superimposed thrust sheets creates a downward flexure of the lithosphere (Price and Mountjoy, 1970; Beaumont, 1981). This basin is filled with sediment derived from the adjacent mountains and the sediment load itself causes further crustal depression. The load of each thrust sheet during crustal shortening results in structurally lower thrusts becoming locked by expulsion of pore waters, leading to an increase in friction. Continued shortening then results in the development of new folds and thrusts in the hitherto undeformed continental basement and cover rocks, so that deformation gradually steps out from the magmatic arc. In this way early basin fill sediments are themselves deformed and become the source of cannibalized or recycled sediment contributing to the younger basin fill. Sediment is usually also fed into the basin from the continental side.

This type of sedimentary basin is a retroarc basin, in Dickinson's (1974) terminology. Bally and Snelson (1980) refer to the crustal shortening and underthrusting of continental basement as A-subduction, after the Austrian geologist Ampferer, who described these processes in the early part of the century. The term is perhaps misleading because it is doubtful if much actual subduction of continental crust occurs.

Alternative (older) terms for this type of sedimentary basin are foreland or foredeep basins. The terms have been applied indiscriminately to many basins flanking orogenic belts, and it is now apparent that they are the result of at least two different plate tectonic processes. Major retroarc basins may develop behind a magmatic arc subducting oceanic crust. Thus the Sevier and Laramide Orogénies and the growth of the central Rocky Mountains foreland basin of Wyoming, Utah and Colorado are attributed to subduction of the Pacific plate during the Jurassic to Paleogene (Dickinson, 1981). The Great Valley forearc basin and Franciscan subduction complex are related to the same plate kinematics, as discussed in an earlier section. Andean foreland basins, such as the headwaters region of the Amazon, are similarly generated. However some of the largest and most well-developed foreland basins owe their origin to continental collision. For example the Alberta foreland basin developed during collision of the western margin of the North American plate with various exotic terranes between the Jurassic and the Paleogene (Monger and Price, 1979; Eisbacher, 1981). Similarly the Appalachian foreland basin of Ordovician to Carboniferous age developed by progressive closure of Iapetus Ocean and associated marginal basins (Graham et al., 1975). More recently the Indus-Ganga Basin has formed following the Cenozoic collision of India with Asia (Graham et al., 1975; Parkash et al., 1980) and the Alpine foreland basin of Germany and Switzerland is a product of microplate accretion in the Cenozoic (Dewey et al., 1973).

All these basins are floored by continental crust, but in the case of foreland basins formed during continental collision it may not be immediately apparent which was the subducted plate and which was the overriding plate until arc polarity can be determined by examining a great deal of complex geology. The Indus–Ganga and Alpine basins are both located on the subducting plate, in a forearc position—a quite different plate setting than the Rocky Mountains and Andean foreland basins. The term retroarc basin should be restricted to those foreland basins whose position can be clearly interpreted with respect to magmatic arc location and arc polarity. These we discuss here. The collision-related type of foreland basin is discussed in section 9.3.4.

As before, Indonesia provides some of the most well-defined examples of developing retroarc basins. Those of Sumatra and Java are illustrated in Figure 9.34. The geology of this area was described by Ben-Avraham and Emery (1973) and Hamilton (1979). The basement consists of an assemblage of Paleozoic to early Tertiary cratonic rocks and arc complexes. Linear basins and ridges of Sunda shelf are mainly controlled by these old alignments. Those basins whose origin can be related directly to Cenozoic retroarc subsidence behind the Java–Sumatra arc are the Madura, East Java, Tjeribon, West Java and South Sumatra basins. Sediment thicknesses reach maxima of about 3 to 6 km in these basins. The Sunda and part of the West Java Basin are graben basins bounded by north–south faults. These faults were active during the Cenozoic but their trend suggests basement control. The West Java–Tjeribon and Madura Basins are separated by a basement high. Sedimentation there began with a marine transgression during the Oligocene. Bathyal conditions persisted until the later Neogene in the east but thick deltaic clastics accumulated in the West Java Basin in the Oligocene. A widespread reef limestone developed in the Miocene. Regression occurred everywhere during the Neogene, and these basins are now largely continental.Basement rocks contributed sediment to the basins at first, but contemporaneous volcanics became the main sediment source during the younger Neogene. Throughout this period the magmatic arc gradually migrated northward. Syndepositional folding commenced first in the south and is probably continuing, resulting in deformation that decreases in intensity northward. Hamilton (1979) attributed the folding to lateral pressures of magmatic intrusion. The sedimentary basins probably owe their origin to downbowing of the crust beneath the weight of volcanic and plutonic rocks in the magmatic belt. A fold-thrust belt has not yet developed here. The geology of the Sumatra foreland is similar. Basement control by north to north–northwest trending structures had an important influence on sediment thicknesses and facies changes. Miocene strata include quartzose deltaic sandstones derived from the Malay Peninsula and Sunda Shelf to the north.

Cretaceous strata in the Rocky Mountain region of Utah, Wyoming, Colorado and Nebraska represent the fill of a more mature foreland basin developed during the Laramide Orogeny. Weimer (1970) showed that these undergo a crude upward coarsening and a facies transition from marine mudstone and limestone to nonmarine sandstone and conglomerate. The same facies transition occurs in reverse toward the east, away from the mountainous source (Fig. 9.35). Syndepositional synclines developed progressively further out from the source, in part as a response to sediment loading in shifting depocentres.

In this foreland basin basement block faulting profoundly affected sedimentation patterns during the latest Cretaceous and Paleogene. Precambrian and Paleozoic strata punch through the Cretaceous and younger detrital cover, dividing the foreland into a series of mountain ranges, such as the Wind River, Uinta and Laramie Ranges, with intervening basins, including the Wind River, Green River, Uinta and Powder River Basins, and so on. Seismic data collected by COCORP (Smithson et al., 1978, 1979; Brewer et al., 1980; Oliver, 1982) have resolved some of the long-standing controversy about the structural style of these faults (summarized by Bally and Snelson, 1980), showing that the Wind River and Laramie Ranges in Wyoming are bounded by steep thrust faults that extend to considerable depths. The Wind River thrust plunges to the base of the crust, whereas the adjacent fold-thrust belt is dominated by listric thrust faults that bottom out at the base of the Paleozoic (Fig. 9.36). Faulting began following a broad upwarping in the Campanian. During the Paleogene sedimentation consisted entirely of local fluvial and lacustrine depositional systems confined within the boundaries of each basin. Marginal clastic wedges con-

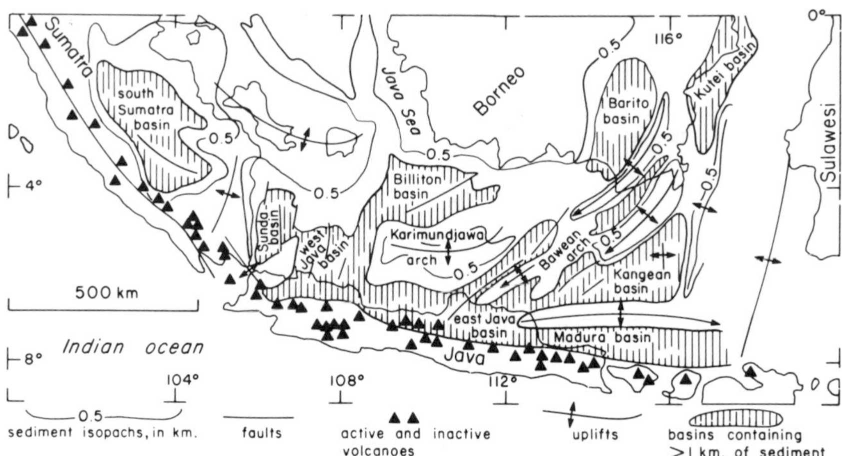

Fig. 9.34. Tectonic elements of Sunda Shelf (Ben-Avraham and Emery, 1973).

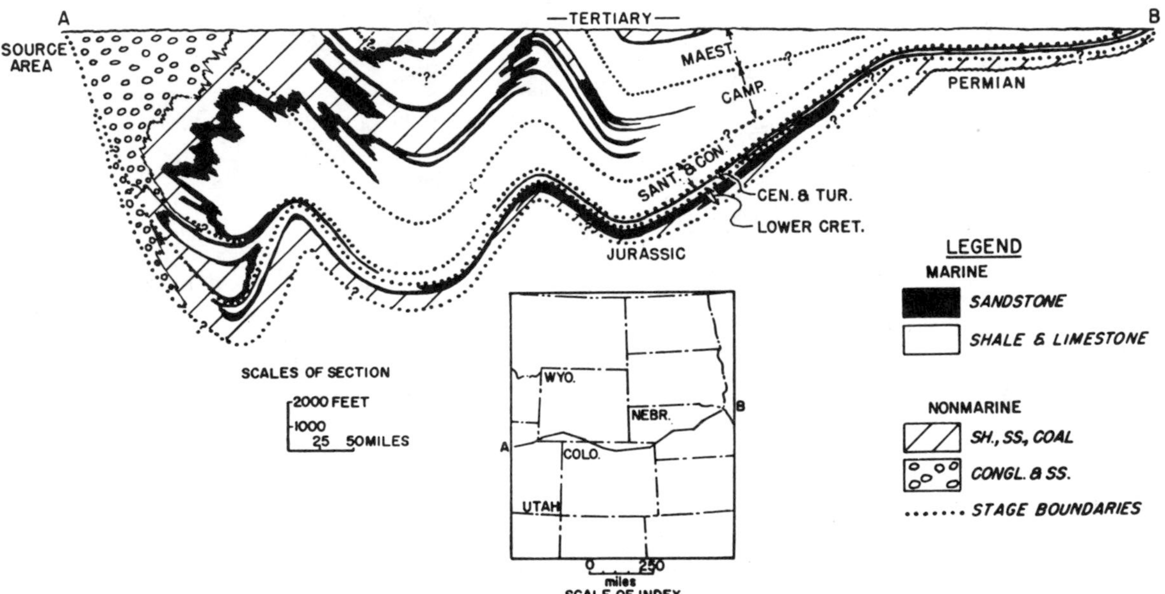

Fig. 9.35. Reconstructed stratigraphic cross section through the Cretaceous fill of the Rocky Mountain retroarc basin (Weimer, 1970).

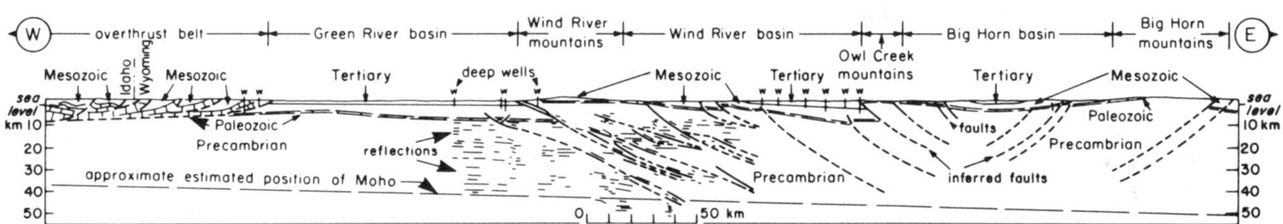

Fig. 9.36. Cross section through Powder River and Wind River Basins, based on deep seismic reflection data. Note that structural style is dominated by deep-seated thrust faults (Bally and Snelson, 1980).

Table 9.6. Plate tectonic settings of strike-slip faults

Plate setting	Examples
1. Plate Boundary Transform Faults	
Intracontinental	San Andreas, California Alpine, New Zealand Dead Sea, Middle East
Intraoceanic	El Pilar-Oca, Caribbean Greater Antillean-Cayman, Caribbean Magellan-North Scotia, Scotia Sea Ninetyeast Ridge, Indian Ocean
Oceanic with continental margin or fragment	Bismark, Bismark Sea, New Guinea Molucca-Sorong, New Guinea Fairweather-Queen Charlotte, western Canada
2. Divergent Margin Transform Faults	
(spreading ridge offsets and fracture zones)	Spitsbergen (Hornsund Fault), Svalbard, Romanche, Atlantic Ocean Falkland-Agulhas, Atlantic Ocean Owen, Indian Ocean Mendocino, Pacific Pioneer, Pacific
3. Convergent Margin Transcurrent Faults	
(arc-parallel)	Sumatra, Sunda Arc Philippine, Philippine Arc Atacama, Chile Median Tectonic Line, Japan
4. Suture Zone Transcurrent Faults	
(oblique collision)	Hornelen Basin faults, Norway Cabot, Maritime Canada Cobequid-Chedabucto, Maritime Canada, Rocky Mountain Trench-Tintina, Canada Kaltag-Porcupine-Alaska Altyn Tagh, Tibet Kunlun, Tibet Red River, Indochina Quetta-Chaman, Pakistan-Afghanistan North Anatolian, Turkey

taining intraformational unconformities, attesting to rapid syndepositional uplift, are typical (eg. Andersen and Picard, 1974).

To summarize, we can begin to establish the elements of a foreland basin model, although additional details are provided in the discussion of collisional foreland basins in section 9.3.4:

1. a basement of normal continental crust, the structural grain of which markedly affects isopach and facies patterns as a result of structural rejuvenation
2. a dominantly shallow marine to nonmarine clastic fill which may contain an upward coarsening transition from marine to nonmarine
3. sediments may be derived from an adjacent magmatic arc or growing fold-thrust belt, or from basement uplifts, or from a cratonic hinterland
4. a structural style dominated by compressional folds and faults, with local intraformational unconformities attesting to syndepositional tectonism.

9.3.3 Transform margins and strike-slip fault basins

Introduction. In simple terms there are three types of plate boundary—convergent, divergent and transform. However, as we have seen, convergent and divergent margins show simple, orthogonal plate trajectories only along short margin segments. Oblique motions are the rule, and the more oblique they are the more important are strike-slip motions at plate junctions and in adjacent plate interiors.

A classification of transcurrent faults is given in Table 9.6. They represent a wide range of plate tectonic environments and vary in importance from major plate boundaries to short faults accommodating local oblique intraplate stresses. Ballance and Reading (1980) used the term **oblique-slip mobile zone** for this entire class of tectonic styles. Fault movement may be convergent–oblique or divergent–oblique. In the first case evidence of shear will be accompanied by compressional structures such as local or regional uplift, thrust faults, nappes and generation of local unconformities in contemporaneous sediments. Harland (1971) coined the term **transpression** for this type of stress. Divergent–oblique stress leads to the development of extensional faults and a variety of different types of sedimentary basin, and was termed **transtension** by Harland. In both cases the presence of a shear component is much more difficult to prove than that of simple compression or extension, which accounts for the fact that geologists have been slow to recognize that strike-slip movements have been important in almost all orogenic belts. The terms transcurrent fault, strike-slip fault and wrench fault are used more or less synonymously, although some authors suggest that only fundamental structures rupturing the entire lithosphere should be termed transcurrent faults (Reading, 1980; Harland, in discussion of Norris et al., 1978).

Transcurrent faults and oblique slip zones in general commonly change style along strike, even though plate trajectories remain constant. This is because the regional stress pattern must accommodate itself to the details of local structural geology. Thus the San Andreas—Gulf of California system (Fig. 9.37) is both transtensional and transpressional in different segments. Crowell (1974a, 1979) pointed out that only three reaches of the fault are parallel to Pacific–North American plate motions. These are regions of pure strike-slip. Elsewhere, as at the Big Bend (Fig. 9.37), the fault zone is under severe compression. Conversely, where the fault passes out into the Gulf of California it becomes extensional (Crowell, 1974a; Kelts, 1981). The initial Gulf rift was slightly oblique to dextral plate motion, and has opened along a series of short spreading centers linked by transform faults (Fig. 9.37). With a subtle change in relative rotation poles regional compressional–extensional relationships on transcurrent faults could drastically change. This has happened on both the San Andreas (Nardin and Henyey, 1978; Blake et al., 1978) and Alpine Fault systems (Carter and Norris, 1976; Norris et al., 1978; Spörli, 1980).

Wrench fault tectonics have been examined by Wilcox et al., (1973), Harding (1974) and Harding and Lowell (1979). They showed that, in areas of regional shear, stress may be accommodated by a single master fault, by several en echelon faults throughout a broad shear zone or by different fault strands across a shear zone that are active at different times. The San Andreas transform system is an example of the latter, as discussed later. The structural style of strike-slip fault zones is distinctive. In cross section they commonly show a splayed or "flower" pattern, the main fault splaying upward into several radiating fractures. In plan view careful mapping may reveal several kinds of associated structure, as summarized in Figure 9.38. Synthetic strike-slip faults occur at an angle of 10 to 30° to the shear direction. Antithetic faults are oriented at 70 to 90° to the shear direction, and have an opposite sense of displacement. Normal faults may occur in early stages of fault movement, and perpendicular to these (15 to 45° to shear direction) are the axes of associated compressional features, including thrust faults and/or folds. All these features may rotate into the direction of shear. The normal faults, in particular, are susceptible to distortion and destruction, and the synthetic faults tend to merge into one master fault, possibly marked by a broad crush zone.

Transcurrent faults are associated with a variety of different types of sedimentary basin, as shown diagrammatically in Figure 9.39. Diagrams *a, b* and *c* illustrate the pattern of braided faults that is so characteristic of major shear zones. Depending on the orientation of individual faults relative to shear direction some may be compressional and others extensional. Uplifted fault blocks may be located adjacent to foundered blocks, leading to the development of local sediment traps.

Diagram *d* shows conditions at the termination of a strike-slip fault. At one end compression occurs and no significant sedimentary deposits will occur. At the other end extension occurs on one

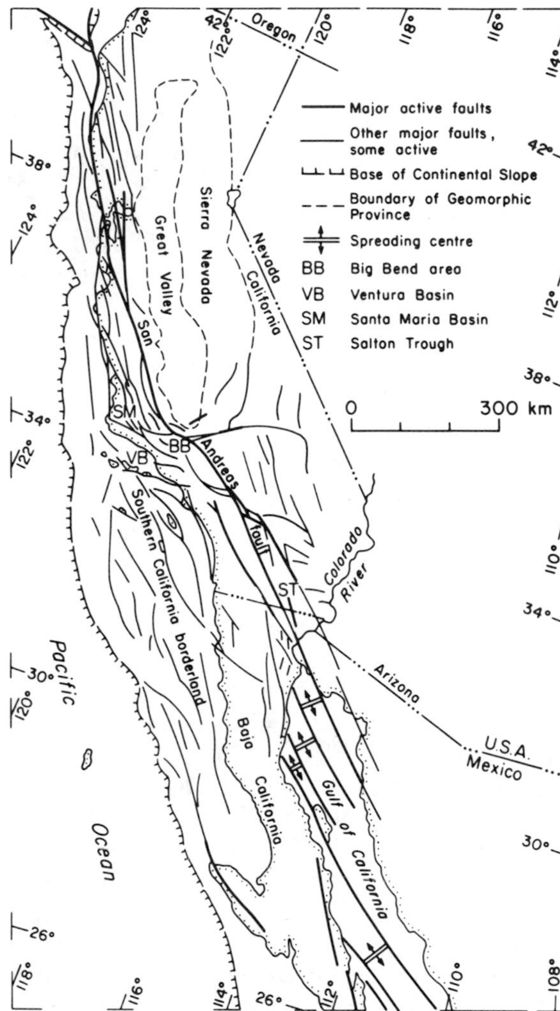

Fig. 9.37. Fault map of California, showing relationship to Gulf of California spreading centers (based on Crowell, 1979; Kelts, 1981).

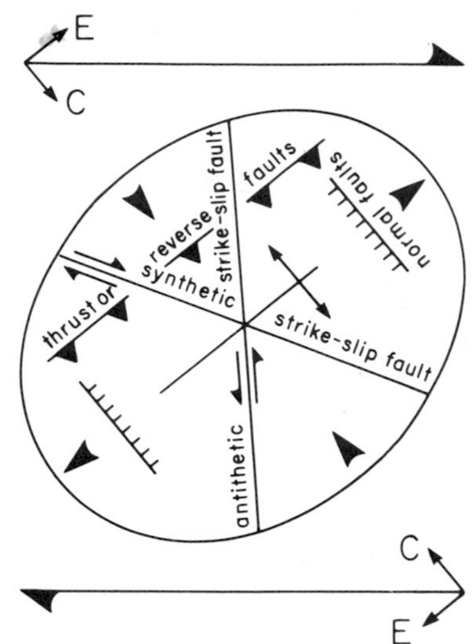

Fig. 9.38. Structural pattern resulting from simple dextral shear (Harding, 1974; Reading, 1980).

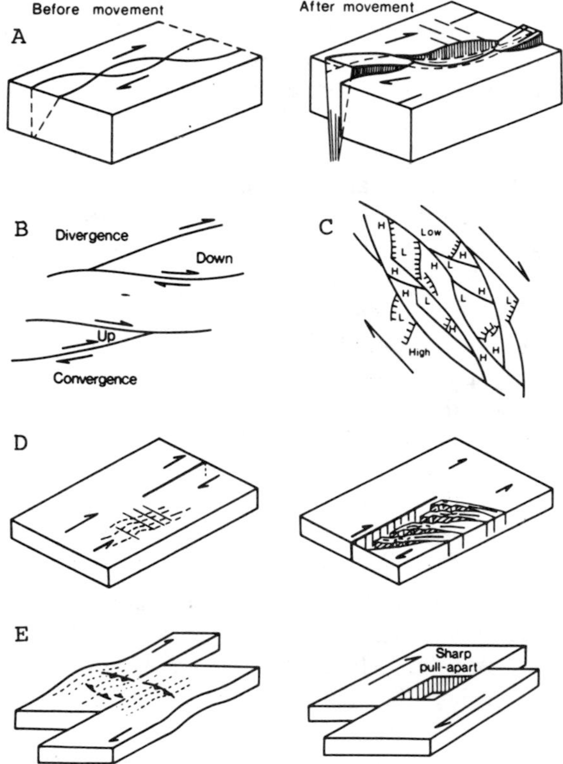

Fig. 9.39. Types of strike-slip fault pattern and resulting sedimentary basins; **A**, **B**, **C** braided faults, **D** fault termination, **E** en echelon faults (Reading, 1980).

side of the fault, and a variety of tensional basins is formed. Lastly diagram e illustrates the generation of a simple pull-apart basin where strike-slip displacement is offset. Initially these may be only a few hundred meters, or less, in length and width, but continued faulting enlarges them into significant sedimentary basins. Aydin and Nur (1982) showed that basins tens of kilometers in width and length could be produced by coalescence of pull-apart structures on adjacent active fault strands.

Basin models for the transcurrent fault setting have the following elements in common:

1. Most basins are only a few tens of kilometers across and are marked by evidence of marked

local syndepositional relief, such as fault–flank conglomeratic wedges.

2. A range of distinctive internal structures with predictable orientations, as summarized in Figure 9.38, is present.
3. Basin fill sediments are characterized by numerous localized facies changes.
4. Movement on individual faults may be local and spasmodic, so that the sediments in adjacent basins may have different stratigraphies.
5. There is usually ample evidence of syndepositional tectonism, including intraformational folds and local unconformities.
6. Basin fill sediments may be offset from their source area, as demonstrable from studies of detrital petrography or from juxtaposition of large depositional systems with small source areas.
7. The basin may itself be dissected by later transcurrent faults, resulting in the juxtaposition of different facies assemblages and sediment thicknesses.
8. In modern basins offset of geomorphic elements such as rivers, alluvial fans or submarine canyons may be demonstrable; sag ponds and pressure ridges are common.
9. Sedimentation rates are rapid, possibly exceeding 1 m/1000 a over periods of several million years.

There are no distinctive depositional environments or facies assemblages, because transcurrent faults occur in all possible environmental settings.

In the next sections the first three of the categories of fault listed in Table 9.6 are described. Transcurrent faults in suture zones are described in section 9.3.4 but do not, in any case, show any significant differences in terms of structural style or sedimentation. For additional data and ideas (some of which are referred to here) the reader is referred to the recent book edited by Ballance and Reading (1980).

Sedimentary basins associated with plate boundary transform faults. Transform faults are a special class of fault associated with sea-floor spreading centers. The actual slip is opposite from the apparent displacement across the fault (Wilson, 1965). Transform faults terminate either at a spreading ridge or a subduction zone. Figure 9.40 illustrates some of the possible configurations by reference to five actual examples. All are hundreds of kilometers in length and the geology of the continental and oceanic plates affected by them is very distinctive. In fact the term "California-type" is commonly used for the geology of continental plates dominated by fundamental transcurrent faults, based on the relatively well-documented examples of the San Andreas transform system (eg. Bally and Snelson, 1980; Dickinson, 1981).

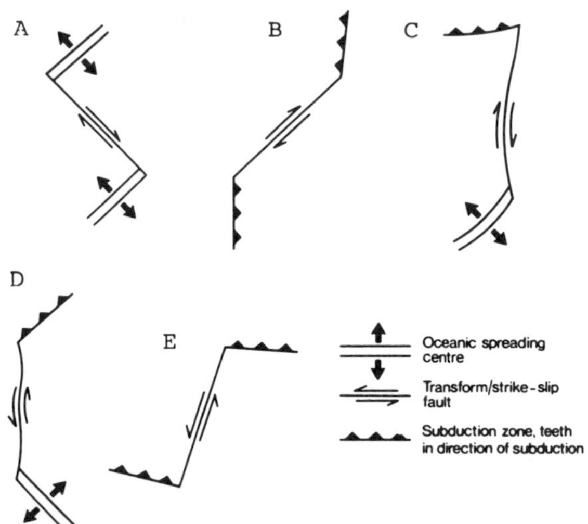

Fig. 9.40. Examples of plate boundary transform faults. **A** San Andreas Fault, **B** Alpine Fault, New Zealand **C** Chugach, Fairweather, Queen Charlotte Fault, **D** Dead Sea Fault, **E** Kirthar–Sulaiman Fault. Various scales (Reading, 1980).

Not shown in Figure 9.40 are the transform faults located entirely within oceanic crust that link spreading centers. These are dealt with in the next section.

The best known intracontinental transform faults are those listed in Table 9.6. The San Andreas system is over 1200 km long and 500 km wide, (Crowell, 1974a, 1979; Howell et al., 1980). It marks the boundary between the Pacific and North American plates, but this is a very diffuse boundary, because there are numerous active faults and many others now inactive that moved in earlier times (Fig. 9.37). The San Andreas Fault, however, is the most continuous and the most seismically active at the present day. Plate tectonic reconstructions based on analysis of the Pacific sea floor indicate that about 1000 km of dextral displacement should have occurred on the transform system since the Oligocene (Atwater

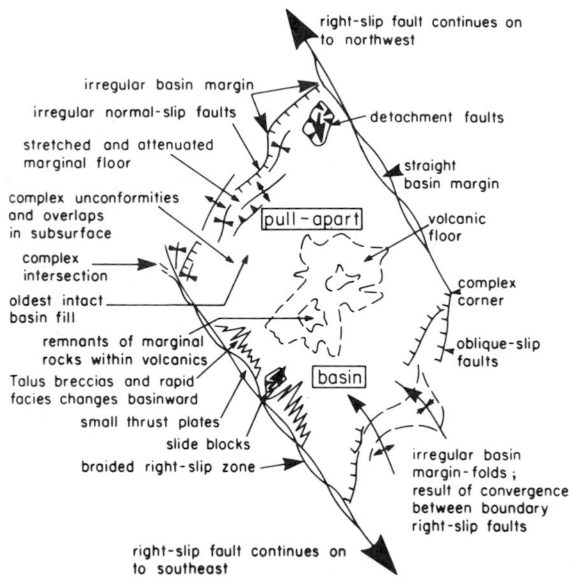

Fig. 9.41. A general model of pull-apart basins (Crowell, 1974b).

and Molnar, 1973), and geological evidence from the San Andreas fault alone indicates at least 300 km of Miocene to recent movement (possibly as much as 600 km, depending on interpretations of offsets). Displacement on other, now inactive faults must be added to this (Crowell, 1979).

The Ridge Basin, near Los Angeles, is a well-exposed example of a pull-apart basin, and excellent studies by Crowell (1974b), Link and Osborne (1978) and Crowell and Link (1982) have turned it into a classic of its kind. The basin was active during Miocene and Pliocene time, when the San Gabriel Fault was the master fault (it is now inactive). The Salton Trough, at the head of the Gulf of California, is another example of a pull-apart basin. It was initiated during the Miocene and is still active, so it has not been uplifted and dissected for our examination as has the Ridge Basin. Volcanism, heat flow data and tectonism in the Salton Trough suggest that it is an incipient spreading center (Crowell, 1974a) and

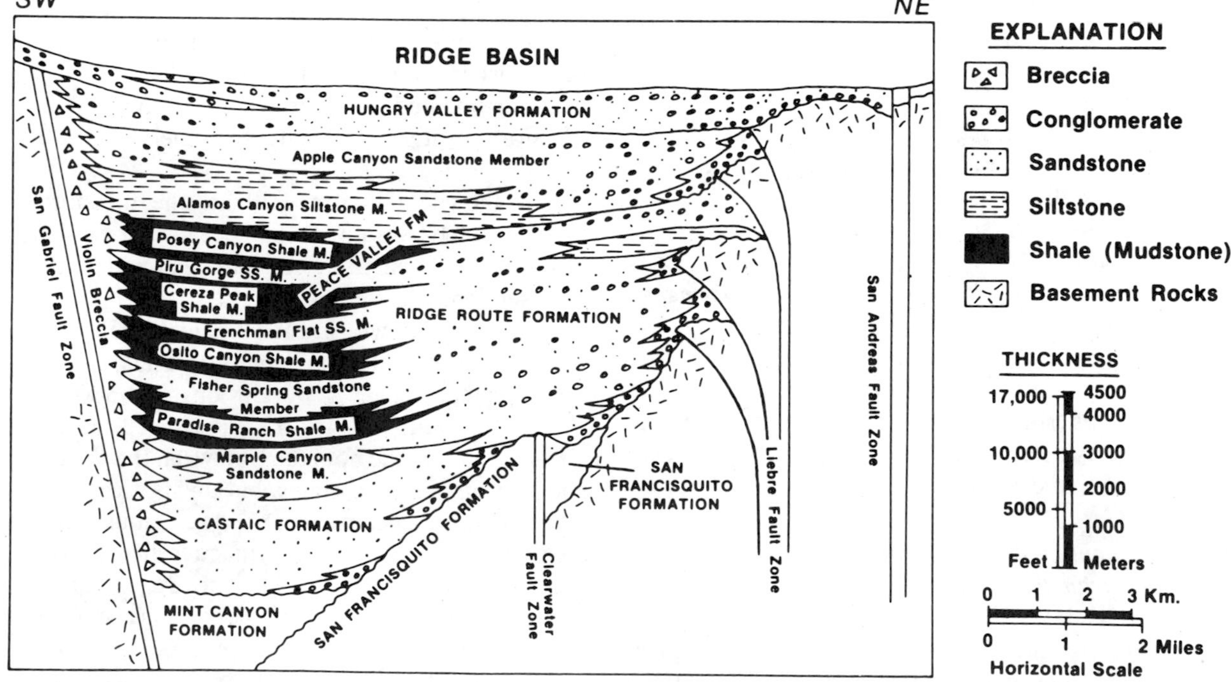

Fig. 9.42. Stratigraphic cross section through Ridge Basin (Crowell, J.C. and Link, M.H., 1982, Ridge Basin, Southern California: Introduction, *in* Crowell, J.C. and Link, M.H., eds., Geological History of Ridge Basin, Southern California: Pacific Section Society of Economic Paleontologists and Mineralogists, p. 1–4.)

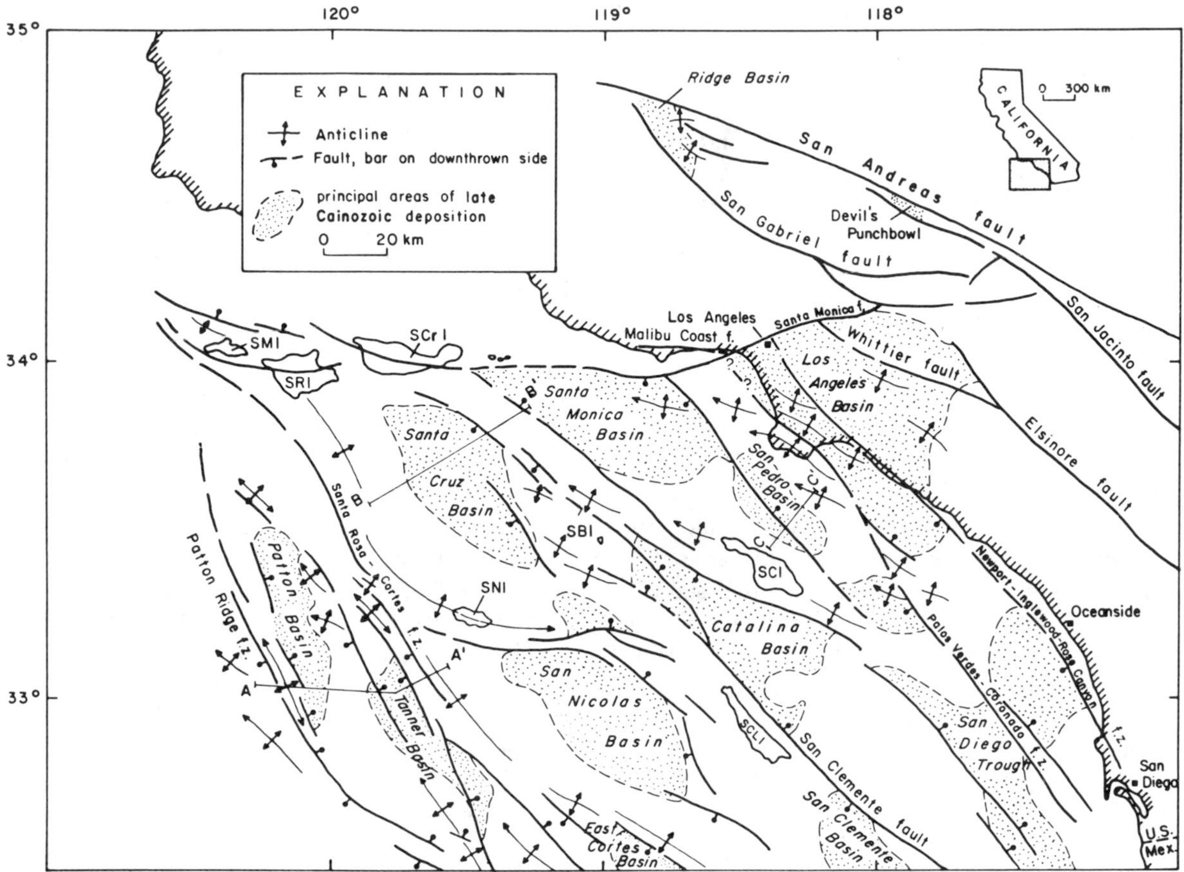

Fig. 9.43. Strike-slip faults and resulting sedimentary basins in the southern California borderland (Howell et al., 1980).

may eventually open fully to expose new oceanic crust, as in the Gulf of California (Kelts, 1981). This would accord with what appears to be a pattern of oblique spreading in the area (a "leaky transform"). Based on these and other Californian basins Crowell (1974b) proposed a general model of pull-apart structure and stratigraphy as shown in Figure 9.41.

The Ridge Basin contains more than 12 km of marine, lacustrine and fluvial sediment that show numerous rapid local facies changes (Fig. 9.42). The San Gabriel Fault bounded the basin to the southwest. It was the main active fault of the San Andreas transform system during sedimentation, and is flanked by the Violin Breccia, an alluvial fan wedge 11 km thick. Much of this great thickness is only apparent, and reflects offlapping of the basin fill to the southeast as the basin subsided and enlarged in that direction. The Liebre Fault Zone, on the northeast side, shows the typical splayed or "flower" pattern of strike-slip fault zones (Fig. 9.42). Uplift on this fault was intermittent and localized, and generated local, short-lived basement sediment sources. The basin center is filled with lacustrine mudstones and with fluvial–deltaic sandstones and minor conglomerates deposited by longitudinal and transverse depositional systems flowing from the north and northwest.

In central and southern California, lying mainly to the west of the San Andreas fault, lies the largely subsea continental borderland (Fig. 9.43). This area is underlain by an arc complex formed during subduction of the Pacific plate in the Mesozoic and early Cenozoic (Howell and Vedder, 1981). Extensive submarine fan complexes formed in small, tectonically active basins along the coast during the Paleogene (Nilsen and Clarke, 1975; Howell and Link, 1979). Commencing in the Oligocene the area was then fragmented by dextral strike-slip faulting and has developed a topography of "en echelon linear to lens-shaped ridges and rhomboid-shaped basins" (Fig. 9.44; Howell et al., 1980) underlain by a structure of braided faults such as those illustrated in Figure 9.39a, b, c. Sedimentation in the basins has been very rapid (up to 1 m/1000 a) and in those basins close to continental source areas (Ventura, Los Angeles) as much as 8 km of Neogene sediment are present (Blake et al., 1978; Howell et al., 1980). Tectonic activity has been almost continuous in these basins during the last 10 to 15 Ma, so that there are numerous syndepositional folds and faults and intraformational unconformities (Fig. 9.44). Fold axes are oriented slightly more westerly than fault trends, reflecting the dextral shear couple (Fig. 9.43). Sediments filling the basins are mainly turbidites, with coastal and nonmarine sediments along the margins. Paleocurrent directions shifted rapidly as local sediment sources were uplifted and then eroded or subsided. Similar sedimentary patterns persist at the present day (Malouta et al., 1981) and the entire borderland area seems likely to remain in a very active condition as long it is controlled by transform tectonics.

The Alpine Fault in New Zealand demonstrates a very similar tectonic and stratigraphic history to that of California, although the evolution of the fault zone has not yet been clearly established (Carter and Norris, 1976; Norris et al., 1978; Spörli, 1980). During the Oligocene to mid-Miocene it functioned as a transtensile or neutral strike-slip fault, with the development of fault-bounded turbidite basins. Toward the end of the Miocene rotation poles between the Australia–India and Pacific plates changed slightly and the fault became transpressive. Many of the sedimentary basins were deformed and uplifted to form the present mountainous backbone of South Island, and nonmarine gravels and sands were shed to fill new basins such as the broad Canterbury Plains. This phase is continuing, and at least 480 km of dextral displacement has accumulated to the present. At its north and south ends the fault merges into opposite-facing subduction zones (Fig. 9.40b). In North Island the Hikurangi arc and accretionary complex provides an excellent example of oblique subduction, well described by Lewis (1980).

The Dead Sea transform is an example of a transtensional fault zone. Sinistral displacement of 105 km has been estimated on this fault (Quennell, 1959; Garfunkel et al., 1981), although this was disputed by Mart and Horowitz (1981). Movement commenced in the Miocene as the Red Sea opened, and is continuing at a rate of 7–10 mm/a. The fault zone shows the typical features of local compression and extension, depending on the orientation of the fault with respect to shear directions. The Dead Sea is a classic example of a pull-apart basin. It contains at least 10 km of Miocene to Recent fluvial clastics and lacustrine limestones and evaporites. Zak and Freund (1981) showed how the depocenters have shifted progressively northward by more than 50 km (Fig. 9.45). This has occurred because the floor of the basin appears to be coupled more strongly to the western margin (Levantine plate) and is being "left behind" by the northward-moving Arabian plate. The pattern of offlapping sedimentary fill (if not the precise mechanism of its formation) is very similar to that of Ridge Basin.

Miall (1981b) suggested that alluvial basin fill models 1, 2 and 6 (Fig. 6.5) would be most characteristic of strike–slip fault basins, and this is borne out by the Ridge Basin and Dead Sea examples. Similar paleocurrent patterns are to be expected in deep sea basins: submarine fans prograding transversly from the margins with sediment gravity flows turning to flow longitudinally down the basin axis. Hsü et al. (1980) demonstrated this pattern for Ventura Basin.

Sedimentary basins associated with divergent margin transform faults. Classifications such as that in Table 9.6 are convenient for mental tidiness, but they can obscure relationships between arbitrarily separated groups. Thus there is a complete gradation between pure transform motion and pure orthogonal spreading. Transforms may be offset by short spreading axes; these have been called "leaky transforms". The Bismark transform north of New Guinea is a good example (Hamilton, 1979). The Atlantic spreading center shows regions of long spreading ridges with few transform offsets, particularly in the North Atlantic, and, between Brazil and Nigeria, an area with long transform offsets and short spreading ridges (Fig. 9.46). During the early stages of spreading these transforms functioned as intracontinental plate boundaries and structural and stratigraphic styles of resulting sedimentary basins were exactly as described in the preceding section. These basins were called "sheared-margin basins" by Wilson and Williams (1979; Fig. 9.47). Initially

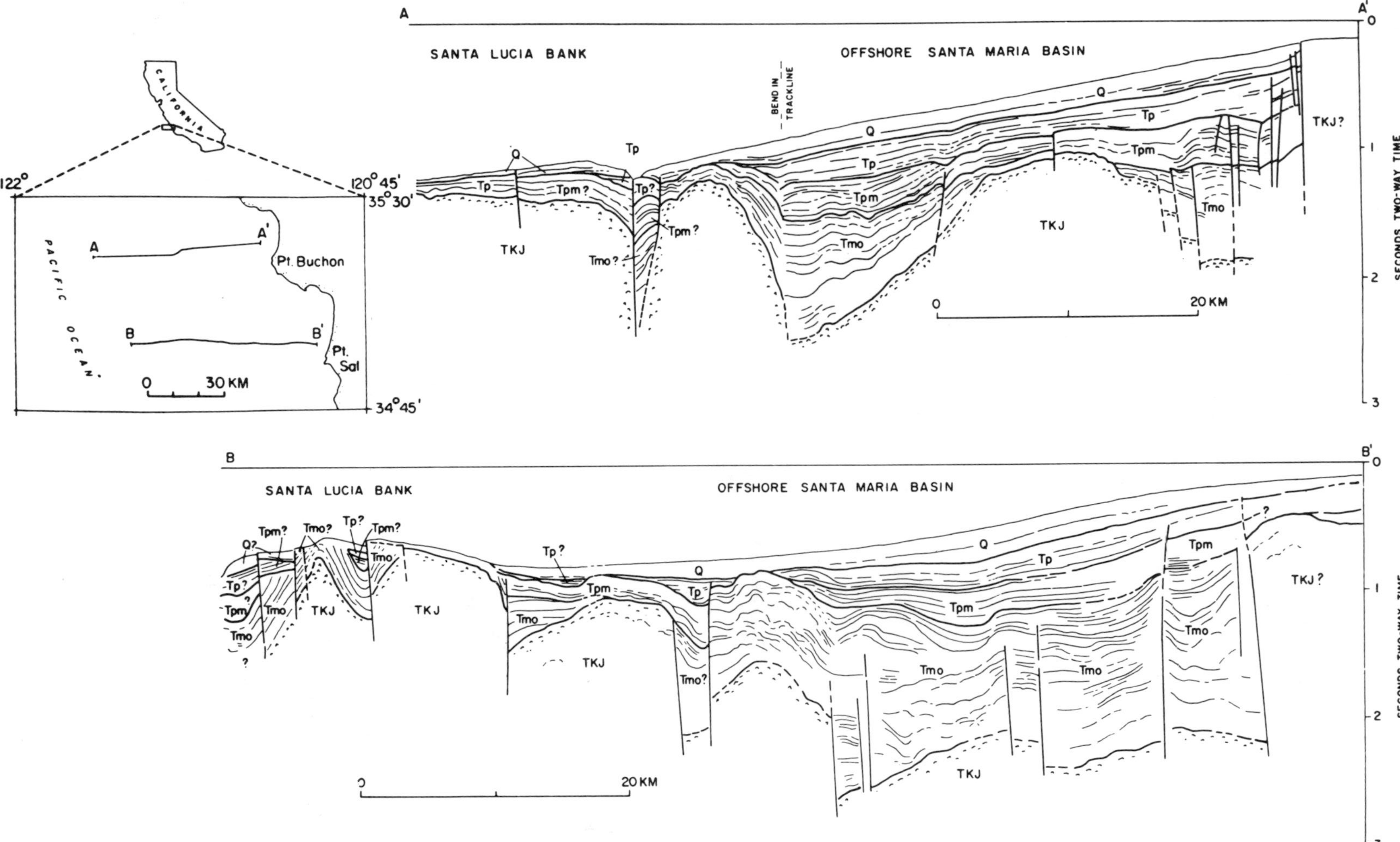

Fig. 9.44. Interpreted seismic reflection profiles across the Santa Maria Basin, southern California Borderland (Howell et al., 1980).

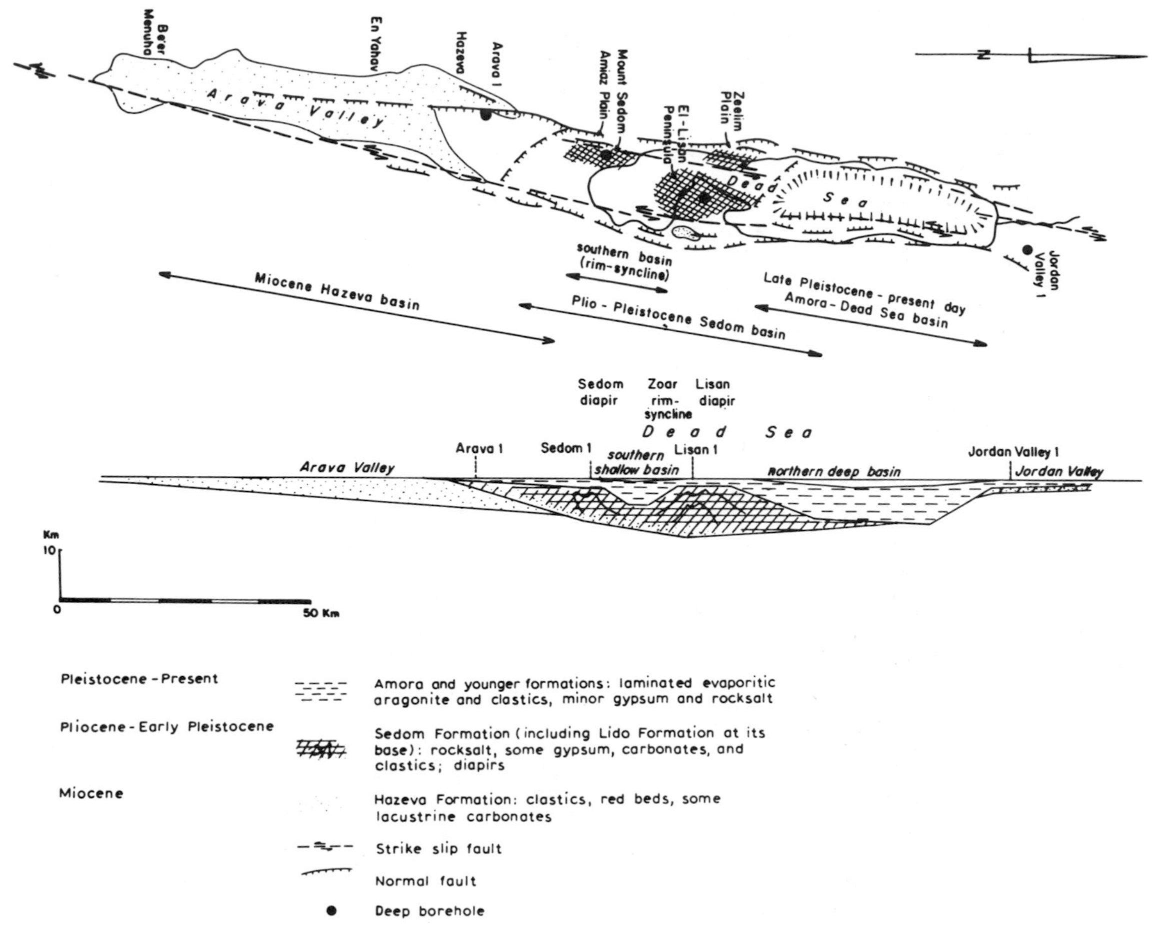

Fig. 9.45. Diachronous fill of Dead Sea Basin during Miocene to Recent strike-slip faulting (Zak and Freund, 1981).

they were two-sided intracontinental basins and may have been either marine or nonmarine. Conditions varied between transpressive and transtensile depending on the positions of local rotation poles. Once the continental margins passed each other the basins faced oceanic crust and continued to develop as divergent margins. Good examples include:

1. Central Tertiary Basin, Spitsbergen, now bounded to the west by the Hornsund Fault, and initially formed along a transform fault separating Svalbard from Greenland (Steel et al., 1981).
2. Barrierinhas and Ceara Basins, Brazil, originally bounded by the West African coast (Francheteau and Le Pichon, 1972; Ponte and Asmus, 1976; Kumar, 1981).
3. Agulhas Basin, off the south coast of south Africa, separated from the Falkland Plateau of the South American plate along the Falkland transform (Dingle, 1973; Du Toit, 1979).

The two-stage evolution of these basins is illustrated in Figures 9.48 and 9.49. Initially sedimentary sources on both sides of the basin were active (Fig. 9.48). Following separation the marginal fracture ridge may have acted as a steep seaward-facing escarpment or as a sediment dam that ponded the sediment until it was buried (Fig. 9.49). In the case of Agulhas Basin dextral shearing reactivated old WNW–ESE structures of the early Mesozoic Cape Fold Belt, and Cretaceous isopachs show a series of depocenters and intervening sediment sources oriented oblique to the continental margin (Fig. 9.50).

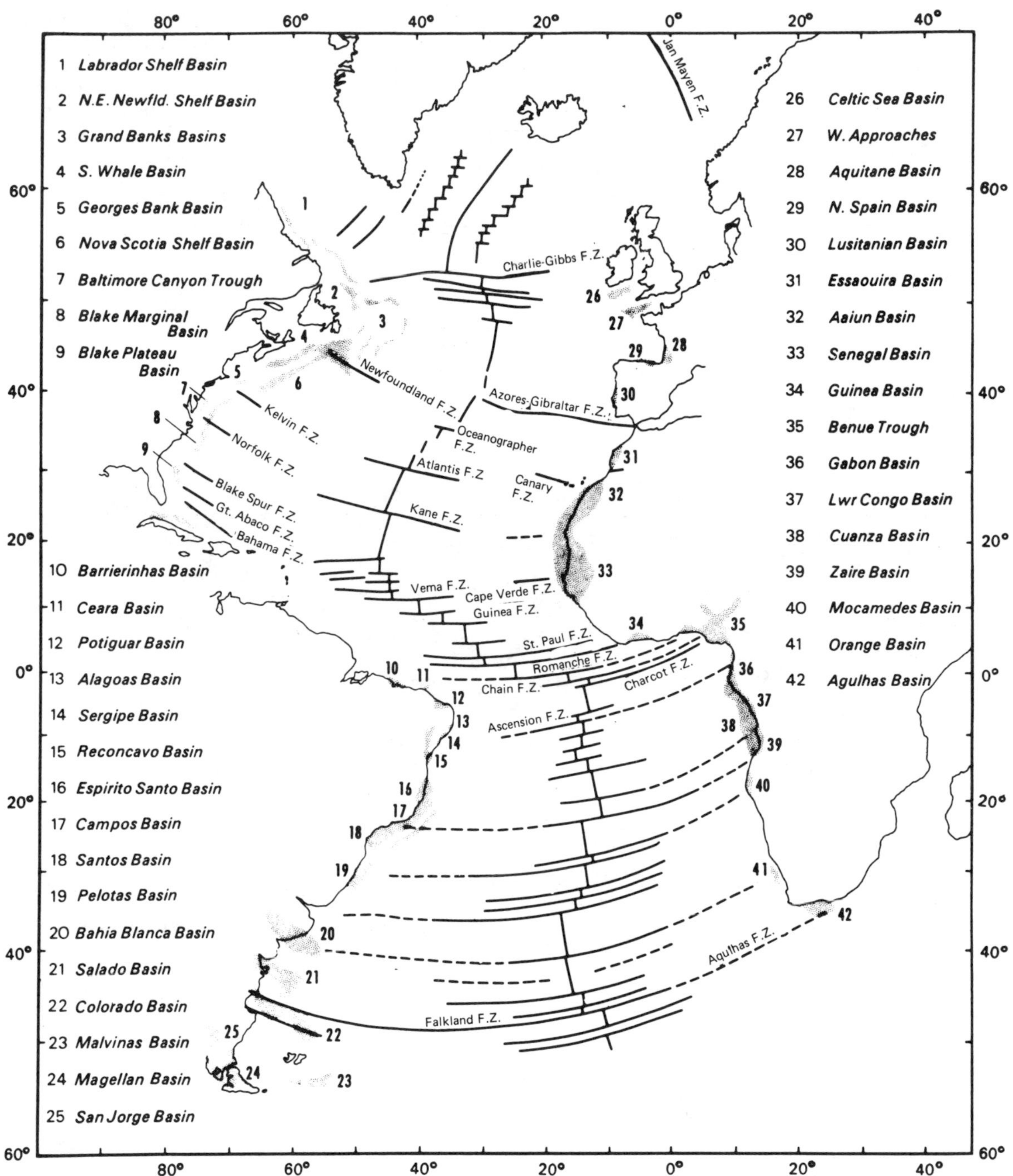

Fig. 9.46. Fracture zones and continental margin basins bordering the Atlantic Ocean (Wilson and Williams, 1979).

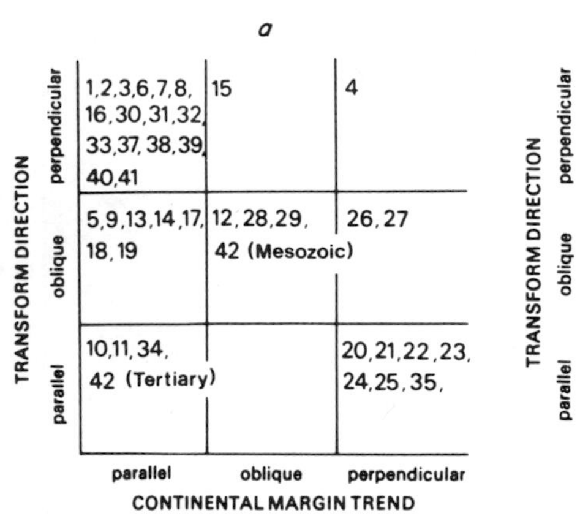

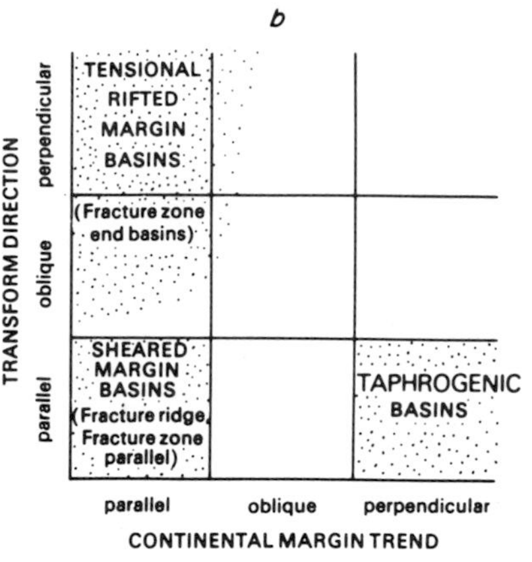

Fig. 9.47. Classification of continental margin basins. Orientation of basin axis is categorized according to relative orientation of transform fault trend and continental margin. Basins are numbered as in Fig. 9.46 (Wilson and Williams, 1979).

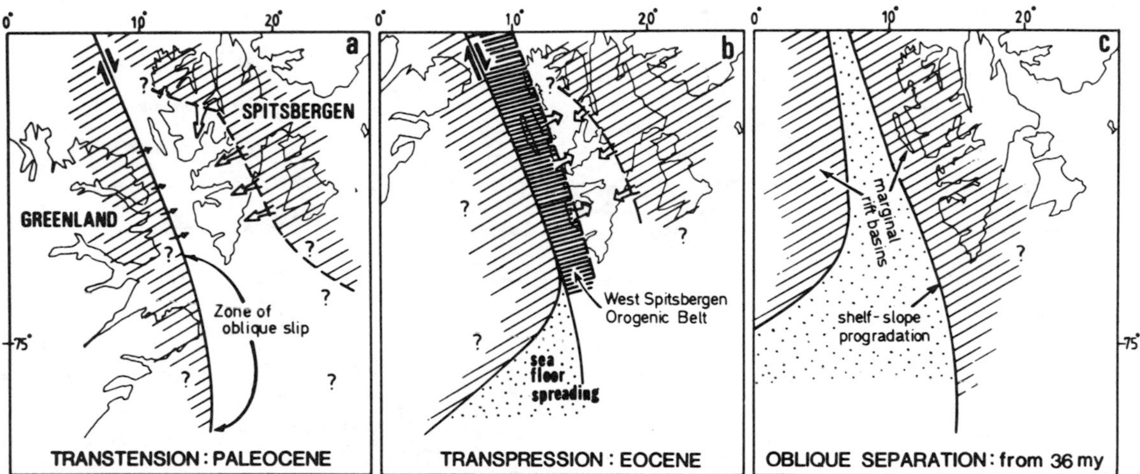

Fig. 9.48. Evolution of the Central Tertiary Basin of Spitsbergen, adjacent to an active transform fault. Sediment was derived from Greenland until this source was moved away along the fault (Steel et al., 1981).

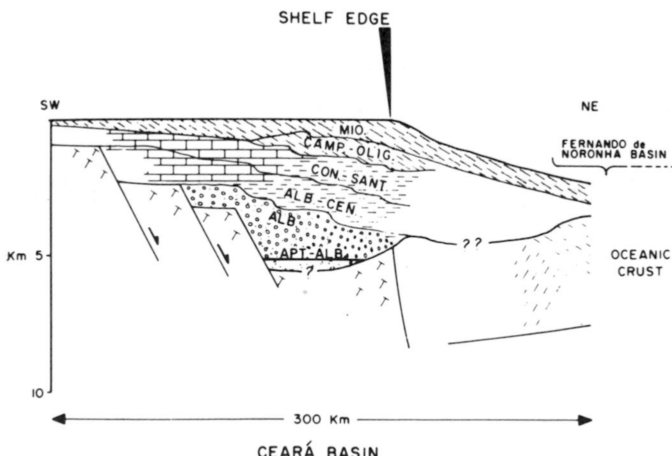

Fig. 9.49. Ceara Basin, Brazil, showing position of transform fault beneath the shelf edge, and its burial during the Cretaceous (Kumar, 1981).

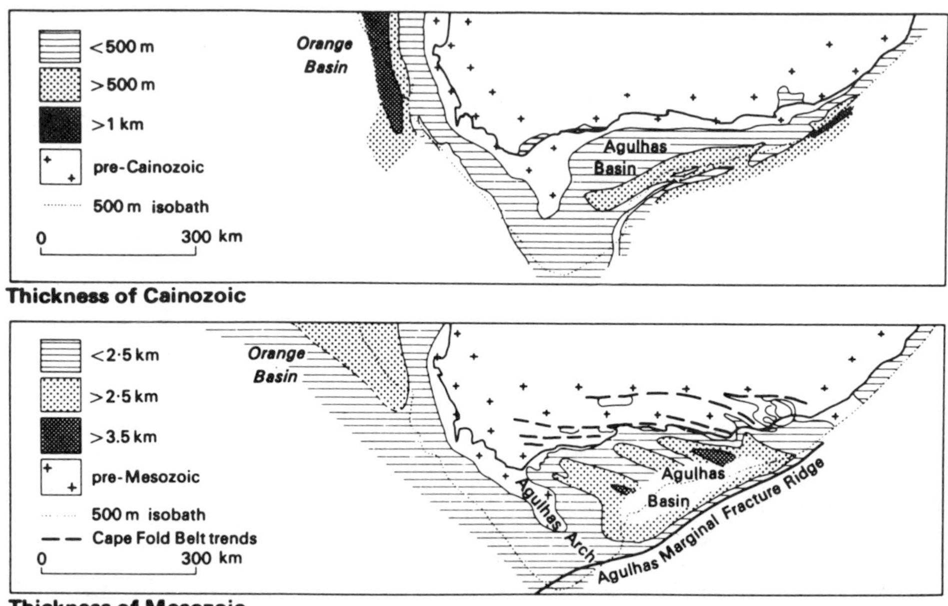

Fig. 9.50. Sediment distribution in Agulhas Basin, off South Africa. The Agulhas Marginal Fracture Ridge is a major transform fault, active during the Cretaceous (Wilson and Williams, 1979).

Francheteau and Le Pichon (1972) and Wilson and Williams (1979) showed that on divergent margins there is a range of different basin types, of which the simple orthogonal spreading type discussed in section 9.3.1 (Table 9.3) and the sheared margin type, discussed above, are but two of nine possible types. These are tabulated in Figure 9.47, which classifies the orientation of basin axis with respect to the orientation of the continental margin and that of spreading-offset transform faults.

Active transcurrent movement on transform faults only occurs between spreading centers, but minor vertical fault movement continues across the fault as the oceanic crust moves away from the spreading center. This is because the crust on either side of the fault is a different age and therefore is at a different point on the cooling curve (Delong et al., 1979). These old tranform faults form distinctive topographic features on the ocean floor, termed **fracture zones**. They lie on small-circle trajectories about the plate rotation pole, and serve a useful function as markers, preserving the tracks of the divergent margins as they move away from each other. Fracture zones may form transverse structural highs where they reach the continental margins and may influence sediment distribution (isopachs) and facies (Francheteau and Le Pichon, 1972). This is particularly well seen where linear basins have formed parallel to transform faults and oriented at a high angle to the continental margins. Wilson and Williams (1979) termed these taphrogenic basins (not a good term, as the word taphrogenic encompasses any type of extensional faulting). Examples include Benue Trough and all the basins on the Atlantic coast of Argentina (Figs. 9.46, 9.47). Benue Trough has been interpreted in the past as an aulacogen, but it is important to build into this interpretation the effect of transcurrent faults extending into the continent from two or three Atlantic fracture zones (section 9.3.1).

Argentinian coastal basins are shown in Figure 9.51. These are graben controlled by a pattern of WNW–ESE trending faults, and contain up to 9 km of Late Jurassic to Recent sediment (Francheteau and Le Pichon, 1972).

It is clear that in analyzing ancient divergent margin sedimentary basins we must be much more sophisticated than in the past in our application of plate tectonic models. Instead of relying on a single, simple orthogonal spreading model we should look for basins and associated bounding faults that trend at oblique to high angles to the presumed ancient continental margin. Recognition of a particular trend may enable us to reconstruct ancient transform fault orientations

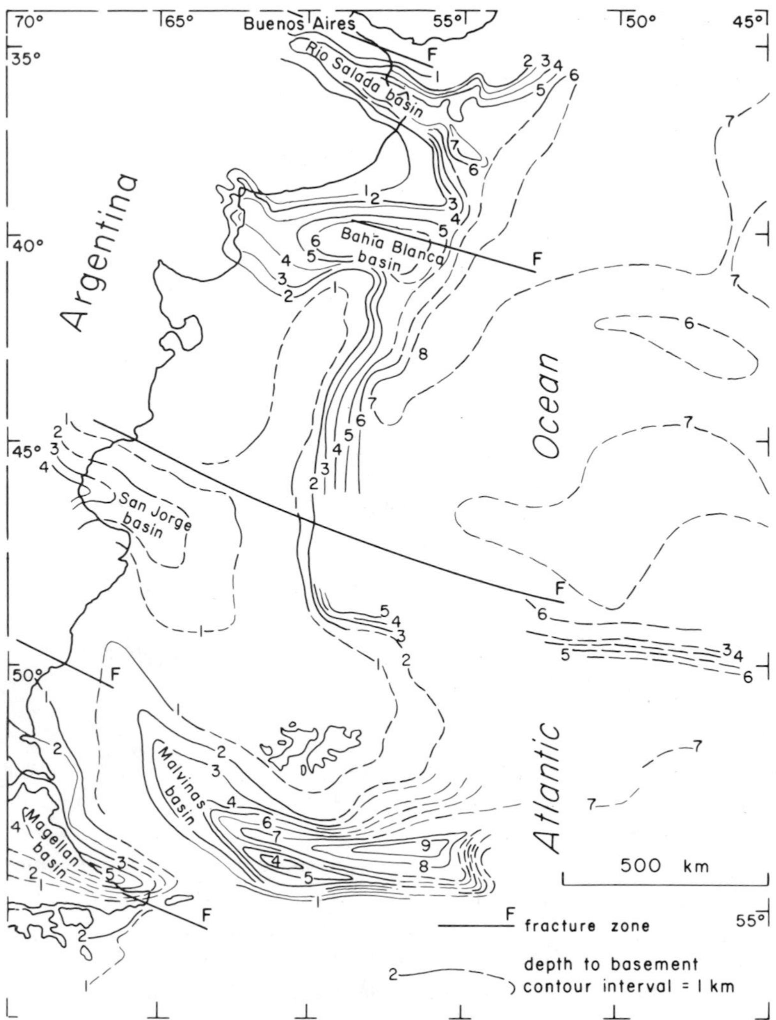

Fig. 9.51. Sediment distribution on Atlantic Margin of Argentina. Note relationship of isopach trends to fracture zones (Francheteau and Le Pichon, 1972).

and plate-spreading trajectories. This would be a particularly instructive exercise where a divergent margin has subsequently undergone a reversal in spreading directions and undergone suturing against another continental plate. However, most of the required evidence is located at the continental margin, a region susceptible to severe deformation by A- or B-subduction during collision.

Sedimentary basins associated with convergent margin transcurrent faults. Many magmatic arcs are cut by major transcurrent faults. These are typically located along the arc itself, where the crust is relatively weak and movement is lubricated by rising magmas. Fitch (1972) and Dewey (1980) showed that these faults are a normal part of arc evolution where convergence is oblique. In the case of neutral or extensional arcs (Dewey, 1980; see section 9.3.2) the faults are transtensile and may be expected to develop extensional sedimentary basins. Even in transpressive systems upthrust fault blocks may be flanked by isostatic depressions that could accumulate sediment.

Few sedimentary basins of any kind have in fact been recorded along transcurrent faults of modern arcs. The Sumatra fault displaces modern alluvium and volcanic rocks and is interpreted as a transtensile rift. Neogene rocks flanking the fault show characteristic en echelon folds at an acute angle to the fault trend (Hamilton, 1979). Neither the Philippine Fault (Hamilton, 1979) nor the Median Tectonic Line of Japan (Uyeda and Miyashiro, 1974) appear to be flanked by syntectonic sedimentary basins.

Fig. 9.52. Block diagram model of tectonic and sedimentary evolution of a suture zone developed between irregular continental margins (Dewey, 1977).

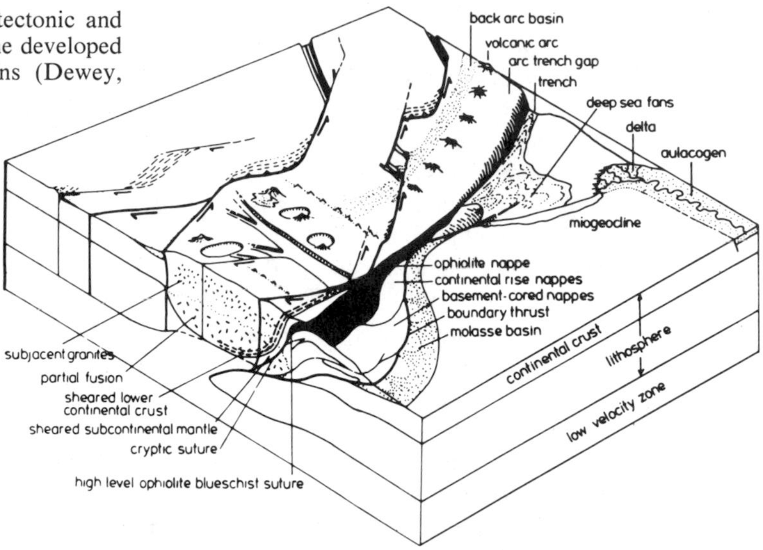

Sedimentary basins associated with suture zone transcurrent faults. These are discussed below (section 9.3.4).

9.3.4 Basins associated with suture zones

Introduction. Sutures are complex orogenic zones where two continental plates have joined by subduction of the intervening oceanic crust. They are characterized by intense structural deformation, regional metamorphism, plutonism and a range of distinctive styles of sedimentary basin. However, a simple classification of suture zone basins is impossible because of the variations inherent in the suturing process. Sutures do not only develop between major continental plates but also occur between continents and island arcs and between microcontinents. Convergence may be orthogonal or, more commonly the case, oblique, with the development of major strike-slip faults or intracontinental transforms. Arcs may be facing toward or away from the colliding continent. In either case a backarc basin may collapse by subduction first, leading to an early orogenic episode predating the terminal collision between major continental masses. Most important of all, collisions between irregular margins can lead to episodic or nearly continuous, diachronous orogeny as the suture zone grinds inexorably shut.

Suture zones are important features of global continental geology. Burke et al. (1977) summarized the geology of 57 of them ranging in age from those still active back to some as old as 2.5 Ga. Examination of the more well-known Phanerozoic sutures such as the Caledonian–Appalachian (Bird and Dewey, 1970; Phillips et al., 1976; Williams and Stevens, 1974; Williams, 1978; Schenk, 1981; Williams and Hatcher, 1982), the Cordilleran (Monger and Price, 1979; Coney et al., 1980) and the Alpine suture (Dewey et al., 1973; Dewey and Sengör, 1979; Biju-Duval et al., 1974, 1978) demonstrates that most suture zones consist of a series of rotated, mangled and partially amalgamated microplates caught between the larger continental plates. Only a few sutures, such as the Urals (Nalivkin 1973; Ager, 1980) and the Indus Suture (Gansser, 1964, 1980; Powell and Conaghan, 1973; LeFort, 1975; Graham et al., 1975; Gupta et al., 1982), appear to have been produced by direct collision of major plates without significant intervening microplate activity. Deciphering microplate kinematics calls for application of the most sophisticated of geological skills, of which sedimentary basin analysis is only one component.

Basic suture model. A collision model for irregular continental margins has been discussed by

Graham et al. (1975), Sengör (1976) and Dewey (1977) and is illustrated in Figure 9.52. Collision between two continental plates requires that subduction must occur on at least one margin (it could occur on both). The overriding plate will therefore have a more or less straight or curving margin, reflecting the presence of the volcanic arc and subduction zone. Collision effects commence as the first projection of the continent on the underriding plate enters the subduction zone. Opposite the point of collision extensional faults, mafic dikes and alkaline volcanism extend into the overriding plate perpendicular to the continental margin. The underriding plate becomes intensely sheared and thrust faulted in the subduction zone, and may be folded back on to itself in the form of giant nappes. Ophiolite slabs and nappes are typical of early collision stages, and may be followed by understacked allochthons of continental cover rocks or basement as the underriding plate is rammed into the subduction zone. Slivers of continental material will be squeezed sideways along transcurrent faults, and a network of transcurrent faults typically also develops in either plate at an angle of about 45° to the suture zone. Crustal shortening may be in the order of several hundred kilometers, leading to crustal thickening, uplift and the development of vast quantities of clastic debris. This sediment is deposited in four types of sedimentary basin.

1. Peripheral or foredeep basins: Detritus shed from the rising orogen at the point of suture is shed onto the underriding plate. The latter becomes depressed into a moat by downbending and, later, by the weight of stacked nappes or thrust sheets. Initially the basin may be under deep water and accumulates thick piles of olisthostromes and sediment gravity flow deposits. Later, as suturing continues, the foredeep becomes shallower, and depositional environments change to nonmarine. The rising orogen may be flanked by giant alluvial fans, which themselves become tilted and deformed. Longitudinal trunk rivers may flow away from the suture along the axis of the foredeep.
2. Intra-suture embayment basins (remnant ocean basins): Along strike the foredeep may pass into embayments between the approaching continents. These serve as sedimentary sinks, and remain undeformed long after suturing has reached an advanced stage elsewhere. In fact they may never be deformed, and may preserve a floor of oceanic crust representing the underriding plate deeply buried beneath a thick blanket of sediment shed into the basin from all sides. The principal sediment source is typically the suture zone itself, from which submarine fans, deltas and alluvial plains prograde in turn longitudinally down the axis of the embayment (Fig. 9.52). Secondary sources may include the arc or subduction complex of the overriding plate, and mature carbonates, clastics or basement rocks from the underriding plate, both of which feed transversly into the embayment. Terminal stages of embayment filling may be lacustrine.
3. Behind the volcanic arc there may be a backarc or retroarc (foreland) basin (section 9.3.2), although this is modified by the graben and transcurrent fault tectonics of the overriding plate that give rise to the fourth type of basin. Backarc basins are typically closed prior to terminal suture, as discussed below.
4. Associated graben and transcurrent fault basins: Strain may be transmitted hundreds or thousands of kilometers into the continental plates on either side of the suture by systems of faults. These may develop graben similar to the rift basins discussed in section 9.3.1 and transcurrent fault basins similar to those described in section 9.3.3.

At this point we should say a few words about the widely used terms **flysch** and **molasse**. Both are old Swiss stratigraphic terms, but acquired a general meaning in early geosynclinal theory. Flysch refers to deep-water clastic sediments deposited under what were described as pre-orogenic or early orogenic conditions. It commonly passes up stratigraphically into molasse, a shallow marine to nonmarine deposit formed under late orogenic to post-orogenic conditions (see Hsü, 1970; Reading, 1972; Van Houten, 1973, 1981). With our knowledge of plate tectonics we can now see that "orogeny" encompasses an enormous variability in plate tectonic settings and processes. The type Swiss flysch and molasse occur in what is now known to be a peripheral basin. However, the sedimentary facies comprising flysch can occur in many tectonic settings, including some that are not orogenic in the original sense (Reading, 1972). Similarly, Miall (1981b) showed that thick alluvial deposits similar to the original Swiss molasse can occur in at least twelve distinct plate tectonic settings, only two or three of which would be recognizable as syn- to post-orogenic geosynclinal settings in pre-plate tectonics terminology. Accordingly most modern workers recommend abandoning the terms flysch and molasse (e.g. Bally and Snelson, 1980).

In the following sections we examine three variations on the basic suture model, with reference to actual ancient examples.

Collapse of outward-facing arcs. Where subduction commences at the edge of a mature ocean, rapid sinking of old, cold oceanic crust favours the development of extensional arcs and the opening of backarc basins. However, with time younger and younger oceanic crust is subducted, and the arc changes to a compressional style (Dewey, 1980). The mechanism is discussed in section 9.3.2 and illustrated in Figure 9.20. The result is that marginal seas first open and then close, with the development of a major orogenic welt, and sedimentation patterns are profoundly altered. It is now widely recognized that this process commonly occurs during the closing-ocean phase of a Wilson cycle. It leads to orogenic episodes that predate by tens or hundreds of millions of years the orogeny of the terminal suture. Ancient examples where this sequence of events has been proposed include the southern Andes (Dalziel et al., 1974), the Western Cordillera of the United States (Churkin, 1974; Burchfiel and Davis, 1972; Dickinson, 1981) and the northern Appalachians (Dewey, 1974; St. Julien and Hubert, 1975).

A single example is illustrated here, that of the Taconic orogen of Quebec. A series of cross sections reconstructing the early Paleozoic history of this region was constructed by St. Julien and Hubert (1975) and is reproduced as Figure 9.53. Cambrian to Lower Ordovician strata constitute a divergent margin shelf–slope–rise assemblage facing the Iapetus Ocean. In the latter part of the Early Ordovician closure of the ocean commenced, with the development of a subduction complex to the southeast. Calc-alkaline volcanics fed tuffaceous material northwestward, and this marked the beginning of a reversal in sediment transport directions. The divergent margin prism became the continental margin of a backarc basin that was progressively filled by northwesterly prograding sediment gravity flows and olisthostromes. The latter were derived by submarine collapse of advancing thrust and nappe sheets consisting of ophiolites, arc volcanics and basin fill sediments. By the end of the Ordovician or Early Silurian closure and uplift of the backarc basin was complete. The Taconic Orogeny, as summarized here, extended over a period of about 30 Ma. It was accompanied by intense structural deformation and regional metamorphism, and the synthesis presented in Figure 9.53 represents the unraveling of a highly complex piece of geology.

St. Julien and Hubert (1975) proposed several changes in arc configuration and polarity during the Ordovician. This is a difficult subject to resolve because of the structural complexity. Many different reconstructions have been made for other parts of the Appalachian–Caledonian orogen, as shown in Figure 9.54. Most of these depend on structural and geochemical variations of the volcanics and obducted ophiolites, a discussion of which is beyond the scope of this book (Dewey, 1974; Strong et al., 1974).

Collisions of continents with oceanic arcs. The islands of Taiwan and Timor are excellent examples of land masses that owe their existence largely to the collision of a continental margin with an oceanic arc. The continental rocks are thrust back on themselves in a much thickened subduction complex, and themselves become the source of clastic detritus feeding a local peripheral basin.

Figure 9.55 is a simplified tectonic map of Taiwan (after Biq, 1964) interpreted in plate tectonic terms (Hamilton, 1979). The volcanic arc is marked only by a few islands off the east coast. The Coastal Range is a forearc basin containing Miocene andesitic lavas and agglomerates and Pliocene turbidites. A thin slice of melange underlies the longitudinal valley, and the main Central and Hsüehshan Ranges consist of a subduction complex of Eocene to Miocene turbidites with a "crystalline core" of migmatites, schists, marbles and argillites of late Paleozoic and Mesozoic age. The core represents deformed rocks of the Chinese continental margin. Along the western edge of the mountains is a fold-thrust belt consisting of Neogene shallow marine to nonmarine sediments floored by Mesozoic shelf sediments and capped by a Pliocene–Pleistocene synorogenic conglomerate that records the climax of the arc–continent collison. This foothills region represents the syndepositionally deformed edge of the peripheral basin. It passes westward into the

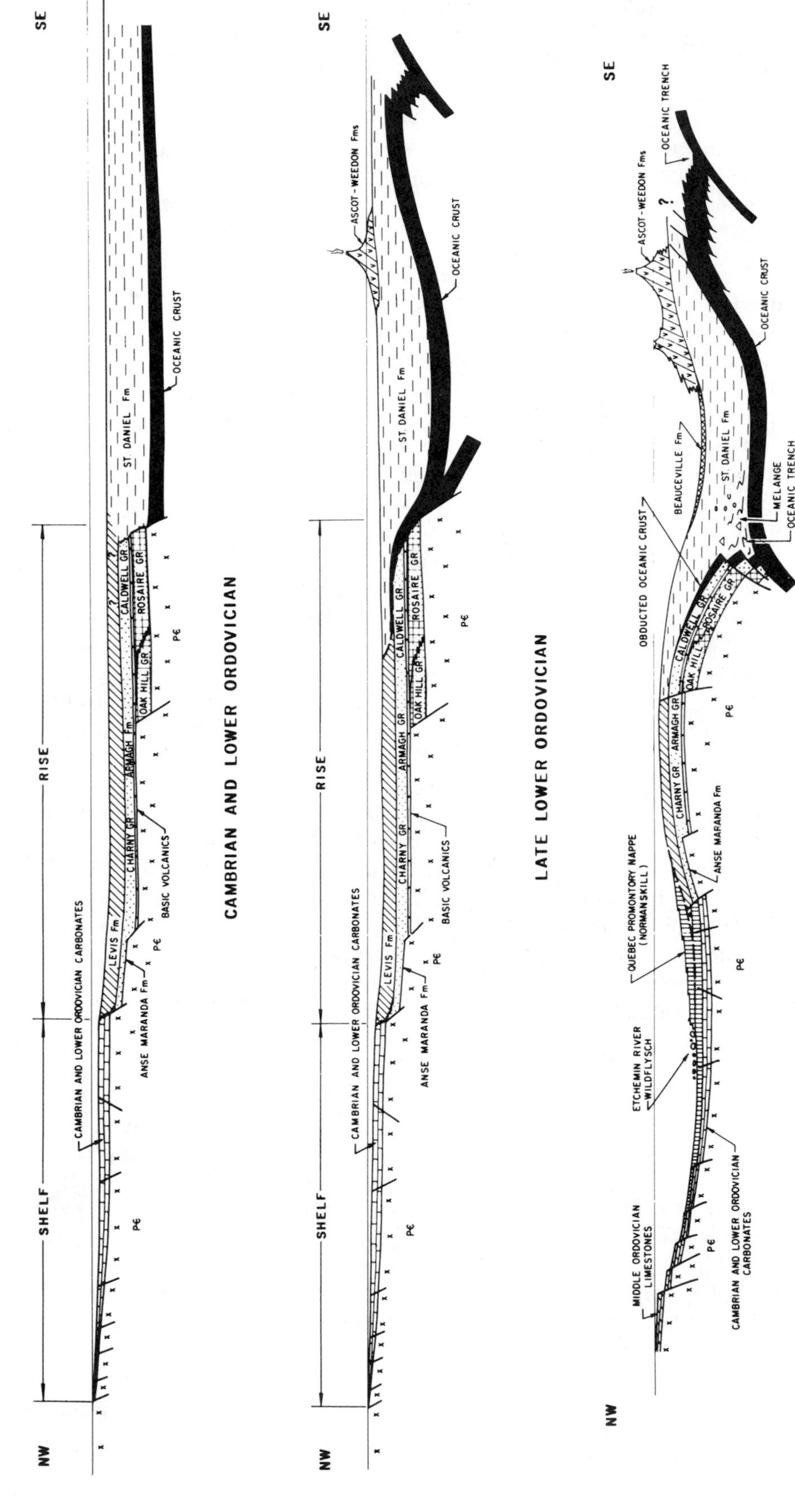
CAMBRIAN AND LOWER ORDOVICIAN
NW
SE
SHELF
RISE
CAMBRIAN AND LOWER ORDOVICIAN CARBONATES
ANSE MARANDA Fm
LEVIS Fm
CHARNY GR
ARMAGH Fm
BASIC VOLCANICS
OAK HILL GR
CALDWELL GR
ROSAIRE GR
ST. DANIEL Fm
OCEANIC CRUST
PЄ
LATE LOWER ORDOVICIAN
NW
SE
SHELF
RISE
CAMBRIAN AND LOWER ORDOVICIAN CARBONATES
ANSE MARANDA Fm
LEVIS Fm
CHARNY GR
ARMAGH GR
BASIC VOLCANICS
OAK HILL GR
CALDWELL GR
ROSAIRE GR
ST. DANIEL Fm
ASCOT - WEEDON Fms
OCEANIC CRUST
PЄ
NW
SE
MIDDLE ORDOVICIAN LIMESTONES
CAMBRIAN AND LOWER ORDOVICIAN CARBONATES
ETCHEMIN RIVER WILDFLYSCH
QUEBEC PROMONTORY NAPPE (NORMANSKILL)
ANSE MARANDA Fm
CHARNY GR.
ARMAGH GR
OAK HILL
CALDWELL GR
ROSAIRE GR
OBDUCTED OCEANIC CRUST
BEAUCEVILLE Fm
ST. DANIEL Fm
MELANGE
OCEANIC TRENCH
OCEANIC CRUST
ASCOT - WEEDON Fms
OCEANIC TRENCH
PЄ

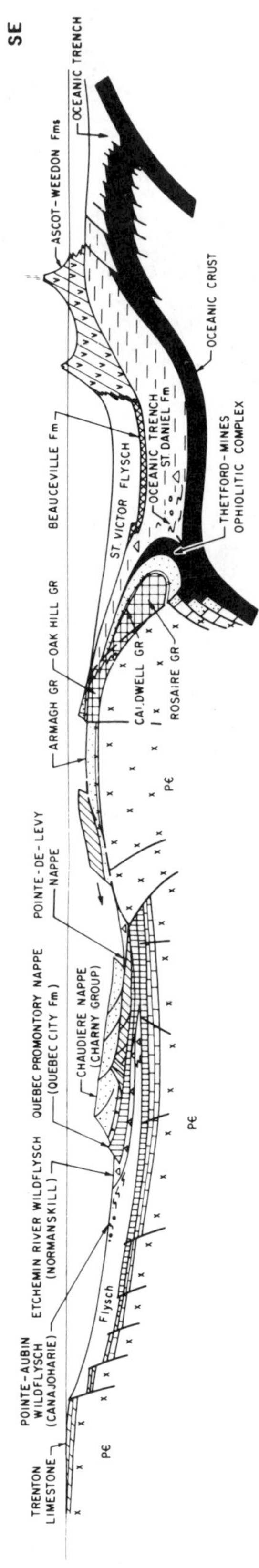

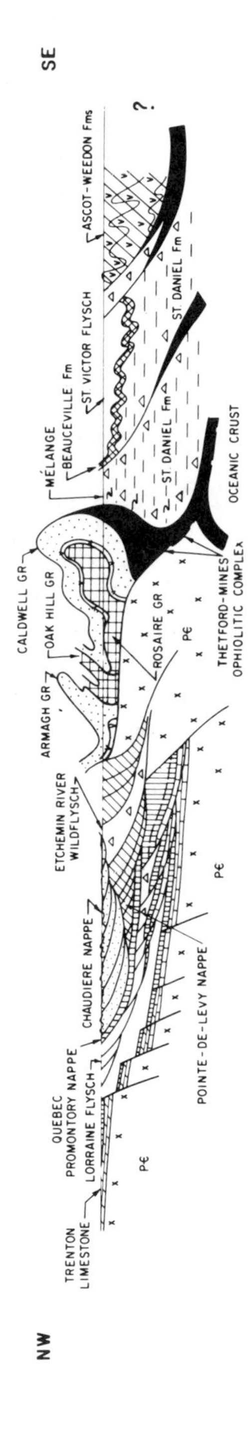

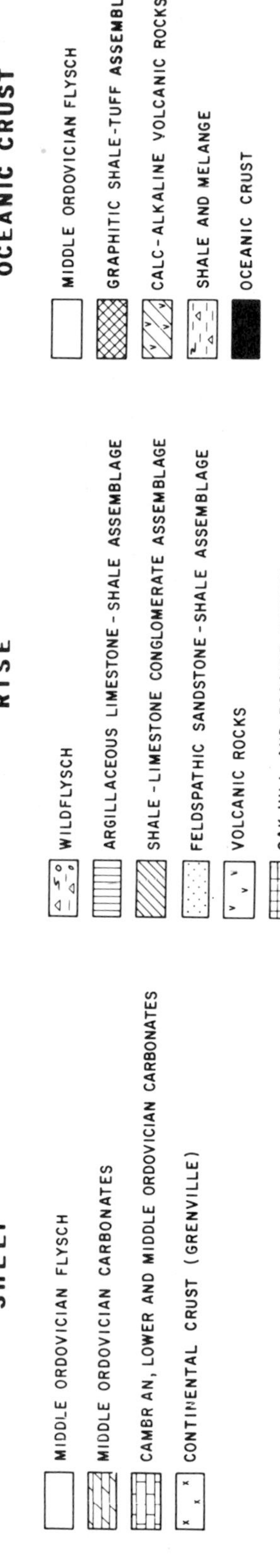

Fig. 9.53. Schematic evolution of the Taconic orogen of the Quebec Appalachians (St. Julien and Hubert, 1975).

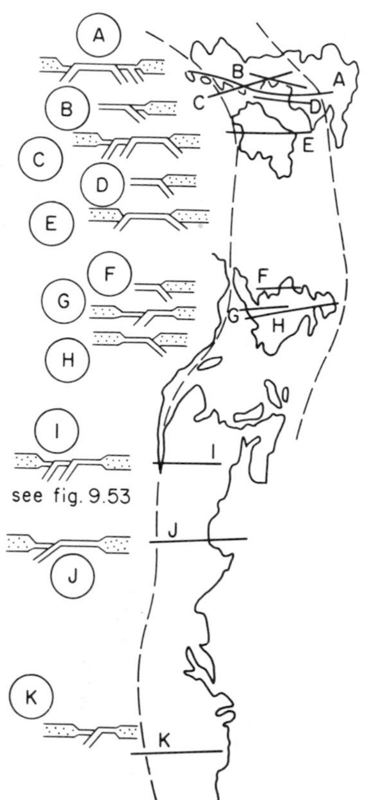

Fig. 9.54. Plate tectonic models suggested for different sections of the Appalachian–British Caledonides suture zone (Windley, 1977). Adapted from B.F. Windley, The Evolving Continents, © 1977, John Wiley and Sons, Ltd. Reprinted by permission of John Wiley and Sons, Ltd.

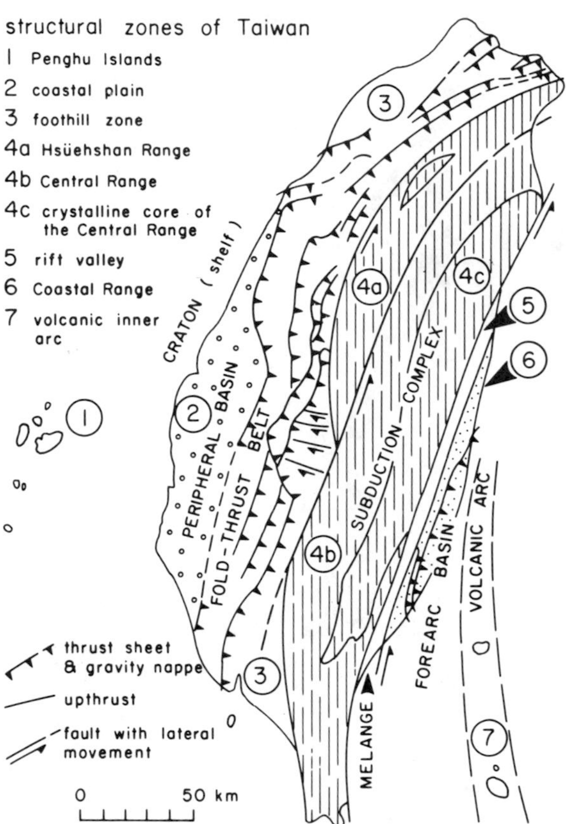

Fig. 9.55. Simplified tectonic map of Taiwan (after Biq, 1964; Hamilton, 1979).

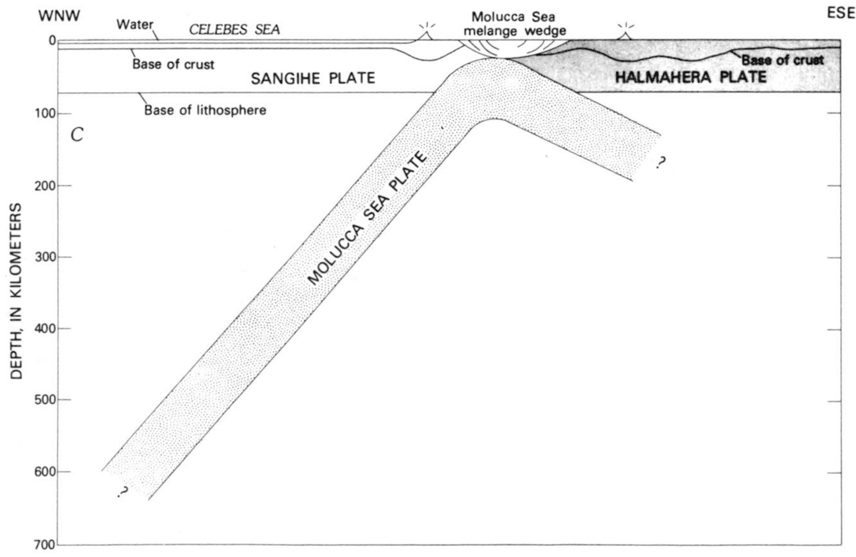

Fig. 9.56. Arc–arc collision in the Molucca Sea (Hamilton, 1979).

coastal plain and Chinese continental shelf, underlain by flat-lying Neogene sediments resting on Mesozoic shelf rocks.

South of Taiwan, in the South China Sea, the arc and subduction complex are still active, and the occurrence of major earthquakes on Taiwan suggests that the collision is not yet complete.

Timor represents a similar plate tectonic configuration, although the local geology is more complex and conflicting interpretations have been published (Carter et al., 1976; Crostella, 1977; Hamilton, 1979). Hamilton's interpretation (see also Dewey, 1977) is that the Australian plate is beginning to be subducted beneath the Sunda Arc. At Timor the first projection of the continental plate entered the north-dipping subduction complex during the Neogene. Most of the island is underlain by imbricate masses and melange of Australian shelf sediments of Permian to early Tertiary age. Faulted into this complex are Miocene to Pleistocene marine marls and shales that were probably deposited on small accretionary basins (Dickinson and Seely, 1979) within the subduction complex. Subduction is still active and Pleistocene reefs have been uplifted to heights of 800 m above sea level. Hamilton (1979) suggested that "choking" of the subduction zone by continental rocks may be leading to a reversal in polarity of the arc, so that a south-dipping zone may be developing in the Banda Sea, to the north of the present volcanic arc. Within about 10 Ma the Timor melange masses may have become the roots of southward-directed nappes sliding southward into a backarc basin developed over the Australian continental shelf—a repeat of a Taconic-style orogeny.

Arc-arc collision. The Molucca Sea, between Indonesia and the Philippines, overlies one of the most curious permutations of arc geology. Here two arcs, facing each other, are both subducting the same oceanic plate (Hamilton, 1979). Figure 9.56 is based on a plot of earthquake activity and shows two downgoing slabs of oceanic crust. Seismic reflection profiles show that the west-dipping Sangihe subduction zone is becoming inactive and that a new trench is developing on the west side of the arc, so that shortly the Sangihe plate will start to be subducted beneath the Halmahera plate.

Recognition of this sequence of events in an ancient orogen would require precise age determinations of tectonic events and sedimentary rocks within three parallel subduction complexes (Halmahera, old Sangihe and new Sangihe).

Terminal sutures. With all the arcs and microcontinents swept into an orogenic welt the stage is set for the most awe-inspiring event in the whole panoply of geological phenomena: the collision between two major continental plates. The Cenozoic collision of India with Asia is the most spectacular example presently visible on earth. Everything about it—tectonic deformation, topographic relief and the scale of consequent depositional systems—is on a gigantic scale. In this section we concentrate on the development of the Himalayan region as the type example of terminal collision (Fig. 9.57; see also Fig. 9.1).

Reconstructions of the northward drift of the Indian continent based on oceanic paleomagnetic evidence (McKenzie and Sclater, 1971; Powell, 1979) indicate that India began to collide with Asia in the latest Cretaceous and Paleocene, and this is confirmed by the age of ophiolites and associated sediments along the Indus Suture (Gansser, 1964, 1980; Powell, 1979; Brookfield and Reynolds, 1981; Andrews-Speed and Brookfield, 1982). North of the suture lie the poorly known mountains of Tibet and Afghanistan, which may have been assembled from an Asian plate and intervening small plates during the Mesozoic (Stocklin, 1977; Powell, 1979). Subduction and underthrusting of the Indian plate has accumulated 300–500 km of crustal shortening (Molnar and Tapponnier, 1975; Powell, 1979), and in the Miocene this led to the beginning of Himalayan uplift that is continuing at the present.

The effects of collision include the following:

1. intense tectonic deformation, metamorphism and igneous activity. Most of the details of this are beyond the scope of this book, except where they are relevant to considerations of sediment age, detrital sources and tectonic position.
2. development and subsequent mangling of island arc complex(es) at the suture zone.
3. development of a gigantic peripheral basin south of the fold–thrust belt, now occupied by the Indus and Ganga rivers and their tributaries.

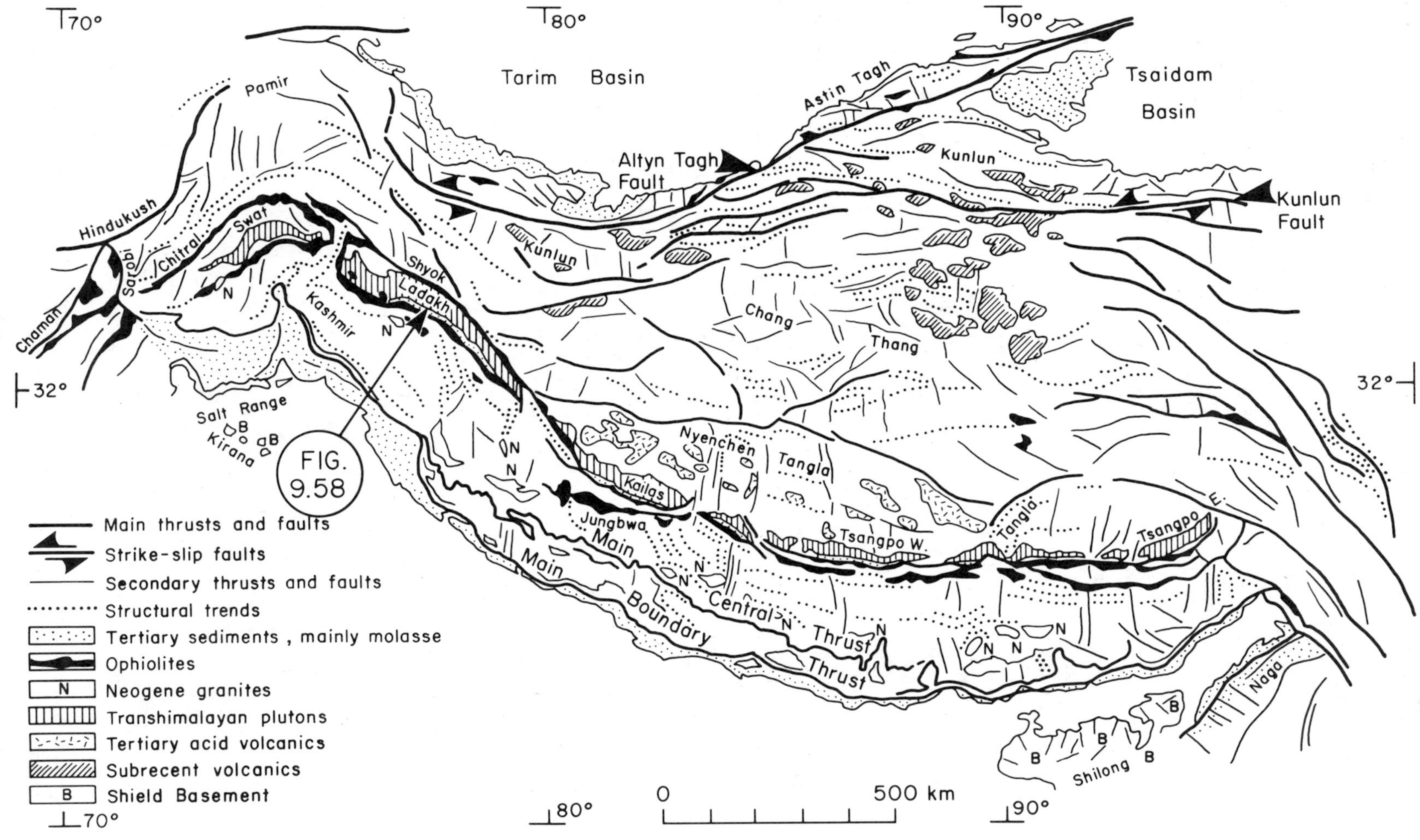

Fig. 9.57. Tectonic sketch map of the Himalayan region (after Molnar and Tapponnier, 1975; Gansser, 1980).

4. spillover of Himalayan detritus into intra-suture embayments southwest (Indus fan) and southeast (Bengal fan) of India.
5. extension of deformation far into China and the Soviet Union along major strike slip faults and graben.

Indus Suture: The geology of the Indus Suture is poorly known, because of topographic and political inaccessibility. Brookfield and Reynolds (1981) and Andrews-Speed and Brookfield (1982) examined cross sections through the suture in Kashmir (Fig. 9.58) and proposed a plate tectonic reconstruction based on a compilation of all available data. They attempted to recognize the various elements of the island arc(s) along the suture, using the kind of approach discussed in section 9.3.2. This was then developed into an evolutionary scheme, using radiometric ages of igneous events and biostratigraphic data from the Spongtang klippe (location in Fig. 9.57), and Figure 9.59 illustrates their two alternative tectonic schemes. Their preference is for alternative two (right column).

Ladakh batholith is interpreted as a magmatic arc that probably was initiated in the Cretaceous with the intrusion of gabbroic plutons. The major intrusions are tonalites and granodiorites of Paleocene–Eocene age. South of the batholith lies the suture zone, now overturned to the north but interpreted as formed originally in front of a south-facing arc (Fig. 9.59c). The complex includes the Lamayuru unit, a Triassic–Jurassic slope and rise deposit of the Indian continental margin; the Dras unit, a volcanic, volcaniclastic and pelagic sequence of mid to Late Cretceous age, interpreted as the product of an island arc; and the "Indus Flysch", a Cretaceous submarine fan deposit. Faulted into these units are sheets of melange, including large ophiolite masses. The melange may partly represent a subduction complex, but probably was further deformed during collision. The Indus Flysch may represent forearc or backarc sediments; relationships in the suture zone are not adequate to determine which.

The cross section (Fig. 9.58) also shows a slice of "Indus Molasse" faulted into the suture zone. Elsewhere the molasse has an unconformable sedimentary contact with the Ladakh batholith. The sediments consist of fluvial conglomerates and sandstones of Eocene to Miocene age. Petrographic and paleocurrent evidence indicate derivation from the batholith to the north and from the arc volcanics and Indian (Tethyan) continental margin rocks to the south, with axial longitudinal flow of large braided streams toward the east. These data are part of the reason why Brookfield and his co-workers proposed a reversal of subduction polarity in the Eocene (Fig. 9.59 *e2*). The two-sided basin is consistent with the expected paleogeography of a backarc basin, and continued arc activity along a suture already choked with continental material (alternative e.1 in Fig. 9.59) seems unlikely (cf. Timor, as discussed earlier). In the Miocene the Ladakh arc closed against the Karakorum massif, generating the Shyok melange and ophiolite belt.

Indian peripheral basin (Fig. 9.60): The structure, stratigraphy and sedimentology of this region have been summarized by LeFort (1975) and Graham et al. (1975) and the sedimentology has been described in more detail by Johnson and Vondra (1972), Visser and Johnson (1978), Johnson et al. (1979) and Parkash et al. (1980).

Cretaceous to Eocene strata of the northern Indian subcontinent are sandstones, mudstones and limestones representing shallow marine shelf environments. Faunal affinities of vertebrates suggest the establishment of a land connection with the Asian plate in the early Eocene, which is confirmed independently by the sequence of events deduced for the developing suture to the north (Fig. 9.59). The Oligocene to Miocene Murree Formation (2000 m) consists of deltaic deposits derived from the Indian Shield and the rising Himalayan orogen. Its relation to the Indus Molasse is unclear.

The Murree passes gradationally up into the Siwaliks, an entirely nonmarine unit up to 5 km thick. Depositional environments were similar to those existing at the present day, namely giant alluvial fans draining transversely from the rising mountains into longitudinal trunk rivers such as the Indus, Ganga and Brahmaputra. Indeed, the major north–south rivers are regarded as type examples of antecedent drainage. They have maintained their course, perhaps since the Miocene, while cutting through ranges that are now higher than their source. The Siwaliks have been divided into three units. The Lower Siwalik comprises a coarsening-upward megacycle consisting of sandstone–mudstone alternations passing up into a predominantly sandy sequence. The Middle Siwalik consists mainly of medium to coarse

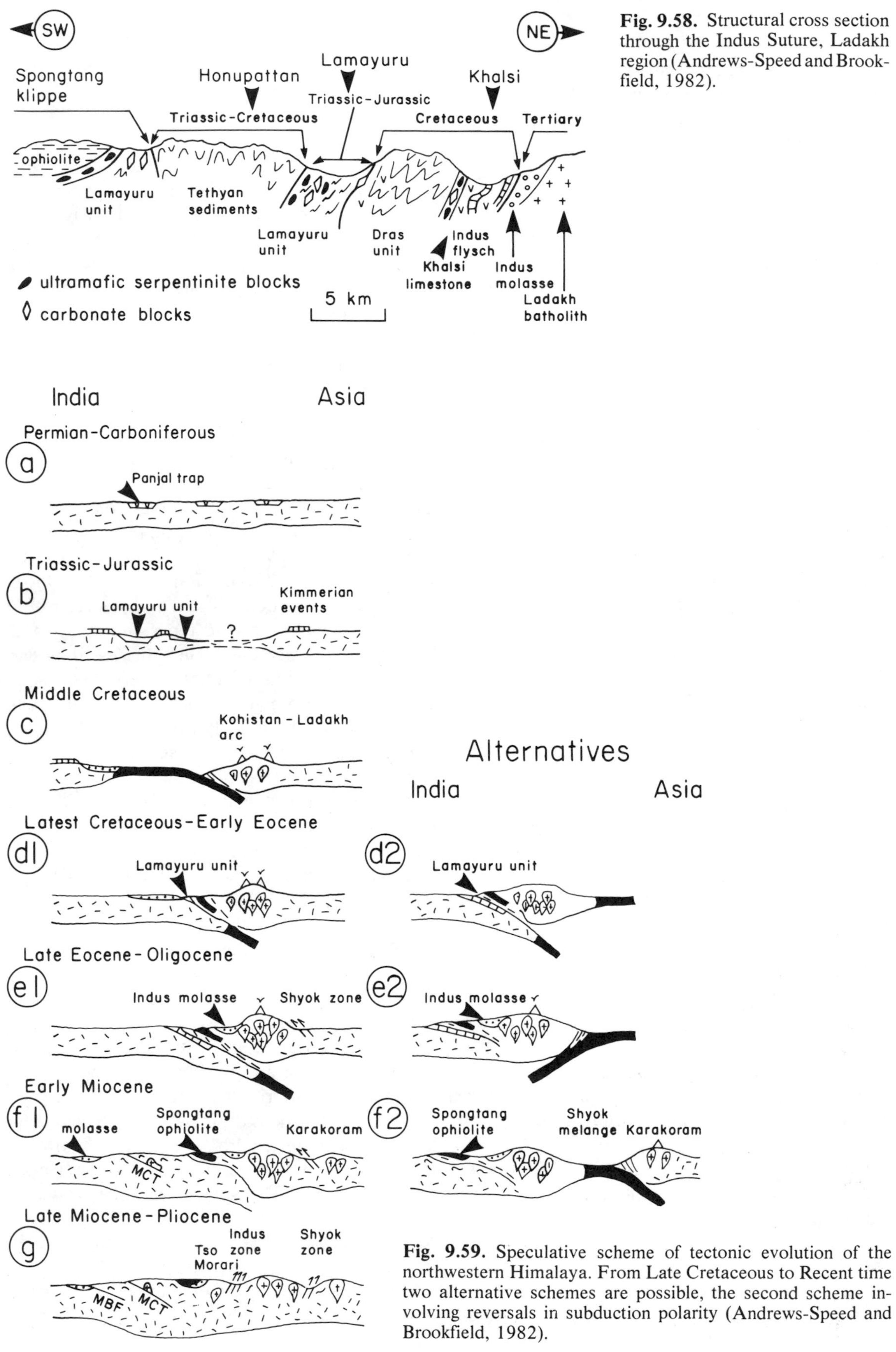

Fig. 9.58. Structural cross section through the Indus Suture, Ladakh region (Andrews-Speed and Brookfield, 1982).

Fig. 9.59. Speculative scheme of tectonic evolution of the northwestern Himalaya. From Late Cretaceous to Recent time two alternative schemes are possible, the second scheme involving reversals in subduction polarity (Andrews-Speed and Brookfield, 1982).

sandstones with interbedded mudstones and it grades up into a largly conglomeratic Upper Siwalik, the Middle and Upper Siwaliks together comprising a second megacycle. Fining-upward cycles 1–35 m thick occur throughout the sequence. The three subdivisions of the Siwaliks are time transgressive, representing southward and southeastward progradation of fluvial depositional systems into and down the axis of the trough. The Siwaliks are folded and faulted near the Main Boundary Fault, but pass out into flat lying beds overlain conformably by modern alluvium. The alluvium itself is tilted near the mountain front, and Landsat images indicte active folds in parts of the trough. Petrographic studies of light and heavy minerals and of the conglomerate clasts document the progressive unroofing of Tethyan rocks that were metamorphosed deep in the collision zone.

Bay of Bengal embayment: Discharge from the delta of the Ganga and Brahmaputra rivers contains the largest sediment load of any river on earth (Stoddart, 1971). It drains into the Bay of Bengal, which functions as an embayment along the Indian–Asian suture (Fig. 9.60). The sediment is deposited on the Bengal Fan, which is probably the world's largest clastic depositional system (the fan is discussed briefly in Chapter 6; see Fig. 6.39). In the Miocene part of the sediment was dispersed southeastward to feed the Nicobar Fan, which extends as far as the Java trench. However, during the Pliocene or Pleistocene northward subduction of the Indian Ocean beneath Sunda arc brought the Ninetyeast Ridge (a volcanic ridge formed along a now extinct leaky transform fault) into contact with the arc, cutting off the sediment supply, and the fan is now inactive (Curray and Moore, 1974).

Effects of collision in Soviet Union and China: Molnar and Tapponnier (1975) showed that the effects of the northward drive of India could be recognized for 3000 km north of the suture. Tibet is being squeezed against Asia, and its response is to slide sideways, generating a series of major sinistral strike slip faults that extend eastward into China. A smaller set of dextral faults trending north–northwest may represent the other pair of a conjugate set. Sengör (1976) and Dewey (1977) predicted that extensional faults and graben would also result from collision of irregular margins and the Baikal Rift zone was probably initiated in this way.

Little is known about the effects, if any, of the strike slip faulting on sedimentation patterns, but the Baikal area has been thoroughly mapped. Logatchev et al. (1978) demonstrated from geophysical data that the zone is underlain by attenuated continental crust, indicating a region of mantle upwelling in response to the regional stress pattern. There are numerous rift valleys in the area, Lake Baikal itself representing onc of the oldest (Eocene), with many others in the process of active formation and initiation (Fig. 9.61). They contain up to 6 km of lacustrine and fluvial sediments. In places the Pleistocene–Holocene section alone, including basalt flows, is up to 400 m thick (Logatchev et al., 1978).

The deformation of central Asia could be classifed as intraplate tectonics, yet it is clearly related to plate margin processes, and demonstrates how far-reaching the effects of these can be. The late Paleozoic Hercynide belt of western Europe is analogous in some respects, showing major strike-slip deformation as a result of the oblique collision of Europe, Africa and intervening microplates (Badham, 1982; Dewey, 1982).

The Caledonian terminal suture: The gradual closure of Iapetus Ocean from Middle Ordovician to Late Devonian time is one of the more well-known Phanerozoic continental collisions. Closure was gradual, in part because of the presence of backarc basins that were the first to collapse (the Taconic orogeny of Quebec is discussed earlier in this section), but in part because subduction and convergence of the opposing continental plates was oblique (Phillips et al., 1976). Terminal collision occurred across northern England in the Late Ordovician, but in southwest Ireland not until the Late Silurian. Convergence continued until the last oceanic "hole", in the Newfoundland area, was closed, probably during the Devonian Acadian orogeny (Williams, 1980). While convergence was continuing, those areas where collision had already occurred continued to deform along several major dextral transcurrent faults.

The resulting Caledonian orogen (Fig. 9.62) is characterized by the famous Devonian Old Red Sandstone, a fluvial and lacustrine unit that occurs in numerous syn- to late-orogenic basins in Britain, Scandinavia, Svalbard, East Greenland and northeast North America (Allen et al., 1968; Friend, 1969, 1981). It is by no means certain how all these basins fit into the plate tectonic

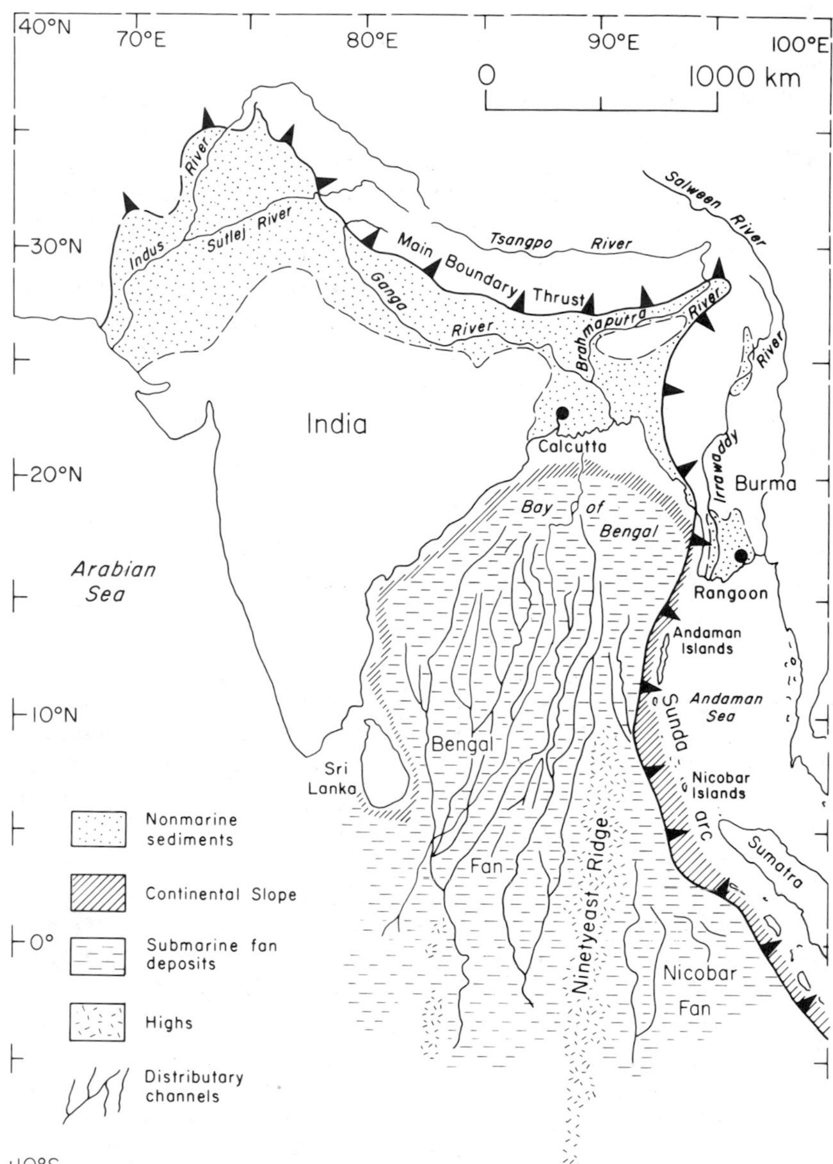

Fig. 9.60. The Ganga–Indus peripheral basin and the Bay of Bengal intra-suture embayment (various sources).

models discussed here, but it does seem likely that many of them were controlled by major strike-slip faults (Friend, 1981). The Hornelen basin in Norway is a classic example of a pull-apart basin, described in several papers by R.J. Steel and his co-workers (eg. Steel, 1976; Steel and Gloppen, 1980). Bluck (1978, 1980) attributed the Devonian basin of the Midland Valley, Scotland, to strike-slip faulting modifying an earlier forearc basin. Other examples are listed by Friend (1981). "External" basins (Allen et al., 1968), namely those with coarse clastic facies passing laterally into marine shelf deposits (e.g. Catskill Delta of New York, Anglo Welsh Cuvette) represent retroarc or peripheral basins.

The Taconic and Acadian suturing episodes represent the closure of a northwestern European plate against eastern North America and Green-

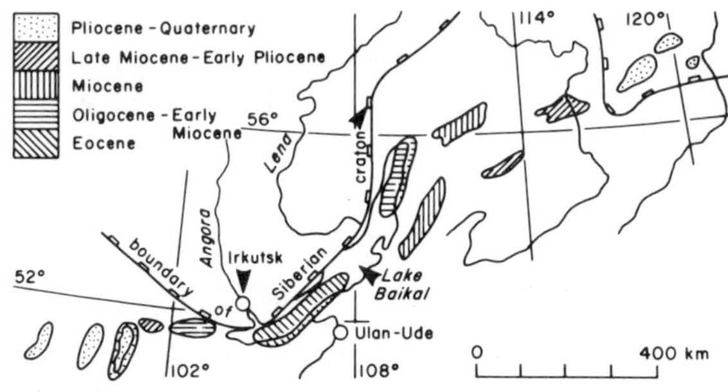

Fig. 9.61. Sequence of rift valley development in the Baikal Rift System (Logatchev et al., 1978).

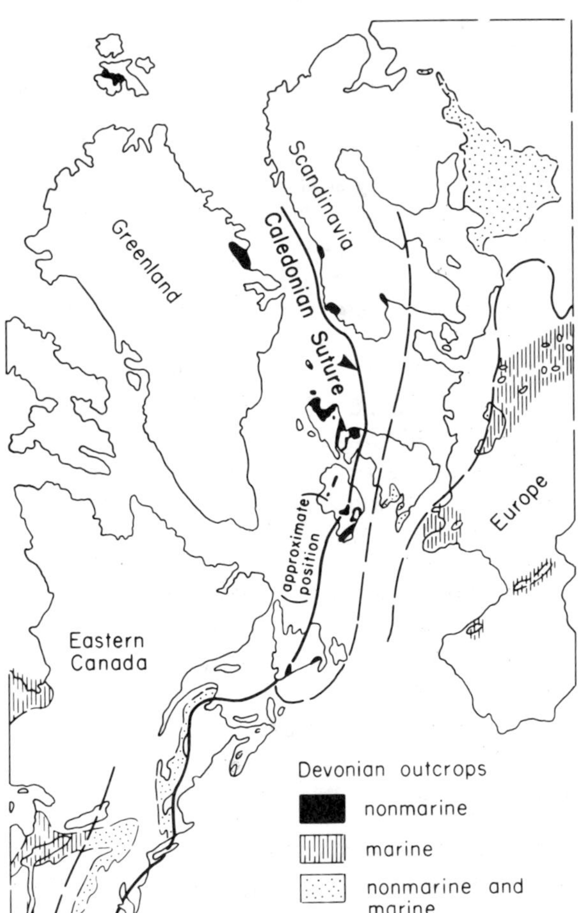

Fig. 9.62. The Caledonian terminal suture and resulting "Old Red Sandstone" basins (after Friend, 1969, and other sources).

land. Paleomagnetic evidence shows that this is only part of a very complex story that extended until Permian time and involved interaction with Africa and several small plates (Kent and Opdyke, 1978; Van der Voo, 1979; Schenk, 1981; Zen, 1981; Cook and Oliver, 1981; Williams and Hatcher, 1982; Dewey, 1982). Interpretations of the Appalachians are rapidly evolving at the time of writing, and it seems advisable to defer further discussion for the present. Graham et al. (1975) used the Cenozoic collision of India with Asia as a model to explain Devonian to Carboniferous clastic wedges derived from the gradual southward zipping up of the Appalachian suture. However, in view of the reevalution of the latter this model needs considerable revision.

9.3.5 Basin analysis of accretionary terranes

Early attempts at plate tectonic reconstructions tended to treat the present continents as major pieces of the puzzle that moved and collided intact, albeit sometimes with ragged tears across earlier sutures (e.g. Wilson's classic analysis of the Atlantic Ocean in 1966). Recently, however, detailed work has demonstrated that many large orogenic areas are composed of unrelated terranes which have become juxtaposed and sutured ("accreted") by complex processes of subduction and transform motion. Clues pointing to such interpretations include the following:

1. Presence of more than one suture assemblage of ophiolites, melange, deformed arc volcanics and metasediments within the orogen
2. Presence of major linear topographic depressions floored by mylonite and possibly representing fossil transform plate junctions
3. Contrasts across a suture of stratigraphic sequence and faunal/floral realm
4. Evidence of laterally offset elements such as plutons, stratigraphic or structural trends, metamorphic isograds, etc.
5. Paleomagnetic evidence of different polar wandering paths for different terranes.

Some terranes may be major microplates floored by Precambrian basement and carrying thick Proterozoic–Phanerozoic sedimentary sequences. Others may consist of arc assemblages

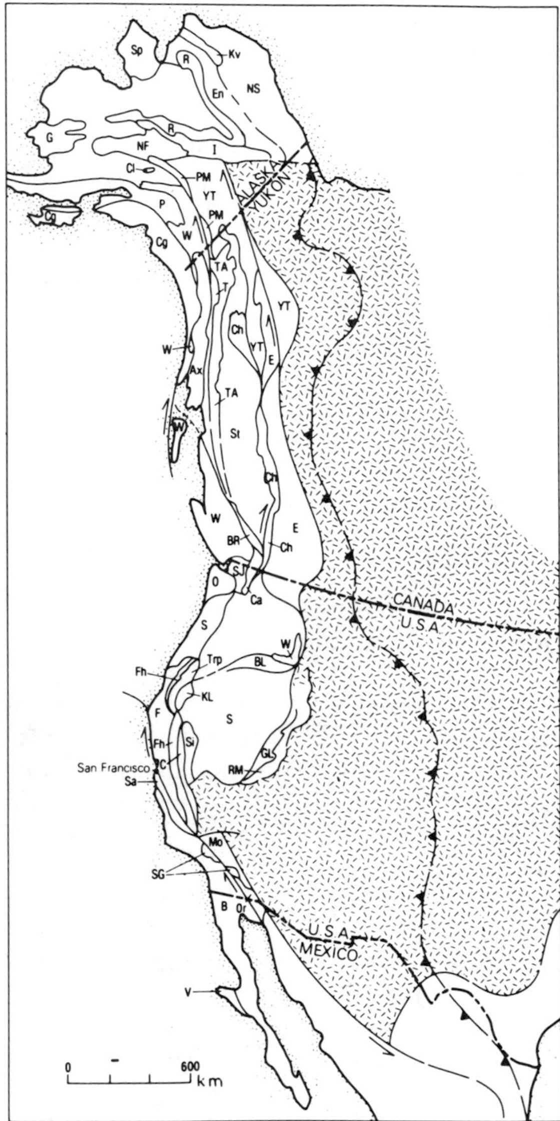

Fig. 9.63. Generalized map of Cordilleran suspect terranes. Dashed pattern is extent of autochthonous North American cratonic basement; thrust fault symbol marks eastern limit of Mesozoic–Cenozoic deformation. Key to terrane symbols is omitted for space reasons but some key terranes are discussed in the text (Coney et al., 1980). Reprinted by permission from Nature, v. 288, pp. 329–333. Copyright © 1980 Macmillan Journals Limited.

and rifted continental fragments without an exposed basement core. Nevertheless, each function as independent continental fragments because they cannot be swallowed up by subduction but accumulate along a suture zone like the candy wrappers and cigarette butts at the top of an escalator.

Geologists have borrowed a term from space exploration, which has advanced concurrently with plate tectonics, and now like to refer to suturing events as "docking". For example, here is a dramatic interpretation of Alaskan geology:

> The long history of successive collisions and terrane accretions around the rim of the northern Pacific suggests that a continental framework existed as a backstop or docking facility for migrating terranes, preventing them from bypassing Alaska and sweeping into the Arctic Basin (Churkin and Trexler, 1981).

Vast areas such as the Cordillera of Canada and Alaska, plus much of Siberia, are composed of accreted terranes or microplates, some of which have travelled hundreds or thousands of kilometres from parts unknown (Churkin and Trexler, 1981; Coney et al., 1980). More than fifty terranes have been identified between Alaska and Mexico, some or many of which were formerly scattered over the paleo-Pacific Ocean and were swept up by the westward advance of North America following the Triassic–Jurassic rifting of this continent away from Africa and Europe (Coney, 1979; Coney et al., 1980).

Similarly Dewey et al. (1973) have shown that the Alpine orogenic belt has been constructed of many small plates welded together between the Jurassic and late Cenozoic.

The British Isles may be a composite of fragments representing four plates: England–Wales–southern Ireland sutured to central and southern Scotland during the Ordovician–Devonian (Phillips et al., 1976); southernmost England sutured on during the Hercynian collision of Europe with Africa (Dewey and Burke, 1973); and northern Scotland emplaced by sinistral transform faulting during the same Carboniferous event (Kent and Opdyke, 1978; Van der Voo and Scotese, 1981). Dewey (1982) has provided an eloquent discussion of this accretionary process.

Basin analysis has a major part to play in the unraveling of microplate accretion because the definition of an exotic terrane depends largely on its stratigraphy and deductions about its original plate setting, source rocks and paleogeographic evolution. Work of this type can also provide clues as to the age and nature of terrane convergence, using the type of analysis discussed in section 9.3.4 and the basin models discussed throughout this chapter. However, as noted else-

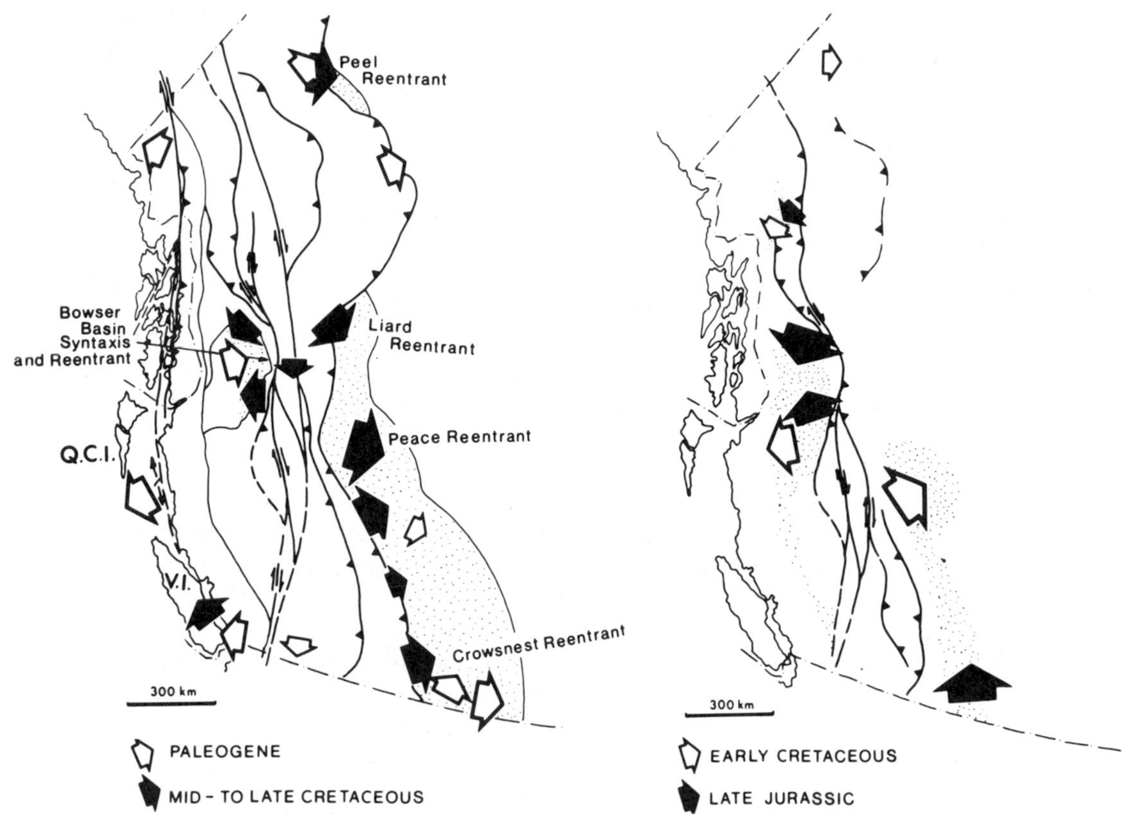

Fig. 9.64. Distribution and timing of major detrital pulses in the Canadian Cordillera (Eisbacher, 1981).

where, a complete analysis of the suturing process calls for the application of a wide range of geological techniques. Paleomagnetic analyses of ancient pole positions and apparent polar wandering curves are essential for any attempt at reconstructing past plate configurations. As this work progresses it generally results in the subdivision of an orogen into a larger number of ever smaller terranes, all with their own independent migration paths and suturing history. The tectonics of suturing are also undergoing a profound re-evaluation as a result of the deep seismic reflection studies of the COCORP team (Oliver, 1982). Both the major North American orogenic belts, the Western Cordillera and Appalachians are undergoing these types of re-interpretation. The history of part of the southern Canadian Cordillera is discussed here as an illustration of basin analysis methods applied to accretionary terranes.

The distribution of the major Cordilleran terranes is shown in Figure 9.63. They range in age from middle Proterozoic to Cenozoic. In general the sutures between the terranes become younger toward the Pacific Ocean, but this is not invariably the case because some terranes were accreted to each other before suturing on to the North American plate.

In Canada the oldest terrane is the composite Eastern Assemblage (*E* in Fig. 9.63), consisting of middle Proterozoic (1.5 Ga) to Middle Devonian sediments, mainly shallow water to nonmarine in origin, overlain by a Mississippian to Triassic chert, argillite, volcaniclastite and mafic volcanic sequence. The Devonian and older rocks have long been regarded as a classic example of a miogeosyncline—the non-volcanic part of a geosyncline—and most such miogeosynclines are now interpreted as divergent margin basins. However Monger and Price (1979) pointed out the length of time represented by the sediments (~1100 Ma) and other complexities such as the presence of coarse lithic arenites and basic volcanics incompatible with a simple divergent margin interpretation. The Eastern Assemblage may represent a composite of more than one rifted margin, or an aborted rift. Much work remains to be done to clarify the problem.

The overlying Mississippian to Triassic sequence is interpreted as the product of backarc spreading and rapid subsidence (Monger and Price, 1979), and records the major change in the tectonic regime of western Canada from a divergent to a convergent environment.

Sutured against the Eastern Assemblage along the Teslin Suture is the Cache Creek or Atlin Terrane (Ch in Fig. 9.63) consisting of Mississippian to Middle or Upper Triassic, highly disrupted radiolarian chert, argillite, basalt and alpine-type ultramafics with local blueschist metamorphism (Coney et al., 1980). Faunas in the sediments are of tropical affinities, similar to those in Tethys and other circum-Pacific regions. This assemblage is a good example of a highly deformed remnant of an ancient ocean floor. It is caught between remnants of Triassic to Jurassic arcs, and the whole complex was accreted to the Cordillera when the Stikine Terrane (St in Fig. 9.63), a major microplate with an exotic Precambrian (?) basement, collided with North America in the Early Jurassic. The kinematics of this convergence are not yet completely understood. Structural evidence in the southern Yukon suggests westerly dipping subduction beneath the Stikine Terrane (Tempelman-Kluit, 1979), but in places there was more than one arc and polarities may have reversed (cf. Indus Suture, section 9.3.4).

The collision climaxed in Middle to Late Jurassic time with regional metamorphism of the Eastern Assemblage, which now began its evolution into the Omineca Crystalline Belt. Uplift followed, producing a major mountain range—the first precursor of the modern Rocky Mountains, and the shedding of floods of detritus both east and west (Eisbacher et al., 1974; Eisbacher, 1981). In Alberta this detritus comprises the first "molasse" pulse of the Rocky Mountains foreland basin (Fig. 9.64) a mainly shallow marine sequence of Upper Jurassic and Lower Cretaceous age (Hamblin and Walker, 1979). Toward the west the sediments comprise the Bowser Lake and Skeena Groups in Bowser Basin (Figs. 9.64, 9.65). This basin had a long and complex history (Eisbacher, 1981). It was initiated on the Stikine microplate following the Jurassic suturing event. Depending on the polarity of the subduction zone along the suture the basin could be interpreted either as a retroarc or peripheral basin at this time. Sedimentary environments ranged from deltaic in the east to basin floor turbidites in the west.

Meanwhile, somewhere out in the Pacific Ocean two other terranes (microplates) were approaching each other, to collide in the Late Jurassic to Early Cretaceous. These are the Wrangellia and Alexander Terranes (W and Ax in Fig. 9.63). Paleomagnetic evidence summarized by Yorath and Chase (1981) shows that both terranes lay several thousand kilometers to the south of their present position during the Triassic. Wrangellia was one of the first allochthonous terranes to be recognized in the Cordillera (Jones et al., 1977). It is represented by five major fragments between the Wrangell Mountains, Alaska, and the Hells Canyon area of Washington–Oregon (Fig. 9.63). Lower Paleozoic to Triassic sequences in each area show marked similarities, but the Triassic rocks have the strongest affinities (Fig. 9.66). A late Paleozoic arc assemblage is overlain by up to 6000 m of tholeiitic basalts representing what must have been a major episode of submarine to subaerial volcanism in the Triassic. In each of the five areas these basalts are then overlain by platform to basinal carbonate sediments containing distinctive invertebrate faunas.

Wrangellia sutured with the Alexander Terrane along a northeast-dipping subduction zone (Fig. 9.67), before its collision with the North American plate in the Early to mid Cretaceous (Coney et al., 1980; Yorath and Chase, 1981). Since then the entire area has been fragmented by dextral strike-slip faulting during oblique subduction of the Pacific and Farallon plates.

Yorath and Chase (1981) examined this sequence of events in some detail in the area of Queen Charlotte Islands, British Columbia, using gravity, magnetic and seismic data, structural analysis, stratigraphy and sedimentology. The suturing of the Wrangellia and Alexander Terranes was accompanied by intrusion of diorite plutons. Subsequent uplift led to the deposition in a forearc basin of conglomeratic detritus containing clasts of the diorite plus andesitic volcanics, both probably derived from the arc along which subduction and collision occurred (Fig. 9.67). The detritus was dispersed by longitudinal flow

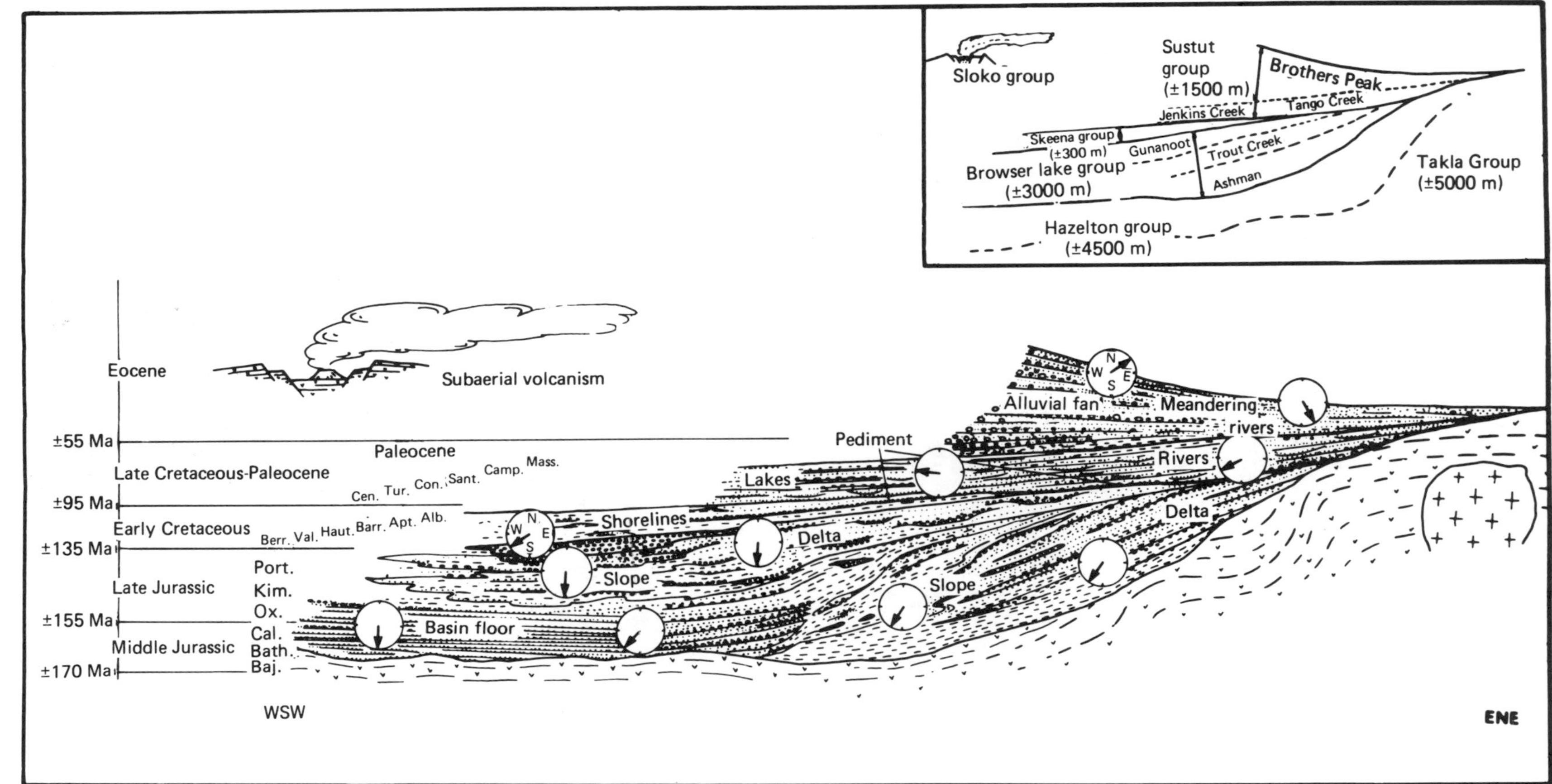

Fig. 9.65. Composite stratigraphic cross section of Bowser Basin, showing major depositional systems and transport directions (Eisbacher, 1981).

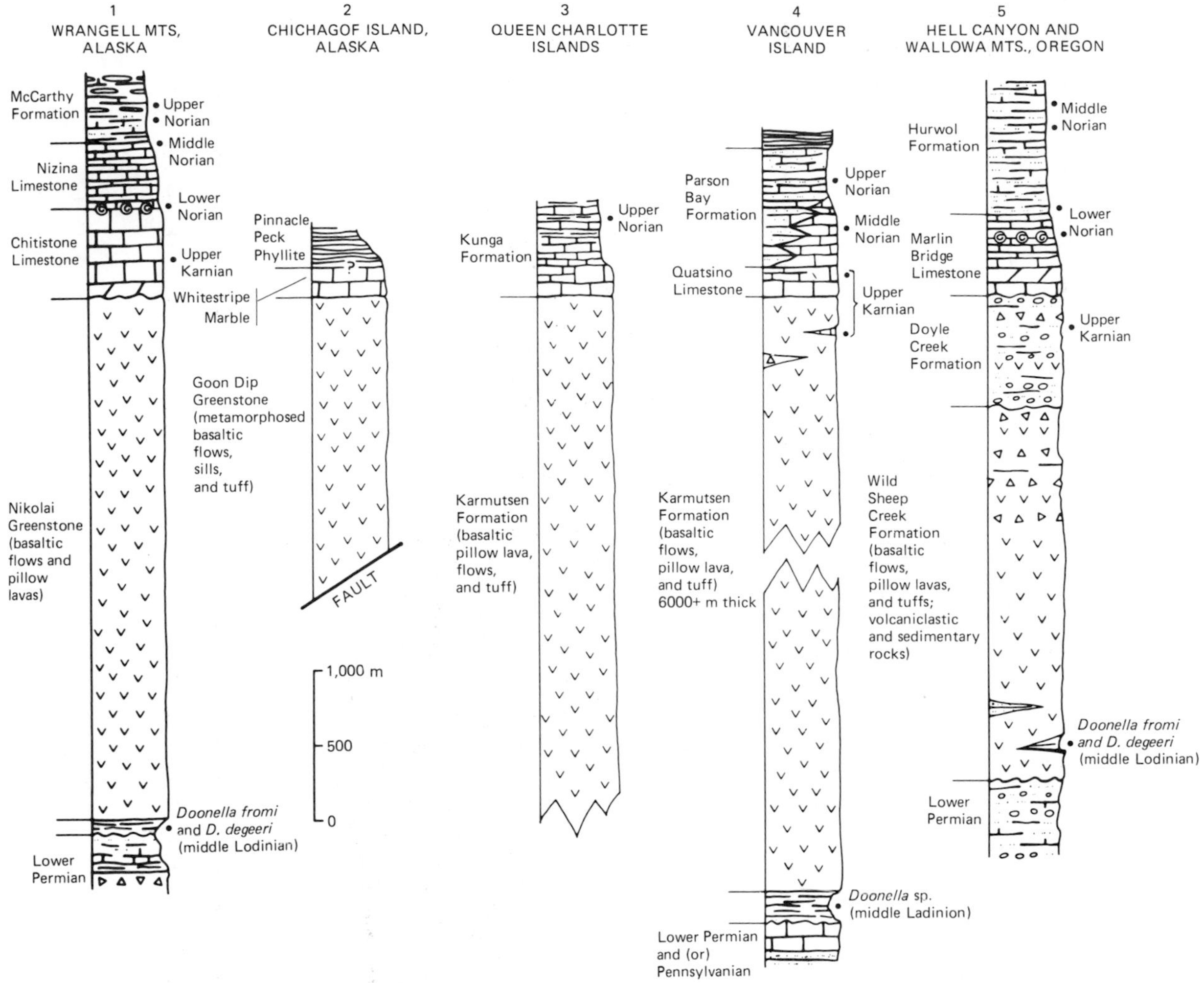

Fig. 9.66. Representative Triassic sections in each of the five fragments of the Wrangellia terrane (shown as W in Fig. 9.63). The stratigraphic similarities between these sections were the main basis for proposing their original combination into a single microplate (Jones et al., 1977).

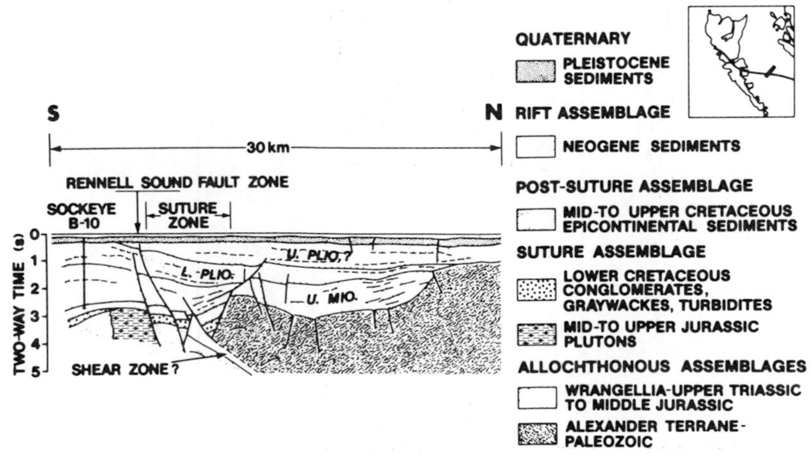

Fig. 9.67. Interpreted seismic reflection profile across the Wrangellia–Alexander suture, east of Queen Charlotte Island (Yorath and Chase, 1981).

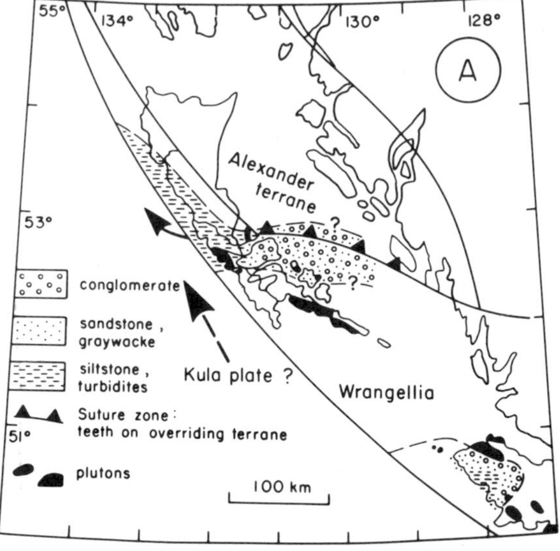

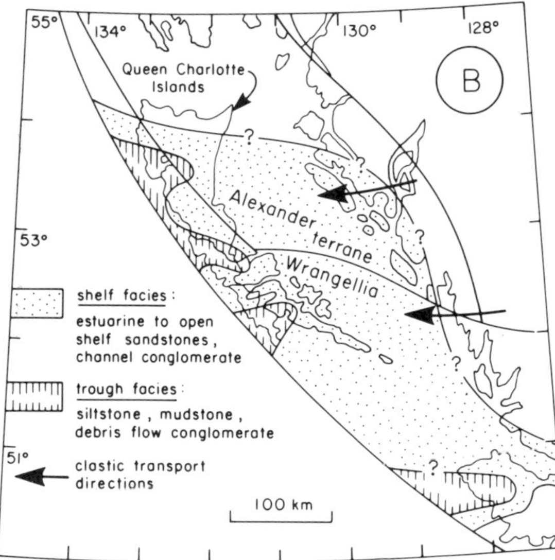

Fig. 9.68. A. Paleogeography of Queen Charlotte Island area during terrane collision and deposition of the "suture assemblage", Early Cretaceous; **B** Paleogeography of "post-suture assemblage", during collision of combined Alexander–Wrangellia block with the mainland in the mid to Late Cretaceous (Yorath and Chase, 1981).

and the conglomerate passes westward into a distal siltstone and turbidite assemblage (Fig. 9.68A).

The collision of Wrangellia–Alexander with the North American plate, beginning in the Early or mid Cretaceous was the start of the most important phase of tectonic deformation, metamorphism and igneous activity in the Cordillera, the Laramide Orogeny, which peaked in the Late Cretaceous and Paleogene. The Coast Plutonic Complex is a magmatic arc that developed at this time, intruding the allochthonous Taku and Tracy Arm terranes (T and TA in Fig. 9.63). Convergence of the Pacific plate probably was oblique, and several major dextral strike-slip faults in the Cordilleras may have been initiated in this phase.

Many important detrital sedimentary sequences formed during the Laramide Orogeny. Some are difficult to classify in terms of the simple basin models discussed earlier in this chapter. The Wrangell–Alexander suture began to subside and was covered by the mid to Upper Cretaceous "post-suture assemblage" of Yorath and Chase (1981). This sequence of mainly shallow marine clastics was derived from the east (Fig. 9.68B). Its paleogeographic situation suggests that this post-suture subsidence should be classified as a peripheral or foredeep basin in front of the Coast Range magmatic arc.

Bowser Basin was deformed and uplifted during the Laramide Orogeny. Active sedimentation was confined to the eastern part of the basin where detritus from the Coast Plutonic complex and older Bowser Basin sediments was deposited in fluvial environments (Brothers Peak Formation of Fig. 9.65). Although both the Bowser and Alberta basins were behind the arc, it is common practice to refer to the Alberta Basin as the foreland (retroarc) basin of the Laramide Orogeny. The Late Cretaceous and Paleogene sediments there comprise the second major "molasse" cycle in the basin (Fig. 9.64).

The Cordillera are cut by numerous dextral strike-slip faults, many of which probably were initiated in the Neogene. They record a change in plate tectonic regime to oblique subduction and transform plate boundaries, just as did the development of the San Andreas system in California. The thick Neogene "rift assemblage" east of Queen Charlotte Islands (Fig. 9.67) includes basic volcanics and pyroclastics overlain by a nonmarine clastic sequence more than 4.5 km thick (Yorath and Chase, 1981), the origin of which may be related to movement on these faults.

Eisbacher (1981) pointed out that restoration of dextral strike-slip movement in the eastern Cordillera places the sediment sources for the first "molasse" pulses in Alberta and the Bowser

Basin (Late Jurassic–Early Cretaceous) more nearly opposite each other (this can be visualized from Fig. 9.64). This is an interesting point as it suggests that the ancestral Rocky Mountains were a much more localized mountain range than the present eastern ranges and owed their origin entirely to the docking event of the Stikine microplate with North America.

This account is necessarily highly simplified, and it ignores the chronology of terrane accretion in Alaska and areas south of the United States border. In fact the story is by no means complete, and large teams of Canadian and American geologists are grappling with the many problems that the terrane model has raised—and probably will be doing so for many years to come.

Space does not permit detailed discussion of other accretionary terranes, but brief mention should be made here of two types of depositional system that may be expected to occur (though not exclusively) in a microplate environment, where plate movements may isolate small areas of ocean or a backarc sea from normal marine circulation patterns. These are euxinic seas giving rise to black, organic and sulfurous muds, and saline seas that may totally desiccate to deposit thick evaporites. Good examples of both occur in the Alpine region. The modern Black Sea is a good example of an euxinic environment (Degens and Ross, 1974). The well-known Messinian (Miocene) evaporites of the Mediterranean exemplify the second style of sedimentation (Hsü et al., 1973; Biju-Duval et al., 1978; see also sections 4.6.6 and 6.6).

9.3.6 Cratonic basins

Thermal fractionation of the mantle to produce continental crust took place relatively rapidly during the Archean because of the greater production of radioactive heat in the early phases of earth evolution. The result was that by the end of the Archean (~2.5 Ga) 50–65% of the present continental mass had been generated (Windley, 1977). This Archean crust now forms the core of a series of shield complexes, around which Proterozoic and Phanerozoic crust has been accreted. Subduction and suturing generate extensive plutonism and regional metamorphism which result in the gradual thickening and stabilization or "cratonization" of the crust. This is the main reason why continents are able to function as rigid plates, transmitting the stresses of a suturing event hundreds or thousands of kilometers beyond the collision zone (e.g. Cenozoic strike-slip faults of Tibet: Molnar and Tapponnier, 1975).

Up to this point we have been discussing what happens at the margins of these plates as they drift about on the earth's surface. However, plate tectonics theory is at present much less helpful in explaining what happens in plate interiors. There a variety of arches, ridges, domes, swells, plateaus and anticlises occur, separating from each other a diversity of basins, troughs, depressions, downwarps, graben, aulacogens and syneclises. Significant thicknesses of sediment have accumulated in some of the latter: more than 3 km of Ordovician to Jurassic section in the Michigan Basin (Sleep and Sloss, 1980), and more than 4 km of Upper Proterozoic to Cenozoic sediment in many of the basins on the Russian Platform (Aleinikov et al., 1980).

Some of these structural features are very long-lived. For example the North American craton is underlain by major depressions, such as the Hudson Bay, Williston, Michigan and Illinois Basins, and by upwarps such as the Peace River and Transcontinental Arches that have been intermittently active throughout much of the Phanerozoic. A detailed analysis of the Russian Platform summarized by Aleinikov et al. (1980) showed that areas of depression and upwarp shifted with time, so that some depressions became broadened and deepened while others were inverted into uplifts (Fig. 9.69).

Undoubtedly some cratonic basins can be explained in terms of plate margin processes. In particular, aulacogens are discussed in section 9.3.1 in terms of "failed" arms of triple-point spreading junctions, and divergent margins develop rim basins that extend for several hundred kilometers inland from the rifted margin arch (Veevers, 1981). Veevers (1980) demonstrated that much of the history of the Australian platform could be explained in this way. Plate tectonic events such as the assembly and rifting of Gondwana, and late Paleozoic convergent tectonics of the Tasman Fold Belt had widespread effects on the evolution of the Australian continental interior. Likewise the North Sea Basin owes its Jurassic to Cenozoic evolution mainly to failed-arm development during Atlantic sea-floor spreading. Whether this type of analysis can be extended to the complex

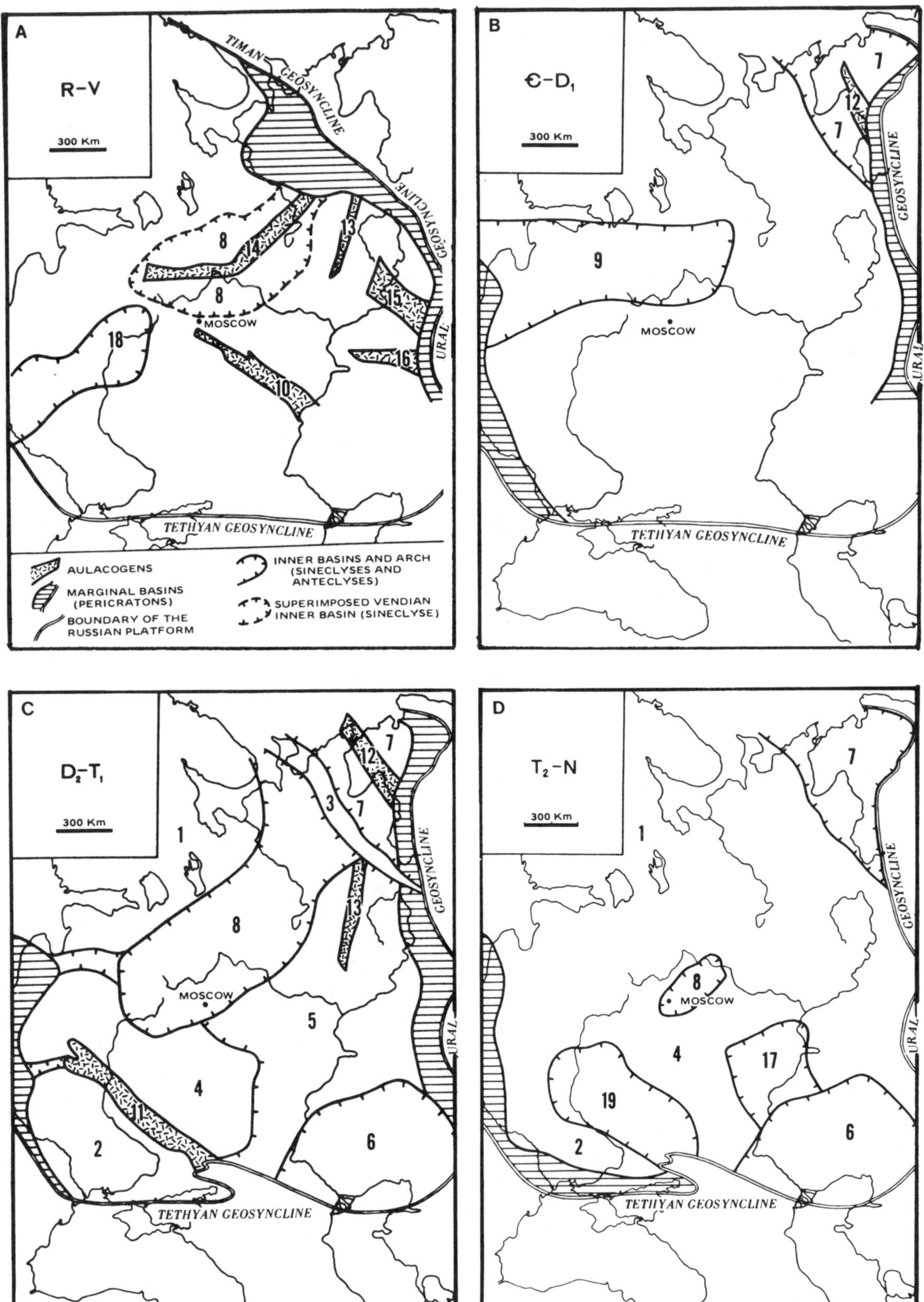

Fig. 9.69. Evolution of basins in the Russian Platform, **A** Late Proterozoic, **B** early Paleozoic, **C** middle and late Paleozoic, **D** Mesozoic and Cenozoic (Aleinikov et al., 1980).

sequence of events recognized in the Russian Platform (Fig. 9.69) is not yet known. The word aulacogen was derived from Russian geology but it had no plate tectonic meaning in its original useage. No attempt has yet been made to interpret the geology of the Russian Platform in plate tectonic terms, but it may well be that the aulacogens there are in fact related to rifted margins that subsequently closed to form the Urals and Alpine Fold Belts.

Bally and Snelson (1980) suggested that one class of cratonic basins may represent subsidence of earlier backarc basins, and they proposed the West Siberian Basin as a type example. This contains a mainly Mesozoic fill overlying Triassic rifts and small late Hercynian (Carboniferous–Permian) basins. The backarc basin was active during convergent tectonics of the Urals Fold Belt.

There remains a large number of basins and upwarps that cannot be explained in terms of contemporaneous plate margin processes. What caused the development of the circular to ovate basins of the North American craton? Why did they subside so steadily that sedimentation and subsidence were in approximate balance and most of the sediments are shallow marine in origin? Why is part of the craton covered by Phanerozoic platform sediments and part uplifted to expose Precambrian basement in the Canadian Shield? Bally and Snelson (1980) and Bally (1980) noted that most cratonic basins are underlain by an ancient rift system, implying that failure of the rift may cause cratonic subsidence at a much later period. This is a tempting idea. For example the Michigan Basin overlies a linear positive Bouguer anomaly that Fowler and Kuenzi (1978) interpreted as a reflection of proto-oceanic basaltic crust along a Keweenawan (1100–1200 Ma) rift. They suggested that the Michigan Basin developed over the thinned and subsided crust along this rift. Continued subsidence during the Late Cambrian to Jurassic would have been caused partly by the weight of sediment overburden. Bronner et al. (1980) described the Taoudeni Basin of West Africa, consisting of more than 3 km of Upper Proterozoic to Carboniferous sediment overlying an Archean basement. They noted that the deepest part of the basin coincides with a belt of Archean ferruginous gneisses and a strong positive gravity anomaly. The origin of the gneisses is not explained, but it is suggested that their high density could have caused increased subsidence.

Conversely Williston Basin lies across a possible Proterozoic suture that Camfield and Gough (1977) traced into the subsurface by geophysical methods. To the north this anomaly lines up with the Wollaston Fold Belt in the Churchill Province of the Canadian Shield (Donaldson et al., 1976). The Transcontinental Arch lies close to another Keweenawan rift, the Central North American Rift of Iowa, Nebraska and Kansas (Van Schmus and Bickford, 1981).

Even if these alignments are not coincidental they do not fully explain the origin of cratonic basins. Most of the various hypotheses (reviewed by Bally and Snelson, 1980) involve cooling and subsidence following a thermal event, or intrusion of dense material deep in the mantle. However, evidence for such events is sparse.

McGetchin et al. (1980) reviewed the style and mechanism of plateau uplifts and listed fourteen different mechanisms that have been proposed to explain such features as the late Cenozoic uplift of the Colorado Plateau. Thermal events, mantle phase changes, mantle hotspots, shallow subduction and other proposals have been made, none of which seem capable yet of being integrated into a unified theory compatible with plate tectonics.

Burke (1976b) suggested that plates may become stationary relative to the mantle, at which time they become differentiated into high areas overlying mantle plumes ("hot spots") and relatively depressed areas between. Volcanism occurs in the elevated areas and sedimentation in the depressed regions, initiating basin subsidence. These basins may survive following decay of the mantle plumes (or migration past them) whereas evidence of volcanism on the elevated regions would have a low preservation potential. Burke (1976b) suggested that this is what happened to Africa about 25 ± 5 Ma ago, and that the Chad Basin is one result. Bond (1978) demonstrated widespread epeirogenic uplift of Africa in the late Tertiary, which is consistent with the idea.

Sedimentation patterns in cratonic platforms and basins are dominated by broad, shallow depositional systems and are markedly affected by the patterns of regional and global cyclicity discussed in Chapter 8. The evolution of the Silurian reef and evaporite fill of Michigan Basin (Fig. 8.22, 8.23) and mesothemic cyclicity in the Cretaceous of the Western Interior (Fig. 8.29 to

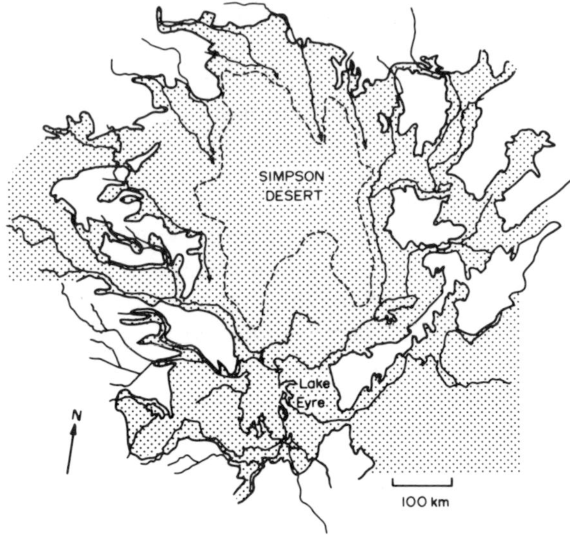

Fig. 9.70. The modern Lake Eyre Basin, a cratonic basin filling with fluvial and eolian sediments (Miall, 1981b) Reproduced with permission from the Geological Association of Canada (1981).

8.32) are excellent examples of this, but the most spectacular examples are the Carboniferous coal-bearing cyclothems, which retain internal evidence of at least two time scales of cyclic eustatic sea level change (Fig. 8.24 to 8.28).

Inland basins may be characterized by broad alluvial plains with centripetal drainage, such as Lake Eyre Basin in Australia (Fig. 9.70) and Lake Chad, Africa. Dispersal directions are sensitive to epeirogenic tilting in response to plate interior and plate margin tectonism (e.g. Wopfner et al., 1974; Doutch, 1976; Ward and McDonald, 1979). Within large continental masses in equatorial regions eolian deserts are to be expected, such as the modern Sahara (Fig. 6.8), Arabian and Simpson deserts, and the Carboniferous to Jurassic deserts of the western United States (Fig. 6.9).

Both marine and nonmarine basins may be characterized by a layer-cake type stratigraphy, with individual lithosomes sometimes traceable for thousands of kilometers. The Lower Paleozoic platform carbonate cover of North America is a

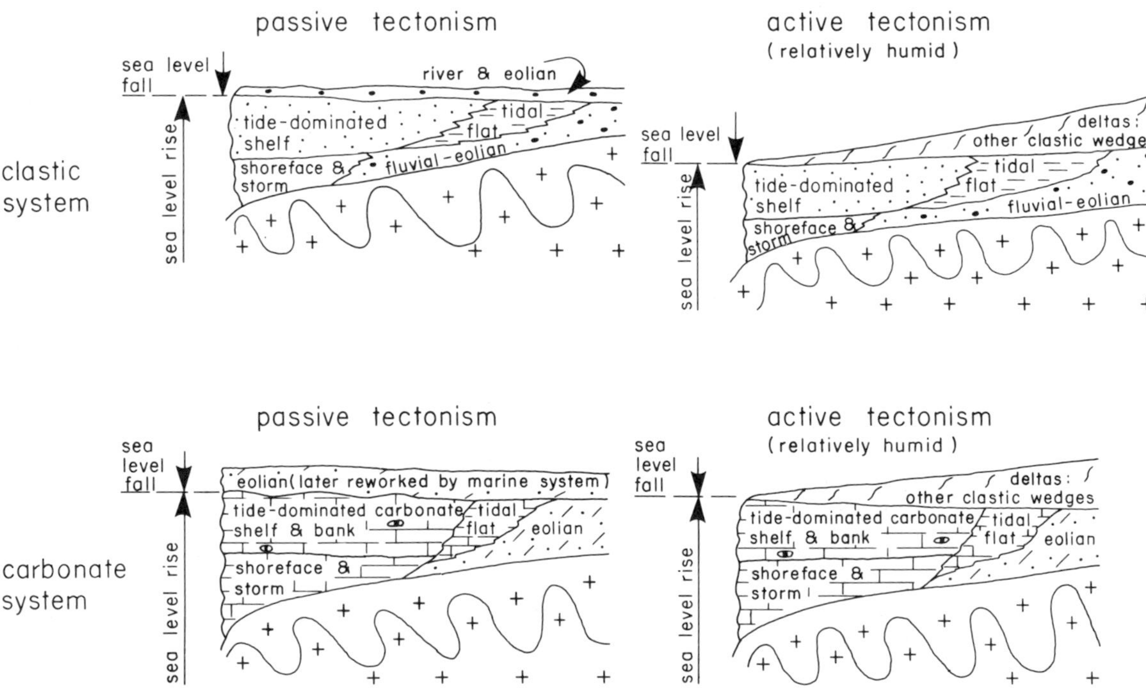

Fig. 9.71. Hypothetical stratigraphic configuration on a craton generated by transgression and regression. Wave-dominated sediments are overlain by tide-dominated facies as shelf-width widens. During regression sediment supply is controlled by rate of marginal uplift; major deltas may grow given adequate sediment supply (Klein, 1982).

good example, as discussed in Chapter 6 (Figs. 6.25, 6.26).

Klein (1982) developed a generalized model for cratonic sedimentation that focuses on the evolution of depositional systems during long term transgressions and regressions. The model could be applied to any of the time scales of cyclicity discussed in Chapter 8. Klein's main thesis is that the balance between wave dominance and tide dominance on continental shelves depends primarily on shelf width. With increased width, wave power (including that of storm waves) becomes attenuated whereas tidal range is amplified. Depositional systems therefore tend to evolve from marginal wave-dominated depositional systems during low sea level stages to broad tidal systems during later stages of marine transgression (Fig. 9.71; also see additional discussion in section 4.6.7). Regressive phases may be characterized by either fluvial, eolian or deltaic depositional systems. The model applies to both clastic and carbonate settings, the differentiation of which depends mainly on climatic and tectonic controls of sediment supply. Klein (1982) quoted numerous examples from the North American Interior, particularly the Illinois Basin, where this sequence of events has been at least partially documented.

9.4 Clastic petrofacies

9.4.1 Introduction

It has long been recognized that the detrital composition of clastic rocks is significantly related to the tectonic setting of their source area. Krynine (1942) interpreted mineral composition in terms of the geosynclinal cycle, which supposedly affected rates of source area uplift and basin subsidence and controlled the geological composition of source terranes. Middleton (1960) reached similar conclusions using chemical composition of sandstones. The geosyncline theory has now been superseded by plate tectonics, but many of Krynine's and Middleton's conclusions have been adapted to plate theory.

Some early attempts to reinterpret detrital mineral and chemical composition in plate tectonics terms were made by Dickinson (1970), Schwab (1971, 1975), Dickinson and Rich (1972) and Crook (1974). The most recent syntheses have been published by Potter (1978), Dickinson and Suczek (1979) and Valloni and Maynard (1981). A brief discussion of modern ideas on the tectonic control of mineral composition is presented here. It should be pointed out that composition is also controlled by transport history (Suttner, 1974; Franzinelli and Potter, 1983), by depositional environment (Davies and Ethridge, 1975; see section 4.5.2) and by paleoclimate (Basu, 1976; Suttner et al., 1981). Also Schwab (1978) demonstrated gross changes in the chemical composition of sediments through time. No attempt has yet been made to integrate the effects of these four controls, although Potter (1978) discussed the interrelationship of climate and tectonics. Dickinson and Suczek (1979) did not deal with environmental, climatic or time variables; and Valloni and Maynard (1981) restricted their discussion to modern deep-sea sands. Ingersoll (1978b) carried out careful statistical tests of the relationship between composition and depositional facies in the Upper Cretaceous Great Valley Sequence of California and showed that in this case the relationship was weak, and was far overshadowed by the control imposed by changes in source area geology. It may be that this is a general rule, but far more study of the problem is required.

The value of petrofacies studies is that they may help elucidate tectonic setting for ancient detrital sequences, particularly those far removed from their original setting by some vagary of the Wilson Cycle. Also the appearance of a particular mineral assemblage in a stratigraphic succession may indicate an important tectonic event such as uplift and erosion of an arc plutonic suite or an ophiolite assemblage along a suture, and this may help date the time of intrusion or of continental collision (e.g. see Ingersoll, 1978b; Eisbacher, 1981; Schwab, 1981. Examples are discussed below). This is an extension of the petrographic mapping technique discussed in section 5.5 in that not only can the data be used to determine source area and paleocurrent directions, but a plate tectonic interpretation can now be made from the same data.

9.4.2 Sandstone composition and plate tectonic setting

Dickinson and Suczek (1979) and Valloni and Maynard (1981) found that the following modal quantities were useful in diagnosing tectonic environment (Dickinson, 1970, discussed the practical details of grain recognition and classification):

- Q: monocrystalline and polycrystalline quartz grains, including chert (Q = Qm + Qp);
- Qm: monocrystalline quartz;
- Qp: polycrystalline quartz, including chert;
- F: monocrystalline feldspar (F = P + K);
- P: plagioclase;
- K: K-feldspar;
- L: unstable polycrystalline lithic fragments (L = Lv + Ls);
- Lv: volcanic and metavolcanic fragments;
- Ls: sedimentary and metasedimentary fragments;
- Lt: L + Qp.

Calcareous grains, ferromagnesian minerals and heavy minerals are not used in this scheme, and care must be taken to distinguish recrystallized detrital matrix (orthomatrix), diagenetic pore-filling matrix (epimatrix) and squashed lithic grains (pseudomatrix), not always an easy matter in immature lithic sandstones (Dickinson, 1970).

Dickinson and Suczek (1979) distinguished nine provenance types, based on tectonic setting:

1. Continental
   - craton interior
   - transitional
   - uplifted basement
2. magmatic arc
   - dissected
   - transitional
   - undissected
3. recyled orogen
   - subducted complex
   - collision orogen
   - foreland uplift

They compiled data from 88 suites of samples representing all these provenance types and ranging in age from Cambrian to modern. Four ternary diagrams were then used to demonstrate the distinctiveness of the detritus derived from the nine provenance types (Fig. 9.72). The QFL and QmFLt plots both show the entire grain population, but differ in the way in which Qp components (primarily chert) are plotted. The QpLvLs and QmPK plots show partial grain populations but help discriminate some of the provenance types, mainly on the basis of the details of lithic and feldspathic composition.

Using all four plots the provenance types are seen to yield quite distinctive detrital suites. The data support many of the ideas of Krynine and the other early workers, once allowance is made for the differences between geosycline theory and modern plate tectonics.

Note that it is not necessary here to use any formal sandstone classification and, indeed, many of the provenance fields straddle the boundaries between some of the classic sandstone types so that the use of such terms as lithic wacke or arkose is not very helpful.

Continental provenances are distinguished mainly on the basis of the presence of abundant quartz and a paucity of lithic fragments. These are rather mature sediments and reflect derivation from crystalline Precambrian shield complexes and overlying platform sediments. The latter are typically polycyclic, and may contain more than 95% monocrystalline quartz. Lithic fragments may be derived from ancient arcs or metamorphic core complexes, the roots of which commonly are exposed in shield areas. K-feldspar is the most resistant of the feldspars to weathering. Its presence reflects ultimate derivation from silicic (granitic) igneous sources, although it can, like quartz, be recycled. The decrease in maturity in continentally derived sandstones as the provenance area becomes uplifted simply reflects the stripping away of the mature platform cover and the exposure of a variety of basement rocks. Sandstones derived from basement uplifts are the most feldspathic of any collected by Dickinson and Suczek (1979). This confirms a long held impression that "arkoses" are typically local deposits related to block faulting or residual deposits above granitic basement (Pettijohn et al., 1972; Potter, 1978; Schwab, 1981).

Arc-derived sandstones are typically immature, in that they are rich in lithic fragments. Undissected arcs yield volcanic detritus containing abundant fine-grained volcanic fragments (Lv) and they tend to be rich in plagioclase. Dissected arcs expose the plutonic core and, in the case of continental (Andean-type) arcs, the local basement. Detrital composition therefore approaches that of sediment derived from uplifted continental basement, although the greater abundance of feldspar, particularly plagioclase, and volcanic fragments remains distinctive.

Recycled orogen sediments are distinguished by an abundance of quartz and sedimentary–metasedimentary lithic fragments. Subduction complexes consist of metasediments and ophiolites, and therefore yield abundant fragments of chert, argillite and lithic arenite. The presence of abundant chert shows up as a marked shift in position of the data points between the QFL and QmFLt plots in Figure 9.72. Suture zones and foreland uplifts expose thrust sheets and nappes of continental margin sediments, together with deformed ophiolites and arc rocks (eg. Figs. 9.1, 9.58). The abundance of sedimentary and metasedimentary rock fragments is therefore to be expected. High Qm contents may reflect derivation from deformed mature platform sediments, whereas high Qp (chert) contents would indicate that the suture zone itself is a major source.

No characteristic detrital mineral assemblages have been proposed for basins in strike-slip fault regimes. It seems intuitively unlikely that they could ever be categorized because they are by their very nature disruptive, juxtaposing source area terranes of all kinds.

Care should be taken to differentiate between the tectonic setting of the provenance area and that of the basin. In some cases a sedimentary basin may not receive the greater part of its detritus from a genetically related terrane. Thus Valloni and Maynard (1981) classified their deep-sea sand samples according to the tectonic setting of the basin itself, and although most of their results confirm the findings of Dickinson and Suczek (1979) there are a few surprises. For example modern offshore basins of Argentina classify as divergent margin basins and would normally be expected to contain mature sands derived from a continental source. However, they proved to be quartz-poor and rich in lithic fragments, falling in the arc field of the QFL plot (Fig. 9.72). This reflects transport by contour currents from the Scotia arc to the south, rather than offshore transport from the adjacent continental margin. Potter (1978) reported on a statistical study of the relationship between tectonic setting of the mouths of major modern rivers and petrographic composition of sands collected therefrom. In general the results support those discussed here but, surprisingly, the Ganga, Brahmaputra and Indus sands classified as "trailing coast" rather than "collision coast" suites. These local anomalies should serve as a warning against too facile an interpretation of petrographic data.

9.4.3 Examples of petrofacies interpretation

Dickinson and Rich (1972) and Ingersoll (1978b) compiled what is probably the most detailed petrographic study of an arc-derived sequence yet attempted. They investigated the Upper Jurassic and Cretaceous Great Valley Sequence of California. In the earlier study Dickinson and Rich (1972) erected six petrofacies which they found to be locally reliable as chronostratigraphic marker units. They attempted to relate variations between volcaniclastic and more quartzofeldspathic petrofacies to magmatic events in the Sierra Nevada source terrane, the sandstones becoming more mature following the temporary cessation of intrusion and volcanism, and consequent dissection of the arc. Ingersoll (1978b) challenged some of these conclusions on the basis of revised radiometric age scales for the igneous rocks and the sediments. However, he was able to show a north–south variation in petrofacies, along the strike of the forearc basin. This can be related to the way in which the Mesozoic arc truncates a late Paleozoic continental margin. Northern segments of the arc and forearc lay across old oceanic crust, whereas the southern part of the arc was intruded into continental crust. Sediments in the north are therefore slightly more lithic, reflecting a greater abundance of andesitic volcanism, whereas in the south sediment sources consisted of felsic volcanics and metamorphic basement rocks. Intermediate plutonic sources were present throughout the arc.

Schwab (1981) sampled sandstones of Carboniferous to Tertiary age across the western French–Italian Alps (Valence to Turin) and related the detrital assemblages to tectonic evolution from a Tethyan rifted margin to a collisional suture (Figs. 9.73, 9.74). Unlike Dickinson and Suczek (1979), Schwab included carbonate sedimentary fragments in his study; they were assigned to the lithic category. His Suite 1 is a typical feldspathic assemblage derived from local fault blocks; Suite 2 is a mature cratonic suite deposited during a period of tectonic quiescence. Suite 3 was derived from the Brianconnais, which

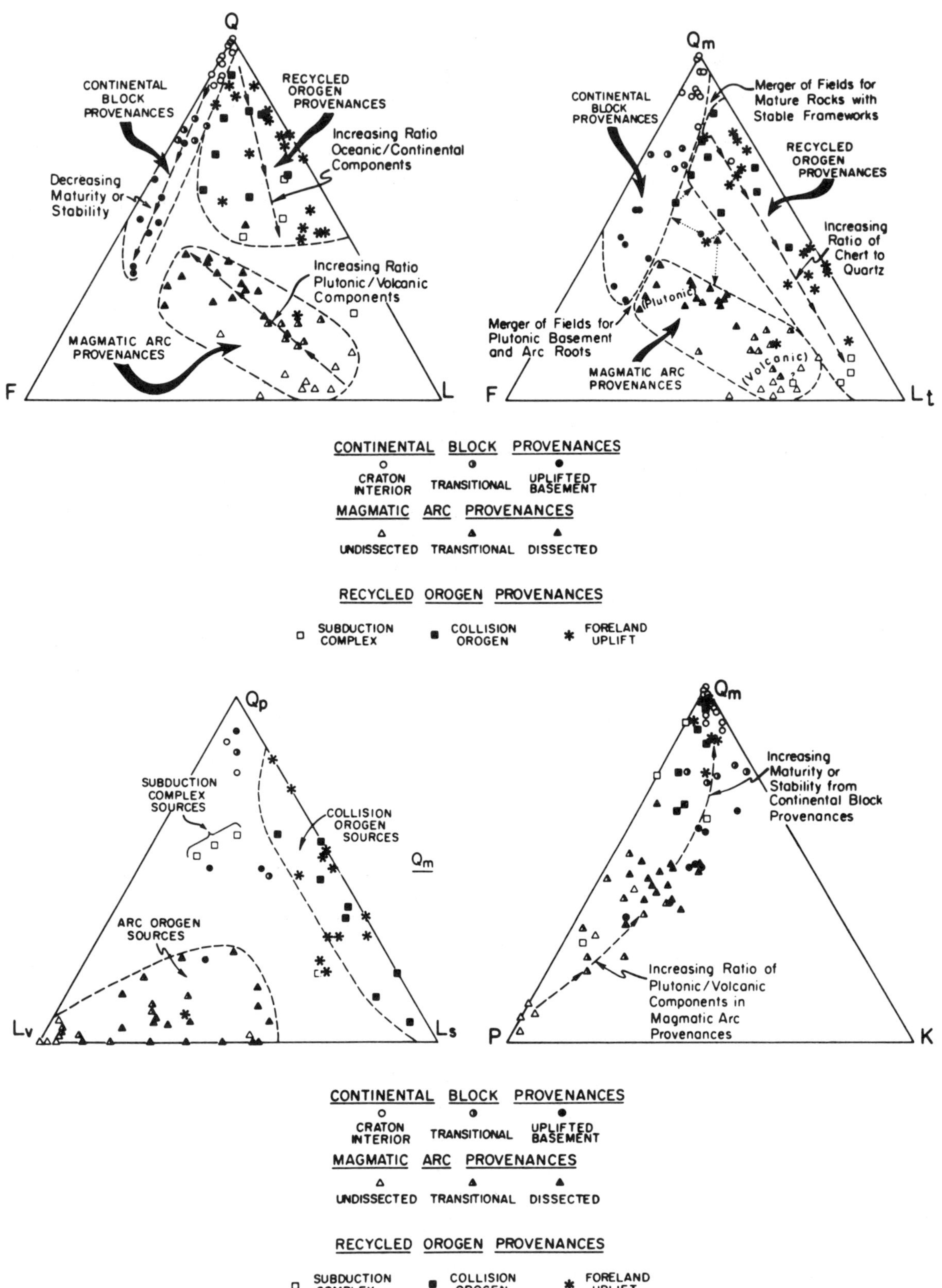

Fig. 9.72. Ternary plots of detrital sandstone composition of 88 sample suites from various tectonic settings. Carbonate grains and heavy minerals omitted. See text for explanation (Dickinson and Suczek, 1979).

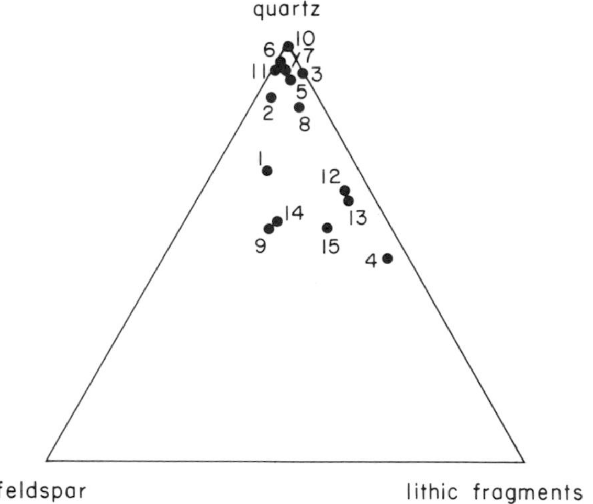

Fig. 9.73. Ternary plot of detrital composition of fifteen sample suites from Carboniferous to Tertiary rocks of the western Alps. Carbonate grains are included with lithic fragments (Schwab, 1981).

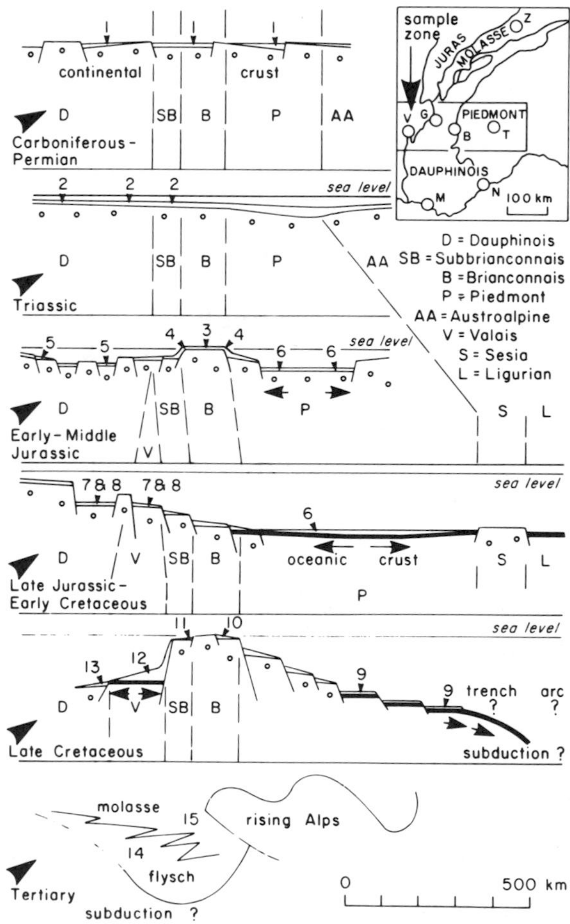

Fig. 9.74. Schematic structural cross sections through the western Alps showing interpreted tectonic evolution and location of sample suites plotted in Fig. 9.73 (Schwab, 1981).

is interpreted as an isolated continental strip that originated much like Blake Plateau on the Atlantic coast (Fig. 9.12). The mineralogy is typical of a mature continental suite. Suite 4 comprises fault breccias deposited during extensional break-up of the Jurassic carbonate platform (cf. Fig. 9.9). The lithic nature of the sands reflects the abundance of carbonate detritus, and because of this the assemblage cannot be compared with the petrofacies classes of Dickinson and Suczek (1979). Suites 6 and 7 are from deep-water starved facies but have the mature characteristics of continentally-derived suites. Suites 7 and 8 are also mature and reflect a similar provenance.

Convergent tectonics commenced during the Late Cretaceous. The resulting deep-water sands (Suite 9) have a composition in the collision orogen provenance class rather than the magmatic arc category. Contemporaneous sands to the west (Suites 10 and 11) are supermature, indicating that their provenance was not affected by the change in tectonic style, and they probably were derived from the same continental sources as Suites 2, 3, 5 and 6. Suites 12 and 13 are deep-water sands and breccias derived from rejuvenated fault blocks. Their relatively lithic nature suggests that they were tapping new (deeper?) sources that had not previously influenced the area.

Suites 14 and 15 represent the great "flysch" and "molasse" wedges produced during closure of the Piedmont Ocean and formation of the present Alpine structures. Their mineralogy compares closely with the collision orogen petrofacies of Dickinson and Suczek (1979). Schwab (1981) discussed the relative paucity of volcaniclastic debris in these sands, which might have been expected to contain abundant volcanic material from adjacent arcs. He pointed out the localized distribution of volcanic detritus adjacent to modern arcs, which may result from local trapping or ponding of the debris.

9.5 Basin models in the Precambrian

9.5.1 Precambrian plate tectonics?

The greater part of this book deals with the Phanerozoic record. Although this represents only about the last one-ninth of geological time since the earth came into being, the record of the Phanerozoic is so much clearer and more complete than that of the Precambrian that it has provided virtually all the data on which modern basin analysis methods are based. Previous sections of this chapter deal exclusively with the Phanerozoic record. This emphasis has been chosen deliberately because it is now generally agreed that the processes of plate tectonics, as deduced from modern and very recent events, can be recognized at least as far back as the early Paleozoic. For example the Ordovician Taconic Orogeny of the Appalachians is discussed in section 9.3.4 as a good example of the evolution of a backarc region during continental collision.

The question of whether or not we can apply the basin models developed here to the Precambrian is part of a larger question concerning long-term crustal evolution. To pose the question in its simplest form, when did plate tectonics "begin" and from what did it evolve? Answers to the question comprise what is probably the most complex and vigorous debate taking place in geology today. Many of the elements of the discussion are far beyond the scope of this book. As Fyfe (1981) has pointed out, the sample of unmodified early crust available to us for analysis is very small, many modern geophysical techniques such as paleomagnetism, seismology and heat flow studies are unavailable for use in the Archean, and radiometric dating methods may have a precision of only about ± 100 Ma. "It is clear that ancient tectonics must involve primarily the mineralogist, petrologist, geochemist and structural geologist" (Fyfe, 1981, p. 549). Nevertheless, as discussed below, basin analysis does have a crucial contribution to make because the basin models discussed in section 9.3 carry definite implications about tectonic style, and clastic petrofacies data tell us much about crustal composition and tectonic setting.

In the remainder of this section we discuss very briefly some of the current ideas on Precambrian crustal evolution relevant to our topic of basin studies. This is drawn from several books (Windley, 1976, 1977; Strangway, 1980; Kröner, 1981a; Moorbath and Windley, 1981) and review articles (Young, 1978; Goodwin, 1981a; Condie, 1982; Sleep and Windley, 1982) to which the reader is referred for complete details.

Few doubt that plate tectonics processes similar to those operating today were already operating during the Proterozoic. The question is to determine how far back into the Proterozoic, and whether in fact they began in the Archean. Some authors, such as Hoffman (1980), have successfully applied plate tectonic models to rocks of early Proterozoic age (see also several papers in Campbell, 1981) and there is one school of thought which maintains that some form of seafloor spreading and subduction mechanism was in operation throughout the Archean. Windley (1981) and Sleep and Windley (1982) discussed the differences that might be expected between modern and Archean plates and plate margin processes based on the widely held idea that heat flow was higher in the Archean. They argued that the crust probably was thicker than at present, plate movements were faster and individual plates smaller. They suggested that along convergent margins both Cordilleran-type compressional arcs (with metamorphic core complexes and arc plutonism) and west Pacific-type extensional arcs (with backarc basins) could have occurred (cf. Dewey, 1980; see section 9.3.2). They summarized examples extending back to the early Proterozoic where deeply eroded elements of these types of margin have been tentatively identified, and they also pointed to possible ophiolites and suture zones (but see conflicting views below).

The two distinctive types of Archean tectonic regime are greenstone–granite belts and granulite–gneiss belts. Windley (1981) compared them to modern backarc basin and plutonic arc complexes, respectively. Greenstone belts (Fig. 9.75) are linear to curvilinear synformal structures up to 800 km long, containing ultramafic and andesitic volcanics and a variety of sedimentary facies. Of the sediments turbidites are the most common, but shale, banded iron formation, chert and limestone, and local thick accumulations of fluvial clastics also occur, most of which (except the latter) indicate deep-water environments (Goodwin, 1981b). Most Archean sediments are associated with greenstone belts and, with the exception of the Kaapvaal Craton of South Africa (discussed below), their petrography and depositional environments suggest that stable cratons or platform areas at or near sea level had a very limited distribution (Kröner, 1981b).

Many other models of Archean tectonics have been proposed, most involving small plates devel-

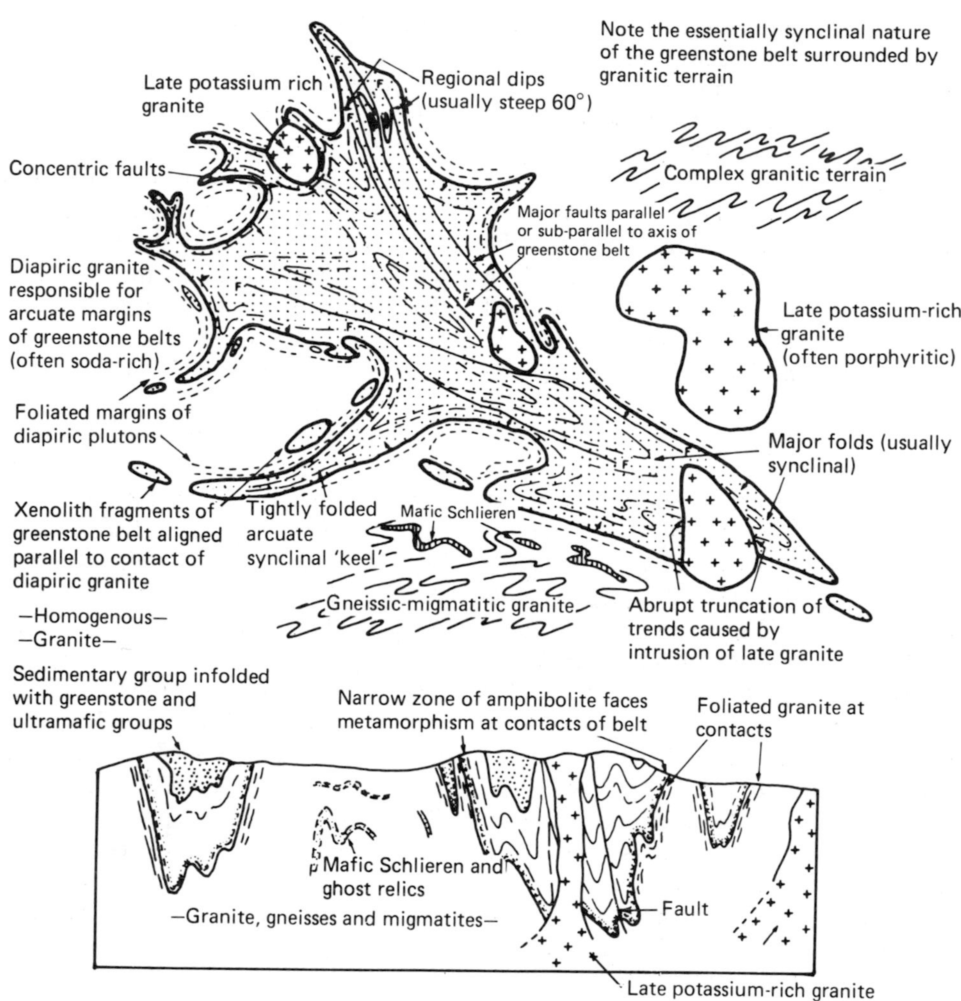

Fig. 9.75. Schematic map and section of a typical Archean greenstone-granite terrain, based on the Barberton Mountain area, Swaziland–South Africa (Anhaeusser et al., 1969).

oped over closely spaced mantle plumes, limited lateral movement, and differentiation into different igneous–metamorphic–sedimentary regimes by doming and sagging ("sagduction") rather than by spreading and subduction (Young, 1978; West, 1980; Kröner, 1981b; Anhaeusser, 1981; Goodwin, 1981b). An example is illustrated in Fig. 9.76. Proponents of these models point to the absence from the Archean record of well-documented subduction zones or sutures with convincing examples of ophiolites (sheeted dikes, etc.) and blueschist associations. Gneissic belts which border the greenstones commonly show structural continuity from one belt to another, which does not fit a drifting plate model.

Another tectonic model is that of ensialic orogeny characterized by A-subduction and thin-skinned tectonics (Kröner, 1981b). Dimroth (1981) applied this model to the Labrador Trough, an early Proterozoic fold belt filled with basic volcanics, turbidites and shelf sediments. He suggested that the trough formed over reactivated Archean crust, with limited lateral movement except that required (100–200 km) by "delamination" and underplating of the crust and generation of a fold-thrust belt.

Kröner (1981b) suggested that some of the differences of opinion with regard to Precambrian tectonics may have arisen from an overly uniformitarian application of plate tectonic principles to the early record. He stated:

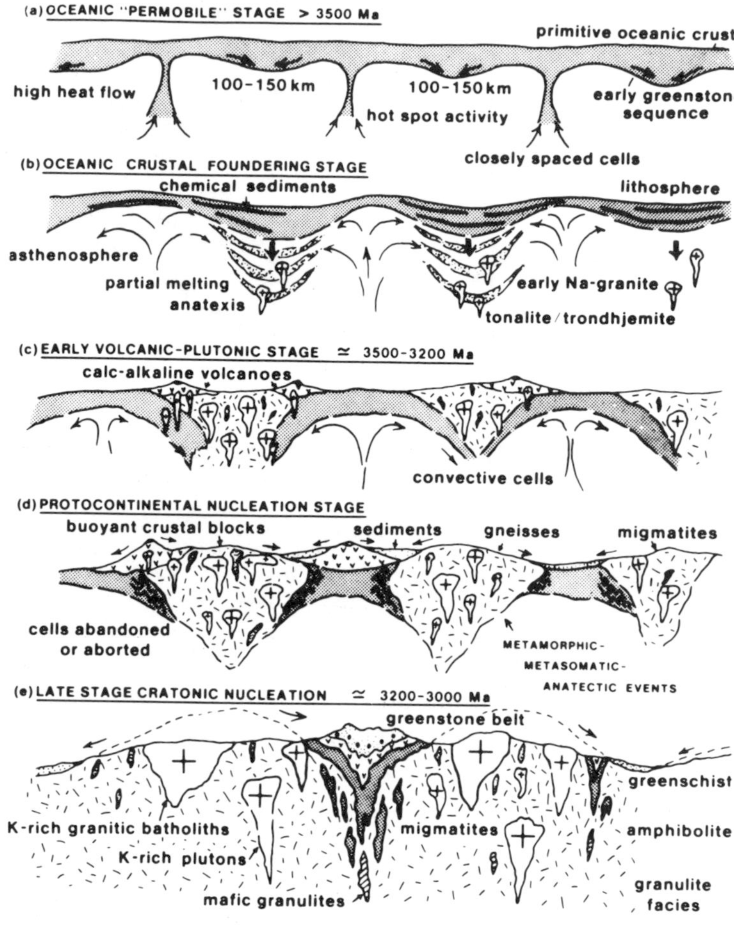

Fig. 9.76. One of many similar models proposed to explain the generation of greenstone belts and cratons in the Archean. The timing of each stage is based on Kaapvaal Craton, southern Africa, and is several hundred million years earlier than in other shield ares (Anhaeusser, 1981).

There is a deplorable lack of detailed stratigraphic and paleogeographic analyses on the basis of physical volcanology and sedimentology in most Archaean greenstone belts (E. Dimroth, pers. commun., 1980) from which depositional environments could be deduced. Instead geochemical data are frequently employed as a substitute for laborious fieldwork. For example Archaean subduction zones and even polarities have sometimes been postulated solely on the basis of a few analyses apparently displaying 'trends' in certain elements such as K, Rb, Sr, Ti. Yet it is well known that chemical composition of modern and ancient submarine volcanic rocks are often, if not always, affected by extensive low-temperature alteration, even within individual flows or pillows . . ., with variations greatly exceeding the apparent trends established.

There is a moral here for basin analysts, as well as a clear warning that plate tectonics may already have become a simplistic orthodoxy as far as the early Precambrian record is concerned.

Goodwin (1981a), Windley (1981) and Kröner (1981b) attempted syntheses of Precambrian tectonic and sedimentary evolution which seem capable of absorbing many of the problems now undergoing debate. Goodwin (1981a) stated:

The Precambrian record is viewed as moving in the direction of increasing continental stability from an early highly mobile microplate tectonics phase, through a more stable, largely intracratonic, ensialic, mobile belt phase, to the modern macroplate tectonics phase involving large, rigid lithospheric plates with modern continents and ocean basins.

These and other evolutionary changes are summarized in Figure 9.77. The microplate stage from about 3.8 to 2.6 Ga resulted in the development of the characteristic shield pattern of cratonic blocks bordered by mobile belts. It was terminated by a widespread cratonization event involving massive intrusion of granitoid bodies apparently on a worldwide scale. This event is the basis for the Archean–Proterozoic division of Precambrian time at 2.5 Ga that has now been adopted virtually worldwide (section 3.6.7).

As indicated in Figure 9.77 cratonization was a

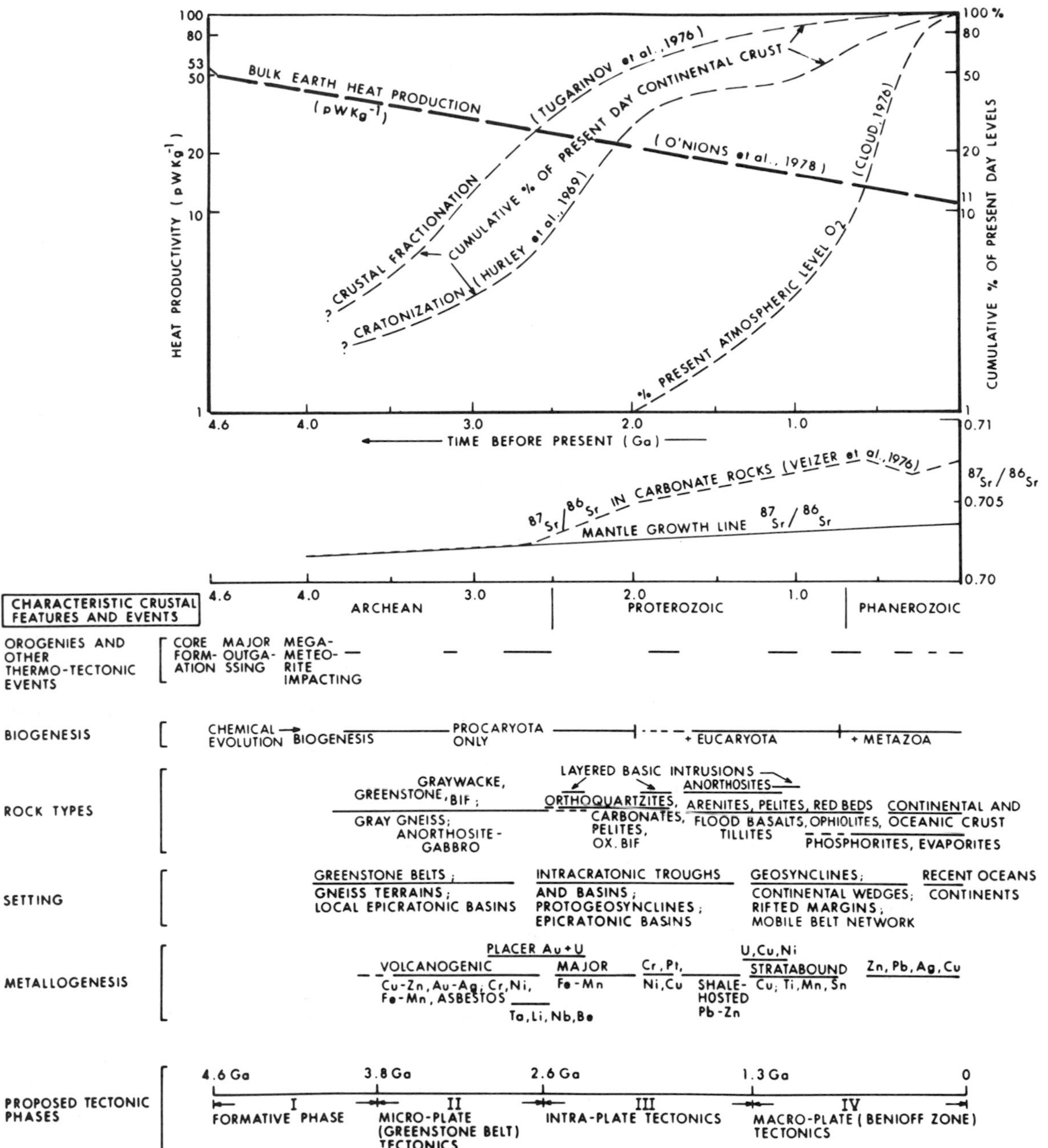

Fig. 9.77. A summary model for the evolution of the Precambrian earth. Reproduced, with permission, from A.M. Goodwin, Science, v. 213, p. 55–61, 1981.

continuous process (the suggested peak at about 2.5 Ga does not appear in some estimates illustrated in Fig. 9.77), and in some areas the development of Proterozoic-style crust and associated sediments took place during the Archean. The prime example of this is the Kaapvaal Craton of southern Africa. There the Swaziland Supergroup (approximately 3.0–3.5 Ga) contains the oldest example of continental margin style sedimentation. It is followed by the Pongola Supergroup (2.9–3.1 Ga), the oldest cratonic cover suite (Anhaeusser, 1981; Tankard et al., 1982). These sediments are described briefly in the next section.

The ensialic type of orogeny discussed above seems to have been typical of the 2.6 to 1.3 Ga time period. Examples in North America include Labrador Trough (Dimroth, 1981) and the Great Lakes tectonic zone (Sims et al., 1980, 1981). Baragar and Scoates (1981) tied both these regions into their supposed circum-Superior belt, a system of continental margin basins developed around the margin of the Superior Province of the Canadian Shield, and characterized by limited sea-floor spreading and subduction.

Macroplate tectonics, which we would recognize as modern plate tectonics, may have become predominant in the Middle Proterozoic (Fig. 9.77), although it began earlier in some areas if Hoffman's (1980) analysis of the Wopmay Orogen (1.8–2.1 Ga) is correct.

The evolutionary scheme summarized here seems an elegant compromise, and yet there remains a doubt. The cartoon approach to Archean tectonics (e.g. Fig. 9.76) is reminiscent of the "fixist" style of geosyncline theory, in its attempt to summarize all known developments in a series of two-dimensional cross sections. This contrasts strongly with Phanerozoic plate tectonics which, we now know, build orogens by unique combinations of macroplate and microplate tectonics, with orthogonal, oblique and strike-slip convergence and diachronous suturing. Geologists, when dealing with complex problems, tend to talk about "fitting together the pieces of a jigsaw puzzle", but jigsaw puzzles have only one solution. For the Phanerozoic, at least, a better analogy would be a Leggo set—a kit of parts that can be put together in an almost infinite number of ways. Can we be sure that many Precambrian orogens are not as complex as the Cordilleran, Appalachian or Alpine belts? Or is there really a fundamental difference in style?

9.5.2 The evidence from sedimentary basins

Basin analysis can contribute to Precambrian tectonic studies in three main ways:

1. Stratigraphic and sedimentological analysis reveals the nature of depositional systems, which tells us much about crustal style, elevation and paleoslope (Chaps. 4, 5, 6).
2. Structural analysis of the sediments yields information on local contemporaneous tectonic control (Harding and Lowell, 1979).
3. Clastic petrofacies data indicate the crustal composition of provenance areas, which can commonly be interpreted in plate tectonic terms (section 9.4).

Although many Precambrian sediments are severely deformed and metamorphosed, others are not and are readily amenable to the types of sedimentological analysis discussed in this book. An excellent example is the Swaziland Supergroup, an Archean sequence (3.0–3.5 Ga) the upper part of which yet appears remarkably fresh. It has been studied sedimentologically by Lowe and Knauth (1977) and by K.A. Eriksson in a series of papers summarized in Eriksson (1982) and Tankard et al. (1982). The sequence is subdivided as follows:

3. Moodies Group (3500 m): polymict conglomerate, feldspathic and quartzose arenite, siltstone, mudstone, minor iron formation.
2. Fig Tree Group (2000 m): lithic arenite, tuff, agglomerate, minor shale, conglomerate and iron formation.
1. Onverwacht Group (16,370 m): ultrabasic rocks, basalt, andesite, dacite, tuff, agglomerate, minor chert, volcaniclastic arenite.

Eriksson was able to recognize fluvial, deltaic and barrier–beach environments in the Moodies Group and discussed them in terms of modern facies model concepts using vertical profile and paleocurrent analyses. The Fig Tree sediments represent a submarine fan complex, and the Onverwacht Group is a typical greenstone belt. The sediments of the Fig Tree and Moodies Groups demonstrate an upward-coarsening character and are reminiscent of the prograding (clinoform) wedges of sediment on a rifted and divergent continental margin (section 9.3.1). Petrographic and geochemical studies indicate an upward change from volcaniclastic to quartzofeldspathic sources, indicating the uplifting and/or unroofing of a sialic plutonic or metamorphic terrane. In terms of modern plate tectonics the Swaziland Supergroup could represent the generation, decay and filling of a backarc basin, with the Fig Tree and Moodies Groups representing the sediment wedge accumulated on the continental side.

Anhaeusser (1981) has interpreted this region in terms of a microplate dome and sag model (Fig. 9.76). The Fig Tree and Moodies Groups are represented by the dotted area of sediments in cartoon d. Their vertical environmental and petrographic changes are attributed to the beginning of a major phase of granitic batholith intrusion that culminated about 3.0 Ga and was responsible for generating the first area of (preserved) cratonic crust, the Kaapvaal Craton. The evidence for this model is beyond the scope of this book, except for the obvious fact that a convergent margin plate tectonic model seems to be precluded by the absence of subduction complexes and by major differences between the Onverwacht Group and Phanerozoic ophiolites. If each greenstone belt represents a separate backarc basin it also implies a veritable festoon of bounding arcs, for which there is no evidence.

Condie (1982) explored the evidence for tectonic styles in the Proterozoic volcanic–sedimentary record. He examined 79 successions around the world and subdivided them into three assemblages, as summarized in Table 9.7. The data clearly show that Phanerozoic-style cratons were widespread by Early Proterozoic time, because stratigraphically, sedimentologically and petrographically assemblages I and II can readily be compared with Phanerozoic cratonic and rift successions from the North American interior, Russia, Siberia and elsewhere. A few Archean assemblages could also be fitted into the first two categories. For example the major cratonization event of the Kaapvaal Craton at about 3.0 Ga, noted above, was followed by deposition of the Pongola Supergroup (2.9–3.1 Ga). This contains a basal volcanic unit up to 7000 m thick which Tankard et al. (1982) compared to Phanerozoic arc volcanics. It passes vertically and laterally into a clastic succession up to 1800 m thick similar to Condie's (1982) assemblage I. Assemblage III is dominated by volcanic rocks and Condie (1982) compares it both to Archean greenstone belts and to Phanerozoic backarc basins. Some basins contain representatives of more than one of Condie's (1982) assemblages. This would be fully in accord with plate tectonic models of sedimentation which, as noted at the beginning of this chapter, may result in the juxtaposition and structural superposition in suture zones of three or four (or more) entirely different basin styles.

Each Precambrian basin should clearly be examined with care because the diachronous development of sialic cratons may imply a similarly

Table 9.7. Proterozoic lithologic assemblages*

Assemblage	% of total	Age range (Ga)	Thickness of individual successions (km)	Lithologies	Interpretation
I	60	1.3–2.5	>15	crossbedded quartzose arenite, dolomite, shale, siltstone; locally iron formation and lithic sandstone important; volcanics minor	stable continental margin or interior, commonly in small basins <500 km across
II	20	1.0–2.5	1–10	feldspathic and quartzose arenite, shale, conglomerate; locally evaporites and carbonates important; bimodal basalt-rhyolite volcanics	intracratonic rift basins, aulacogens
III	20	1.3–2.4	>15	tholeiitic or calc-alkaline volcanics, pyroclastics, lithic sandstone, argillite, minor conglomerate, quartz arenite, iron formation, carbonate	rift developed in backarc basin or over mantle plume; cf. Archean greenstone belts.

*data from Condie, 1982.

diachronous evolution of tectonic styles. It is possible that the change from the relatively fixed microplate style to the freely moving macroplate style took place at different times in different parts of the world. In this case a meticulous basin analysis of the volcanic–sedimentary record may contribute much to the clarification of the problem.

Direct comparison with the Phanerozoic is complicated for the Archean and Early Proterozoic record by the controversy over the composition of the atmosphere at that time. Cloud (1973) and others have pointed out that sedimentary iron formations, which are the world's greatest source of iron ore, were formed mainly between about 1.8 and 2.6 Ga (although in lesser amounts they span most of geological time). Precambrian formations consist of iron oxides, carbonates, silicates and sulfides, commonly interbanded with chert and other sediments, and were deposited in shallow to deep marine environments. The "Superior" type is typical of Proterozoic iron formations. It is associated with shelf sediments and can be compared to Condie's (1982) assemblage I. Cloud (1973) reasoned that the iron must have been transported in the ferrous state, implying a complete lack of oxygen in the atmosphere. Free oxygen is thought to have become abundant in the atmosphere between about 2.2 and 1.8 Ga as a result of the rapid spread of primitive photosynthesizing organisms. Other observations supporting this idea are: (1) the abundance of uranium as a placer mineral in the Witwatersrand, Africa (2.3–2.8 Ga) and lower Huronian, Canada (2.1–2.7 Ga) successions, compared with the virtual absence of uranium placers in younger rocks; and (2) the lack of red, hematite–limonite stained sandstones in rocks older than about 2.2 Ga. These facts complicate our facies analysis of the early Precambrian record, particularly as not all geologists agree with the interpreted change in atmospheric composition. Dimroth and Kimberley (1976) and Clemmey and Badham (1982) presented arguments that from the time of the earliest dated rocks the earth had an oxygenated atmosphere. For example many of the features ascribed to a reducing environment could be explained by a reducing diagenetic environment.

Other difficulties with facies analysis of the distant past are discussed in section 4.4.4.

References

AGER, D.V., 1980: The geology of Europe; McGraw-Hill, London, 535 p.

ALEINIKOV, A.L., BELLAVIN, O.V., BULASHEVICH, Y.P., TAVRIN, I.F., MAKSIMOV, E.M., RUDKEVICH, M.Y., NALIVKIN, V.D., SHABLINSKAYA, N.V. and SURKOV, V.S., 1980: Dynamics of the Russian and West Siberian Platforms, *in* A.W. Bally, P.L. Bender, T.R. McGetchin and R.I. Walcott, eds., Dynamics of plate interiors; Am. Geophys. Union and Geol. Soc. Am., Geodynamics Ser., v.1, p. 53–71.

ALLEN, J.R.L., DINELEY, D.L., and FRIEND, P.F., 1968: Old Red Sandstone basins of North America and Northwest Europe, in D.H. Oswald, ed., Internat. Symp. Devonian System; Alta. Soc. Petrol. Geol. Calgary, Canada, v.1, p. 69–98.

ANDERSEN, D.W., and PICARD, M.D., 1974: Evolution of synorogenic clastic deposits in the intermontane Uinta Basin of Utah, *in* W.R. Dickinson, ed., Tectonics and Sedimentation; Soc. Econ. Paleont. Mineral. Spec. Publ. 22, p. 167–189.

ANDREWS-SPEED, C.P., and BROOKFIELD, M.E., 1982; Middle Paleozoic to Cenozoic geology and tectonic evolution of the northwestern Himalaya; Tectonophysics, v. 82, p. 253–276.

ANHAEUSSER, C.R., 1981: Geotectonic evolution of the Archaean successions in the Barberton Mountain land, South Africa, *in* A. Kröner, ed., Precambrian plate tectonics, Develop. in Precamb. Geol. 4, p. 137–160.

ANHAEUSSER, C.R., MASON, R., VILJOEN, M.J. and VILJOEN, R.P., 1969: A reappraisal of some spects of Precambrian Shield geology; Geol. Soc. Am. Bull., v. 80, p. 2175–2200.

ARTHUR, M.A., and SCHLANGER, S.O., 1979: Cretaceous "Oceanic anoxic events" as causal factors in development of reef reservoired giant oil fields; Am. Assoc. Petrol. Geol. Bull., v. 63, p. 870–885.

ATWATER, T., and MOLNAR, P., 1973: Relative motion of the Pacific and North American plates deduced from sea-floor spreading in the Atlantic Indian and South Pacific Oceans, in R.L. Kovach and A. Nur, eds., Proc. Conf. on Tectonic Problems of the San Andreas Fault System; Stanford Univ. Pub. Geol. Sci., v. 13, p. 136–148.

AUBOUIN, J., 1965; Geosynclines; Developments in Geotectonics 1, Elsevier, Amsterdam, 335 p.

AYDIN, A., and NUR, A., 1982: Evolution of pull-apart basins and their scale independence; Tectonics, v. 1, p. 91–105.

BADHAM, J.P.N., 1982: Strike-slip orogens—an explanation for the Hercynides; J. Geol. Soc. London, v. 139, p. 495–506.

BAKER, B.H. and MORGAN, P., 1981: Continental rifting: progress and outlook; EOS, v. 62. p. 585–586.

BALKWILL, H.R., 1978: Evolution of Sverdrup Basin, Arctic Canada; Am. Assoc. Petrol. Geol. Bull., v. 62, p. 1004–1028.

BALLANCE, P.F., 1980: Models of sediment distribution in nonmarine and shallow marine environments in oblique-slip fault zones, *in* P.F. Ballance and H.G. Reading, eds., Sedimentation in oblique-slip mobile zones; Internat. Assoc. Sedimentologists Spec. Publ. 4, p. 229–236.

BALLANCE, P.F., and READING, H.G., eds., 1980: Sedimentation in oblique-slip mobile zones, Internat. Assoc. Sedimentologists Spec. Publ. 4, 265 p.

BALLY, A.W., 1980: Basins and subsidence—a summary, *in* A.W. Bally, P.L. Bender, T.R. McGetchin and R.I. Walcott, eds., Dynamics of plate interiors; Am. Geophys. Union and Geol. Soc. Am., Geodynamics Ser., v. 1, p. 5–20.

BALLY, A.W. and SNELSON, S., 1980: Realms of subsidence, *in* A.D. Miall, ed., Facts and principles of world petroleum occurence; Can. Soc. Petrol. Geol. Mem. 6, p. 9–94.

BANNERT, D., and HELMCKE, D., 1981: The evolution of the Asian plate in Burma; Geol. Rundschau, v. 70, p. 446–458.

BARAGAR, W.R.A., and SCOATES, R.F.J., 1981: The circum-Superior belt: a Proterozoic plate margin?, *in* A. Kroner, ed., Precambrian plate tectonics; Develop. in Precamb. Geol. 4, Elsevier, Amsterdam, p. 297–330.

BASU, A., 1976: Petrology of Holocene fluvial sand derived from plutonic source rocks: Implications to paleoclimatic interpretation; J. Sediment. Petrol., v. 46, p. 694–709.

BEAUMONT, C., 1981: Foreland basins; Geophys. J. Roy. Astron. Soc., v. 65, p. 291–329.

BEN-AVRAHAM, Z., and EMERY, K.O., 1973: Structural framework of Sunda Shelf; Am. Assoc. Petrol. Geol. Bull., v. 57, p. 2323–2366.

BENKHELIL, J., 1982: Benue Trough and Benue Chain; Geol. Mag., v. 119, p. 155–168.

BERNOULLI, D., and JENKYNS, H.C., 1974: Alpine, Mediterranean, and central Atlantic Mesozoic facies in relation to the early evolution of the Tethys, *in* R.H. Dott and R.H. Shaver, eds., Modern and ancient geosynclinal sedimentation, Soc. Econ. Paleont. Mineral. Spec. Publ. 19, p. 129–160.

BIJU-DUVAL, B., LETOUZEY, J., MONTADERT, L., COURRIER, P., MUGNIOT, J.F., and SANCHO, J., 1974: Geology of the Mediterranean Sea Basins, *in* C.A. Burk and C.L. Drake, eds., The Geology of Continental margins; Springer-Verlag, New York, p. 695–721.

BIJU-DUVAL, B., LETOUZEY, J., and MONTADERT, L., 1978: Variety of margins and deep basins in the Mediterranean, *in* J.S. Watkins, L. Montadert and P.W. Dickerson, eds., Geological and geophysical investigations of continental margins; Am. Assoc. Petrol. Geol. Mem. 29, p. 293–317.

BIQ, CHINGCHANG, 1964: Taiwan and the Alps: an attempt at comparison of island and mountain arcs; 22nd. Internat. Geol. Congr., India, p. 220–238.

BIRD, J.M., and DEWEY, J.F., 1970, Lithosphere plate-continental margin tectonics and the evolution of the Appalachian Orogen; Geol. Soc. Amer. Bull., v. 81, p. 1031–1060.

BLAKE, M.C., Jr., CAMPBELL, R.H., DIBBLEE, T.W., Jr., HOWELL, D.G., NILSEN, T.H., NORMARK, W.R., VEDDER, J.C., and SILVER, E.A., 1978: Neogene basin formation in relation to Plate tectonic evolution of San Andreas fault system, California; Am. Assoc. Petrol. Geol., Bull. v. 62, p. 344–372.

BLAKE, M.C., Jr. and JONES, D.L., 1974: Origin of Franciscan melanges in northern California, *in* R.H. Dott Jr. and R.H. Shaver, eds., Modern and ancient geosynclinal sedimentation; Soc. Econ. Paleont. Mineral. Spec. Publ. 19, p. 345–357.

BLAKE, M.C., Jr., and JONES, D.L., 1981: The Franciscan assemblage and related rocks in northern California: A reinterpretation, *in* W.G. Ernst, ed., The geotectonic development of California; Prentice-Hall, New Jersey, p. 307–328.

BLUCK, B.J., 1978: Sedimentation in a late orogenic basin: the Old Red Sandstone of the Midland Valley of Scotland, *in* D.R. Bowes and B.E. Leake, eds., Crustal evolution in northwestern Britain and adjacent regions; Geol. J. Special Issue 10, p. 249–278.

BLUCK, B.J., 1980: Evolution of a strike-slip fault-controlled basin, Upper Old Red Sandstone, Scotland, *in* P.F. Ballance and H.G. Reading, eds., Sedimentation in oblique-slip mobile zones, Internat. Assoc. Sed., Spec. Publ. 4, p. 63–78.

BOEUF, M.G., and DOUST, H., 1975: Structure and development of the southern margin of Australia; Australia Petrol. Expl. Assoc. J., v. 15, p. 33–43.

BOND, G., 1978: Evidence for late Tertiary uplift of Africa relative to North America, South America, Australia and Europe; J. Geol., v. 86, p. 47–65.

BOTT, M.H.P., 1978: Subsidence mechanisms at passive continental margins, *in* J.S. Watkins, L. Montadert, and P.W. Dickerson, eds., Geological and geophysical investigations of continental margins; Am. Assoc. Petrol. Geol. Mem. 29, p. 3–10.

BOTT, M.H.P., 1982: The mechanism of continental splitting; Tectonophysics, v. 81, p. 301–309.

BREWER, J.A., GOOD, R., OLIVER, J.E., BROWN, L.D., and KAUFMAN, S., 1983: COCORP profiling across the Southern Oklahoma aulacogen: overthrusting of the Wichita Mountains and compression within the Anadarko Basin; Geology, v. 11, p. 109–114.

BREWER, J., SMITHSON, S.B., OLIVER, J., KAUFMAN, S., BROWN, L.D., 1980: The Laramide Orogeny: evidence from COCORP deep crustal seismic profiles in the Wind River Mountains, Wyoming; Tectonophysics, v. 62, p. 165–187.

BRONNER, G., ROUSSEL, J., and TROMPETTE, R., 1980: Genesis and geodynamic evolution of the Taoudeni cratonic basin (Upper Precambrian and Paleozoic), western Africa, *in* A.W. Bally, P.L. Bender, T.R. McGetchin and R.I. Walcott, eds., Dynamics of plate interiors; Am. Geophys. Union and Geol. Soc. Am., Geodynamics Ser., v. 1, p. 81–90.

BROOKFIELD, M.E., and REYNOLDS, P.H., 1981: Late Cretaceous emplacement of the Indus suture zone ophiolitic melanges and an Eocene–Oligocene magmatic arc on the northern edge of the Indian Plate; Earth. Plan. Sci. Letters, v. 55, p. 157–162.

BROWN, R.N., 1980: History of exploration and discovery of Morgan, Ramadan and July oilfields, Gulf of Suez, Egypt, *in* A.D. Miall, ed., Facts and principles of world petroleum occurrence; Can. Soc. Petrol. Geol. Mem. 6, p. 733–764.

BURCHFIEL, B.C. and DAVIS, G.A., 1972: Structural framework and evolution of the southern part of the Cordilleran orogen, western United States; Am. J. Sci., v. 272, p. 97–118.

BURCHFIEL, B.C. and ROYDEN, L., 1982: Carpathian fold and thrust belt and its relation to Pannonian and other basins; Am. Assoc. Petrol. Geol. Bull., v. 66, p. 1179–1195.

BURKE, K., 1976a: Development of graben associated with the initial ruptures of the Atlantic Ocean; Tectonophysics, v. 36, p. 93–112.

BURKE, K., 1976b: The Chad basin: an active intracontinental basin; Tectonophysics, v. 36, p. 197–206.

BURKE, K., 1977: Aulacogens and continental breakup; Ann. Rev. Earth Planet. Sci. v. 5, p. 371–396.

BURKE, K. and DEWEY, J.F., 1973: Plume generated triple junctions: key indicators in applying plate tectonics to old rocks; J. Geol., v. 81, p. 406–433.

BURKE, K., DEWEY, J.F., and KIDD, W.S.F., 1977: World distribution of sutures—the sites of former oceans; Tectonophysics, v. 40, p. 69–99.

BURKE, K., and KIDD, W.S.F., 1980: Volcanism on earth through time , *in* D.W. Strangway, ed., The continental crust and its mineral deposits; Geol. Assoc. Can. Spec. Paper 20, p. 503–522.

CAMFIELD, P.A., and GOUGH, D.I., 1977: A possible Proterozoic plate boundary in North America; Can. J. Earth Sci., v. 14, p. 1229–1238.

CAMPBELL, F.H.A., ed., 1981: Proterozoic basins of Canada; Geol. Survey Canada Paper 81–10.

CARTER, D.J., AUDLEY-CHARLES, M.G., and BARBER, A.J., 1976: Stratigraphical analysis of island arc-continental margin collision in eastern Indonesia; J. Geol. Soc. London, v. 132, p. 179–198.

CARTER, R.M., and NORRIS, R.J., 1976: Cainozoic history of southern New Zealand: an accord between geological observations and plate-tectonic predictions; Earth Planet. Sci. Letters, v. 31, p. 85–94.

CHURKIN, M. Jr., 1974: Paleozoic marginal ocean basin-volcanic arc systems in the Cordilleran foldbelt, *in* R.H. Dott Jr., and R.H. Shaver, eds., Modern and ancient geosynclinal sedimentation; Soc. Econ. Paleont. Mineral. Spec. Publ. 19, p. 174–192.

CHURKIN, M. Jr., and TREXLER, J.H., Jr., 1981: Continental plates and accreted oceanic terranes in the Arctic, *in* A.E.M. Nairn, M. Churkin Jr., and F.G. Stehli, eds., The Ocean basins and margins, v. 5, The Arctic Ocean; Plenum Press, New York, p. 1–20.

CLEMMEY, H., and BADHAM, N., 1982: Oxygen in the Precambrian atmosphere: an evaluation of the geological evidence; Geology, v. 10, p. 141–146.

CLOUD, P., 1973: Paleoecological significance of banded iron-formation; Econ. Geol., v. 68, p. 1135–1143.

CONDIE, K.C., 1982: Early and middle Proterozoic supracrustal successions and their tectonic settings; Am. J. Sci., v. 282, p. 341–357.

CONEY, P.J., 1979: Mesozoic–Cenozoic Cordilleran plate tectonics, *in* R.B. Smith and G.P. Eaton, eds., Cenozoic tectonics and regional geophysics of the Western Cordillera; Geol. Soc. Am. Mem. 152, p. 33–50.

CONEY, P.J., JONES, D.L., and MONGER, J.W.H., 1980: Cordilleran suspect terranes; Nature, v. 288, p. 329–333.

COOK, F.A., and OLIVER, J.E., 1981: The Late Precambrian—Early Paleozoic continental edge in the Appalachian orogen; Am. J. Sci., v. 281, p. 993–1008.

CROOK, K.A.W., 1974: Lithogenesis and geotectonics: the significance of compositional variations in flysch arenites (graywackes), *in* R.H. Dott and R.H. Shaver, eds., Modern and ancient geosynclinal sedimentation; Soc. Econ. Paleont. Mineral. Spec. Publ. 19, p. 304–310.

CROSS, T.A., and PILGER, R.H., Jr., 1982: Controls of subduction geometry, location of magmatic arcs, and tectonics of arc and back-arc regions; Geol. Soc. Am. Bull., v. 93, p. 545–562.

CROSSLEY, R., 1979: Structure and volcanism in the S. Kenya Rift, *in* Geodynamic evolution of the Afro-Arabian Rift System; Accademia Nazionale dei Lincei, Roma, p. 89–98.

CROSTELLA, A., 1977: Geosynclines and plate tectonics in Banda Arcs, eastern Indonesia; Am. Assoc. Petrol. Geol. Bull., v. 61, p. 2063–2081.

CROWELL, J.C., 1974a, Sedimentation along the San Andreas Fault, California, *in* R.H. Dott and R.H. Shaver, eds., Modern and ancient geosynclinal sedimentation; Soc. Econ. Paleont. Mineral. Spec. Publ. 19, p. 292–303.

CROWELL, J.C., 1974b: Origin of late Cenozoic basins in southern California, *in* W.R. Dickinson, ed., Tectonics and Sedimentation; Soc. Econ. Paleont. Mineral. Spec. Publ. 22, p. 190–204.

CROWELL, J.C., 1979. The San Andreas Fault System through time; J. Geol. Soc. London, v. 136, p. 293–302.

CROWELL, J.C. and LINK, M.H., 1982, Ridge Basin, Southern California: Introduction, *in*: Crowell, J.C. and Link, M.H.., eds., Geologic History of Ridge Basin, Southern California: Pacific Section Society of Economic Paleontologists and Mineralogists, p. 1–4.

CURRAY, J.R., and MOORE, D.G., 1974: Sedimentary and tectonic processes in the Bengal deep-sea fan and geosyncline, *in* C.A. Burk and C.L. Drake, eds., The geology of continental margins; Springer-Verlag, New York, p. 617–627.

CURRAY, J.R., MOORE, D.G., LAWVER, L.A., EMMEL, F.J., RAITT, R.W., HENRY, M. and KIECKHEFER, R., 1978: Tectonics of the Andaman Sea and Burma, *in* J.S. Watkins, L. Montadert and P.W. Dickerson, eds., Geological and geophysical investigations of continental margins; Am. Assoc. Petrol. Geol. Mem. 29, p. 189–198.

DALZIEL, I.W.D., DE WIT, M.J. and PALMER, K.F., 1974: A fossil marginal basin in the Southern Andes; Nature, v. 250, p. 291–294.

DAVIES, D.K., and ETHRIDGE, F.G., 1975: Sandstone composition and depositional environments; Am.

Assoc. Petrol. Geol. Bull., v. 59, p. 239–264.

DEGENS, E.T., and ROSS, D.A., eds., 1974: The Black Sea—geology, chemistry, and biology; Am. Assoc. Petrol. Geol. Mem. 20, 633 p.

DELONG, S.E., DEWEY, J.F., and FOX, P.J., 1979: Topographic and geologic evolution of fracture zones; J. Geol. Soc. London, v. 136, p. 303–310.

DEWEY, J.F., 1974: Continental margins and ophiolite obduction: Appalachian–Caledonian system, *in* C.A. Burk and C.L. Drake, eds., The geology of continental margins; Springer-Verlag, New York, p. 933–950.

DEWEY, J.F., 1977: Suture zone complexities: A review; Tectonophysics, v. 40, p. 53–67.

DEWEY, J.F., 1980: Episodicity, sequence, and style at convergent plate boundaries, *in* D.W. Strangway, ed., The continental crust and its mineral deposits; Geol. Assoc. Can. Spec. Paper 20, p. 553–573.

DEWEY, J.F., 1982: Plate tectonics and the evolution of the British Isles; J. Geol. Soc. London, v. 139, p. 371–414.

DEWEY, J.F., and BURKE, K.C.A., 1973: Tibetan, Variscan, and Precambrian basement reactivation; Products of continental collision; J. Geol., v. 81, p. 683–692.

DEWEY, J.F., PITMANN, W.C., III., RYAN, W.B.F., and BONNIN, J., 1973: Plate tectonics and the evolution of the Alpine system; Geol. Soc. Am. Bull., v. 84, p. 3137–3180.

DEWEY, J.F., and SENGÖR, A.M.C., 1979: Aegean and surrounding regions: complex multiplate and continuum tectonics in a convergent zone; Geol. Soc. Am. Bull., v. 90, Pt. I, p. 84–92.

DICKINSON, W.R., 1970: Interpreting detrital modes of graywacke and arkose; J. Sediment. Petrol., v. 40, p. 695–707.

DICKINSON, W.R., 1974: Plate tectonics and sedimentation, *in* W.R. Dickinson, ed., Tectonics and sedimentation; Soc. Econ. Paleont. Mineral. Spec. Publ. 22, p. 1–27.

DICKINSON, W.R., 1980: Plate tectonics and key petrologic associations, *in* D.W. Strangway, ed., The continental crust and its mineral deposits; Geol. Assoc. Can. Spec. Paper 20, p. 341–360.

DICKINSON, W.R., 1981: Plate tectonics and the continental margin of California, *in* W.G. Ernst, ed., The geotectonic development of California; Prentice-Hall Inc., Englewood Cliffs, New Jersey, p. 1–28.

DICKINSON, W.R., and RICH, E.I., 1972: Petrologic intervals and petrofacies in the Great Valley sequence, Sacramento Valley, California; Geol. Soc. Am. Bull., v. 83, p. 3007–3024.

DICKINSON, W.R. and SEELY, D.R., 1979: Structure and stratigraphy of forearc regions; Am. Assoc. Petrol. Geol. Bull., v. 63, p. 2–31.

DICKINSON, W.R., and SUCZEK, C.A., 1979, Plate tectonics and sandstone compositions; Am. Assoc. Petrol. Geol. Bull., v. 63, p. 2164–2182.

DIMROTH, E., 1981: Labrador geosynclinal: type example of Early Proterozoic cratonic reactivation, *in* A. Kröner, ed., Precambrian plate tectonics; Develop. in Precamb. Geol. 4, Elsevier, Amsterdam, p. 331–352.

DIMROTH, E., and KIMBERLEY, M.M., 1976: Precambrian atmospheric oxygen: evidence in the sedimentary distributions of carbon, sulfur, uranium and iron; Can. J. Earth Sci., v. 13, p. 1161–1185.

DINGLE, R.V., 1973: Regional distribution and thickness of post-Paleozoic sediments on the continental margin of southern Africa; Geol. Mag., v. 110, p. 97–102.

DINGLE, R.V., 1980: Large allochthonous sediment masses and their role in the construction of the continental slope and rise off southwestern Africa, Mar. Geol., v. 37, p. 333–354.

DONALDSON, J.A., IRVING, E., TANNER, J., and McGLYNN, J., 1976: Stockwell symposium on the Hudsonian orogeny and plate tectonics, Geosci. Canada, v. 3, p. 285–291.

DOTT, R.H. Jr., 1974: The geosynclinal concept, *in* R.H. Dott, Jr. and R.H. Shaver, eds., Modern and ancient geosynclinal sedimentation; Soc. Econ. Paleont. Mineral. Spec. Publ. 19, p. 1–13.

DOTT, R.H., Jr., 1978: Tectonics and sedimentation a century later; Earth Sci. Rev., v. 14, p. 1–34.

DOUGLAS, R.J.W., GABRIELSE, H., WHEELER, J.O., SCOTT, D.F., and BELYEA, H.R., 1970: Geology of Western Canada, *in* R.J.W. Douglas, ed., Geology and Economic Minerals of Canada; Geol. Surv. Can. Econ. Geol. Rept. 1, p. 366–488.

DOUTCH, H.F., 1976: The Karumba Basin, northeastern Australia and southern New Guinea; Bur. Miner. Resour. J. Aust. Geol. Geophys., v. 1, p. 131–140.

DU TOIT, S.R., 1979: The Mesozoic history of the Agulhas Bank in terms of the plate tectonic theory; Geol. Soc. S. Africa Spec. Publ. 6, p. 197–203.

EISBACHER, G.H., 1981: Late Mesozoic—Paleogene Bowser Basin molasse and Cordilleran tectonics, Western Canada, *in* A.D. Miall, ed., Sedimentation and tectonics in alluvial basins; Geol. Assoc. Can. Spec. Paper 23, p. 125–151.

EISBACHER, G.H., CARRIGY, M.A., and CAMPBELL, R.B., 1974: Paleodrainage pattern and late orogenic basins of the Canadian Cordillera, *in* W.R. Dickinson, ed., Tectonics and sedimentation; Soc. Econ. Paleont. Mineral. Spec. Publ. 22, p. 143–166.

ERIKSSON, K.A., 1982: Sedimentation patterns in the Barberton Mountain land, South Africa, and the Pilbara Block, Australia: evidence for Archean rifted continental margins; Tectonophysics, v. 81, p. 179–194.

FAHRIG, W.F., IRVING, E., and JACKSON, J.D., 1973: Test of nature and extent of continental drift as provided by study of Proterozoic dike swarms of the Canadian Shield, *in* M.G. Pitcher, ed., Arctic Geology; Am. Assoc. Petrol. Geol. Mem. 19, p. 583–586.

FIELD, M.E., and CLARKE, S.H. Jr., 1979: Small-scale slumps and slides and their significance for basin slope processes, southern California borderland, *in* L.T. Doyle and O.H. Pilkey Jr., eds., Geology of continental slopes; Soc. Econ. Paleont. Mineral. Spec. Publ. 27, p. 223–230.

FISHER, M.A., and MAGOON, L.B., 1978: Geologic framework of Lower Cook Inlet, Alaska; Am. Assoc. Petrol. Geol. Bull., v. 62, p. 373–402.

FISHER, M.A., and VON HUENE, R., 1980, Structure of Upper Cenozoic strata beneath Kodiak Shelf, Alaska; Am. Assoc. Petrol. Geol. Bull. v. 64, p. 1014–1033.

FITCH, T.J., 1972: Plate convergence, transcurrent faults and internal deformation adjacent to southeast Asia and the western Pacific; J. Geophys. Res., v. 77, p. 4432–4460.

FOWLER, J.H., and KUENZI, W.D., 1978: Keweenawan turbidites in Michigan (deep borehole red beds): a foundered basin sequence developed during evolution of a protoceanic rift system; J. Geophys. Res., v. 83, p. B5833–B5843.

FRANCHETEAU, J., and LE PICHON, X., 1972: Marginal fracture zones as structural framework of continental margins in South Atlantic Ocean; Am. Assoc. Petrol. Geol. Bull., v. 56, p. 991–1007.

FRANZINELLI, E., and POTTER, P.E., 1983: Petrology, chemistry and texture of modern river sands, Amazon River system; J. Geol., v. 91, p. 23–40.

FRIEND, P.F., 1969: Tectonic features of Old Red sedimentation in North Atlantic borders, *in* M. Kay, ed., North Atlantic geology and continental drift; Am. Assoc. Petrol. Geol., Mem. 12, p. 703–710.

FRIEND, P.F., 1981: Devonian sedimentary basins and deep faults of the northernmost Atlantic borderlands, *in* J.W. Kerr and A.J. Fergusson eds., Geology of the North Atlantic borderlands; Can. Soc. Petrol. Geol. Mem. 7, p. 149–165.

FYFE, W.S., 1981: How do we recognize plate tectonics in very old rocks?, *in* A. Kröner, ed., Precambrian plate tectonics; Develop. in Precamb. Geol. 4, Elsevier, Amsterdam, p. 549–560.

GANSSER, A., 1964: Geology of the Himalayas; Interscience, London, 289 p.

GANSSER, A., 1980: The significance of the Himalayan suture zone; Tectonophysics, v. 62, p. 37–52.

GARFUNKEL, Z., ZAK, I., and FREUND, R., 1981: Active faulting in the Dead Sea Rift; Tectonophysics, v. 80, p. 1–26.

GOODWIN, A.M., 1981a: Precambrian perspectives; Science, v. 213, p. 55–61.

GOODWIN, A.M., 1981b: Archean plates and greenstone belts, *in* A. Kröner, ed., Precambrian plate tectonics; Develop. in Precamb. Geol. 4., Elsevier, Amsterdam, p. 105–135.

GRAHAM, S.A., DICKINSON, W.R., INGERSOLL, R.V., 1975: Himalayan Bengal model for flysch dispersal in the Appalachian Ouachita System; Geol. Soc. Am. Bull., v. 86, p. 273–286.

GUPTA, H.K., RAO, V.D., and SINGH, J., 1982: Continental collision tectonics: evidence from the Himalaya and the neighbouring regions; Tectonophysics, v. 81, p. 213–238.

HALL, J.M., 1981: The Thulean volcanic line, *in* J.W. Kerr and A.J. Fergusson, eds., Geology of the North Atlantic borderlands; Can. Soc. Petrol. Geol. Mem. 7, p. 231–244.

HALLAM, A.M., 1980: A reassessment of the fit of Pangaea components and the time of their initial breakup, *in* D.W. Strangway, ed., The continental crust and its mineral deposits; Geol. Assoc. Can. Spec. Paper 20, p. 375–387.

HAMBLIN, A.P., and WALKER, R.G., 1979: Storm-dominated shallow marine deposits: the Fernie–Kootenay (Jurassic) transition, Southern Rocky Mountains; Can. J. Earth Sci., v. 16, p. 1673–1690.

HAMILTON, W., 1977: Subduction in the Indonesian region, *in* Island arcs, deep sea trenches and backarc basins, Maurice Ewing Series, v. 1, Am. Geophys. Union, p. 15–31.

HAMILTON, W., 1979: Tectonics of the Indonesian Region, U.S. Geol. Survey Prof. Paper 1078.

HARDING, T.P., 1974: Petroleum traps associated with wrench faults; Am. Assoc. Petrol. Geol. Bull., v. 58, p. 1290–1304.

HARDING, T.P., and LOWELL, J.D., 1979: Structural styles, their plate-tectonic habitats, and hydrocarbon traps in petroleum provinces; Am. Assoc. Petrol. Geol. Bull., v. 63, p. 1016–1058.

HARLAND, W.B., 1971: Tectonic transpression in Caledonian Spitsbergen; Geol. Mag., v. 108, p. 27–42.

HAWKINGS, T.J., and HATLELID, W.G., 1975: The regional setting of the Taglu Field, *in* C.J. Yorath, E.R. Parker and D.J. Glass, eds., Canada's continental margins and offshore petroleum exploration; Can. Soc. Petrol. Geol. Mem. 4, p. 633–647.

HAYES, J.B., HARMS, J.C. and WILSON, T.W., 1976: Contrasts between braided and meandering stream deposits, Beluga and Sterling formations (Tertiary), Cook Inlet, Alaska, *in* Recent and ancient sedimentary environments in Alaska; Alaska Geol. Soc. Proc., p. J1–J27.

HAYES, M.O., and MICHEL, J., 1982: Shoreline sedimentation within a forearc embayment, Lower Cook Inlet, Alaska; J. Sediment. Petrol., v. 52, p. 251–264.

HERMAN, B.M., ANDERSON, R.N. and TRUCHAN, M., 1978: Extensional tectonics in the Okinawa Trough, *in* J.S. Watkins, L. Montadert and P.W. Dickerson, eds., Geological and geophysical investigations of continental margins; Am. Assoc. Petrol. Geol. Mem. 29, p. 199–208.

HOBDAY, D.K., and VON BRUNN, V., 1979: Fluvial sedimentation and paleogeography of an early Paleozoic failed rift, southeastern margin of Africa; Palaeogeog., Palaeoclim., Palaeoec., v. 28, p. 169–184.

HOFFMAN, P.F., 1973: Evolution of an early Proterozoic continental margin: the Coronation Geosyncline and associated aulacogens of the NW Canadian Shield; Roy Soc. London Phil. Trans., v. A273, p. 547–581.

HOFFMAN, P.F., 1980: Wopmay orogen: a Wilson cycle of early Proterozoic age in the northwest of the Canadian Shield, *in* D.W. Strangway, ed., The continental crust and its mineral deposits; Geol. Assoc. Can. Spec. Paper 20, p. 523–549.

HOFFMAN, P.F., DEWEY, J.F., and BURKE, K., 1974: Aulacogens and their genetic relation to geosynclines, with a Proterozoic example from Great Slave Lake, Canada, *in* R.H. Dott, Jr., and R.H. Shaver, eds., Modern and ancient geoosynclinal sedimentation; Soc. Econ. Paleont. Mineral. Spec. Publ. 19, p. 38–55.

HOWELL, D.G., CROUCH, J.K., GREENE, H.G., McCULLOCH, D.S., and VEDDER, J.G., 1980: Basin development along the late Mesozoic and Cainozoic California margin: a plate tectonic margin of subduction, oblique subduction and transform tectonics, *in* P.F. Ballance and H.G. Reading, eds., Sedimentation in oblique-slip mobile zones; Internat. Assoc. Sediment. Spec. Pub. 4, p. 43–62.

HOWELL, D.G., and LINK, M.H., 1979: Eocene conglomerate sedimentology and basin analysis, San Diego and the southern California borderland; J. Sediment. Petrol., v. 49, p. 517–540.

HOWELL, D.G., and VEDDER, J., 1981: Structural implications of stratigraphic discontinuities across the southern California borderland, *in* W.G. Ernst, ed., the Geotectonic development of California; Prentice Hall Inc., Englewood Cliffs, New Jersey, p. 535–558.

HSÜ, K.J., 1970: The meaning of the word flysch—a short historical search, *in* J. Lajoie, ed., Flysch sedimentology in North America; Geol. Assoc. Can. Spec. Paper 7, p. 1–11.

HSÜ, K.J., 1974: Melanges and their distinction from olistostromes, *in* R.H. Dott Jr. and R.H. Shaver, eds., Modern and ancient geosynclinal sedimentation; Soc. Econ. Paleont. Mineral. Spec. Publ. 19, p. 321–333.

HSÜ, K.J., CITA, M.B., and RYAN, W.B.F., 1973: The origin of the Mediterranean evaporites, *in* W.B.F. Ryan, K.J. Hsü et. al., Initial Reports of the Deep Sea Drilling Project, v. 13, p. 1203–1231.

HSÜ, K.J., KELTS, K., and VALENTINE, J.W., 1980: Resedimented facies in Ventura Basin, California, and model of longitudinal transport of turbidity currents; Am. Assoc. Petrol. Geol. Bull., v. 64, p. 1034–1051.

HUBERT, J.F., REED, A.A., and CAREY, P.J., 1976: Paleogeography of the East Berlin Formation, Newark Group, Connecticut Valley; Am. J. Sci., v. 276, p. 1183–1207.

HYDEN, G. and TANNER, P.W.G., 1981: Late Palaeozoic–early Mesozoic fore-arc basin sedimentary rocks at the Pacific margin in western Antarctica; Geol. Rundschau., v. 70, p. 529–541.

INGERSOLL, R.V., 1978a: Submarine fan facies of the Upper Cretaceous Great Valley Sequence, northern and central California; Sediment. Geol., v. 21, p. 205–230.

INGERSOLL, R.V., 1978b: Petrofacies and petrologic evolution of the late Cretaceous fore-arc basin, northern and central California; J. Geol., v. 86, p. 335–352.

INGERSOLL, R.V., 1979: Evolution of the Late Cretaceous forearc basin, northern and central California; Geol. Soc. Am. Bull., v. 90, Pt. 1, p. 813–826.

INGERSOLL, R.V., 1982: Triple-junction instability as a cause for late Cenozoic extension and fragmentation of the western United States; Geology, v. 10, p. 621–624.

INGERSOLL, R.V., and DICKINSON, W.R., 1981: Great Valley Group (Sequence), Sacramento Valley, California, *in* V. Frizzell, ed., Upper Mesozoic Franciscan rocks and Great Valley Sequence, Central Coast Ranges, California; Pacific Sect., Soc. Econ. Paleont. Mineral. Guidebook, p. 1–33.

JOHNSON, G.D., JOHNSON, N.M., OPDYKE, N.D., and TAHIRKHELI, R.A.K., 1979: Magnetic reversal stratigraphy and sedimentary tectonic history of the Upper Siwalik Group, eastern Salt Range and southwestern Kashmir, *in* A. Farah and K.A. DeJong, eds., Geodynamics of Pakistan; Geol. Surv. Pakistan, Quetta, Pakistan, p. 149–165.

JOHNSON, G.D. and VONDRA, C.F., 1972: Siwalik sediments in a portion of the Punjab re-entrant: the sequence at Haritalyangar, District Bilaspur, H.P.; Himalayan Geol., v. 2, p. 118–144.

JONES, D.L., SILBERLING, N.J., and HILLHOUSE, J., 1977: Wrangellia—a displaced terrane in northwestern North America; Can. J. Earth Sci., v. 14, p. 2565–2577.

KARIG, D.E., 1971a: Origin and development of marginal basins in the western Pacific; J. Geophys. Res., v. 76, p. 2542–2561.

KARIG, D.E., 1971b: Structural history of the Mariana Island arc system: Geol. Soc. Amer. Bull., v. 82, p. 323–344.

KARIG, D.E., and MOORE, G.F., 1975: Tectonically controlled sedimentation in marginal basins; Earth Plan. Sci. Letters, v. 26, p. 233–238.

KARIG, D.E., and SHARMAN, G.F., III, 1975: Subduction and accretion in trenches; Geol. Soc. Amer. Bull., v. 86, p. 377–389.

KARIG, D.E., SUPARKA, S., MOORE, G.F., and HEHANUSSA, P.E., 1978: Structure and Cenozoic evolution of the Sunda Arc in the central Sumatra region, *in* J.S. Watkins, L. Montadert and P.W. Dickerson, eds., Geological and geophysical investigations of continental margins; Am. Assoc. Petrol. Geol. Mem. 29, p. 223–237.

KATZ, H.R., 1971: Continental margin in Chile—is tectonic style compressional or extensional?; Am. Assoc. Petrol. Geol. Bull., v. 55, p. 1753–1758.

KAY, M., 1951: North American geosynclines; Geol. Soc. Amer. Mem. 48.

KEEN, C.E., 1979: Thermal history and subsidence of rifted continental margins—evidence from wells on the Nova Scotia and Labrador Shelves; Can. J. Earth Sci., v. 16, p. 505–522.

KELTS, K., 1981: A comparison of some aspects of sedimentation and translational tectonics from the Gulf of California and the Mesozoic Tethys, Northern Penninic Margin; Eclog. Geol. Helv., v. 74, p. 317–338.

KENT, D.V., and OPDYKE, N.D., 1978: Paleomagnetism of the Devonian Catskill red beds: evidence for motion of the coastal New England—Canadian Maritime region relative to cratonic North America; J. Geophys. Res., v. 83, p. 4441–4450.

KLEIN, G. deV., 1982: Probable sequential arrangement of depositional systems on cratons; Geology, v. 10, p. 17–22.

KLEIN, G. deV., and LEE, Y.I., 1983: A preliminary assessment of geodynamic controls on depositional systems and sandstone diagenesis in back-arc basins, western Pacific Ocean, Tectonophysics, in press.

KRÖNER, A., ed., 1981a: Precambrian plate tectonics; Develop. in Precamb. Geol. 4, Elsevier, Amsterdam, 781 p.

KRÖNER, A., 1981b: Precambrian plate tectonics, *in* A. Kroner, ed., Precambrian plate tectonics; Develop. in Precamb. Geol. 4, Elsevier, Amsterdam, p. 57–90.
KRUMBEIN, W.C., and SLOSS, L.L., 1963: Stratigraphy and sedimentation; W.H. Freeman, San Francisco, 660 p.
KRYNINE, P.D., 1942: Differential sedimentation and its products during one complete geosynclinal cycle; Ann. Prim. Congr. Panamericano de Ingen. de Minas Geolog. (Chile), Geologia 1st Pt., v. 2, p. 537–561.
KUENZI, W.D., HORST, O.H., and McGEHEE, R.V., 1979: Effect of volcanic activity on fluvial—deltaic sedimentation in a modern arc-trench gap, southwestern Guatemala; Geol. Soc. Amer. Bull., v. 90, Pt. 1, p. 827–838.
KULM, L.D., and FOWLER, G.A., 1974: Cenozoic sedimentary framework of the Gorda–Juan de Fuca Plate and adjacent continental margin—a review, *in* R.H. Dott Jr. and R.H. Shaver, eds., Modern and ancient geosynclinal sedimentation; Soc. Econ. Paleont. Mineral. Spec. Publ. 19, p. 212–239.
KUMAR, N., 1981: Geologic history of north and northeastern Brazilian margin: controls imposed by seafloor spreading on the continental structures, *in* J.W. Kerr and A.J. Fergusson, eds., Geology of the North Atlantic borderlands, Can. Soc. Petrol. Geol. Mem. 7, p. 527–542.
LAJOIE, J., ed., 1970: Flysch sedimentology in North America; Geol. Assoc. Can. Spec. Paper 7.
LEEDER, M.R., 1976: Sedimentary facies and the origins of basin subsidence along the northern margin of the supposed Hercynian Ocean; Tectonophysics, v. 36, p. 167–180.
LEEDER, M.R., 1982: Upper Paleozoic basins of the British Isles—Caledonide inheritance versus Hercynian plate margin processes; J. Geol. Soc. London, v. 139, p. 479–491.
LeFORT, P., 1975: Himalayas: the collided range, present knowledge of the continental arc; Am. J. Sci., v. 275-A, p. 1–44.
LEGGETT, J.K., 1980: The sedimentological evolution of a Lower Paleozoic accretionary fore-arc in the Southern Uplands of Scotland; Sedimentology, v. 27, p. 401–418.
LE PICHON, X., ANGELIER, J., and SIBUET, J-C., 1982: Plate boundaries and extensional tectonics; Tectonophysics, v. 81, p. 239–256.
LEWIS, K.B., 1980: Quaternary sedimentation on the Hikurangi oblique-subduction and transform margin, New Zealand, *in* P.F. Ballance and H.G. Reading, eds., Sedimentation in oblique-slip mobile zones, Internat. Assoc. Sed. Spec. Publ. 4, p. 171–189.
LINK, M.H., and OSBORNE, R.H., 1978: Lacustrine facies in the Pliocene Ridge Basin Group: Ridge Basin, California, *in* A. Matter and M.E. Tucker, eds., Modern and ancient lake sediments; Internat. Assoc. Sediment. Spec. Publ. 2, p. 169–187.
LOGATCHEV, N.A., ROGOZHINA, V.A., and SOLONENKO, V.P., 1978: Deep structure and evolution of the Baikal Rift Zone, *in* I.B. Ramberg and E.R. Neumann, eds., Tectonics and geophysics of continental rifts; D. Reidel Pub. Co., Dordrecht, Holland, p. 49–61.
LOWE, D.R., and KNAUTH, L.P., 1977: Sedimentology of the Onverwacht Group (3.4 billion years) Transvaal, South Africa, and its bearing on the characteristics and evolution of the early earth; J. Geol., v. 85, p. 699–723.
LOWELL, J.D., and GENIK, G.J., 1972: Sea-floor spreading and structural evolution of southern Red Sea; Am. Assoc. Petrol. Geol. Bull., v. 56, p. 247–259.
MALOUTA, D.N., GORSLINE, D.S., and THORNTON, S.E., 1981: Processes and rates of recent (Holocene) basin filling in an active transform margin: Santa Monica Basin, California continental borderland; J. Sediment. Petrol., v. 51, p. 1077–1096.
MART, Y., and HOROWITZ, A., 1981: The tectonics of the Timna region in southern Israel and the evolution of the Dead Sea rift; Tectonophysics, v. 79, p. 165–200.
MASCLE, A., and BISCARRAT, P.A., 1978: The Sulu Sea: a marginal basin in southeast Asia, *in* J.S. Watkins, L. Montadert and P.W. Dickerson, eds., Geological and geophysical investigations of continental margins, Am. Assoc. Petrol. Geol. Mem. 29, p. 373–381.
McCONNELL, R.B., 1977: East African Rift System dynamics in view of Mesozoic apparent polar wander; J. Geol. Soc. London, v. 134, p. 33–39.
McCONNELL, R.B., 1980: A resurgent taphrogenic lineament of Precambrian origin in eastern Africa; J. Geol. Soc. London, v. 137, p. 483–489.
McGETCHIN, T.R., BURKE, K.C., THOMPSON, G.A., and YOUNG, R.A., 1980: Mode and mechanisms of plateau uplifts, *in* A.W. Bally, P.L. Bender, T.R. McGetchin and R.I. Walcott, eds., Dynamics of plate interiors; Am. Geophys. Union and Geol. Soc. Am., Geodynamics Ser., v. 1, p. 99–110.
McKENZIE, D., 1978: Active tectonics of the Alpine–Himalayan belt: the Aegean Sea and surrounding regions; Geophys. J. Roy. Astron. Soc., v. 55, p. 217–254.
McKENZIE, D. and SCLATER, J.G., 1971: The evolution of the Indian Ocean Since the Late Cretaceous: Geophys. J. Roy. Astron. Soc., v. 25, p. 437–528.
McWHAE, J.R.H., 1981: Structure and spreading history of the northwestern Atlantic region from the Scotian Shelf to Baffin Bay, *in* J.W. Kerr and A.J. Fergusson, eds., Geology of the North Atlantic borderlands, Can. Soc. Petrol. Geol. Mem. 7, p. 299–332.
MIALL, A.D., ed., 1981a: Sedimentation and tectonics in alluvial basins; Geol. Assoc. Can. Spec. Paper 23.
MIALL, A.D., 1981b: Alluvial sedimentary basins: tectonic setting and basin architecture, *in* A.D. Miall, ed., Sedimentation and tectonics in alluvial basins; Geol. Assoc. Can. Spec. Paper 23, p. 1–33.
MIALL, A.D., BALKWILL, H.R., and HOPKINS, W.S., Jr., 1980: Cretaceous and Tertiary sediments of Eclipse Trough, Bylot Island area, Arctic Canada, and their regional setting; Geol. Surv. Can. Paper, 79–23.
MIDDLETON, G.V., 1960: Chemical composition of

sandstones; Geol. Soc. Am. Bull., v. 71, p. 1011–1026.

MITCHELL, A.H.G. and McKERROW, W.S., 1975: Analogous evolution of the Burma Orogen and the Scottish Caledonides; Geol. Soc. Am. Bull., v. 86, p. 305–315.

MITCHELL, A.H.G., and READING, H.G., 1978: Sedimentation and tectonics, *in* H.G. Reading, ed., Sedimentary environments and facies; Elsevier, p. 439–476.

MIYASHIRO, A., 1980: Metamorphism and plate convergence, *in* D.W. Strangway, ed., The Continental crust and its mineral deposits; Geol. Assoc. Can. Spec. Paper 20, p. 591–605.

MOLNAR, P., and ATWATER, T., 1978: Interarc spreading and Cordilleran tectonics as alternates related to the age of subducted oceanic lithosphere; Earth Planet. Sci. Letters, v. 41, p. 330–340.

MOLNAR, P. and TAPPONNIER, P., 1975: Cenozoic tectonics of Asia: effects of a continental collision; Science, v. 189, p. 419–426.

MONGER, J.W.H. and PRICE, R.A., 1979: Geodynamic evolution of the Canadian Cordillera—progress and problems; Can. J. Earth Sci., v. 16, p. 770–791.

MOORBATH, S., and WINDLEY, B.F., eds., 1981: The origin and evolution of the earth's continental crust; Roy. Soc. London Phil. Trans., v. A301, p. 183–487.

MOORE, D.G., CURRAY, J.R., and EMMEL, F.J., 1976: Large submarine slide (olistostrome) associated with Sunda Arc subduction zone, northeast Indian Ocean; Marine Geol., v. 21, p. 211–226.

MOORE, G.F., and KARIG, D.E., 1976: Development of sedimentary basins on the lower trench slope; Geology, v. 4, p. 693–697.

MOORE, G.F., and KARIG, D.E., 1980: Structural geology of Nias Island, Indonesia: implications for subduction zone tectonics; Am. J. Sci., v. 280, p. 193–223.

MOORE, J.C. and KARIG, D.E., 1976: Sedimentology, structural geology, and tectonics of the Shikiku subduction zone, southwestern Japan; Geol. Soc. Am. Bull., v. 87, p. 1259–1268.

MOORE, J.C., and WHEELER, R.L., 1978: Structural fabric of a melange, Kodiak Islands, Alaska; Am. J. Sci., v. 278, p. 739–765.

MORGAN, W.J., 1971: Convection plumes in the lower mantle; Nature, v. 230, p. 42–43.

MORGAN, W.J., 1972: Deep mantle convection plumes and plate motions; Am. Assoc. Petrol. Geol. Bull., v. 56, p. 203–213.

NALIVKIN, D.V., 1973: Geology of the USSR; Univ. Toronto Press, 855 p. (English translation by N. Rast).

NARDIN, T.R., and HENYEY, T.L., 1978: Pliocene–Pleistocene diastrophism of Santa Monica and San Pedro Shelves, California continental borderland; Am. Assoc. Petrol. Geol. Bull., v. 62, p. 247–272.

NILSEN, T.H., and CLARKE, S.H., Jr., 1975: Sedimentation and tectonics in the Early Tertiary continental borderland of central California; U.S. Geol. Surv. Prof. Paper 925.

NORRIS, R.J., CARTER, R.M., and TURNBULL, I.M., 1978: Cainozoic sedimentation in basins adjacent to a major continental transform boundary in southern New Zealand; J. Geol. Soc. London, v. 135, p. 191–206.

OLIVER, J., 1982: Tracing surface features to great depths: a powerful means for exploring the deep crust; Tectonophysics, v. 81, p. 257–272.

PARKASH, B., SHARMA, R.P., and ROY, A.K., 1980: The Siwalik Group (molasse) sediments shed by collision of continental plates; Sediment. Geol., v. 25, p. 127–159.

PAUTOT, G., RENARD, V., DANIEL, J., and DUPONT, J., 1973: Morphology, limits, origin and age of salt layer along South Atlantic African margin; Am. Assoc. Petrol. Geol. Bull., v. 57, p. 1658–1671.

PETTIJOHN, F.J., POTTER, P.E., and SIEVER, R., 1972: Sand and sandstone; Springer-Verlag, New York, 618 p.

PHILLIPS, W.E.A., STILLMAN, C.J., and MURPHY,T., 1976: A Caledonian plate tectonic model; J. Geol. Soc. London, v. 132, p. 579–609.

PONTE, F.C., and ASMUS, H.E., 1976: The Brazilian marginal basins: current state of knowledge, *in* Proc. Internat. Symp. on the continental margins of Atlantic Type, Sao Paulo, Brazil, Oct. 1975; Academia Brasileria de Ciencias, Anais, 48, Suppl., p. 215–240.

PONTE, F.C., FONSECA, J.dR., and CAROZZI, A.V., 1980: Petroleum habitats in the Mesozoic–Cenozoic of the continental margin of Brazil, *in* A.D. Miall, ed., Facts and principles of world petroleum occurrence; Can. Soc. Petrol. Geol. Mem. 6, p. 857–886.

POTTER, P.E., 1978: Petrology and composition of modern big-river sands, J. Geol:, v. 86, p. 423–449.

POWELL, C. McA., 1979: A speculative tectonic history of Pakistan and surroundings: some constraints from the Indian Ocean, *in* A. Farah and K.A. DeJong, eds., the Geodynamics of Pakistan; Geol. Surv. Pakistan, Quetta, Pakistan, p. 5–24.

POWELL, C.McA., and CONAGHAN, P.J., 1973: Plate tectonics and the Himalayas; Earth Plan. Sci. Letters, v. 20, p. 1–12.

PRICE, R.A. and MOUNTJOY, E.W., 1970: Geologic structure of the Canadian Rocky Mountains between Bow and Athabasca Rivers—a progress report; Geol. Assoc. Can. Spec. Paper 6, p. 7–26.

QUENNELL, A.M., 1959: Tectonics of the Dead Sea rift; 20th Int. Geol. Congr. Mexico; Assoc. Serv. Geol. Afr., p. 385–405.

READING, H.G., 1972: Global tectonics and the genesis of flysch successions; 24th Internat. Geol. Congr. Proc., Sect. 6, p. 59–66.

READING, H.G., 1980: Characteristics and recognition of strike-slip fault systems, *in* P.F. Ballance and H.G. Reading, eds., Sedimentation in oblique-slip mobile zones; Internat. Assoc. Sediment. Spec. Publ. 4, p. 7–26.

RONA, P.A., 1982: Evaporites at passive margins, *in* R.A. Scutton, ed., Dynamics of passive margins; Am. Geophys. Union and Geol. Soc. Am., Geodynamics Ser. v. 6, p. 116–132.

SCHENK, P.E., 1981: The Meguma zone of Nova Scotia—A remnant of western Europe, South America, or Africa, *in* J.W. Kerr and A.J. Fergusson, eds., Geology of the North Atlantic borderlands; Can. Soc. Petrol. Geol. Mem. 7, p. 119–148.

SCHOLZ, C.H., BARAZANGI, M., and SBAR, M.L., 1971: Late Cenozoic evolution of the Great Basin, western United States, as an ensialic interarc basin; Geol. Soc. Am. Bull., v. 82, p. 2979–2990.

SCHWAB, F.L., 1971: Geosynclinal compositions and the new global tectonics; J. Sediment. Petrol., v. 41, p. 928–938.

SCHWAB, F.L., 1975: Framework mineralogy and chemical composition of continental margin-type sandstone; Geology, v. 3, p. 487–490.

SCHWAB, F.L., 1978: Secular trends in the composition of sedimentary rock assemblages—Archean through Phanerozoic time; Geology, v. 6, p. 532–536.

SCHWAB, F.L., 1981: Evolution of the western continental margin, French–Italian Alps: sandstone mineralogy as an index of plate tectonic setting; J. Geol., v. 89, p. 349–368.

SCLATER, J.G., ANDERSON, R.N., and BELL, M.L., 1971: The elevation of ridges and the evolution of the central eastern Pacific; J. Geophys. Res., v. 76, p. 7888–7915.

SCLATER, J.G., and CHRISTIE, P.A.F., 1980: Continental stretching: an explanation of the post-mid Cretaceous subsidence of the central North Sea Basin; J. Geophys. Res., v. 85, p. 3711–3739.

SEARLE, M.P., and GRAHAM, G.M., 1982: "Oman Exotics"—Oceanic carbonate build-ups associated with the early stages of continental rifting; Geology, v. 10, p. 43–49.

SENGÖR, A.M.C., 1976: Collision of irregular continental margins: implications for foreland deformation of Alpine-type orogens; Geology, v. 4, p. 779–782.

SENGÖR, A.M.C., and BURKE, K., 1978: Relative timing of rifting and volcanism on earth and its tectonic implications; Geophys. Res. Letters, v. 5, p. 419–421.

SHERIDAN, R.E., 1981: Recent research on passive continental margins, *in* J.E. Warme, R.G. Douglas and E.L. Winterer, eds., The Deep Sea Drilling Project: a decade of progress; Soc. Econ. Paleont. Mineral. Spec. Publ. 32, p. 39–55.

SHERIDAN, R.E., CROSBY, J.T., KENT, K.M., DILLON, W.P., and PAULL, C.K., 1981: The geology of the Blake Plateau and Bahamas regions, *in* J.W. Kerr and A.J. Fergusson, eds., Geology of the North Atlantic borderlands; Can. Soc. Petrol. Geol. Mem. 7, p. 487–502.

SIEVER, R. and HAGER, J.L., 1981: Paleogeography, tectonics and thermal history of some Atlantic margin sediments, *in* J.W. Kerr and A.J. Fergusson, eds., Geology of the North Atlantic borderlands; Can. Soc. Petrol. Geol. Mem. 7, p. 95–117.

SIMPSON, E.S.W., 1977: Evolution of the South Atlantic; Geol. Soc. S. Afr. Trans., Annexure to v. 80, 15 p.

SIMS, P.K., CARD, K.D., and LUMBERS, S.B., 1981: Evolution of early Proterzoic basins of the Great Lakes region, *in* F.H.A. Campbell, ed., Proterozoic basins of Canada; Geol. Survey Canada, Paper 81–10, p. 379–397.

SIMS, P.K., CARD, K.D., MOREY, G.B. and PETERMAN, Z.E., 1980: The Great Lakes tectonic zone—a major Precambrian crustal structure in central North America; Geol. Soc. Amer. Bull., v. 91, p. 690–698.

SLEEP, N.H., 1971: Thermal effects of the formation of Atlantic continental margins by continental breakup; Geophys. J. Roy. Astron. Soc., v. 24, p. 325–350.

SLEEP, N.H., and SLOSS, L.L., 1980: The Michigan Basin, *in* A.W. Bally, P.L. Bender, T.R. McGetchin and R.I. Walcott, eds., Dynamics of plate interiors, Am. Geophys. Union. and Geol. Soc. Am., Geodynamics Ser., v. 1, p. 93–98.

SLEEP, N.H., and WINDLEY, B.F., 1982: Archean plate tectonics: constraints and inferences; J. Geol., v. 90, p. 363–380.

SMELLIE, J.L., 1981: A complete arc-trench system recognized in Gondwana sequences of the Antarctic Peninsula region; Geol. Mag., v. 118, p. 139–159.

SMITHSON, S.B., BREWER, J.A., KAUFMAN, S., and OLIVER, J., 1979: Structure of the Laramide Wind River Uplift, Wyoming, from COCORP deep reflection data and from gravity data; J. Geophys. Res., v. 84, p. 5955–5972.

SMITHSON, S.B., BREWER, J.A., KAUFMAN, S., OLIVER., J., and HURICH, C., 1978: Nature of the Wind River thrust, Wyoming, from COCORP deep reflection and gravity data; Geology, v. 6, p. 648–652.

SPÖRLI, K.B., 1980: New Zealand and oblique-slip margins: tectonic development up to and during the Cainozoic, *in* P.F. Ballance and H.G. Reading, eds., Sedimentation in oblique-slip mobile zones; Internat. Assoc. Sedimentologists Special Publ. 4, p. 147–170.

STEEL, R.J., 1976: Devonian basins of western Norway—sedimentary response to tectonism and varying tectonic context; Tectonophysics, v. 36, p. 207–224.

STEEL, R.J., DALLAND, A., KALGRAFF, K. and LARSEN, V., 1981: The Central Tertiary Basin of Spitsbergen: sedimentary development of a sheared-margin basin, *in* J.W. Kerr and A.J. Fergusson, eds., Geology of the North Atlantic borderlands; Can. Soc. Petrol. Geol. Mem. 7, p. 647–664.

STEEL, R.J., and GLOPPEN, T.G., 1980: Late Caledonian (Devonian) basin formation, western Norway: signs of strike-slip tectonics during infilling, *in* P.F. Ballance and H.G. Reading, eds. Sedimentation in oblique-slip mobile zones; Internat. Assoc. Sedimentologists Spec. Publ. 4, p. 79–103.

STEWART, J.H., 1972: Initial deposits in the Cordilleran Geosyncline: evidence of a late Precambrian (<850 m.y.) separation; Geol. Soc. Amer. Bull., v. 83, p. 1345–1360.

STEWART, J.H., 1978: Basin-range structure in western North America: a review, *in* R.B. Smith and G.P. Eaton, eds., Cenozoic tectonics and regional geophysics of the Western Cordillera; Geol. Soc. Amer. Mem. 152, p. 1–31.

ST. JULIEN, P., and HUBERT, C., 1975: Evolution

of the Taconian orogen in the Quebec Appalachians; Am. J. Sci., v. 275–A, p. 337–362.
STOCKLIN, J., 1977: Structural correlation of the Alpine ranges between Iran and central Asia; Mem. h. Ser. Soc. Geol. France, 8, p. 333–353.
STODDART, D.R., 1971: World erosion and sedimentation, *in* R.J. Chorley, ed., Introduction to fluvial processes; Methuen and Co., p. 8–29.
STRANGWAY, D.W., ed., 1980: The continental crust and its mineral deposits; Geol. Assoc. Can. Spec. Paper 20, 804 p.
STRONG, D.F., DICSON, W.L., O'DRISCOLL, C.F., KEAN, B.F., and STEVENS, R.K., 1974: Geochemical evidence for an east-dipping Appalachian subduction zone in Newfoundland; Nature, v. 248, p. 37–39.
SURLYK, F., CLEMMENSEN, L.B., and LARSEN, H.C., 1981: Post-Paleozoic evolution of the East Greenland continental margin, *in* J.W. Kerr and A.J. Fergusson, eds., Geology of the north Atlantic borderlands; Can., Soc. Petrol. Geol. Mem. 7, p. 611–646.
SUTTNER, L.J., 1974: Sedimentary petrographic provinces: an evaluation, *in* C.A. Ross, ed., Paleogeographic provinces and provinciality; Soc. Econ. Paleont. Mineral. Spec. Publ. 21, p. 75–84.
SUTTNER, L.J., BASU, A., and MACK, G. H., 1981: Climate and the origin of quartz arenites; J. Sediment. Petrol., v. 51, p. 1235–1246.
TALWANI, M., MUTTER, J., HOUTZ, R., and KONIG, M., 1978: The crustal structure and evolution of the area underlying the magnetic quiet zone on the margin south of Australia, *in* J.S. Watkins, L. Montadert and P.W. Dickerson, eds., Geological and geophysical investigations of continental margins; Am. Assoc. Petrol. Geol., Mem. 29, p. 151–175.
TANKARD, A.J., JACKSON, M.P.A., ERIKSSON, K.A., HOBDAY, D.K., HUNTER, D.R. and MINTER, W.E.L., 1982: Crustal evolution of Southern Africa; Springer-Verlag, New York, 523 p.
TEMPELMAN-KLUIT, D.J., 1979: Transported cataclasite, ophiolite and granodiorite in Yukon: evidence of arc-continent collision; Geol. Surv. Can. Paper 79–14.
TIERCELIN, J.J. and FAURE, H., 1978: Rates of sedimentation and vertical subsidence in neorifts and paleorifts, *in* I.B. Ramberg and E.R. Neumann, eds., Tectonics and geophysics of continental rifts; D. Reidel Pub. Co., Dordrecht, Holland, p. 41–47.
TOKSÖZ, M.N., and BIRD, P., 1977: Formation and evolution of marginal basins and continental plateaus, *in* M. Talwani and W.C. Pitman III, eds., Island arcs, deep sea trenches and back-arc basins; Maurice Ewing Ser. 1, Am. Geophys. Union., p. 379–393.
TRETTIN, H.P., FRISCH, T.O., SOBCZAK, L.W., WEBER, J.R., NIBLETT, E.R., LAW, L.K., DELAURIER, J.M., and WHITHAM, K., 1972: The Innuitian Province, *in* R.A. Price and R.J.W. Douglas, eds., Variations in tectonic styles in Canada; Geol. Assoc. Canada Spec. Paper 11, p. 83–179.
UYEDA, S., 1981: Subduction zones and back arc basins—a review, Geol. Rundschau., v. 70, p. 552–569.
UYEDA, S., 1982: Subduction zones: an introduction to comparative subductology; Tectonophysics, v. 81, p. 133–159.
UYEDA, S., and MIYASHIRO, A., 1974: Plate tectonics and the Japanese Islands: a synthesis; Geol. Soc. Amer. Bull., v. 85, p. 1159–1170.
VALLONI, R. and MAYNARD, J.B., 1981: Detrital modes of recent deep–sea sands and their relation to tectonic setting: a first approximation; Sedimentology, v. 28, p. 75–84.
VAN DER VOO, R., 1979: Paleozoic assembly of Pangea: a new plate–tectonic model for the Taconic, Caledonian, and Hercynian orogenies; Eos, v. 60, p. 241.
VAN DER VOO, R., and SCOTESE, C., 1981: Paleomagnetic evidence for a large (~2,000 km) sinistral offset along the Great Glen fault during Carboniferous time; Geology, v. 9, p. 583–589.
VAN HOUTEN, F.B., 1973: Meaning of molasse; Geol. Soc. Am. Bull., v. 84, p. 1973–1975.
VAN HOUTEN, F.B., 1981: The odyssey of molasse, *in* A.D. Miall, ed., Sedimentation and tectonics in alluvial basins; Geol. Assoc. Can. Spec. Paper 23, p. 35–48.
VAN SCHMUS, W.R., and BICKFORD, M.E., 1981: Proterozoic chronology and evolution of the Mid-Continent region; North America, *in* A. Kröner, ed., Precambrian plate-tectonics; Develop. in Precambrian Geol. 4, Elsevier, p. 261–296.
VEEVERS, J.J., 1980: Basins of the Australian craton and margin, *in* A.W. Bally, P.L. Bender, T.R. McGetchin and R.I. Walcott, eds., Dynamics of plate interiors; Am. Geophys. Union and Geol. Soc. Am., Geodynamics Ser. 1, p. 73–80.
VEEVERS, J.J., 1981: Morphotectonics of rifted continental margins in Embryo (East Africa), youth (Africa-Arabia), and maturity (Australia); J. Geol., v. 89, p. 57–82.
VISSER, C.F., and JOHNSON, G.D., 1978: Tectonic control of Late Pliocene molasse sedimentation in a portion of the Jhelum re-entrant, Pakistan; Geol. Rundschau., v. 67, p. 15–37.
VON HUENE, R., 1978: Structure of the outer convergent margin off Kodiak Island, Alaska, from multichannel seismic records, *in* J.S. Watkins, L. Montadert and P.W. Dickerson, eds., Geological and geophysical investigations of continental margins; Am. Assoc. Petrol. Geol. Mem. 29, p. 261–272.
VON HUENE, R., 1981: Review of early results from drilling of the IPOD-1 active margin transects across the Japan, Mariana, and middle-America convergent margins, *in* J.E. Warme, R.G. Douglas and E.L. Winterer, eds., The Deep Sea Drilling Project: A decade of progress; Soc. Econ. Paleont. Mineral. Spec. Publ. 32, p. 57–66.
WADE, J.A., Geology of the Canadian Atlantic margin from Georges Bank to the Grand Banks, *in* J.W. Kerr and A.J. Fergusson, eds., Geology of the North Atlantic borderlands; Can. Soc. Petrol. Geol. Mem. 7, p. 447–460.
WARD, W.C., and McDONALD, K.C., 1979: Nubia Formation of central eastern desert, Egypt—major subdivisions and depositional setting; Am. Assoc. Petrol.

Geol. Bull., v. 63, p. 975–983.
WEIMER, R.J., 1970: Rates of deltaic sedimentation and intrabasin deformation, Upper Cretaceous of Rocky Mountain region, *in* J.P. Morgan, ed., Deltaic sedimentation modern and ancient, Soc. Econ. Paleont. Mineral. Spec. Publ. 15, p. 270–292.
WEST, G.F., 1980: Formation of continental crust, *in* D.W. Strangway, ed., The continental crust and its mineral deposits; Geol. Assoc. Can. Spec. Paper 20, p. 117–148.
WILCOX, R.E., HARDING, T.P. and SEELY, D.R., 1973: Basic wrench tectonics; Am. Assoc. Petrol. Geol. Bull., v. 57, p. 74–96.
WILLIAMS, H. (compiler), 1978: Tectonic lithofacies map of the Appalachian Orogen; Memorial Univ. Newfoundland, Map 1.
WILLIAMS, H., 1980: Structural telescoping across the Appalachian orogen and the minimum width of the Iapetus Ocean, *in* D.W. Strangway, ed., The continental crust and its mineral deposits; Geol. Assoc. Can. Spec. Paper 20, p. 421–440.
WILLIAMS, H. and HATCHER, R.D., Jr., 1982: Suspect terranes and accretionary history of the Appalachian orogen; Geology, v. 10, p. 530–536.
WILLIAMS, H., and STEVENS, R.K., 1974: The ancient continental margin of eastern North America, *in* C.A. Burk and C.L. Drake, eds., The geology of continental margins; Springer-Verlag, New York, p. 781–796.
WILSON, J.T., 1963: Evidence from islands on the spreading of ocean floors; Nature, v. 197, p. 536–538.
WILSON, J.T., 1965: A new class of faults and their bearing on continental drift; Nature, v. 207, p. 343–347.
WILSON, J.T., 1966: Did the Atlantic close and then re-open?; Nature, v. 211, p. 676–681.
WILSON, J.T., and BURKE, K., 1972: Two types of mountain building; Nature, v. 239, p. 448–449.
WILSON, L.M., 1981: Circum-North Atlantic tectono-stratigraphic reconstruction, *in* J.W. Kerr and A.J. Fergusson, eds., Geology of the North Atlantic borderlands; Can. Soc. Petrol. Geol. Mem. 7, p. 167–184.
WILSON, R.C.L., and WILLIAMS, C.A., 1979: Oceanic transform structures and the development of Atlantic continental margin sedimentary basins—a review; J. Geol. Soc. London, v. 136, p. 311–320.
WINDLEY, B.F., ed., 1976: The early history of the Earth; John Wiley, 619 p.
WINDLEY, B.F., 1977: The evolving continents; John Wiley and Sons, New York, 385 p.
WINDLEY, B.F., 1981: Precambrian rocks in the light of the plate-tectonic concept, *in* A. Kröner, ed., Precambrian plate tectonics, Develop. in Precambrian Geol. 4, Elsevier, Amsterdam, p. 1–20.
WOODBURY, H.O., MURRAY, I.B. Jr., and OSBORNE, R.E., 1980: Diapirs and their relation to hydrocarbon accumulation, *in* A.D. Miall, ed., Facts and principles of world petroleum occurrence; Can. Soc. Petrol. Geol. Mem. 6, p. 119–142.
WOPFNER, H., CALLEN, R. and HARRIS, W.K., 1974: The Lower Tertiary Eyre Formation of the southwestern Great Artesian Basin; J. Geol. Soc. Austr., v. 21, p. 17–51.
WORZEL, J.L., and BURK, C.A., 1979: The margins of the Gulf of Mexico, *in* J.S. Watkins, L. Montadert and P.W. Dickerson, eds., Geological and geophysical investigations of continental margins; Am. Assoc. Petrol. Geol. Mem. 29, p. 403–419.
YORATH, C.J., and CHASE, R.L., 1981: Tectonic history of the Queen Charlotte Islands and adjacent areas—a model; Can. J. Earth Sci., v. 18, p. 1717–1739.
YOUNG, G.M., 1978: Some aspects of the evolution of the Archean crust; Geosci. Canada, v. 5, p. 140–149.
ZAK, I., and FREUND, R., 1981: Asymmetry and basin migration in the Dead Sea rift; Tectonophysics, v. 80, p. 27–38.
ZEN, E-AN, 1981: An alternative model for the development of the allochthonous southern Appalachian Piedmont; Am. J. Sci., v. 281, p. 1153–1163.
ZIEGLER, P.A., 1980: Hydrocarbon provinces of the Northwest European Basin, *in* A.D. Miall, ed., Facts and principles of world petroleum occurrence; Can. Soc. Petrol. Geol. Mem. 6, p. 653–706.

CHAPTER 10

Conclusions

One of the main messages of this book is the need for synthesis of many types of data. Thus I believe it is useful at the conclusion to review the structure of the book and re-emphasize some of the ideas I have tried to convey.

The organization of the chapters is intended to reflect an increasing complexity of information and sophistication, corresponding to more advanced levels of basin analysis. However, it by no means reflects the way in which all basin analyses are carried out. In mature basins the analyst may be faced at the start with a proliferation of old stratigraphic names. Sorting out the validity of these, their relative ages, facies interpretations and paleogeographic evolution will consume the first round of analysis (Chapters 3 to 6). Proceeding to interpretations of global cycles (Chap. 8) and plate tectonic setting (Chap. 9) may come much later and, for many basins, may seem irrelevant. For example those working on the Alberta Oil Sands or the coal-bearing sequences of the Illinois Basin may regard plate tectonics as a very remote concern. Yet this is not a wise approach to take. We have seen how plate margin tectonism may be transmitted for thousands of kilometers across a cratonic interior, and we have seen how large-scale stratigraphic architecture can be controlled by eustatic sea level changes that depend largely on spreading rates in faraway mid-oceanic ridges. It is a mistake to define too small an area as one's focus of interest. This can lead to serious problems, ranging from questions of stratigraphic correlation to erroneous interpretations of plate history. This point was convincingly argued by Dewey (1982) in his discussion of the evolution of the British Isles. His interpretive maps extend from Mexico to Russia.

Marine geologists and those involved with frontier petroleum exploration, particularly in offshore regions, tend to take the large-scale approach to basin analysis. They start with regional geophysical data: magnetic anomaly maps for sea-floor spreading history, regional crustal sections based on gravity and seismic, and detailed seismic stratigraphic interpretations. These data provide a plate tectonics framework and a broad structural–stratigraphic interpretation on which to base a drilling program. Interpretations of depositional systems and applications of facies model concepts come much later and only at advanced stages of exploration is there either the time to spend on or the data to justify the luxuries of refined lithostratigraphic nomenclature and formal biostratigraphic zonation. For these types of basins the chapters in this book would be used more or less in the reverse order. In much frontier exploration the formalities of stratigraphy (Chap. 3) are dealt with in a rather relaxed way, as new material is constantly being added to the data base. Informal stratigraphic names and biostratigraphic subdivisions provide adequate working tools until a late stage of basin exploration.

We can see, therefore, that the organization of a basin analysis must reflect the concerns of the analyst and the nature of the problem at hand. In many cases different specialists will be working on various aspects of the work at the same time, either as an industry project team or as a government–academic group (e.g. the teams assembled for the DSDP legs) or as separate individuals who may or may not communicate with each other.

One of the initial plans of this book was to include a series of complete worked examples to illustrate the evolution of a basin synthesis. This idea was quickly abandoned as it became obvious that any such series could only cover a few of the enormous variety of combinations of plate setting, stratigraphic architecture and depositional system. To deal with this complexity it is necessary to touch on a few important details about many basins around the world, thus subjecting the reader to a surfeit of geography. Readers can amplify this treatment for themselves by studing the examples in the advanced chapters (8 and 9). Details of stratigraphic correlation, facies analysis, etc., are dealt with very briefly here (for example, arguments about correlation and locations of data points are omitted), but enough information is provided that the reader should be able to see between the lines, and ample references are provided.

Basin analysis methods have been in a stage of rapid development for at least the last twenty years, and continue to evolve at a furious pace. The five major developments outlined in Chapter 1 are by no means complete—in fact refinements are appearing so frequently that it has been difficult to decide when to stop writing this book. It seems likely that particularly significant progress may be anticipated in three main areas:

1. Depositional systems of shelf, slope and rise areas. These are the most inaccessible for sedimentological study, and much remains to be learned. Shelf dynamics and submarine fan evolution are undergoing intense study, both in modern environments and in appropriate ancient sequences.

2. Documentation and interpretation of global stratigraphic cycles. The Vail et al. (1977) curves of changing sea level are being widely reproduced as geologists attempt to document their implications for local geological events or for broader stratigraphic problems. There is as yet incomplete agreement on their cause or their global correlation, and this has prompted a renaissance of detailed stratigraphic studies.

3. Plate tectonic interpretations. Geophysical models of crustal behaviour are rapidly improving in their ability to predict geological events. At the same time the stratigraphic–structural basin models outlined in Chapter 9 are providing increasingly sophisticated templates for the interpretation of ancient basins. New plate models for most areas of the globe will therefore continue to appear, as incorporation of these new ideas becomes necessary.

However, as Dewey (1982) has pointed out, the development of plate tectonics has in many respects made our work more difficult. We can no longer build *ad hoc* geosyclinal models, but are severely constrained by a complex theoretical framework. It may never be possible to unravel completely the kinematics of many areas of repeated suturing and rifting or microplate accretion, where the spreading histories of large and small oceans have been destroyed by subduction. Even where these can be deduced, perhaps from motions of adjacent larger plates, rational interpretation may be severely hampered by the overprint of successive plate processes at different orientations. These each develop their own crustal anisotropy, which may be reactivated later under a different plate regime, with complex results. Good examples are the interaction of Caledonide and Hercynide trends in Britain and the Hercynide–Alpine overprint through the Mediterranean area. Even the structurally and stratigraphically simple Alberta Basin shows the effects of rejuvenation of Precambrian tectonic elements periodically throughout the Phanerozoic. Resolving the historical geology of areas such as these will continue to provide the greatest challenges for basin analysts in the years to come.

References

DEWEY, J.F., 1982: Plate tectonics and the evolution of the British Isles; J. Geol. Soc. London, v. 139, p. 371–414.

VAIL, P.R., MITCHUM, R.M. Jr., and THOMPSON, S., III, 1977: Seismic stratigraphy and global changes of sea level, Part four: global cycles of relative changes of sea level; Am. Assoc. Petrol. Geol. Mem. 26, p. 83–98.

Subject index

D

E

Q

R

Authors index